7.1
7.4

Thermal Physics

Thermal Physics

Thermodynamics and Statistical Mechanics for Scientists and Engineers

Robert F. Sekerka

Carnegie Mellon University
Pittsburgh, PA 15213, USA

ELSEVIER

AMSTERDAM • BOSTON • HEIDELBERG • LONDON • NEW YORK • OXFORD
PARIS • SAN DIEGO • SAN FRANCISCO • SINGAPORE • SYDNEY • TOKYO

Elsevier
Radarweg 29, PO Box 211, 1000 AE Amsterdam, Netherlands
The Boulevard, Langford Lane, Kidlington, Oxford OX5 1GB, UK
225 Wyman Street, Waltham, MA 02451, USA

Library of Congress Cataloging-in-Publication Data
A catalog record for this book is available from the Library of Congress

British Library Cataloguing in Publication Data
A catalogue record for this book is available from the British Library

For information on all Elsevier publications
visit our website at http://store.elsevier.com/

ISBN: 978-0-12-803304-3

Dedication

To Care ....

who cared about every word
and helped me write what I meant to say
rather than what I had written

Table of Contents

About the Cover

To represent the many scientists who have made major contributions to the foundations of thermodynamics and statistical mechanics, the cover of this book depicts four significant scientists along with some equations and graphs associated with each of them.

- James Clerk Maxwell (1831-1879) for his work on thermodynamics and especially the kinetic theory of gases, including the Maxwell relations derived from perfect differentials and the Maxwell-Boltzmann Gaussian distribution of gas velocities, a precursor of ensemble theory (see Sections 5.2, 19.4, and 20.1).
- Ludwig Boltzmann (1844-1906) for his statistical approach to mechanics of many particle systems, including his Eta function that describes the decay to equilibrium and his formula showing that the entropy of thermodynamics is proportional to the logarithm of the number of microscopic realizations of a macrosystem (see Chapters 15–17).
- J. Willard Gibbs (1839-1903) for his systematic theoretical development of the thermodynamics of heterogeneous systems and their interfaces, including the definition of chemical potentials and free energy that revolutionized physical chemistry, as well as his development of the ensemble theory of statistical mechanics, including the canonical and grand canonical ensembles. The contributions of Gibbs are ubiquitous in this book, but see especially Chapters 5–8, 12–14, 17, 20, and 21.
- Max Planck (1858-1947, Nobel Prize 1918) for his quantum hypothesis of the energy of cavity radiation (hohlraum blackbody radiation) that connected statistical mechanics to what later became quantum mechanics (see Section 18.3.2); the Planck distribution of radiation flux versus frequency for a temperature 2.725 K describes the cosmic microwave background, first discovered in 1964 as a remnant of the Big Bang and later measured by the COBE satellite launched by NASA in 1989.

The following is a partial list of many others who have also made major contributions to the field, all deceased. Recipients of a Nobel Prize (first awarded in 1901) are denoted by the letter "N" followed by the award year. For brief historical introductions to thermodynamic and statistical mechanics, see Cropper [11, pp. 41-136] and Pathria and Beale [9, pp. xxi-xxvi], respectively. The scientists are listed in the order of their year of birth:

Sadi Carnot (1796-1832); Julius von Mayer (1814-1878); James Joule (1818-1889); Hermann von Helmholtz (1821-1894); Rudolf Clausius (1822-1888); William Thomson, Lord Kelvin (1824-1907); Johannes van der Waals (1837-1923, N1910); Jacobus van't Hoff (1852-1911, N1901); Wilhelm Wien (1864-1928, N1911); Walther Nernst (1864-1941, N1920); Arnold Sommerfeld (1868-1951); Théophile de Donder (1872-1957); Albert

Einstein (1879-1955, N1921); Irving Langmuir (1881-1957, N1932); Erwin Schrödinger (1887-1961, N1933); Satyendra Bose (1894-1974); Pyotr Kapitsa (1894-1984, N1978); William Giauque (1895-1982, N1949); John van Vleck (1899-1980, N1977); Wolfgang Pauli (1900-1958, N1945); Enrico Fermi (1901-1954, N1938); Paul Dirac (1902-1984, N1933); Lars Onsager (1903-1976, N1968); John von Neumann (1903-1957); Lev Landau (1908-1968, N1962); Claude Shannon (1916-2001); Ilya Prigogine (1917-2003, N1977); Kenneth Wilson (1936-2013, N1982).

Preface

This book is based on lectures in courses that I taught from 2000 to 2011 in the Department of Physics at Carnegie Mellon University to undergraduates (mostly juniors and seniors) and graduate students (mostly first and second year). Portions are also based on a course that I taught to undergraduate engineers (mostly juniors) in the Department of Metallurgical Engineering and Materials Science in the early 1970s. It began as class notes but started to be organized as a book in 2004. As a work in progress, I made it available on my website as a pdf, password protected for use by my students and a few interested colleagues.

It is my version of what I learned from my own research and self-study of numerous books and papers in preparation for my lectures. Prominent among these sources were the books by Fermi [1], Callen [2], Gibbs [3, 4], Lupis [5], Kittel and Kroemer [6], Landau and Lifshitz [7], and Pathria [8, 9], which are listed in the bibliography. Explicit references to these and other sources are made throughout, but the source of much information is beyond my memory.

Initially it was my intent to give an integrated mixture of thermodynamics and statistical mechanics, but it soon became clear that most students had only a cursory understanding of thermodynamics, having encountered only a brief exposure in introductory physics and chemistry courses. Moreover, I believe that thermodynamics can stand on its own as a discipline based on only a few postulates, or so-called laws, that have stood the test of time experimentally. Although statistical concepts can be used to motivate thermodynamics, it still takes a bold leap to appreciate that thermodynamics is valid, within its intended scope, independent of any statistical mechanical model. As stated by Albert Einstein in Autobiographical Notes (1946) [10]:

> *"A theory is the more impressive the greater the simplicity of its premises is, the more different kinds of things it relates, and the more extended is its area of applicability. Therefore the deep impression which classical thermodynamics made on me. It is the only physical theory of universal content concerning which I am convinced that within the framework of the applicability of its basic concepts, it will never be overthrown."*

Of course thermodynamics only allows one to relate various measurable quantities to one another and must appeal to experimental data to get actual values. In that respect, models based on statistical mechanics can greatly enhance thermodynamics by providing values that are independent of experimental measurements. But in the last analysis, any model must be compatible with the laws of thermodynamics in the appropriate limit of sufficiently large systems. Statistical mechanics, however, has the potential to treat smaller systems for which thermodynamics is not applicable.

Consequently, I finally decided to present thermodynamics first, with only a few connections to statistical concepts, and then present statistical mechanics in that context. That allowed me to better treat reversible and irreversible processes as well as to give a thermodynamic treatment of such subjects as phase diagrams, chemical reactions, and anisotropic surfaces and interfaces that are especially valuable to materials scientists and engineers.

The treatment of statistical mechanics begins with a mathematical measure of disorder, quantified by Shannon [48, 49] in the context of information theory. This measure is put forward as a candidate for the entropy, which is formally developed in the context of the microcanonical, canonical, and grand canonical ensembles. Ensembles are first treated from the viewpoint of quantum mechanics, which allows for explicit counting of states. Subsequently, classical versions of the microcanonical and canonical ensembles are presented in which integration over phase space replaces counting of states. Thus, information is lost unless one establishes the number of states to be associated with a phase space volume by requiring agreement with quantum treatments in the limit of high temperatures. This is counter to the historical development of the subject, which was in the context of classical mechanics. Later in the book I discuss the foundation of the quantum mechanical treatment by means of the density operator to represent pure and statistical (mixed) quantum states.

Throughout the book, a number of example problems are presented, immediately followed by their solutions. This serves to clarify and reinforce the presentation but also allows students to develop problem-solving techniques. For several reasons I did not provide lists of problems for students to solve. Many such problems can be found in textbooks now in print, and most of their solutions are on the internet. I leave it to teachers to assign modifications of some of those problems or, even better, to devise new problems whose solutions cannot yet be found on the internet.

The book also contains a number of appendices, mostly to make it self-contained but also to cover technical items whose treatment in the chapters would tend to interrupt the flow of the presentation.

I view this book as an intermediate contribution to the vast subjects of thermodynamics and statistical mechanics. Its level of presentation is intentionally more rigorous and demanding than in introductory books. Its coverage of statistical mechanics is much less extensive than in books that specialize in statistical mechanics, such as the recent third edition of Pathria's book, now authored by Pathria and Beale [9], that contains several new and advanced topics. I suspect the present book will be useful for scientists, particularly physicists and chemists, as well as engineers, particularly materials, chemical, and mechanical engineers. If used as a textbook, many advanced topics can be omitted to suit a one- or two-semester undergraduate course. If used as a graduate text, it could easily provide for a one- or two-semester course. The level of mathematics needed in most parts of the book is advanced calculus, particularly a strong grasp of functions of several variables, partial derivatives, and infinite series as well as an elementary knowledge of differential equations and their solutions. For the treatment of anisotropic surfaces and interfaces, necessary relations of differential geometry are presented in an appendix. For the statistical mechanics part, an appreciation of stationary quantum states, including degenerate states, is essential, but the calculation of such states is not needed. In a few places, I use the notation of the Dirac vector space, bras and kets, to represent quantum states, but always with reference to other representations; the only exceptions are Chapter 26, Quantum Statistics, where the Dirac notation is used to treat the density operator, and Appendix I, where creation and annihilation operators are treated.

I had originally considered additional information for this book, including more of my own research on the thermodynamics of inhomogeneously stressed crystals and a few more chapters on the statistical mechanical aspects of phase transformations. Treatment of the liquid state, foams, and very small systems were other possibilities. I do not address many-body theory, which I leave to other works. There is an introduction to Monte Carlo simulation at the end of Chapter 27, which treats the Ising model. The renormalization group approach is described briefly but not covered in detail. Perhaps I will address some of these topics in later writings, but for now I choose not to add to the already considerable bulk of this work.

Over the years that I shared versions of this book with students, I received some valuable feedback that stimulated revision or augmentation of topics. I thank all those students. A few faculty at other universities used versions for self-study in connection with courses they taught, and also gave me some valuable feedback. I thank these colleagues as well. I am also grateful to my research friends and co-workers at NIST, where I have been a consultant for nearly 45 years, whose questions and comments stimulated a lot of critical thinking; the same applies to many stimulating discussions with my colleagues at Carnegie-Mellon and throughout the world. Singular among those was my friend and fellow CMU faculty member Prof. William W. Mullins who taught me by example the love, joy and methodologies of science. There are other people I could thank individually for contributing in some way to the content of this book but I will not attempt to present such a list. Nevertheless, I alone am responsible for any misconceptions or outright errors that remain in this book and would be grateful to anyone who would bring them to my attention.

In bringing this book to fruition, I would especially like to thank my wife Carolyn for her patience and encouragement and her meticulous proofreading. She is an attorney, not a scientist, but the logic and intellect she brought to the task resulted in my rewriting a number of obtuse sentences and even correcting a number of embarrassing typos and inconsistent notation in the equations. I would also like to thank my friends Susan and John of Cosgrove Communications for their guidance with respect to several aesthetic aspects of this book. Thanks are also due to the folks at my publisher Elsevier: Acquisitions Editor Dr. Anita Koch, who believed in the product and shepherded it through technical review, marketing and finance committees to obtain publication approval; Editorial Project Manager Amy Clark, who guided me though cover and format design as well as the creation of marketing material; and Production Project Manager Paul Prasad Chandramohan, who patiently managed to respond positively to my requests for changes in style and figure placements, as well as my last-minute corrections. Finally, I thank Carnegie Mellon University for providing me with an intellectual home and the freedom to undertake this work.

Robert F. Sekerka
Pittsburgh, PA

PART I

Thermodynamics

1 Introduction

Thermal physics deals with the quantitative physical analysis of macroscopic systems. Such systems consist of a very large number, $\mathcal{N}$, of atoms, typically $\mathcal{N} \sim 10^{23}$. According to classical mechanics, a detailed knowledge of the microscopic state of motion (say, position $\mathbf{r}_i$ and velocity $\mathbf{v}_i$) of each atom, $i = 1, 2, \ldots, \mathcal{N}$, at some time t, even if attainable, would constitute an overwhelmingly huge database that would be practically useless. More useful quantities would be averages, such as the average kinetic energy of an atom in the system, which would be independent of time if the system were in equilibrium. We might also be interested in knowing such things as the volume V of the system or the pressure p that it exerts on the walls of a containing vessel. In other words, a useful description of a macroscopic system is necessarily statistical and consists of knowledge of a few macroscopic variables that describe the system to our satisfaction.

We shall be concerned primarily with macroscopic systems in a state of equilibrium. An equilibrium state is one whose macroscopic parameters, which we shall call state variables, do not change with time. We accept the proposition, in accord with our experience, that any macroscopic system subject to suitable constraints, such as confinement to a volume and isolation from external forces or sources of matter and energy, will eventually come to a state of equilibrium. Our concept, or model, of the system will dictate the number of state variables that constitute a complete description—a complete set of state variables—of that system. For example, a gas consisting of a single atomic species might be described by three state variables, its energy U, its volume V, and its number of atoms $\mathcal{N}$. Instead of its number of atoms, we usually avoid large numbers and specify its number of moles, $N := \mathcal{N}/\mathcal{N}_A$ where $\mathcal{N}_A = 6.02 \times 10^{23}$ molecules/mol is Avogadro's number.[1] The state of a gas consisting of two atomic species, denoted by subscripts 1 and 2, would require four variables, U, V, N_1, and N_2. A simple model of a crystalline solid consisting of one atomic species would require eight variables; these could be taken to be U, V, N, and five more variables needed to describe its state of shear strain.[2]

1.1 Temperature

A price we pay to describe a macroscopic system is the introduction of a state variable, known as the temperature, that is related to statistical concepts and has no counterpart in simple mechanical systems. For the moment, we shall regard the temperature to be an

[1]The notation $A := B$ means A is defined to be equal to B, and can be written alternatively as $B =: A$.

[2]This is true if the total number of unit cells of the crystal is able to adjust freely, for instance by means of vacancy diffusion; otherwise, a total of nine variables is required because one must add the volume per unit cell to the list of variables. More complex macroscopic systems require more state variables for a complete description, but usually the necessary number of state variables is small.

Thermal Physics. http://dx.doi.org/10.1016/B978-0-12-803304-3.00001-6

empirical quantity, measured by a thermometer, such that temperature is proportional to the expansion that occurs whenever energy is added to matter by means of heat transfer. Examples of thermometers include thermal expansion of mercury in a long glass tube, bending of a bimetallic strip, or expansion of a gas under the constraint of constant pressure. Various thermometers can result in different scales of temperature corresponding to the same physical states, but they can be calibrated to produce a correspondence. If two systems are able to freely exchange energy with one another such that their temperatures are equal and their other macroscopic state variables do not change with time, they are said to be in equilibrium.

From a theoretical point of view, the most important of these empirical temperatures is the temperature θ measured by a gas thermometer consisting of a fixed number of moles N of a dilute gas at volume V and low pressure p. This temperature θ is defined to be proportional to the volume at fixed p and N by the equation

$$\theta := \frac{p}{RN} V, \tag{1.1}$$

where R is a constant. For variable p, Eq. (1.1) also embodies the laws of Boyle, Charles, and Gay-Lussac. Provided that the gas is sufficiently dilute (small enough N/V), experiment shows that θ is independent of the particular gas that is used. A gas under such conditions is known as an ideal gas. The temperature θ is called an absolute temperature because it is proportional to V, not just linear in V. If the constant $R = 8.314\,\mathrm{J/(mol\,K)}$, then θ is measured in degrees Kelvin, for which one uses the symbol K. On this scale, the freezing point of water at one standard atmosphere of pressure is 273.15 K. Later, in connection with the second law of thermodynamics, we will introduce a unique thermodynamic definition of a temperature, T, that is independent of any particular thermometer. Fermi [1, p. 42] uses a Carnot cycle that is based on an ideal gas as a working substance to show that $T = \theta$, so henceforth we shall use the symbol T for the absolute temperature.[3]

Example Problem 1.1. The Fahrenheit scale °F, which is commonly used in the United States, the United Kingdom, and some other related countries, is based on a smaller temperature interval. At one standard atmosphere of pressure, the freezing point of water is 32 °F and the boiling point of water is 212 °F. How large is the Fahrenheit degree compared to the Celsius degree?

The Rankine scale R is an absolute temperature scale but based on the Fahrenheit degree. At one standard atmosphere of pressure, what are the freezing and boiling points of water on the Rankine scale? What is the value of the triple point of water on the Rankine scale, the Fahrenheit scale and the Celsius scale? What is the value of absolute zero in °F?

[3]The Kelvin scale is defined such that the triple point of water (solid-liquid-vapor equilibrium) is exactly 273.16 K. The Celsius scale, for which the unit is denoted °C, is defined by $T(°\mathrm{C}) = T(\mathrm{K}) - 273.15$.

Solution 1.1. The temperature interval between the boiling and freezing points of water at one standard atmosphere is 100 °C or 212 − 32 = 180 °F. Therefore, 1 °F = 100/180 = 5/9 °C = (5/9) K. The freezing and boiling points of water are 273.15 × (9/5) = 491.67 R and 373.15 × (9/5) = 671.67 R. The triple point of water is 273.16 × (9/5) = 491.688 R = 32.018 °F = 0.01 °C. The value of absolute zero in °F is −(491.67 − 32) = −459.67 °F.

In the process of introducing temperature, we alluded to the intuitive concept of heat transfer. At this stage, it suffices to say that if two bodies at different temperatures are brought into "thermal contact," a process known as heat conduction can occur that enables energy to be transferred between the bodies even though the bodies exchange no matter and do no mechanical work on one another. This process results in a new equilibrium state and a new common temperature for the combined body. It is common to say that this process involves a "transfer of heat" from the hotter body (higher initial temperature) to the colder body (lower initial temperature). This terminology, however, can be misleading because a conserved quantity known as "heat" does not exist.[4] We should really replace the term "transfer of heat" by the longer phrase "transfer of energy by means of a process known as heat transfer that does not involve mechanical work" but we use the shorter phrase for simplicity, in agreement with common usage. The first law of thermodynamics will be used to quantify the amount of energy that can be transferred between bodies without doing mechanical work. The second law of thermodynamics will then be introduced to quantify the maximum amount of energy due to heat transfer (loosely, "heat") that can be transformed into mechanical work by some process. This second law will involve a new state variable, the entropy S, which like the temperature is entirely statistical in nature and has no mechanical counterpart.

1.2 Thermodynamics Versus Statistical Mechanics

Thermodynamics is the branch of thermal physics that deals with the interrelationship of macroscopic state variables. It is traditionally based on three so-called laws (or a number of postulates that lead to the same results, see Callen [2, chapter 1]). Based on these laws, thermodynamics is independent of detailed models involving atoms and molecules. It results in criteria involving state variables that must be true of systems that are in equilibrium with one another. It allows us to develop relationships among measurable quantities (e.g., thermal expansion, heat capacity, compressibility) that can be represented by state variables and their derivatives. It also results in inequalities that must be obeyed by any naturally occurring process. It does not, however, provide *values* of the quantities with which it deals, only their interrelationship. Values must be provided by experiments or by models based on statistical mechanics. For an historical introduction to thermodynamics, see Cropper [11, p. 41].

[4]Such a quantity was once thought to exist and was called caloric.

Statistical mechanics is based on the application of statistics to large numbers of atoms (or particles) that obey the laws of mechanics, strictly speaking quantum mechanics, but in limiting cases, classical mechanics. It is based on postulates that relate certain types of averages, known as ensemble averages, to measurable quantities and to thermodynamic state variables, such as entropy mentioned above. Statistical mechanics can be used to rationalize the laws of thermodynamics, although it is based on its own postulates which were motivated by thermodynamics. By using statistical mechanics, specific models can be analyzed to provide values of the quantities employed by thermodynamics and measured by experiments. In this sense, statistical mechanics appears to be more complete; however, it must be borne in mind that the validity of its results depends on the validity of the models. Statistical mechanics can, however, be used to describe systems that are too small for thermodynamics to be applicable. For an excellent historical introduction to statistical mechanics, see Pathria and Beale [9, pp. xxi-xxvi].

A crude analogy with aspects of mathematics may be helpful here: thermodynamics is to statistical mechanics as Euclidean geometry is to analytic geometry and trigonometry. Given the few postulates of Euclidean geometry, which allow things such as lengths and angles to be compared but never measured, one can prove very useful and general theorems involving the interrelationships of geometric forms, for example, congruence, similarity, bisections, conditions for lines to be parallel or perpendicular, and conditions for common tangency. But one cannot assign numbers to these geometrical quantities. Analytic geometry and trigonometry provide quantitative measures of the ingredients of Euclidean geometry. These measures must be compatible with Euclidean geometry but they also supply precise information about such things as the length of a line or the size of an angle. Moreover, trigonometric identities can be quite complicated and transcend simple geometrical construction.

1.3 Classification of State Variables

Much of our treatment will be concerned with homogeneous bulk systems in a state of equilibrium. By bulk systems, we refer to large systems for which surfaces, either external or internal, make negligible contributions. As a simple example, consider a sample in the shape of a sphere of radius R and having volume $V = (4/3)\pi R^3$ and surface area $A = 4\pi R^2$. If each atom in the sample occupies a volume a^3, then for $a \ll R$, the ratio of the number of surface atoms to the number of bulk atoms is approximately

$$r = \frac{4\pi (R/a)^2}{(4/3)\pi (R/a)^3 - 4\pi (R/a)^2} \sim 3(a/R) \ll 1. \tag{1.2}$$

For a sufficiently large sphere, the number of surface atoms is completely negligible compared to the number of bulk atoms, and so presumably is their energy and other properties. More generally, for a bulk sample having $\mathcal{N}$ atoms, roughly $\mathcal{N}^{2/3}$ are near the surface, so the ratio of surface to bulk atoms is roughly $r \sim \mathcal{N}^{-1/3}$. For a mole of atoms, we have $\mathcal{N} \sim 6 \times 10^{23}$ and $r \sim 10^{-8}$. In defining bulk samples, we must be careful to exclude samples such as thin films or thin rods for which one or more dimension is small compared to others. Thus, a thin film of area L^2 and thickness $H \ll L$ contains roughly $\mathcal{N} \sim L^2H/a^3$ atoms, but about $2L^2/a^2$ of these are on its surfaces. Thus, the ratio of surface to bulk atoms is $r \sim a/H$ which will not be negligible for a sufficiently thin film. We must also exclude samples that are finely subdivided, such as those containing many internal cavities.

From the considerations of the preceding paragraph, atoms of bulk samples can be regarded as being equivalent to one another, independent of location. It follows that certain state variables needed to describe such systems are proportional to the number of atoms. For example, for a homogeneous sample, total energy $U \propto \mathcal{N}$ and total volume $V \propto \mathcal{N}$, provided we agree to exclude from consideration small values of $\mathcal{N}$ that would violate the idealization of a bulk sample.[5] State variables of a homogeneous bulk thermodynamic system that are proportional to its number of atoms are called **extensive** variables. They are proportional to the "extent" or "size" of the sample. For a homogeneous gas consisting of three atomic species, a complete set of extensive state variables could be taken to be U, V, N_1, N_2, and N_3, where the N_i are the number of moles of atomic species i.

There is a second kind of state variable that is independent of the "extent" of the sample. Such a variable is known as an **intensive** variable. An example of such a variable would be a ratio of extensive variables, say U/V, because both numerator and denominator are proportional to $\mathcal{N}$. Another example of an intensive variable would be a derivative of some extensive variable with respect to some other extensive variable. This follows because a derivative is defined to be a limit of a ratio, for example,

$$\frac{\mathrm{d}U}{\mathrm{d}V} = \lim_{\Delta V \to 0} \frac{U(V + \Delta V) - U(V)}{\Delta V}. \tag{1.3}$$

If other quantities are held constant during this differentiation, the result is a partial derivative $\partial U/\partial V$, which is also an intensive variable, but its value will depend on which other variables are held constant. It will turn out that the pressure p, which is an intensive state variable, can be expressed as

$$p = -\frac{\partial U}{\partial V} \tag{1.4}$$

provided that certain other variables are held constant; these variables are the entropy S, an extensive variable alluded to previously, as well as all other extensive variables of a remaining complete set. Another important intensive variable is the absolute temperature T, which we shall see can also be expressed as a partial derivative of U with respect to the entropy S while holding constant all other extensive variables of a remaining complete set.

Since the intensive variables are ratios or derivatives involving extensive variables, we will not be surprised to learn that the total number of *independent intensive* variables is one less than the total number of *independent extensive* variables. The total number of

[5]The symbol $\propto$ means "proportional to."

independent intensive variables of a thermodynamic system is known as its number of **degrees of freedom**, usually a small number which should not be confused with the huge number of microscopic degrees of freedom $6\mathcal{N}$ for $\mathcal{N}$ particles that one would treat by classical statistical mechanics.

In Chapter 5, we shall return to a systematic treatment of extensive and intensive variables and their treatment via Euler's theorem of homogeneous functions.

1.4 Energy in Mechanics

The concept of energy is usually introduced in the context of classical mechanics. We review such considerations briefly in order to shed light on some aspects of energy that will be important in thermodynamics.

1.4.1 Single Particle in One Dimension

A single particle of mass m moving in one dimension, x, obeys Newton's law

$$m\frac{\mathrm{d}^2x}{\mathrm{d}t^2} = F, \tag{1.5}$$

where t is the time and $F(x)$ is the force acting on the particle when it is at position x. We introduce the potential energy function

$$V(x) = -\int_{x_0}^{x} F(u)\,\mathrm{d}u, \tag{1.6}$$

which is the negative of the work done by the force on the particle when the particle moves from some position x_0 to position x. Then the force $F = -\mathrm{d}V/\mathrm{d}x$ can be written in terms of the derivative of this potential function. We multiply Eq. (1.5) by $\mathrm{d}x/\mathrm{d}t$ to obtain

$$m\frac{\mathrm{d}x}{\mathrm{d}t}\frac{\mathrm{d}^2x}{\mathrm{d}t^2} + \frac{\mathrm{d}V}{\mathrm{d}x}\frac{\mathrm{d}x}{\mathrm{d}t} = 0, \tag{1.7}$$

which can be rewritten as

$$\frac{\mathrm{d}}{\mathrm{d}t}\left[\frac{1}{2}mv^2 + V\right] = 0, \tag{1.8}$$

where the velocity $v := \mathrm{d}x/\mathrm{d}t$. Equation (1.8) can then be integrated to obtain

$$\frac{1}{2}mv^2 + V = E, \tag{1.9}$$

where E is independent of time and known as the total energy. The first term in Eq. (1.9) is known as the kinetic energy and the equation states that the sum of the kinetic and potential energy is some constant, independent of time. It is important to note, however, that the value of E is undetermined up to an additive constant. This arises as follows: If some constant V_0 is added to the potential energy $V(x)$ to form a new potential $\tilde{V} := V + V_0$, the same force results because

$$-\frac{d\tilde{V}}{dx} = -\frac{d}{dx}(V + V_0) = -\frac{dV}{dx} = F. \tag{1.10}$$

Thus, Eq. (1.9) could equally well be written

$$\frac{1}{2}mv^2 + \tilde{V} = \tilde{E}, \tag{1.11}$$

where $\tilde{E}$ is a new constant. Comparison of Eq. (1.11) with Eq. (1.9) shows that $\tilde{E} = E + V_0$, so the total energy shifts by the constant amount V_0. Therefore, only differences in energy have physical meaning; to obtain a numerical value of the energy, one must always measure energy relative to some well-defined state of the particle or, what amounts to the same thing, adopt the *convention* that the energy in some well-defined state is equal to zero. In view of Eq. (1.6), the potential energy $V(x)$ will be zero when $x = x_0$, but the choice of x_0 is arbitrary.

In classical mechanics, it is possible to consider more general force laws such as $F(x, t)$ in which case the force at point x depends *explicitly* on the time that the particle is at point x. In that case, we can obtain $(d/dt)(1/2)mv^2 = Fv$ where Fv is the power supplied by the force. Similar considerations apply for forces of the form $F(x, v, t)$ that can depend explicitly on velocity as well as time. In such cases, one must solve the problem explicitly for the functions $x(t)$ and $v(t)$ before the power can be evaluated. In these cases, the total energy of the system changes with time and it is not possible to obtain an energy integral as given by Eq. (1.9).

1.4.2 Single Particle in Three Dimensions

The preceding one-dimensional treatment can be generalized to three dimensions with a few modifications. In three dimensions, where we represent the position of a particle by the vector $\mathbf{r}$ with Cartesian coordinates x, y, and z, Eq. (1.5) takes the form

$$m\frac{d^2\mathbf{r}}{dt^2} = \mathbf{F}, \tag{1.12}$$

where $\mathbf{F}(\mathbf{r})$ is now a vector force at the point $\mathbf{r}$. The mechanical work done by the force on the particle *along a specified path* leading from $\mathbf{r}_A$ to $\mathbf{r}_B$ is now given by

$$W_{\mathbf{r}_A \text{ to } \mathbf{r}_B} = \int_{\text{path}} \mathbf{F} \cdot d\mathbf{r}. \tag{1.13}$$

According to the theorem of Stokes, one has

$$\int (\nabla \times \mathbf{F}) \cdot d\mathbf{A} = \oint \mathbf{F} \cdot d\mathbf{r}, \quad \text{closed loop}, \tag{1.14}$$

where the integral on the right is a line integral around a closed loop and the integral on the left is over an area that subtends that loop. For a force such that $\nabla \times \mathbf{F} = \mathbf{0}$, we see that the line integral around any closed loop is equal to zero. Thus, if we integrate from A to B along path 1 and from B back to A along some other path 2 we get zero. But the latter integral is just the negative of the integral from A to B along path 2, so the integral from A to B is the same along path 1 as along path 2. For such a force, it follows that the work

$$W_{AB} = \int_{\mathbf{r}_A}^{\mathbf{r}_B} \mathbf{F} \cdot \mathrm{d}\mathbf{r}, \quad \text{any path,} \tag{1.15}$$

is independent of path and depends only on the end points. Such a force is called a conservative force and may be represented as the gradient of a potential

$$V(\mathbf{r}) = -\int_{\mathbf{r}_0}^{\mathbf{r}} \mathbf{F}(\mathbf{r}') \cdot \mathrm{d}\mathbf{r}' \tag{1.16}$$

such that $\mathbf{F} = -\nabla V$. In this case, it follows that the work

$$W_{AB} = -\int_{\mathbf{r}_A}^{\mathbf{r}_B} \nabla V \cdot \mathrm{d}\mathbf{r} = -\int_{\mathbf{r}_A}^{\mathbf{r}_B} \mathrm{d}V = V(\mathbf{r}_A) - V(\mathbf{r}_B). \tag{1.17}$$

For such a conservative force, we can dot the vector $\mathbf{v} := \mathrm{d}\mathbf{r}/\mathrm{d}t$ into Eq. (1.12) to obtain

$$m\frac{\mathrm{d}\mathbf{r}}{\mathrm{d}t} \cdot \frac{\mathrm{d}^2\mathbf{r}}{\mathrm{d}t^2} + \frac{\mathrm{d}\mathbf{r}}{\mathrm{d}t} \cdot \nabla V = 0. \tag{1.18}$$

Then by noting that

$$\frac{\mathrm{d}}{\mathrm{d}t}(1/2)m\mathbf{v} \cdot \mathbf{v} = m\frac{\mathrm{d}\mathbf{r}}{\mathrm{d}t} \cdot \frac{\mathrm{d}^2\mathbf{r}}{\mathrm{d}t^2} \quad \text{and} \quad \frac{\mathrm{d}V}{\mathrm{d}t} = \frac{\mathrm{d}\mathbf{r}}{\mathrm{d}t} \cdot \nabla V, \tag{1.19}$$

we are led immediately to Eq. (1.8) and its energy integral Eq. (1.9) just as in one dimension, except now $v^2 = \mathbf{v} \cdot \mathbf{v}$ in the kinetic energy.

1.4.3 System of Particles

We next consider a system of particles, $k = 1, 2, \ldots, \mathcal{N}$, having masses m_k, positions $\mathbf{r}_k$, and velocities $\mathbf{v}_k = \mathrm{d}\mathbf{r}_k/\mathrm{d}t$. Each particle is assumed to be subjected to a conservative force

$$\mathbf{F}_k = -\nabla_k V(\mathbf{r}_1, \mathbf{r}_2, \ldots, \mathbf{r}_{\mathcal{N}}), \tag{1.20}$$

where ∇_k is a gradient operator that acts only on $\mathbf{r}_k$. Then by writing Newton's equations in the form of Eq. (1.12) for each value of k, summing over k and proceeding as above, we obtain

$$\frac{\mathrm{d}}{\mathrm{d}t}[\mathcal{T} + V] = 0, \tag{1.21}$$

where the total kinetic energy

$$\mathcal{T} := \sum_{k=1}^{\mathcal{N}} \frac{1}{2} m_k \mathbf{v}_k \cdot \mathbf{v}_k \tag{1.22}$$

and

$$\frac{\mathrm{d}V}{\mathrm{d}t} = \sum_{k=1}^{\mathcal{N}} \frac{\mathrm{d}\mathbf{r}_k}{\mathrm{d}t} \cdot \nabla_k V. \tag{1.23}$$

Furthermore, we can suppose that the forces on each particle can be decomposed into internal forces $\mathbf{F}^{\mathrm{i}}$ due to the other particles in the system and to external forces $\mathbf{F}^{\mathrm{e}}$, that is,

$\mathbf{F} = \mathbf{F}^{\mathrm{i}} + \mathbf{F}^{\mathrm{e}}$. Since these forces are additive, we also have a decomposition of the potential, $V = V^{\mathrm{i}} + V^{\mathrm{e}}$, into internal and external parts. The integral of Eq. (1.21) can therefore be written in the form

$$\mathcal{T} + V^{\mathrm{i}} + V^{\mathrm{e}} = E, \tag{1.24}$$

where E is the total energy constant. This suggests a related decomposition of $\mathcal{T}$ which we proceed to explore.

We introduce the position vector of the center of mass of the system of particles, defined by

$$\mathbf{R} := \frac{1}{M}\sum_{k=1}^{\mathcal{N}} m_k \mathbf{r}_k, \tag{1.25}$$

where $M := \sum_{k=1}^{\mathcal{N}} m_k$ is the total mass of the system. The velocity of the center of mass is

$$\mathbf{V} := \frac{\mathrm{d}\mathbf{R}}{\mathrm{d}t} = \frac{1}{M}\sum_{k=1}^{\mathcal{N}} m_k \mathbf{v}_k. \tag{1.26}$$

The kinetic energy relative to the center of mass, namely $\mathcal{T}^{\mathrm{i}}$, can be written

$$\mathcal{T}^{\mathrm{i}} := \frac{1}{2}\sum_{k=1}^{\mathcal{N}} m_k (\mathbf{v}_k - \mathbf{V}) \cdot (\mathbf{v}_k - \mathbf{V}) = \mathcal{T} - \frac{1}{2} M V^2. \tag{1.27}$$

Eq. (1.27) may be verified readily by expanding the left-hand side to obtain four terms and then using Eq. (1.26). The term $(1/2)MV^2$ is recognized as the kinetic energy associated with motion of the center of mass of the system. Equation (1.24) can therefore be written

$$\mathcal{T}^{\mathrm{i}} + V^{\mathrm{i}} + \frac{1}{2} M V^2 + V^{\mathrm{e}} = E. \tag{1.28}$$

The portion of this energy exclusive of the kinetic energy of the center of mass and the external forces, namely $U = \mathcal{T}^{\mathrm{i}} + V^{\mathrm{i}}$, is an internal energy of the system of particles and is the energy usually dealt with in thermodynamics. Thus, when energies of a thermodynamic system are compared, they are compared under the assumption that the state of overall motion of the system, and hence its overall motional kinetic energy, $(1/2)MV^2$, is unchanged. This is equivalent to supposing that the system is originally at rest and remains at rest. Moreover, it is usually assumed that there are no external forces so the interaction energy V^{e} is just a constant. Thus, the energy integral is usually viewed in the form

$$U =: \mathcal{T}^{\mathrm{i}} + V^{\mathrm{i}} = E - \frac{1}{2} M V^2 - V^{\mathrm{e}} =: U_0, \tag{1.29}$$

where U_0 is a new constant. If such a system does interact with its environment, U is no longer a constant. Indeed, if the system does work or if there is heat transfer from its environment, U will change according to the first law of thermodynamics, which is taken up in Chapter 2.

Sometimes one chooses to include conservative external forces in the energy used in thermodynamics. Such treatments require the use of a generalized energy that includes potential energy due to conservative external forces, such as those associated with gravity or an external electric field. In that case, one deals with the quantity

$$\tilde{U} =: \mathcal{T}^{\mathrm{i}} + V^{\mathrm{i}} + V^{\mathrm{e}} = E - \frac{1}{2}MV^2. \tag{1.30}$$

In terms of chemical potentials, which we shall discuss in Chapter 12, such external forces give rise to gravitational chemical potentials and electrochemical potentials that play the role [6, p. 122] of intrinsic chemical potentials when external fields are present. It is also possible to treat uniformly rotating coordinate systems by including in the thermodynamic energy the effective potential associated with fictitious centrifugal forces [7, p. 72].

1.5 Elementary Kinetic Theory

More insight into the state variables temperature T and pressure p can be gained by considering the elementary kinetic theory of gases. We consider a monatomic ideal gas having particles of mass m that do not interact and whose center of mass remains at rest. Its kinetic energy is

$$\mathcal{T} = \frac{1}{2}\sum_{k=1}^{\mathcal{N}} m\frac{\mathrm{d}\mathbf{r}_k}{\mathrm{d}t}\cdot\frac{\mathrm{d}\mathbf{r}_k}{\mathrm{d}t} = \frac{1}{2}\sum_{k=1}^{\mathcal{N}} m\mathbf{v}_k^2. \tag{1.31}$$

If the gas is in equilibrium, the time average $\overline{\mathcal{T}}$ of this kinetic energy is a constant. This kinetic energy represents the vigor of motion of the atoms, so it is natural to suppose that it increases with temperature because temperature can be increased by adding energy due to heat transfer. A simple and fruitful assumption is to assume that $\overline{\mathcal{T}}$ is proportional to the temperature. In particular, we postulate that the time average kinetic energy per atom is related to the temperature by[6]

$$\frac{1}{\mathcal{N}}\overline{\mathcal{T}} = \frac{1}{2\mathcal{N}}\overline{\sum_{k=1}^{\mathcal{N}} m\mathbf{v}_k^2} = \frac{3}{2}k_B T, \tag{1.32}$$

where k_{B} is a constant known as Boltzmann's constant. In fact, $k_{\mathrm{B}} = R/\mathcal{N}_A$ where R is the gas constant introduced in Eq. (1.1) and $\mathcal{N}_A$ is Avogadro's number. We shall see that Eq. (1.32) makes sense by considering the pressure of an ideal gas.

The pressure p of an ideal gas is the force per unit area exerted on the walls of a containing box. For simplicity, we treat a monatomic gas and assume for now that each atom of the gas has the same speed v, although we know that there is really a distribution of speeds given by the Maxwell distribution, to be discussed in Chapter 19. We consider

[6] If the center of mass of the gas were not at rest, Eq. (1.27) would apply and $\overline{\mathcal{T}}$ would have to be replaced by $\overline{\mathcal{T}^{\mathrm{i}}}$. In other words, the kinetic energy $(1/2)MV^2$ of the center of mass makes no contribution to the temperature.

an infinitesimal area $\mathrm{d}A$ of a wall perpendicular to the x direction and gas atoms with velocities that make an angle of θ with respect to the positive x direction. In a time $\mathrm{d}t$, all atoms in a volume $v\,\mathrm{d}t\,\mathrm{d}A\cos\theta$ will strike the wall at $\mathrm{d}A$, provided that $0 < \theta < \pi/2$. Each atom will collide with the wall with momentum $m\,v\cos\theta$ and be reflected with the same momentum,[7] so each collision will contribute a force $(1/\mathrm{d}t)2m\,v\cos\theta$, which is the time rate of change of momentum. The total pressure (force per unit area) is therefore

$$p = \frac{1}{2}\left\langle \frac{n(v\,\mathrm{d}t\,\mathrm{d}A\cos\theta)(2m\,v\cos\theta)}{\mathrm{d}A\,\mathrm{d}t} \right\rangle = nm\langle v^2\cos^2\theta\rangle = nm\langle v_x^2\rangle, \tag{1.33}$$

where n is the number of atoms per unit volume and the angular brackets denote an average over time and all θ. The factor of $1/2$ arises because of the restriction $0 < \theta < \pi/2$. Since the gas is isotropic, $\langle v_x^2\rangle = \langle v_y^2\rangle = \langle v_z^2\rangle = (1/3)\langle v^2\rangle$. Therefore,[8]

$$p = \frac{1}{3}nm\langle v^2\rangle = \frac{2}{3}n\frac{\overline{\mathcal{T}}}{\mathcal{N}} = nk_BT = \frac{NR}{V}T, \tag{1.34}$$

where Eq. (1.32) has been used. Equation (1.34) is the well-known ideal gas law, in agreement with Eq. (1.1) if the absolute temperature is denoted by T. In the case of an ideal gas, all of the internal energy is kinetic, so the total internal energy is $U = \overline{\mathcal{T}}$. Eq. (1.34) therefore leads to $p = (2/3)(U/V)$, which is also true for an ideal monatomic gas.

These simple relations from elementary kinetic theory are often used in thermodynamic examples and are borne out by statistical mechanics.

[7]Reflection with the same momentum would require specular reflection from perfectly reflecting walls, but irrespective of the nature of actual walls, one must have reflection with the same momentum *on average* to avoid a net exchange of energy.

[8]If we had accounted for a Maxwell distribution of speeds, this result would still hold provided that we interpret $\langle v^2\rangle$ to be an average of the square of the velocity with respect to that distribution. See Eqs. (20.28-20.30) for details.

2 First Law of Thermodynamics

The first law of thermodynamics extends the concept of energy from mechanical systems to thermodynamic systems, specifically recognizing that a process known as heat transfer can result in a transfer of energy to the system in addition to energy transferred by mechanical work. We first state the law and then discuss the terminology used to express it. As stated below, the law applies to a **chemically closed system**, by which we mean that the system can exchange energy with its environment by means of heat transfer and work but cannot exchange mass of any chemical species with its environment. This definition is used by most chemists; many physicists and engineers use it as well but it is not universal. Some authors, such as Callen [2] and Chandler [12], regard a closed system as one that can exchange nothing with its environment. In this book, we refer to a system that can exchange nothing with its environment as an **isolated system**.

2.1 Statement of the First Law

For a thermodynamic system, there exists an extensive function of state, U, called the **internal energy**. Every equilibrium state of a system can be described by a complete set of (macroscopic) *state variables*. The number of such state variables depends on the complexity of the system and is usually small. For now we can suppose that U depends on the temperature T and additional extensive state variables needed to form a complete set.[1] Alternatively, any equilibrium state can be described by a complete set of extensive state variables that includes U. For a chemically closed system, the change ΔU from an initial to a final state is equal to the heat, Q, added *to the system* minus the work, $\mathcal{W}$, done *by the system*, resulting in[2]

$$\Delta U = Q - \mathcal{W}. \tag{2.1}$$

Q and $\mathcal{W}$ are not functions of state because they depend on the path taken during the process that brings about the change, not on just the initial and final states. Eq. (2.1)

[1] There are other possible choices of a complete set of state variables. For example, a homogeneous isotropic fluid composed a single chemical component can be described by three extensive variables, the internal energy U, the volume V, and the number of moles N. One could also choose state variables T, V, and N and express U as a function of them, and hence a function of state. Alternatively, U could be expressed as a function of T, the pressure p, and N. In Chapter 3, we introduce an extensive state variable S, the entropy, in which case U can be expressed as a function of a complete set of extensive variables including S, known as a fundamental equation.

[2] In agreement with common usage, we use the terminology "heat transferred to the system" or "heat added to the system" in place of the longer phrase "energy transferred to the system by means of a process known as heat transfer that does not involve mechanical work."

Thermal Physics. http://dx.doi.org/10.1016/B978-0-12-803304-3.00002-8

actually defines Q, since ΔU and $\mathcal{W}$ can be measured independently, as will be discussed in detail in Section 2.1.1.

If there is an infinitesimal amount of heat δQ transferred to the system and the system does an infinitesimal amount of work $\delta\mathcal{W}$, the change in the internal energy is

$$dU = \delta Q - \delta\mathcal{W}, \quad \text{infinitesimal change.} \tag{2.2}$$

For an isolated system, $\Delta U = 0$, and for such a system, the internal energy is a constant.

2.1.1 Discussion of the First Law

As explained in Chapter 1, the term internal energy usually excludes kinetic energy of motion of the center of mass of the entire macroscopic system, as well as energy associated with overall rotation (total angular momentum). The internal energy also usually excludes the energy due to the presence of external fields, although it is sometimes redefined to include conservative potentials. We will only treat thermodynamic systems that are at rest with respect to the observer (zero kinetic energy due to motion of the center of mass or total angular momentum). For further discussion of this point, see Landau and Lifshitz [7, p. 34].

We emphasize that $\mathcal{W}$ is positive if work is done by the system on its environment. Many authors, however, state the first law in terms of the work $W = -\mathcal{W}$ done by the environment on the system by some external agent. In this case, the first law would read $\Delta U = Q + W$. This is especially common[3] in Europe [14] and Russia [7].

The symbol Δ applied to any **state function** means the value of that function in the final state (after some process) minus the value of that function in the initial state. Specifically, $\Delta U := U(\text{final state}) - U(\text{initial state})$. As mentioned above, Q and $\mathcal{W}$ are *not* state functions, although their difference is a state function. As will be illustrated below, Q and $\mathcal{W}$ depend on the details of the *process* used to change the state function U. In other words, Q and $\mathcal{W}$ depend on the *path* followed during a process. Therefore, it makes no sense to apply the Δ symbol or the differential symbol d to Q or $\mathcal{W}$. We use δQ and $\delta\mathcal{W}$ to denote *infinitesimal transfers* of energy to remind ourselves that Q and $\mathcal{W}$ are not state functions. Some authors [6, 12] use a d with a superimposed strikethrough (đ) instead of δ.

The first law of thermodynamics is a theoretical generalization based on many experiments. Particularly noteworthy are the experiments of Joule who found that for two states of a closed thermodynamic system, say A and B, it is always possible to cause a transition that connects A to B by a process in which the system is thermally insulated, so $\delta Q = 0$ at every stage of the process. This also means that $Q = 0$ for the whole process.

[3]Fermi [1] uses the symbol L for the work done by the system; note that the Italian word for work is 'lavoro' (cognate labor). The introductory physics textbook by Young and Freedman [13] also states the first law of thermodynamics in terms of the work done by the system. Landau and Lifshitz [7] use the symbol $R \equiv -\mathcal{W}$ ('rabota') to denote the work done on the system. Chandler [12] and Kittel and Kroemer [6] use $W \equiv -\mathcal{W}$ to denote the work done on the system. This matter of notation and conventions can cause confusion, but we have to live with it.

Thus by work alone, either the transformation $A \to B$ or the transformation $B \to A$ is possible. Since the energy change due to work alone is well defined in terms of mechanical concepts, it is possible to establish either the energy difference $U_A - U_B$ or its negative $U_B - U_A$. The fact that one of these transformations might be impossible is related to concepts of irreversibility, which we will discuss later in the context of the second law of thermodynamics.

According to the first law, as recognized by Rudolf Clausius in 1850, **heat transfer** accounts for energy received by the system in forms other than work. Since ΔU can be measured and $\mathcal{W}$ can be determined for any mechanical process, Q is actually *defined* by Eq. (2.1). It is common to measure the amount of energy due to heat transfer in units of calories. One calorie is the amount of heat necessary to raise the temperature of one gram (10^{-3} kg) of water from 14 °C to 15 °C at standard atmospheric pressure. The mechanical equivalent of this heat is 1 calorie = 4.184 J = 4.184×10^7 erg. The amount of heat required to raise the temperature by ΔT of an arbitrary amount of water is proportional to its mass.

It was once believed that heat was a conserved quantity called caloric, and hence the unit calorie, but no such conserved quantity exists. This discovery is usually attributed to Count Rumford who noticed that water used to cool a cannon during boring would be brought to a boil more easily when the boring tool became dull, resulting in even less removal of metal. Thus, "heat" appears to be able to be produced in virtually unlimited amounts by doing mechanical work, and thus cannot be a conserved quantity. Therefore, we must bear in mind that heat transfer refers to a *process* for energy transfer and that there is actually no identifiable quantity, "heat," that is transported. From an atomistic point of view, we can think of conducted heat as energy transferred by means of microscopic atomic or molecular collisions in processes that occur without the transfer of matter and without changing the macroscopic physical boundaries of the system under consideration. Heat can also be transferred by radiation that is emitted or absorbed by a system.

We can enclose a system of interest and a heat source of known heat capacity (see Section 2.3) by insulation to form a calorimeter, assumed to be an isolated system, and allow the combined system to come to equilibrium. The temperature change of the heat source will allow determination of the amount of energy transferred from it (or to it) by means of heat transfer and this will equal the increase (or decrease) in energy of the system of interest.[4]

2.2 Quasistatic Work

If a thermodynamic system changes its volume V by an amount dV and does work against an external pressure p_{ext}, it does an infinitesimal amount of work

$$\delta \mathcal{W} = p_{\text{ext}}\, dV. \tag{2.3}$$

[4]If the heat source changes volume, it could exchange work with its environment and this would have to be taken into account.

This external pressure can be established by purely mechanical means. For example, an external force $F^{\rm ext}$ acting on a piston of area A would give rise to an external pressure $p^{\rm ext} = F^{\rm ext}/A$. Note that Eq. (2.3) is valid for a fluid system even if the process being considered is so rapid and violent that an internal pressure of the system cannot be defined during the process. This equation can also be generalized for a more complex system as long as one uses actual mechanical external forces and the distances through which they displace portions of the system, for example, pushing on part of the system by a rod or pulling on part of a system by a rope.

If an isotropic system (same in all directions, as would be true for a fluid, a liquid or a gas) expands or contracts sufficiently slowly (hence the term "quasistatic") that the system is practically in equilibrium at each instant of time, it will have a well-defined internal pressure p. Under such conditions, $p \approx p_{\rm ext}$ and the system will do an infinitesimal amount of work

$$\delta W = p\,\mathrm{d}V, \quad \text{quasistatic work.} \tag{2.4}$$

Note that δW and $\mathrm{d}V$ are positive if work is done by the system and both are negative if work is done on the system by an external agent.

Eq. (2.4) applies only to an idealized process. For an actual change to take place, we need p to be at least slightly different from $p_{\rm ext}$ to provide a net force in the proper direction. This requires $(p - p_{\rm ext})\,\mathrm{d}V > 0$. Thus $p_{\rm ext}\,\mathrm{d}V < p\,\mathrm{d}V$ which, in view of Eq. (2.3), may be written

$$\delta W < p\,\mathrm{d}V, \quad \text{actual process.} \tag{2.5}$$

For the case of quasistatic work, it will be necessary for p to be slightly greater than $p_{\rm ext}$ for the system to expand ($\mathrm{d}V > 0$); conversely, it will be necessary for p to be slightly less than $p_{\rm ext}$ for the system to contract. These small differences are assumed to be second order and are ignored in writing Eq. (2.4). Consistent with this idealization, a process of quasistatic expansion can be reversed to a process of quasistatic contraction by making an infinitesimal change in p. Therefore, **quasistatic work** is also called **reversible work**.[5] We can combine Eq. (2.4) with Eq. (2.5) to obtain

$$\delta W \le p\,\mathrm{d}V \tag{2.6}$$

with the understanding that the inequality applies to all actual processes (which are irreversible) and the equality applies to the idealized process of reversible quasistatic work.

For a finite change of V, the quasistatic work can be computed by integration:

$$W = \int_{\rm path} p\,\mathrm{d}V, \quad \text{quasistatic work.} \tag{2.7}$$

To evaluate this integral, we *must* specify the path that connects the initial and final states of the system. It makes no sense to write this expression with lower and upper limits of

[5]A process involving quasistatic work will be reversible only if all other processes that go on in the system are reversible. For example, an irreversible chemical reaction would be forbidden.

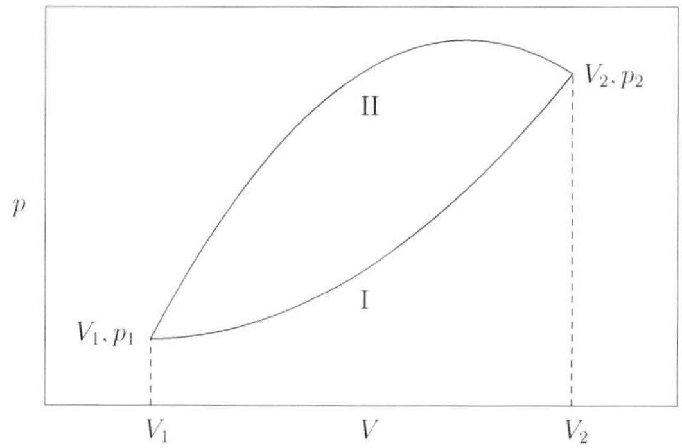

FIGURE 2–1 Illustration of quasistatic work for a system whose states can be represented by points in the V, p plane. The system makes a quasistatic transition from a state at V_1, p_1 to a state V_2, p_2 by two different paths, I and II. According to Eq. (2.7), the quasistatic work is the area under each curve and is obviously greater for path II. The difference in work is the area between the paths. Since ΔU for the two paths is the same, the difference in the heat Q for the two paths is also equal to the area between the paths.

integration unless the path is clearly specified. For a system whose equilibrium states can be represented by points in the V, p plane, the quasistatic work is represented by the area under the curve that represents the path that connects the initial and final states, as illustrated in Figure 2–1. Since the areas under two curves that connect the same two end points can be different, the quasistatic work $\mathcal{W}$ clearly depends on the path. Since $Q = \Delta U + \mathcal{W}$ and ΔU is independent of path, Q also depends on path.

If work and heat are exchanged with a system, it is important to recognize that the internal energy of the system will not be partitioned in any way that allows part of it to be associated with heat and part with work. That is because work and heat refer to *processes* for changing the energy of a system and lose their identity once equilibrium is attained and the energy of the system is established. On the other hand, other state variables of the system can differ depending on the relative amounts of heat and work that bring about the same change of internal energy. For example, consider two alternative processes in which the internal energy of an ideal gas is increased by exactly the same amount, the first by means of only work done by a constant external pressure p_{ext} and the second by means of only heat transfer. In the case of only work, the volume of the gas will necessarily be decreased but in the case of only heat transfer, the volume of the system will not be changed. Therefore, the two processes result in different thermodynamic states, even though both result in the same internal energy.

2.3 Heat Capacities

We can define heat capacities for changes in which the work done by the system is the quasistatic work given by Eq. (2.4). In that case, the first law takes the form

$$dU = \delta Q - p\,dV. \tag{2.8}$$

The heat capacity at constant volume, C_V, is defined to be the ratio of the infinitesimal amount of heat δQ needed to raise the temperature by an infinitesimal amount $\mathrm{d}T$ while holding the volume constant, namely

$$C_V := \left(\frac{\delta Q}{\mathrm{d}T}\right)_V = \left(\frac{\partial U}{\partial T}\right)_V, \tag{2.9}$$

where the last expression, a partial derivative at constant volume, follows from Eq. (2.8). The heat capacity is an extensive quantity and should not be confused with the specific heat, which is the heat capacity per unit mass, which is intensive.[6]

The heat capacity at constant pressure, C_p, is defined to be the ratio of the infinitesimal amount of heat δQ needed to raise the temperature by an infinitesimal amount $\mathrm{d}T$ while holding the pressure constant, namely

$$C_p := \left(\frac{\delta Q}{\mathrm{d}T}\right)_p = \left(\frac{\partial U}{\partial T}\right)_p + p\left(\frac{\partial V}{\partial T}\right)_p, \tag{2.10}$$

where the last expression again follows from Eq. (2.8). Note that the partial derivatives of U in Eqs. (2.9) and (2.10) are not the same because different variables are held constant. Thus

$$\left(\frac{\partial U}{\partial T}\right)_p = \left(\frac{\partial U}{\partial T}\right)_V + \left(\frac{\partial U}{\partial V}\right)_T \left(\frac{\partial V}{\partial T}\right)_p. \tag{2.11}$$

Example Problem 2.1. The specific heat of silver at 20 °C is 0.0558 cal g^{-1} K^{-1}. Here we ignore the small difference between constant volume and constant pressure for this condensed phase. What is the heat capacity of 3 kg of silver? How many Joules of energy are needed to raise the temperature of 3 kg of silver from 15 °C to 25 °C?

Solution 2.1. The heat capacity of 3 kg of silver is $3000 \times 0.0558 = 167$ cal K^{-1}. The temperature interval is 10 K so the energy required is $1670\ \text{cal} \times 4.184\ \text{J/cal} = 6990\ \text{J}$. We only keep three significant figures because the specific heat was only given to three figures.

2.3.1 Heat Capacity of an Ideal Gas

One mole of an ideal gas obeys the equation of state

$$pV = RT, \quad \text{one mole of ideal gas,} \tag{2.12}$$

where T is the absolute temperature and R is the gas constant. Equation (2.12) is essentially a definition of an ideal gas, based on experiments for real dilute gases that obey Eq. (1.1) that was used to define the empirical temperature θ. For such a real dilute gas, Joule conducted experiments in which the gas was confined originally to a subvolume V_1 of an insulated rigid container having overall volume V_2. The remainder of the volume, $V_2 - V_1$ was initially evacuated (see Figure 3–2). In these experiments, $Q = 0$ because the

[6]Analogous intensive quantities such as the heat capacity per atom, per molecule, or per mole are often used.

container is insulated and $\mathcal{W} = 0$ because the overall container having volume V_2 is rigid. Therefore, by the first law, the internal energy U remains constant. In the experiments, the gas was allowed to expand *internally* from V_1 to V_2. Joule observed that the temperature T of the gas remained practically unchanged during the process. More accurate experiments were performed later by Thomson (Lord Kelvin) and Joule by causing the gas to expand through a porous plug until a steady state is reached and measuring the temperature of the exiting gas directly. For hydrogen, there was hardly appreciable change in temperature; see the treatise by Planck [15, p. 50] for details. Therefore, the internal energy of such a dilute gas is practically independent of its volume. For an ideal gas, we shall assume that U is strictly independent of its volume, V, and therefore only a function of T. We shall see later that this conclusion can be derived by applying the second law of thermodynamics for a gas that obeys Eq. (2.12).[7]

Therefore, for an ideal gas, the second term on the right-hand side of Eq. (2.11) is zero. The second term on the right of Eq. (2.10) can be evaluated by means of Eq. (2.12), resulting in[8]

$$C_p = C_V + R, \quad \text{one mole of ideal gas.} \tag{2.13}$$

We observe that C_p is larger than C_V by an amount needed to supply the work $p\,dV$ done by the gas as it expands at constant pressure. The value of C_V depends on the type of gas under consideration and can be derived by means of statistical mechanics. For a mole of gas, we shall see that each translational or rotational degree of freedom of a gas molecule, made up of atoms that are considered to be point particles, contributes an amount $R/2$ to C_V. For a monatomic gas, each atom has three translational degrees of freedom, translation along x, y, and z, so $C_V = 3R/2$. A diatomic gas molecule would have six total degrees of freedom (three translational degrees for each atom) but the distance of separation of the two atoms remains practically constant due to strong chemical bonds. The atoms of a diatomic gas can execute vibrations along the line joining them, but these vibrations are hardly excited except at very high temperatures.[9] Thus only five degrees of freedom are usually active (three translational and two rotational) and $C_V = 5R/2$ for a diatomic gas. Similarly, if we neglect vibrational degrees of freedom for polyatomic gases, six degrees of freedom are usually active (three translational and three rotational) and $C_V = 3R$. This leads to the values listed in Table 2–1.

[7]By calculating derivatives of the entropy, it can be shown that $dU = C_V dT + (T\alpha/\kappa_T - p)dV$, where α is the isobaric compressibility and κ is the isothermal compressibility. From the ideal gas law, $\alpha = 1/T$ and $\beta = 1/p$, so the coefficient of dV vanishes and U depends only on T.

[8]As defined by Eqs. (2.9) and (2.10), the heat capacities C_V and C_p are extensive. Thus they depend not only on the substance under consideration but also on the amount of that substance. One can obtain intensive quantities by dividing by the number of moles or the mass. These intensive quantities depend only on the substance under consideration. Here we deal with one mole, which is equivalent to dividing the extensive heat capacities by the number of moles being considered.

[9]If partially excited, the contribution of a vibrational degree of freedom would depend on temperature. If fully excited, a vibrational degree of freedom would contribute $R/2$ for kinetic energy and $R/2$ for potential energy, for a total of R. Polyatomic gases with linear molecules behave somewhat like diatomic molecules insofar as rotational degrees of freedom are concerned. See Section 21.3 for a detailed discussion of ideal gases with internal structure.

Table 2–1 Heat Capacities per Mole of Ideal Gases

Molecule	C_V	C_p	$\gamma = C_p/C_V$
monatomic	$3R/2$	$5R/2$	$5/3 \approx 1.67$
diatomic	$5R/2$	$7R/2$	$7/5 = 1.40$
polyatomic	$3R$	$4R$	$4/3 \approx 1.33$

It is assumed that the atoms are point particles, translational and rotational degrees of freedom are totally excited, and vibrational degrees of freedom of diatomic and polyatomic gases are not excited.

2.3.2 General Relationship of C_p to C_V

By means of a general result of thermodynamics, it will turn out that C_p is always larger than C_V. For the moment, we state this result without proof but will derive it after we cover the second and third laws of thermodynamics. First, we need to define two other measurable quantities:

isobaric coefficient of thermal expansion:

$$\alpha := \frac{1}{V}\left(\frac{\partial V}{\partial T}\right)_p, \tag{2.14}$$

isothermal compressibility:

$$\kappa_T := -\frac{1}{V}\left(\frac{\partial V}{\partial p}\right)_T. \tag{2.15}$$

The signs in Eqs. (2.14) and (2.15) have been chosen so that κ_T is positive and α is usually positive.[10] The general result (see Eq. (5.32)) is

$$C_p = C_V + \frac{TV\alpha^2}{\kappa_T}. \tag{2.16}$$

From the form of Eq. (2.16), we observe that $C_p \ge C_V$ for any substance, which is not obvious from their definitions. From stability considerations, it will be shown in Section 7.4 that $C_p \ge C_V \ge 0$. For an ideal gas, we readily calculate from Eq. (2.12) that $\alpha = 1/T$ and $\kappa_T = 1/p$, in which case Eq. (2.16) becomes Eq. (2.13) for $N = 1$ mole of gas. For condensed phases, $|\alpha| \ll 1/T$ and $\kappa_T \ll 1/p$, but the second term in Eq. (2.16) is quadratic in α so the difference between C_p and C_V is very small. Thus, the difference between C_p and C_V is very important for gases but small and often negligible for liquids and solids.

[10]This agrees with our intuition and with experiment. It can be proven from general thermodynamic stability considerations (see Chapter 7) that κ_T is positive. α is usually positive but negative values of α are possible, for example, for water below about 4 °C.

Example Problem 2.2. The equation of state for one mole of a van der Waals fluid is

$$(p + a/v^2)(v - b) = RT,$$

where p is the pressure, v is the volume per mole, T is the temperature, and a and b are constants. Calculate the following quantities and show that they agree with the results for an ideal gas in the limit $a = b = 0$:

(a) The isothermal compressibility, $\kappa_T = -(1/v)(\partial v/\partial p)_T$
(b) The isobaric coefficient of thermal expansion, $\alpha = (1/v)(\partial v/\partial T)_p$
(c) The molar heat capacity difference, $(C_p - C_V)/N$
(d) Show directly that $(\partial p/\partial T)_v = \alpha/\kappa_T$. Why is this true?

Solution 2.2. We first take the differential of the given equation to obtain

$$\mathrm{d}p\ (v-b) + \left(p - \frac{a}{v^2} + \frac{2\,a\,b}{v^3}\right)\mathrm{d}v = R\,\mathrm{d}T.$$

(a)

$$\kappa_T = -\frac{1}{v}\left(\frac{\partial v}{\partial p}\right)_T = \frac{1}{v}(v-b)\left(p - \frac{a}{v^2} + \frac{2\,a\,b}{v^3}\right)^{-1} \to \frac{1}{p} \quad \text{for an ideal gas,}$$

(b)

$$\alpha = \frac{1}{v}\left(\frac{\partial v}{\partial T}\right)_p = \frac{R}{v}\left(p - \frac{a}{v^2} + \frac{2\,a\,b}{v^3}\right)^{-1} \to \frac{R}{pv} = \frac{1}{T} \quad \text{for an ideal gas,}$$

(c)

$$(C_p - C_V)/N = \frac{Tv\alpha^2}{\kappa_T} = \frac{R^2 T}{(v-b)}\left(p - \frac{a}{v^2} + \frac{2\,a\,b}{v^3}\right)^{-1} \to \frac{R^2 T}{pv} = R \quad \text{for an ideal gas,}$$

(d)

$$\left(\frac{\partial p}{\partial T}\right)_v = \frac{R}{v-b} = -\left(\frac{\partial v}{\partial T}\right)_p \Big/ \left(\frac{\partial v}{\partial p}\right)_T = \frac{\alpha}{\kappa_T}.$$

This relation is generally true, not just true for the van der Waals fluid, as can be seen from the differential

$$\mathrm{d}v = \left(\frac{\partial v}{\partial T}\right)_p \mathrm{d}T + \left(\frac{\partial v}{\partial p}\right)_T \mathrm{d}p.$$

2.4 Work Due to Expansion of an Ideal Gas

We calculate the work due to expansion of one mole of an ideal gas that obeys the equation of state Eq. (2.12). For simplicity, we will further assume that the gas has a constant heat capacity C_V at constant volume. According to Eq. (2.9), this results in

$$dU = C_V\,dT; \quad U = C_V\,T + \text{constant}. \tag{2.17}$$

2.4.1 Reversible Isothermal Process

For a reversible isothermal process, the path in the V, p plane is an equilateral hyperbola, $pV = \text{constant}$, where the value of the constant depends on T. We assume that this path joins two states that satisfy $p_1V_1 = p_2V_2$, so the quasistatic work is

$$\mathcal{W} = \int_{T=\text{ constant}} p\,dV = RT\int_{V_1}^{V_2}\frac{dV}{V} = RT\ln(V_2/V_1), \quad \text{one mole.} \tag{2.18}$$

For $V_2 > V_1$ the gas expands and does positive work, as shown in Figure 2–2. For the reverse transformation from V_2 to V_1, the gas contracts and does negative work; in this case, the environment of the gas does positive work on the gas. Since U depends only on T, we have $U_1 = U_2$ so $\Delta U = 0$. Therefore, by the first law, $Q = \mathcal{W}$ for this process.

2.4.2 Reversible Isobaric Expansion Followed by Isochoric Transformation

We assume the path to be a reversible expansion from V_1 to V_2 at constant pressure p_1 (**isobaric** expansion) followed by lowering the pressure to p_2 at constant volume V_2 (**isochoric** transformation). This is illustrated by the dashed line in Figure 2–3. The quasistatic work is

$$\mathcal{W} = p_1\int_{V_1}^{V_2} dV + \int_{V_2}^{V_2} p\,dV = p_1(V_2 - V_1) \tag{2.19}$$

because the second integral is zero. The temperature will change throughout this process. In general, the end points will have different temperatures, $T_1 = p_1V_1/R$ and $T_2 = p_2V_2/R$, and the change in internal energy will be $\Delta U = C_V(T_2 - T_1)$. If the end points happen to satisfy $p_1V_1 = p_2V_2$, then $T_1 = T_2$, but *during* the process T will not be constant. In general, $Q = \Delta U + \mathcal{W}$, but if $T_1 = T_2$, then $\Delta U = 0$ and $Q = \mathcal{W}$. Then the work given by Eq. (2.19) can also be written as $RT_1(V_2 - V_1)/V_1 = RT_2(V_2 - V_1)/V_1$ and is greater than that given by Eq. (2.18) with $T = T_1 = T_2$. The reader is invited to prove this statement mathematically.

2.4.3 Isochoric Transformation Followed by Reversible Isobaric Expansion

We assume the path to consist of lowering the pressure to p_2 at constant volume V_1 followed by reversible expansion from V_1 to V_2 at constant pressure p_2. This is illustrated by the dot-dashed line in Figure 2–3. The quasistatic work is

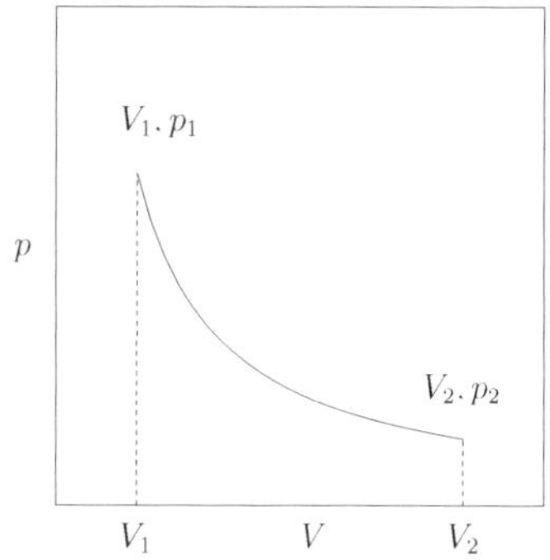

FIGURE 2–2 Illustration of quasistatic work for isothermal expansion of one mole of an ideal gas. The system makes a quasistatic transition from a state at V_1, p_1 to a state V_2, p_2 such that $pV = RT$. The work done by the gas is equal to the area under the curve.

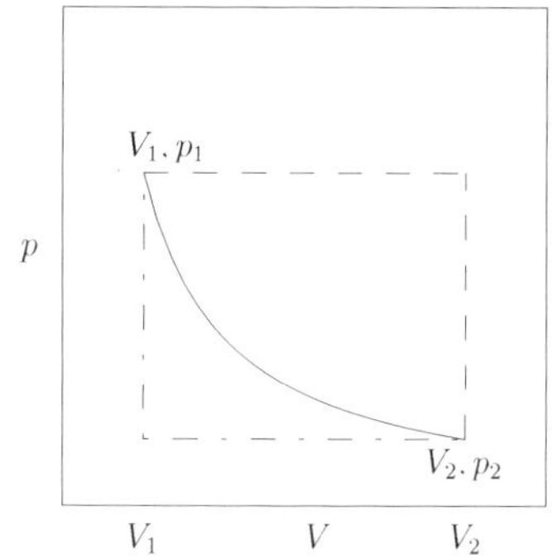

FIGURE 2–3 Illustration of quasistatic work for one mole of an ideal gas. The dashed line represents an isobaric expansion at pressure p_1 followed by an isochoric transformation at V_2. The dot-dashed line represents an isochoric transformation at V_1 followed by an isobaric transformation at pressure p_2. The full line represents an isothermal transformation from V_1, p_1 to V_2, p_2, which is only possible if $V_1 p_1 = V_2 p_2$.

$$\mathcal{W} = \int_{V_1}^{V_1} p\,\mathrm{d}V + p_2 \int_{V_1}^{V_2} \mathrm{d}V = p_2(V_2 - V_1), \tag{2.20}$$

which is clearly smaller than that given by Eq. (2.19). If the end points happen to be at the same temperature, the work given by Eq. (2.20) can be written $RT_1(V_2 - V_1)/V_2 = RT_2(V_2 - V_1)/V_2$ and is less than that given by Eq. (2.18) with $T = T_1 = T_2$.

2.4.4 Reversible Adiabatic Expansion

We assume that the gas is perfectly insulated from its surroundings so that $\delta Q = 0$ at each stage of the process. Such processes are called **adiabatic** processes.[11] We allow the gas to expand quasistatically, and therefore reversibly, from a state V_2, p_2 to a state V_3, p_3. Applying the first law to each stage of this process gives

$$C_V\,\mathrm{d}T = -p\,\mathrm{d}V, \tag{2.21}$$

which by Eq. (2.12) may be rewritten in the form

$$C_V \frac{\mathrm{d}T}{T} + R\frac{\mathrm{d}V}{V} = 0, \quad \text{one mole of ideal gas.} \tag{2.22}$$

[11]Some authors use the word adiabatic to mean that $\delta Q = 0$ *and* that the process is reversible, but we use adiabatic to mean only $\delta Q = 0$. An irreversible adiabatic process is illustrated in Section 2.4.5. See Eq. (3.13) for the entropy change of an adiabatic process.

Taking the logarithmic derivative of Eq. (2.12) gives $\mathrm{d}T/T = \mathrm{d}p/p + \mathrm{d}V/V$ which allows Eq. (2.22) to be recast in the form

$$C_V \frac{\mathrm{d}p}{p} + (C_V + R)\frac{\mathrm{d}V}{V} = 0. \tag{2.23}$$

Eq. (2.23) is a differential equation for the path in the V, p plane. It can be integrated to give

$$\ln p + \gamma \ln V = \text{constant}, \tag{2.24}$$

where $\gamma := (C_V + R)/C_V = C_p/C_V > 1$. Exponentiating Eq. (2.24) gives a more usual form

$$pV^\gamma = p_2 V_2^\gamma = p_3 V_3^\gamma = \mathcal{K} = \text{constant}. \tag{2.25}$$

The path of the system is represented by the solid line in Figure 2–4. The quasistatic work is therefore

$$\mathcal{W} = \mathcal{K}\int_{V_2}^{V_3} V^{-\gamma}\,\mathrm{d}V = \mathcal{K}\left.\frac{V^{1-\gamma}}{1-\gamma}\right|_{V_2}^{V_3} = \frac{(p_3V_3 - p_2V_2)}{1-\gamma} = C_V(T_2 - T_3). \tag{2.26}$$

We have labored to produce this result which, however, could have been derived simply by applying the first law with $Q = 0$ to give $\mathcal{W} = -\Delta U = -C_V(T_3 - T_2)$. Nevertheless, we see clearly how the quasistatic work integral depends on path.

Just as Eq. (2.23) is a differential equation for the path in the V, p plane, Eq. (2.22) is the differential equation of the path in the T, V plane. It could be integrated directly, but the same result can be obtained by substitution of Eq. (2.12) into Eq. (2.25) to obtain

$$TV^{\gamma-1} = \text{constant}. \tag{2.27}$$

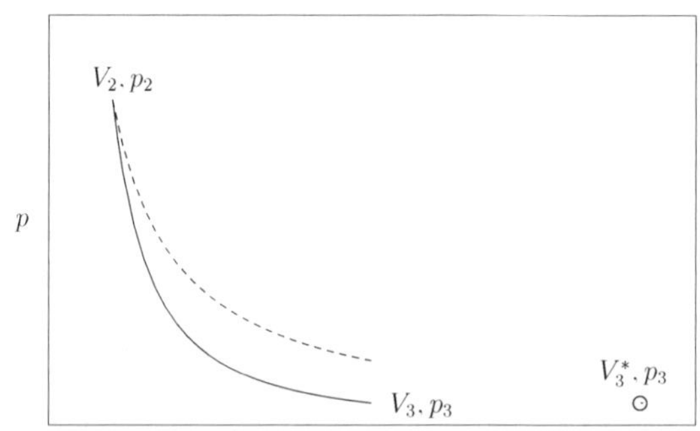

FIGURE 2–4 The solid line represents a reversible adiabatic expansion for one mole of an ideal gas for $\gamma = 5/3$. The dotted line represents, for the sake of comparison, an isothermal expansion from the same initial state. The point at V_3^*, p_3 is the final state for an irreversible adiabatic expansion at constant *external* pressure p_3. In this irreversible case, the system starts in the state V_2, p_2, "leaves the page" as it progresses through non-equilibrium states, and "reenters the page," ultimately coming to equilibrium at the state V_3^*, p_3, which is represented by a circled point.

Note that $\gamma - 1 = R/C_V$, consistent with Eq. (2.22). Similarly,

$$\frac{T}{p^{(\gamma-1)/\gamma}} = \text{constant}. \tag{2.28}$$

2.4.5 Irreversible Adiabatic Expansion

Here again we assume that the gas is perfectly insulated from its surroundings so that $\delta Q = 0$ at each stage of the process. We start out at the same state V_2, p_2 as for the reversible adiabatic process treated above, but we allow the gas to expand suddenly against a constant reduced *external* pressure p_3 that is chosen to have the same value as p_3 for the final state of the reversible adiabatic expansion considered above. During this expansion, the pressure of the gas is not well-defined, so we cannot represent this process by a path in Figure 2–4. Because this process is irreversible, it will come to equilibrium in a state having temperature T_3^* and volume V_3^* different from those for the reversible case. The work done will be $\mathcal{W} = p_3(V_3^* - V_2)$ and the change in internal energy will be $\Delta U = C_V(T_3^* - T_2)$. Since $Q = 0$ we will have $\mathcal{W} = -\Delta U$, which becomes

$$p_3(V_3^* - V_2) = C_V(T_2 - T_3^*). \tag{2.29}$$

By using Eq. (2.12), we can write $p_3 V_3^* = RT_3^*$ and $p_3 V_2 = RT_2 p_3/p_2$, in which case Eq. (2.29) can be written in the form

$$\frac{T_3^*}{T_2} = \frac{C_V + R p_3/p_2}{C_V + R} = 1 - q + qr, \tag{2.30}$$

where $r := p_3/p_2$ and $q := (\gamma - 1)/\gamma$. In this same notation, Eq. (2.28) for the reversible adiabatic expansion leads to

$$\frac{T_3}{T_2} = r^q. \tag{2.31}$$

We shall see that $T_3^* > T_3$. This is illustrated in Figure 2–5.

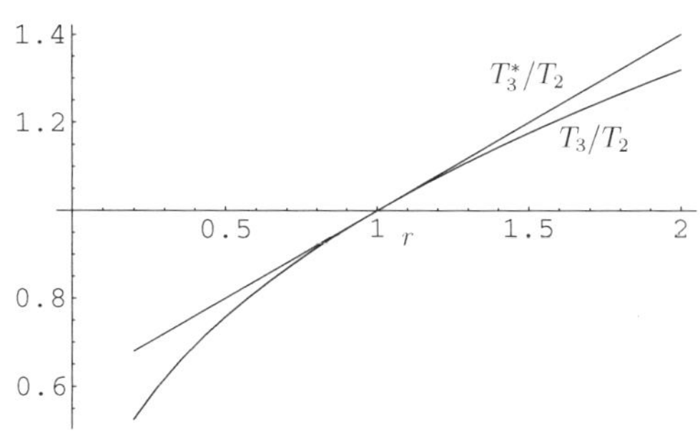

FIGURE 2–5 Graphs of T_3/T_2 for a reversible adiabatic process, Eq. (2.30), and T_3^*/T_2 for an irreversible adiabatic process, Eq. (2.31), versus $r = p_3/p_2$ for $q = 2/5$, which corresponds to $\gamma = 5/3$. The straight line corresponds to T_3^*/T_2 and shows that $T_3^* > T_3$ for $r \neq 1$.

We first note for $r = 1$ that $T_3^* = T_3 = T_2$ as expected. Then we take derivatives with respect to r to obtain

$$\frac{\mathrm{d}}{\mathrm{d}r}\left(\frac{T_3^*}{T_2}\right) = q; \quad \frac{\mathrm{d}}{\mathrm{d}r}\left(\frac{T_3}{T_2}\right) = qr^{q-1}. \tag{2.32}$$

These derivatives are also equal for $r = 1$, so the curve represented by Eq. (2.31) is tangent to the line represented by Eq. (2.30) at $r = 1$. Since $q - 1 = -1/\gamma$ is negative, we see that the slope of a graph of T_3^* versus r is less than that of T_3 versus r for any $r < 1$. Moreover, the slope of a graph of T_3^* versus r is greater than that of T_3 versus r for any $r > 1$. Hence, $T_3^* > T_3$ for any $r \neq 1$, which means that the irreversible adiabatic expansion results in a final state with greater temperature than the reversible adiabatic expansion. The same would be true for contraction, in which case $V_3 < V_2$ and $r > 1$. For the end points of the two processes, Eq. (2.12) can be written $p_3 V_3 = RT_3$ and $p_3 V_3^* = RT_3^*$. Taking the ratio of these equations gives

$$\frac{V_3^*}{V_3} = \frac{T_3^*}{T_3}. \tag{2.33}$$

From this result, we see that $V_3^* > V_3$. In summary, irreversible adiabatic expansion or contraction against a constant external pressure p_3 results in a different final state (larger temperature and volume) than a reversible adiabatic expansion to a final state[12] having pressure p_3.

2.5 Enthalpy

The **enthalpy** (sometimes called the **heat function**) is defined by

$$H := U + pV. \tag{2.34}$$

Since U, p, and V are all functions of state, H is also a function of state. In general,

$$\mathrm{d}H = \mathrm{d}U + p\,\mathrm{d}V + V\,\mathrm{d}p. \tag{2.35}$$

For quasistatic work such that Eq. (2.8) holds, Eq. (2.35) becomes

$$\mathrm{d}H = \delta Q + V\,\mathrm{d}p. \tag{2.36}$$

By combining Eqs. (2.10) and (2.36), we obtain

$$C_p := \left(\frac{\delta Q}{\mathrm{d}T}\right)_p = \left(\frac{\partial H}{\partial T}\right)_p. \tag{2.37}$$

[12]The final state depends on the details of the irreversible process. Here we have considered only a specific case and demonstrated that the final state is different from that for a reversible adiabatic process. Later we shall introduce a new state variable S, the entropy, in which case it can be shown that the entropy change for a reversible adiabatic process is zero but that for an irreversible adiabatic process is positive.

Comparison of Eq. (2.37) with Eq. (2.9) shows that H plays the same role at constant p as U does at constant V. We will see that this role is very general after developing the second law and studying Legendre transformations. In essence, the dependence of U on V is replaced by the dependence of H on p. Thus, if $Q = 0$ and $\mathcal{W} = 0$, we have $\Delta U = 0$, so energy is conserved and U is a constant. From Eq. (2.36) we see that for $\delta Q = 0$ and constant p we have $\mathrm{d}H = 0$, so H is a constant. Actually, a less restrictive condition than constant p suffices for finite changes. If $Q = 0$ and the *only* work done by the system is against a pressure reservoir with constant pressure p_r, the first law gives $\Delta U = -p_r \Delta V$ which can be written in the form

$$\Delta(U + p_r V) = 0. \tag{2.38}$$

Then if $p = p_r$ in the initial and final states of the system, Eq. (2.38) becomes

$$\Delta H = 0; \quad Q = 0 \text{ and } p = p_r \text{ in initial and final states.} \tag{2.39}$$

Example Problem 2.3. We saw above that the internal energy, U, of an ideal gas was independent of volume, V, and therefore only a function of temperature, T. Use this information together with the definition of H to show that $(\partial H/\partial p)_T = 0$, which means that the enthalpy of an ideal gas is a function of only the temperature, T.

Solution 2.3. We take the partial derivative of Eq. (2.34) while holding T constant to obtain

$$\left(\frac{\partial H}{\partial p}\right)_T = \left(\frac{\partial U}{\partial V}\right)_T \left(\frac{\partial V}{\partial p}\right)_T + V + p\left(\frac{\partial V}{\partial p}\right)_T. \tag{2.40}$$

For an ideal gas, $(\partial U/\partial V)_T = 0$ so the first term on the right vanishes. From the ideal gas law $V = NRT/p$ we obtain

$$\left(\frac{\partial V}{\partial p}\right)_T = -\frac{NRT}{p^2} = -\frac{V}{p}. \tag{2.41}$$

Hence the last two terms on the right of Eq. (2.40) cancel, and we are left with

$$\left(\frac{\partial H}{\partial p}\right)_T = 0, \quad \text{ideal gas.} \tag{2.42}$$

Actually, Eq. (2.42) follows from more elementary considerations. Substitution of the ideal gas law into Eq. (2.34) for one mole gives $H = U(T) + RT$, so we see immediately that H depends only on T. Differentiation with respect to T gives our former result $C_p = C_V + R$.

Example Problem 2.4. As heat is supplied to ice at temperature 0 °C and atmospheric pressure, the ice melts to become water, still at its melting point 0 °C, until all of the ice has melted. The heat needed to melt the ice is 80 cal/g. How much does the enthalpy change if one mole of ice is melted? Show that this is equivalent to an effective heat capacity that is a Dirac delta function at the melting point.

What would you have to know to calculate the corresponding change of the internal energy U and what would that change be?

Solution 2.4. Integration of Eq. (2.36) at constant p gives an enthalpy change $\Delta H = Q$. One mole of ice has a mass of 18 g, so $\Delta H = 18\,\mathrm{g/mol} \times 80\,\mathrm{cal/g} = 1440\,\mathrm{cal/mol}$. Since the temperature does not change during melting, an effective heat capacity can be defined formally by

$$C_p^{\mathrm{eff}} = \Delta H\,\delta(T - T_{\mathrm{M}}), \tag{2.43}$$

where T_{M} is the melting point and $\delta(T - T_{\mathrm{M}})$ is the Dirac delta function. Equation (2.43) can be justified by integration from $T_{\mathrm{M}} - \epsilon$ to $T_{\mathrm{M}} + \epsilon$. From a different perspective, a graph of H versus T has a discontinuous step at T_{M} whose formal derivative is a delta function. See Section 3.4.1 for a more thorough discussion.

From Eq. (2.15) at constant p we obtain $\Delta U = \Delta H - p\Delta V$ so we would have to know ΔV to evaluate ΔU. We can estimate ΔV as follows: The volume of ice shrinks about 9% on melting and its density is about $1\,\mathrm{g/cm^3}$. So for one mole, $\Delta V \approx -0.09 \times 1\,\mathrm{cm^3/g} \times 18\,\mathrm{g/mol} = -1.6\,\mathrm{cm^3/mol} = -1.6 \times 10^{-6}\,\mathrm{m^3/mol}$. One standard atmosphere is $p = 1.01 \times 10^5\,\mathrm{N/m^2}$. Thus $p\Delta V = -0.16\,\mathrm{J/mol} = -0.04\,\mathrm{cal/mol}$, which is a negligible correction. So $\Delta U \approx \Delta H$ for melting of ice. This is typical for melting of condensed phases. On the other hand, for the water-steam transition, $\Delta V \approx 2.24 \times 10^{-2}\,\mathrm{m^3/mol}$, roughly 1000 times larger in magnitude than for melting. So for evaporation, $p\Delta V \approx 6\,\mathrm{cal/mol}$. But for the evaporation transition, $\Delta H = 9720\,\mathrm{cal/mol}$ so the difference between ΔU and ΔH is larger but still practically negligible.

3 Second Law of Thermodynamics

Even though the first law of thermodynamics is obeyed, there are additional limitations on processes that can occur naturally. The second law of thermodynamics deals quantitatively with these limitations and is expressed in terms of an inequality that is obeyed by changes of a new state function, the **entropy** S, which is postulated to exist. These limitations are due to the fact that all natural processes in thermodynamic systems are irreversible. The boundary between natural processes and processes that are forbidden by thermodynamics can be characterized in terms of idealized processes that are reversible. For an idealized reversible process, which is hypothetical, the entropy change obeys an equality and this allows the entropy change to be calculated. If a system, by virtue of suitable constraints, is such that all natural processes are forbidden by the second law, it is in a state of thermodynamic equilibrium. This leads to a criterion for thermodynamic equilibrium in terms of the entropy.

Historically, the entropy function was discovered by studying limitations that occur during the process of transformation of heat into work, even though energy is conserved. Theoretically, these processes were imagined to be accomplished by engines that exchange heat with external heat sources, do mechanical work, and return to their original thermodynamic state after each cycle. These processes were assumed to obey the following postulates [1, p. 30]:

> ***Postulate of Kelvin:*** *"A transformation whose only final result is to transfer into work heat extracted from a source which is at the same temperature throughout is impossible."*

> ***Postulate of Clausius:*** *"A transformation whose only final result is to transfer heat from a body at a given temperature to a body at a higher temperature is impossible."*

These historical postulates forbid the existence of a process in which a virtually infinite amount of work can be obtained by extracting with 100% efficiency heat from a huge thermal source (e.g., the ocean). An engine that would accomplish such a process is sometimes called a perpetual motion machine of the second kind.[1] In fact, many people have come up with clever ideas and claims of such perpetual motion machines and have attempted to patent them, but careful analysis has always shown that some irreversible

[1]A perpetual motion machine of the first kind is one that would violate the conservation of energy itself, which is already ruled out by the first law of thermodynamics.

Thermal Physics. http://dx.doi.org/10.1016/B978-0-12-803304-3.00003-X

process occurs such that their efficiency cannot exceed the theoretical efficiency (see Eq. (3.27)) allowed by the second law.

Fermi [1, pp. 31-34] has shown that the postulates of Kelvin and Clausius are equivalent. The key phrase in each of them is "only final result." One can certainly transfer heat from a refrigerator to a room at higher temperature, but other things must change in the process, for example, work must be expended by a motor. Based on these postulates, a Carnot engine, which is a hypothetical reversible engine, and other imagined irreversible engines can be used [1, chapter IV] to develop a logical process that leads to a classical formula for the entropy (see Eq. (3.33)).

Rather than dwell on this historical justification of the second law, we shall state it as a postulate in very general terms and then relate it to its historical roots.

3.1 Statement of the Second Law

For a thermodynamic system, there exists a function of state, S, called the entropy. S is a function of a complete set of extensive state variables that includes the internal energy, U. For all *other* extensive variables held fixed, S is a monotonically increasing function of the internal energy U. For a homogeneous system, S is an extensive function and its slope $\partial S/\partial U = 1/T$, where the positive quantity T is the absolute temperature.[2] If the system is a composite system, S is the sum of the entropies of its constituent subsystems.

An **isolated system** is a chemically closed system for which $\delta Q = 0$ and $\delta \mathcal{W} = 0$, so $\mathrm{d}U = 0$ and U is a constant. Therefore also $Q = 0$, $\mathcal{W} = 0$, and $\Delta U = 0$. For an isolated system, changes of S obey the inequality

$$\Delta S \geq 0, \quad \text{isolated system, allowed changes,} \tag{3.1}$$

where the inequality corresponds to a natural irreversible process and the equality corresponds to a hypothetical idealized reversible process.

If the entropy of an isolated system is a maximum subject to its internal and external constraints, all natural irreversible processes are forbidden by Eq. (3.1) so the system is in a state of equilibrium. This leads to the following equilibrium criterion:

Entropy criterion for equilibrium: The criterion for an isolated thermodynamic system to be in internal equilibrium is that its total entropy be a maximum with respect to variation of its *internal extensive* parameters, subject to external constraints and any remaining internal constraints. Isolation constitutes the external constraints of chemical closure, perfect thermal insulation and zero external work, which require the internal energy to be constant.

For example, consider an isolated composite system consisting of two subsystems having different temperatures and separated by an insulating wall (internal constraint). If

[2] For a homogeneous system, the absolute thermodynamic temperature is *defined* by a partial derivative $1/T := \partial S/\partial U$ or alternatively by $T = \partial U/\partial S$, where all other members of the complete set of extensive variables are held constant. Thus T exists independent of any particular measuring device (thermometer). See Fermi [1, p. 45] for a related discussion in terms of the Carnot cycle.

the wall is then allowed to conduct heat (removal of an internal constraint), the energies of the two systems will change until the temperatures are equalized and a new equilibrium, corresponding to a state of higher entropy, is established.

In Chapter 6 we will discuss the application of this entropy criterion for equilibrium and deduce from it several alternative and useful criteria for equilibrium.

3.1.1 Discussion of the Second Law

The second law of thermodynamics is a postulate. The fact that it is believed to be true is based on extensive experimental testing. It can be rationalized on the basis of statistical mechanics, which of course is based on its own postulates. It can also be derived, as is done in classical thermodynamics for chemically closed systems, from other postulates of Kelvin or Clausius, as stated above. In order to make contact with the historical development of the second law and to derive equations that allow calculation of the entropy, we first digress to apply Eq. (3.1) to a composite system consisting of sources of heat and work.

We consider an isolated composite system having total entropy $S_{\rm tot}$ and apply Eq. (3.1) in the form

$$\Delta S_{\rm tot} \geq 0, \quad \text{isolated system, allowed changes.} \tag{3.2}$$

We assume that our composite system consists of a chemically closed system of interest having entropy S, a heat source having entropy S_s, and a purely mechanical system capable only of exchanging work. By definition, there is no entropy associated with this purely mechanical system, so the total entropy of our composite system is

$$S_{\rm tot} = S + S_s. \tag{3.3}$$

The heat source is assumed to be a homogeneous thermodynamic system whose only function is to exchange heat; it does no work, has a fixed number of moles of each chemical component, a temperature T_s and an internal energy U_s. Thus $\mathrm{d}S_s = (1/T_s)\mathrm{d}U_s$ by definition of the absolute temperature of the heat source. We denote by δQ a small amount of heat *extracted* from the source.[3] From the first law we have $-\delta Q = \mathrm{d}U_s$, so $\mathrm{d}S_s = -\delta Q/T_s$. Thus $\mathrm{d}S_{\rm tot} = \mathrm{d}S - \delta Q/T_s$ and for infinitesimal changes, Eq. (3.2) becomes

$$\mathrm{d}S \geq \frac{\delta Q}{T_s}, \quad \text{chemically closed system, allowed changes.} \tag{3.4}$$

In Eq. (3.4), the term chemically closed system pertains to the system of interest, having entropy S. The inequality pertains to a natural irreversible process and the equality pertains to an idealized reversible process. Thus

$$\mathrm{d}S > \frac{\delta Q}{T_s}, \quad \text{chemically closed system, natural irreversible changes.} \tag{3.5}$$

[3] δQ is assumed to be so small and the heat source has, by definition, a sufficiently large heat capacity that it remains practically unchanged during this process.

For reversible heat flow, which is an idealization that separates irreversible heat flow from forbidden heat flow, T_s can differ only infinitesimally from T, the temperature of the system, so we have

$$dS = \frac{\delta Q}{T}, \quad \text{chemically closed system, idealized reversible changes.} \tag{3.6}$$

Equations (3.5) and (3.6) are sometimes offered as a statement of the second law, although the distinction between T_s and T is not always made.[4]

If our system of interest were simply another heat source capable of no other change, we would have $dS = dU/T$ by definition of its absolute temperature. Then $\delta \mathcal{W} = 0$ so $dU = \delta Q$ from the first law and we would have $dS = \delta Q/T$. For spontaneous heat conduction, a natural irreversible process, we would need

$$dS_{\text{tot}} = dS + dS_s = \delta Q\left(\frac{1}{T} - \frac{1}{T_s}\right) > 0, \tag{3.7}$$

which results in $\delta Q(T_s - T) > 0$. This means that spontaneous heat conduction, with no other change, occurs only from a higher temperature to a lower temperature, in agreement with our intuition and the postulate of Clausius stated above.

For finite changes, we can integrate Eq. (3.4) to obtain

$$\Delta S \geq \int \frac{\delta Q}{T_s}, \quad \text{chemically closed system, allowed changes,} \tag{3.8}$$

where the equality sign is for a reversible process and requires $T_s = T$. Our system of interest can do work (on the mechanical subsystem) of amount

$$\mathcal{W} = -\Delta U + \int \delta Q, \tag{3.9}$$

provided that Eq. (3.8) is satisfied. We emphasize that our system of interest is not isolated, so its entropy can be made to *decrease* by extracting heat reversibly. Therefore, if a chemically closed system is not isolated, its entropy can increase or decrease, and the process that brings about this change can be either reversible or irreversible, depending on the relationship of ΔS to $\int \delta Q/T_s$ for that process.

In classical thermodynamics, one often speaks of **heat reservoirs**. A heat reservoir is a heat source with such a large heat capacity that its temperature remains constant.[5] If the heat source in Eq. (3.8) is replaced by a heat reservoir of temperature T_r from which an amount of heat Q_r is extracted, we obtain

$$\Delta S \geq \frac{Q_r}{T_r}, \quad \text{chemically closed system, allowed changes.} \tag{3.10}$$

[4]See the footnote on page 48 of Fermi [1] for further discussion of T_s. Some books [5, 16] write $dS > \delta Q/T$ which is more restrictive than Eq. (3.5); such an equation applies to a process in which the heat conduction between the heat source and the system of interest is reversible but other processes that take place within the system of interest are irreversible.

[5]For example, if a heat source has a constant heat capacity C_r and an amount of heat Q_r is extracted from it, its temperature would change by $\Delta T_r = -Q_r/C_r$. For a reservoir, C_r is assumed to be so large that ΔT_r can be made arbitrarily small, and therefore zero for all practical purposes.

If the heat source consists of a number of such reservoirs, Eq. (3.8) becomes

$$\Delta S \geq \sum_r \frac{Q_r}{T_r}, \quad \text{chemically closed system, allowed changes} \tag{3.11}$$

and Eq. (3.9) is replaced by

$$\mathcal{W} = -\Delta U + \sum_r Q_r. \tag{3.12}$$

If the amounts of heat Q_r in Eqs. (3.11) and (3.12) are very small, the sums can be replaced by integrals, and the result is essentially the same as Eqs. (3.8) and (3.9).

A system surrounded by perfectly insulating walls requires $\delta Q = 0$ and is said to be **adiabatic**. For an adiabatic system, Eq. (3.8) becomes

$$\Delta S \geq 0, \quad \text{chemically closed adiabatic system, allowed changes.} \tag{3.13}$$

But Eq. (3.9) yields $\mathcal{W} = -\Delta U$, so such a system is not isolated and can still do work. Chandler [12, p. 8] states the second law by means of Eq. (3.13) which applies to transformations that are adiabatically accessible, those corresponding to the inequality being irreversible and those corresponding to the equality being reversible.

For a cyclic process, the system returns to its original state after each cycle. Since S is a state function, $\Delta S = 0$ for a cyclic process and Eq. (3.11) becomes

$$0 \geq \sum_r \frac{Q_r}{T_r}, \quad \text{cyclic process, chemically closed system, allowed changes.} \tag{3.14}$$

For a continuous distribution of reservoirs,

$$0 \geq \oint \frac{\delta Q}{T_r}, \quad \text{cyclic process, chemically closed system, allowed changes.} \tag{3.15}$$

For an adiabatic cyclic process, $\delta Q = 0$, so Eq. (3.15) becomes $0 \geq 0$ and compatibility would require the equality sign to hold, consistent with the fact that an adiabatic cyclic process is reversible.

3.2 Carnot Cycle and Engines

In classical thermodynamics, the second law of thermodynamics is usually rationalized by considering processes involving the conversion of work to heat by engines that return to their original thermodynamic state after one cycle. Comparison is made to a hypothetical engine, known as a Carnot engine, which is imagined to execute a reversible cycle. The **Carnot cycle** pertains to an idealized engine in which the working substance is one mole[6] of an ideal gas. There are four segments to the cycle, as depicted in Figure 3–1. All segments involve reversible processes, so the whole cycle is reversible. Segment AB is a reversible isothermal expansion in which an amount of heat $|Q_2| = Q_2$ is extracted from a heat source at a high temperature T_2. Segment BC is a reversible adiabatic expansion. Segment CD

[6]We could also consider any fixed number of moles of an ideal gas.

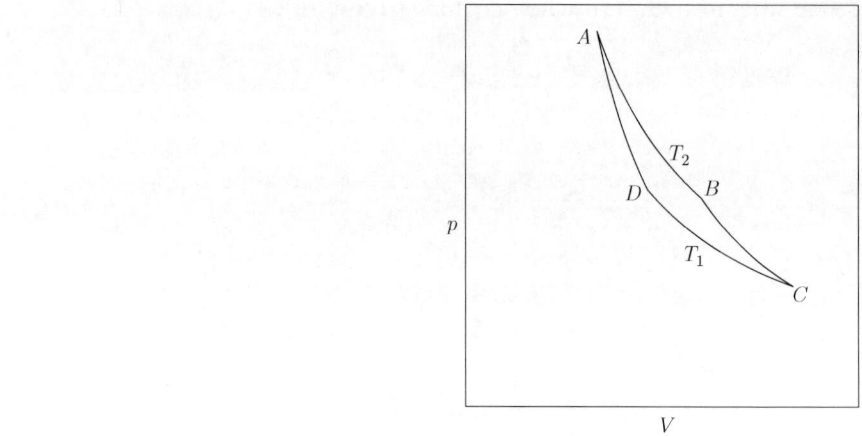

FIGURE 3–1 The Carnot cycle in the V,p plane. The working substance is an ideal gas and the cycle consists of four reversible segments. *AB* is isothermal expansion at temperature T_2, *BC* is adiabatic expansion, *CD* is isothermal compression at temperature T_1, and *DA* is adiabatic compression. The figure is drawn for $\gamma = 5/3$.

is a reversible isothermal compression in which an amount of heat $|Q_1| = -Q_1$ is given up to a heat sink at temperature T_1. Both the source and the sink are assumed to be heat reservoirs, so their temperatures do not change. Finally, segment *DA* is a reversible adiabatic compression. In order for these segments to form a closed cycle, we can apply Eq. (2.27) to each of the adiabatic segments to obtain

$$T_2 V_B^{\gamma-1} = T_1 V_C^{\gamma-1}; \quad T_2 V_A^{\gamma-1} = T_1 V_D^{\gamma-1}. \tag{3.16}$$

Division of one of these equations by the other and extraction of the $\gamma - 1$ root gives

$$\frac{V_A}{V_B} = \frac{V_D}{V_C}. \tag{3.17}$$

Combining Eq. (3.17) with the ideal gas law gives

$$\frac{p_A}{p_B} = \frac{p_D}{p_C}, \tag{3.18}$$

so the geometry of the cycle is completely known and simple to express.

On the adiabatic segment *BC*, $\delta Q = 0$ so we have $\mathcal{W}_{BC} = -\Delta U_{BC} = C_V(T_2 - T_1)$. This exactly cancels the work $C_V(T_1 - T_2)$ done by the gas on the other adiabatic segment. The work done by the gas on the isothermal expansion segment *AB* is $RT_2 \ln(V_B/V_A)$. Recalling that U depends only on T for an ideal gas means that $\Delta U_{AB} = 0$ for that segment, so

$$|Q_2| = RT_2 \ln(V_B/V_A). \tag{3.19}$$

Similarly, for the isothermal compression segment *CD*, we obtain

$$|Q_1| = -RT_1 \ln(V_D/V_C) = RT_1 \ln(V_B/V_A), \tag{3.20}$$

where Eq. (3.17) has been used in the last step. Dividing Eq. (3.19) by Eq. (3.20) we obtain

$$\frac{|Q_1|}{T_1} = \frac{|Q_2|}{T_2}. \tag{3.21}$$

For the entire cycle, $\Delta U = 0$ so the total work done by the gas during the cycle is $\mathcal{W} = |Q_2| - |Q_1|$. The efficiency of the cycle is therefore

$$\eta := \frac{\mathcal{W}}{|Q_2|} = 1 - \frac{|Q_1|}{|Q_2|} = 1 - \frac{T_1}{T_2}. \tag{3.22}$$

This efficiency is always less than unity except for a heat sink at absolute zero, which is deemed to be impossible.

Let us examine the meaning of Eq. (3.21) in terms of the second law. Since the entropy is a function of state, we have $\Delta S = 0$ for a cycle. Applying Eq. (3.11) with the equality, for our reversible cycle, we obtain

$$0 = \frac{Q_2}{T_2} + \frac{Q_1}{T_1} = \frac{|Q_2|}{T_2} - \frac{|Q_1|}{T_1} \tag{3.23}$$

in agreement with Eq. (3.21).

Beginning with the Carnot cycle, Fermi [1, chapter IV] proves a number of other things based on the Kelvin/Clausius postulates. These are used to rationalize the existence of the entropy and to formulate the second law. Here, we take the opposite approach by quoting the main results and demonstrating how they follow from the second law.

- *Any reversible engine working between the same two temperatures T_2 and T_1 has the same efficiency as a Carnot engine.* We follow the same procedure as we did in deriving Eq. (3.23) except that the amounts of heat are now $|Q_2'|$ and $|Q_1'|$ which might differ from those for a Carnot engine. Thus we obtain

$$0 = \frac{Q_2'}{T_2} + \frac{Q_1'}{T_1} = \frac{|Q_2'|}{T_2} - \frac{|Q_1'|}{T_1}. \tag{3.24}$$

It follows that the *ratio* $|Q_1'|/|Q_2'| = T_1/T_2$ is the same as for a Carnot engine. From Eq. (3.12) with $\Delta U = 0$, the amount of work done in the cycle is now $\mathcal{W}' = |Q_2'| - |Q_1'|$, so

$$\eta' := \frac{\mathcal{W}'}{|Q_2'|} = 1 - \frac{|Q_1'|}{|Q_2'|} = 1 - \frac{T_1}{T_2} = \eta. \tag{3.25}$$

- *Any irreversible engine working between the same two temperatures T_2 and T_1 has a smaller efficiency than a Carnot engine.* This result follows by applying Eq. (3.11) with the inequality to obtain (superscript *i* for irreversible)

$$0 > \frac{Q_2^{\mathrm{i}}}{T_2} + \frac{Q_1^{\mathrm{i}}}{T_1} = \frac{|Q_2^{\mathrm{i}}|}{T_2} - \frac{|Q_1^{\mathrm{i}}|}{T_1}, \tag{3.26}$$

which leads to $|Q_1^{\mathrm{i}}|/|Q_2^{\mathrm{i}}| > T_1/T_2$. The amount of work done in the cycle is now $\mathcal{W}^{\mathrm{i}} = |Q_2^{\mathrm{i}}| - |Q_1^{\mathrm{i}}|$, resulting in

$$\eta^{\mathrm{i}} := \frac{\mathcal{W}^{\mathrm{i}}}{|Q_2^{\mathrm{i}}|} = 1 - \frac{|Q_1^{\mathrm{i}}|}{|Q_2^{\mathrm{i}}|} < \eta. \tag{3.27}$$

- *In a cycle of any reversible engine that receives heat δQ from a number of sources at temperature T,*

$$\oint \frac{\delta Q}{T} = 0. \tag{3.28}$$

This follows from Eq. (3.8) with the equality by recognizing that $\Delta S = 0$ for a cycle. In classical thermodynamics, Eq. (3.28) is deduced by arguing that any reversible cycle can be approximated to arbitrary accuracy by a very large number of small Carnot cycles. It is actually Eq. (3.28) that was used to deduce that a state function, now known as the entropy, exists. By integrating from point A to point B along some reversible path and the back again to A along some other reversible path, we create a reversible cycle. Since the integral from B to A along the return path is the negative of the integral from A to B along that path, it follows that

$$\left(\int_A^B \frac{\delta Q}{T}\right)_{\text{reversible path I}} = \left(\int_A^B \frac{\delta Q}{T}\right)_{\text{reversible path II}}. \tag{3.29}$$

Since the values of the integrals in Eq. (3.29) depend only on their end points, their integrand must be the differential of some function, namely $\mathrm{d}S = \delta Q/T$, which is Eq. (3.6). In mathematics, $1/T$ would be called an integrating factor for δQ.

- *In a cycle of any irreversible engine that receives heat δQ from a number of sources at temperature T_s,*

$$\oint \frac{\delta Q}{T_s} < 0. \tag{3.30}$$

This follows from Eq. (3.11) with the inequality by recognizing that $\Delta S = 0$ for a cycle.

Example Problem 3.1. Analyze a Carnot refrigerator in which heat $|Q_1| = Q_1$ is extracted (from the refrigerator) at a low temperature T_1 and given to a Carnot engine running in reverse; then $|Q_2| = -Q_2$ is extracted from that Carnot engine and given to a sink at higher temperature T_2.

Solution 3.1. The magnitudes $|Q_2|$ and $|Q_1|$ are still given, respectively, by Eqs. (3.19) and (3.20), so Eq. (3.21) still applies. But now an amount of work $W = -\mathcal{W} > 0$ must be done **on** the system, where $\mathcal{W} = Q_1 + Q_2 = |Q_1| - |Q_2| = -W$. Thus

$$\frac{|Q_1|}{W} = \frac{T_1}{T_2 - T_1}. \tag{3.31}$$

We see that only a small amount of work W must be provided to extract $|Q_1|$ from the refrigerator provided that T_1 is not too much lower than T_2. Since an amount of heat $|Q_2| = |Q_1|(T_2/T_1)$ must be given up to the source, the cooling of a refrigerator can result in a large amount of heat given up to the surrounding room. Of course the process that takes place in an actual refrigerator is irreversible, so even a larger ratio of the removed heat to the work W is required than given by Eq. (3.31). Indeed, by using Eq. (3.26) for an irreversible engine, we obtain the inequality $|Q_1^{\mathrm{i}}|/W < T_1/(T_2 - T_1)$.

The considerations that led to Eq. (3.31) can also be applied to analyze a heat pump that adds an incremental amount of heat from an inexpensive source at temperature T_1 to heat a room at temperature T_2. In that case, a more meaningful quantity is

$$\frac{|Q_2|}{W} = \frac{T_2}{T_2 - T_1}. \tag{3.32}$$

Thus the heat pump will require only a small amount of work to provide $|Q_2|$ if the source temperature T_1 is close to T_2. For a real (irreversible) heat pump we would have $|Q_2^{\mathrm{i}}|/W < T_2/(T_2 - T_1)$.

3.3 Calculation of the Entropy Change

From Eq. (3.29) it follows that the change in entropy of a system that begins in state A and ends in state B is given by

$$\Delta S \equiv S(B) - S(A) = \int_A^B \frac{\delta Q}{T}, \quad \text{any reversible path connecting } A \text{ and } B. \tag{3.33}$$

In Eq. (3.33), we emphasize that the path of integration is *any reversible path*. Since S is function of state, the entropy change $S(B)-S(A)$ will be the same no matter how the system changes from A to B, for example by an irreversible process, but it can only be calculated by using a reversible path. In practice, one uses some convenient reversible path to make the computation simple. Equation (3.33) only defines the difference in entropy between states. We could choose some standard state O and then calculate the differences $S(A) - S(O)$ and $S(B) - S(O)$. Later we will encounter the third law of thermodynamics, according to which there is a standard state whose entropy can be taken to be zero.

Example Problem 3.2. The heat capacity at constant volume of a number of substances can be represented empirically by an equation of the form

$$C_V = a + bT + cT^2, \tag{3.34}$$

where a, b, and c are constants. Calculate the change in internal energy and the change in entropy when the temperature changes from T_1 to T_2 at constant volume.

Solution 3.2. At constant volume, we have $\mathrm{d}U = C_V\,\mathrm{d}T$ and $\mathrm{d}U = \delta Q = T\,\mathrm{d}S$. Thus,

$$\Delta U = U_2 - U_1 = \int_{T_1}^{T_2} C_V\,\mathrm{d}T = aT + bT^2/2 + cT^3/3\Big|_{T_1}^{T_2} \tag{3.35}$$

and

$$\Delta S = S_2 - S_1 = \int_{T_1}^{T_2} C_V/T\,\mathrm{d}T = a\ln T + bT + cT^2/2\Big|_{T_1}^{T_2}. \tag{3.36}$$

Example Problem 3.3. Consider an isolated composite system consisting of two subsystems, (1) and (2) respectively, having fixed volumes V_1 and V_2 and heat capacities at constant volume at temperature T of $C_1(T)$ and $C_2(T)$. Suppose that the subsystems are separated initially by an insulating wall and are at equilibrium with initial temperatures $T_1 < T_2$. Then let very small amounts of energy pass very slowly through the wall by heat transfer so that each subsystem passes through a series of equilibrium states until the system comes to a final equilibrium state. Calculate the temperatures of the subsystems at each stage of the process and study the total entropy change until a maximum entropy has been reached.

Solution 3.3. At some intermediate stage of the process, the changes in energy and entropy will be given by

$$0 = \Delta(U) = \int_{T_1}^{T_1^*} C_1(T)\,\mathrm{d}T + \int_{T_2}^{T_2^*} C_2(T)\,\mathrm{d}T \tag{3.37}$$

and

$$\Delta(S) = \int_{T_1}^{T_1^*} \frac{C_1(T)}{T}\,\mathrm{d}T + \int_{T_2}^{T_2^*} \frac{C_2(T)}{T}\,\mathrm{d}T \geq 0. \tag{3.38}$$

Then take differentials of these expressions to obtain

$$0 = C_1(T_1^*)\,\mathrm{d}T_1^* + C_2(T_2^*)\,\mathrm{d}T_2^* \tag{3.39}$$

and

$$\mathrm{d}\Delta(S) = \frac{C_1(T_1^*)}{T_1^*}\,\mathrm{d}T_1^* + \frac{C_2(T_2^*)}{T_2^*}\,\mathrm{d}T_2^* \geq 0. \tag{3.40}$$

Substitution of Eq. (3.39) into Eq. (3.40) gives

$$\mathrm{d}\Delta(S) = C_1(T_1^*)\left(\frac{1}{T_1^*} - \frac{1}{T_2^*}\right)\mathrm{d}T_1^* \geq 0, \tag{3.41}$$

which for positive $\mathrm{d}T_1^*$ requires $(1/T_1^* - 1/T_2^*) \geq 0$. In view of Eq. (3.37), this requires $T_1 < T_1^* \leq T_2^* < T_2$ at each stage of the process. When T_1^* increases to T_2^*, $\mathrm{d}\Delta(S) = 0$ and S will reach its maximum value at some new equilibrium temperature $T_1^* = T_2^* = T_{\mathrm{eq}}$. This can be seen in principle by integrating Eq. (3.41) from T_1 to T_{eq}, but that would require specification of $C_1(T)$ and $C_2(T)$ to enable T_2^* to be expressed as a function of T_1^*. Nevertheless, the final result will satisfy

$$0 = \Delta(U) = \int_{T_1}^{T_{\mathrm{eq}}} C_1(T)\,\mathrm{d}T + \int_{T_2}^{T_{\mathrm{eq}}} C_2(T)\,\mathrm{d}T \tag{3.42}$$

and

$$\Delta(S) = \int_{T_1}^{T_{\mathrm{eq}}} \frac{C_1(T)}{T}\,\mathrm{d}T + \int_{T_2}^{T_{\mathrm{eq}}} \frac{C_2(T)}{T}\,\mathrm{d}T > 0. \tag{3.43}$$

For the simple case when C_1 and C_2 are independent of T, the reader is invited to carry out these calculations explicitly.

3.4 Combined First and Second Laws

For a chemically closed system, the first law gives

$$dU = \delta Q - \delta W. \tag{3.44}$$

For a simple[7] homogeneous isotropic system for which U depends only on S and V,

$$dU = \left(\frac{\partial U}{\partial S}\right)_V dS + \left(\frac{\partial U}{\partial V}\right)_S dV. \tag{3.45}$$

For a reversible transformation in this system, for which the only work is the quasistatic work, we have

$$\delta Q = T\,dS; \quad \delta W = p\,dV; \quad \text{reversible}. \tag{3.46}$$

Substitution of Eq. (3.46) into Eq. (3.44) gives

$$dU = T\,dS - p\,dV. \tag{3.47}$$

We can therefore identify the derivatives

$$T = \left(\frac{\partial U}{\partial S}\right)_V; \quad -p = \left(\frac{\partial U}{\partial V}\right)_S. \tag{3.48}$$

We emphasize that Eq. (3.47) holds for all infinitesimal changes of $U(S, V)$ within the field of equilibrium states. Equation (3.46), which is only true for reversible processes, was only used to identify the derivatives in Eq. (3.45). Equations that give explicit forms of the functions $T(S, V)$ and $p(S, V)$ are known as **equations of state**. If all[8] equations of state are known, Eq. (3.47) can be integrated to recover the function $U(S, V)$, except for an additive constant which has to do with the arbitrary zero of energy. If the second partial derivatives of U are continuous, as we shall assume to be the case for thermodynamic functions, the order of partial differentiation does not matter and we obtain

$$\left(\frac{\partial T}{\partial V}\right)_S = \frac{\partial^2 U}{\partial V \partial S} = \frac{\partial^2 U}{\partial S \partial V} = -\left(\frac{\partial p}{\partial S}\right)_V. \tag{3.49}$$

$(\partial T/\partial V)_S = -(\partial p/\partial S)_V$ is an example of a **Maxwell relation**. In Chapter 5 we will take up Maxwell relations for systems that depend on several variables.

Since Eq. (3.44) holds even for irreversible transformations and Eq. (3.47) is generally true, we can eliminate dU to obtain

$$p\,dV - \delta W = T\,dS - \delta Q. \tag{3.50}$$

[7]Note that Eqs. (3.45) and (3.47) hold only for a chemically closed system in which no chemical reactions are occurring. If chemical reactions are allowed, U would depend on additional variables (progress variables of the reactions). Equation (3.6) would not hold if these reactions were irreversible. See Eq. (5.128) for further clarification.

[8]For open systems, one must include the numbers of moles of each chemical component, $N_1, N_2, \ldots, N_\kappa$ as additional variables in U, in which case there are more equations of state (see Chapter 5). In general, U depends on a complete set of extensive state variables.

For reversible transformations, Eq. (3.46) holds and both sides of Eq. (3.50) are zero. But for an irreversible process, Eq. (3.46) no longer applies. Instead, Eq. (3.5) applies and Eq. (3.50) leads to an interesting inequality. We divide Eq. (3.50) by T and rearrange to obtain

$$\frac{p\,\mathrm{d}V - \delta W}{T} + \frac{\delta Q}{T} = \mathrm{d}S. \tag{3.51}$$

Then we subtract $\delta Q/T_s$ from both sides of Eq. (3.51) and apply Eq. (3.5) to obtain

$$\frac{p\,\mathrm{d}V - \delta W}{T} + \delta Q\left(\frac{1}{T} - \frac{1}{T_s}\right) = \mathrm{d}S - \frac{\delta Q}{T_s} > 0, \quad \text{natural, irreversible.} \tag{3.52}$$

The first term on the left of Eq. (3.52) is due to the process of irreversible work and the second term on the left is due to the irreversible process of heat conduction between the external source and the system. These terms can be regarded [16, pp. 95-95] as representing entropy production during independent irreversible processes and are *separately* positive. A positive value of the first term leads to the inequality $W < p\,\mathrm{d}V$, in agreement with Eq. (2.5). If we substitute $W = p_{\mathrm{ext}}\,\mathrm{d}V$ where p_{ext} is an effective pressure of purely mechanical origin as in Eq. (2.3), this work inequality becomes $(p - p_{\mathrm{ext}})\,\mathrm{d}V > 0$. The second term is the same as in Eq. (3.7), derived for the case in which the system was considered to be a heat source that could do no work.

We can rearrange Eq. (3.47) in the form

$$\mathrm{d}S = \frac{1}{T}\,\mathrm{d}U + \frac{p}{T}\,\mathrm{d}V \tag{3.53}$$

from which it follows that

$$\frac{1}{T} = \left(\frac{\partial S}{\partial U}\right)_V; \quad \frac{p}{T} = \left(\frac{\partial S}{\partial V}\right)_U. \tag{3.54}$$

Equations that give $1/T$ and p/T as functions of U and V are also equations of state. If we know these functions, Eq. (3.53) can be integrated to recover $S(U, V)$. We also have the Maxwell relation $(\partial(1/T)/\partial V)_U = \left(\partial(p/T)/\partial U\right)_V$.

Since the entropy is postulated to be a monotonically increasing function of the internal energy, the internal energy is also a monotonically increasing function of the entropy. The inverse transformation between $S(U, V)$ and $U(S, V)$ is therefore unique, and either of these functional forms can be chosen to give a complete representation of the thermodynamic system.[9] One speaks of the entropy representation $S(U, V)$ or the energy representation $U(S, V)$. Either of these equations can be regarded as a **fundamental equation of the system** and either contains complete information about the system.

[9]For more complicated systems, both S and U depend on an additional set of extensive variables, but these behave just like V.

Example Problem 3.4. For a hypothetical thermodynamic system, $T = (4/A)(U/V)^{3/4}$ and $p = 3U/V$, where A is a constant. Find the fundamental equation in the entropy representation.

Solution 3.4. We readily calculate $1/T = (A/4)(V/U)^{3/4}$ and $p/T = (3A/4)(U/V)^{1/4}$ so Eq. (3.53) takes the form

$$dS = (A/4)(V/U)^{3/4}\, dU + (3A/4)(U/V)^{1/4}\, dV, \tag{3.55}$$

which integrates to give $S = AU^{1/4}V^{3/4} + S_0$, where S_0 is a constant.

Example Problem 3.5. This problem concerns one mole of an ideal monatomic gas that obeys the equation $pV = RT$, where p is the pressure, V is the volume, T is absolute temperature, and R is the universal gas constant. The gas has a heat capacity (per mole) at constant volume of $C_V = (3/2)R$. In its initial state, it is in equilibrium at temperature T_1 and volume V_1 in the left chamber of a box, as shown in Figure 3–2. The right chamber of the box, which has volume $V_2 - V_1$, is initially evacuated. The two chambers are surrounded by exterior walls that are rigid and impenetrable. The chambers are separated *initially* by an interior wall that is rigid, impenetrable, and insulating. Under various conditions detailed below, the gas is allowed to expand and finally comes to equilibrium in the total volume V_2.

Apply the first and second laws of thermodynamics, the definition of C_V, the ideal gas equation of state, and integration to answer the following questions.

(a) Suppose, by whatever means, that the gas expands into the total volume V_2 and comes to equilibrium at temperature T_2. What is the change, ΔS, in entropy of the gas from its initial to its final state?

(b) The entire system is maintained at constant temperature T_1 by contact with a heat reservoir. The gas is allowed to expand by means of an external agent that moves the internal wall separating the chambers very slowly (such that the gas is practically in equilibrium at each stage of the process) until the gas occupies the entire volume V_2. What is the change, ΔU, in its internal energy? How much external work, $\mathcal{W}$, does the system do on the external agent that moves the wall? How much heat, Q, is added to the system during this process? Compare Q to the relevant ΔS and deduce whether this process is reversible or irreversible.

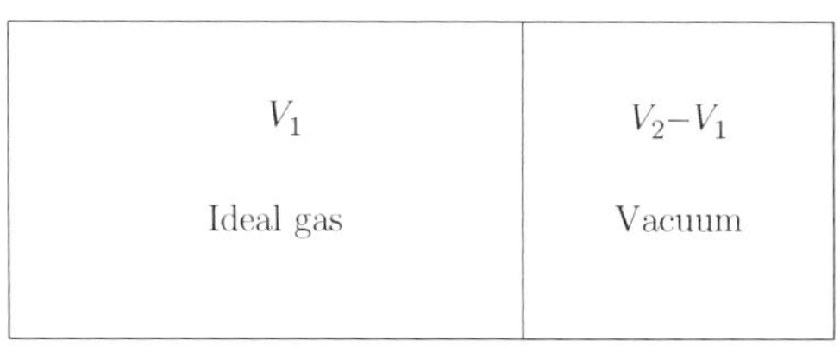

FIGURE 3–2 A monatomic ideal gas at temperature T_1 initially occupies the left chamber of the box. The right chamber of the box, which has volume $V_2 - V_1$, is evacuated. The interior wall that separates the gas from the evacuated chamber is rigid, impenetrable and insulating, but can be moved or ruptured.

(c) The entire system is insulated and the wall separating the chambers is suddenly ruptured, allowing the gas to fill the entire volume V_2. How much external work, $\mathcal{W}$, does the system do? What is the final temperature of the gas? Compare Q to the relevant ΔS and deduce whether this process is reversible or irreversible.

(d) The entire system is insulated. The gas is allowed to expand by means of an external agent which moves the internal wall separating the chambers very slowly (such that the gas is practically in equilibrium at each stage of the process) until the gas occupies the entire volume V_2. What is the final temperature, T_2, of the gas? Compare Q to the relevant ΔS and deduce whether this process is reversible or irreversible.

Solution 3.5.

(a) Since S is a state function, ΔS depends only on the initial and final states of the system, irrespective of how the system gets from the initial state to the final state. We substitute the ideal gas law and the equation $\mathrm{d}U = C_V\,\mathrm{d}T$ into Eq. (3.53) to obtain

$$\mathrm{d}S = C_V \frac{\mathrm{d}T}{T} + R\frac{\mathrm{d}V}{V}, \tag{3.56}$$

which integrates to give

$$\Delta S = C_V \ln(T_2/T_1) + R\ln(V_2/V_1), \quad \text{one mole of ideal gas.} \tag{3.57}$$

(b) U depends only on T for an ideal gas, so $\Delta U = 0$. Thus from the first law, $\mathcal{W} = Q$. Since the work is quasistatic, $\mathcal{W} = \int p\,\mathrm{d}V$ where the integral is to be carried out along an isothermal path $T = T_1$. Therefore we can use $p = RT_1/V$ and take the constants RT_1 outside the integral to obtain

$$Q = \mathcal{W} = RT_1 \int_{V_1}^{V_2} \frac{\mathrm{d}V}{V} = RT_1 \ln(V_2/V_1). \tag{3.58}$$

Since $T_2 = T_1$ for this process, part (a) gives $\Delta S = R\ln(V_2/V_1)$ so

$$\Delta S = Q/T_1 \tag{3.59}$$

and the process is reversible (as expected for quasistatic work). Note that the entropy increases for this reversible process. In this case, entropy increase does not automatically imply irreversibility *because the system is not isolated.* Similarly, for reversible adiabatic contraction, both Q and ΔS are negative, and the entropy of the system decreases. This does not violate Eq. (3.1) *because the system is not isolated.*

(c) $\mathcal{W} = 0$ because the outer wall is rigid and there is no way to do mechanical work on the environment of the system. Since $Q = 0$, we conclude from the first law that $\Delta U = 0$.

Since U depends only on T, we have $T_2 = T_1$. (During the process itself, which we shall see is irreversible, T is at best inhomogeneous and probably undefined.) The change in entropy, from part (a), is again $\Delta S = R\ln(V_2/V_1) > 0$. Therefore, since $\delta Q = 0$ at every stage of the process,

$$\Delta S > \int \frac{\delta Q}{T} = 0, \tag{3.60}$$

so the process is irreversible as expected.

(d) $Q = 0$ because the system is insulated. The work is quasistatic so $\delta W = p\,dV$, and since $\delta Q = 0$ at each stage of the process, the first law gives $dU + p\,dV = 0$. Since $dU = C_V\,dT$, this becomes $C_V\,dT + RT\,dV/V = 0$. Division by T (which is *not* constant in this process) yields $C_V\,dT/T + R\,dV/V = 0$ which integrates to give

$$C_V \ln(T_2/T_1) + R\ln(V_2/V_1) = 0. \tag{3.61}$$

Thus, $\Delta S = 0$ and

$$\Delta S = \int \frac{\delta Q}{T} = 0,$$

so the process is reversible and isentropic, as expected for this quasistatic process with adiabatic walls. By means of Eq. (2.27), the final temperature can be written more succinctly as $T_2 = T_1(V_1/V_2)^{2/3}$, so the temperature drops, as expected, for this reversible adiabatic expansion.

3.4.1 Latent Heat

When a substance melts or evaporates, heat must be supplied to partially or totally break atomic bonds and rearrange structure, and hence to change the phase to a state of higher disorder, which we shall see later is a state of higher entropy. Melting and vaporization processes are generally carried out at constant pressure, for example, atmospheric pressure. The heat needed to change the phase reversibly at constant pressure and temperature is known as **latent heat**. Heat must be supplied when a solid melts to become a liquid (heat of melting); the same amount is given up when a liquid freezes to become a solid (latent heat of fusion). When a liquid becomes a gas, it is necessary to supply heat (heat of vaporization); when a gas condenses to become a liquid, the same amount of heat is given up (latent heat of condensation). These are generally reported as positive quantities, usually per mole or per unit mass.

Consider, for example, the melting of ice, which takes place at atmospheric pressure at a temperature of $0\,°C = 273.15\,K$. As we supply heat to cold ice, it is warmed from below its melting point to 273.15 K where melting occurs and water begins to form. As heat continues to be supplied, the ice-water mixture remains at 273.15 K until all of the ice melts. This requires 80 calories of heat per gram of ice, the latent heat of fusion. Further heating causes the temperature of the water to rise.

Processes such as this, which take place at constant pressure, may be analyzed conveniently in terms of the enthalpy, $H = U + pV$ previously introduced in connection with the first law (see Section 2.5). We saw that $dH = dU + p\,dV + V\,dp$ which in view of Eq. (3.47) becomes

$$dH = T\,dS + V\,dp. \tag{3.62}$$

But at constant pressure we have

$$dH = C_p\,dT, \tag{3.63}$$

where C_p is the heat capacity at constant pressure. Equation (3.63) applies in the absence of phase change, say for $T_{\rm I} \le T < T_{\rm M}$ and also for $T_{\rm M} < T \le T_{\rm W}$, where $T_{\rm I}$ is the initial temperature of the ice, $T_{\rm M}$ is the melting point and $T_{\rm W}$ is the final temperature of the water. At $T = T_{\rm M}$, H increases by the amount $\Delta H_{\rm M}$, the latent heat of fusion. The total change in H is therefore

$$\Delta H = \int_{T_{\rm I}}^{T_{\rm M}} C_p(\text{ice})\, \mathrm{d}T + \Delta H_{\rm M} + \int_{T_{\rm M}}^{T_{\rm W}} C_p(\text{water})\, \mathrm{d}T. \tag{3.64}$$

ΔH as a function of T is shown in Figure 3–3a. Formally, the effective heat capacity at the melting point can be represented as a delta function (the formal derivative of a step function) as shown in Example Problem 2.4.

By combining Eq. (3.62) with Eq. (3.63) at constant pressure, we obtain

$$\mathrm{d}S = \frac{C_p}{T}\, \mathrm{d}T, \tag{3.65}$$

which can be integrated to find the entropy change that occurs prior to melting and subsequent to melting. During the melting itself, we integrate[10] Eq. (3.62) at constant p to obtain $\Delta S_{\rm M} = \Delta H_{\rm M}/T_{\rm M}$, which is called the entropy of fusion. Therefore, the total change of entropy is given by

$$\Delta S = \int_{T_{\rm I}}^{T_{\rm M}} \frac{C_p(\text{ice})}{T}\, \mathrm{d}T + \frac{\Delta H_{\rm M}}{T_{\rm M}} + \int_{T_{\rm M}}^{T_{\rm W}} \frac{C_p(\text{water})}{T}\, \mathrm{d}T. \tag{3.66}$$

ΔS as a function of T is shown in Figure 3–3b.

If the range of temperature is not large, $C_p(\text{ice})$ and $C_p(\text{water})$ can be considered to be practically independent of T, so we have the simplifications

$$\Delta H \approx C_p(\text{ice})(T_{\rm M} - T_{\rm I}) + \Delta H_{\rm M} + C_p(\text{water})(T_{\rm W} - T_{\rm M}) \tag{3.67}$$

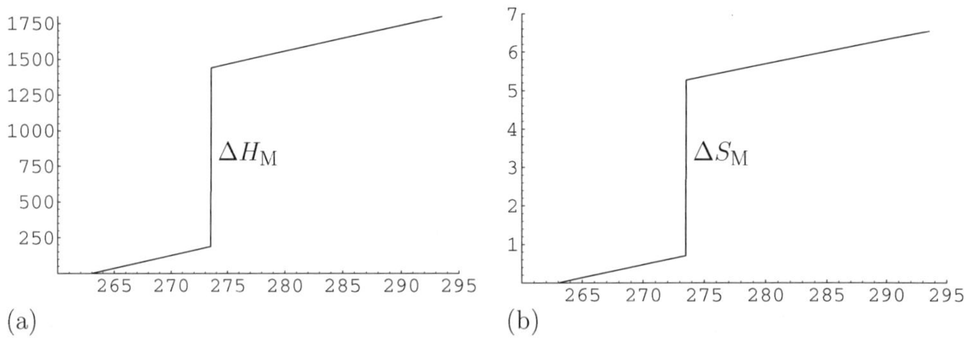

FIGURE 3–3 (a) Enthalpy change ΔH in cal/mol and (b) entropy change ΔS in cal/(mol K) as a function of temperature T in K for melting of ice. The curvature of the logarithms in ΔS is not apparent on this scale. The jumps are related by $\Delta H_{\rm M} = T_{\rm M}\Delta S_{\rm M}$. (a) Enthalpy ΔH versus T and (b) Entropy ΔS versus T.

[10]We assume that the whole process is done slowly and carefully so that it is reversible.

and

$$\Delta S \approx C_p(\text{ice}) \ln \frac{T_\text{M}}{T_\text{I}} + \frac{\Delta H_\text{M}}{T_\text{M}} + C_p(\text{water}) \ln \frac{T_W}{T_\text{M}}. \tag{3.68}$$

To get an idea of the magnitudes involved, we approximate $C_p(\text{ice}) \approx C_p(\text{water}) \approx 1\,\text{cal/g K}$, take $T_\text{I} = -10\,°\text{C}$ and $T_\text{W} = 20\,°\text{C}$. Then for every mole of H_2O (18 g/mol) we have

$$\Delta H = (189 + 1440 + 351)\,\text{cal/mol} = 1980\,\text{cal/mol} \tag{3.69}$$

and

$$\Delta S = (0.67 + 5.27 + 1.27)\,\text{cal/K mol} = 7.21\,\text{cal/K mol}. \tag{3.70}$$

For many monatomic substances, $\Delta S_\text{M} = \Delta H_\text{M}/T_\text{M} \sim R \approx 2\,\text{cal/K mol}$, an empirical rule known as **Richard's rule**. For ice, the entropy of fusion is much larger (5.27 cal/K mol) because of the complexity of the H_2O molecule. For vaporization, a similar empirical rule known as **Trouton's rule** leads to the estimate $\Delta S_V = \Delta H_V/T_V \sim 10.5R \approx 21\,\text{cal/K mol}$, as compared to 26 cal/K mol for water. The fact that the entropy of vaporization is larger than the entropy of fusion is because essentially all atomic bonds must be broken for evaporation and because of the large volume change from liquid to gas.

3.5 Statistical Interpretation of Entropy

The entropy S enters classical thermodynamics as a mysterious state function whose changes can be calculated from Eq. (3.33). Unlike other state variables such as the internal energy U or the pressure p, it has no roots in classical mechanics. Its existence is related to the fact that the absolute temperature T is regarded in thermodynamics to be a state variable, and the entropy S turns out to be its **conjugate variable**.[11] A more thorough understanding of entropy requires a statistical analysis. Later we will discuss entropy in the context of the formal postulates that underlie statistical mechanics. For now, we give a brief statistical interpretation based on a few simple ideas.

3.5.1 Relationship of Entropy to Microstates

In order to understand entropy, we must appreciate that for every macrostate of a system, which corresponds to a fixed energy and other extensive parameters, there are a number Ω of compatible microstates, and the system could be in any one of them.[12] In fact, it could progress through a number of compatible microstates as time evolves. If we assume that the probability of a given microstate is $1/\Omega$, it is reasonable to postulate that the entropy is a function of the number of microstates, that is,

$$S = f(\Omega). \tag{3.71}$$

[11] In the differential $\mathrm{d}U = T\,\mathrm{d}S - p\,\mathrm{d}V$, T is said to be conjugate to S and $-p$ is conjugate to V. For a more general definition of conjugate variables, see Section 5.5.

[12] According to quantum mechanics, the system will have a discrete set of energy eigenstates, which are actually countable. See Chapters 16 and 26 for details.

For an isolated system, natural processes are those that correspond to an increase in S. Moreover, S is defined to be an increasing function of the internal energy and we would expect the number of compatible microstates to increase with energy. We therefore anticipate that $f(\Omega)$ will be a monotonically increasing function of Ω, which turns out to be the case.

Once Eq. (3.71) is accepted, the form of the function $f(\Omega)$ can be determined by considering an isolated composite system S made up of two subsystems having entropies S_1 and S_2. Since S is assumed to be additive for composite systems, we have

$$S = S_1 + S_2. \tag{3.72}$$

If the number of microstates for S_1 is Ω_1 and that for S_2 is Ω_2, then for the total system S the number of microstates is $\Omega_1\Omega_2$. Therefore, Eq. (3.72) may be written

$$f(\Omega_1\Omega_2) = f(\Omega_1) + f(\Omega_2). \tag{3.73}$$

In Eq. (3.73), we first set $\Omega_2 = 1$ to obtain

$$f(\Omega_1) = f(\Omega_1) + f(1), \tag{3.74}$$

from which we conclude that $f(1) = 0$. Then we differentiate Eq. (3.73) partially with respect to Ω_2 to get (the prime denotes the derivative with respect to the argument)

$$\Omega_1 f'(\Omega_1\Omega_2) = f'(\Omega_2) \tag{3.75}$$

and again set $\Omega_2 = 1$ to get

$$f'(\Omega_1) = \frac{k}{\Omega_1}, \tag{3.76}$$

where $k = f'(1)$ is a constant. We then integrate Eq. (3.76) to obtain $f(\Omega_1) = k \ln \Omega_1 + C$ where C is a constant. Since $f(1) = 0$, we conclude that $C = 0$. Therefore, returning to our general notation, we have

$$S = k \ln \Omega. \tag{3.77}$$

In order for S to be a monotonically increasing function of Ω, we must choose $k > 0$.

For an isolated system, Eq. (3.77) is a fundamental equation that relates entropy to statistical mechanical concepts. It states that the entropy is proportional to the logarithm of the number of microstates that are compatible with a given macrostate. The constant of proportionality k depends on the units used to measure S. In order to agree with classical thermodynamics, we need to choose $k = k_B$ which is known as Boltzmann's constant:

$$k_B = 1.381 \times 10^{-16}\,\text{erg/K} = 1.381 \times 10^{-23}\,\text{J/K} = 3.301 \times 10^{-24}\,\text{cal/K}. \tag{3.78}$$

It is related to the gas constant $R = \mathcal{N}_A k_B$ where $\mathcal{N}_A = 6.022\,\text{mol}^{-1}$ is Avogadro's number (also known as Loschmidt-Zahl in the German literature). Hence

$$R = 8.314 \times 10^{-7}\,\text{erg/(mol K)} = 8.314\,\text{J/(mol K)} = 1.987\,\text{cal/(mol K)}. \tag{3.79}$$

For a more rigorous justification of Eq. (3.77) in the context of information theory and the microcanonical ensemble, see Chapter 15, particularly Eq. (15.14), and Chapter 16.

4

Third Law of Thermodynamics

The third law of thermodynamics is the latest of the three laws of thermodynamics to be developed. It insures that the entropy remains well-defined at the absolute zero of temperature and allows one to define a zero of entropy that is consistent with statistical mechanics. This avoids having to deal with entropy differences; instead, we can deal with entropies as absolute quantities, analogous to absolute temperature but unlike energy.

4.1 Statement of the Third Law

The entropy S of a thermodynamic system in internal equilibrium approaches a universal constant S_0, independent of phase, as the absolute temperature T tends to zero. Alternatively, one could say that $S \to S_0$ in a state for which the quantity $(\partial U/\partial S)_{\{\mathrm{ext}\}} \to 0$, where {ext} stands for the remaining members of a complete set of extensive variables. By convention, and in agreement with statistical mechanics, the value of this universal constant $S \to S_0$ is taken to be zero. Since entropy is a monotonically increasing function of temperature, this convention results in the entropy being a positive quantity.

4.1.1 Discussion of the Third Law

According to statistical mechanics, as motivated by Eq. (3.77), the entropy of an isolated system is given by

$$S = k_B \ln \Omega, \tag{4.1}$$

where k_B is Boltzmann's constant and Ω is the number of microstates that correspond to a given macrostate. If at absolute zero only a unique ground state of the system is occupied, then $\Omega = 1$ and $S = 0$. Possibly the ground state could be degenerate, in which case $\Omega \neq 1$ even at $T = 0$. But this degeneracy would have to be massive to make a significant difference in the entropy of a macroscopic system at $T = 0$. Indeed, to get a contribution $S = 10^{-10}R = 10^{-10}k_B\mathcal{N}_A$ for one mole at absolute zero would require the ground state degeneracy Ω_0 to satisfy $10^{-10}\mathcal{N}_A = \ln \Omega_0$, where $\mathcal{N}_A$ is Avogadro's number. This yields $\Omega_0 \sim e^{6\times 10^{13}} \sim 10^{2.6\times 10^{13}}$. But such a huge degeneracy is contrary to experience. As the ground state of a quantum system is approached (as $T \to 0$), the number of accessible quantum states decreases quite rapidly and is no longer of exponential order, even though there could still be a ground state of much smaller degeneracy. An illuminating discussion of this point has been presented by Benjamin Widom [17, chapter 5].

The third "law" is an extension by Max Planck [15, p. 273] of the so-called Nernst postulate [2, p. 277] that was made in an attempt to justify an empirical rule of Thomsen and Berthelot for chemical reactions that take place at constant temperature and pressure.

Thermal Physics. http://dx.doi.org/10.1016/B978-0-12-803304-3.00004-1

Nernst conjectured that their empirical rule for equilibrium, which is equivalent to minimizing the enthalpy change ΔH of the reaction, would be in agreement with the proper thermodynamic criterion obtained by minimizing an appropriate change in free energy[1] of the reaction, provided that the entropy change ΔS tends to zero as $T \to 0$. This can be interpreted to mean that the entropy S itself tends to some constant, independent of the extent of the reaction, as $T \to 0$. For convenience, Planck set this entropy constant to zero, which agrees with the convention used to define entropy in statistical mechanics. Callen [2, p. 30] states the third law as an independent postulate, namely that $S=0$ in a state for which $\partial U/\partial S=0$ (which is true at absolute zero by definition of the thermodynamic temperature). From the point of view of classical thermodynamics, one could deal with entropy differences and it would not be necessary to adopt a state of zero entropy; however, doing so leads to simplicity and builds a strong bridge to statistical mechanics.

4.2 Implications of the Third Law

The third law has certain implications regarding heat capacities and other properties of materials as $T \to 0$. From Eq. (3.47) with $\mathrm{d}V=0$, we obtain $C_V\,\mathrm{d}T=T\,\mathrm{d}S$ where C_V is the heat capacity at constant volume. The change in entropy at constant volume from one temperature to another is given by $\int C_V/T\,\mathrm{d}T$. Thus

$$S(T_1,V)=\int_0^{T_1}\frac{C_V(T,V)}{T}\,\mathrm{d}T. \tag{4.2}$$

In order for this integral to converge, it is necessary for C_V to depend on T in such a way that $C_V \to 0$ as $T \to 0$. Recall that C_V was taken to be a constant for an ideal gas; clearly such an ideal gas becomes impossible as $T \to 0$. For insulating solids, one finds both theoretically and experimentally that $C_V \propto T^3$ as $T \to 0$. For metals, nearly free electrons contribute to the heat capacity and $C_V \propto T$ as $T \to 0$. Similar considerations apply to the heat capacity at constant pressure. From Eq. (3.62) with $\mathrm{d}p=0$, we obtain $C_p\,\mathrm{d}T=T\,\mathrm{d}S$, where C_p is the heat capacity at constant pressure. Thus

$$S(T_1,p)=\int_0^{T_1}\frac{C_p(T,p)}{T}\,\mathrm{d}T \tag{4.3}$$

and it is necessary[2] for $C_p \to 0$ as $T \to 0$.

An interesting experimental verification of the third law has been discussed by Fermi [1, p. 146]. At temperatures below $T_0=292$ K, gray tin (α, diamond cubic) is the stable form and above this temperature, white tin (β, tetragonal) is stable. These are allotropic forms of pure tin. It turns out, however, that white tin can exist (in internal equilibrium) below 292 K, even though it is unstable with respect to transformation to gray tin. It is also

[1]This is the change ΔG of the Gibbs free energy of the reaction, whose definition and properties we explore later.

[2]In order for the integrals in Eqs. (4.2) and (4.3) to converge at $T=0$, it will suffice for C_V or C_p to go to zero very weakly as $T \to 0$, for instance $\propto T^\epsilon$ where $\epsilon > 0$.

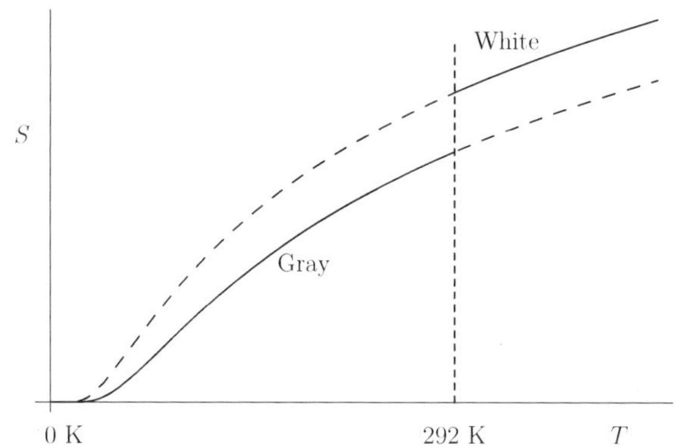

FIGURE 4–1 Entropies S of gray and white tin as a function of absolute temperature T. Below $T_0 = 292$ K, gray tin is stable and above this temperature white tin is stable. The full curves denote stable phases and the dashed curves denote unstable phases. White tin can be supercooled below T_0 so its heat capacity can be measured and its entropy can be calculated. The jump in entropy at T_0 between gray tin and white tin is due to the latent heat of transformation.

possible to measure the heat capacities of both forms of tin down to very low temperatures. One can therefore evaluate the entropy of white tin at 292 K in two different ways, the first by integrating its heat capacity from absolute zero and the second by integrating the heat capacity of gray tin from absolute zero and then adding the entropy associated with transformation to white tin at 292 K. See Figure 4–1 for a graphic illustration. Thus (with subscripts g and w for gray and white), we have

$$S_{\rm w}(292\,K) = \int_0^{292\,K} \frac{C_{\rm w}(T)}{T}\,{\rm d}T = 12.30\,{\rm cal/mol\,K}, \tag{4.4}$$

and

$$S_{\rm g}(292\,K) = \int_0^{292\,K} \frac{C_{\rm g}(T)}{T}\,{\rm d}T = 10.53\,{\rm cal/mol\,K}. \tag{4.5}$$

The heat of transformation from gray to white tin is $\Delta H_{\rm g\to w} = 535$ cal/mol so the entropy of transformation is $\Delta S_{\rm g\to w} = \Delta H_{\rm g\to w}/T_0 = 535/292 = 1.83$ cal/mol K. Adding this to the result of Eq. (4.5) gives 12.36 cal/mol K, in reasonable agreement with Eq. (4.4).

The third law can also shed light on the behavior of the coefficient of thermal expansion, α, and the compressibility, κ_T, as $T \to 0$. Since $S \to 0$ as $T \to 0$ independent of V or p, one has

$$\left(\frac{\partial S}{\partial V}\right)_{T=0} = 0; \quad \left(\frac{\partial S}{\partial p}\right)_{T=0} = 0. \tag{4.6}$$

Through a Maxwell relation (see Eq. (5.90)), it can be shown that

$$\left(\frac{\partial S}{\partial p}\right)_T = -\left(\frac{\partial V}{\partial T}\right)_p = -V\alpha, \tag{4.7}$$

where α is the coefficient of isobaric thermal expansion. Indeed, it has been verified experimentally that $\alpha \to 0$ as $T \to 0$. By means of another Maxwell relation (see Eq. (5.86))

$$\left(\frac{\partial S}{\partial V}\right)_T = \left(\frac{\partial p}{\partial T}\right)_V = \alpha/\kappa_T, \tag{4.8}$$

where κ_T is the coefficient of isothermal compressibility. Thus, κ_T must either remain nonzero as $T \to 0$ or go to zero more slowly than α.

See Lupis [5, pp. 21-23] for further discussion of experimental verification of the third law as well as a discussion of some of its other consequences, particularly consequences concerning chemical reactions. See Fermi [1, p. 150] for an excellent discussion of the entropy of mercury vapor.

5

Open Systems

Until now we have dealt with chemically closed thermodynamic systems in which there is no exchange of chemical components with the environment. Such chemically closed systems can receive heat Q from the environment and do work $\mathcal{W}$ on their environment. Their change in internal energy is given by $\Delta U = Q - \mathcal{W}$, which for infinitesimal changes is $dU = \delta Q - \delta \mathcal{W}$. For reversible changes in a simple isotropic system, the (quasistatic) work is $\delta \mathcal{W} = p\,dV$, where p is the pressure and V is the volume. The heat received in a reversible change is $\delta Q = T\,dS$, where T is the absolute temperature and S is the entropy. If the mole numbers of each chemical component are constant (no chemical reactions), the combined first and second laws (see Chapter 3) lead to

$$dU = T\,dS - p\,dV. \tag{5.1}$$

Open systems can exchange chemical components with their environment. Consequently the number of moles of each chemical component, N_i, for $i = 1, 2, \ldots, \kappa$, are variables. This requires several modifications. The first law must be amended to read

$$\Delta U = Q - \mathcal{W} + E_{\text{ch}}, \tag{5.2}$$

where E_{ch} is the energy (sometimes called chemical heat) that is added to the system when chemical components are exchanged with its environment. Moreover, U now becomes a function of S, T and all of the N_i, so additional terms are needed in Eq. (5.1). This also sets the stage for changes of N_i due to chemical reactions within the system, which can even occur for a chemically closed system for which $E_{\text{ch}} = 0$. We shall first treat an open system having a single component and then go on to treat multicomponent systems.

5.1 Single Component Open System

If the simple chemically closed isotropic system discussed above has only one chemical component and is now opened to allow exchange of that component with the environment, U must be regarded as a function of S, V, and N, the number of moles[1] of that component. Then the differential of the internal energy becomes

$$dU = T\,dS - p\,dV + \mu\,dN, \tag{5.3}$$

where now

$$T = \left(\frac{\partial U}{\partial S}\right)_{V,N}; \quad -p = \left(\frac{\partial U}{\partial V}\right)_{S,N}; \quad \mu = \left(\frac{\partial U}{\partial N}\right)_{S,V}. \tag{5.4}$$

[1] Instead of N, we could use the mass, M or the number of atoms $\mathcal{N}$. Then the resulting chemical potentials would be per unit mass or per atom instead of per mole.

Thermal Physics. http://dx.doi.org/10.1016/B978-0-12-803304-3.00005-3

The quantity μ, introduced originally by Gibbs [3], is called the **chemical potential**, by analogy to the thermal potential T. We shall see later that μ can be expressed as a function of only p and T. It is the internal energy per mole[2] of component added reversibly to the system at constant S and V. Nevertheless, Eq. (5.3) is a relationship among state functions and always holds for infinitesimal changes within the field of equilibrium states.

5.1.1 Ideal Gas

The chemical potential of a monocomponent ideal gas can be written in the form

$$\mu(T, p) = \mu^*(T) + RT \ln p = RT \ln \frac{p}{p^*(T)}, \tag{5.5}$$

where $p^*(T)$ is a function of temperature with dimensions of pressure. In this standard but misleading notation, $\mu^*(T)$ is a function of temperature only but does not have the dimensions of an energy. This can be fixed by writing $\mu^*(T) + RT \ln p = \mu^*(T) + RT \ln p_0 + RT \ln p/p_0$, where p_0 is a reference pressure, usually taken to be one atmosphere. Then the quantity $\mu^0(T, p_0) = \mu^*(T) + RT \ln p_0$ is the chemical potential of this ideal gas at p_0 and we can write Eq. (5.5) in the form

$$\mu(T, p) = \mu^0(T, p_0) + RT \ln p/p_0. \tag{5.6}$$

Moreover, if p_0 is equal to one atmosphere, it is often omitted from formulas with the understanding that all pressures are expressed in atmospheres. In this case, the term $RT \ln p_0 = 0$, so numerically $\mu^*(T) = \mu^0(T, p_0)$, even though the dimensions do not agree. We avoid this shortcut in the interest of clarity, so pressures can be measured in any units. Even though each term on the right of Eq. (5.6) depends on p_0, their sum is independent of p_0. Even for a real gas, liquid, or solid, $\mu(T, p)$ must be independent of any reference pressure such as p_0.

Example Problem 5.1. For N moles of an ideal gas, the equation of state Eq. (2.12) takes the form $pV = NRT$. Show that its chemical potential can be expressed as a function of only its concentration $c = N/V$ and the temperature. At the standard temperature $T_0 = 25\,°\text{C}$ and a pressure of one standard atmosphere (1.01325×10^5 Pa), the volume of one mole of an ideal gas is 22.4 l. If one mole of gas remains at temperature T_0 but is compressed so that it occupies only 2 l, how much does its chemical potential change compared to that at standard temperature and pressure?

Solution 5.1. We substitute $p = cRT$ into Eq. (5.5) to obtain

$$\mu = \mu^*(T) + RT \ln cRT = \mu^*(T) + RT \ln RT + RT \ln c. \tag{5.7}$$

[2] Since U is only defined up to an additive constant, μ is similarly only defined up to a compatible additive constant. In practice, one usually adopts so-called standard states and deals with the quantities $\mu - \mu^0$, where μ^0 refers to the standard state.

The change in chemical potential at temperature T_0 is $\Delta\mu = RT_0 \ln c/c_0 = RT_0 \ln(22.4/2) = 2.416RT_0$. We have $T_0 = 25 + 273.15 = 298.15$ K, so $RT_0 = 2479$ J and $\Delta\mu = 5868$ J. Alternatively, we could evaluate the new pressure, which would be $p = 11.2$ atmospheres, and use Eq. (5.6) to get the same answer. At a given temperature, we see that the chemical potential of an ideal gas is just a measure of concentration, or pressure, on a logarithmic scale.

Example Problem 5.2. According to statistical mechanics, the chemical potential per atom of a monatomic gas having atoms of mass m is given by $k_B T \ln[n/n_Q(T)]$, where k_B is Boltzmann's constant, n is the number of atoms per unit volume, and $n_Q(T)$ is the quantum concentration given by $n_Q(T) = (mk_BT/2\pi\hbar^2)^{3/2}$, where $\hbar = h/2\pi$ and h is Planck's constant. This result is based on a convention for the zero of energy used to deduce the quantum states of a free particle; see Section 19.3.1 for details. Find explicit expressions for $p^*(T)$ and $\mu^*(T)$ in Eq. (5.5).

Solution 5.2. To obtain the chemical potential per mole, we simply multiply the given chemical potential per atom by Avogadro's number $\mathcal{N}_A$ and recall that $\mathcal{N}_A k_B = R$. The ideal gas law can similarly be converted to obtain $p = nk_BT$. Therefore,

$$\mu = RT \ln \frac{p}{n_Q(T)k_BT} \tag{5.8}$$

from which we identify

$$p^*(T) = n_Q(T)k_BT = (mk_BT/2\pi\hbar^2)^{3/2}k_BT. \tag{5.9}$$

Of course $\mu^*(T) = -RT \ln p^*(T)$. Formally, numerical evaluation of $\mu^*(T)$ involves taking the logarithm of a quantity with dimensions of pressure, but the units in which pressures are expressed will cancel when μ is evaluated, as illustrated by Eq. (5.8).

5.2 Multicomponent Open Systems

The generalization of Eq. (5.3) to open multicomponent systems is straightforward. U now depends on the variable set $S, V, N_1, N_2, \ldots, N_\kappa$ for a system of κ chemical components. Then

$$dU = T\,dS - p\,dV + \sum_{i=1}^{\kappa} \mu_i\,dN_i, \tag{5.10}$$

where

$$T = \left(\frac{\partial U}{\partial S}\right)_{V,\{N_i\}}; \quad -p = \left(\frac{\partial U}{\partial V}\right)_{S,\{N_i\}}; \quad \mu_i = \left(\frac{\partial U}{\partial N_i}\right)_{S,V,\{N_i'\}}. \tag{5.11}$$

Here, $\{N_i\}$ stands for the entire set $N_1, N_2, \ldots, N_\kappa$, and $\{N_i'\}$ stands for that same set but with N_i missing. Since this notation is cumbersome, we will often omit these subscripts, but they should always be borne in mind to avoid confusion. Equation (5.11) defines $\kappa + 2$ intensive variables (p, T, and κ chemical potentials, one for each chemical component), although we shall see that only $\kappa + 1$ of them are independent (see Eq. (5.45)). One says that such a thermodynamic system has $\kappa + 1$ **degrees of freedom**. These thermodynamic degrees of freedom should not be confused with the number of degrees of freedom (typically of order 10^{23}) of the underlying microscopic system. Equation (5.11) is valid for all infinitesimal changes of $S, V, \{N_i\}$ within the field of equilibrium states and these changes are reversible.

5.2.1 Maxwell Relations for Open Systems

In general, Maxwell relations are obtained by equating the mixed second derivatives of a function of two or more variables. Suppose that we have some function f of three variables, x, y, and z. Then

$$\mathrm{d}f = \left(\frac{\partial f}{\partial x}\right)_{y,z} \mathrm{d}x + \left(\frac{\partial f}{\partial y}\right)_{z,x} \mathrm{d}y + \left(\frac{\partial f}{\partial z}\right)_{x,y} \mathrm{d}z \tag{5.12}$$

and[3]

$$\frac{\partial^2 f}{\partial x \partial y} = \frac{\partial^2 f}{\partial y \partial x}; \quad \frac{\partial^2 f}{\partial y \partial z} = \frac{\partial^2 f}{\partial z \partial y}; \quad \frac{\partial^2 f}{\partial z \partial x} = \frac{\partial^2 f}{\partial x \partial z}. \tag{5.13}$$

Equations (5.12) and (5.13) can be extended to any number of dependent variables. If we apply Eq. (5.13) to the first two members of Eq. (5.10), we obtain

$$\left(\frac{\partial T}{\partial V}\right)_{S,\{N_i\}} = -\left(\frac{\partial p}{\partial S}\right)_{V,\{N_i\}}. \tag{5.14}$$

Since all N_i are held constant in Eq. (5.14), it would also hold for a chemically closed system in which there are no chemical reactions. For an open system, we have additional Maxwell relations such as

$$\left(\frac{\partial T}{\partial N_i}\right)_{S,V,\{N_i'\}} = \left(\frac{\partial \mu_i}{\partial S}\right)_{V,\{N_i\}}, \tag{5.15}$$

$$-\left(\frac{\partial p}{\partial N_i}\right)_{S,V,\{N_i'\}} = \left(\frac{\partial \mu_i}{\partial V}\right)_{S,\{N_i\}}, \tag{5.16}$$

and for $i \neq j$

$$\left(\frac{\partial \mu_i}{\partial N_j}\right)_{S,V,\{N_j'\}} = \left(\frac{\partial \mu_j}{\partial N_i}\right)_{S,V,\{N_i'\}}. \tag{5.17}$$

For a system having κ chemical components, the number of these Maxwell relations is $(\kappa + 2)(\kappa + 1)/2$.

Additional Maxwell relations may be obtained by solving Eq. (5.10) for the differential of another variable, for example,

[3]These relations are true if the derivatives exist and are continuous, which we will assume to be the case for thermodynamic functions.

$$\mathrm{d}S = \frac{1}{T}\,\mathrm{d}U + \frac{p}{T}\,\mathrm{d}V - \sum_{i=1}^{\kappa} \frac{\mu_i}{T}\,\mathrm{d}N_i. \tag{5.18}$$

Then

$$\left(\frac{\partial(1/T)}{\partial V}\right)_{U,\{N_i\}} = \left(\frac{\partial(p/T)}{\partial U}\right)_{V,\{N_i\}}, \tag{5.19}$$

$$\left(\frac{\partial(1/T)}{\partial N_i}\right)_{U,V\{N_i'\}} = -\left(\frac{\partial(\mu_i/T)}{\partial U}\right)_{V,\{N_i\}}, \tag{5.20}$$

$$\left(\frac{\partial(p/T)}{\partial N_i}\right)_{U,V\{N_i'\}} = -\left(\frac{\partial(\mu_i/T)}{\partial V}\right)_{U,\{N_i\}}, \tag{5.21}$$

and for $i \neq j$

$$\left(\frac{\partial(\mu_i/T)}{\partial N_j}\right)_{U,V,\{N_j'\}} = \left(\frac{\partial(\mu_j/T)}{\partial N_i}\right)_{U,V,\{N_i'\}}. \tag{5.22}$$

Maxwell relations may be used to simplify thermodynamic expressions and also to derive formulae for desired quantities in terms of experimentally measurable quantities.

Example: relationship of C_p to C_V: A useful result of Maxwell relations is the general formula (previously quoted in Eq. (2.16) without proof) that connects the heat capacity at constant pressure C_p to that at constant volume, C_V. Here, we deal with a chemically closed system in the absence of chemical reactions, so we drop the subscripts $\{N_i\}$ for the sake of simplicity. From the definitions $C_V := (\delta Q/\mathrm{d}T)_V$ and $C_p := (\delta Q/\mathrm{d}T)_p$ and $\delta Q = \mathrm{d}U + p\,\mathrm{d}V$ for quasistatic work, we have

$$C_V = \left(\frac{\partial U}{\partial T}\right)_V \quad \text{and} \tag{5.23}$$

$$C_P = \left(\frac{\partial U}{\partial T}\right)_p + p\left(\frac{\partial V}{\partial T}\right)_p = \left(\frac{\partial U}{\partial T}\right)_V + \left(\frac{\partial U}{\partial V}\right)_T\left(\frac{\partial V}{\partial T}\right)_p + p\left(\frac{\partial V}{\partial T}\right)_p. \tag{5.24}$$

Therefore,

$$C_p = C_V + \left[\left(\frac{\partial U}{\partial V}\right)_T + p\right]\left(\frac{\partial V}{\partial T}\right)_p. \tag{5.25}$$

To get the derivative $(\partial U/\partial V)_T$, we make use of the fact that the entropy is a function of state with differential

$$\mathrm{d}S = \frac{1}{T}\,\mathrm{d}U + \frac{p}{T}\,\mathrm{d}V = \frac{1}{T}\left(\frac{\partial U}{\partial T}\right)_V \mathrm{d}T + \left[\frac{1}{T}\left(\frac{\partial U}{\partial V}\right)_T + \frac{p}{T}\right]\mathrm{d}V, \tag{5.26}$$

where we now regard S to be a function of T and V. Therefore

$$\left\{\frac{\partial}{\partial V}\left[\frac{1}{T}\left(\frac{\partial U}{\partial T}\right)_V\right]\right\}_T = \left\{\frac{\partial}{\partial T}\left[\frac{1}{T}\left(\frac{\partial U}{\partial V}\right)_T + \frac{p}{T}\right]\right\}_V, \tag{5.27}$$

which becomes

$$\frac{1}{T}\frac{\partial^2 U}{\partial V \partial T} = \frac{1}{T}\frac{\partial^2 U}{\partial T \partial V} - \frac{1}{T^2}\left[\left(\frac{\partial U}{\partial V}\right)_T + p\right] + \frac{1}{T}\left(\frac{\partial p}{\partial T}\right)_V. \tag{5.28}$$

After cancellation of the mixed partial derivatives, Eq. (5.28) gives[4]

$$\left[\left(\frac{\partial U}{\partial V}\right)_T + p\right] = T\left(\frac{\partial p}{\partial T}\right)_V, \tag{5.29}$$

which may be substituted into Eq. (5.25) to give

$$C_p = C_V + T\left(\frac{\partial p}{\partial T}\right)_V \left(\frac{\partial V}{\partial T}\right)_p. \tag{5.30}$$

Finally, we use the relation[5]

$$\left(\frac{\partial p}{\partial T}\right)_V \left(\frac{\partial T}{\partial V}\right)_p \left(\frac{\partial V}{\partial p}\right)_T = -1 \tag{5.31}$$

to eliminate $(\partial p/\partial T)_V$. Then the definitions of the coefficient of expansion $\alpha := (1/V)(\partial V/\partial T)_p$ and the compressibility $\kappa_T := -(1/V)(\partial V/\partial p)_T$ lead to

$$C_p = C_V + \frac{TV\alpha^2}{\kappa_T}, \tag{5.32}$$

which is the same as Eq. (2.16). The isothermal compressibility κ_T is always positive whereas the coefficient of thermal expansion α is usually positive but can be negative or even zero, as it is for water near 4 °C. Since Eq. (5.32) depends on α^2, we see that $C_p - C_V \geq 0$. For gases, C_p can be considerably larger than C_V but for condensed phases the difference between them is relatively small.

Example: relationship of C_p to C_V, alternative method: For a reversible change we have $\delta Q = T\,\mathrm{d}S$, so we can write

$$C_V = T\left(\frac{\partial S}{\partial T}\right)_V \quad \text{and} \tag{5.33}$$

$$C_p = T\left(\frac{\partial S}{\partial T}\right)_p = T\left(\frac{\partial S}{\partial T}\right)_V + T\left(\frac{\partial S}{\partial V}\right)_T \left(\frac{\partial V}{\partial T}\right)_p. \tag{5.34}$$

Thus

$$C_p = C_V + T\left(\frac{\partial S}{\partial V}\right)_T \left(\frac{\partial V}{\partial T}\right)_p. \tag{5.35}$$

[4]This is called the Helmholtz equation and sometimes written $(\partial U/\partial V)_T = T^2(\partial(p/T)/\partial T)_V$.

[5]This relation follows immediately from the differential $\mathrm{d}V = (\partial V/\partial T)_p\,\mathrm{d}T + (\partial V/\partial p)_T\,\mathrm{d}p$ by setting $\mathrm{d}V = 0$ and solving for the remaining partial derivative.

To find an expression for $(\partial S/\partial V)_T$, we invent a new state function[6] $F := U - TS$ so that

$$dF = dU - T\,dS - S\,dT = -S\,dT - p\,dV. \tag{5.36}$$

Regarding F to be a function of T and V, we see that

$$\left(\frac{\partial S}{\partial V}\right)_T = \left(\frac{\partial p}{\partial T}\right)_V, \tag{5.37}$$

so Eq. (5.35) becomes Eq. (5.30) and Eq. (5.32) follows as above.

5.2.2 Other Maxwell Relations

Maxwell relations can be obtained by equating the mixed second partial derivatives of any state function. This is usually done by defining other functions that are related to U and S by means of Legendre transformations. A number of specific examples are presented in Section 5.5.1. Tables of some Maxwell relations and a mnemonic diagram for remembering them are given by Callen [2, chapter 7] but many others exist and can be derived as needed.

5.3 Euler Theorem of Homogeneous Functions

A function $f(x, y, z)$ is said to be a homogeneous function of degree n with respect to the variables x, y, and z if

$$f(\lambda x, \lambda y, \lambda z) = \lambda^n f(x, y, z), \tag{5.38}$$

where λ is some parameter. Note that Eq. (5.38) requires a very special type of function and that many functions are not homogeneous. For a homogeneous function of degree n, the Euler theorem states that

$$x\left(\frac{\partial f}{\partial x}\right)_{y,z} + y\left(\frac{\partial f}{\partial y}\right)_{z,x} + z\left(\frac{\partial f}{\partial z}\right)_{x,y} = nf(x, y, z). \tag{5.39}$$

This theorem, illustrated for three dependent variables, holds for any number of dependent variables.

Proof. We differentiate Eq. (5.38) partially with respect to λ to obtain

$$\frac{\partial f(\lambda x, \lambda y, \lambda z)}{\partial(\lambda x)}\frac{\partial(\lambda x)}{\partial\lambda} + \frac{\partial f(\lambda x, \lambda y, \lambda z)}{\partial(\lambda y)}\frac{\partial(\lambda y)}{\partial\lambda} + \frac{\partial f(\lambda x, \lambda y, \lambda z)}{\partial(\lambda z)}\frac{\partial(\lambda z)}{\partial\lambda} = n\lambda^{n-1}f(x, y, z) \tag{5.40}$$

and note that $\partial(\lambda x)/\partial\lambda = x$, $\partial(\lambda y)/\partial\lambda = y$ and $\partial(\lambda z)/\partial\lambda = z$. After the differentiation is done, we set $\lambda = 1$ in Eqs. (5.40) and (5.39) results. Note especially that if the function f depends on additional variables, say u and v, such that

$$f(\lambda x, \lambda y, \lambda z, u, v) = \lambda^n f(x, y, z, u, v), \tag{5.41}$$

[6]This state function is actually the Helmholtz free energy, a useful thermodynamic potential that we shall define later. For now, it is just a convenient state function that will allow us to get the desired result.

Eq. (5.39) still holds, with no corresponding terms for u and v. In other words, Eq. (5.41) should be interpreted to mean that f is homogeneous in the variables x, y, and z; the variables u and v are held constant during differentiation and are simply irrelevant insofar as homogeneity with respect to x, y, and z is concerned.

Examples: The function $\phi(x,y,z) := x^2 + y^2 + z^4/(x^2+y^2)$ is a homogeneous function of degree 2 in x, y, and z. We have $\partial\phi/\partial x = 2x - 2xz^4/(x^2+y^2)^2$, $\partial\phi/\partial y = 2y - 2yz^4/(x^2+y^2)^2$, and $\partial\phi/\partial z = 4z^3/(x^2+y^2)$. Thus

$$x\frac{\partial\phi}{\partial x} + y\frac{\partial\phi}{\partial y} + z\frac{\partial\phi}{\partial z} = 2x^2 - \frac{2x^2z^4}{(x^2+y^2)^2} + 2y^2 - \frac{2y^2z^4}{(x^2+y^2)^2} + \frac{4z^3}{x^2+y^2} = 2\phi.$$

The function $\psi(x,y,z) := \sin(x/y) + z^2/x^2$ is a homogeneous function of degree zero in x, y, and z. We have $\partial\psi/\partial x = (1/y)\cos(x/y) - 2z^2/x^3$, $\partial\psi/\partial y = -(x/y^2)\cos(x/y)$ and $\partial\psi/\partial z = 2z/x^2$, which yields $x\partial\psi/\partial x + y\partial\psi/\partial y + z\partial\psi/\partial z = 0$.

The function $\eta(x,y,z) := x^3 + y^3 + z^2$ is not a homogeneous function with respect to the variables x, y, and z. The function $\zeta(x,y,z) := x^3z + y^3z^2$ is not a homogeneous function with respect to the variables x, y, and z, but it is a homogeneous function of degree 3 in x and y with z held constant. Then $x(\partial\zeta/\partial x)_{y,z} + y\big(\partial\zeta/\partial y\big)_{x,z} = 3\zeta$.

Note that it is not necessary for n to be an integer, and that n can even be negative. Thus, the function $\phi(x,y,z) := (x/yz)^{1/3} + (1/x)^{1/3}$ is a homogeneous function of degree $n = -1/3$ in x, y, and z and Eq. (5.39) holds, as the reader may verify.

5.3.1 Euler Theorem Applied to Extensive Functions

We note that U, which is extensive, is a homogeneous function of degree one in the extensive variables $S, V, N_1, N_2, \ldots, N_\kappa$. Thus,

$$U(\lambda S, \lambda V, \lambda N_1, \lambda N_2, \ldots, \lambda N_\kappa) = \lambda U(S, V, N_1, N_2, \ldots, N_\kappa). \tag{5.42}$$

For example, if we double all of the extensive variables on which U depends, we will obtain a system that is twice as large but whose *intensive* variables are unchanged,[7] so we will have twice as much of the same thermodynamic state. Applying the Euler theorem for $n = 1$, we obtain

$$U = TS - pV + \sum_{i=1}^{\kappa} \mu_i N_i. \tag{5.43}$$

We call this the **Euler equation** for U. By taking its differential, we obtain

$$dU = T\,dS + S\,dT - p\,dV - V\,dp + \sum_{i=1}^{\kappa} \mu_i\,dN_i + \sum_{i=1}^{\kappa} N_i\,d\mu_i. \tag{5.44}$$

[7]This follows because the intensive variables are partial derivatives of the extensive variable U with respect to the extensive variables on which U depends, so any constant multiple such as 2 will cancel.

By comparing with Eq. (5.10), we deduce that

$$0 = S\,\mathrm{d}T - V\,\mathrm{d}p + \sum_{i=1}^{\kappa} N_i\,\mathrm{d}\mu_i. \tag{5.45}$$

Eq. (5.45) is the **Gibbs-Duhem equation** for a multicomponent system. It shows that T, p, and the μ_i are not independent intensive variables because changes in them are related. Thus for a κ component system, there are only $\kappa + 1$ independent intensive variables.

For a monocomponent system, there are two independent intensive variables, say p and T, and $\mu(p, T)$ can be regarded to be a function of them.[8] In that case, Eq. (5.45) can be divided by N and written in the form

$$\mathrm{d}\mu = -s\,\mathrm{d}T + v\,\mathrm{d}p, \tag{5.46}$$

where $s := S/N$ is the entropy per mole and $v := V/N$ is volume per mole (molar volume). If the functions $s(T, p)$ and $v(T, p)$ are known (these are the two equations of state of the system), Eq. (5.46) can be integrated to determine μ up to an additive constant that results from the arbitrary zero of energy. For a two component system, there would be three independent intensive variables, say p, T, and μ_1, and then $\mu_2(p, T, \mu_1)$. For a three component system, there would be four independent intensive variables, etc.

Example Problem 5.3. By using the chemical potential of an ideal gas given by Eqs. (5.5) and (5.46), determine its equations of state. From these results, calculate the enthalpy per mole h and the internal energy per mole u and comment on their dependence on pressure. Deduce the relationship between the molar heat capacities c_V and c_p at constant volume and constant pressure and compare with Eq. (2.13).

Solution 5.3. We have $v = (\partial\mu/\partial p)_T = RT/p$, which just reproduces the ideal gas law, which is one equation of state. Also, $s = -(\partial\mu/\partial T)_p = -\mathrm{d}\mu^*(T)/\mathrm{d}T - R\ln p$, which is the other equation of state. Thus, by dividing the Euler equation Eq. (5.43) for a single component by N, we obtain $u = Ts - pv + \mu$, so the molar enthalpy

$$h = u + pv = \mu + Ts = \mu^*(T) - T\frac{\mathrm{d}\mu^*(T)}{\mathrm{d}T}, \tag{5.47}$$

which is a function of only the temperature, independent of pressure. Moreover,

$$u = h - pv = \mu^*(T) - T\frac{\mathrm{d}\mu^*(T)}{\mathrm{d}T} - RT, \tag{5.48}$$

which is also a function of only the temperature, independent of v. We readily compute $c_p = \mathrm{d}h/\mathrm{d}T = -T\mathrm{d}^2\mu^*(T)/\mathrm{d}T^2$ and $c_v = \mathrm{d}u/\mathrm{d}T = c_p - R$ in agreement with Eq. (2.13), even if c_p and c_v depend on T.

[8] We could make other choices, such as regarding T and μ as the independent variables, and then writing $p(T, \mu)$.

Composition: For multicomponent systems, one often regards the independent intensive variables as being p, T, and **composition**, where composition[9] is designated by the $\kappa - 1$ independent **mole fractions**

$$X_i := N_i/N, \tag{5.49}$$

where $N = \sum_{i=1}^{\kappa} N_i$ is the total number of moles. Note that the X_i are intensive variables, because they are ratios of extensive variables. Moreover, we have

$$\sum_{i=1}^{\kappa} X_i = 1 \tag{5.50}$$

so only $\kappa - 1$ of them are independent, as already stated. Taking the differential of Eq. (5.50) gives

$$\sum_{i=1}^{\kappa} \mathrm{d}X_i = 0 \tag{5.51}$$

so we note that it is *impossible* to take a partial derivative with respect to one of the X_i while holding *all* of the others constant. In particular, we cannot calculate chemical potentials by taking a single partial derivative with respect to an X_i, that is,

$$\mu_i \neq \left(\frac{\partial U}{\partial X_i}\right)_{S,V,\{X_i'\}} \tag{5.52}$$

because the right-hand side is meaningless.

For a system with two components, we could take a set of the independent intensive variables to be p, T, and X_1, in which case $\mu_1(p, T, X_1)$ and $\mu_2(p, T, X_1)$. For three components, independent intensive variables could be p, T, X_1, and X_2, in which case $\mu_1(p, T, X_1, X_2)$, $\mu_2(p, T, X_1, X_2)$, and $\mu_3(p, T, X_1, X_2)$. To recover an extensive description, we could add to these variable sets N or any one of the N_i.

Enthalpy of a multicomponent system: Recall that we defined the enthalpy $H := U + pV$. For a multicomponent system

$$\mathrm{d}H = \mathrm{d}U + p\,\mathrm{d}V + V\,\mathrm{d}p = T\,\mathrm{d}S + V\,\mathrm{d}p + \sum_{i=1}^{\kappa} \mu_i\,\mathrm{d}N_i, \tag{5.53}$$

[9]For a mass based description, we can describe composition by the mass fractions $\omega_i := M_i/M$, where M_i is the mass of the ith component and $M = \sum_{i=1}^{\kappa} M_i$ is the total mass. The relationship of the ω_i to the X_i is nonlinear and depends on the molecular masses, m_i. Specifically, $\omega_i = m_i X_i / \sum_j m_j X_j$.

where Eq. (5.10) has been used. Thus, H is a **natural function**[10] of S, p, and the N_i and we have

$$T = \left(\frac{\partial H}{\partial S}\right)_{p,\{N_i\}} ; \quad V = \left(\frac{\partial H}{\partial p}\right)_{S,\{N_i\}} ; \quad \mu_i = \left(\frac{\partial H}{\partial N_i}\right)_{S,p,\{N_i'\}} . \tag{5.54}$$

Furthermore, with regard to homogeneity, we see that

$$H(\lambda S, p, \lambda N_1, \lambda N_2, \ldots, \lambda N_\kappa) = \lambda H(S, p, N_1, N_2, \ldots, N_\kappa) \tag{5.55}$$

because p, being intensive, does not participate. Thus, application of the Euler theorem gives

$$H = TS + \sum_{i=1}^{\kappa} \mu_i N_i \tag{5.56}$$

which is in agreement with Eq. (5.43) once the definition of H is used. So we actually get nothing new (except self-consistency) and Eq. (5.45) follows as well.

5.3.2 Euler Theorem Applied to Intensive Functions

Intensive functions are homogeneous functions of degree zero with respect to extensive variables. For example, the energy $u := U/N$ per mole or the energy $u_V := U/V$ per unit volume are intensive. They are therefore homogeneous functions of degree zero in the variables S, V, N_1, $N_2, \ldots, N_\kappa$, which means that they can depend only on *ratios* of these variables, which ratios are themselves intensive. To see this formally, note that

$$u = \frac{U(S, V, N_1, N_2, \ldots, N_\kappa)}{N} = \frac{U(\lambda S, \lambda V, \lambda N_1, \lambda N_2, \ldots, \lambda N_\kappa)}{\lambda N} \tag{5.57}$$

and then choose $\lambda = 1/N$ to deduce

$$u = U(s, v, X_1, X_2, \ldots, X_\kappa), \tag{5.58}$$

where $s = S/N$ is the entropy per mole and $v = V/N$ is the volume per mole. But since the X_i are not all independent we can omit the last of them and write, in terms of $\kappa + 1$ independent variables,

$$u = u(s, v, X_1, X_2, \ldots, X_{\kappa-1}) \tag{5.59}$$

whose differential turns out to be

$$\mathrm{d}u = T\,\mathrm{d}s - p\,\mathrm{d}v + \sum_{i=1}^{\kappa-1} (\mu_i - \mu_\kappa)\,\mathrm{d}X_i. \tag{5.60}$$

[10]A natural function is a thermodynamic potential that contains information equivalent to a fundamental equation (for U or S) and whose independent variables are either members of the original complete set of extensive variables (on which U or S depends) or their conjugate variables (which are the partial derivatives of U or S with respect to their extensive variables).

Equation (5.60) can be verified by taking the differential of Eq. (5.57) and using Eqs. (5.10) and (5.43) to simplify the result. Thus

$$du = \frac{dU}{N} - \frac{U}{N^2}\,dN = \frac{1}{N}\left[T\,dS - p\,dV + \sum_{i=1}^{\kappa}\mu_i\,dN_i\right] - \frac{1}{N^2}\left[TS - pV + \sum_{i=1}^{\kappa}\mu_i N_i\right]dN. \tag{5.61}$$

But $ds = dS/N - (S/N^2)\,dN$, $dv = dV/N - (V/N^2)\,dN$ and $dX_i = dN_i/N - (N_i/N^2)\,dN$, so Eq. (5.61) becomes

$$du = T\,ds - p\,dv + \sum_{i=1}^{\kappa}\mu_i\,dX_i, \tag{5.62}$$

which reduces to Eq. (5.60) after dX_κ is eliminated. Note in this derivation that the total number of moles N was treated as a variable, even though the result appears as if we just treated it as a constant and divided by it.

In a similar way, we can deduce that

$$u_V = u(s_V, c_1, c_2, \ldots, c_\kappa), \tag{5.63}$$

where $s_V := S/V$ is the entropy per unit volume and the $c_i := N_i/V$ are the **concentrations** of each component (in moles per unit volume). The corresponding differential is

$$du_V = T\,ds_V + \sum_{i=1}^{\kappa}\mu_i\,dc_i. \tag{5.64}$$

Similar considerations apply to other intensive variables, such as the enthalpy $h := H/N$ per mole, which is a function of s, p, $X_1, X_2, \ldots, X_{\kappa-1}$ and whose differential is

$$dh = T\,ds + v\,dp + \sum_{i=1}^{\kappa-1}(\mu_i - \mu_\kappa)\,dX_i. \tag{5.65}$$

5.4 Chemical Potential of Real Gases, Fugacity

As a further application of Eq. (5.46), we treat the dependence of the chemical potential of a pure non-ideal gas on temperature and pressure by means of a function known as the fugacity (see, for example, Denbigh [18, p. 125]). To do this, we replace Eq. (5.5) by

$$\mu(T,p) = \mu^*(T) + RT\ln f; \quad f \to p \text{ as } p \to 0. \tag{5.66}$$

The **fugacity** $f(T,p)$ is an effective pressure[11] that replaces the pressure p of an ideal gas. Equation (5.66) is based on the idea that all gases will tend toward ideal gas behavior if sufficiently dilute, which will be the case for fixed temperature at sufficiently low pressure. Therefore, the function $\mu^*(T)$ is precisely the same function of T as for the corresponding ideal gas.

[11]One can also employ a dimensionless fugacity $f^D = f/p_0$, where p_0 is a reference pressure, by adding a term $RT\ln p_0$ to $\mu^*(T)$. Then if $p_0 = 1$ atmosphere and pressures are measured in atmospheres, the term $RT\ln p_0 = 0$ numerically.

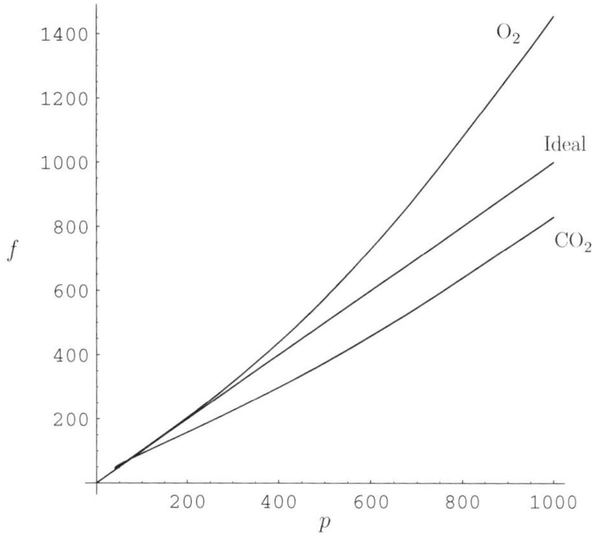

FIGURE 5–1 Fugacity f (atmospheres) versus pressure p (atmospheres) of O_2 and CO_2 at $T = 200\,^\circ\text{C}$ based on a cubic spline fit of data of Darken and Gurry [19, p. 210]. The middle line is for an ideal gas for which $f = p$. The lowest data points are for $p = 50$ atmospheres for which $f = 50.5$ atmospheres for O_2 and 47.8 atmospheres for CO_2.

The general dependence of the fugacity on pressure may be deduced by integrating the equation

$$v(T,p) = \left(\frac{\partial \mu(T,p)}{\partial p}\right)_T = RT\left(\frac{\partial \ln f}{\partial p}\right)_T; \quad f \to p \text{ as } p \to 0. \tag{5.67}$$

In order to avoid a singularity and to incorporate the condition on f at low pressures, we rewrite Eq. (5.67) in the form

$$\left(\frac{\partial \ln(f/p)}{\partial p}\right)_T = \frac{v(T,p)}{RT} - \frac{1}{p}. \tag{5.68}$$

Then integration on pressure at constant temperature from 0 to p gives

$$\ln(f/p) = \int_0^p \left[\frac{v(T,p')}{RT} - \frac{1}{p'}\right] \mathrm{d}p'. \tag{5.69}$$

If the gas is ideal, the integrand vanishes and one obtains simply $f = p$. Depending on the temperature, many gases behave like ideal gases at atmospheric pressure p_0, but at high pressures the deviations from ideality can be quite significant. Figure 5–1 shows a plot of fugacity versus pressure for the gases O_2 and CO_2 at a temperature of $200\,^\circ\text{C}$, as well as for an ideal gas, for which $f = p$ at all temperatures. Note the opposite deviations from ideality.

Example Problem 5.4. Suppose that a non-ideal gas has an expansion (called a virial expansion) in terms of pressure of the form

$$\frac{v(T,p)}{RT} = \frac{1}{p}\left[1 + \tilde{B}p + \tilde{C}p^2 + \cdots\right] = \frac{1}{p} + \tilde{B} + \tilde{C}p + \cdots, \tag{5.70}$$

where $\tilde{B}(T)$ and $\tilde{C}(T)$ are virial coefficients that depend on the temperature. Calculate the fugacity of this gas. For a simple model based on a potential consisting of a hard repulsive core of diameter σ_1, an attractive potential well of constant depth ε in the annular region between a sphere of diameter σ_2 and the repulsive core, and zero potential beyond, the first virial coefficient is given by

$$\tilde{B}(T) = \frac{2\pi}{3}\left[\frac{\sigma_1^3 - (e^{\varepsilon/k_B T} - 1)(\sigma_2^3 - \sigma_1^3)}{k_B T}\right]. \tag{5.71}$$

If this is the only important virial coefficient, discuss briefly the effect of temperature on the fugacity.

Solution 5.4. Equation (5.69) becomes

$$\ln[f(T,p)/p] = \int_0^p [\tilde{B} + \tilde{C}p' + \cdots]dp' = \tilde{B}(T)p + (\tilde{C}(T)/2)p^2 + \cdots. \tag{5.72}$$

Thus,

$$f(T,p) = p\,\exp[\tilde{B}(T)p + (\tilde{C}(T)/2)p^2 + \cdots]. \tag{5.73}$$

The first virial coefficient given by Eq. (5.71) becomes $\tilde{B}(T) = -(2\pi/3)(\sigma_2^3 - \sigma_1^3)(e^{\varepsilon/k_B T}/k_B T)$ at low temperatures and $\tilde{B}(T) = (2\pi/3)(\sigma_1^3/k_B T)$ at high temperatures. It therefore changes sign from negative to positive as the temperature increases. If this is the only important virial coefficient, $f < p$ and varies strongly with temperature for low temperatures and $f > p$ and varies weakly with temperature for high temperatures.

Example Problem 5.5. For the previous example, compare the chemical potential difference $\mu(T,p) - \mu(T,p_0)$ for a real gas with that for a condensed phase (solid or liquid) for which the molar volume is given approximately by $v(T,p) = v(T,p_0)[1-\kappa_T(p-p_0)]$, where the isothermal compressibility κ_T is evaluated at p_0.

Solution 5.5. For the real gas,

$$\mu(T,p) - \mu(T,p_0) = RT\left[\ln(p/p_0) + \tilde{B}(T)(p-p_0) + (\tilde{C}(T)/2)(p^2 - p_0^2) + \cdots\right]. \tag{5.74}$$

The term $RT\ln(p/p_0)$ is very important and the other terms represent a small correction unless the pressure is very large.

For a condensed phase, the integral of $(\partial\mu(T,p)/\partial p)_T = v(T,p)$ at constant temperature yields

$$\mu(T,p) - \mu(T,p_0) = v(T,p_0)[(p-p_0) - (\kappa_T/2)(p-p_0)^2], \quad \text{solid or liquid.} \tag{5.75}$$

Except for very large pressure differences, this difference is small compared to RT.

Therefore, for gases the chemical potential has a significant dependence on pressure but for condensed phases it is practically independent of pressure.

5.5 Legendre Transformations

Legendre transformations are frequently used in thermodynamics to define new functions that depend on a convenient variable set. An example is the enthalpy function $H = U + pV$ discussed in Chapter 2 with differential given by Eq. (5.53) for a multicomponent system. In Chapter 6 we will show how some of these functions can be used to formulate useful criteria for thermodynamic equilibrium. Here we cover some formal aspects of such transformations.

We treat a system having κ chemical components and for which $\mathrm{d}U$ is given by Eq. (5.10) which we write in the schematic form

$$\mathrm{d}U = \sum_{i=1}^{\kappa+2} p_i \, \mathrm{d}E_i. \tag{5.76}$$

The extensive variables $E_1 = S$, $E_2 = V$, and $E_{i+2} = N_i$ for $i = 1, 2, \ldots, \kappa$. Evidently

$$p_i = \left(\frac{\partial U}{\partial E_i}\right)_{\{E_i'\}}, \tag{5.77}$$

so the corresponding (intensive) potentials are $p_1 = T$, $p_2 = -p$, and $p_{i+2} = \mu_i$ for $i = 1, 2, \ldots, \kappa$. The variables p_i and E_i are called **conjugate variables**. We now define a **Legendre transform** by means of the function

$$L_j := U - p_j E_j = U - E_j \left(\frac{\partial U}{\partial E_j}\right)_{\{E_j'\}}, \tag{5.78}$$

obtained by subtracting from U the product of the conjugate variables p_j and E_j. We obtain

$$\mathrm{d}L_j = \mathrm{d}U - p_j \, \mathrm{d}E_j - E_j \, \mathrm{d}p_j = -E_j \, \mathrm{d}p_j + \sum_{i \neq j}^{\kappa+2} p_i \, \mathrm{d}E_i. \tag{5.79}$$

We can regard L_j to be a function of p_j and the remaining E_i for $i \neq j$. In other words, L_j depends on the slope of the function U with respect to E_j. Moreover, L_j itself is the $E_j = 0$ intercept of a graph of U versus E_j. Since a curve may be defined by the envelope of its tangent lines, a knowledge of intercept L_j as a function of slope p_j is equivalent to a knowledge of U as a function of E_j, regarding all of the remaining E_i, for $i \neq j$ to be fixed. See Callen [2, p. 140] for an extended discussion of this equivalence. Given the function L_j, Eq. (5.79) yields

$$E_j = -\left(\frac{\partial L_j}{\partial p_j}\right)_{\{E_j'\}}. \tag{5.80}$$

We can therefore write the inverse of Eq. (5.78) in the form

$$U = L_j + p_j E_j = L_j - p_j \left(\frac{\partial L_j}{\partial p_j}\right)_{\{E_j'\}}. \tag{5.81}$$

We note the reciprocity (with appropriate sign changes) of the relationship between U and L_j, so they can be regarded as Legendre transforms of one another.

We next obtain a useful relation between second derivatives of Legendre transforms. We have

$$\frac{\partial^2 U}{\partial E_j^2} = \frac{\partial p_j}{\partial E_j} = -\frac{\partial p_j}{\partial(\partial L_j/\partial p_j)} = -1/\frac{\partial^2 L_j}{\partial p_j^2}. \tag{5.82}$$

Thus if U is a convex function (positive second derivative) of E_j, then L_j will be a concave function (negative second derivative) of p_j.

Simple example: For a single variable E, suppose that $U = A + BE^2$, where A and B are constants. Then $p = \partial U/\partial E = 2BE$ and $L = U - pE = A - BE^2 = A - p^2/4B$. For the inverse transformation, we start with $L(p)$ and obtain $E = -\partial L/\partial p = p/2B$. Then $U = L + pE = A + p^2/4B = A + BE^2$. We also have $\partial^2 U/\partial E^2 = 2B$ and $\partial^2 L/\partial p^2 = -1/(2B)$.

We could make an additional Legendre transformation by selecting a second pair of conjugate variables, say $p_k E_k$, and subtracting from L_j. This produces a function

$$L_{jk} := L_j - p_k E_k = U - p_j E_j - p_k E_k = L_k - p_j E_j := L_{kj}, \tag{5.83}$$

which can be thought of as a double Legendre transform of the original U. We will then have

$$\mathrm{d}L_{jk} = -E_j\,\mathrm{d}p_j - E_k\,\mathrm{d}p_k + \sum_{i\neq j,k}^{\kappa+2} p_i\,\mathrm{d}E_i \tag{5.84}$$

and we can regard L_{jk} to be a function of p_j, p_k and the remaining E_i, where $i \neq j, k$. This process can be continued up to $\kappa + 1$ successive Legendre transforms. Since the Euler equation is $U = \sum_{i=1}^{\kappa+2} p_i E_i$, we see that $\kappa + 2$ Legendre transforms of U would lead identically to zero. The total number of possible transforms is therefore $2^{\kappa+2} - 2$.

5.5.1 Specific Legendre Transforms

We end this section by identifying several specific Legendre transforms that play an important role in thermodynamics and statistical mechanics.

Helmholtz free energy *F*: We define the Helmholtz free energy by the Legendre transformation

$$F := U - TS \tag{5.85}$$

with differential

$$\mathrm{d}F = -S\,\mathrm{d}T - p\,\mathrm{d}V + \sum_{i=1}^{\kappa} \mu_i\,\mathrm{d}N_i. \tag{5.86}$$

Effectively, the dependence of U on S is replaced by the dependence of F on T, whereas both U and F depend on V and $\{N_i\}$. Thus, F is useful in situations where T is a control variable. We note that $S = -(\partial F/\partial T)_{V,N_i}$ and $\partial^2 U/\partial S^2 = -1/(\partial^2 F/\partial T^2)$. A number of Maxwell relations can be deduced from dF, one being $(\partial S/\partial V)_{T,\{N_i\}} = \left(\partial p/\partial T\right)_{V,\{N_i\}}$.

Enthalpy *H*: We have previously mentioned the enthalpy defined by

$$H := U + pV \tag{5.87}$$

with differential

$$dH = T\,dS + V\,dp + \sum_{i=1}^{\kappa} \mu_i\,dN_i. \tag{5.88}$$

Effectively, the dependence of U on V is replaced by the dependence of H on p. We note that $V = \left(\partial H/\partial p\right)_{S,N_i}$ and $\partial^2 U/\partial V^2 = -1/(\partial^2 H/\partial p^2)$. A Maxwell relation is $\left(\partial T/\partial p\right)_{S,\{N_i\}} = (\partial V/\partial S)_{p,\{N_i\}}$.

Gibbs free energy *G*: The Gibbs free energy is a double Legendre transformation from U or a single Legendre transformation from F or H and is defined by

$$G := U - TS + pV = F + pV = H - TS. \tag{5.89}$$

It has a differential

$$dG = -S\,dT + V\,dp + \sum_{i=1}^{\kappa} \mu_i\,dN_i. \tag{5.90}$$

The control variables for G are T and p as opposed to S and V for U. G is especially important for the study of chemical reactions that take place at various temperatures and atmospheric pressure. We note that $S = -(\partial G/\partial T)_{p,N_i}$ and $V = \left(\partial G/\partial p\right)_{T,N_i}$ as well as $\partial^2 H/\partial S^2 = -1/(\partial^2 G/\partial T^2)$ and $\partial^2 F/\partial V^2 = -1/(\partial^2 G/\partial p^2)$. One Maxwell relation is $\left(\partial S/\partial p\right)_{T,\{N_i\}} = -(\partial V/\partial T)_{p,\{N_i\}}$. Other useful Maxwell relations involving the chemical potentials are $\left(\partial \mu_i/\partial p\right)_{T,\{N_i\}} = (\partial V/\partial N_i)_{T,p,\{N_i'\}} =: \bar{V}_i$ and $(\partial \mu_i/\partial T)_{p,\{N_i\}} = -(\partial S/\partial N_i)_{T,p,\{N_i'\}} =: \bar{S}_i$. The quantities $\bar{V}_i$ and $\bar{S}_i$ are known respectively as the partial molar volume and the partial molar entropy and are examples of partial molar quantities discussed in Section 5.6.

Kramers potential *K*: The Kramers potential, also known as the grand potential and often denoted by Ω, is obtained by transforming all variables in U except V. It is defined by

$$K := U - TS - \sum_{i=1}^{\kappa} \mu_i N_i \tag{5.91}$$

and has a differential

$$dK = -S\,dT - p\,dV - \sum_{i=1}^{\kappa} N_i\,d\mu_i. \tag{5.92}$$

K depends on T, V, and all of the chemical potentials μ_i. We note that $S = -(\partial K/\partial T)_{V,\{\mu_i\}}$ and $N_i = -(\partial K/\partial \mu_i)_{T,V,\{\mu_i'\}}$, where $\{\mu_i'\}$ stands for the entire set $\{\mu_i\}$ of chemical potentials but with μ_i missing. The potential K is closely related to the grand canonical ensemble of statistical mechanics and is also useful in problems involving surfaces and interfaces.

Massieu functions: The functions F, H, G, and K, which are known as thermodynamic potentials, are all Legendre transforms of the internal energy U. One can also begin with the entropy whose differential is given by Eq. (5.10), namely

$$dS = (1/T)\,dU + (p/T)\,dV - \sum_{i=1}^{\kappa}(\mu_i/T)\,dN_i. \tag{5.93}$$

Its Legendre transforms are known as Massieu functions. For example,[12]

$$M_1(1/T, V, \{N_i\}) := S - (1/T)U \equiv S[1/T], \tag{5.94}$$

where the last notation has been used by Callen [2, p. 151]. It has a differential

$$dM_1 = -U\,d(1/T) + (p/T)\,dV - \sum_{i=1}^{\kappa}(\mu_i/T)\,dN_i. \tag{5.95}$$

This differential leads to the Maxwell relation

$$\left(\frac{\partial U}{\partial V}\right)_{1/T,\{N_i\}} = -\left(\frac{\partial (p/T)}{\partial (1/T)}\right)_{V,\{N_i\}} = -p + T\left(\frac{\partial p}{\partial T}\right)_{V,\{N_i\}}, \tag{5.96}$$

which is the same as Eq. (5.29). For an ideal gas, $p/T = NR/V$ and Eq. (5.96) yields $(\partial U/\partial V)_{T,\{N_i\}} = 0$. Thus, the fact that U depends only on T for an ideal gas, which was deduced on the basis of experiments on dilute gases, follows from the ideal gas equation of state and the second law. Some other Massieu functions are

$$M_2(U, p/T, \{N_i\}) : = S - (p/T)V \equiv S[p/T]; \tag{5.97}$$

$$dM_2 = (1/T)\,dU - V\,d(p/T) - \sum_{i=1}^{\kappa}(\mu_i/T)\,dN_i$$

and[13]

$$M_3(1/T, p/T, \{N_i\}) : = S - (1/T)U - (p/T)V \equiv S[(1/T, p/T]; \tag{5.98}$$

$$dM_3 = -U\,d(1/T) - V\,d(p/T) - \sum_{i=1}^{\kappa}(\mu_i/T)\,dN_i.$$

One could also add quantities such as μ_1/T to S to obtain a function $S[\mu_1/T]$ that depends on μ_1/T instead of N_1. The total number of possible transforms of the entropy is $2^{\kappa+2} - 2$, just as for the transforms of the thermodynamic potentials.

[12] M_1 is sometimes denoted by Φ and called the Helmholtz free entropy.

[13] M_3 is sometimes denoted by Ξ (or Ψ by Planck) and is called the Gibbs free entropy.

Natural variables: The natural variables of a thermodynamic potential are the set of independent variables that give complete information about the system under consideration. For isotropic multicomponent fluids, the natural variables of the entropy are the set of extensive variables U, V, $\{N_i\}$, where $\{N_i\} = N_1, N_2, \ldots, N_\kappa$. For the internal energy, the natural variables are S, V, and $\{N_i\}$. Since S is a monotonically increasing function of U with other extensive variables held constant, one can always transform uniquely from $S(U, V, \{N_i\})$ to $U(S, V, \{N_i\})$ and vice versa for these fundamental equations. For the thermodynamic potentials, the natural variables are those independent variables that result from Legendre transformation. For example, any of the functions $F(T, V, \{N_i\})$, $H(U, p, \{N_i\})$, $G(T, p, \{N_i\})$, $K(T, V, \{\mu_i\})$, $M_3(1/T, p/T, \{N_i\})$ contain complete information about a system. It is possible and sometimes useful to express these functions in terms of other variable sets, as discussed in the next section.

5.6 Partial Molar Quantities

This section applies to any extensive state function that can be expressed in terms of the complete variable set $T, p, \{N_i\}$ for a homogeneous system having κ chemical components. For example, we could consider the internal energy $U(T, p, \{N_i\})$, even though the natural variables for U are the set $S, V, \{N_i\}$. We could also consider the entropy $S(T, p, \{N_i\})$ or the enthalpy $H(T, p, \{N_i\})$, etc. Of course a transformation of variables is necessary to convert from the set of natural variables of a function to the set $T, p, \{N_i\}$, except for $G(T, p, \{N_i\})$ where these are also its natural variables. As we shall see in Chapter 6, the temperature T and the pressure p are uniform for phases in mutual equilibrium, so functions expressed in terms of these intensive variables are particularly important.

For the generic extensive function $Y(T, p, N_1, N_2, \ldots, N_\kappa)$, the partial molar quantities $\bar{Y}_i$ are defined as derivatives[14]

$$\bar{Y}_i := \left(\frac{\partial Y}{\partial N_i}\right)_{T,p,\{N_i'\}}. \tag{5.99}$$

Since Y is an extensive function in the variables $N_1, N_2, \ldots, N_\kappa$, we have

$$Y(T, p, \lambda N_1, \lambda N_2, \ldots, \lambda N_\kappa) = \lambda Y(T, p, N_1, N_2, \ldots, N_\kappa) \tag{5.100}$$

so the Euler theorem gives

$$Y = \sum_{i=1}^{\kappa} \bar{Y}_i N_i. \tag{5.101}$$

Since T and p are held constant in the definition Eq. (5.99), we can differentiate the equations that define H, F, and G to obtain $\bar{H}_i = \bar{U}_i + p\bar{V}_i$, $\bar{F}_i = \bar{U}_i - T\bar{S}_i$, and

[14] Instead of the mole numbers N_i, we could use the masses M_i of each chemical component. Then one could develop a parallel treatment in terms of partial specific quantities defined by $\tilde{Y}_i := (\partial Y/\partial M_i)_{T,p,\{M_i'\}}$.

$\bar{G}_i = \bar{U}_i - T\bar{S}_i + p\bar{V}_i$. Therefore, these partial molar quantities obey the same algebra as their definitions. Since the natural variables for G are $T, p, \{N_i\}$, we observe that $\bar{G}_i \equiv \mu_i$, which is just a special symbol for this very important partial molar quantity. By the same reasoning as for Y, the corresponding Euler equations in terms of partial molar quantities are $H = \sum_i \bar{H}_i N_i$, $F = \sum_i \bar{F}_i N_i$, and $G = \sum_i \bar{G}_i N_i = \sum_i \mu_i N_i$. Given the algebra of the partial molar quantities just mentioned, the first two of these are in agreement with the Euler equations $H = TS + \sum_i \mu_i N_i$ and $F = -pV + \sum_i \mu_i N_i$.

For our further development, we take the volume $V(T, p, N_1, N_2, \ldots, N_\kappa)$ as a specific example, but the procedure is quite general and applies to any extensive function $Y(T, p, N_1, N_2, \ldots, N_\kappa)$. If the partial molar volumes $\bar{V}_i$ were constants, Eq. (5.101) for V would have the obvious interpretation that $\bar{V}_i$ was the volume per mole actually occupied by species i, in which case the total volume would be a linear function of the N_i. But Eq. (5.101) for $Y = V$ is true even when the $\bar{V}_i$ vary with composition as well as T and p. From the form of Eq. (5.99), it is clear that the $\bar{V}_i$ are intensive variables, so they can only depend on the ratios of the N_i. Thus, they can be expressed as functions of the independent variable set $T, p, X_1, X_2, \ldots, X_{\kappa-1}$. Written out in full, we have

$$V(T, p, N_1, N_2, \ldots, N_\kappa) = \sum_{i=1}^{\kappa} \bar{V}_i(T, p, X_1, X_2, \ldots, X_{\kappa-1}) N_i. \tag{5.102}$$

The differential of V is

$$dV = V\alpha \, dT - V\kappa_T \, dp + \sum_{i=1}^{\kappa} \bar{V}_i \, dN_i, \tag{5.103}$$

but from $V = \sum_i \bar{V}_i N_i$ it can also be written

$$dV = \sum_{i=1}^{\kappa} \bar{V}_i \, dN_i + \sum_{i=1}^{\kappa} N_i \, d\bar{V}_i. \tag{5.104}$$

Comparison of Eqs. (5.103) and (5.104) shows that

$$\sum_{i=1}^{\kappa} N_i \, d\bar{V}_i = V\alpha \, dT - V\kappa_T \, dp, \tag{5.105}$$

which is an equation of Gibbs-Duhem type. We can divide Eq. (5.101) by N to obtain an equation for the molar volume

$$v := \frac{V}{N} = \sum_{i=1}^{\kappa} \bar{V}_i X_i. \tag{5.106}$$

For a single component material there is only one partial molar volume, $\bar{V}_1 = (\partial V/\partial N)_{T,p}$ and it depends only on T and p. In that case, Eq. (5.106) takes the form

$$v = \frac{V}{N} = \left(\frac{\partial V}{\partial N}\right)_{T,p}, \quad \text{single component.} \tag{5.107}$$

In this simple case, the derivative with respect to N becomes just the ratio V/N. Equation (5.105) becomes $\mathrm{d}v = v\alpha\,\mathrm{d}T - v\kappa_T\,\mathrm{d}p$ which can be rewritten

$$\mathrm{d}\ln v = \alpha\,\mathrm{d}T - \kappa_T\,\mathrm{d}p, \quad \text{single component.} \tag{5.108}$$

Thus

$$\alpha = \left(\frac{\partial \ln v}{\partial T}\right)_p; \quad \kappa_T = -\left(\frac{\partial \ln v}{\partial p}\right)_T, \quad \text{single component.} \tag{5.109}$$

For a multicomponent material, we see from Eq. (5.106) that Eq. (5.107) must be replaced by

$$v = \frac{V}{N} = \left(\frac{\partial V}{\partial N}\right)_{T,p,\{X_i\}}, \quad \text{multicomponent.} \tag{5.110}$$

To obtain Eq. (5.110), we hold the composition constant, in addition to T and p, in Eq. (5.102) and just allow the total number of moles N to vary, so $\mathrm{d}N_i = X_i\,\mathrm{d}N$. On the other hand, Eq. (5.108) is insufficient and must be replaced by

$$\begin{aligned}\mathrm{d}v = \frac{\mathrm{d}V}{N} - \frac{V}{N^2}\,\mathrm{d}N &= v\alpha\,\mathrm{d}T - v\kappa_T\,\mathrm{d}p + \sum_{i=1}^{\kappa} \bar{V}_i\,\mathrm{d}X_i \\ &= v\alpha\,\mathrm{d}T - v\kappa_T\,\mathrm{d}p + \sum_{i=1}^{\kappa-1} (\bar{V}_i - \bar{V}_\kappa)\,\mathrm{d}X_i,\end{aligned} \tag{5.111}$$

where the second form is written in terms of the differentials of $\kappa + 1$ independent variables. Instead of Eq. (5.109), we now have

$$\alpha = \left(\frac{\partial \ln v}{\partial T}\right)_{p,\{X_i\}}; \quad \kappa_T = -\left(\frac{\partial \ln v}{\partial p}\right)_{T,\{X_i\}}. \tag{5.112}$$

5.6.1 Method of Intercepts

The method of intercepts provides a useful graphical representation of partial molar quantities. We illustrate it for partial molar volumes, but it applies to any partial molar quantities. We first illustrate it for a binary system and then derive the general formulae for a multicomponent system.

Binary system: For a binary system, there are only two chemical components, so we choose an independent variable set p, T, X_2. Since X_1 is not a member of this set, X_2 is allowed to vary freely, so we can take partial derivatives with respect to T, p, or X_2 while holding the other pair constant. For a binary system, Eq. (5.106) becomes

$$v = \bar{V}_1 X_1 + \bar{V}_2 X_2 = \bar{V}_1(1 - X_2) + \bar{V}_2 X_2 \tag{5.113}$$

and from Eq. (5.111), with $\mathrm{d}X_1 = -\mathrm{d}X_2$, we obtain

$$\left(\frac{\partial v}{\partial X_2}\right)_{T,p} = \bar{V}_2 - \bar{V}_1. \tag{5.114}$$

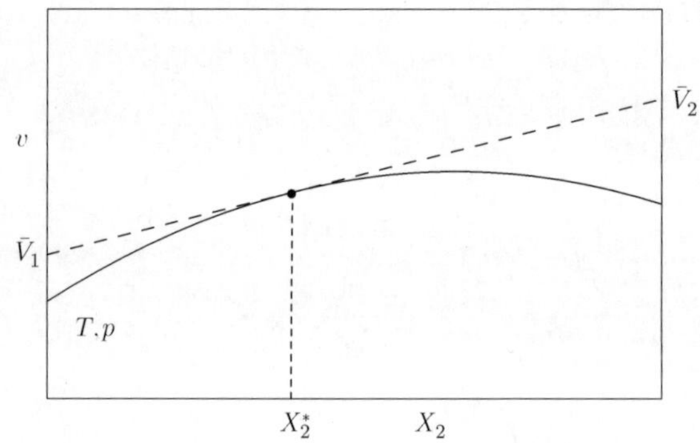

FIGURE 5–2 Illustration of the method of intercepts to calculate partial molar volumes for a system of two components. We plot a graph of v versus X_2 at fixed p and T. Then the partial molar volumes for some composition X_2^* are given by the intercepts of the tangent to v at X_2^*. $\bar{V}_1(T, p, X_2^*)$ is the intercept at $X_2 = 0$ which corresponds to pure component 1. $\bar{V}_2(T, p, X_2^*)$ is the intercept at $X_2 = 1$ which corresponds to pure component 2.

We solve Eqs. (5.113) and (5.114) simultaneously for $\bar{V}_1$ and $\bar{V}_2$ to obtain

$$\bar{V}_1 = v - X_2 \left(\frac{\partial v}{\partial X_2}\right)_{T,p}; \tag{5.115}$$

$$\bar{V}_2 = v + (1 - X_2) \left(\frac{\partial v}{\partial X_2}\right)_{T,p}. \tag{5.116}$$

Equations (5.115) and (5.116) are illustrated in Figure 5–2. On a graph of v versus X_2 (at fixed T and p) we see that the partial molar volumes for some composition X_2^* are given by the intercepts, at $X_2 = 0$ and $X_2 = 1$, of the tangent to v at X_2^*. This graphic construction allows one to see immediately how $\bar{V}_1$ and $\bar{V}_2$ vary with composition X_2^*.

For example, if V is a linear function of X_2, its tangent is coincident with V itself and $\bar{V}_1$ and $\bar{V}_2$ are independent of X_2. In that case, one can imagine that each component of the solution has a fixed physical volume. Moreover, if the curve V versus X_2 is convex, instead of concave as in Figure 5–2, a partial molar volume could be negative! In that case, it makes no sense to think of a partial molar volume as a physical volume; instead, it is only a manifestation of the slope of the V versus X_2 curve, even though Eq. (5.113) still holds.

Multicomponent system: For multicomponent systems, we use the second form of Eq. (5.111) which depends on the set of independent variables $p, T, X_1, X_2, \ldots, X_{\kappa-1}$. Within this reduced variable set, we can take the partial derivative with respect to X_i to obtain

$$\left(\frac{\partial v}{\partial X_i}\right)_{T,p,\{X_i'\}} = (\bar{V}_i - \bar{V}_\kappa); \quad i = 1, 2, \ldots, \kappa - 1. \tag{5.117}$$

We multiply Eq. (5.117) by X_i and sum to get

$$\sum_{i=1}^{\kappa-1} X_i \left(\frac{\partial v}{\partial X_i}\right)_{T,p,\{X_i'\}} = \sum_{i=1}^{\kappa-1} X_i \bar{V}_i - \bar{V}_\kappa \sum_{i=1}^{\kappa-1} X_i = \sum_{i=1}^{\kappa-1} X_i \bar{V}_i + X_\kappa \bar{V}_\kappa - \bar{V}_\kappa. \tag{5.118}$$

Then by using Eq. (5.106), Eq. (5.118) becomes

$$\bar{V}_\kappa = v - \sum_{i=1}^{\kappa-1} X_i \left(\frac{\partial v}{\partial X_i}\right)_{T,p,\{X_i'\}}. \tag{5.119}$$

Equation (5.119) can be interpreted geometrically by imagining v to be plotted as a hypersurface in the coordinates $X_1, X_2, \ldots, X_{\kappa-1}$. The quantity on its right-hand side is then seen to be the intercept on the v axis, at the origin of the X_i, of a hyperplane that is tangent to v at composition $X_1, X_2, \ldots, X_{\kappa-1}$. Unlike the case of two components, this is not particularly easy to visualize.

Example Problem 5.6. A solution of A and B atoms at constant temperature and pressure has a molar volume $v = 3 + 2X_B - X_B^2\ \mathrm{cm}^3/\mathrm{mol}$, where X_B is the mole fraction of B atoms.

(a) Use the method of intercepts to calculate the partial molar volumes $\bar{V}_A$ and $\bar{V}_B$.
(b) Show *explicitly* from your results that $v = X_A\bar{V}_A + X_B\bar{V}_B$, where $X_A = 1 - X_B$ is the mole fraction of A atoms. Why is such a relation true?
(c) Show *explicitly* from your results that $0 = X_A(\mathrm{d}\bar{V}_A/\mathrm{d}X_B) + X_B(\mathrm{d}\bar{V}_B/\mathrm{d}X_B)$. Why is such a relation true?

Solution 5.6.

(a) We calculate $\bar{V}_A = v(X_B) - X_B\mathrm{d}v/\mathrm{d}X_B = X_B^3 - 3$ and $\bar{V}_B = v(X_B) + (1 - X_B)\mathrm{d}v/\mathrm{d}X_B = X_B^2 - 2X_B + 5$.
(b) We can easily check that $v = X_A\bar{V}_A + X_B\bar{V}_B$, where $X_A = 1 - X_B$. This is just a special case of Eq. (5.106).
(c) We readily compute $\partial\bar{V}_A/\partial X_B = 2X_B$ and $\partial\bar{V}_B/\partial X_B = 2X_B - 2 = -2X_A$ so $X_A\partial\bar{V}_A/\partial X_B + X_B\partial\bar{V}_B/\partial X_B = 0$. This result follows from Eq. (5.105) for constant T and p after division by N.

5.7 Entropy of Chemical Reaction

Before leaving this chapter, we show how the formalism developed for open systems can be used to treat chemically closed systems in which the mole numbers can vary by means of chemical reactions. Then we proceed to calculate the entropy due to a chemical reaction. See Chapter 12 for a more complete treatment of chemical reactions that includes heats of reaction and detailed conditions for equilibrium.

We begin with Eq. (5.10) and write

$$dN_i = d^{\text{int}}N_i + d^{\text{ext}}N_i, \tag{5.120}$$

where $d^{\text{ext}}N_i$ denotes changes in N_i due to exchanges of chemical species with the external environment and $d^{\text{int}}N_i$ denotes changes due to chemical reactions internal to the system. For simplicity, we treat only one chemical reaction, which we write in the symbolic form

$$\sum_i \nu_i A_i = 0, \tag{5.121}$$

where A_i is the symbol (such as C, CO, CO_2, H, H_2, etc.) of the chemical species i and ν_i is its stoichiometric coefficient in the reaction. We regard ν_i to be negative for reactants and positive for products. For example, reaction of carbon and oxygen to form carbon monoxide, namely

$$\text{C} + (1/2)\text{O}_2 \rightarrow \text{CO} \tag{5.122}$$

could be written in the form of Eq. (5.121) with $A_1 = \text{C}$, $A_2 = \text{O}_2$, $A_3 = \text{CO}$ and $\nu_1 = -1$, $\nu_2 = -1/2$, $\nu_3 = 1$. We can therefore write

$$d^{\text{int}}N_i = \nu_i \, d\tilde{N}, \tag{5.123}$$

where $\tilde{N}$ is a progress variable that represents the extent to which the reaction has taken place. Equation (5.10) therefore becomes

$$dU = T\,dS - p\,dV + \sum_{i=1}^{\kappa} \mu_i \nu_i \, d\tilde{N} + \sum_{i=1}^{\kappa} \mu_i \, d^{\text{ext}}N_i. \tag{5.124}$$

A special case of Eq. (5.124) is a chemically closed system for which $d^{\text{ext}}N_i = 0$, in which case it becomes

$$dU = T\,dS - p\,dV + \sum_{i=1}^{\kappa} \mu_i \nu_i \, d\tilde{N}. \tag{5.125}$$

Equation (5.125) replaces Eq. (3.47) when there is a chemical reaction. Combining Eq. (5.125) with the first law $dU = \delta Q - \delta W$ and eliminating dU, we obtain

$$\frac{\delta Q}{T} + \frac{p\,dV - \delta W}{T} - \sum_{i=1}^{\kappa} \frac{\mu_i \nu_i}{T} \, d\tilde{N} = dS. \tag{5.126}$$

Subtracting $\delta Q/T_s$ from both sides of Eq. (5.126) and applying the second law in the form of Eq. (3.4), we obtain

$$\delta Q\left(\frac{1}{T} - \frac{1}{T_s}\right) + \frac{p\,dV - \delta W}{T} - \sum_{i=1}^{\kappa} \frac{\mu_i \nu_i}{T} \, d\tilde{N} = dS - \frac{\delta Q}{T_s} \geq 0, \tag{5.127}$$

where the inequality holds for natural irreversible processes and the equal sign holds for an idealized reversible process. Comparison with Eq. (3.52) reveals an additional term that can represent irreversible entropy production due to chemical reaction.

If only quasistatic work is done so that $\delta W = p\,\mathrm{d}V$, and $T = T_s$ so there is no entropy production due to irreversible heat transfer, Eq. (5.127) becomes

$$\mathrm{d}S = \frac{\delta Q}{T} - \sum_{i=1}^{\kappa} \frac{\mu_i \nu_i}{T}\,\mathrm{d}\tilde{N} \geq 0. \tag{5.128}$$

For a reversible process, the equal sign holds in Eq. (5.128) and Eq. (3.6) also holds, so $\mathrm{d}S = \delta Q/T$, which would require the second term on the right-hand side of Eq. (5.128) to vanish. For $\mathrm{d}\tilde{N} \neq 0$, this would require $\sum_i^{\kappa} \mu_i \nu_i = 0$, which turns out to be the condition that the reaction is in equilibrium. For an irreversible process, the inequality sign in Eq. (5.128) holds, so

$$\mathrm{d}S - \frac{\delta Q}{T} = -\sum_{i=1}^{\kappa} \frac{\mu_i \nu_i}{T}\,\mathrm{d}\tilde{N} > 0, \tag{5.129}$$

which results in entropy production due to an irreversible chemical reaction. In that case, Eq. (3.6) would no longer hold. Such a reaction will continue until equilibrium is reached or until at least one of the reactants in the system is used up, which will occur when $\mathrm{d}\tilde{N} = 0$.

In their book *Modern Thermodynamics*, Kondepudi and Prigogine [16] break the entropy change $\mathrm{d}S$ into external and internal parts by writing[15] $\mathrm{d}S = \mathrm{d}^{\mathrm{ext}}S + \mathrm{d}^{\mathrm{int}}S$, where $\mathrm{d}^{\mathrm{ext}}S = \delta Q/T$ and $\mathrm{d}^{\mathrm{int}}S \geq 0$. The inequality applies to a natural irreversible process and the equality applies to an idealized reversible process. This leads to

$$\mathrm{d}^{\mathrm{int}}S = -\sum_{i=1}^{\kappa} \frac{\mu_i \nu_i}{T}\,\mathrm{d}\tilde{N} \geq 0. \tag{5.130}$$

This interpretation is consistent with our more general Eqs. (5.127) and (5.128) in the special case of $T_s = T$ (no irreversible heat flow) and no irreversible work.

For a cyclic process,

$$0 = \oint \mathrm{d}S = \oint \mathrm{d}^{\mathrm{ext}}S + \oint \mathrm{d}^{\mathrm{int}}S, \tag{5.131}$$

which requires

$$-\oint \mathrm{d}^{\mathrm{ext}}S = -\oint \frac{\delta Q}{T} = \oint \mathrm{d}^{\mathrm{int}}S \geq 0. \tag{5.132}$$

Equation (5.132) is in agreement with Eq. (3.15) for a cyclic process during which $T = T_r$. When Eq. (5.130) holds, we also have

$$\oint \mathrm{d}^{\mathrm{int}}S = -\oint \sum_{i=1}^{\kappa} \frac{\mu_i \nu_i}{T}\,\mathrm{d}\tilde{N} \geq 0. \tag{5.133}$$

[15] As shown below, $\mathrm{d}^{\mathrm{ext}}S$ and $\mathrm{d}^{\mathrm{int}}S$ are not exact differentials because their integrals around a closed path are not necessarily equal to zero.

Since S depends on U, V, and $\tilde{N}$ for this system, these quantities must return to their original values for a cyclic process. This means that any chemical reaction that takes place during part of a cycle must be reversed during another part of the cycle. If the inequality holds in Eq. (5.133), the chemical reaction is irreversible and entropy is produced; this requires heat to be exchanged with the system in such a way that Eq. (5.132) holds, so an equal amount of entropy is extracted from the system.

6

Equilibrium and Thermodynamic Potentials

In Chapter 3 we introduced the criterion for thermodynamic equilibrium for an isolated system in terms of the entropy. We now develop alternative criteria for equilibrium in terms of the internal energy and other thermodynamic potentials, the latter being related to the internal energy by Legendre transformations. Each of these potentials depends on a specific set of natural variables. The various resulting equilibrium criteria are useful for a situation in which a particular variable set is subject to control in an experiment. For example, many experiments on gases are conducted in a fixed volume V at constant temperature T. In this case, heat must be exchanged with the environment to keep the temperature constant. Experiments on liquids or solids are often conducted at fixed T and fixed pressure p, in which case both heat and work must be exchanged with the environment to insure that these quantities remain constant. Therefore, our alternative equilibrium criteria will generally pertain to systems that are not isolated.

6.1 Entropy Criterion

We first review the criterion for equilibrium in terms of the entropy, S. This criterion is based on the second law for an isolated system, according to which

$$\Delta S \geq 0, \quad \text{isolated system, allowed changes.} \tag{6.1}$$

For an isolated system, we have chemical closure, $\delta Q = 0$ and $\delta \mathcal{W} = 0$, so $\mathrm{d}U = 0$. The inequality in Eq. (6.1) pertains to a natural irreversible process and the equality corresponds to a hypothetical idealized process that is reversible. Thus for natural irreversible processes, we have

$$\Delta S > 0, \quad \text{isolated system, natural irreversible changes.} \tag{6.2}$$

Equilibrium pertains to a situation in which all natural irreversible processes are forbidden.

Suppose that a composite system, which consists of a number of parts, is initially in equilibrium by virtue of some internal constraints, such as rigid, insulating, and impenetrable walls that separate its parts. When some of these constraints are removed, transformations to which Eq. (6.2) applies can occur, and the entropy can continue to increase as much as allowed by any remaining constraints until S achieves a maximum value. When this maximum value is reached, the system will no longer be able to undergo irreversible changes, and it will be in a new equilibrium state. We need not worry about

Thermal Physics. http://dx.doi.org/10.1016/B978-0-12-803304-3.00006-5

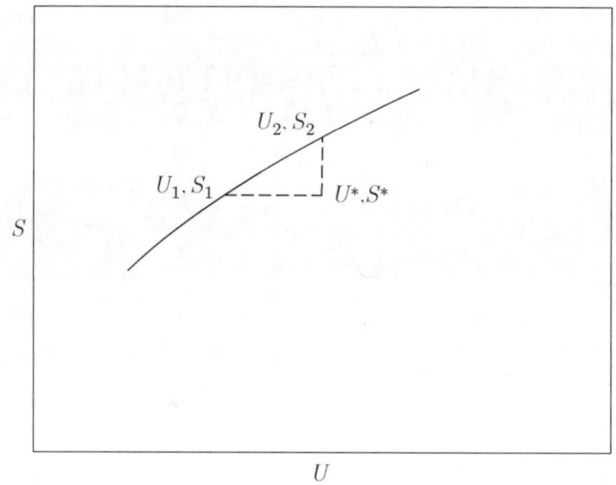

FIGURE 6–1 Curve of entropy S versus internal energy U for a system in internal equilibrium. The state U^*, S^* is an equilibrium state of the same system but with constraints on some internal extensive variables. $U^* = U_2$ and $S^* < S_2$ since, according to the entropy criterion, the equilibrium state of the unconstrained system is higher. But this implies the existence of an equilibrium state U_1, S_1 with the same entropy $S_1 = S^*$ as the constrained state but lower internal energy, $U_1 < U^*$, in agreement with the energy criterion.

the equality in Eq. (6.1) which corresponds to idealized reversible changes, because there are no driving forces for such changes to occur. This approach to equilibrium can be understood with reference to Figure 6–1 in which the curved line represents entropy S as a function of internal energy U for the equilibrium state of the system, with other extensive variables fixed. We recall that S is a monotonically increasing function of U, in agreement with the way that the curve is drawn. We focus on the equilibrium state U_2, S_2. The state U^*, S^* is also an equilibrium state for the same system except that some of its internal extensive variables are constrained to have different values from those of the state U_2, S_2. It has the same energy $U^* = U_2$ but a lower entropy $S^* < S_2$ as compared to the equilibrium state. As constraints are removed, natural irreversible processes occur, the internal extensive variables change, and the entropy increases toward S_2. After all internal constraints are removed, except for those present for the state U_2, S_2, the entropy rises to its final value S_2 and the internal extensive variables reach their final values.

The foregoing considerations suggest the following test to find the equilibrium state. We select a state having fixed energy and other constraints on its extensive variables that are necessary for an isolated system. The selected state corresponds to some fixed values of the internal extensive variables of the system. We then imagine the internal extensive variables of the selected state to vary, resulting in a varied state. If any varied state has higher entropy than the original selected state, the selected state is not the correct equilibrium state. But if all such varied states have lower entropy, the selected state has the maximum possible entropy and is the equilibrium state. When a varied state has lower entropy than the selected state, that varied state cannot be reached from the selected state by means of a natural irreversible process. We can therefore think of the variations that lower the entropy as virtual variations, since they are allowed by the constraints but forbidden by the second law of thermodynamics.

This approach leads to the following criterion, already stated in Section 3.1:

Entropy criterion: The criterion for an isolated thermodynamic system to be in internal equilibrium is that its total entropy be a maximum with respect to variation of its *internal extensive* parameters, subject to external constraints and any remaining internal constraints. Isolation constitutes the external constraints of chemical closure, perfect thermal insulation and zero external work, which require the internal energy to be constant. In this chapter, we will discuss the application of this criterion and deduce from it other useful and equivalent criteria for equilibrium.

For a system sufficiently simple that constant total volume V guarantees that there is no external work, and in which there are no chemical reactions such that constant values of $\{N_i\}$ guarantee that the system is chemically closed, the entropy criterion can be based on the relation

$$(\Delta S)_{U,V,\{N_i\}} \geq 0, \quad \text{isolated system, allowed changes.} \tag{6.3}$$

Such a thermodynamic system will be in internal equilibrium if its entropy is a maximum subject to the constraints of constant internal energy, constant volume, and constant mole numbers.

6.1.1 Conditions for Equilibrium, Multicomponent Subsystems

We can apply the entropy criterion for equilibrium to a composite system consisting of two subsystems, I and II, having respective entropies $S^{\mathrm{I}}(U^{\mathrm{I}}, V^{\mathrm{I}}, \{N_i^{\mathrm{I}}\})$, and $S^{\mathrm{II}}(U^{\mathrm{II}}, V^{\mathrm{II}}, \{N_i^{\mathrm{II}}\})$ with differentials[1]

$$\mathrm{d}S^{\mathrm{I}} = (1/T^{\mathrm{I}})\,\mathrm{d}U^{\mathrm{I}} + (p^{\mathrm{I}}/T^{\mathrm{I}})\,\mathrm{d}V^{\mathrm{I}} - \sum_{i=1}^{\kappa}(\mu_i^{\mathrm{I}}/T^{\mathrm{I}})\,\mathrm{d}N_i^{\mathrm{I}}; \tag{6.4}$$

$$\mathrm{d}S^{\mathrm{II}} = (1/T^{\mathrm{II}})\,\mathrm{d}U^{\mathrm{II}} + (p^{\mathrm{II}}/T^{\mathrm{II}})\,\mathrm{d}V^{\mathrm{II}} - \sum_{i=1}^{\kappa}(\mu_i^{\mathrm{II}}/T^{\mathrm{II}})\,\mathrm{d}N_i^{\mathrm{II}}. \tag{6.5}$$

Equation (6.3) applies to finite entropy changes that we designate by ΔS. Of course it also applies to infinitesimal changes[2] that we designate by dS. For such an infinitesimal change of the total entropy $S = S^{\mathrm{I}} + S^{\mathrm{II}}$, allowed changes must obey

$$0 \leq \mathrm{d}S = \mathrm{d}S^{\mathrm{I}} + \mathrm{d}S^{\mathrm{II}}, \quad \text{constraints } U, V, \{N_i\} \text{ held constant.} \tag{6.6}$$

[1]These apply to bulk systems in the absence of chemical reactions.

[2]Examination of infinitesimal entropy changes will lead to an extremum of the entropy, but not necessarily a maximum. We must examine finite entropy changes, sometimes possible by examining higher derivatives, to *guarantee* that we have a maximum of entropy and hence that the equilibrium is stable. This leads to stability conditions we shall examine in Chapter 7.

These constraints require

$$dU^{\rm I} = -dU^{\rm II}; \quad dV^{\rm I} = -dV^{\rm II}; \quad dN_i^{\rm I} = -dN_i^{\rm II}, \quad \text{for } i = 1, 2, \ldots, \kappa. \tag{6.7}$$

Thus Eq. (6.6) becomes

$$0 \le (1/T^{\rm I} - 1/T^{\rm II})\, dU^{\rm I} + (p^{\rm I}/T^{\rm I} - p^{\rm II}/T^{\rm II})\, dV^{\rm I} - \sum_{i=1}^{\kappa}(\mu_i^{\rm I}/T^{\rm I} - \mu_i^{\rm II}/T^{\rm II})\, dN_i^{\rm I}. \tag{6.8}$$

The key to extracting detailed **conditions for equilibrium** from Eq. (6.8) is to recognize that it must be true for *arbitrary* and *independent* changes $dU^{\rm I}$, $dV^{\rm I}$ and $dN_i^{\rm I}$ of either sign[3] including zero. We can therefore get information about equilibrium by considering a number of *special* variations such as

$$dU^{\rm I} = \text{arbitrary} \pm; \quad dV^{\rm I} = 0; \quad dN_i^{\rm I} = 0 \text{ for } i = 1, 2, \ldots, \kappa, \tag{6.9}$$

which leads to

$$0 \le (1/T^{\rm I} - 1/T^{\rm II})\, dU^{\rm I}, \quad \text{allowed changes.} \tag{6.10}$$

In view of Eq. (6.10), the only way to achieve equilibrium is to prevent an actual irreversible process (which obeys the inequality) in which a change $dU^{\rm I}$ of either sign can occur, and this requires

$$T^{\rm I} = T^{\rm II}. \tag{6.11}$$

If $T^{\rm I} > T^{\rm II}$, then a natural irreversible process $dU^{\rm I} < 0$ can occur; whereas for $T^{\rm I} < T^{\rm II}$, a natural irreversible process $dU^{\rm I} > 0$ can occur. These processes are consistent with the notion that there will be spontaneous heat transfer from hot to cold, and their prevention leads to the equilibrium condition of equal temperatures.

We now use Eq. (6.11) to recast Eq. (6.8) in the form

$$0 \le (1/T^{\rm I})(p^{\rm I} - p^{\rm II})\, dV^{\rm I} - (1/T^{\rm I})\sum_{i=1}^{\kappa}(\mu_i^{\rm I} - \mu_i^{\rm II})\, dN_i^{\rm I}. \tag{6.12}$$

We then apply the special variation

$$dV^{\rm I} = \text{arbitrary} \pm; \quad dN_i^{\rm I} = 0 \text{ for } i = 1, 2, \ldots, \kappa, \tag{6.13}$$

to Eq. (6.12) to obtain

$$0 \le (1/T^{\rm I})(p^{\rm I} - p^{\rm II})\, dV^{\rm I}, \tag{6.14}$$

from which we deduce the equilibrium condition

$$p^{\rm I} = p^{\rm II}. \tag{6.15}$$

[3]One can consider more general constraints that allow one-way changes only, for example $dV^{\rm I} \ge 0$. These lead to equilibrium conditions that are inequalities, for example, $p^{\rm II} \ge p^{\rm I}$ instead of $p^{\rm II} = p^{\rm I}$ which would result if $dV^{\rm I}$ could have either sign. Similarly, one-way constraints on the $dN_i^{\rm I}$ correspond to semipermeable membranes.

If $p^{\mathrm{I}} > p^{\mathrm{II}}$, an irreversible process in which $\mathrm{d}V^{\mathrm{I}} > 0$ can occur, and if $p^{\mathrm{I}} < p^{\mathrm{II}}$, an irreversible process in which $\mathrm{d}V^{\mathrm{I}} < 0$ can occur, in agreement with the notion that the volume corresponding to the system having higher pressure will expand.

Proceeding in this manner, we consider a variation in which only $\mathrm{d}N_j^{\mathrm{I}} \neq 0$, which leads to

$$0 \leq -(1/T^{\mathrm{I}})(\mu_j^{\mathrm{I}} - \mu_j^{\mathrm{II}})\,\mathrm{d}N_j^{\mathrm{I}}, \tag{6.16}$$

from which we deduce the equilibrium conditions

$$\mu_j^{\mathrm{I}} = \mu_j^{\mathrm{II}} \text{ for each } j = 1, 2, \ldots, \kappa. \tag{6.17}$$

If $\mu_j^{\mathrm{I}} > \mu_j^{\mathrm{II}}$, a natural irreversible process in which $\mathrm{d}N_j^{\mathrm{I}} < 0$ can occur, and if $\mu_j^{\mathrm{I}} < \mu_j^{\mathrm{II}}$, a natural irreversible process in which $\mathrm{d}N_j^{\mathrm{I}} > 0$ can occur, in agreement with the notion that there will be diffusion from high chemical potential to low chemical potential for the jth chemical component.

This leads to the following conditions for equilibrium:

Conditions for equilibrium: The conditions for thermodynamic equilibrium for two systems capable of freely exchanging energy, volume, and chemical components with one another are: equality of temperature, equality of pressure, and equality of chemical potential of each chemical component. For a system of κ components, these conditions are expressed by $\kappa + 2$ equations, namely Eqs. (6.11), (6.15), and (6.17). Note that these conditions imply uniformity of temperature, pressure, and chemical potential of each chemical component *within* any system. This follows because any two portions of a system can be regarded as subsystems that must be in equilibrium with one another.

6.1.2 Phase Rule

If there are more than two subsystems in a given system, we can consider their equilibria in pairs and eventually arrive at the same conclusions for all of them. If each subsystem having κ chemical components corresponds to a different phase (e.g., solid with crystal structure α, solid with crystal structure β, liquid, vapor) we can count the number of independent intensive variables, subtract from it the number of equations needed to specify equilibrium, and get the number of free variables (if any) that remain. Requiring the number of free variables f to be positive or zero puts a limitation on the number of phases, n, that can exist in equilibrium. The result is the **Gibbs phase rule** which can be derived as follows:

$$\text{number of independent intensive variables} = n(\kappa + 1),$$

$$\text{number of equilibrium equations} = (n - 1)(\kappa + 2),$$

$$\text{number of free variables} = f = n(\kappa + 1) - (n - 1)(\kappa + 2) = \kappa + 2 - n. \tag{6.18}$$

Thus, for a monocomponent system, $\kappa = 1$ and the only possibilities are $n = 1, 2$, and 3. $n = 1$ corresponds to a single phase region for which the number of free variables is $f = 2$, so the pressure p and temperature T can be chosen independently. $n = 2$ corresponds to a coexistence curve (say between solid and liquid) and on such a curve, $f = 1$ so p is a function of T. $n = 3$ corresponds to a triple point where, for example, solid, liquid, and vapor are at equilibrium; since $f = 0$, both p and T are fixed. See Chapter 8 and especially Figure 8–1 for more detail. There could be more than one triple point, for example, one where two solid phases having different crystal structures and a liquid phase are at equilibrium.

For a binary system, $\kappa = 2$ and the only possibilities are $n = 1, 2, 3$, and 4 corresponding to $f = 3, 2, 1$, and 0, respectively. The possibilities become more numerous as κ increases.

6.2 Energy Criterion

From the entropy criterion for equilibrium, one can derive an equilibrium criterion in terms of the internal energy with the entropy held constant. Such a criterion is suggested by Figure 6–1 by consideration of the state U_1, S_1 which is also an equilibrium state of the system. This state has the same entropy $S_1 = S^*$ as the internally constrained state U^*, S^* but a lower energy, $U_1 < U^*$. According to the entropy criterion, *all* constrained states having the same entropy S^* lie *below* the equilibrium curve. Therefore, as internal constraints are removed at constant entropy S^*, the system can lower its energy to U_1 but no lower, and we see that the equilibrium state U_1, S_1 has the minimum possible energy at fixed entropy. The state U_1, S_1 is a different equilibrium state from U_2, S_2 which was found by applying the entropy criterion beginning with the state U^*, S^*, but this is only because $U^* = U_2$. Had we begun with a state U^{**}, S^{**} with $S^{**} = S_2$ and $U^{**} > U_2$, then as internal constraints are removed at constant entropy S_2, the system can lower its energy to U_2 but no lower. The equilibrium state U_2, S_2 found in this manner is the same as that found by applying the entropy criterion starting from U^*, S^*.

This leads to the following criterion for equilibrium:

Energy criterion: The criterion for a chemically closed thermodynamic system to be in internal equilibrium is that its total internal energy be a minimum with respect to variation of its *internal extensive* parameters, subject to any remaining internal constraints and the constraint of constant total entropy and no external work.

Paradox: The entropy criterion applies to an isolated system, for which the internal energy cannot change. The equilibrium state U_1, S_1 can therefore be found by applying the entropy criterion at fixed energy U_1, so it is an equilibrium state for an isolated system. On the other hand, application of the energy criterion to find the state U_1, S_1 requires the energy to change, which is impossible for an isolated system! So how do we apply the energy criterion? We have no choice but to deal with a system that is not isolated. In principle, we must put our system in contact with a hypothetical external system that has the unusual capability of exchanging heat in such a way as to maintain a constant entropy of our system. Physically, it is difficult if not impossible to imagine such a system, but mathematically the result of applying the energy criterion gives the desired result.

To see this in more detail, we develop a general inequality that applies to a chemically closed system that can exchange both heat and work at constant entropy. If S is constant during a process, then $\mathrm{d}S = 0$ at every infinitesimal stage of the process. We can then combine the differential forms of the first and second laws, Eqs. (2.2) and (3.4), to obtain

$$\delta\mathcal{W} + \mathrm{d}U = \delta Q \le T_s\,\mathrm{d}S = 0, \quad \text{constant } S, \tag{6.19}$$

where T_s is the temperature of an external heat source. Equation (6.19) can be integrated over the path of the process to obtain

$$\mathcal{W} + \Delta U = Q \le 0, \tag{6.20}$$

from which it follows that

$$\mathcal{W} \le -\Delta U, \quad \text{constant } S. \tag{6.21}$$

The maximum amount of work that can be done in such a process is equal to the decrease in internal energy and occurs for the reversible process for which the equality holds in Eqs. (6.19) and (6.21). In that case, $\delta Q = 0$ and $Q = 0$, so no heat is exchanged with the system and $\mathcal{W} = -\Delta U$. This would be true for a purely mechanical system. For an irreversible isentropic process, Eq. (6.21) shows that the actual amount of work done is $\mathcal{W} < -\Delta U$; in such a case, $Q < 0$ so heat was extracted from the system to keep its entropy constant. If $\mathcal{W} = 0$ in Eq. (6.21), we have

$$\Delta U \le 0, \quad \mathcal{W} = 0 \text{ and constant } S, \tag{6.22}$$

and equilibrium corresponds to a minimum of U, compatible with constraints.

We pause here to emphasize a subtle point. Constant S certainly guarantees $\Delta S = 0$ but $\Delta S = 0$ does not guarantee constant S, because S can still vary throughout the process. If we only know that $\Delta S = 0$ for a process, it is possible to have $\delta Q > 0$ for some parts of the process which can lead to $Q > 0$ and violation of Eq. (6.20). This fact is easy to illustrate for a process in which the system exchanges heat with only two reservoirs. In that case, we must only satisfy Eq. (3.11) for $\Delta S = 0$, which results in

$$\frac{Q_1}{T_1} + \frac{Q_2}{T_2} \le 0. \tag{6.23}$$

If $Q_1 \le 0$ and $Q_2 \le 0$, then $Q \le 0$ and Eq. (6.20) is not violated. But suppose $T_2 > T_1$, $Q_2 = |Q_2| > 0$, and $Q_1 = -|Q_1| < 0$. If then we choose $|Q_2| = (T_1 + T_2)/(2T_1)|Q_1|$, a little algebra shows that

$$\frac{Q_1}{T_1} + \frac{Q_2}{T_2} = \frac{T_1 - T_2}{2T_1T_2}|Q_1| < 0; \quad Q = Q_1 + Q_2 = \frac{T_2 - T_1}{2T_1}|Q_1| > 0. \tag{6.24}$$

Thus Eq. (6.23) will be satisfied but Eq. (6.20) will be violated.

In the next two sections, we proceed to give further motivation and finally an indirect proof, due to Gibbs, that the energy criterion and the entropy criterion are equivalent.

6.2.1 Local Energy Criterion

We first follow closely a calculation by Callen [2, p. 134] to show that a local maximum of the entropy S at constant internal energy U implies a local minimum of U at constant S. To simplify the notation, we consider S to depend on U and some internal extensive variable Ξ and suppress all of the other extensive variables on which S depends. Then if S is a local maximum at constant U when $\Xi = \Xi_0$, we have

$$\left(\frac{\partial S}{\partial \Xi}\right)_U = 0; \quad \Xi = \Xi_0 \tag{6.25}$$

and

$$\left(\frac{\partial^2 S}{\partial \Xi^2}\right)_U < 0; \quad \Xi = \Xi_0. \tag{6.26}$$

Since S is a monotonically increasing function of U at constant Ξ (and constant values of the suppressed parameters as well), it has a unique inverse function $U(S, \Xi)$. Such a functional relationship among three variables implies that

$$\left(\frac{\partial U}{\partial \Xi}\right)_S \left(\frac{\partial \Xi}{\partial S}\right)_U \left(\frac{\partial S}{\partial U}\right)_\Xi = -1, \tag{6.27}$$

which can be rewritten

$$\left(\frac{\partial U}{\partial \Xi}\right)_S = -\left(\frac{\partial S}{\partial \Xi}\right)_U \Big/ \left(\frac{\partial S}{\partial U}\right)_\Xi =: P(\Xi, U). \tag{6.28}$$

The fact that S is a monotonically increasing function of U requires $(\partial S/\partial U)_\Xi > 0$, so evaluation of Eq. (6.28) at $\Xi = \Xi_0$ shows that

$$\left(\frac{\partial U}{\partial \Xi}\right)_S = 0; \quad \Xi = \Xi_0. \tag{6.29}$$

We proceed to examine the second derivative of U, namely

$$\left(\frac{\partial^2 U}{\partial \Xi^2}\right)_S = \left(\frac{\partial P}{\partial \Xi}\right)_U + \left(\frac{\partial P}{\partial U}\right)_\Xi \left(\frac{\partial U}{\partial \Xi}\right)_S. \tag{6.30}$$

By using Eq. (6.29), we see that the second term in Eq. (6.30) vanishes at $\Xi = \Xi_0$. The first term on the right-hand side can be written

$$\left(\frac{\partial P}{\partial \Xi}\right)_U = -\left(\frac{\partial^2 S}{\partial \Xi^2}\right)_U \Big/ \left(\frac{\partial S}{\partial U}\right)_\Xi + \left(\frac{\partial S}{\partial \Xi}\right)_U \frac{\partial^2 S}{\partial U \partial \Xi} \Big/ \left(\frac{\partial S}{\partial U}\right)_\Xi^2. \tag{6.31}$$

By using Eq. (6.25), we see that the second term in Eq. (6.31) also vanishes at $\Xi = \Xi_0$, resulting in

$$\left(\frac{\partial^2 U}{\partial \Xi^2}\right)_S = -\left(\frac{\partial^2 S}{\partial \Xi^2}\right)_U \Big/ \left(\frac{\partial S}{\partial U}\right)_\Xi; \quad \Xi = \Xi_0. \tag{6.32}$$

Since $(\partial S/\partial U)_\Xi > 0$ as discussed above, the use of Eq. (6.26) in Eq. (6.32) shows that

$$\left(\frac{\partial^2 U}{\partial \Xi^2}\right)_S > 0; \quad \Xi = \Xi_0. \tag{6.33}$$

From Eqs. (6.29) and (6.33), we conclude that U has a local minimum at $\Xi = \Xi_0$.

This local analysis suggests that the energy criterion is true. A general proof requires one to prove that a global maximum of S at constant U corresponds to a global minimum of U at constant S.

6.2.2 Equivalence of Entropy and Energy Criteria

We shall prove that the entropy criterion and the energy criterion are equivalent. Both $S(U, \{\Lambda\})$ and $U(S, \{\Lambda\})$ (where $\{\Lambda\}$ stands for all other extensive variables of a complete set) are fundamental equations; if one is known, the other can be found because S is a monotonically increasing function of U and vice versa. The procedure to obtain detailed conditions for equilibrium of systems by minimizing U is analogous to that for maximizing S, and the resulting conditions (uniformity of temperature, pressure, and chemical potential of each chemical component) are the same.

This equivalence was recognized and emphasized by Gibbs [3, p. 56] who stated

> *"That these two theorems [of entropy maximization at constant energy and energy minimization at constant entropy] are equivalent will appear from the consideration that it is always possible to increase both the energy and entropy of the system, or to decrease both together, viz., by imparting heat to any part of the system or by taking it away."*

A key word in this statement is "both" and this relates to the fact that S is a monotonically increasing function of U and vice versa, as already stated. As had been stated previously by Gibbs [3, p. 55]:

> *"For by mechanical and thermodynamic contrivances, supposed theoretically perfect, any supply of work and heat may be transformed into any other which does not differ from it either in the amount of work and heat taken together [which is equal to ΔU] or in the value of the integral $\int \delta Q/T$."*

Based on these statements of Gibbs, one can prove the equivalence of the entropy criterion and the energy criterion as follows:

- First, suppose that for the equilibrium state, the entropy criterion is true but that the energy criterion is not true, that is, the entropy is a maximum at constant energy but the internal energy is not a minimum at constant entropy. Then there exists a state of the system with lower energy and the same entropy. We can therefore use a combination of heat and work to *raise both* the internal energy and the entropy of this state, and thus achieve a state having the original internal energy but higher entropy. This contradicts the fact that the entropy is a maximum at constant internal energy.

- Second, suppose that for the equilibrium state, the energy criterion is true but that the entropy criterion is not true, that is, the internal energy is a minimum at constant entropy but the entropy is not a maximum at constant internal energy. Then there exists a state of the system with higher entropy and the same internal energy. We can therefore use a combination of heat and work to *lower both* the internal energy and the entropy of this state, and thus achieve a state having the original entropy but lower internal energy. This contradicts the fact that the internal energy is a minimum at constant entropy.

6.3 Other Equilibrium Criteria

We have shown that the entropy criterion and the internal energy criterion are equivalent. By means of Legendre transformations, one can use other so-called "thermodynamic potentials" (such as Helmholtz free energy, Gibbs free energy, enthalpy) for which an equilibrium criterion of minimization exists, but with other variables (some intensive) held constant. This is taken up in the following sections.

6.3.1 Helmholtz Free Energy Criterion

If a chemically closed thermodynamic system is in contact with a heat reservoir having constant temperature T_r, then Eq. (3.10) becomes $Q_r \le T_r \Delta S$. Combining this with the first law, we obtain

$$\Delta U + \mathcal{W} = Q_r \le T_r \Delta S, \tag{6.34}$$

which may be rewritten

$$\mathcal{W} \le -(\Delta U - T_r \Delta S). \tag{6.35}$$

Equation (6.35) is a formula for the maximum work that a system in contact with a heat reservoir at constant T_r can do. We define the Helmholtz free energy[4] by the Legendre transformation

$$F := U - TS. \tag{6.36}$$

If $T = T_r$ in the initial and final states of a process,[5] Eq. (6.35) can be written

$$\mathcal{W} \le -\Delta F; \quad T = T_r \text{ in initial and final states.} \tag{6.37}$$

[4]Many books denote the Helmholtz free energy by the symbol A and use F for the Gibbs free energy. We denote the Gibbs free energy by $G := U - TS + pV = H + pV$.

[5]Note that Eqs. (6.37) and (6.38) hold even if the temperature of the system is undefined *during* the process. Of course they also hold if $T = T_r$ throughout the process, which is the case treated in most books. Fermi [1, p. 78] gives a careful discussion of this more general treatment, which is also mentioned by Landau and Lifshitz [7, p. 59].

Hence the name "free" energy because the decrease in F is the energy that is free to do work for a system that exchanges heat with a heat reservoir at constant temperature T_r. If $\mathcal{W} = 0$, Eq. (6.37) becomes

$$\Delta F \le 0; \quad \mathcal{W} = 0 \text{ and } T = T_r \text{ in initial and final states.} \tag{6.38}$$

For a chemically closed system that does no external work and is held at constant temperature $T = T_r$ in its initial and final states, the Helmholtz free energy can only decrease, and equilibrium is achieved when F reaches its minimum, compatible with constraints. This leads to the following equilibrium criterion:

Helmholtz free energy criterion: The criterion for a chemically closed thermodynamic system held at constant temperature $T = T_r$ in its initial and final states and which does no external work to be in internal equilibrium is that its Helmholtz free energy be a minimum with respect to variations of its internal extensive parameters.

If there are no chemical reactions such that constant values of $\{N_i\}$ guarantee that the system is chemically closed, the system is sufficiently simple that constant total volume V guarantees that there is no external work, and the system temperature T is held constant by an external source, Eq. (6.38) reduces to

$$(\Delta F)_{T,V,\{N_i\}} \le 0, \quad \text{allowed changes.} \tag{6.39}$$

Such a thermodynamic system will be in internal equilibrium if its Helmholtz free energy is a minimum subject to the constraints of constant temperature, constant volume, and constant mole numbers.

6.3.2 Gibbs Free Energy Criterion

If a chemically closed thermodynamic system is in contact with a heat reservoir having constant temperature T_r and a pressure reservoir having pressure p_r and against which it does work $p_r\,\Delta V$, then Eq. (6.34) becomes

$$\Delta U + \mathcal{W}^{\rm xs} + p_r\,\Delta V = Q_r \le T_r \Delta S, \tag{6.40}$$

where $\mathcal{W}^{\rm xs}$ is any excess external work that the system can do in addition to that done on the reservoir.[6] Equation (6.40) can be rewritten

$$\mathcal{W}^{\rm xs} \le -[\Delta U - T_r\,\Delta S + p_r\,\Delta V]. \tag{6.41}$$

We define the Gibbs free energy[7] by the Legendre transformation

$$G := U - TS + pV = H - TS, \tag{6.42}$$

[6] In order to have $\mathcal{W}^{\rm xs} \neq 0$, the system must be sufficiently complex to do work by means other than just expanding against an external pressure.

[7] Note that G has the same relationship to H as F does to U. Consequently, G is sometimes called the free enthalpy, rather than the Gibbs free energy.

where T and p are temperature and pressure of the *system*. If $T = T_r$ and $p = p_r$ in the initial and final states of a process,[8] then Eq. (6.41) can be written

$$\mathcal{W}^{\mathrm{xs}} \le -\Delta G, \quad T = T_r \text{ and } p = p_r \text{ in initial and final states.} \tag{6.43}$$

Equation (6.43) gives the maximum excess work (useful work) that a system in contact with a pressure reservoir can do at constant temperature. If $\mathcal{W}^{\mathrm{xs}} = 0$, Eq. (6.43) becomes

$$\Delta G \le 0, \quad \mathcal{W}^{\mathrm{xs}} = 0, T = T_r \text{ and } p = p_r \text{ in initial and final states.} \tag{6.44}$$

For a chemically closed system held in its initial and final states at constant temperature $T = T_r$ and constant pressure $p = p_r$ that does external work only on a pressure reservoir at pressure p_r, the Gibbs free energy can only decrease, and equilibrium is achieved whenever G reaches its minimum, compatible with constraints. This leads to the following equilibrium criterion:

Gibbs free energy criterion: The criterion for a chemically closed thermodynamic system held in its initial and final states at constant temperature $T = T_r$ and constant pressure $p = p_r$ which only does external work $p_r\,\Delta V$ to be in internal equilibrium is that its Gibbs free energy be a minimum with respect to variations of its internal extensive parameters. If there are no chemical reactions such that constant values of $\{N_i\}$ guarantee that the system is chemically closed, Eq. (6.44) becomes

$$(\Delta G)_{T,p,\{N_i\}} \le 0, \quad \text{allowed changes.} \tag{6.45}$$

Such a thermodynamic system will be in internal equilibrium if its Gibbs free energy is a minimum subject to the constraints of constant temperature, constant pressure, and constant mole numbers.

6.3.3 Enthalpy Criterion

If we apply Eq. (6.21) to a chemically closed system in contact with a pressure reservoir at pressure p_r and against which it does work $p_r\Delta V$, we obtain

$$\mathcal{W}^{\mathrm{xs}} + p_r\Delta V \le -\Delta U, \quad \text{constant } S, \tag{6.46}$$

where $\mathcal{W}^{\mathrm{xs}}$ is any excess external work that the system can do in addition to that done on the reservoir. Equation (6.46) can be rewritten

$$\mathcal{W}^{\mathrm{xs}} \le= -\Delta(U + p_r\Delta V), \quad \text{constant } S. \tag{6.47}$$

The enthalpy is defined by the Legendre transformation

$$H := U + pV, \tag{6.48}$$

[8]Note that Eqs. (6.43) and (6.44) hold even if the temperature and pressure of the system are undefined *during* the process. Of course they also hold if $T = T_r$ and $p = p_r$ throughout the process, which is the case treated in most books.

where p is the pressure of the *system*. If $p = p_r$ in the initial and final state of the system, Eq. (6.47) becomes

$$\mathcal{W}^{\rm xs} \le -\Delta H, \quad \text{constant } S \text{ and } p = p_r \text{ in initial and final states.} \tag{6.49}$$

Thus the maximum excess work that can be done under these conditions is given by the decrease in the enthalpy. If $\mathcal{W}^{\rm xs} = 0$, we obtain

$$\Delta H \le 0, \quad \mathcal{W}^{\rm xs} = 0, \text{ constant } S \text{ and } p = p_r \text{ in initial and final states.} \tag{6.50}$$

Under these conditions, the enthalpy can only decrease, and equilibrium is achieved when H reaches its minimum, compatible with constraints. We are therefore led to the following equilibrium criterion:

Enthalpy criterion: The criterion for a chemically closed thermodynamic system held at constant pressure $p = p_r$ in its initial and final states which only does external work $p_r\,\Delta V$ to be in internal equilibrium is that its enthalpy be a minimum with respect to variations of its internal extensive parameters, subject to the constraint of constant entropy.

If there are no chemical reactions such that constant values of $\{N_i\}$ guarantee that the system is chemically closed, Eq. (6.50) becomes

$$(\Delta H)_{S,p,\{N_i\}} \le 0, \quad \text{allowed changes.} \tag{6.51}$$

Such a thermodynamic system will be in internal equilibrium if its enthalpy is a minimum subject to the constraints of constant entropy, constant pressure, and constant mole numbers.

6.3.4 Kramers Potential Criterion

A somewhat different criterion for equilibrium can be obtained in terms of the Kramers potential (also known as the grand potential, and often denoted by Ω),

$$K = F - \sum_{i=1}^{\kappa} \mu_i N_i, \tag{6.52}$$

introduced by Eq. (5.91). We consider a set[9] of chemical reservoirs, each having fixed temperature and volume and respective chemical potential μ_{ri} for chemical component i. We apply Eq. (6.38) to a composite system having total Helmholtz free energy $F_{\rm tot}$ and consisting of the system of interest and all of the reservoirs. The total system is chemically closed and we forbid chemical reactions, so that $\Delta N_i + \Delta N_{ri} =$ constant, where N_{ri} is the number of moles of component i in its reservoir. Then

$$\Delta F_{\rm tot} = \Delta F + \sum_{i=1}^{\kappa} \mu_{ri} \Delta N_{ri} = \Delta F - \sum_{i=1}^{\kappa} \mu_{ri} \Delta N_i, \tag{6.53}$$

[9]The reservoirs need not be separate systems. In fact, this criterion is often used where the system of interest is a surface and the bulk of the system is the reservoir.

where for each reservoir, $dF_{ri} = \mu_{ri}\, dN_{ri}$, has been integrated. If $\mu_i = \mu_{ir}$, at least in the initial and final states, we have $\Delta F_{\text{tot}} = \Delta K$, so minimization of F_{tot} at constant T and no external work is the same as minimization of K at constant T, no external work and constant chemical potentials equal to those of the reservoirs. This leads to the following criterion:

Kramers potential criterion: The criterion for a thermodynamic system to be in equilibrium at constant temperature and a constant value of each of its chemical potentials is that its Kramers potential be a minimum with respect to variations of its internal extensive parameters under the constraints of no external work and no chemical reactions.

If there are no chemical reactions, the system is sufficiently simple that constant total volume guarantees no external work, and if the system temperature T and its chemical potentials $\{\mu_i\}$ are held constant by external reservoirs, the equilibrium criterion for the Kramers potential reduces to a minimization of the Kramers potential at constant temperature, constant volume, and constant chemical potentials.

6.4 Summary of Criteria

For cases in which $\Delta V = 0$ guarantees $\mathcal{W} = 0$, or for p constant in which the only external work is $p\,\Delta V$ (so $\mathcal{W}^{\text{xs}} = 0$), and no chemical reactions such that constant values of $\{N_i\}$ guarantee that the system is chemically closed, or for constant $\{\mu_i\}$ imposed by external chemical reservoirs, the criteria for equilibrium can be summarized by first noting the natural variable set[10] on which the various thermodynamic functions depend. For the entropy and the thermodynamic potentials discussed above, these variable sets are summarized in Table 6–1.

Then for internal equilibrium, S is a maximum, and each thermodynamic potential is a minimum, with respect to variations of its *internal extensive variables*, with all designated

Table 6–1 Natural Variable Sets of Thermodynamic Functions

Function	Variable →	S	U	V	$\{N_i\}$	T	p	$\{\mu_i\}$
Entropy	S		×	×	×			
Internal Energy	U	×		×	×			
Helmholtz Free Energy	F			×	×	×		
Gibbs Free Energy	G				×	×	×	
Enthalpy	H	×			×		×	
Kramers (Grand) Potential	K			×		×		×

[10]This is the variable set that gives complete information about the system, namely extensive variables for U and S and variables obtained by Legendre transformations in the case of F, G and H. For further discussion of this point in the context of Legendre transformations, see Callen [2, pp. 137-145].

variables held constant for the overall system. We emphasize that these are *alternative* criteria for equilibrium, each applicable for different constraints.

6.4.1 Equilibrium Conditions

No matter which of these criteria are applied, the conditions for mutual equilibrium of subsystems of a composite system will be the same as those derived from the entropy condition in Section 6.1.1. For a composite system containing more than two subsystems, the systems may be considered in pairs. These conditions are uniformity, throughout the entire system, of the temperature T, the pressure p, and each chemical potential μ_i. For the potentials, this can be seen by carrying out the same kind of variations as in Section 6.1.1 for only the subset of variables that are unconstrained. For the Gibbs free energy, for example, T and p are already uniform and assumed to be held constant by external reservoirs, so only exchanges of the $\{N_i^{\text{sub}}\}$ among the subsystems need to be considered. This leads to uniformity of the μ_i.

6.4.2 Extension to Chemical Reactions

In event that chemical reactions are allowed, one must revert to an equilibrium criterion that allows variations of the $\{N_i\}$ due to those reactions. Thus, to apply the entropy criterion for a single chemical reaction, one would have to vary the progress variable $\tilde{N}$ that appears in Eq. (5.125). Then according to the discussion of Eq. (5.128), there would be an additional condition $\sum_i \mu_i \nu_i = 0$ that the chemical potentials must satisfy for that chemical reaction to be in equilibrium. That same condition would apply to all subsystems because the chemical potentials must be uniform at equilibrium. This additional condition would lower the number of degrees of freedom in the phase rule, Eq. (6.18), by one. If there were c independent chemical reactions, the phase rule would take the modified form

$$f = (\kappa - c) + 2 - n, \tag{6.54}$$

where $\kappa - c \geq 1$ is the number of*independent* chemical components. See Darken and Gurry [19, p. 287] for a discussion of the phase rule for a variety of conditions, including "frozen reactions," in the context of the thermochemistry of metals.

7 Requirements for Stability

In Chapter 6 we discussed the criterion for thermodynamic equilibrium of an isolated system, namely that its entropy, S, be a maximum with respect to variations of its internal extensive variables. If Ξ is such an internal extensive variable, then $\mathrm{d}S/\mathrm{d}\Xi = 0$ at equilibrium. But this condition could correspond to a maximum, a minimum or a horizontal point of inflection in a graph of S versus Ξ. We must therefore examine higher derivatives in order to insure that S is a local maximum, and finite changes $\Delta\Xi$ to ascertain if S is a global maximum. In this chapter, we examine the requirements for stable equilibrium, particularly with respect to the stability of homogeneous systems. We pose the question of whether a homogeneous system is stable with respect to breakup into a composite system consisting of two (or more) subsystems, each of which is homogeneous. This will lead to requirements concerning the functional dependence of S on its complete set of extensive variables.

In Chapter 6 we also discussed equilibrium criteria in terms of minimization of the internal energy, U, and its Legendre transforms, subject to suitable overall constraints on the system. Here again, criteria such as $\mathrm{d}U=0$ can lead to an extremum, but not necessarily a minimum, and we must examine higher derivatives or finite changes in order to ascertain requirements for stability. Similar considerations apply to stability criteria based on minimization of other thermodynamic potentials such as F, G, and H, but some of the natural variables on which these potentials depend are intensive, so their behavior with respect to stability must be ascertained by relating to extensive variables by means of Legendre transforms.

Examination of these requirements will also result in useful information about the signs of various physical quantities, such as heat capacities, and compressibilities, as well as inequalities that restrict the relative magnitudes or ratios of these quantities.

7.1 Stability Requirements for Entropy

For simplicity, we consider a homogeneous system having entropy $S(U, V, N)$ and assume that constant values of U, V, and N will guarantee isolation. We first follow Callen [2, p. 203] based on an analysis by Griffiths [20] and pose the question of whether this system is stable with respect to breakup into two homogeneous subsystems, each having a volume $V/2$ and number of moles $N/2$, one having energy $(U - \Delta U)/2$ and the other having energy $(U + \Delta U)/2$. The energy of the combined subsystems is $(1/2)(U - \Delta U) + (1/2)(U + \Delta U) = U$. Since S is a homogeneous function of degree one in these extensive variables, the corresponding entropies of the subsystems are $(1/2)S(U - \Delta U, V, N)$ and

Thermal Physics. http://dx.doi.org/10.1016/B978-0-12-803304-3.00007-7

$(1/2)S(U + \Delta U, V, N)$. Therefore, the homogeneous system will be stable with respect to this breakup by an irreversible process if

$$(1/2)S(U - \Delta U, V, N) + (1/2)S(U + \Delta U, V, N) \leq S(U, V, N). \tag{7.1}$$

This requirement is represented graphically in Figure 7–1. By rewriting the left-hand side of Eq. (7.1) in the form

$$S(U - \Delta U, V, N) + (1/2)[S(U + \Delta U, V, N) - S(U - \Delta U, V, N)] \leq S(U, V, N), \tag{7.2}$$

we verify that the entropy of the composite system lies on the straight line (chord) joining $(1/2)S(U - \Delta U, V, N)$ and $(1/2)S(U + \Delta U, V, N)$ at the value U, midway between $U - \Delta U$ and $U + \Delta U$. Thus, stability for all values of U requires S to be a **concave function** of U (as viewed from below). Thus, the situation in Figure 7–1a is stable, and that in Figure 7–1b is unstable. The equal sign in Eq. (7.2) would correspond to a situation of neutral stability that would involve a hypothetical reversible process. We will discuss this possibility in Chapters 9 and 10 in connection with phase transformations.

For infinitesimal changes $\Delta U \to \delta U$, we can expand the entropies in Eq. (7.1) to obtain

$$S(U \pm \delta U, V, N) = S(U, V, N) \pm S_U(U, V, N)\delta U + (1/2)S_{UU}(\delta U)^2 + \cdots, \tag{7.3}$$

where the subscripts U represent partial differentiation.[1] Then neglecting terms of the third order and higher, Eq. (7.1) becomes, after division by $(\delta U)^2/2$,

$$S_{UU} \equiv \left(\frac{\partial^2 S}{\partial U^2}\right)_{V,N} \leq 0. \tag{7.4}$$

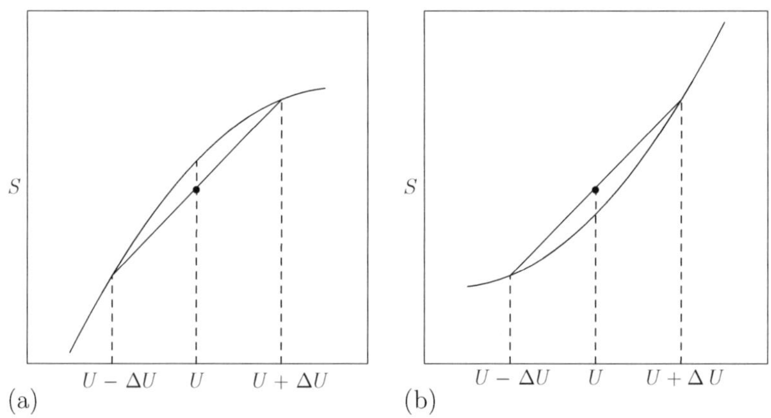

FIGURE 7–1 Conditions for $S(U, V, N)$, represented by the solid curves, for stability (a) or instability (b). To be stable, $S(U, V, N)$ must be a concave function of U at fixed V and N. A composite system having the same values of U, V, and N would have an entropy represented by the intersection of the chord with the vertical line at U. (a) Stable (concave) and (b) Unstable (convex).

[1] In this chapter, subscripts that indicate partial derivatives imply the natural variable sets for each function, explicitly $S(U, V, N)$, $U(S, V, N)$, $H(S, p, N)$, $F(T, V, N)$, and $G(T, p, N)$.

Equation (7.4) is a requirement for *local* stability because it corresponds to infinitesimal changes. If $S_{UU} = 0$, we could examine higher derivatives. For example, we would need $S_{UUU} = 0$ and $S_{UUUU} < 0$, but such a requirement would still be local.

The situation depicted in Figure 7–2 is more complicated because the second derivative S_{UU} changes sign at the so-called spinodal points U_{S1} and U_{S2}. The region between points U_{S1} and U_{S2} is clearly unstable with respect to infinitesimal variations δU. The remainder of the curve is stable with respect to infinitesimal variations. The states between U_1 and U_{S1} and between U_{S2}, and U_2, where U_1 and U_2 are points of common tangency, are more difficult to analyze because the above analysis requires values of $U - \Delta U$ and $U + \Delta U$ that are symmetrically situated and can span distant portions of the curve.

We therefore resort to the following modified analysis. We represent the entropy, internal energy, and volume per mole by the lower case letters *s*, *u*, and v, respectively. The original system has N moles, entropy $S(U, V, N) = Ns(u, v)$, internal energy Nu, and volume Nv. We consider breakup onto a composite system consisting of two homogeneous systems, one having $(1-f)N$ moles and intensive parameters $u_1, v, s(u_1, v)$, and the other having fN moles and intensive parameters $u_2, v, s(u_2, v)$, where

$$1 - f = \frac{u_2 - u}{u_2 - u_1}; \quad f = \frac{u - u_1}{u_2 - u_1}. \tag{7.5}$$

Without loss of generality we take $u_2 > u_1$. The volume of the composite system is $N(1-f)v + Nfv = Nv = V$ and its number of moles is $N(1-f) + Nf = N$. It has energy

$$N(1-f)u_1 + Nfu_2 = \frac{N}{u_2 - u_1}[(u_2 - u)u_1 + (u - u_1)u_2] = Nu = U. \tag{7.6}$$

The entropies of the subsystems are $(1-f)Ns(u_1, v)$ and $fNs(u_2, v)$. After division by N, the requirement for stability becomes

$$(1-f)s(u_1, v) + fs(u_2, v) \leq s(u, v), \tag{7.7}$$

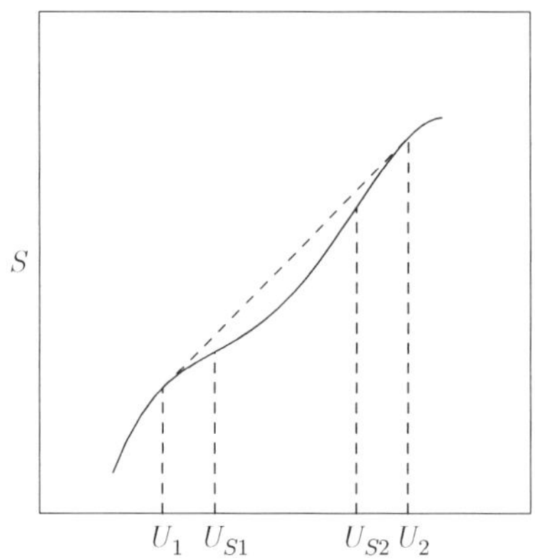

FIGURE 7–2 $S(U, V, N)$ versus U under conditions for which some states are locally stable and others are locally unstable. The states between the spinodal points U_{S1} and U_{S2} are locally unstable and states outside these points are locally stable. But states between U_1 and U_{S1} and between U_{S2} and U_2 are globally unstable, so they are metastable.

which can be rewritten

$$s(u_1, v) + \frac{u - u_1}{u_2 - u_1}[s(u_2, v) - s(u_1, v)] \le s(u, v). \tag{7.8}$$

The requirement represented by Eq. (7.8) is shown in Figure 7–3, from which we see that the entropy per mole of the composite system is represented by the intersection of a vertical line at u with a chord joining *any* points $s(u_1, v)$ and $s(u_2, v)$, as long as $u_2 > u > u_1$ is satisfied. This criterion shows that the general requirement for stability is concavity of $s(u, v)$ as a function of u at fixed v. Since $S(U, V, N) = Ns(u, v) = Ns(U/N, V/N)$, we see for stability that $S(U, V, N)$ is a concave function of U at fixed V and N. Thus the states in Figure 7–2 between U_1 and U_{S1} and between U_{S2} and U_2, although locally stable, are globally unstable and are termed metastable. By letting $u_1 = u - \delta u$, $u_2 = u + \delta u$ and expanding Eq. (7.8) for small δu, one obtains $\partial^2 s/\partial u^2 \le 0$ as a local stability condition, consistent with Eq. (7.4).

Returning to the general analysis of $S(U, V, N)$, we can inquire about stability against breakup into two homogeneous subsystems, each having the same energy $U/2$ and mole numbers $N/2$, but different volumes $(V - \Delta V)/2$ and $(V + \Delta V)/2$. By the same reasoning as above, stability requires

$$(1/2)S(U, V - \Delta V, N) + (1/2)S(U, V + \Delta U, N) \le S(U, V, N). \tag{7.9}$$

For infinitesimal changes δV

$$S_{VV} \equiv \left(\frac{\partial^2 S}{\partial V^2}\right)_{U,N} \le 0. \tag{7.10}$$

The same reasoning applies to changes of N or to any other extensive variables on which S could depend.

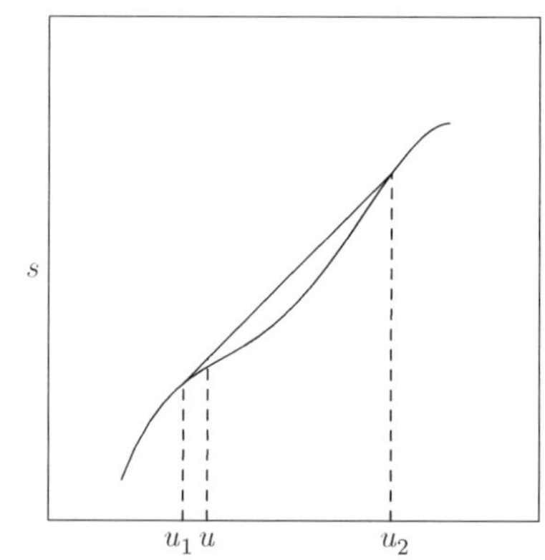

FIGURE 7–3 $s(u, v)$ versus u under conditions for which some states are locally stable and others are locally unstable. At constant v, we test the state at u against breakup into a composite system consisting of states having molar energies u_1 and u_2 that are not equidistant from u. The entropy per mole of the composite system lies on the straight line at position u and exceeds $s(u, v)$ which lies on the curve. Therefore, the state at u is globally unstable, even though it is locally stable.

If both U and V are different for the subsystems, we obtain

$$(1/2)S(U-\Delta U, V-\Delta V, N)+(1/2)S(U+\Delta U, V+\Delta U, N) \leq S(U, V, N). \tag{7.11}$$

For infinitesimal changes in U and V, Eq. (7.11) becomes

$$S_{UU}(\delta U)^2+2S_{UV}\delta U\delta V+S_{VV}(\delta V)^2 \leq 0, \tag{7.12}$$

where the derivatives are evaluated at U, V, N. Testing Eq. (7.12) for $\delta V = 0$ or $\delta U = 0$ recovers $S_{UU} \leq 0$ and $S_{VV} \leq 0$ as above. But for general δV and δU, a new condition emerges. We can write Eq. (7.12) in the matrix form

$$\begin{pmatrix} \delta U & \delta V \end{pmatrix}\begin{pmatrix} S_{UU} & S_{UV} \\ S_{UV} & S_{VV} \end{pmatrix}\begin{pmatrix} \delta U \\ \delta V \end{pmatrix} \leq 0, \tag{7.13}$$

which involves a real symmetric matrix that can be diagonalized. Its eigenvalues λ satisfy

$$\det\begin{pmatrix} S_{UU}-\lambda & S_{UV} \\ S_{UV} & S_{VV}-\lambda \end{pmatrix} = 0, \tag{7.14}$$

which leads to a quadratic equation with roots

$$\begin{aligned}\lambda_{\pm} &= \frac{S_{UU}+S_{VV}}{2} \pm \sqrt{\left(\frac{S_{UU}+S_{VV}}{2}\right)^2+S_{UV}^2-S_{UU}S_{VV}} \\ &= \frac{S_{UU}+S_{VV}}{2} \pm \sqrt{\left(\frac{S_{UU}-S_{VV}}{2}\right)^2+S_{UV}^2}.\end{aligned} \tag{7.15}$$

From the second form, we see that both roots are real, which is a general property for the eigenvalues of any real symmetric matrix. From the first form, and recalling that $S_{UU} \leq 0$ and $S_{VV} \leq 0$, we see that there are no positive roots provided that

$$S_{UU}S_{VV}-S_{UV}^2 \geq 0. \tag{7.16}$$

After diagonalization, Eq. (7.13) can be rewritten in the form

$$\lambda_{+}(\delta X_1)^2+\lambda_{-}(\delta X_2)^2 \leq 0, \tag{7.17}$$

where $\lambda_{\pm} \leq 0$ and δX_1 and δX_2 are linear combinations of δU and δV that can be found by calculating the eigenvectors of the matrix. Thus, $S_{UU} \leq 0$ and $S_{VV} \leq 0$ together with Eq. (7.16) guarantee that Eq. (7.12) is satisfied.[2] They insure *locally* that the surface S will not lie above its *local* tangent plane. Callen [2, p. 206] refers to Eq. (7.16) as a "fluting condition."

By a procedure similar to that used to derive Eq. (7.8), we can test a system with entropy $Ns(u, v)$ with respect to breakup into a composite of three systems having entropies $Nf_1s(u_1, v_1)$, $Nf_2s(u_2, v_2)$, and $Nf_3s(u_3, v_3)$, where f_1, f_2, and f_3 are positive fractions that sum to unity, chosen such that total energy and total volume are conserved. This leads to a stability criterion of the form

$$f_1(u, v)s(u_1, v_1)+f_2(u, v)s(u_2, v_2)+f_3(u, v)s(u_3, v_3) \leq s(u, v), \tag{7.18}$$

[2]For an alternative procedure that would lead to Eq. (7.16), see Section 7.2.

where the f_i satisfy the following linear equations:

$$\begin{aligned} f_1 + f_2 + f_3 &= 1, \\ f_1 u_1 + f_2 u_2 + f_3 u_3 &= u, \\ f_1 v_1 + f_2 v_2 + f_3 v_3 &= v. \end{aligned} \tag{7.19}$$

We could use Cramer's rule to solve Eq. (7.19) by means of determinants, but the actual expressions are cumbersome and not needed as long as we note the following properties. A solution is only possible if the determinant of the coefficients of the f_i is not zero, which will be true if the points (u_1, v_1), (u_2, v_2), and (u_3, v_3) lie at the vertices of a non-degenerate triangle in the u, v plane. We shall refer to these vertices as 1, 2, and 3, respectively, in which case that determinant is equal to $2A_{123}$, where $A_{123} > 0$ is the area of that triangle. As shown below, the point (u, v) where $s(u, v)$ is to be tested for stability must be chosen within or on that triangle. With (u_1, v_1), (u_2, v_2), and (u_3, v_3) fixed, the f_i will be linear functions of u and v that can be written in the form $f_i(u, v)$, as already indicated in Eq. (7.19); furthermore, they will satisfy

$$f_i(u_j, v_j) = \delta_{ij}; \quad f_i(u_0, v_0) = A_{0jk}/A_{123}, \tag{7.20}$$

where δ_{ij} is the Kronecker delta, i, j, k, are cyclic permutations of 123, and the quantities A_{0jk} are areas of triangles defined below. The first member of Eq. (7.20) follows from Cramer's rule because the determinant of a matrix having two identical columns is zero. If the point (u_0, v_0) is referred to as point zero, Cramer's rule can also be used to show that A_{0jk} is the area of triangle $0jk$. Consistent with $A_{123} > 0$, the areas $A_{0jk} \geq 0$ are positive as long as (u_0, v_0) lies inside or on triangle 123. If (u_0, v_0) were to lie outside triangle 123, at least one of the f_i will be negative, which is unacceptable. See Figure 8–11 that pertains to an isomorphous problem.

From these properties of the $f_i(u, v)$, it follows that the left-hand side of Eq. (7.18) represents a plane that passes through the points $s(u_1, v_1)$, $s(u_2, v_2)$, and $s(u_3, v_3)$. Therefore, geometrically, the global stability criterion represented by Eq. (7.18) states that $s(u, v)$ lies above or on *any* such plane. In other words, for stability $s(u, v)$ must be a **concave function** of u and v. If $s(u, v)$ violates Eq. (7.18) for any such plane, that state will be globally unstable, but would be locally stable if Eq. (7.12) were satisfied.

7.2 Stability Requirements for Internal Energy

We can establish similar requirements for stability in terms of the internal energy U since at equilibrium it is a minimum at constant entropy and other extensive variables. For example, for $U(S, V, N)$ we have the stability requirement

$$(1/2)U(S - \Delta S, V, N) + (1/2)U(S + \Delta S, V, N) \geq U(S, V, N), \tag{7.21}$$

which for infinitesimal changes in S gives the local condition

$$\left(\frac{\partial^2 U}{\partial S^2}\right)_{V,N} \geq 0. \tag{7.22}$$

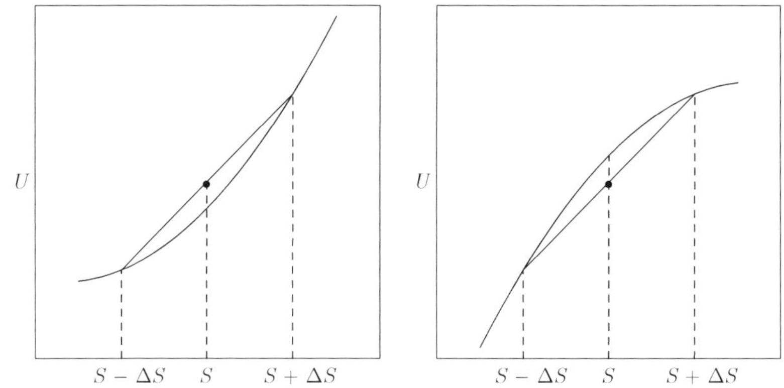

(a) (b)

FIGURE 7–4 Conditions for *U*(*S*, *V*, *N*), represented by the solid curves, for stability (a) or instability (b). To be stable, *U*(*S*, *V*, *N*) must be a convex function of *S* at fixed *V* and *N*. A composite system having the same values of *S*, *V*, and *N* would have an energy represented by the intersection of the chord with the vertical line at *S*. (a) Stable (convex) and (b) Unstable (concave).

Similar equations would apply for the other extensive variables V and N on which U depends. Thus, for stability, U is a **convex function** of S, V, and N (and of its other extensive variables for more complicated systems). This requirement is represented graphically in Figure 7–4.

If both S and V are different for members of the composite system, stability requires

$$(1/2)U(S-\Delta S, V-\Delta V, N) + (1/2)U(S+\Delta S, V+\Delta V, N) \geq U(S, V, N). \tag{7.23}$$

For infinitesimal changes, Eq. (7.23) yields the stability requirement

$$U_{SS}(\delta S)^2 + U_{VV}(\delta V)^2 + 2U_{SV}\delta S\delta V \geq 0. \tag{7.24}$$

We can proceed as in the case of Eq. (7.12) to examine eigenvalues and to find the condition that both are non-negative. In addition to $U_{SS} \geq 0$ and $U_{VV} \geq 0$, this leads to the fluting condition

$$\mathcal{D} \equiv U_{SS}U_{VV} - U_{SV}^2 \geq 0, \tag{7.25}$$

which has the *same* sense of the inequality as Eq. (7.16). We can also deduce Eq. (7.25) by another method as follows. We multiply Eq. (7.24) by the non-negative quantity U_{SS} to deduce

$$(U_{SS}\delta S + U_{SV}\delta V)^2 + \mathcal{D}(\delta V)^2 \geq 0. \tag{7.26}$$

For given δV, the first term can be made equal to zero by choice of δS, so the second term must be non-negative, thus resulting in Eq. (7.25). Moreover, if Eq. (7.25) holds, Eq. (7.26) is always satisfied, so Eq. (7.25) is both necessary and sufficient. A similar technique can be applied to analyze Eq. (7.12); in that case, one multiplies first by the non-positive quantity S_{UU} which reverses the sense of the inequality. Thus,

$$(S_{UU}\Delta U + S_{UV}\Delta V)^2 + (S_{UU}S_{VV} - S_{UV}^2)(\Delta V)^2 \geq 0, \tag{7.27}$$

which results in Eq. (7.16). For the internal energy we could also carry out the same procedure that led to Eq. (7.18), resulting in the stability requirement

$$g_1(s, v)u(s_1, v_1) + g_2(s, v)u(s_2, v_2) + g_3(s, v)s(u_3, v_3) \geq u(s, v). \tag{7.28}$$

Here, the fractions $g_i(s, v)$ are linear functions of s and v that satisfy $g_i(s_j, v_j) = \delta_{ij}$. Equation (7.28) shows that $u(s, v)$ must lie below any plane represented by its left-hand side, so $u(s, v)$ must be a **convex function** for global stability.

7.3 Stability Requirements for Other Potentials

We can also obtain stability requirements for other potentials, such as H, F, and G, which are Legendre transforms of U. An important distinction arises, however, because some of the natural variables on which these functions depend are intensive.

7.3.1 Enthalpy

For the enthalpy $H(S, p, N)$, stability requires

$$(1/2)H(S - \Delta S, p, N) + (1/2)H(S + \Delta S, p, N) \geq H(S, p, N). \tag{7.29}$$

For infinitesimal changes δS, the local stability requirement is

$$H_{SS} := \left(\frac{\partial^2 H}{\partial S^2}\right)_{p,N} \geq 0. \tag{7.30}$$

But there is no equation analogous to Eq. (7.29) involving changes Δp because p is intensive and therefore must be the same in each member of the composite system that we compare to $H(S, p, N)$. We therefore deduce an inequality for H_{pp} by relating to a partial derivative of its Legendre transform U. As shown in Section 5.5, we have

$$H_{pp} := \left(\frac{\partial^2 H}{\partial p^2}\right)_{S,N} = -\frac{1}{U_{VV}} \leq 0. \tag{7.31}$$

Thus for local stability, H is a locally convex function of the extensive variable S but a locally concave function of the intensive variable p. As a result of this, the fluting condition $H_{SS}H_{pp} - H_{Sp}^2 \leq 0$ is true by default because both terms are non-positive. The fact that this inequality has the correct sense can also be seen as follows. We suppress N for simplicity of notation. Then $(\partial U/\partial S)_V = T = (\partial H/\partial S)_p$, so

$$\left(\frac{\partial^2 U}{\partial S^2}\right)_V = \left[\frac{\partial}{\partial S}\left(\frac{\partial H}{\partial S}\right)_p\right]_V = H_{SS} + H_{Sp}\left(\frac{\partial p}{\partial S}\right)_V. \tag{7.32}$$

But

$$\left(\frac{\partial p}{\partial S}\right)_V = -\frac{(\partial V/\partial S)_p}{(\partial V/\partial p)_S} = -\frac{H_{Sp}}{H_{pp}}. \tag{7.33}$$

Therefore

$$U_{SS} = \frac{H_{SS}H_{pp} - H_{Sp}^2}{H_{pp}}. \tag{7.34}$$

Since $U_{SS} \geq 0$, $H_{SS} \geq 0$, and $H_{pp} \leq 0$, we see consistently that $H_{SS}H_{pp} - H_{Sp}^2 \leq 0$.

In a similar manner, we can show that

$$H_{SS} = \frac{U_{SS}U_{VV} - U_{SV}^2}{U_{VV}} = \frac{\mathcal{D}}{U_{VV}}, \tag{7.35}$$

so the fact that $\mathcal{D} \geq 0$ could have been deduced from $H_{SS} \geq 0$ and $U_{VV} \geq 0$. It is generally the case that all fluting conditions can be deduced from conditions on non-mixed second derivatives provided that appropriate Legendre transforms are considered.

7.3.2 Helmholtz Free Energy

For the Helmholtz free energy $F(T, V, N)$, we have an equation analogous to Eq. (7.29) but involving ΔV and this leads directly to the local requirement $F_{VV} \geq 0$. We also have $F_{TT} = -1/U_{SS} \leq 0$. So for local stability, F is a locally convex function of the extensive variable V and a locally concave function of the intensive variable T. By methods similar to those discussed for the enthalpy, we have the local stability requirement

$$U_{VV} = \frac{F_{VV}F_{TT} - F_{VT}^2}{F_{TT}} \tag{7.36}$$

so $F_{VV}F_{TT} - F_{VT}^2 \leq 0$, which is no contest because $F_{TT} \leq 0$ so both terms are non-positive. We also have

$$F_{VV} = \frac{\mathcal{D}}{U_{SS}} \geq 0, \tag{7.37}$$

another redundancy.

7.3.3 Gibbs Free Energy

For the Gibbs free energy $G(T, p, N)$, both T and p are intensive, so local stability requirements involving their derivatives must be obtained indirectly from their Legendre transforms. We have $G_{TT} = -1/H_{SS} \leq 0$ and $G_{pp} = -1/F_{VV} \leq 0$ as anticipated for both principal second partial derivatives with respect to intensive variables. In this case, the fluting condition is not trivial. It is most easily related to derivatives of F or H, which differ from it by a single Legendre transform. Thus we can use either

$$F_{TT} = \frac{G_{TT}G_{pp} - G_{Tp}^2}{G_{pp}} \tag{7.38}$$

or

$$H_{pp} = \frac{G_{TT}G_{pp} - G_{Tp}^2}{G_{TT}}, \tag{7.39}$$

either of which shows that

$$G_{TT}G_{pp} - G_{Tp}^2 \geq 0. \tag{7.40}$$

A somewhat more involved calculation[3] shows that $G_{pp} = -U_{SS}/\mathcal{D}$, $G_{TT} = -U_{VV}/\mathcal{D}$, and $G_{Tp} = -U_{SV}/\mathcal{D}$ which results in

$$G_{TT}G_{pp} - G_{Tp}^2 = \frac{1}{U_{SS}U_{VV} - U_{SV}^2} \geq 0, \tag{7.41}$$

so the two non-trivial fluting conditions are just reciprocals of one another.

7.3.4 Summary of Stability Requirements

By means similar to those discussed above, we can extend the stability requirements to any number of variables. For stability of a homogeneous system:

- The entropy, S, must be a concave function of its natural extensive variables.
- The internal energy, U, must be a convex function of its natural extensive variables.
- Legendre transforms of U, such as H, F, and G, must be convex functions of their natural extensive variables and concave functions of their natural intensive variables.

We did not discuss the Massieu functions, which are Legendre transforms of the entropy, but they must be concave functions of their extensive variables and convex functions of their intensive variables.

Fluting conditions involve mixed partial derivatives, but are always redundant with requirements on non-mixed second partial derivatives of S, U, or some Legendre transform of U.

It is possible to consider thermodynamic functions, perhaps derived from some model, for which the requirements for local stability are true for some range of variables but for which the requirements for global stability are violated. Such situations can occur when different phases of a composite system are in equilibrium but in which phase transitions can occur. We shall illustrate this in Chapter 9 by means of the van der Waals model.

In applying the above requirements, it is extremely important to note that they only apply to the extensive thermodynamic functions and the natural variables, extensive and intensive, on which they depend. Moreover, if one uses a "density" of some extensive variable, such as the Helmholtz free energy per mole, $f = F/N$, one finds that $\mathrm{d}f = -s\,\mathrm{d}T - p\,\mathrm{d}v$ where $v = V/N$ is also a "density," namely the volume per mole. Although f and v are certainly intensive, they still behave from the point of view of stability like the extensive variables F and V from which they originate. In other words, $(\partial^2 f/\partial v^2)_T \geq 0$ for local stability, corresponding to f being a convex function of v, just as F is a convex function of V. But T is not a "density" so $(\partial^2 f/\partial T^2)_v \leq 0$ for local stability, meaning that f is a concave function of T. This peculiarity arises because the local stability condition

[3]For instance, $G_{TT} = -1/(\partial U_S/\partial S)_p$, $(\partial U_S/\partial S)_p = U_{SS} + U_{SV}(\partial V/\partial S)_p$, and $(\partial V/\partial S)_p = -U_{VS}/U_{VV}$.

for an intensive variable such as T is derived from a Legendre transformation, rather than splitting a system into parts having different values of T, as was done for V.

7.4 Consequences of Stability Requirements

By using the stability requirements previously derived, we can deduce several useful relationships about the signs and relative magnitudes of some measurable physical properties of stable homogeneous phases. Thus

$$U_{SS} = \left(\frac{\partial T}{\partial S}\right)_{V,N} = T/C_V \geq 0 \tag{7.42}$$

so the heat capacity at constant volume $C_V \geq 0$. Similarly

$$H_{SS} = \left(\frac{\partial T}{\partial S}\right)_{p,N} = T/C_p \geq 0 \tag{7.43}$$

so the heat capacity at constant pressure $C_p \geq 0$. We also have

$$F_{VV} = -\left(\frac{\partial p}{\partial V}\right)_{T,N} = 1/(V\kappa_T) \geq 0 \tag{7.44}$$

so the isothermal compressibility $\kappa_T \geq 0$. From Eq. (5.32) we have

$$C_p - C_V = TV\alpha^2/\kappa_T \tag{7.45}$$

so

$$C_p \geq C_V \geq 0. \tag{7.46}$$

We can define a compressibility at constant entropy[4] by the relation

$$\kappa_S := -\frac{1}{V}\left(\frac{\partial V}{\partial p}\right)_{S,N}. \tag{7.47}$$

Since $U_{VV} = -(\partial p/\partial V)_{S,N} \geq 0$, we see that $\kappa_S \geq 0$. It can be related to κ_T as follows. We have (with constant N suppressed for simplicity)

$$-V\kappa_S = \left(\frac{\partial V}{\partial p}\right)_T + \left(\frac{\partial V}{\partial T}\right)_p \left(\frac{\partial T}{\partial p}\right)_S \tag{7.48}$$

so

$$\kappa_S = \kappa_T - \alpha\left(\frac{\partial T}{\partial p}\right)_S. \tag{7.49}$$

Then from $(\partial T/\partial p)_S(\partial p/\partial S)_T(\partial S/\partial T)_p = -1$ we deduce

$$\left(\frac{\partial T}{\partial p}\right)_S = -\frac{(\partial S/\partial p)_T}{(\partial S/\partial T)_p} = \frac{V\alpha}{C_p/T}, \tag{7.50}$$

[4]This is sometimes called the adiabatic compressibility, but strictly speaking it is isentropic.

where the Maxwell relation $-(\partial S/\partial p)_T = (\partial V/\partial T)_p$ from the differential dG has been used. Combining these relations gives

$$\kappa_T - \kappa_S = TV\alpha^2/C_P. \tag{7.51}$$

We therefore see that

$$\kappa_T \geq \kappa_S \geq 0. \tag{7.52}$$

In fact, combination of Eqs. (7.45) and (7.51) gives the interesting relation

$$\kappa_S/\kappa_T = C_V/C_p. \tag{7.53}$$

For an alternative derivation of Eq. (7.53) that involves Jacobians, see Appendix B.

7.5 Extension to Many Variables

A number of other relationships can be derived in the same manner as illustrated above. We illustrate these beginning with the internal energy as a function of many extensive variables. If we write dU in the form

$$\mathrm{d}U = \sum_{j=1}^{n} p_j\,\mathrm{d}E_j, \tag{7.54}$$

where the E_j are extensive variables and the p_j are their conjugate intensive variables, local stability with respect to a single variable will require

$$\left(\frac{\partial^2 U}{\partial E_i^2}\right)_{\{E_i'\}} = \left(\frac{\partial p_i}{\partial E_i}\right)_{\{E_i'\}} \geq 0, \tag{7.55}$$

where $\{E_i'\}$ is the set $\{E_i\}$ with E_i missing. If all of the extensive variables are allowed to change by infinitesimal amounts, the generalization of the local stability condition Eq. (7.24) is

$$\sum_{i,j} \delta U_i\, U_{ij}\, \delta U_j \geq 0; \quad U_{ij} = \frac{\partial^2 U}{\partial E_i \partial E_j} = U_{ji}. \tag{7.56}$$

We could proceed to diagonalize the real symmetric matrix $\mathcal{U} = \{U_{ij}\}$, in which case Eq. (7.56) would become

$$\sum_i \lambda_i\,(\delta X_i)^2 \geq 0, \tag{7.57}$$

where λ_i are its eigenvalues and δX_i are linear combinations of the δU_j that depend on the eigenvectors of $\mathcal{U}$. The condition for all eigenvalues of $\mathcal{U}$ to be positive definite is that the determinants of all of its principal minors be positive definite. Its principal minor of order r is the square symmetric matrix obtained from $\{U_{ij}\}$ by eliminating all of its rows for $i > r$ and all of its columns for $j > r$. If $\mathcal{U}$ is an $n \times n$ matrix, there are n of these principal minors, the largest being the entire matrix $\mathcal{U}$. For the simple case in which only δU_1 and

δU_2 are non-zero, the minor of order $r = 1$ gives $U_{11} > 0$ and the minor of order $r = 2$ gives $U_{11}U_{22} - U_{12}^2 > 0$, in agreement with Eq. (7.25).

For the entropy as a function of many extensive variables, the corresponding local stability criterion is a little trickier. In that case, one wants the eigenvalues of the matrix $\{S_{ij}\}$ to be non-positive. In order for such eigenvalues to be negative definite, one needs the determinants of the principal minors of odd order to be negative and those of even order to be positive. Thus, if only δS_1 and δS_2 are non-zero, one needs $S_{11} < 0$ but $S_{11}S_{22} - S_{12}^2 > 0$, in agreement with Eq. (7.16).

If we consider the Legendre transform

$$\mathcal{L} = U - \sum_{k=r+1}^{n} p_k E_k \tag{7.58}$$

with differential

$$\mathrm{d}\mathcal{L} = \sum_{j=1}^{r} p_j \,\mathrm{d}E_j - \sum_{k=r+1}^{n} E_k \,\mathrm{d}p_k, \tag{7.59}$$

we get either a stability requirement of the type

$$\frac{\partial \mathcal{L}}{\partial E_j} = \left(\frac{\partial p_j}{\partial E_j}\right)_{\{E'_j\},\{p_k\}} \geq 0; \quad j = 1, \ldots, r; \quad k = r+1, \ldots, n, \tag{7.60}$$

or

$$\frac{\partial \mathcal{L}}{\partial p_k} = -\left(\frac{\partial E_k}{\partial p_k}\right)_{\{E_j\},\{p'_k\}} \leq 0; \quad j = 1, \ldots, r; \quad k = r+1, \ldots, n. \tag{7.61}$$

Comparison of Eq. (7.60) with Eq. (7.55) shows that a partial derivative of an intensive variable with respect to its conjugate extensive variable is non-negative, but different variables can be held constant in the partial differentiations. For instance, $(\partial \mu_j/\partial N_j)_{S,V,\{N'_j\}} \geq 0$ but also $(\partial \mu_j/\partial N_j)_{S,p,\{N'_j\}} \geq 0$, $(\partial \mu_j/\partial N_j)_{T,V,\{N'_j\}} \geq 0$, and $(\partial \mu_j/\partial N_j)_{T,p,\{N'_j\}} \geq 0$ follow from consideration of U, H, F, and G, respectively.

7.6 Principles of Le Chatlier and Le Chatlier-Braun

Before leaving the subject of stability, we mention some general principles that govern the approach of systems to equilibrium. The first, due to Le Chatlier, states that if some extensive variable fluctuates from its equilibrium value, its conjugate intensive variable will change in such a way as to restore that extensive variable to its equilibrium value. The second, due to Le Chatlier-Braun, states that if some extensive variable fluctuates and also produces changes in non-conjugate intensive variables, secondary induced processes occur in such a way as to oppose the change in the conjugate intensive variable associated with the original extensive variable. Thus, any fluctuations of a stable state will tend to decay in such a way as to restore equilibrium values. For formal treatments, see Landau and Lifshitz [7, p. 63] or Callen [2, p. 212].

8

Monocomponent Phase Equilibrium

In this Chapter, we examine equilibrium for a monocomponent system for the simple case in which the solid phase has only a single crystal structure. The situation can be described by means of a **phase diagram** in the T, p plane, such as sketched in Figure 8–1. This diagram divides the plane into regions where the phases solid (S), liquid (L), and vapor[1] (V) are stable. Therefore, the only lines that appear on the diagram are curves where pairs of these phases are in equilibrium. These are called coexistence curves and we shall proceed to develop equations that describe them.

According to the thermodynamics of open monocomponent systems, the conditions for phases to be in equilibrium (see Chapter 6) are for them to have the same temperature T, the same pressure p, and the same chemical potential μ. But according to Eq. (5.45), the Gibbs-Duhem equation, these variables are not independent and one can regard the chemical potential $\mu(T,p)$ to be a function of temperature and pressure. This function is not the same for different phases, so the **coexistence curves** are given by the following equations:

$$\mu_{\mathrm{S}}(T,p) = \mu_{\mathrm{L}}(T,p), \quad \text{solid-liquid coexistence curve,} \tag{8.1}$$

$$\mu_{\mathrm{S}}(T,p) = \mu_{\mathrm{V}}(T,p), \quad \text{solid-vapor coexistence curve,} \tag{8.2}$$

$$\mu_{\mathrm{L}}(T,p) = \mu_{\mathrm{V}}(T,p), \quad \text{liquid-vapor coexistence curve.} \tag{8.3}$$

According to the Gibbs phase rule for a monocomponent system, the number of thermodynamic degrees of freedom is $3 - n$ where n is the number of phases. A single phase region, such as the solid, is represented by an area; accordingly, $n=1$ and there are two degrees of freedom, p and T, that may be chosen independently throughout this area. Along each of the coexistence curves, $p=2$ so there is one degree of freedom along these curves. Thus, if T is specified, p is known from the curve. For either solid-vapor or solid-liquid equilibrium, the corresponding pressure of the vapor for a given value of T is known as the vapor pressure. If $n=3$, there are no degrees of freedom; this happens at a point known as the **triple point** where solid, liquid, and vapor are in mutual equilibrium with each other. Thus we have

$$\mu_{\mathrm{S}}(T,p) = \mu_{\mathrm{L}}(T,p) = \mu_{\mathrm{V}}(T,p), \quad \text{triple point.} \tag{8.4}$$

Equation (8.4) represents two equations in two unknowns; their solution determines T_t and p_t, the unique coordinates of the triple point. It turns out that the liquid-vapor coexistence curve actually ends at a point T_c and p_c known as the **critical point**. Thus,

[1]A vapor is a gaseous phase that can be condensed to form a liquid or solid. Sometimes the word "gas" is used interchangeably with "vapor," but an ideal gas cannot undergo a phase transformation.

Thermal Physics. http://dx.doi.org/10.1016/B978-0-12-803304-3.00008-9

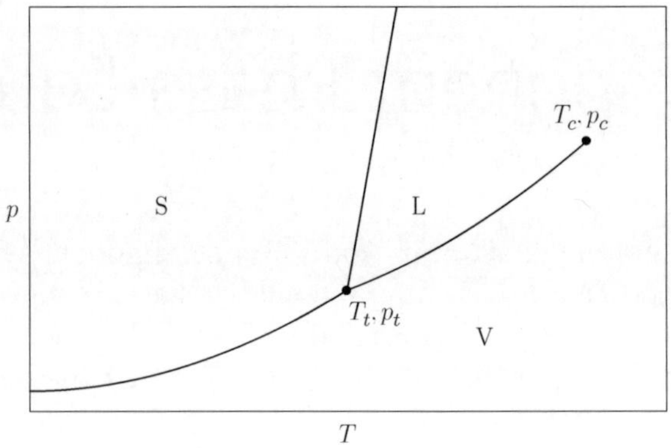

FIGURE 8–1 Sketch (not to scale) of a phase diagram for a monocomponent system. The curves are coexistence curves for pairs of the phases solid (S), liquid (L), and vapor (V). All three phases coexist in mutual equilibrium at the triple point T_t, p_t. The liquid-vapor coexistence curve ends at the critical point T_c, p_c. This diagram pertains to the usual case in which the molar volume of the solid is less than that of the liquid from which it freezes. See Figure 8–3 for the unusual case.

for $T > T_c$ or $p > p_c$, liquid and vapor become indistinguishable. In Chapter 9 we will see how such a behavior follows from the van der Waals model of a fluid.

Phase diagrams for monocomponent systems can have great variety because the crystalline solids can have different crystal structures, each considered to be a phase. For example, if the solid can have two crystal structures, say α and β, there can be more than one triple point, for example, for equilibrium among (α, L, V) and (α, β, L). See deHoff [21, chapter 7] for some specific examples as well as geometrical details of chemical potential surfaces.

8.1 Clausius-Clapeyron Equation

We proceed to find a differential equation for one of the coexistence curves; we choose the liquid-vapor coexistence curve as a specific example. We take the differential of Eq. (8.3) to obtain

$$\left(\frac{\partial \mu_{\rm L}}{\partial T}\right)_p \mathrm{d}T + \left(\frac{\partial \mu_{\rm L}}{\partial p}\right)_T \mathrm{d}p = \left(\frac{\partial \mu_{\rm V}}{\partial T}\right)_p \mathrm{d}T + \left(\frac{\partial \mu_{\rm V}}{\partial p}\right)_T \mathrm{d}p. \tag{8.5}$$

The derivatives in Eq. (8.5) can be identified by noting for a monocomponent system that the chemical potential μ is equal to $g := G/N$, the Gibbs free energy per mole. This follows because the Euler equation is just $G = \mu N$ for a monocomponent system. Since $\mathrm{d}G = -S\,\mathrm{d}T + V\,\mathrm{d}p + \mu\,\mathrm{d}N$, we readily verify that

$$\mathrm{d}\mu = \mathrm{d}g = -s\,\mathrm{d}T + v\,\mathrm{d}p, \quad \text{monocomponent system,} \tag{8.6}$$

where s is the entropy per mole and v is the volume per mole. Thus,

$$\left(\frac{\partial \mu}{\partial T}\right)_p = -s; \quad \left(\frac{\partial \mu}{\partial p}\right)_T = v. \tag{8.7}$$

Therefore, Eq. (8.5) becomes

$$\frac{\mathrm{d}p}{\mathrm{d}T} = \frac{s_V - s_L}{v_V - v_L}. \tag{8.8}$$

We can further transform Eq. (8.8) by recalling that $G = H - TS$ so that $g = h - Ts$ where h is the enthalpy per mole. Thus

$$\mu = h - Ts, \quad \text{monocomponent system,} \tag{8.9}$$

and Eq. (8.3) leads to

$$s_V - s_L = \frac{h_V - h_L}{T} \tag{8.10}$$

along the coexistence curve. The quantity $h_V - h_L$ is the latent heat of vaporization per mole from liquid to vapor. Similarly, the quantity $s_V - s_L$ is the entropy of vaporization per mole from liquid to vapor. According to Eq. (8.3), μ is continuous at a coexistence curve. But its first partial derivatives $-s$ and v are not continuous. They have jumps from liquid to vapor that are related by Eq. (8.10). Thus, it turn out that both $s_V - s_L$ and $h_V - h_L$ are positive quantities.[2] Substitution of Eq. (8.10) into Eq. (8.8) leads to

$$\frac{\mathrm{d}p}{\mathrm{d}T} = \frac{h_V - h_L}{T(v_V - v_L)}, \tag{8.11}$$

which is known as the **Clausius-Clapeyron equation.**[3] It is a differential equation for the liquid-vapor coexistence curve. It is generally more useful than Eq. (8.3) because the quantities on the right-hand side of Eq. (8.11) are better understood than μ itself and can be measured experimentally. Since $v_V - v_L > 0$, the vapor pressure curve of p versus T has a positive slope, so vapor pressure clearly increases with increasing T. To get the actual shape of the vapor pressure curve, we must know how $h_V - h_L$ and $v_V - v_L$ depend on T and v. Equations of the same form apply to the other coexistence curves.

8.1.1 Approximate Vapor Pressure Curve

We can integrate Eq. (8.11) by making the following approximations:

- The latent heat $\Delta h := h_V - h_L$ is a positive constant.
- The molar volume of the vapor is much greater than that of the liquid, so $v_V - v_L \approx v_V$.
- We can approximate the volume of the vapor by using the ideal gas law, $v_V \approx RT/p$.

These approximations are terrible near the critical point, but otherwise they are not too bad over a limited range of T. Of course, an ideal vapor will not condense to form a liquid, but the ideal gas law can still give a reasonable estimate of the molar volume of a real vapor. With these approximations, Eq. (8.11) becomes

[2]Experiment as well as elementary considerations of statistical mechanics lead to the fact that a mole of vapor has a higher entropy (more disorder) than a mole of liquid. For similar reasons, $s_V - s_S$, $h_V - h_S$, $s_L - s_S$, and $h_L - h_S$ are all positive quantities.

[3]According to Planck [15, p. 149] this equation was deduced by Clapeyron from Carnot's incorrect theory, but first rigorously proved by Clausius.

$$\frac{dp}{dT} = \frac{p\Delta h}{RT^2}. \tag{8.12}$$

The variables separate to give

$$\frac{dp}{p} = \frac{\Delta h}{R}\frac{dT}{T^2}, \tag{8.13}$$

which integrates to yield

$$\ln p = -\frac{\Delta h}{RT} + \ln C, \tag{8.14}$$

where C is a constant. We can exponentiate Eq. (8.14) to obtain

$$p = C \exp\left(-\frac{\Delta h}{RT}\right). \tag{8.15}$$

The constant C can be determined by relating to one point, T_0, p_0, on the coexistence curve, resulting in

$$p = p_0 \exp\left[-\frac{\Delta h}{R}\left(\frac{1}{T} - \frac{1}{T_0}\right)\right]. \tag{8.16}$$

The exponential form of Eq. (8.15) indicates that the vapor pressure p increases very rapidly as T increases. Consequently, it is often represented graphically by reverting to Eq. (8.14) and plotting $\ln p$ as a function of $1/T$, which yields a straight line of slope $-\Delta h/R$, as illustrated in Figure 8–2. Such a plot of vapor pressure data could be used to determine experimentally a value of Δh. Any process that obeys an equation of the general form of Eq. (8.14) is known as an **activated process** and is said to have **Arrhenius form**. The quantity Δh is often referred to as an **activation energy**, although it is really an enthalpy difference. The reason that many processes are activated will become apparent from statistical mechanics.

The same approximations can be made for the vapor pressure along the solid-vapor coexistence curve, resulting in Eq. (8.16) with $\Delta h = h_V - h_S$. The process of formation

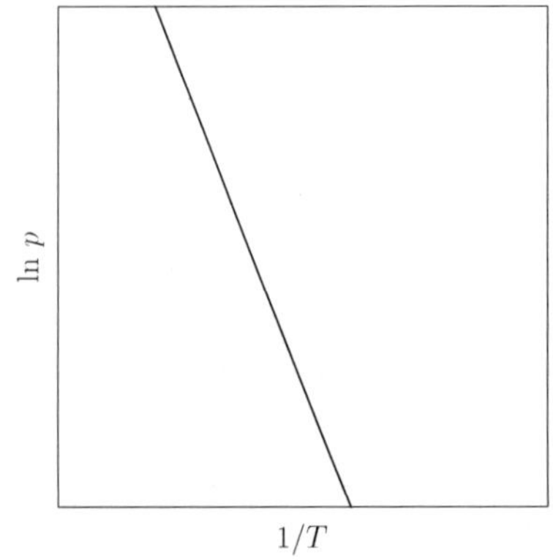

FIGURE 8–2 Plot of the logarithm of the vapor pressure p versus $1/T$ according to Eq. (8.14). The slope of the line is $-\Delta h/R$. Quantities that depend on temperature in this way are said to have Arrhenius form with an activation energy of Δh.

of a vapor directly from a solid is called sublimation, so this could also be called the sublimation pressure, but vapor pressure is a more common usage.

8.1.2 Approximate Solid-Liquid Coexistence Curve

For the solid-liquid coexistence curve, the Clausius-Clapeyron equation becomes

$$\frac{\mathrm{d}p}{\mathrm{d}T} = \frac{h_{\mathrm{L}} - h_{\mathrm{S}}}{T(v_{\mathrm{L}} - v_{\mathrm{S}})} \tag{8.17}$$

and a different set of approximations applies as follows:

- The entropy of fusion $\Delta s = \Delta h/T := (h_{\mathrm{L}} - h_{\mathrm{S}})/T$ is a positive constant.
- The molar volume of the liquid is comparable to that of the solid, typically only a few percent different. But $v_{\mathrm{L}} - v_{\mathrm{S}}$ can have either sign. For most materials, $v_{\mathrm{L}} - v_{\mathrm{S}}$ is positive, resulting in p increasing with T. But for some materials, including H_20, and semi-metals such as antimony and bismuth, $v_{\mathrm{L}} - v_{\mathrm{S}}$ is negative and p decreases as T increases.
- $\Delta v := v_{\mathrm{L}} - v_{\mathrm{S}}$ is constant.

With these approximations, Eq. (8.17) becomes

$$\frac{\mathrm{d}p}{\mathrm{d}T} \approx \frac{\Delta s}{\Delta v}, \tag{8.18}$$

which integrates to give

$$p - p_0 = \frac{\Delta s}{\Delta v}(T - T_0). \tag{8.19}$$

Thus the solid-liquid coexistence curve is nearly a straight line with steep slope. In the case for which Δv is positive, the phase diagram looks like Figure 8–1, but when it is negative, as it is for H_2O, the phase diagram resembles Figure 8–3.

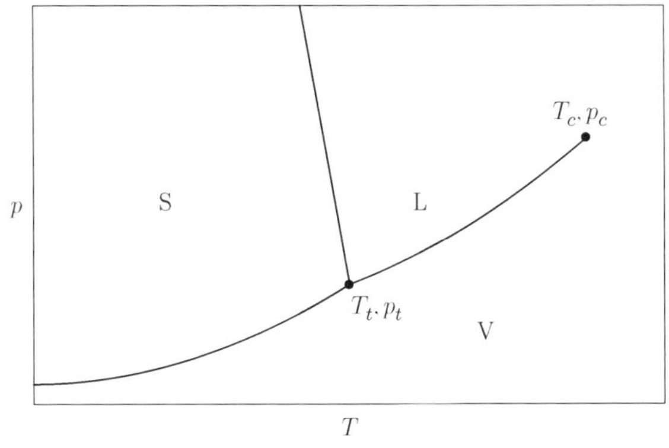

FIGURE 8–3 Sketch (not to scale) of a phase diagram for a monocomponent system for the unusual case for which the molar volume of the solid exceeds that of the liquid from which it freezes. The curves are coexistence curves for pairs of the phases solid (S), liquid (L), and vapor (V). See Figure 8–1 for the usual case and other notation.

Example Problem 8.1. At atmospheric pressure, silver melts at $T = 1235\,\mathrm{K}$ and its volume expands about 4%, the actual volume change being about $0.4\,\mathrm{cm^3/mol}$. Its latent heat of fusion is 11,950 J/mol. How much must the pressure increase to raise its melting point by 1 K?

Solution 8.1. Inserting this data into Eq. (8.17), we obtain

$$\frac{\mathrm{d}p}{\mathrm{d}T} = \frac{11{,}950}{1235 \times 0.4 \times 10^{-6}} = 2.38 \times 10^{7}\,\mathrm{Pa/K} = 2.38 \times 10^{2}\,\mathrm{atm/K}. \tag{8.20}$$

Thus, an enormous pressure of about 240 atmospheres would be required to raise the melting point of by 1 K. We conclude that the melting point of silver is practically insensitive to pressure, which is typical of other substances as well. On the other hand, as will be shown below, boiling points are quite sensitive to pressure because the molar volumes of gaseous phases depend strongly on pressure and are many times larger than the molar volumes of condensed phases.

8.1.3 Approximate Relative Magnitudes

The approximations used to obtain Eqs. (8.16) and (8.19) are rather crude and only meant to be illustrative. Although they lead to results that resemble the phase diagrams for real systems, they are no substitute for accurate experimental data. We can, however, gain some insight into the relative magnitudes of the slopes $\mathrm{d}p/\mathrm{d}T$ by using empirical rules to estimate the latent heats. At atmospheric pressure, for many simple metals, we have **Trouton's rule** that estimates $\Delta h/(RT) = 10.5$ for vaporization and **Richard's rule** that estimates $\Delta h/(RT) = 1.0$ for melting. By using these rules, Eq. (8.11) becomes

$$\frac{\mathrm{d}p}{\mathrm{d}T} \approx \frac{10.5R}{v_{\mathrm{V}}}, \quad \text{vaporization at atmospheric pressure,} \tag{8.21}$$

and Eq. (8.17) becomes

$$\frac{\mathrm{d}p}{\mathrm{d}T} \approx \frac{1.0R}{\Delta v}, \quad \text{melting at atmospheric pressure.} \tag{8.22}$$

By taking the ratio of Eq. (8.21) to Eq. (8.22) we obtain

$$\left(\frac{\mathrm{d}p}{\mathrm{d}T}\right)_{\text{vaporization}} \left| \left(\frac{\mathrm{d}p}{\mathrm{d}T}\right)_{\text{melting}} \right|^{-1} \approx 10.5\frac{|\Delta v|}{v_{\mathrm{V}}} \ll 1, \tag{8.23}$$

where the inequality applies because v_{V} is typically many orders of magnitude larger than $|\Delta v|$ for melting. Therefore, the slope of the solid-liquid coexistence curve is much steeper for melting than for vaporization. For vaporization of water at $373.1\,\mathrm{K} = 100\,°\mathrm{C}$, Fermi [1, p. 67] estimates $\mathrm{d}p/\mathrm{d}T = 2.7\,\mathrm{cm\,Hg/K} = 0.036\,\mathrm{atm/K}$, whereas for melting of ice at $273.1\,\mathrm{K} = 0\,°\mathrm{C}$ he estimates[4] $\mathrm{d}p/\mathrm{d}T = -134\,\mathrm{atm/K}$. The ratio of these slopes is -2.7×10^{-4}.

[4]This temperature is 100 K lower than for vaporization, but $\mathrm{d}p/\mathrm{d}T$ is nearly constant along the line of solid-liquid coexistence.

If we compute this ratio using Eq. (8.23), we get -5.5×10^{-4}. But the latent heats of H_2O deviate significantly from those given by Trouton's rule and Richard's rule as given above because of the complexity of the water molecule and the structure of ice. For H_2O, 10.5 for Trouton's rule should be replaced by 13.0 and for 1.0 for Richard's rule should be replaced by 2.64. This has the net effect of replacing 10.5 in Eq. (8.23) by $13.0/2.64 = 4.92$, so the corrected value of the slope ratio for H_2O is -2.6×10^{-4}, in reasonable agreement.

8.2 Sketches of the Thermodynamic Functions

We can gain more insight into monocomponent systems by sketching the thermodynamic functions μ, h, and s as functions of p and T. For a phase diagram of the form of Figure 8–1, we choose three constant pressures, p_1, p_2, and p_3 as indicated in Figure 8–4, and then discuss μ, h, and s as a function T at each of these pressures.

Along a line of constant p, μ is a continuous function of T. According to Eq. (8.7), its slope is $-s$. But s is discontinuous at a coexistence curve, so μ has a discontinuity of slope as T crosses a coexistence curve. Figure 8–5 shows a sketch of μ as a function of T along the line p_1 in Figure 8–4.

To quantify the behavior of s and h, we must view them as functions of T and p. Within a bulk phase,

$$ds = \left(\frac{\partial s}{\partial T}\right)_p \mathrm{d}T + \left(\frac{\partial s}{\partial p}\right)_T \mathrm{d}p = \frac{c_p}{T}\,\mathrm{d}T + \left(\frac{\partial s}{\partial p}\right)_T \mathrm{d}p, \tag{8.24}$$

where c_p is the heat capacity per mole at constant pressure. From Eq. (8.7) we have the Maxwell relation

$$\left(\frac{\partial s}{\partial p}\right)_T = -\left(\frac{\partial v}{\partial T}\right)_p = -v\alpha, \tag{8.25}$$

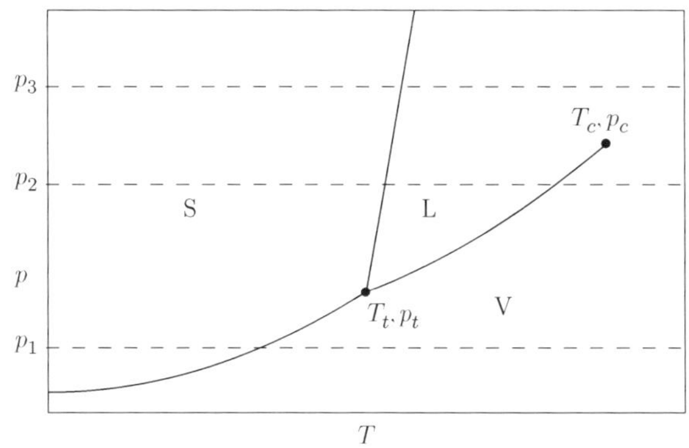

FIGURE 8–4 Constant pressure paths p_1, p_2, and p_3 on a phase diagram for the monocomponent system of Figure 8–1. The chemical potential μ is continuous along these paths, but its slope, $-s$, changes as T crosses a coexistence curve.

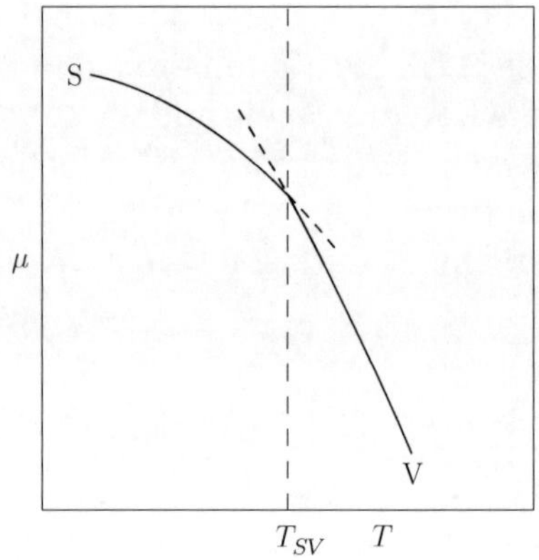

FIGURE 8–5 Sketch of the chemical potential μ as a function of T along the line $p = p_1$ in Figure 8–4. The full line corresponds to the stable solid and vapor phases. The dashed lines are extrapolations into unstable regions of superheated solid and supercooled vapor, intended to emphasize the discontinuity of slope of the full line at the phase transition. The stable phase is solid (S) for $T \le T_{SV}$ and vapor (V) for $T \ge T_{SV}$.

where α is the coefficient of expansion. Thus

$$ds = \frac{c_p}{T}\,dT - v\alpha\,dp. \tag{8.26}$$

For the enthalpy per mole, we have

$$dh = T\,ds + v\,dp = T\left(\frac{\partial s}{\partial T}\right)_p dT + \left[T\left(\frac{\partial s}{\partial p}\right)_T + v\right] dp. \tag{8.27}$$

Thus

$$dh = c_p\,dT + v(1 - T\alpha)\,dp. \tag{8.28}$$

For the sake of consistency of Eqs. (8.26) and (8.28), note that substitution into $d\mu = dh - s\,dT - T\,ds$ leads back to Eq. (8.6). Thus within a single phase at constant pressure we have

$$h(T_2) - h(T_1) = \int_{T_1}^{T_2} c_p\,dT \approx c_p(T_2 - T_1); \tag{8.29}$$

$$s(T_2) - s(T_1) = \int_{T_1}^{T_2} \frac{c_p}{T}\,dT \approx c_p \ln(T_2/T_1), \tag{8.30}$$

where the approximate expressions hold if c_p is a constant. Figure 8–6 shows sketches of h and s as a function of T along the line $p = p_1$ in Figure 8–4.

Along the line $p = p_2$, there are two phase transitions, from S to L and from L to V, so μ has a discontinuity of slope at each transition, and h and s have jumps at each transition. Along the line $p = p_3$, there is only one phase transition, because $p_3 > p_c$ and there is no distinction between liquid and vapor above the critical pressure.

Next, we choose three constant temperatures T_1, T_2, and T_3, as indicated in Figure 8–7. Along a line of constant T, μ is continuous and within a single phase, according to Eq. (8.7), it has a slope of v. Figure 8–8 is a sketch of μ versus p at $T = T_1$. We observe the discontinuity of slope as the vapor-solid coexistence curve is crossed.

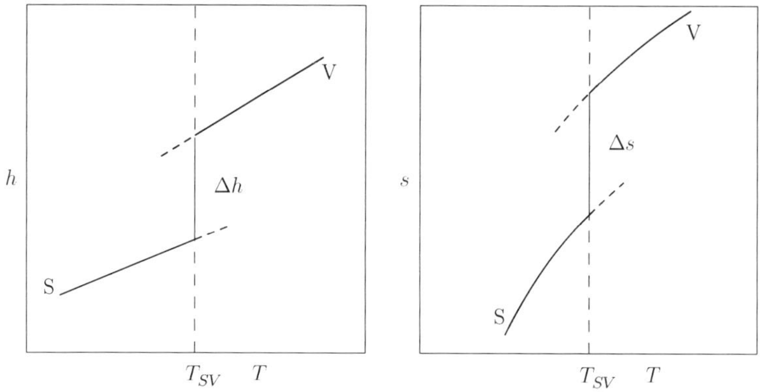

FIGURE 8–6 Sketches of the enthalpy h per mole and entropy s per mole as a function of T along the line $p = p_1$ in Figure 8–4. The full line corresponds to the stable solid and vapor phases. The stable phase is solid (S) for $T \le T_{SV}$ and vapor (V) for $T \ge T_{SV}$. The dashed lines are extrapolations into unstable regions of superheated solid and supercooled vapor. The jump Δh in enthalpy is the latent heat of vaporization per mole and the jump Δs is the entropy of vaporization per mole. These jumps are related by $\Delta h = T\Delta s$ so there is no jump in μ, consistent with Figure 8–5.

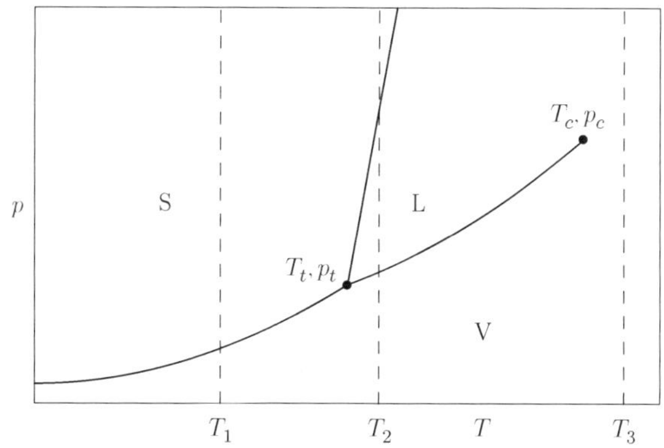

FIGURE 8–7 Constant temperature paths T_1, T_2, and T_3 on a phase diagram for the monocomponent system of Figure 8–1. The chemical potential μ is continuous along these paths, but its slope, v, changes as p crosses a coexistence curve.

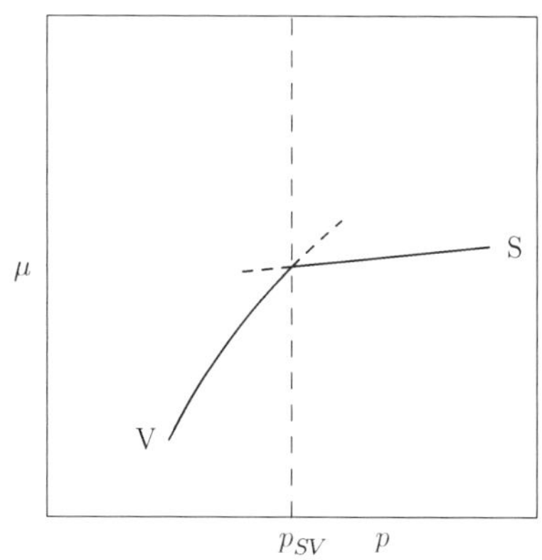

FIGURE 8–8 Sketch of the chemical potential μ as a function of p along the line $T = T_1$ in Figure 8–7. The full line corresponds to the stable solid and vapor phases. The dashed lines are extrapolations into unstable regions of superheated solid and supercooled vapor, intended to emphasize the discontinuity of slope of the full line at the phase transition. The stable phase is vapor (V) for $p \le p_{SV}$ and solid (S) for $p \ge p_{SV}$.

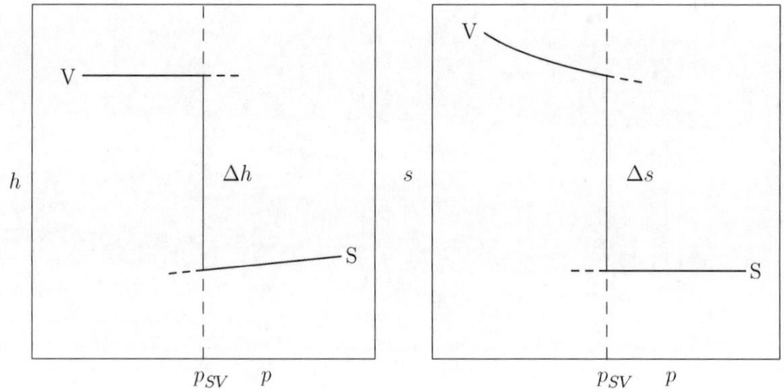

FIGURE 8–9 Sketches of the enthalpy h per mole and the entropy s per mole as a function of p along the line $T = T_1$ in Figure 8–7. The full line corresponds to the stable solid and vapor phases. The stable phase is vapor (V) for $p \le p_{SV}$ and solid (S) for $p \ge p_{SV}$. The dashed lines are extrapolations into unstable regions of superheated solid and supercooled vapor. The jump Δh is the latent heat of vaporization and the jump Δs is the entropy of vaporization. These jumps are related by $\Delta h = T\Delta s$ so there is no jump in μ.

The behaviors of h and s versus p within a single phase can be ascertained from Eqs. (8.26) and (8.28) which along a line of constant T lead to

$$h(p_2) - h(p_1) = \int_{p_1}^{p_2} v(1 - T\alpha)\,dp; \tag{8.31}$$

$$s(p_2) - s(p_1) = \int_{p_1}^{p_2} -v\alpha\,dp. \tag{8.32}$$

For an ideal vapor, $T\alpha = 1$ which gives

$$h(p_2) - h(p_1) = 0; \quad s(p_2) - s(p_1) = \int_{p_1}^{p_2} -\frac{v}{T}\,dp = -R\ln(p_2/p_1). \tag{8.33}$$

For the solid phase, $T\alpha \ll 1$, so for constant v we have[5]

$$h(p_2) - h(p_1) \approx v(p_2 - p_1); \quad s(p_2) - s(p_1) \approx 0. \tag{8.34}$$

Figure 8–9 shows sketches of h and s as functions of p along a line $T = T_1$ in Figure 8–7.

Along the line $T = T_2$ in Figure 8–7, there are two phase transitions, from V to L and from L to S, so μ has a discontinuity of slope at each transition and h and s have jumps at each transition. Along the line $T = T_3$, the liquid-vapor phase transition is absent because $T_3 > T_c$ and there is no distinction between liquid and vapor above the critical temperature.

8.3 Phase Diagram in the v, p Plane

In the v, p plane, the phase diagram of a monocomponent system is sketched in Figure 8–10. The regions where single phases are stable are separated by miscibility gaps.

[5]The same approximation would be true for a liquid.

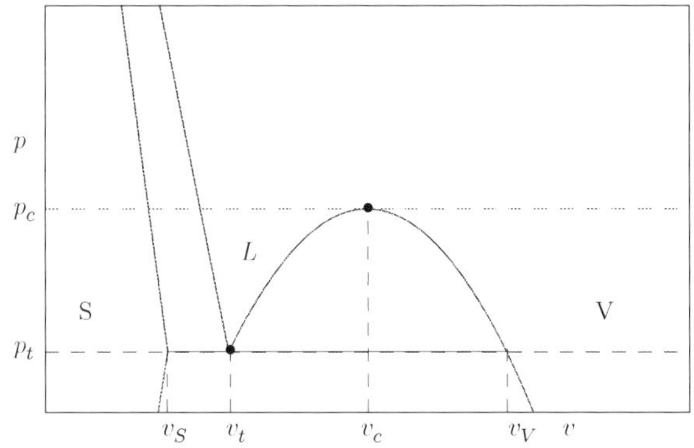

FIGURE 8–10 Sketch of the phase diagram for a monocomponent system in the v, p plane. The solid S is stable to the left of all lines. The liquid L is stable in the region of distorted triangular shape, the bottom vertex of which is at the triple point v_t, p_t. The triple point of the p, T phase diagram actually becomes a triple line on the v, p diagram and extends from v_S to v_V. Thus v_t also represents the molar volume v_L of the liquid phase that is in equilibrium with solid of molar volume v_S and vapor of molar volume v_V. The vapor V is stable to the right of all lines. The critical point v_c, p_c is at the top of the miscibility gap that separates L from V. For $p > p_c$, there is no distinction between liquid and vapor. The regions of stable phases are separated by miscibility gaps. A point within a miscibility gap can represent a composite made up of stable phases at its boundaries having the same pressure. This diagram is not to scale. Typically, the difference in molar volume of L and S is a few percent, whereas the molar volume of a vapor in equilibrium with S or L can be thousands of times larger.

These gaps occur because of the jumps in molar volumes between phases that are in equilibrium with one another. A point within a miscibility gap could correspond to an unstable single phase, for example, a supersaturated vapor that, for kinetic reasons, has not yet transformed to precipitate some liquid. At equilibrium, a point within a miscibility gap can represent a composite made up of stable phases at its boundaries having the same pressure. Most points within a miscibility gap correspond to only two stable phases that lie along a coexistence curve of a T, p phase diagram. But along the line $p = p_t$, three phases can coexist in equilibrium. The amounts of these three phases cannot be determined by specifying the overall molar volume v alone, except for the special cases $v = v_S$ and $v = v_V$ which correspond to the ends of the triple line. However, the three phases have different molar enthalpies, h_S, h_L, and h_V, so we could also specify the overall molar enthalpy, h. If we denote the phases S, L, V by the indices $1, 2, 3$, then their mole fractions f_i must satisfy

$$\begin{aligned} f_1 + f_2 + f_3 &= 1, \\ f_1 v_1 + f_2 v_2 + f_3 v_3 &= v, \\ f_1 h_1 + f_2 h_2 + f_3 h_3 &= h. \end{aligned} \tag{8.35}$$

But Eq. (8.12) is isomorphous with Eq. (8.35), so the $f_i(v, h)$ will have analogous properties to the $f_i(s, v)$ discussed in Chapter 7. In particular, to get positive values of $f_i(v, h)$, the point v, h must be chosen within or on the non-degenerate triangle with vertices (v_S, h_S), (v_L, h_L), and (v_V, h_V). The values of the $f_i(v, h)$ are given by the triangle construction in Figure 8–11.

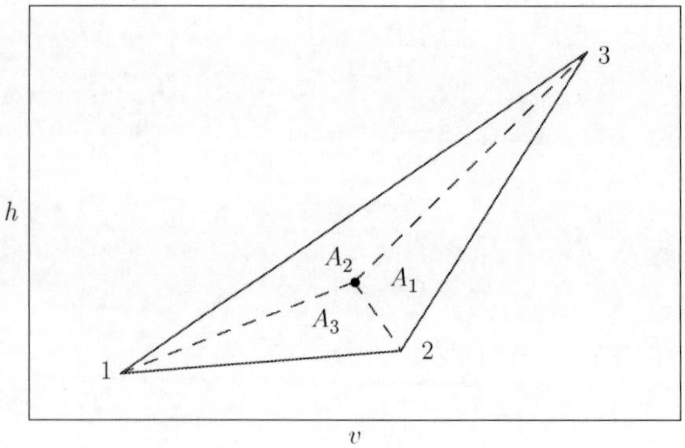

FIGURE 8–11 Triangle construction for solution of the $f_i(v, h)$ in Eq. (8.35). The vertices (1,2,3) are located at the states (v_S, h_S), (v_L, h_L), and (v_V, h_V), respectively. The point within triangle 123 where the dashed lines meet has coordinates (h, v) where the total molar volume and total molar enthalpy are specified. If we designate the area of triangle 123 by A_{123}, then $f_i(u, v) = A_i/A_{123}$, where A_i denotes the area of the inner triangle opposite to the vertex i. The diagram is not to scale because typically $v_V - v_S \gg v_L - v_S$ and $h_V - h_S \gg h_L - h_S$. If u, v lies on one of the sides of triangle 123, only two phases will be present.

9

Two-Phase Equilibrium for a van der Waals Fluid

In this chapter, we use the van der Waals model of a fluid to develop the methods that enable one to analyze the thermodynamics of two-phase equilibrium for a monocomponent system. This model will also serve to illustrate why the Helmholtz and Gibbs free energies are useful thermodynamic functions. We will focus particular attention on two graphical constructions, the common tangent and the chord, that will enable us to see easily the conditions under which two phases can exist in equilibrium as well as identify regions of stability and metastability. We will also derive Maxwell's construction that allows one to determine the miscibility gap in the v, p plane. Although we have used the simple model of a van der Waals fluid, the methods developed in this chapter are general and apply to more realistic models or data.

9.1 van der Waals Equation of State

A simple model for a monocomponent system that exhibits a liquid-vapor phase transition and a critical point is based on a generalized[1] equation of state, due to van der Waals, of the form

$$(p + \frac{a}{v^2})(v - b) = RT, \tag{9.1}$$

which holds for one mole of a van der Waals fluid. In Eq. (9.1), p is the pressure, v is the molar volume, T is the absolute temperature, R is the gas constant, and a and b are positive constants. Equation (9.1) can be rewritten in the form

$$p = \frac{RT}{v - b} - \frac{a}{v^2}, \tag{9.2}$$

which becomes the equation of state for one mole of an ideal gas for $a = 0$ and $b = 0$. The constant b accounts for the finite size of vapor molecules, so $v - b$ is the volume per mole that is free for occupancy. The constant a accounts for an attractive force between vapor molecules, which for sufficiently low temperatures will lead to condensation to a liquid. The explicit form of the term $-a/v^2$ in the pressure can be justified on the basis of mean field theory, but we postpone this connection until Section 9.2.

[1]Strictly speaking, an equation of state expresses a partial derivative of a fundamental equation (for S or U) with respect to one of its dependent extensive variables as a function of its complete set of extensive variables. In this generalized equation of state, we have a relation among intensive variables which gives the partial derivative, $-p$, of the Helmholtz free energy per mole, f, with respect to the molar volume v as a function of T and v.

Thermal Physics. http://dx.doi.org/10.1016/B978-0-12-803304-3.00009-0

The van der Walls fluid is a useful model because it is tractable and gives rise to an approximate phase diagram that exhibits many of the features of real phase diagrams. Nevertheless, it is wrong in detail, especially near the critical point where correlations become important and mean field models fail. We shall examine this model with these shortcomings in mind, but with the aim of illustrating important constructions that allow one to analyze graphs of the Helmholtz and Gibbs free energies.

9.1.1 Isotherms

Insight about the van der Waals fluid can be gained by using Eq. (9.2) to plot isotherms in the v, p plane, as sketched below in Figure 9–1. In doing this, we make the restriction $v > b$ in order to avoid infinite values of p. For high T, the term in a is negligible and the isotherms resemble those for an ideal gas, except shifted to the right by b. For sufficiently low T, p is not a monotonically decreasing function of v and there are three values of v for a given p (see Figure 9–1b which shows such an isotherm on an exaggerated scale). These values are roots of the cubic equation

$$pv^3 - (pb + RT)v^2 + av - ab = 0, \tag{9.3}$$

which is equivalent to Eq. (9.2). For T sufficiently low, one root of Eq. (9.3) can be small (of the order of b) and will be associated with a liquid; another can be large (of the order of RT/p) and can be associated with a vapor, and a middle sized root is spurious and can be associated with an unstable phase.

To find out when the isotherms display a non-monotonic behavior, we look for a maximum and minimum of p by examining

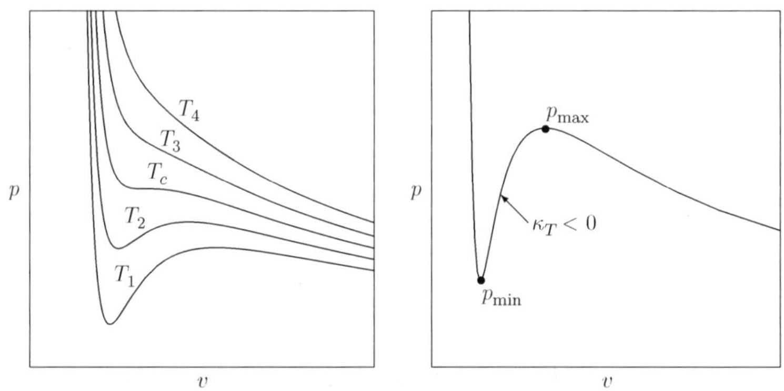

FIGURE 9–1 (a) Sketch of isotherms in the v, p plane according to Eq. (9.2), for $T_4 > T_3 > T_c > T_2 > T_1$. For sufficiently high temperatures, the isotherms are monotonically decreasing functions of v, as they would be for an ideal gas. T_c is the critical temperature and its isotherm has a horizontal point of inflection. For sufficiently low temperatures, the isotherms display multiple values of v for the same value of p. (b) A low temperature isotherm on an exaggerated scale, illustrating a maximum and a minimum value of p. The curve between the maximum and minimum values, p_{max} and p_{min}, corresponds to unstable states having negative compressibility, $\kappa_T < 0$.

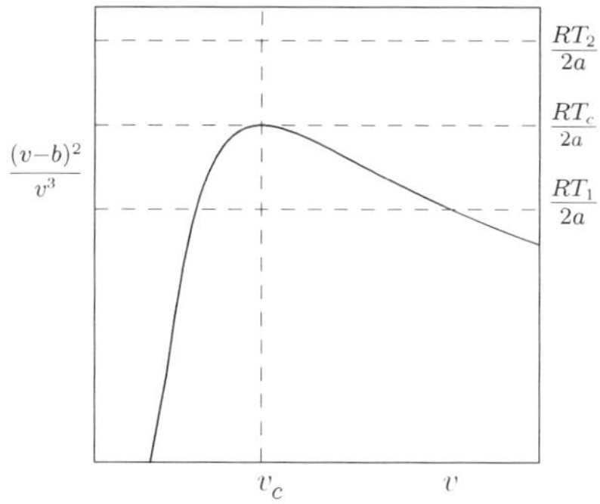

FIGURE 9–2 Graphical solution to Eq. (9.5). The straight horizontal lines represent values of $RT/(2a)$. For values of $RT/(2a)$ above the maximum of the curve, there are no real roots. For $RT_c/(2a)$ corresponding to the maximum of the curve, there is one real root, and this defines the critical temperature T_c. Below the critical temperature, there are two real roots, and these lie on the spinodal curve of Figure 9–3.

$$\left(\frac{\partial p}{\partial v}\right)_T = -\frac{RT}{(v-b)^2} + \frac{2a}{v^3} = 0. \tag{9.4}$$

Eq. (9.4) may be rewritten in the form

$$\frac{(v-b)^2}{v^3} = \frac{RT}{2a}, \tag{9.5}$$

which admits a graphic solution depicted in Figure 9–2. For $T > T_c$ where T_c is a critical value of temperature, there are no real roots, so p versus v is monotonic; for $T = T_c$, there is one real root at the critical molar volume v_c; and for $T < T_c$ there are two real roots, the smaller corresponding to a minimum of p and the larger to a maximum of p. By setting the derivative of $(v-b)^2/v^3$ to zero, its maximum is found to occur at $v_c = 3b$ and has the value $(v_c - b)^2/v_c^3 = 4/(27b)$. Therefore

$$T_c = \frac{8a}{27bR}, \tag{9.6}$$

and the corresponding critical pressure is[2]

$$p_c = \frac{RT_c}{v_c - b} - \frac{a}{v_c^2} = \frac{a}{27b^2}. \tag{9.7}$$

Returning to Eq. (9.4), we note that the partial derivative $(\partial p/\partial v)_T = -1/(v\kappa_T)$ where $\kappa_T := -(1/v)(\partial v/\partial p)_T$ is the isothermal compressibility. Therefore, the maximum and minimum of p as a function of v correspond to points of infinite compressibility, and the values of v in between to a region of negative compressibility. As discussed in Chapter 7, this region of negative compressibility corresponds to an unstable phase, which is an artifact of the van der Waals model.

[2]These results are the same as those obtained by Fermi [1, p. 73] by using the clever method of finding a triple root of v for Eq. (9.3) when $p = p_c$ and $T = T_c$. Thus Eq. (9.3) can be written $p_c(v - v_c)^3 = 0$ and comparison of coefficients of powers of v gives three simultaneous equations.

Another consideration of the van der Walls model is the need to restrict T to prevent negative pressures. Setting $p = 0$ in Eq. (9.2) and solving the resulting quadratic equation for v yields

$$v = \frac{a \pm \sqrt{a^2 - 4abRT}}{2RT}, \tag{9.8}$$

which has a double root, corresponding to the minimum of a p versus v curve just touching zero, whenever $a^2 - 4abRT = 0$. This gives $T = 27T_c/32$. The restriction $T > 27T_c/32$ would seem to allow only a narrow range of temperature, but we must recall that we are dealing here with absolute temperatures. For $H_2 0$, for example, $T_c = 647$ K, so $27T_c/32 = 546$ K, allowing a range of 100 K. If one restricts the model to temperatures above which *stable* phases have positive pressure, even lower temperatures are possible.

9.1.2 Spinodal Curve

The locus in the v, p plane of the maximum and minimum of p as a function of v is known as the **spinodal**. It separates the (unstable) region of negative compressibility from that of positive compressibility (where states can be either stable or metastable, as we shall see later). A simple equation for this spinodal curve can be obtained by substituting Eq. (9.5) into Eq. (9.2) to eliminate T and thus obtaining

$$p = \frac{a(v - 2b)}{v^3}, \quad \text{spinodal curve,} \tag{9.9}$$

which yields positive values for $v > 2b$, which is equivalent to $v > 2v_c/3$. The value $v = 2v_c/3$ corresponds to $T = 27T_c/32$, obtained above. In terms of the dimensionless pressure $y := p/p_c$ and the dimensionless volume $\nu := v/v_c$, Eq. (9.9) can be written

$$y = \frac{3\nu - 2}{\nu^3}, \quad \text{spinodal curve,} \tag{9.10}$$

which is depicted by the dashed curve in Figure 9–3. Note the asymmetry of the plot relative to its maximum. This asymmetry is due to the fact that the liquid always has a molar volume of the order of v_c, whereas the vapor has a large volume as well as a large range of volume, depending on its pressure p.

9.2 Thermodynamic Functions

We now calculate thermodynamic functions for the van der Waals fluid. Since Eq. (9.2) gives p as a function of v and T, the most natural function to deal with is the Helmholtz free energy per mole, $f := F/N$. Its differential is

$$df = -s\,dT - p\,dv, \tag{9.11}$$

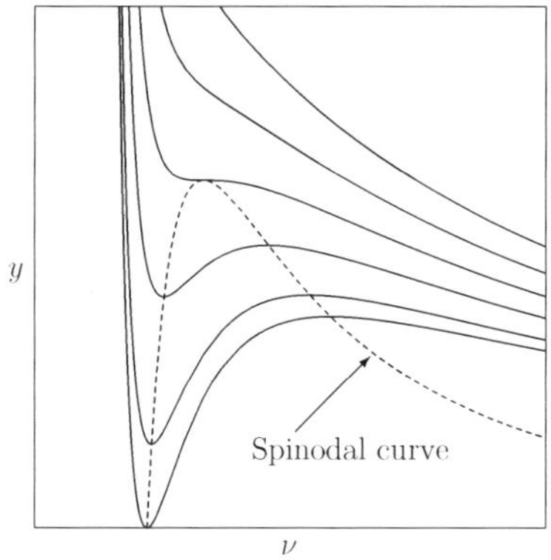

FIGURE 9–3 Plot of dimensionless pressure $y = p/p_c$ as a function of dimensionless molar volume $\nu = v/v_c$. The dashed curve is the spinodal given by Eq. (9.10) and the solid curves are isotherms. From top to bottom, $T/T_c = 32/28, 32/30, 1, 30/32, 28/32, 27/32$. The lowest isotherm touches zero pressure at $\nu = 2/3$, which is the same as $v = 2b$.

so

$$-p = \left(\frac{\partial f}{\partial v}\right)_T = \frac{a}{v^2} - \frac{RT}{v-b}. \tag{9.12}$$

We can therefore integrate Eq. (9.12) at constant T to obtain

$$f = -\frac{a}{v} - RT\ln(v/b - 1) + f_0(T), \tag{9.13}$$

where the function ("constant" as far as v is concerned) of integration $f_0(T)$ depends on T and we have used the degree of freedom provided by it to make the argument of the logarithm dimensionless. Then by differentiation we can calculate the entropy

$$s = -\left(\frac{\partial f}{\partial T}\right)_v = R\ln(v/b - 1) - f_0'(T), \tag{9.14}$$

where $f_0'(T)$ denotes the derivative of $f_0(T)$. From Eq. (9.14) it follows that the heat capacity at constant volume

$$c_v = T\left(\frac{\partial s}{\partial T}\right)_v = -Tf_0''(T) \tag{9.15}$$

so it depends only on T. The internal energy per mole is

$$u = f + Ts = -\frac{a}{v} + f_0(T) - Tf_0'(T). \tag{9.16}$$

Note from Eq. (9.16) that u depends on both v and T, whereas for an ideal gas, $a = 0$ and u is a function of only T, as we know.

In the following, we shall be concerned with the behavior of f as a function of v at various fixed values of T, so the unknown function $f_0(T)$ will either just shift the origin of f or drop out entirely when values of f are compared at fixed T for different v.

Example Problem 9.1. In many treatments of the van der Waals fluid, c_v is taken to be a constant. In that case, find an explicit form of $f_0(t)$ by integrating Eq. (9.15) and introducing any necessary constants of integration. Then calculate the corresponding values of s and u.

Solution 9.1. We integrate $f_0''(T) = -c_v/T$ once to obtain $f_0'(T) = -c_v \ln T + c_1$ where c_1 is a constant of integration. Then we integrate again to obtain $f_0(T) = -c_v T \ln T + c_v T + c_1 T - u_{00}$ where u_{00} is another constant of integration. For convenience we choose the form of c_1 so that $f_0(T) = -c_v T \ln(T/T_c) - s_{00}T + u_{00}$ where s_{00} is a new constant with the dimensions of entropy. Then $s = R\ln(v/b - 1) + c_v[\ln(T/T_c) + 1] + s_{00}$ and $u = -a/v + c_v T + u_{00}$.

9.2.1 Origin of the Constant *a*

As mentioned above, the constant a accounts for an attractive force between vapor molecules. We proceed to explain this interpretation on the basis of mean field theory. We assume the potential energy for interaction of a vapor molecule, located at the origin $\mathbf{r} = 0$, with another molecule of the vapor at distance r to be given by a function $\varphi(r)$ such as sketched in Figure 9–4. For a system of $\mathcal{N}$ molecules, there are $\mathcal{N}(\mathcal{N} - 1)/2 \approx \mathcal{N}^2/2$ distinct pairs, so the attractive energy associated with these pairs is

$$\Delta U_a \approx \frac{\mathcal{N}^2}{2V} \int_{r_b}^{\infty} \varphi(r)\, 4\pi r^2 \, dr, \tag{9.17}$$

where V is the volume and we have taken the integral to infinity provided the potential cuts off sufficiently rapidly. The mean field approximation is implicit in Eq. (9.17) because the factor $\mathcal{N}^2/V$ is taken outside the integral as a constant, whereas in reality, there are correlations among the vapor molecules and their density is not uniform. Introducing the

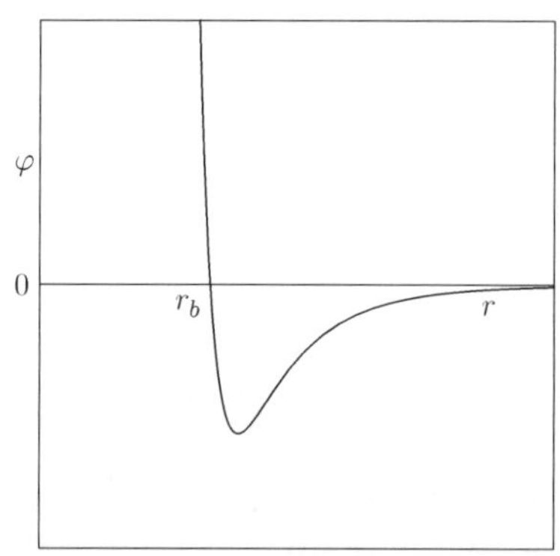

FIGURE 9–4 Sketch of the potential function $\varphi(r)$ as a function of r. For small r, say $r = r_b$, corresponding roughly to the excluded volume b, $\varphi(r)$ is large and positive, resulting in strong repulsion, while for larger values of r, $\varphi(r)$ is negative, resulting in mild attraction.

number of moles $N = \mathcal{N}/\mathcal{N}_{\mathrm{A}}$, where $\mathcal{N}_{\mathrm{A}}$ is Avogadro's number, the molar volume $v = V/N$, and the molar energy change $\Delta u_a = \Delta U_a/N$, Eq. (9.17) can be rewritten in the form

$$\Delta u_a = -\frac{a}{v} \quad \text{with} \quad a = -\frac{\mathcal{N}_{\mathrm{A}}^2}{2}\int_{r_b}^{\infty} \varphi(r)\, 4\pi r^2\, \mathrm{d}r > 0. \tag{9.18}$$

Equation (9.18) has the same form as the first term in Eq. (9.13). Some values of a and b for a variety of van der Walls fluids are given by Callen [2, p. 77].

9.3 Phase Equilibrium and Miscibility Gap

Armed with a knowledge of f, we shall now use several methods to examine the conditions for phase equilibrium, particularly the representation in the v, p plane of the coexistence curve (in the T, p plane) for these phases. Coexistence in the v, p plane is represented by two regions, one in which the equilibrium state is a single phase and the other, known as the **miscibility gap**, in which the equilibrium state is a composite system consisting of two phases. The spinodal curve derived above lies entirely within the miscibility gap except at the critical point where the two intersect. Outside the miscibility gap, the fluid is stable; between the miscibility gap and the spinodal, the fluid is metastable; and within the spinodal, it is unstable. We shall use several methods to illustrate these points.

9.3.1 Common Tangent Construction

The **common tangent construction** is a useful method that provides a graphical solution to phase equilibrium problems. We develop it in general, and then apply it specifically to the van der Waals fluid.

We consider a composite system at uniform temperature T and consisting of two homogeneous phases of the same substance, one having mole number N_1, volume V_1, and molar volume $v_1 = V_1/N_1$ and the other having N_2, V_2 and $v_2 = V_2/N_2$. The total Helmholtz free energy of the system is

$$F = N_1 f(T, v_1) + N_2 f(T, v_2). \tag{9.19}$$

If these phases are in equilibrium, F must be a minimum with respect to changes of the *internal extensive* variables N_1, V_1, N_2, V_2 *subject to the constraints* $N_1 + N_2 =$ constant, and $V_1 + V_2 =$ constant.

We first hold N_1 and N_2 constant and set the differential $\mathrm{d}F = 0$ to obtain

$$N_1 \left(\frac{\partial f(T, v_1)}{\partial v_1}\right)_T \left(\frac{\partial v_1}{\partial V_1}\right)_{N_1} \mathrm{d}V_1 + N_2 \left(\frac{\partial f(T, v_2)}{\partial v_2}\right)_T \left(\frac{\partial v_2}{\partial V_2}\right)_{N_2} \mathrm{d}V_2 = 0. \tag{9.20}$$

Since $(\partial v_1/\partial V_1)_{N_1} = 1/N_1$, $(\partial v_2/\partial V_2)_{N_2} = 1/N_2$, and $\mathrm{d}V_1 = -\mathrm{d}V_2$ from a constraint, Eq. (9.20) becomes

$$\left(\frac{\partial f(T, v_1)}{\partial v_1}\right)_T = \left(\frac{\partial f(T, v_2)}{\partial v_2}\right)_T. \tag{9.21}$$

In view of the left-hand equality in Eq. (9.12), Eq. (9.21) is recognized as equality of pressure for two phases at the same temperature, one having molar volume v_1 and the other having molar volume v_2. Thus Eq. (9.21) could be rewritten

$$p(T, v_1) = p(T, v_2). \tag{9.22}$$

This result is not to be unexpected!

Next, we hold V_1 and V_2 constant and set the differential $dF = 0$ to obtain

$$\left[f(T, v_1) + N_1\left(\frac{\partial f(T, v_1)}{\partial v_1}\right)_T\left(\frac{\partial v_1}{\partial N_1}\right)_{V_1}\right]\mathrm{d}N_1 + \left[f(T, v_2) + N_2\left(\frac{\partial f(T, v_2)}{\partial v_2}\right)_T\left(\frac{\partial v_2}{\partial N_2}\right)_{V_2}\right]\mathrm{d}N_2 = 0. \tag{9.23}$$

Since $(\partial v_1/\partial N_1)_{V_1} = -V_1/N_1^2$, $(\partial v_2/\partial N_2)_{V_2} = -V_2/N_2^2$, and $\mathrm{d}N_1 = -\mathrm{d}N_2$ from a constraint, Eq. (9.23) becomes

$$f(T, v_1) - \left(\frac{\partial f(T, v_1)}{\partial v_1}\right)_T v_1 = f(T, v_2) - \left(\frac{\partial f(T, v_2)}{\partial v_2}\right)_T v_2. \tag{9.24}$$

We identify the members on the left-hand and right-hand sides of Eq. (9.24) as chemical potentials, that is,

$$f(T, v) - \left(\frac{\partial f(T, v)}{\partial v}\right)_T v = f(T, v) + p(T, v)v =: \mu(T, v). \tag{9.25}$$

This enables Eq. (9.24) to be rewritten

$$\mu(T, v_1) = \mu(T, v_2). \tag{9.26}$$

From general considerations, Eq. (9.26) is also not to be unexpected!

We seem to have labored to obtain what amounts to Eqs. (9.22) and (9.26), which we might have just written down from general considerations. Nevertheless, the chemical potential μ, which for a monocomponent system is equal to the Gibbs free energy per mole, g, is ordinarily regarded as a function of T and p, not T and v. The variables T and v are the natural variables of f, not μ. We therefore return to Eqs. (9.21) and (9.24) and establish the following geometrical interpretation: According to Eq. (9.21), a graph of f versus v has the same slope at two values of v, namely at v_1 and v_2. There can be many pairs of v_1 and v_2 for which this is true. But either the left or the right member of Eq. (9.24) can be interpreted as the *intercept*, on the f axis ($v = 0$), of a tangent to a graph of f versus v at v_1 and v_2. So Eq. (9.21) requires parallel tangents and Eq. (9.24) requires equal intercepts. It follows that the simultaneous solution to Eqs. (9.21) and (9.24) requires a *common tangent* at v_1 and v_2, as illustrated in Figure 9–5a.

Another important feature of Figure 9–5a is noteworthy. A composite system consisting of the two phases having molar volumes v_1 and v_2 at its common tangent has a total molar free energy that lies along the tangent line joining the points of tangency. To see this, consider first a homogeneous system consisting of N moles and having molar volume v^*. For our composite system to have the same volume, we would need $N_1 + N_2 = N$ and

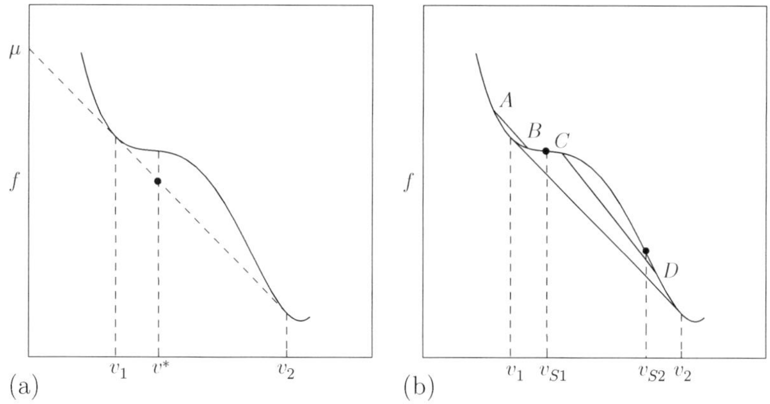

FIGURE 9–5 (a) Common tangent construction and (b) chord construction. The curves represent the Helmholtz free energy per mole, $f = F/N$, versus the molar volume v. For the common tangent construction, the dashed line is tangent to the curve at v_1 and v_2. Its slope is the negative of the common pressure and its intercept is μ, the common value of the chemical potential. The free energy per mole of a composite state having total molar volume v^* lies along the tangent line at v^* and is lower than the free energy of a single phase having v^*. Hence, the composite system is more stable than a homogeneous system. The chord construction can be used to investigate the local or global stability of a homogeneous phase. If the chord lies above the curve, as for the chord *AB*, the homogeneous phases along the curve *AB* are stable with respect to a composite, whereas if it lies below the curve, as for the chord *CD*, the homogeneous phases along the curve *CD* are unstable. v_{S1} and v_{S2} mark the spinodal points S_1 and S_2 at which $\partial^2 f/\partial v^2 = 0$.

$N_1 v_1 + N_2 v_2 = N v^*$. This results in $N_1/N = (v_2 - v^*)/(v_2 - v_1)$ and $N_2/N = (v^* - v_1)/(v_2 - v_1)$ which is known as the **lever rule**. Inserting these values in Eq. (9.19) gives

$$
\begin{aligned}
F/N &= [(v_2 - v^*)/(v_2 - v_1)]f(T, v_1) + [(v^* - v_1)/(v_2 - v_1)]f(T, v_2) \\
&= f(T, v_1) + [(v^* - v_1)/(v_2 - v_1)][f(T, v_2) - f(T, v_1)], \quad \text{composite system.}
\end{aligned}
\tag{9.27}
$$

Comparing this value with that for the homogeneous system having molar volume v^*, we see that the free energy of the composite system is *lower* for all v^* between v_1 and v_2. Thus, for these values of v^*, the *composite system is the equilibrium state.* Put another way, given the opportunity, the homogeneous system will decompose to form the equilibrium composite system consisting of two phases that are in equilibrium with each other.

9.3.2 Chord Construction

We can also use the reasoning that led to Eq. (9.27) to establish another valuable construction which we shall call the **chord construction**. Indeed, Eq. (9.27) is still valid if v_1 and v_2 correspond to any two points along the curve, provided only that $v_1 < v_2$. We can therefore apply it to various points along the curve, such as the pair *AB* or the pair *CD*, as illustrated in Figure 9–5b. For chord *AB*, the free energy of a composite lies along the chord, which is above the curve *AB*, so a homogeneous phase along the curve *AB* is stable with respect to a composite consisting of its end points. But for *CD*, the free energy of a composite lies on a chord of f that is below f, so the single phase is unstable with respect to that particular composite. Any homogeneous state that lies above the chord corresponding to the common tangent is unstable, but some of those states are locally stable. Local stability or instability requires the chord construction to be applied to neighboring points. The spinodal points where $-\partial p/\partial v = \partial^2 f/\partial v^2 = 0$ separate locally stable states from locally unstable states. States that are locally stable but globally unstable are said to be **metastable**.

9.3.3 Summary for $f(v)$ Curves

We can summarize this situation as follows: With respect to a curve of f versus v for a monocomponent system, portions of the curve that are convex (as viewed from below) correspond to locally stable states; portions of the curve that are concave (as viewed from below) correspond to locally unstable states. All states that lie above the common tangent between v_1 and v_2 are ultimately unstable. Thus there are three kinds of states:

unstable locally unstable (locally concave) and also above the common tangent between v_1 and v_2; therefore, locally unstable and globally unstable;

metastable locally stable (locally convex) but above the common tangent between v_1 and v_2; therefore, locally stable but globally unstable;

stable locally stable (locally convex) but outside the common tangent region, i.e., $v \leq v_1$ or $v \geq v_2$; therefore, locally stable and globally stable.

The concave and convex regions are separated by points, S_1 and S_2, at which the second partial derivative $(\partial^2 f/\partial v^2)_T = 0$. But since $(\partial^2 f/\partial v^2)_T = -(\partial p/\partial v)_T$, these points also correspond to maxima and minima of p, which means that they correspond to the spinodal curve in the v, p plane. Thus the spinodal curve separates the metastable region from the unstable region. On the other hand, the locus of the points of common tangency in the v, p plane separate the stable region from the metastable region; the region inside this curve is called the **miscibility gap** because within it a composite *mixture* of phases located at the ends of the common tangent, rather than a homogeneous phase, is globally stable.

9.3.4 Explicit Equations for van der Waals Miscibility Gap

For the van der Waals fluid, the explicit forms of Eqs. (9.21) and (9.24) are

$$\frac{RT}{v_1 - b} - \frac{a}{v_1^2} = \frac{RT}{v_2 - b} - \frac{a}{v_2^2} \tag{9.28}$$

and

$$-\frac{2a}{v_1} - RT\ln(v_1/b - 1) + \frac{RTv_1}{v_1 - b} = -\frac{2a}{v_2} - RT\ln(v_2/b - 1) + \frac{RTv_2}{v_2 - b}. \tag{9.29}$$

For a given value of T, the simultaneous equations (9.28) and (9.29) can be solved for v_1 and v_2 and then p can be evaluated by using Eq. (9.2). In principle, such a solution determines the shape of the miscibility gap, but these equations would have to be solved numerically because they are not tractable analytically. A graphical representation of their

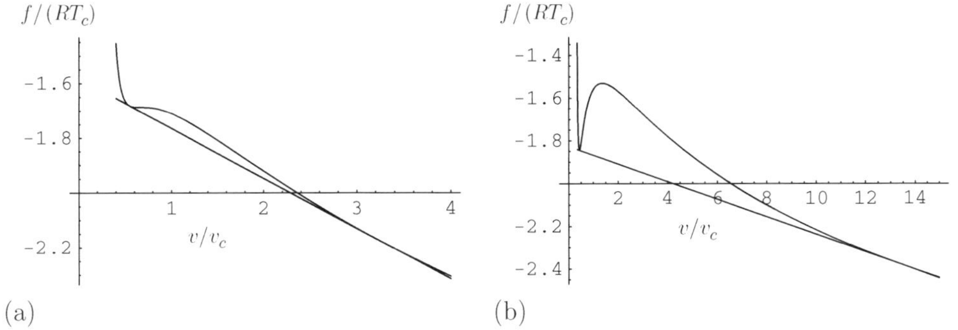

FIGURE 9–6 Graphical representation of simultaneous solution of Eqs. (9.28) and (9.29). (a) $T/T_c = 27/32$, $v/v_c = 0.548$ for the liquid and 3.241 for the vapor. The pressure is $p/p_c = 0.183$. (b) $T/T_c = 20/32$, $v/v_c = 0.440$ for the liquid and 13.585 for the vapor. The pressure is $p/p_c = 0.0411$.

solution is presented in Figure 9–6. Since the liquid and vapor have quite different molar volumes, the curve of f becomes quite flat at large values of v and such a solution is not very practical. We therefore turn to other methods to demonstrate the nature of these solutions.

9.4 Gibbs Free Energy

For the van der Waals fluid, the Gibbs free energy per mole is $g = f + pv$, which is also equal to the chemical potential μ for this monocomponent system. It can be written in the form

$$g = -\frac{2a}{v} - RT\ln(v/b - 1) + \frac{vRT}{v - b} + f_0(T), \tag{9.30}$$

where Eqs. (9.2) and (9.13) have been used. Equation (9.30) gives g as a function of T and v, but this is not very useful information because the equilibrium criterion for g is that it be a minimum for given values of T and p. Equation (9.2) for p could be solved for $v(p)$. But since v is a root of the cubic Eq. (9.3), one would have to write analytical expressions for its three roots by using the cubic formula and then substitute into Eq. (9.30). This would result in unwieldy expressions for g whose functional behavior would not be easy to analyze. A much more useful procedure is to regard Eqs. (9.30) and (9.2) to be a *parametric* representation[3] of $g(T, p)$, with v as a parameter. This can be done by assigning a fixed value to T and then allowing v to range through a set of numerical values. For each numerical value of v, one can calculate a pair of numerical values of $g - f_0(T)$ and p, and ultimately construct a graph of g versus p, such as shown in Figure 9–7a. Note that

[3]A parametric representation is a very powerful way of representing a function, especially if it is multiple valued. A simple example would be the ellipse $x^2/a^2 + y^2/b^2 = 1$ which can be represented parametrically by $x = a\cos\psi$ and $y = b\sin\psi$, where ψ is a parameter that ranges from 0 to 2π as the ellipse is traversed once in the counterclockwise direction. Powerful software packages, such as ParametricPlot of Mathematica®, can be used to construct plots of such functions with ease.

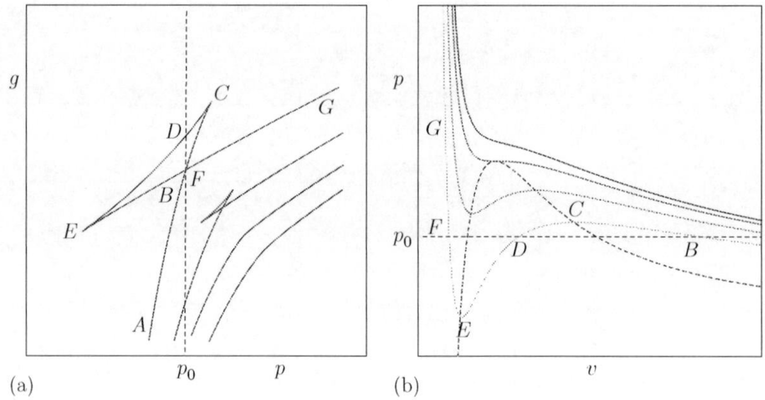

FIGURE 9–7 (a) Isotherms in the p, g plane for two temperatures below T_c, one at T_c and one above T_c. The labeled curve is for the lowest temperature, $T/T_c = 27/32$. (b) Isotherms in the v, p plane for temperatures corresponding to those in (a). The isotherm with labeled points is for $T/T_c = 27/32$. The dashed curve is the spinodal. Point A is outside the figure to the right on the labeled isotherm in (b). The segment AB is stable vapor, BC is metastable vapor, CDE is unstable fluid, EF is metastable liquid, and FG is stable liquid. The pressure p_0 intersects the $T/T_c = 27/32$ isotherm at points F and B, corresponding to the miscibility gap, and also at point D on the unstable branch. Points C and E lie along the spinodal curve.

there are three values of g for a range of p between the maximum and minimum values of p that correspond to an isotherm in the p, v plane. These triple roots end at cusps of g that correspond to points on the spinodal curve. Points along the curve EDC correspond to unstable[4] fluid states inside the spinodal. Points along BC or EF represent metastable states; they are outside the spinodal region but inside the miscibility gap (gap separating stable phases, to be defined more precisely in the following section). States along AB and FG are stable and lie outside the boundary of the miscibility gap.

At constant T, $\mathrm{d}g = v\,\mathrm{d}p$, and since $v > 0$, the isotherms of g are monotonically increasing functions of p. But for $T < T_c$, v is a multiple valued function of p, so the corresponding isotherms of g have three monotonic branches that join as shown in Figure 9–7. For values of $T \geq T_c$, there is a single branch. For $T \gg T_c$, the isotherms begin to resemble those for an ideal gas,

$$g(p, T) = RT \ln(p/p_0) + g(p_0, T), \quad \text{ideal gas,} \tag{9.31}$$

where p_0 is some reference pressure. For the van der Waals fluid at any fixed T, one could integrate $\mathrm{d}g = v\,\mathrm{d}p$ at constant T by parts, which would lead to the already-known Eq. (9.30).

[4]We can draw this conclusion based on the fact that EDC in Figure 9–7a corresponds to the convex region of a curve of f as a function of v for fixed T. Note, however, that the curves ABC and EFG are concave. This is because p is an intensive variable. By the methods of Chapter 7, we know that g is a convex function of p for an unstable state but a concave function of p for an unstable state. Specifically, $\partial^2 f/\partial v^2 = -1/\partial^2 g/\partial p^2$.

9.4.1 Maxwell Construction

A miscibility gap is a two-phase region that separates stable phases. For the van der Walls fluid, this gap separates stable liquid from stable vapor. A point within that miscibility gap does not represent a homogeneous phase but instead a mixture of liquid and vapor in proportions given by the lever rule discussed immediately above Eq. (9.27). A simple way to determine the miscibility gap graphically is the equal area construction of Maxwell. According to this construction, the horizontal line $p = p_0$ that makes equal areas with an isotherm in the v, p plane intersects that isotherm at the boundaries, v_1 and v_2, of the miscibility gap, as illustrated in Figure 9–8. To prove this construction, we rewrite Eq. (9.13) in the form

$$f(v,T) = -\int_{v_0}^{v} p(T,v')\,\mathrm{d}v' + f(v_0,T), \tag{9.32}$$

where v_0 is some reference molar volume and the function $p(T, v)$ is given by Eq. (9.2). For a *fixed* pressure p_0, equality of the molar Gibbs free energies at two different volumes, v_1 and v_2, that lie on the miscibility gap, gives

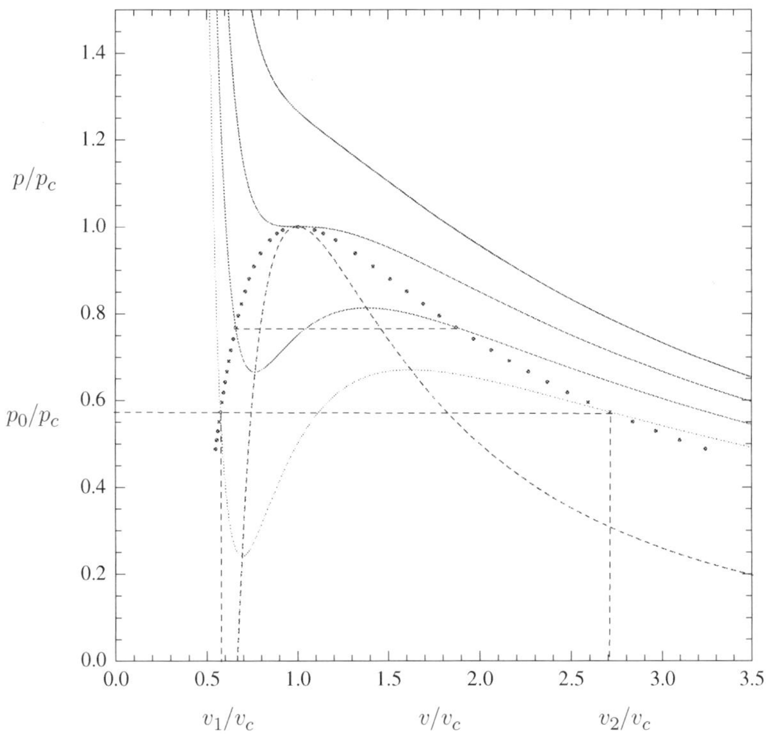

FIGURE 9–8 Maxwell's equal area construction for determining the miscibility gap. The isotherms from top to bottom are $T/T_c = 32/30$, 1, 30/32, 28/32. The dashed curve is the spinodal and the dotted curve is the miscibility gap, computed numerically as discussed in the example problem. The dashed horizontal line at p_0/p_c illustrates the equal-area construction for the lowest isotherm and the shorter horizontal line illustrates the equal area construction for the next lowest isotherm.

$$-\int_{v_0}^{v_1} p(T,v)\,\mathrm{d}v + f(v_0,T) + p_0 v_1 = -\int_{v_0}^{v_2} p(T,v)\,\mathrm{d}v + f(v_0,T) + p_0 v_2. \tag{9.33}$$

After canceling $f(v_0, T)$, Eq. (9.33) can be rewritten in the form

$$\int_{v_1}^{v_2} p(T,v)\,\mathrm{d}v - p_0(v_2 - v_1) = 0, \tag{9.34}$$

where we shall take $v_2 > v_1$ to correspond to Figure 9–8. But $p_0(v_1 - v_2) = \int_{v_1}^{v_2} p_0\,\mathrm{d}v$ because p_0 is a constant, so Eq. (9.34) becomes

$$\int_{v_1}^{v_2} [p(T,v) - p_0]\,\mathrm{d}v = 0. \tag{9.35}$$

In Eq. (9.35), regions of integration where $p(T,v) > p_0$ give positive contributions whereas regions where $p(T,v) < p_0$ give negative contributions. If p_0 is chosen to satisfy Eq. (9.35), the areas between the $p(T,v)$ curve and p_0 will be equal. This proves the Maxwell construction which holds for the van der Waals fluid.

However, the **Maxwell construction** holds generally for any fluid for which the isotherms in the v, p plane have the qualitative features of the van der Waals fluid. This can be seen in a simple way from Fermi's observation [1, p. 71] that the reversible work $\mathcal{W}$ done by the system in a cyclic reversible isothermal process is zero. That observation follows from the fact that Eq. (3.28) (which is the same as Eq. (3.33) for a reversible cycle) holds for such a process and T can be taken outside the integral to give $\oint \delta Q = 0$, which requires $Q = 0$. But for a cyclic process, $\Delta U = 0$, so by the first law, $\mathcal{W} = 0$. Thus

$$\oint p\,\mathrm{d}v = \mathcal{W} = 0. \tag{9.36}$$

One can therefore integrate from one side of the miscibility gap to the other side along the curve $p(v,T)$ and then return to the first side along the line $p = p_0$, thus proving the Maxwell construction generally.

Example Problem 9.2. Although the Maxwell construction allows one to visualize the miscibility gap, it does not provide an accurate quantitative determination. Solve simultaneously Eqs. (9.28) and (9.29) to determine the miscibility gap and discuss the relationship of pressure to temperature on the miscibility gap.

Solution 9.2. This can only be done numerically because of the transcendental form of Eq. (9.29). Rather than choosing specific values of the constants a and b, we introduce the dimensionless variables $t = T/T_c$, $\mathrm{v}_1 = v_1/v_c$, and $\mathrm{v}_2 = v_2/v_c$ and the corresponding dimensionless pressure $y = p/p_c$. The equivalent dimensionless equations are

$$\frac{8t}{3\mathrm{v}_1 - 1} - \frac{3}{\mathrm{v}_1^2} = \frac{8t}{3\mathrm{v}_2 - 1} - \frac{3}{\mathrm{v}_2^2} = y \tag{9.37}$$

and

$$-\frac{6}{\nu_1}-\frac{8t}{3}\ln(3\nu_1-1)+\frac{8t\nu_1}{3\nu_1-1}=-\frac{6}{\nu_2}-\frac{8t}{3}\ln(3\nu_2-1)+\frac{8t\nu_2}{3\nu_2-1}. \tag{9.38}$$

Forgetting about y for the moment, we specify a value of t which leads to two simultaneous equations in ν_1 and ν_2 that can be solved numerically, for example by using a procedure such as FindRoot in Mathematica®. To do this, one must specify guesses for ν_1 and ν_2 which serve as a starting point for an iterative procedure; the Maxwell construction is useful in this respect. Then one can compute the value of y for that temperature and repeat the whole process for a number of temperatures. The result is the dotted curve in Figure 9–8. Along the miscibility gap, T is an increasing function of p so there will also be a miscibility gap in the v, T plane where the corresponding spinodal curve will be given by Eq. (9.5), as illustrated in Figure 9–2.

10

Binary Solutions

One commonly thinks of solutions as liquids, such as salt or sugar dissolved in water. In thermodynamics, we regard a solution more generally as being a homogeneous phase consisting of more than one chemical species, intermixed on an atomic scale, and thus mutually soluble. Solutions can be solids, liquids, or gases, and can contain any number of chemical components. Binary solutions consist of two chemical components, A and B, which for simplicity we will refer to as atoms, even though they may actually be molecules that do not dissociate. In a solution, chemical components can interact but cannot form new molecules. A solution should not be confused with a mixture of more than one phase.

In this chapter we consider binary solutions and their phase equilibria. A given amount of a solution of chemical components A and B can characterized by a set of state variables consisting of the temperature T, the pressure p and the composition that can be expressed by one of the mole fractions, say X_B. Although vapor phases could be considered, we shall confine ourselves to the important case of condensed phases (solids and liquids) for which the thermodynamic functions, in particular the Gibbs free energy g per mole, are insensitive to the pressure except for very large applied pressures of many atmospheres. This is true because $\partial g/\partial p = v$, the molar volume, which is quite small relative to that of a gaseous phase. Therefore, the problem reduces to one in which g effectively depends on only two important variables, T and X_B, similar to the case of monocomponent systems for which the molar Helmholtz free energy f depends on only the two variables T and v. As a result, we will be able to analyze two-phase equilibria in terms of a common tangent construction and a chord construction, analogous to those for monocomponent systems. We will illustrate our treatment with simple models, namely ideal solutions and so-called regular solutions, recognizing that the subject of binary phase diagrams for real materials is huge and quite complex. The interested reader is referred to the materials science literature for a thorough analysis of real systems.

10.1 Thermodynamics of Binary Solutions

We consider a binary solution made up of chemical components A and B. The internal energy U of such a solution is a function of the entropy S, the volume V, and the mole numbers N_A and N_B. Its differential is

$$dU = T\,dS - p\,dV + \mu_A\,dN_A + \mu_B\,dN_B, \tag{10.1}$$

where T is the temperature, p is the pressure, μ_A is the chemical potential of component A and μ_B is the chemical potential of component B. $U = \tilde{U}(S, V, N_A, N_B)$ is a fundamental

Thermal Physics. http://dx.doi.org/10.1016/B978-0-12-803304-3.00010-7

equation,[1] so it contains complete information about the system. The Euler equation for U is

$$U = TS - pV + \mu_A N_A + \mu_B N_B \tag{10.2}$$

and the Gibbs-Duhem equation is

$$0 = S\,\mathrm{d}T - V\,\mathrm{d}p + N_A\,\mathrm{d}\mu_A + N_B\,\mathrm{d}\mu_B. \tag{10.3}$$

There are four equations of state, which give T, p, μ_A, and μ_B as functions of S, V, N_A, and N_B. Thus

$$\begin{aligned} T &= \tilde{T}(S, V, N_A, N_B); \quad p = \tilde{p}(S, V, N_A, N_B); \\ \mu_A &= \tilde{\mu}_A(S, V, N_A, N_B); \quad \mu_B = \tilde{\mu}_B(S, V, N_A, N_B). \end{aligned} \tag{10.4}$$

Note, however, that T, p, μ_A, and μ_B are intensive variables so they can only depend on the ratios of the extensive variables S, V, N_A, and N_B, of which only three are independent.

A convenient way to make the reduction to three independent intensive variables is to introduce the per mole quantities $u := U/N$, $s := S/N$, $v := V/N$, $X_A = N_A/N$, and $X_B = N_B/N$. Since the mole fractions satisfy $X_A + X_B = 1$, we only need to keep one of them, so we retain the three independent variables s, v, and X_B. Then Eqs. (10.1)–(10.3) become

$$\mathrm{d}u = T\,\mathrm{d}s - p\,\mathrm{d}v + (\mu_B - \mu_A)\,\mathrm{d}X_B; \tag{10.5}$$

$$u = Ts - pv + \mu_A(1 - X_B) + \mu_B X_B; \tag{10.6}$$

$$0 = s\,\mathrm{d}T - v\,\mathrm{d}p + (1 - X_B)\,\mathrm{d}\mu_A + X_B\,\mathrm{d}\mu_B. \tag{10.7}$$

The equations of state can be written

$$\begin{aligned} T &= \tilde{T}(s, v, X_B); \quad p = \tilde{p}(s, v, X_B); \\ \mu_A &= \tilde{\mu}_A(s, v, X_B); \quad \mu_B = \tilde{\mu}_B(s, v, X_B). \end{aligned} \tag{10.8}$$

Since s, v, and X_B are independent, it is possible to take partial derivatives with respect to one of them while holding the other two constant. For example, from Eq. (10.5) we obtain[2]

$$\left(\frac{\partial u}{\partial X_B}\right)_{s,v} = (\mu_B - \mu_A). \tag{10.9}$$

To obtain μ_A and μ_B separately, we would have to solve Eq. (10.9) simultaneously with Eq. (10.6). The result is

$$\mu_A = u - Ts + pv - X_B\left(\frac{\partial u}{\partial X_B}\right)_{s,v}; \quad \mu_B = u - Ts + pv + (1 - X_B)\left(\frac{\partial u}{\partial X_B}\right)_{s,v}. \tag{10.10}$$

[1]Here, we use the more elaborate functional notation $U = \tilde{U}(S, V, N_A, N_B)$ to distinguish the value U from its functional form $\tilde{U}$. In cases where there is no confusion between these quantities, we use the abbreviated functional notation $U = U(S, V, N_A, N_B)$.

[2]Note well that X_A is *not* held constant in this differentiation.

10.1.1 Molar Gibbs Free Energy

In working with binary solutions, we shall often be concerned with situations in which T and p are specified and uniform throughout the system. The natural function to use to discuss this situation is the Gibbs free energy $G := U - TS + pV$ or the Gibbs free energy per mole $g := G/N = u - Ts + pv$. This is because complete information about the system is contained in the function $G = \tilde{G}(T, p, N_A, N_B)$ or, for one mole, in the function $g = \tilde{g}(T, p, X_B)$. Since g is related to u by Legendre transformations, we obtain

$$\mathrm{d}g = -s\,\mathrm{d}T + v\,\mathrm{d}p + (\mu_B - \mu_A)\,\mathrm{d}X_B; \tag{10.11}$$

$$g = \mu_A(1 - X_B) + \mu_B X_B; \tag{10.12}$$

$$0 = s\,\mathrm{d}T - v\,\mathrm{d}p + (1 - X_B)\,\mathrm{d}\mu_A + X_B\,\mathrm{d}\mu_B, \tag{10.13}$$

and the equations of state

$$s = \tilde{s}(T, p, X_B); \quad v = \tilde{v}(T, p, X_B); \tag{10.14}$$
$$\mu_A = \tilde{\mu}_A(T, p, X_B); \quad \mu_B = \tilde{\mu}_B(T, p, X_B).$$

Note that Eq. (10.13) is the same as Eq. (10.7) but in Eqs. (10.12) and (10.13) the independent variable set is T, p, and X_B rather than s, v, and X_B in Eqs. (10.5) and (10.6).

By partial differentiation of Eq. (10.11) we obtain

$$\left(\frac{\partial g}{\partial X_B}\right)_{T,p} = (\mu_B - \mu_A), \tag{10.15}$$

which resembles Eq. (10.9) except that g replaces u and the variables T and p are now held constant in the differentiation. Simultaneous solution of Eqs. (10.15) and (10.12) gives

$$\mu_A = g - X_B\left(\frac{\partial g}{\partial X_B}\right)_{T,p}; \quad \mu_B = g + (1 - X_B)\left(\frac{\partial g}{\partial X_B}\right)_{T,p}. \tag{10.16}$$

10.1.2 Intercept and Common Tangent Constructions

Unlike Eq. (10.10), Eq. (10.16) contains the same function g and its partial derivative with respect to X_B, so these equations can be interpreted geometrically. This is illustrated in Figure 10–1a from which it can be seen that μ_A is just the intercept at $X_B = 0$ of the tangent to a graph of g versus X_B at constant T and p. Similarly, μ_B is the intercept at $X_B = 1$ of that same tangent. As the point of tangency slides along the curve, one obtains from the intercepts of the tangent an appreciation for μ_A and μ_B as a function of the value of X_B at the point of tangency. This procedure is known as **the method of intercepts**.[3]

[3]This famous construction, known as the method of intercepts, works for the molar value $y := Y/N$ of any extensive function $Y = \tilde{Y}(T, p, N_A, N_B)$. The partial derivatives $\bar{Y}_A := (\partial Y/\partial N_A)_{T,p,N_B}$ and $\bar{Y}_B := (\partial Y/\partial N_B)_{T,p,N_A}$, which are analogous to the chemical potentials, are known as partial molar values of Y. $\bar{Y}_A$ and $\bar{Y}_B$ are then given by the intercepts to the tangent of a graph of y as a function of X_B for fixed T and p. In fact, the method can be generalized to multicomponent systems for which the relevant intercepts are those of tangent planes or hyperplanes. See Section 5.6.1 for more detail.

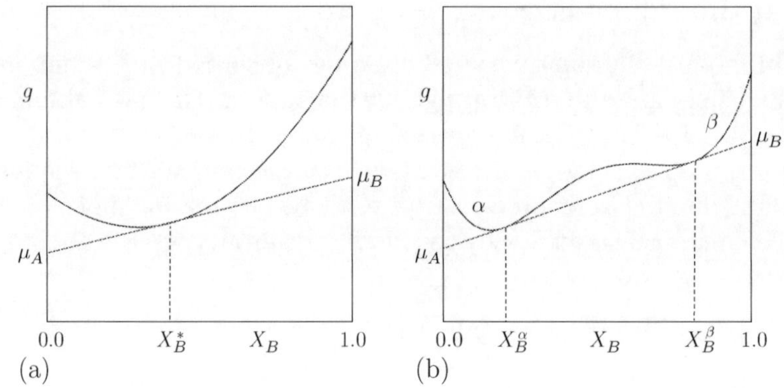

FIGURE 10–1 (a) Sketch of $g(T, p, X_B)$, in arbitrary units, as a function of X_B for fixed T and p illustrating the method of intercepts. The $X_B = 0$ and $X_B = 1$ intercepts of the tangent give the values of μ_A and μ_B, respectively, at the point of tangency X_B^*. (b) Common tangent construction. The chemical potentials of *each* component are equal for α and β phases having compositions X_B^α and X_B^β at points of common tangency. The α phase is stable for $X_B < X_B^\alpha$ and the β phase is stable for $X_B > X_B^\beta$. The region $X_B^\alpha < X_B < X_B^\beta$ is a miscibility gap.

From Figure 10–1b, we note that the graph of g versus X_B is not convex, as will be the case when an unstable phase and more than one stable phase are involved.[4] In such a case, the common tangent intersects the curve at values of X_B that satisfy

$$\mu_A(T, p, X_B^\alpha) = \mu_A(T, p, X_B^\beta); \quad \mu_B(T, p, X_B^\alpha) = \mu_B(T, p, X_B^\beta), \quad \text{common tangent,} \tag{10.17}$$

where X_B^α and X_B^β are the values of X_B at which common tangency occurs. This **common tangent** construction solves the equilibrium problem for phases of a binary system. It is the analog of the common tangent construction for the molar Helmholtz free energy, f, treated in Chapter 9. The α phase is stable for $X_B < X_B^\alpha$ and the β phase[5] is stable for $X_B > X_B^\beta$. The region $X_B^\alpha < X_B < X_B^\beta$ is a miscibility gap where no homogeneous phase is stable; it is a two-phase region where a mixture of α with composition X_B^α and β with composition X_B^β is globally stable.

Example Problem 10.1. Show that the values of X_B that correspond to a common tangent construction of a non-convex g, such as depicted in Figure 10–1b, are unchanged if a linear function of X_B is added to g.

Solution 10.1. For the function $g(x)$, the values of X_B subtended by a common tangent are solutions to the simultaneous equations

[4]In Chapter 7 we showed for stability that the extensive Gibbs free energy G must be a convex function of each of its extensive variables. For a binary solution we could regard G to depend on the variable set T, p, N_B, N, so with N constant, $g = G/N$ must be a convex function of $X_B = N_B/N$ for stability.

[5]The fact that the present g-curve is continuous and has a continuous derivative implies that there is not a change of structure as X_B increases from 0 to 1. Therefore, one could rename the β phase α' since it only differs from α by composition. If there were a change in structure, there could be two separate curves that cross.

$$g'(x) = g'(y); \quad g(x) - xg'(x) = g(y) - yg'(y), \tag{10.18}$$

where the prime denotes differentiation and fixed values of p and T have been suppressed. Let $\tilde{g}(x) = g(x) + ax + b$ where a and b are constants. Then

$$\tilde{g}'(x) = g'(x) + a; \quad \tilde{g}(x) - x\tilde{g}'(x) = g(x) - xg'(x) + b. \tag{10.19}$$

Therefore, if $\tilde{g}$ is substituted for g in Eq. (10.18), the constants a and b cancel. This happens because a straight line is its own common tangent.

10.1.3 Chord Construction

We also have a **chord construction** for a binary system. Consider a composite system consisting of one mole and made up of a mole fraction f_1 of material having composition X_{B1} and a mole fraction f_2 of material having composition X_{B2}. Suppose that the total composite has X_B^* moles of component B, where $X_{B1} \le X_B^* \le X_{B2}$. This requires $f_1 + f_2 = 1$ and $f_1 X_{B1} + f_2 X_{B2} = X_B^*$. The fractions f_1 and f_2 are therefore given by the **lever rule**:

$$f_1 = \frac{X_{B2} - X_B^*}{X_{B2} - X_{B1}}; \quad f_2 = \frac{X_B^* - X_{B1}}{X_{B2} - X_{B1}}; \quad \text{lever rule.} \tag{10.20}$$

The molar Gibbs free energy of the composite is

$$\begin{aligned} g_c &= f_1\, g(T, p, X_{B1}) + f_2\, g(T, p, X_{B2}) \\ &= g(T, p, X_{B1}) + \frac{X_{B2} - X_B^*}{X_{B2} - X_{B1}} [g(T, p, X_{B1}) - g(T, p, X_{B2})]. \end{aligned} \tag{10.21}$$

From Eq. (10.21), we see that g_c lies on the chord connecting $g(T, p, X_{B1})$ with $g(T, p, X_{B2})$ on a graph of g versus X_B at fixed T and p, as illustrated in Figure 10–2a. Since that chord is below the curve, the homogeneous state with $g(T, p, X_B^*)$ is unstable with respect to the composite, which has lower energy g_c. If a chord is above the curve, the homogeneous states below it are at least locally stable. They are only globally stable if they lie outside the common tangent chord BE. A state that is locally stable but globally unstable is said to be metastable and can exist for significant periods of time if kinetics of nucleation of a new phase are slow. Thus, the g-curve can be divided into stable, metastable, and unstable regions, as illustrated in Figure 10–2b. The situation is similar to the analysis of $f(T, v)$ as a function of v for fixed T, but there are some differences. For one thing, the phase rule for a binary system allows up to four phases to coexist at equilibrium, not just three as for a monocomponent system. Second, we now have three independent variables, rather than two. As a consequence, one often confines analysis of condensed binary systems to atmospheric pressure and studies phase stability as T and X_B are varied. Such a procedure is fairly general because condensed phases have sufficiently small molar volumes and are practically incompressible, resulting in a very weak dependence of g on pressure. For gaseous phases, which have relatively large molar volumes that are approximately inversely proportional to pressure at constant temperature, the dependence of g on pressure is large and cannot be ignored.

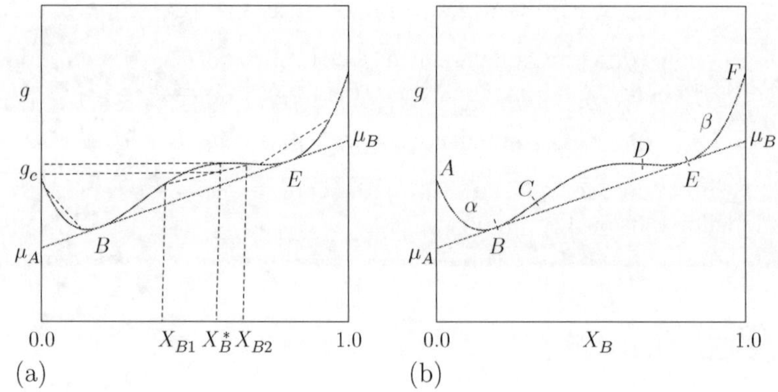

FIGURE 10–2 (a) Graph of g versus X_B at fixed T and p illustrating the chord construction. The energy g_c of a composite system having compositions X_{B1} and X_{B2} lies along the chord where it intersects the composition X_B^* and is lower than $g(T, p, X_B^*)$ of homogeneous material, which is unstable. The chord to the left of one point B of common tangency lies above the curve, so the states below it are stable. The chord near the other point E of common tangency lies above the curve, so the states below it are locally stable, but those to the left of E are only metastable because a composite along BE would have lower energy. (b) C and D are points of inflection where $\partial^2 g/\partial X_B^2 = 0$. Branch AB is stable phase α, BC is metastable α, CD is an unstable branch, DE is metastable β and EF is stable phase β.

10.2 Ideal Solutions

We shall first give a brief description of an ideal binary solution from the point of view of elementary statistical mechanics.[6] Let N_A be the number of moles of A, N_B the number of moles of B, and $N = N_A + N_B$ the total number of moles. We could equally well describe the system in terms of the numbers, $\mathcal{N}_A$ and $\mathcal{N}_B$, of atoms of A and B, respectively, and $\mathcal{N} = \mathcal{N}_A + \mathcal{N}_B$ the total number of atoms. This latter description relates better to statistical mechanics, which we can see in the following way. Suppose we have an ideal solution, in which A and B atoms are mixed but do not interact chemically. The number of ways that we can arrange these atoms is

$$\Omega = \frac{\mathcal{N}!}{\mathcal{N}_A!\mathcal{N}_B!}. \tag{10.22}$$

According to Eq. (3.77), this leads to an entropy of mixing

$$\Delta S^{\text{ideal}} = k_B \ln \frac{\mathcal{N}!}{\mathcal{N}_A!\mathcal{N}_B!} \sim k_B[\mathcal{N} \ln \mathcal{N} - \mathcal{N}_A \ln \mathcal{N}_A - \mathcal{N}_B \ln \mathcal{N}_B], \tag{10.23}$$

where Stirling's approximation $\ln \mathcal{N} \sim \mathcal{N} \ln \mathcal{N} - \mathcal{N}$, valid for large numbers, has been used.[7] The symbol Δ is used here to remind us that this entropy is in addition to the entropy of the same number of A and B atoms in their undissolved states. By introducing

[6]For a detailed treatment of the ideal entropy of mixing, see Section 16.5.1.
[7]See Appendix A.

the mole (or atom) fractions $X_A = N_A/N = \mathcal{N}_A/\mathcal{N}$ and $X_B = N_B/N = \mathcal{N}_B/\mathcal{N}$, Eq. (10.2) becomes

$$\Delta S^{\text{ideal}} = -\mathcal{N} k_B [X_A \ln X_A + X_B \ln X_B] = -NR[X_A \ln X_A + X_B \ln X_B], \tag{10.24}$$

which is known as the **ideal entropy of mixing**.

In terms of extensive variables, Eq. (10.24) can be written in the form

$$\Delta S^{\text{ideal}} = -R\left[N_A \ln(N_A/N) + N_B \ln(N_B/N)\right], \tag{10.25}$$

which could be taken as a definition, irrespective of its derivation from statistical mechanics. An **ideal solution** is one in which ΔS^{ideal} is the only entropy in addition to that of the undissolved components, and for which the enthalpy of mixing ΔH, also known as the "heat of mixing," is zero. This requires that there be no chemical interaction between the components A and B. Thus

$$\Delta H^{\text{ideal}} = 0. \tag{10.26}$$

If there is chemical interaction in addition to mechanical mixing of A and B, $\Delta H \neq 0$ and the solution is not ideal. We shall deal with a model of such a solution in Section 10.4.

The entire Gibbs free energy of an ideal solution is therefore

$$\begin{aligned} G &= N_A \mu_A^0(T,p) + N_B \mu_B^0(T,p) - T\Delta S^{\text{ideal}} \\ &= N_A \mu_A^0(T,p) + N_B \mu_B^0(T,p) + RT\left[N_A \ln(N_A/N) + N_B \ln(N_B/N)\right], \end{aligned} \tag{10.27}$$

where $\mu_A^0(T,p) = g_A(T,p)$ is the Gibbs free energy per mole of pure A and $\mu_B^0(T,p) = g_B(T,p)$ is the Gibbs free energy per mole of pure B.[8] The chemical potentials are

$$\mu_A = \left(\frac{\partial G}{\partial N_A}\right)_{T,p,N_B} = \mu_A^0(T,p) + RT\ln(N_A/N), \tag{10.28}$$

$$\mu_B = \left(\frac{\partial G}{\partial N_B}\right)_{T,p,N_A} = \mu_B^0(T,p) + RT\ln(N_B/N). \tag{10.29}$$

For any Gibbs free energy, we have

$$\left(\frac{\partial(G/T)}{\partial(1/T)}\right)_{p,N_A,N_B} = H, \tag{10.30}$$

which can be proven by noting that $G/T = H/T - S$ and carrying out the required differentiation. Applying Eq. (10.30) to Eq. (10.27) shows that ΔS^{ideal} does not contribute to H, consistent with Eq. (10.26), so we obtain

$$H = N_A \left(\frac{\partial(\mu_A^0/T)}{\partial(1/T)}\right)_{p,N_A,N_B} + N_B \left(\frac{\partial(\mu_B^0/T)}{\partial(1/T)}\right)_{p,N_A,N_B} = N_A h_A + N_B h_B, \tag{10.31}$$

where h_A and h_B are the molar enthalpies of A and B, respectively.

[8]The chemical potentials $\mu_A^0(T,p)$ and $\mu_B^0(T,p)$ refer to so-called "reference states" which we have chosen to be pure A and pure B, respectively. One can also make solutions whose constituents are molecules, such as AB or AB_2, or even by combining other solutions. For a general discussion, see Lupis [5, pp. 179-184].

On a per mole basis, the ideal entropy of mixing is

$$\Delta s^{\text{ideal}} = -R\left[X_A \ln X_A + X_B \ln X_B\right] = -R\left[(1 - X_B)\ln(1 - X_B) + X_B \ln X_B\right] \quad (10.32)$$

and the corresponding molar Gibbs free energy is

$$g = (1 - X_B)\mu_A^0(T, p) + X_B\mu_B^0(T, p) + RT\left[(1 - X_B)\ln(1 - X_B) + X_B \ln X_B\right]. \quad (10.33)$$

Figure 10–3a shows a plot of g as a function of X_B. We could obtain the chemical potentials graphically from the method of intercepts. We could also use the analytic formulae Eq. (10.16), which are the basis of the common tangent construction, to obtain

$$\mu_A = \mu_A^0(T, p) + RT \ln(1 - X_B); \quad \mu_B = \mu_B^0(T, p) + RT \ln X_B. \quad (10.34)$$

The chemical potentials in Eq. (10.34) are the same as in Eqs. (10.28) and (10.29) except for notation. Note that $\mu_B = -\infty$ for $X_B = 0$. This can be traced back to the fact that ΔS^{ideal} has an infinite slope at $\mathcal{N}_B = 0$. More precisely, the first member on the right-hand side of Eq. (10.23) shows that $\Delta S^{\text{ideal}} = 0$ for $\mathcal{N}_B = 0$ and $\Delta S^{\text{ideal}} = k_B \ln \mathcal{N}$ for $\mathcal{N}_B = 1$. Its slope at $\mathcal{N}_B = 0$ is therefore $k_B \ln \mathcal{N}$ which is actually finite but diverges logarithmically as $\mathcal{N} \to \infty$ in the thermodynamic limit. Similarly, $\mu_A = -\infty$ for $X_B = 1$.

From Figure 10–3a we see that g is a convex function of X_B, so there is a stable homogeneous phase for every value of X_B; in other words, there is a complete range of mutual solubility. In particular, such a homogeneous solution is stable with respect to phase separation to a composite state. At temperatures for which pure A and pure B are both crystalline solids, they can only form solid solutions for all X_B if they have the same crystal structure. Examples are silicon and germanium (Si, Ge) which are both diamond cubic, or nickel and copper (Ni, Cu) which are both face centered cubic. The solid solutions of these pairs of elements formed by substituting one atom for the other on the same crystalline lattice are ideal solutions to a first approximation. Similarly, if T is above the melting points of both pure A and pure B, one could possibly have an ideal liquid.

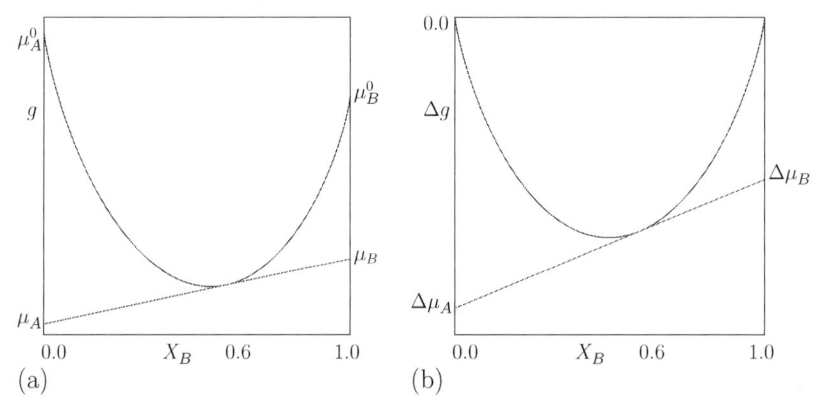

FIGURE 10–3 (a) Plot of g versus X_B for an ideal solution according to Eq. (10.33). The chemical potentials of the pure components are μ_A^0 and μ_B^0. At $X_B = 0.6$, intercepts of the tangent give the chemical potentials μ_A and μ_B. (b) Plot of Δg versus X_B for an ideal solution according to Eq. (10.35). At $X_B = 0.6$, intercepts of the tangent give the chemical potential differences $\Delta\mu_A = \mu_A - \mu_A^0$ and $\Delta\mu_B = \mu_B - \mu_B^0$.

It is often convenient to eliminate consideration of unmixed constituents by defining

$$\Delta g := g - (1 - X_B)\mu_A^0(T,p) - X_B\mu_B^0(T,p) = RT\left[(1 - X_B)\ln(1 - X_B) + X_B \ln X_B\right]. \qquad (10.35)$$

Figure 10–3b shows a plot of Δg as a function of X_B. Its minimum value occurs for $X_B = 0.5$ and has the value $-RT\ln 2 = -0.693RT$. Applying the method of intercepts to Δg gives the chemical potential *differences* $\Delta\mu_A = \mu_A - \mu_A^0$ and $\Delta\mu_B = \mu_B - \mu_B^0$.

10.3 Phase Diagram for an Ideal Solid and an Ideal Liquid

As we have seen above, g or Δg for an ideal binary solution is a convex function of X_B, and this could happen either below or above the melting points of the pure components. The interesting question is: What happens for temperatures between the melting points of the pure elements? For example, Si melts at 1685 K and Ge melts at 1210.4 K. Moreover, Ni melts at 1736 K and Cu melts at 1356.5 K.

The answer is that there is a miscibility gap and phase separation into a composite, part solid and part liquid. We analyze this situation as an application of the ideal solution model, and arbitrarily take A to be the component having the higher melting point, T_A. Thus $T_A > T_B$, where T_B is the melting point of pure B. For the ideal liquid, we have

$$\mu_A^{\rm L} = \mu_A^{\rm 0L}(T,p) + RT\ln(1 - X_B); \quad \mu_B^{\rm L} = \mu_B^{\rm 0L}(T,p) + RT\ln X_B, \qquad (10.36)$$

and for the ideal solid

$$\mu_A^{\rm S} = \mu_A^{\rm 0S}(T,p) + RT\ln(1 - X_B); \quad \mu_B^{\rm S} = \mu_B^{\rm 0S}(T,p) + RT\ln X_B, \qquad (10.37)$$

where the superscripts L and S denoting liquid and solid have been added to Eq. (10.34). For a temperature T that satisfies $T_A > T > T_B$, we know that $\mu_A^{\rm 0L}(T,p) > \mu_A^{\rm 0S}(T,p)$ and $\mu_B^{\rm 0L}(T,p) < \mu_B^{\rm 0S}(T,p)$. Therefore, graphs of $g^{\rm L} = \mu_A^{\rm L}(1 - X_B) + \mu_B^{L}X_B$ and $g^{\rm S} = \mu_A^{\rm S}(1 - X_B) + \mu_B^{\rm S}X_B$ result in two curves that cross, as shown in Figure 10–4. By applying the common

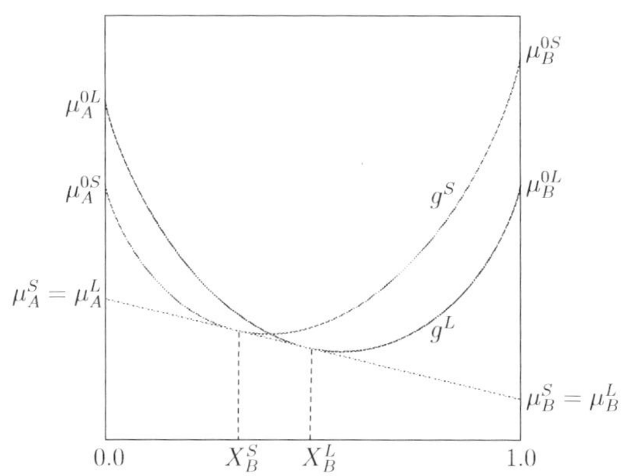

FIGURE 10–4 Curves of $g^{\rm L}$ for an ideal liquid solution and $g^{\rm S}$ for an ideal solid solution versus X_B for a temperature T between the melting points of pure A and pure B. The common tangent construction applies, with tangency at $X_B^{\rm S}$ and $X_B^{\rm L}$. As the temperature changes, the curves shift, resulting in a change of the points of common tangency.

tangent construction to Figure 10–4, we see that there is a miscibility gap for $X_B^S < X_B < X_B^L$. Here, X_B^S and X_B^L are the compositions at which common tangency occurs for the value of T corresponding to the figure. As T varies, these curves shift and we can trace out the miscibility gap in the X_B, T plane. The result is a "lens type" binary phase diagram, such as plotted in Figure 10–5.

10.3.1 Equations for the Miscibility Gap

Equations to determine the values of X_B^S and X_B^L that correspond to the boundary of the miscibility gap in the X_B, T plane can be determined by equating chemical potentials:

$$\mu_A^L(T,p,X_B^L) = \mu_A^S(T,p,X_B^S); \quad \mu_B^L(T,p,X_B^L) = \mu_B^S(T,p,X_B^S). \tag{10.38}$$

Substitution of Eqs. (10.36) and (10.37) into Eq. (10.38) gives:

$$\mu_A^{0L}(T,p) - \mu_A^{0S}(T,p) = RT \ln\left(\frac{1-X_B^S}{1-X_B^L}\right);$$
$$\mu_B^{0L}(T,p) - \mu_B^{0S}(T,p) = RT \ln\left(\frac{X_B^S}{X_B^L}\right). \tag{10.39}$$

In order to proceed further, we need a model to evaluate the chemical potential differences on the left of Eq. (10.39). We write these in the forms

$$\Delta\mu_A := \mu_A^{0L}(T,p) - \mu_A^{0S}(T,p) = \Delta h_A - T\Delta s_A;$$
$$\Delta\mu_B := \mu_B^{0L}(T,p) - \mu_B^{0S}(T,p) = \Delta h_B - T\Delta s_B, \tag{10.40}$$

where Δh_A and Δh_B are enthalpy differences (liquid minus solid) and Δs_A and Δs_B are the corresponding entropy differences. We assume as an approximation that these enthalpy and entropy differences are constants that we relate by requiring $\Delta\mu_A = 0$ at $T = T_A$ and $\Delta\mu_B = 0$ at $T = T_B$. This gives $\Delta h_A = T_A\Delta s_A$ and $\Delta h_B = T_B\Delta s_B$, so Eqs. 10.40 become approximately

$$\Delta\mu_A = \Delta h_A(1 - T/T_A); \quad \Delta\mu_B = \Delta h_B(1 - T/T_B). \tag{10.41}$$

We recognize Δh_A and Δh_B as the respective latent heats of fusion per mole and observe that Eqs. (10.41) have the expected algebraic signs, that is, $\Delta\mu_A > 0$ and $\Delta\mu_B < 0$ for $T_A > T > T_B$. An alternative view of Eqs. (10.41) is to assume that they are the leading terms in expansions in the variables $(T - T_A)/T_A$ and $(T - T_B)/T_B$.

Substitution of Eqs. (10.41) into Eqs. (10.39) shows that $X_B^S < X_B^L$ as expected and

$$\frac{1-X_B^L}{1-X_B^S} = \exp\left[-\frac{\Delta h_A}{R}\left(\frac{1}{T} - \frac{1}{T_A}\right)\right] \equiv E_A(T);$$
$$\frac{X_B^S}{X_B^L} = \exp\left[\frac{\Delta h_B}{R}\left(\frac{1}{T} - \frac{1}{T_B}\right)\right] \equiv E_B(T). \tag{10.42}$$

For $T_A > T > T_B$, we note that $0 < E_A(T) < 1$ and $0 < E_B(T) < 1$. Solving Eq. (10.42) for X_B^S and X_B^L gives

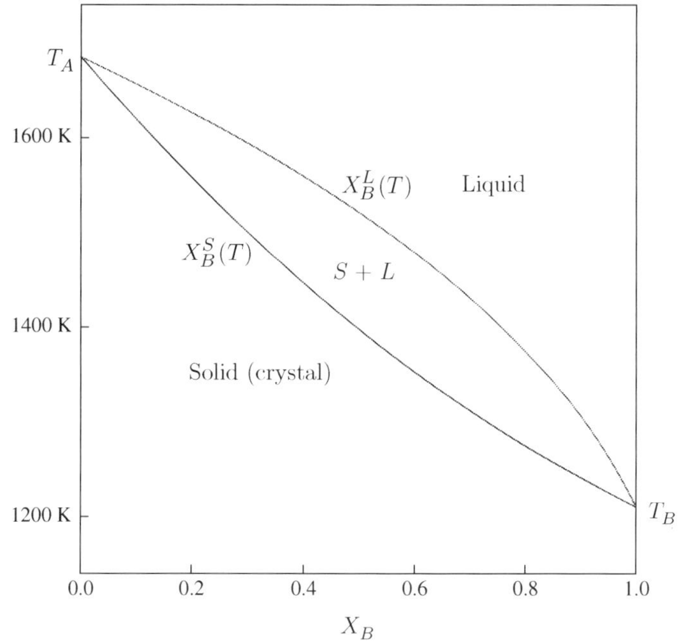

FIGURE 10–5 Computed phase diagram for an ideal solid solution and an ideal liquid solution for physical constants that resemble A = Si and B = Ge. The plot shows X_B^S and X_B^L as functions of T according to Eq. (10.43). $T_A = 1685$ K and $T_B = 1210.4$ K.

$$X_B^S = E_B(T)\frac{1 - E_A(T)}{1 - E_A(T)E_B(T)}; \quad X_B^L = \frac{1 - E_A(T)}{1 - E_A(T)E_B(T)}. \tag{10.43}$$

We observe from Eq. (10.43) that X_B^S and X_B^L increase from 0 to 1 with $X_B^S \leq X_B^L$ as T decreases from T_A to T_B.

In order to make a plot of Eq. (10.43) we need some numerical values of the physical constants. If A were Si and B were Ge, then $\Delta h_A/(RT_A) = 3.59$ and $\Delta h_B/(RT_B) = 3.14$. Figure 10–5 shows a plot of the resulting phase diagram. There is a **miscibility gap** with a "lens shape" connecting the melting points of pure A and B. Above the miscibility gap, the liquid solution is stable, and below it the solid solution is stable. The curve $X_B^S(T)$ is called the **solidus** and $X_B^L(T)$ is called the **liquidus**. For a point within the miscibility gap, a homogeneous solution is unstable, so the corresponding equilibrium state is a composite, consisting of part liquid solution and part solid solution. The amounts of solid and liquid in this composite are governed by the lever rule.

Example Problem 10.2. For the phase diagram depicted in Figure 10–5, what is the mole fraction of solid in equilibrium with liquid at $T = 1600$ K if the overall composition is $X_B = 0.22$ mole fraction? By how much does the chemical potential of A in this solid differ from that of pure solid A at $T = 1600$ K?

Solution 10.2. At $T = 1600$ K, the compositions at the solidus and liquidus are estimated to be 0.13 and 0.28. By the lever rule, the mole fraction of solid is $(0.28-0.22)/(0.28-0.13) = 0.4$. From the first of Eqs. (10.37) we obtain $\mu_A^S - \mu_A^{0S} = RT\ln(1 - X_B) = 3200\ln(1 - 0.13) = -456$ cal/mol.

Example Problem 10.3. For a very dilute ideal solution $X_B \ll 1$, develop approximate formulae for the distribution coefficient $k := X_B^S/X_B^L$ and the slope dX_B^L/dT.

Solution 10.3. From the second of Eqs. (10.42) we can approximate $T = T_A$ to obtain $k = E_B(T_A) = \text{constant} < 1$. Then we take the derivative of the first of Eqs. (10.42) with respect to T and evaluate the result at $X_B^S = X_B^L = 0$ and $T = T_A$ to obtain

$$\frac{d(X_B^L - X_B^S)}{dT} = -\frac{\Delta h_A}{RT_A^2}. \tag{10.44}$$

Then use of k to eliminate X_B^S gives

$$\frac{dX_B^L}{dT} = -\frac{\Delta h_A}{(1-k)RT_A^2}. \tag{10.45}$$

This result is related to J. H. van't Hoff's law of freezing point lowering for a dilute solid solution [22, p. 235]. Similar formulae can be obtained at the other end of the phase diagram for $X_A \ll 1$. The results are $k' := X_A^S/X_A^L = 1/E_A(T_B) = \text{constant} > 1$ and $dX_A^L/dT = \Delta h_B/[(k'-1)RT_B^2]$.

10.4 Regular Solution

A so-called **regular solution** is a solution having an ideal entropy of mixing, but also a heat of mixing of the form

$$\Delta H^{\text{reg}} = \tilde{\Omega} X_A X_B = \tilde{\Omega}(1 - X_B)X_B, \tag{10.46}$$

where $\tilde{\Omega}$ is a constant. In a quasichemical approximation [23], this heat of mixing arises from interactions among A and B atoms in a mean field approximation. If correlations among atoms are neglected, the probabilities of AA, AB, and BB interactions are just the terms on the right-hand side of the expression

$$1 = (X_A + X_B)^2 = X_A^2 + 2X_AX_B + X_B^2. \tag{10.47}$$

If E_{AA}, E_{AB}, and E_{BB} are the respective interaction energies,[9] then the energy of formation of a solution of A and B from pure A and pure B is proportional to

$$E_{AA}X_A^2 + 2E_{AB}X_AX_B + E_{BB}X_B^2 - E_{AA}X_A - E_{BB}X_B = 2X_AX_B[E_{AB} - (1/2)(E_{AA} + E_{BB})]. \tag{10.48}$$

To get the actual energy for N moles, we need to multiply by $(1/2)Nz$, where z is the coordination number (number of significant, probably nearest) of neighbors, which we

[9] These energies are negative for attraction and positive for repulsion. We deal here with condensed phases at constant pressure, so the distinction between energy and enthalpy is not important.

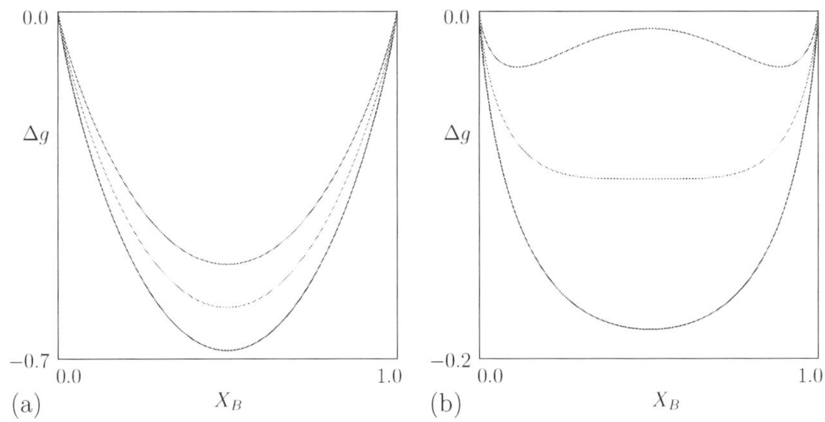

FIGURE 10–6 Plots of Δg versus X_B for a regular solution according to Eq. (10.51). (a) For $RT/\omega = -3/8, -1/2, -5/8$ for top to bottom curves, there is relative attraction of A and B, and Δg is convex. (b) For $RT/\omega = 3/8, 1/2, 5/8$ for top to bottom curves, there is relative repulsion of A and B. Δg is convex at high T but develops a concave portion at low T, resulting in a miscibility gap given by the common tangent construction. The critical temperature satisfies $RT_c/\omega = 1/2$. Note the different vertical scales (units of $|\omega|$). (a) Δg, regular solution attractive and (b) Δg, regular solution repulsive.

assume to be the same for the pure components and the solution. The factor of (1/2) arises to avoid double counting. This results in

$$\tilde{\Omega} \approx Nz[E_{AB} - (1/2)(E_{AA} + E_{BB})]. \tag{10.49}$$

The main thing we learn from Eq. (10.49) is that $\tilde{\Omega}$ will be negative if A atoms are attracted to B atoms more than these atoms are attracted to their own kind. Otherwise, there will be net repulsion of A and B atoms, and $\tilde{\Omega}$ will be positive. If $\tilde{\Omega}$ is positive, we anticipate that a miscibility gap will form at sufficiently low temperatures.

We therefore regard $\tilde{\Omega}$ as an empirical parameter of the regular solution model and proceed to analyze the thermodynamics. The total free energy of mixing is

$$\Delta G = \Delta H^{\text{reg}} - T\Delta S^{\text{ideal}} = \tilde{\Omega}\, X_A X_B + RT\left[N_A \ln(N_A/N) + N_B \ln(N_B/N)\right]. \tag{10.50}$$

For one mole, this becomes

$$\Delta g = \omega\, X_B(1 - X_B) + RT\left[(1 - X_B)\ln(1 - X_B) + X_B \ln X_B\right], \tag{10.51}$$

where $\omega := \tilde{\Omega}/N$ is the value of the interaction parameter per mole.

Figure 10–6 shows some plots of Δg versus X_B for a few temperatures and values of ω. For $\omega < 0$ (relative attraction of A and B) Δg is convex. In this case, A and B are mutually soluble for all X_B, just as for an ideal solution. For $\omega > 0$ (relative repulsion of A and B) Δg is convex at high T but develops a concave portion at low T. Therefore a miscibility gap develops at sufficiently low T and its boundary is given by the common tangent construction.

Within the miscibility gap is a spinodal curve, which is the locus in the X_B, T plane of the points where Δg changes from convex to concave. To compute the spinodal curve, we solve

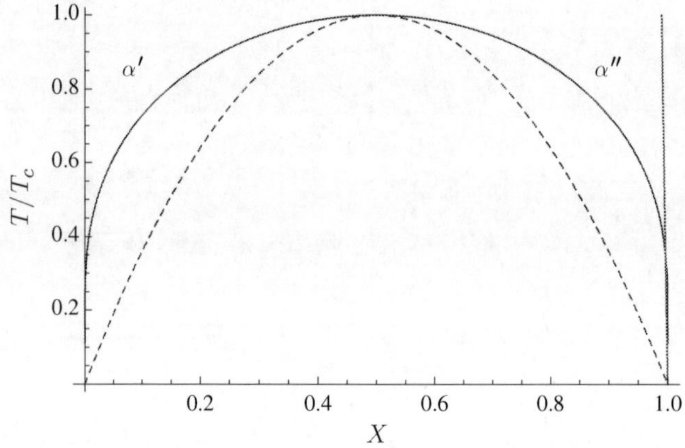

FIGURE 10–7 Miscibility gap boundary (solid curve) and spinodal curve (dashed curve) for a regular solution. For $T > T_c$ there is a stable solid phase α with a complete range of solid solubility. For $T < T_c$, the stable phases are α' and α'' which have the same crystal structure as α but different compositions. A point between the miscibility gap and the spinodal curve represents either metastable α' for $X_B < 1/2$ or metastable α'' for $X_B > 1/2$. A point within the spinodal curve represents an unstable phase. An unstable or metastable phase will eventually transform to a composite consisting of α' and α'' phases that lie on the miscibility gap at the same temperature and in proportion given by the lever rule.

$$\frac{\partial^2 \Delta g}{\partial X_B^2} = -2\omega + RT\left(\frac{1}{X_B} + \frac{1}{1-X_B}\right) = 0, \tag{10.52}$$

which yields

$$X_B(1-X_B) = RT/(2\omega); \quad \text{spinodal.} \tag{10.53}$$

The maximum value of $X_B(1-X_B)$ is $1/4$ and occurs at $X_B = 1/2$. The top of the spinodal curve occurs at the critical temperature

$$T_c = \omega/(2R) \tag{10.54}$$

because for higher temperatures Eq. (10.53) has no allowable roots. From the form of Eq. (10.53), we see that the spinodal curve is symmetric with respect to $X_B = 1/2$. By making the substitution $X_B = 1/2 - X$ we can write the equation of the spinodal in the form

$$T/T_c = 1 - 4X^2, \tag{10.55}$$

which is a parabola that ranges from $X = -1/2$ to $X = 1/2$ with its maximum at $T = T_c$. The spinodal is represented by the dashed curve in Figure 10–7.

To compute the boundary of the miscibility gap, we need the chemical potentials. These can be obtained by differentiation of the extensive[10] ΔG given by Eq. (10.50) with

[10]Note that one must write $\tilde{\Omega} X_A X_B = \omega N_A N_B/(N_A + N_B)$ before performing the differentiation.

respect to N_A and N_B or by using the intensive Δg given by Eq. (10.51) and the method of intercepts. The results are

$$\mu_A - \mu_A^0(T,p) = RT\ln(1-X_B) + \omega X_B^2; \quad \mu_B - \mu_B^0(T,p) = RT\ln X_B + \omega(1-X_B)^2. \tag{10.56}$$

Note that by working with ΔG or Δg rather than G or g, we get $\Delta\mu_A = \mu_A - \mu_A^0(T,p)$ and $\Delta\mu_B = \mu_B - \mu_B^0(T,p)$. Equating chemical potentials for A and B at X_{B1} and X_{B2} gives

$$\begin{aligned} RT\ln(1-X_{B1}) + \omega X_{B1}^2 &= RT\ln(1-X_{B2}) + \omega X_{B2}^2; \\ RT\ln X_{B1} + \omega(1-X_{B1})^2 &= RT\ln X_{B2} + \omega(1-X_{B2})^2. \end{aligned} \tag{10.57}$$

Solving Eq. 10.57 appears to be formidable at first sight, but study of their symmetry reveals that the boundary of the miscibility gap is symmetric with respect to $X_B = 1/2$. This can be demonstrated by making the substitutions $X_{B1} = 1/2 - X$ and $X_{B2} = 1/2 + X$, in which case they both become[11]

$$RT\ln\left(\frac{1}{2}+X\right) + \omega\left(\frac{1}{2}-X\right)^2 = RT\ln\left(\frac{1}{2}-X\right) + \omega\left(\frac{1}{2}+X\right)^2. \tag{10.58}$$

Equation (10.58) can be rearranged to yield

$$\ln\left[\frac{\frac{1}{2}+X}{\frac{1}{2}-X}\right] = 4(T_c/T)X, \tag{10.59}$$

where Eq. (10.54) has been used. The function on the left-hand side of Eq. (10.59) is sketched in Figure 10–8 along with three possibilities for the right-hand side. For $T = T_c$, the full line is tangent to the curve at $X = 0$, which corresponds to the top of the miscibility gap at $X_B = 1/2$. For $T > T_c$ there is just one root at $X = 0$ which corresponds to the stable state $X_B = 1/2$ on a convex Δg curve. For $T < T_c$ there are two unequal roots, being of equal magnitude but opposite sign, and corresponding to distinct values of X_{B1} and X_{B2} on the miscibility gap. The root at $X = 0$ for $T < T_c$ corresponds to an unstable state at $X_B = 1/2$ on the upper curve of Figure 10–6b. Thus for $T < T_c$ the equilibrium state of the system is a composite consisting of one phase[12] α' having composition $1/2 - |X|$ and another phase α'' having composition $1/2 + |X|$, where X is a root of Eq. (10.59). The boundary of the resulting miscibility gap is shown in Figure 10–7.

Note from Figure 10–7 that the top of the miscibility gap is flatter than the spinodal curve. Expansion of Eq. (10.59) in powers of X shows for small X that

$$\frac{T}{T_c} = 1 - (4/3)X^2 - (64/45)X^4 \cdots, \tag{10.60}$$

[11]A shorter determination of the miscibility gap can be made by noting from Figure 10–6b that the common tangent has a slope of zero. Thus we could solve $\partial^2\Delta g/\partial X_B^2 = 0$ which would also lead to Eq. (10.59). We follow a more general procedure, however, because other models of solutions do not have such high symmetry and the slope of the common tangent is not zero.

[12]As mentioned above, the regular solution model only makes sense for solid phases if they have the same crystal structure. When a miscibility gap develops, the phases in the equilibrium composite still have the same crystal structure, but different composition. Hence we denote one by α' and the other by α'', reserving the notation α and β for models in which different crystal structures occur. For systems having liquid miscibility gaps, one usually uses L' and L''.

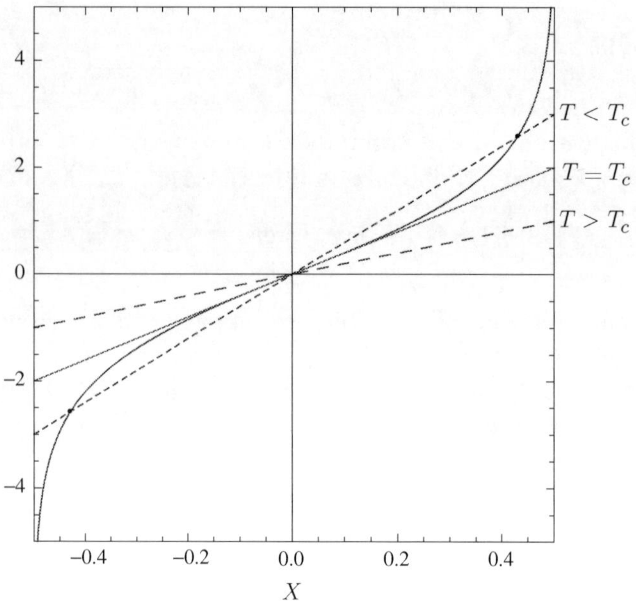

FIGURE 10–8 Plot of $\ln\left[\left(\frac{1}{2}+X\right)/\left(\frac{1}{2}-X\right)\right]$ versus X and comparison to $4(T_c/T)X$. The full line is for $T = T_c$ and is tangent to the curve at $X = 0$, which corresponds to the top of the miscibility gap. For $T > T_c$, illustrated by the line with large dashes for $T = 2T_c$, there is only a root at $X = 0$, which corresponds to a stable state on a convex Δg curve. For $T < T_c$, illustrated by the line with small dashes for $T = (2/3)T_c$, Eq. (10.59) has two non-zero roots that lie on the miscibility gap; its root at $X = 0$ corresponds to an unstable state at $X_B = 1/2$.

which should be compared to Eq. (10.55). For the regular solution, the top of the miscibility gap occurs at temperature T_c and composition $X_B = 1/2$ and the first three partial derivatives of Δg with respect to X_B are equal to zero there. This is not general, however, because the vanishing of the first derivative is only due to the symmetry of the regular solution model.

For a general solution model, we know that the spinodal curve is the locus of the points of inflection of Δg and is therefore given by

$$\frac{\partial^2 \Delta g}{\partial X_B^2} = 0. \tag{10.61}$$

At the top of the spinodal, two such inflection points coalesce, which requires

$$\frac{\partial^3 \Delta g}{\partial X_B^3} = 0. \tag{10.62}$$

Therefore, at the top of the miscibility gap, both the second and third partial derivatives of Δg with respect to X_B vanish simultaneously. This determines T_c and the corresponding X_c. Since Δg differs from g only by a linear function of X_B, Eqs. (10.61) and (10.62) also hold if Δg is replaced by g.

10.5 General Binary Solutions

We have barely scratched the surface of the subject of binary solutions and their phase diagrams. In general, binary phase diagrams are much more complicated and display intricate topologies including eutectics, peritectics, and the occurrence of several intermediate phases. The interested reader is referred to Lupis [5, chapter VIII] for a very thorough discussion of binary phase diagrams, with particular attention to their relationship to free energy curves. The book by DeHoff [21] devotes several detailed chapters to this subject and goes on to treat multicomponent solutions, which are of great practical importance to the understanding of commercial alloys. Nevertheless we have covered all of the essential physics and the most important constructions (method of intercepts, common tangent, lever rule, chord construction) that allow analysis of more general models. For a compendium of information about the phase diagrams of real materials, the reader is referred to three volumes edited by Massalski [24].

11

External Forces and Rotating Coordinate Systems

In Chapter 6 we developed criteria for equilibrium of a thermodynamic system under various conditions. In the absence of external forces, those criteria included the minimization of the internal energy U for a system that does no work at constant entropy S and the minimization of the Helmholtz free energy F for a system that does no work at constant temperature, T. In this chapter, we generalize these equilibrium conditions to include external forces, such as gravity and electromagnetic forces, that can exert body forces and do work on a system.

At equilibrium, such systems will usually be inhomogeneous. For example, in the case of gravity, the pressure, and the chemical potentials will be functions of position within a sample.[1] We shall restrict our development to conservative external forces that can be derived from a potential function. Then the equilibrium conditions can be expressed conveniently in terms of new potentials, such as gravitational chemical potentials and electrochemical potentials that are uniform at equilibrium.

We examine in detail the equilibrium conditions for multicomponent ideal gases and binary liquids in a uniform gravitational field. Then we treat rotating systems by means of a potential that relates to centrifugal force. Finally, we give a brief treatment of applied electric fields.

11.1 Conditions for Equilibrium

We begin with the inequality Eq. (6.19) which we rewrite in the form

$$dU + \delta\mathcal{W} < 0, \quad \text{constant } S, \text{ natural process, chemically closed,} \tag{11.1}$$

where we have written a strict inequality to confine our attention to actual processes that are always natural and irreversible. This eliminates hypothetical reversible processes for which there is really no driving force. To have equilibrium, all such natural irreversible processes must be prevented. A state will therefore be in equilibrium if all *virtual* variations of the system, which we indicate by δ applied to U, violate Eq. (11.1), that is,

$$\delta U + \delta\mathcal{W} \geq 0, \quad \text{constant } S, \text{ chemically closed, all virtual variations.} \tag{11.2}$$

[1]In this chapter, we make use of methods of the calculus of variations to treat inhomogeneous systems. This subject is treated in many standard references, for example [25, p. 198] and [26, p. 164]. The main results, however, can be appreciated and in many cases applied without a complete understanding of their derivation.

Thermal Physics. http://dx.doi.org/10.1016/B978-0-12-803304-3.00011-9

By a **virtual variation**, we mean any imagined variation that is compatible with the constraints of the system but does not necessarily satisfy the laws of thermodynamics. If all virtual variations satisfy Eq. (11.2), they all violate the laws of thermodynamics, so no natural irreversible processes are possible and the system is in equilibrium.

We next confine ourselves to the case in which the *only* work is done against conservative external forces. Thus, there exists a potential function Φ such that $\delta\mathcal{W} = \delta\Phi$ and in which the δ applied to Φ denotes a *change* in Φ due to a virtual variation, unlike the general meaning of the symbol $\delta\mathcal{W}$ which means a small amount of work, not a change of work. We shall assume that the overall volume of the system is constant. Our equilibrium criterion Eq. (11.2) therefore becomes

$$\delta(U + \Phi) \geq 0, \quad \text{constant } S \text{ and } V, \text{ chemically closed, all virtual variations.} \tag{11.3}$$

Although Eq. (11.3) is a valid criterion for equilibrium, it would be difficult if not impossible to realize it experimentally because one would have to devise a way to hold the entropy constant. Theoretically, however, one can satisfy the constraint of constant entropy by means of a Lagrange multiplier λ and consider virtual variations of the form $\delta(U + \Phi - \lambda S)$, but we must also insure that the system is chemically closed. For simplicity we first forbid chemical reactions. Therefore, we also introduce additional Lagrange multipliers λ_i for each chemical component and obtain

$$\delta(U + \Phi - \lambda S - \sum_i \lambda_i N_i) \geq 0, \quad \text{constant } V, \text{ all virtual variations.} \tag{11.4}$$

As is usual with Lagrange multipliers, we could choose λ and the λ_i later to satisfy the constraints of constant entropy and constant mole numbers.[2] By following this methodology, one finds, not surprisingly, that $\lambda = T$ at equilibrium, so the absolute temperature of the system is a constant.

An alternative approach is to assume at the outset that the system in question is in contact with a heat reservoir having constant temperature T and that the system is maintained at temperature T throughout any process under consideration. Then returning to Eq. (6.19) with $T_s = T$ but allowing S to vary gives

$$\mathrm{d}F + \delta\mathcal{W} < 0, \quad \text{constant } T, \text{ natural process, chemically closed,} \tag{11.5}$$

where $F = U - TS$ is the Helmholtz free energy. Then if there exists a potential function Φ such that the only work done is $\delta\mathcal{W} = \delta\Phi$, as discussed above, one can use Lagrange multipliers λ_i to insure the chemical closure. This results in the equilibrium criterion

$$\delta\left(F + \Phi - \sum_i \lambda_i N_i\right) \geq 0, \quad \text{constant } T \text{ and } V, \text{ all virtual variations.} \tag{11.6}$$

[2]We could equally well constrain the masses M_i of each chemical component since $M_i = m_i N_i$, where m_i are molecular weights. This would change the values of the λ_i but not the fact that they are just constants. This is often done when the force under consideration is due to gravity, which acts on the masses, and was preferred by Gibbs [3, p. 144].

We proceed to illustrate the use of the equilibrium criteria, Eq. (11.4) or Eq. (11.6), extensively for the case of gravitational forces and then discuss several other forces.

11.2 Uniform Gravitational Field

We first consider the case of a uniform gravitational field that exerts a constant downward acceleration g per unit mass on any chemical species. We take the z axis of a cartesian coordinate system to be vertically upward, "up" defined as antiparallel to the acceleration due to gravity. In that case,

$$\Phi = \int_V \rho g z \, d^3x + \text{constant}, \tag{11.7}$$

where $\rho = \sum_i \rho_i$ is the local mass density and the integral is over the constant volume V of the system. We use masses rather than moles to simplify the form of the potential for gravity.[3] The total internal energy, entropy, and masses of each species are given by

$$U = \int_V u_V \, \mathrm{d}^3x; \quad S = \int_V s_V \, \mathrm{d}^3x; \quad M_i = \int_V \rho_i \, \mathrm{d}^3x, \tag{11.8}$$

where the internal energy density, entropy density, and mass densities are denoted by u_V, s_V, and ρ_i, respectively.

We treat a multicomponent fluid for which $\mathrm{d}u_V$ is given by Eq. (5.64) except we use partial densities ρ_i instead of the mole concentrations c_i, resulting in

$$\mathrm{d}u_V = T\,\mathrm{d}s_V + \sum_{i=1}^{\kappa} \mu_i^m \, \mathrm{d}\rho_i. \tag{11.9}$$

Then the general equilibrium criterion Eq. (11.4) becomes

$$\int \left[\frac{\partial u_V}{\partial s_v} \delta s_V + \sum_{i=1}^{\kappa} \frac{\partial u_V}{\partial \rho_i} \delta \rho_i + g z \delta \rho - \lambda \delta s_V - \sum_{i=1}^{\kappa} \lambda_i^m \delta \rho_i \right] \mathrm{d}^3x \ge 0. \tag{11.10}$$

By identifying the partial derivatives, using $\delta\rho = \sum_{i=1}^{\kappa} \delta\rho_i$ and grouping terms, Eq. (11.10) becomes

$$\int \left[(T - \lambda)\delta s_V + \sum_{i=1}^{\kappa} (\mu_i^m + gz - \lambda_i^m)\delta\rho_i \right] \mathrm{d}^3x \ge 0. \tag{11.11}$$

Then by requiring Eq. (11.11) to be true for all independent and arbitrary virtual variations δs_V and $\delta\rho_i$, both positive and negative, the only possibility is for their coefficients to be zero. Thus we obtain the $\kappa + 1$ conditions

$$T = \lambda \tag{11.12}$$

[3]The corresponding chemical potentials μ_i^m are per unit mass and are related to those per mole by $\mu_i^m = \mu_i / m_i$, where the m_i are molecular weights. Consistent with this change, we use total masses $M_i = \int \rho_i d^3x$ instead of total mole numbers N_i, so the Lagrange multipliers λ_i^m will have corresponding units.

and[4]

$$\mu_i^m + gz = \lambda_i^m; \quad i = 1, 2, \ldots, \kappa. \tag{11.13}$$

Therefore, at equilibrium, the temperature is constant and uniform, as anticipated in the discussion of Eq. (11.4), but the chemical potentials μ_i^m of each chemical component are no longer uniform. Instead, the quantities $\mu_i^m + gz$, which are often referred to as **gravitational chemical potentials**, are uniform. For this reason, the μ_i^m are often referred to as **intrinsic chemical potentials** because they are the same as those in the absence of external forces. Of course one could also incorporate the potential Φ with U to form a new potential $\tilde{U} = U + \Phi$ whose density would have a differential

$$\mathrm{d}\tilde{u}_V = T\,\mathrm{d}s_V + \sum_{i=1}^{\kappa} (\mu_i^m + gz)\,\mathrm{d}\rho_i \tag{11.14}$$

in which the gravitational chemical potentials appear directly.

According to Eq. (11.13), values of the intrinsic chemical potentials will depend linearly on z at equilibrium. Through the Gibbs-Duhem equation, we can relate this to a dependence of the pressure on z. The Euler equation (see Eq. (5.43)) per unit volume can be written

$$u_V = Ts_V - p + \sum_{i=1}^{\kappa} \mu_i^m \rho_i. \tag{11.15}$$

By taking the differential of Eq. (11.15) and subtracting Eq. (11.9), the required Gibbs-Duhem equation is found to be

$$s_V\,\mathrm{d}T - \mathrm{d}p + \sum_{i=1}^{\kappa} \rho_i\,\mathrm{d}\mu_i^m = 0. \tag{11.16}$$

However, we already know that the temperature is constant, so

$$\mathrm{d}p = \sum_{i=1}^{\kappa} \rho_i\,\mathrm{d}\mu_i^m. \tag{11.17}$$

From Eq. (11.13), $\mathrm{d}\mu_i^m = -g\,\mathrm{d}z$, so Eq. (11.17) yields

$$\mathrm{d}p = -\sum_{i=1}^{\kappa} \rho_i g\,\mathrm{d}z = -\rho g\,\mathrm{d}z, \tag{11.18}$$

where the total density ρ depends on pressure and the *local* composition. Of course Eq. (11.17) can be invoked on the basis of mechanical equilibrium by using the mechanical interpretation of pressure, but here it arises naturally as a consequence of applying the laws of thermodynamics. For a single component liquid with negligible compressibility, ρ can often be treated as a constant and Eq. (11.18) can be integrated to give $p = \rho gz + p_0$,

[4]In terms of chemical potentials per mole, Eq. (11.13) would be $\mu_i + m_i gz = \lambda_i$.

where p_0 is the pressure at $z=0$. In the case of a single component ideal gas of molecular weight m, $\rho = mp/RT$ and Eq. (11.18) can be integrated to give

$$p = p_0 \exp\left(-mgz/RT\right), \tag{11.19}$$

which is often called the **law of atmospheres**. Of course the atmosphere of the Earth is not a single component ideal gas and its temperature is not uniform, so Eq. (11.19) would only provide a crude approximation to the decrease of its pressure with height.

Had we used Eq. (11.6) at the outset, with T presumed to be imposed and uniform from the start, we could use the differential

$$\mathrm{d}f_V = \sum_{i=1}^{\kappa} \mu_i^m \, \mathrm{d}\rho_i \tag{11.20}$$

and obtain

$$\int \left[\sum_{i=1}^{\kappa} (\mu_i^m + gz - \lambda_i^m)\delta\rho_i \right] \mathrm{d}^3x \geq 0, \tag{11.21}$$

which for arbitrary $\delta\rho_i$ of either sign leads immediately to Eq. (11.13).

Example Problem 11.1. If a single chemical reaction is allowed, show that the conditions for equilibrium in a uniform gravitational field are the same as in the absence of such a reaction but there is an additional condition for the reaction to be in equilibrium.

Solution 11.1. When a chemical reaction is allowed in a homogeneous system, the differential of the number of moles of component i is given by combining Eqs. (5.120) and (5.123) to obtain $\mathrm{d}N_i = \nu_i \, \mathrm{d}\tilde{N} + \mathrm{d}N_i^{\mathrm{ext}}$, where ν_i is the stoichiometric coefficient for the reaction, $\mathrm{d}\tilde{N}$ is the change in the progress variable of the reaction, and $\mathrm{d}N_i^{\mathrm{ext}}$ is the number of moles of component i that enter the system from its exterior. For an inhomogeneous system having constant volume, we have

$$\delta N_i^{\mathrm{ext}} = \int_V [\delta c_i - \nu_i \delta\tilde{c}] \, \mathrm{d}^3x, \tag{11.22}$$

where c_i is the concentration of component i in moles per unit volume and $\tilde{c}$ is the progress variable per unit volume. In order to assure chemical closure of our system, Eq. (11.4) must be replaced by

$$\delta(U + \Phi - \lambda S - \sum_i \lambda_i N_i^{\mathrm{ext}}) \geq 0, \quad \text{constant } V, \text{ all virtual variations.} \tag{11.23}$$

Given that the chemical reaction is expressed in terms of moles, it is expedient to express all other quantities in terms of the c_i instead of the ρ_i. For example, the density in Eq. (11.7) for Φ should be replaced by $\rho = \sum_i^{\kappa} m_i c_i$. Thus Eq. (11.23) becomes

$$\int \left[(T-\lambda)\delta s_V + \sum_{i=1}^{\kappa} (\mu_i + m_i gz - \lambda_i)\delta c_i + \sum_{i=1}^{\kappa} \lambda_i \nu_i \delta\tilde{c} \right] \mathrm{d}^3x \geq 0. \tag{11.24}$$

Arbitrary variation of s_V gives Eq. (11.12) whereas arbitrary variation of c_i gives $\mu_i + m_i gz = \lambda_i$, which is the same as Eq. (11.13) divided by m_i. Arbitrary variation of $\tilde{c}$ gives an additional condition

$$0 = \sum_{i=1}^{\kappa} \lambda_i \nu_i = \sum_{i=1}^{\kappa} [\mu_i + m_i gz]\nu_i, \tag{11.25}$$

which is the condition for the chemical reaction to be in equilibrium. But since mass is conserved in a chemical reaction, we have $\sum_{i=1}^{\kappa} m_i \nu_i = 0$, so Eq. (11.25) reduces to $\sum_{i=1}^{\kappa} \mu_i \nu_i = 0$, which is the same as obtained in Section 5.7 (and also later in Chapter 12, Eq. (12.29)) in the absence of gravity. Nevertheless, the quantities $\mu_i + m_i gz$ in Eq. (11.25) are uniform in equilibrium, which shows clearly that the chemical reaction is in equilibrium at every height z, even though each μ_i varies linearly with z.

11.2.1 Multicomponent Ideal Gas in Gravity

A **multicomponent ideal gas** is an ideal solution (see Section 10.2) of chemical components, each of which obeys the ideal gas law, $p_i = n_i RT$, where p_i is the partial pressure of gas i and n_i is its concentration in moles per unit volume. Its total pressure is $p = \sum_{i=1}^{\kappa} p_i = NRT/V$. By a Maxwell relation readily obtained from $\mathrm{d}G$, one has

$$\left(\frac{\partial \mu_i}{\partial p}\right)_{T,\{N_i\}} = (\partial V/\partial N_i)_{T,p,\{N_i'\}} = \frac{RT}{p}, \tag{11.26}$$

where $\{N_i\}$ denotes the entire set of mole numbers and $\{N_i'\}$ denotes the same set with N_i missing. If the N_i are constant, the compositions X_i are constant so we can integrate Eq. (11.26) at constant composition and temperature to obtain

$$\mu_i = RT \ln p + w(T, \{X_i\}), \tag{11.27}$$

where $w(T, \{X_i\})$ is a function of integration. But the chemical potential of an ideal solution has the form $\mu_i = \mu_i^0(p, T) + RT \ln X_i$, where $X_i = N_i/N$ is the mole fraction of component i. Therefore, this chemical potential has the form

$$\mu_i = RT \ln p + RT \ln X_i + q_i(T), \tag{11.28}$$

where $q_i(T)$ is a function of only the temperature. In fact, Denbigh [18, p. 115] takes Eq. (11.28) to be the definition of an ideal gas mixture.

Since $p_i = pX_i$, Eq. (11.28) divided by the molecular weight m_i gives

$$\mu_i^m = \frac{RT}{m_i} \ln p_i + q_i^m(T) \tag{11.29}$$

for the chemical potential per unit mass of an ideal gas. Here, $q_i^m(T) = q_i(T)/m_i$. We take its differential at constant T and substitute into the differential of Eq. (11.13) to obtain

$$\frac{RT}{m_i} \ln \frac{\mathrm{d}p_i}{p_i} = -g\,\mathrm{d}z. \tag{11.30}$$

Equation (11.30) can then be integrated to give

$$p_i = p_{i0} \exp(-m_i gz/RT), \tag{11.31}$$

where p_{i0} is the partial pressure in the plane $z = 0$. The total pressure is therefore

$$p = \sum_{i=1}^{\kappa} p_{i0} \exp(-m_i gz/RT). \tag{11.32}$$

The composition at z is given by

$$X_i = \frac{p_i}{p} = \frac{p_{i0} \exp(-m_i gz/RT)}{\sum_{j=1}^{\kappa} p_{j0} \exp(-m_j gz/RT)}. \tag{11.33}$$

In terms of the composition $X_{i0} = p_{i0}/p_0$, where p_0 is the pressure in the plane $z=0$, Eq. (11.33) can be written in the form[5]

$$X_i = \frac{X_{i0} \exp(-m_i gz/RT)}{\sum_{j=1}^{\kappa} X_{j0} \exp(-m_j gz/RT)} \tag{11.34}$$

and the total pressure can therefore be written

$$p = p_0 \sum_{i=1}^{\kappa} X_{i0} \exp(-m_i gz/RT). \tag{11.35}$$

Equation (11.32) should be compared to the simple result Eq. (11.19) for the monocomponent ideal gas. The molar density is given by $n = p/(RT)$ but the total mass density is a little more complicated because

$$\rho = \sum_{i=1}^{\kappa} m_i n_i = n \sum_{i=1}^{\kappa} m_i X_i = \frac{p_0}{RT} \sum_{i=1}^{\kappa} m_i X_{i0} \exp(-m_i gz/RT). \tag{11.36}$$

The reader is invited to verify that Eq. (11.18) is satisfied.

Although Eq. (11.34) correctly describes the gravitational segregation of the chemical components of an ideal gas that have different molecular weights, it has been expressed in terms of composition in the plane $z=0$. If the overall mole numbers N_{i00} of a sample were known, one would have to integrate $n_i = p_i/(RT)$ over z with due respect to the dependence of cross sectional area $A(z)$ on z and solve to determine the quantities X_{i0}. For a sample of height H, this gives

$$N_{i00} = \frac{p_0}{RT} \int_0^H A(z) X_{i0} \exp(-m_i gz/RT)\, dz. \tag{11.37}$$

If $A(z)$ is independent of z, it may be factored out and the integral performed to yield

$$N_{i00} = \frac{A p_0 X_{i0}}{m_i g} \left[1 - \exp(-m_i gH/RT)\right]. \tag{11.38}$$

[5]For future reference, we remark that the structure of Eq. (11.34) is exactly what one would expect from the canonical ensemble of statistical mechanics (see Chapters 18 and 19) in which case the denominator is analogous to a canonical partition function for degenerate states.

Then by using $\sum_{i=1}^{\kappa} X_{i0} = 1$, we can solve for the pressure in the plane $z = 0$ to obtain

$$p_0 = \frac{1}{A}\sum_{i=1}^{\kappa} \frac{m_i N_{i00}\, g}{[1 - \exp(-m_i gH/RT)]}. \tag{11.39}$$

This expression for p_0 can be substituted into Eq. (11.38) to obtain

$$X_{i0} = \frac{m_i N_{i00}}{[1 - \exp(-m_i gH/RT)]}\left\{\sum_{j=1}^{\kappa} \frac{m_j N_{j00}}{[1 - \exp(-m_j gH/RT)]}\right\}^{-1}. \tag{11.40}$$

Then by substitution of Eqs. (11.39) and (11.40) into Eqs. (11.34) and (11.35), one can determine exactly the composition and the pressure as a function of z.

For samples of laboratory size, however, it is important to recognize that the effect of gravitational segregation on gases is extremely small. For example, for $H = 1\,\text{m}$ and $T = 300\,\text{K}$ one has $gH/RT = 3.93 \times 10^{-3}\,\text{mol/kg}$. Thus for N_2 ($m_i = 28\,\text{g/mol}$) and O_2 ($m_i = 32\,\text{g/mol}$) one would have $m_i gH/RT$ equal to 1.1×10^{-4} and 1.26×10^{-4}, respectively. Even for a heavy gas such as uranium hexafluoride UF_6 ($m_i = 352\,\text{g/mol}$) one would have $m_i gH/RT = 1.4 \times 10^{-3}$. Therefore, for samples of laboratory size, one has $m_i gH/RT \ll 1$ and the above expressions for ideal gases can be expanded in these small quantities. If we keep only linear terms in $m_i gH/RT$ and $m_i gz/RT$, the pressure and the compositions become linear functions of z. The results can be written in terms of a few new symbols: the total number of moles $N_{00} := \sum_{i=1}^{\kappa} N_{i00}$, the total mass $M_{00} := \sum_{i=1}^{\kappa} m_i N_{i00}$, the average molecular weight $\bar{m} = M_{00}/N_{00}$, and the volume of the system $V = AH$. Then some algebra yields

$$p \approx \frac{N_{00}RT}{V} + \frac{M_{00}g}{A}\left(\frac{1}{2} - \frac{z}{H}\right) = \frac{N_{00}RT}{V}\left[1 + \frac{\bar{m}gH}{RT}\left(\frac{1}{2} - \frac{z}{H}\right)\right]; \tag{11.41}$$

$$X_i \approx \frac{N_{i00}}{N_{00}}\left[1 + (m_i - \bar{m})\frac{gH}{RT}\left(\frac{1}{2} - \frac{z}{H}\right)\right]. \tag{11.42}$$

Equation (11.41) shows that the pressure, largest at the bottom and smallest at the top, is equal to its value in the absence of gravity plus a linear correction related to the mass of the gas. According to Eq. (11.42), the composition is given by the overall composition times a linear function that either decreases or increases with height depending on whether the molecular weight is smaller or larger than the average molecular weight.

11.2.2 Binary Liquid in Gravity

A binary liquid in a uniform gravitational field will also undergo segregation of its A and B species but the situation is different from that of a gas because a liquid is much denser and comparatively incompressible. We carry out the calculation for an ideal solution for which

$$\mu_A^m = \mu_A^{m0}(p, T) + \frac{RT}{m_A}\ln X_A; \tag{11.43}$$

$$\mu_B^m = \mu_B^{m0}(p, T) + \frac{RT}{m_B}\ln X_B, \tag{11.44}$$

where $\mu_A^{m0}(p,T)$ and $\mu_B^{m0}(p,T)$ correspond to standard states of pure A and B, respectively. We substitute into Eq. (11.13) and identify the Lagrange multipliers by setting $z = 0$, where we denote the pressure by p_0 and the compositions by X_{A0} and X_{B0}, to obtain

$$\mu_A^{m0}(p,T) - \mu_A^{m0}(p_0,T) + \frac{RT}{m_A}\ln(X_A/X_{A0}) + gz = 0; \tag{11.45}$$

$$\mu_B^{m0}(p,T) - \mu_B^{m0}(p_0,T) + \frac{RT}{m_B}\ln(X_B/X_{B0}) + gz = 0. \tag{11.46}$$

From the differential of the Gibbs free energy, we note that the partial specific volumes are given by $\partial\mu_A^m/\partial p = \partial\mu_A^{m0}/\partial p = 1/\rho_A$ and similarly for B. Although the quantities ρ_A and ρ_B depend on p and T, the temperature T is constant and the dependence on p is very weak because liquids have such small compressibilities. We shall therefore treat ρ_A and ρ_B as constants in order to obtain a tractable problem. This results in

$$\frac{1}{\rho_A}(p - p_0) + \frac{RT}{m_A}\ln(X_A/X_{A0}) + gz = 0; \tag{11.47}$$

$$\frac{1}{\rho_B}(p - p_0) + \frac{RT}{m_B}\ln(X_B/X_{B0}) + gz = 0. \tag{11.48}$$

Since $X_B = 1 - X_A$ we could solve Eqs. (11.47) and (11.48) simultaneously for X_A and p as functions of z but the results are cumbersome so we take advantage immediately of the fact that mgz/RT is very small, where m characterizes m_A or m_B or the combination of them given by Eq. (11.52). Thus we expand the logarithms to first order to obtain the linear equations

$$(p - p_0) + \frac{\rho_A RT}{m_A}\frac{(X_A - X_{A0})}{X_{A0}} + \rho_A gz = 0; \tag{11.49}$$

$$(p - p_0) - \frac{\rho_B RT}{m_B}\frac{(X_A - X_{A0})}{X_{B0}} + \rho_B gz = 0. \tag{11.50}$$

We then subtract to eliminate $p - p_0$ and obtain

$$(X_A - X_{A0})\left[\frac{f}{X_{A0}} + \frac{1-f}{X_{B0}}\right] = -\frac{m^* gz}{RT}, \tag{11.51}$$

where

$$f = \frac{\rho_A/m_A}{\rho_A/m_A + \rho_B/m_B}; \quad m^* = \frac{\rho_A - \rho_B}{\rho_A/m_A + \rho_B/m_B}. \tag{11.52}$$

If $\rho_A > \rho_B$, we observe that X_A decreases with increasing z as would be expected. Finally, we can solve Eqs. (11.49) and (11.50) for $p - p_0$ to obtain

$$p - p_0 = -\rho^* gz, \tag{11.53}$$

where

$$\rho^* = \frac{\rho_A\rho_B(m_A X_{A0} + m_B X_{B0})}{\rho_A m_B X_{B0} + \rho_B m_A X_{A0}}. \tag{11.54}$$

Thus the pressure always decreases linearly with height z, but the weighting of densities is not obvious. Incidentally, if $\rho_A = \rho_B$, there is no segregation and the pressure increases with ρ^* equal to their common density. As compared to a binary ideal gas, the magnitude of the segregation is comparable but the magnitude of the pressure change is much larger for the binary liquid because the effective density is much larger than for the gas.

11.3 Non-Uniform Gravitational Field

For a non-uniform gravitational field, Eq. (11.7) can be written

$$\Phi = \int_V \rho\,\varphi(\mathbf{r})\,\mathrm{d}^3x + \text{constant}, \tag{11.55}$$

where $\varphi(\mathbf{r})$ is the gravitational potential (potential energy per unit mass) at position $\mathbf{r}$. For example, the gravitational potential due to attraction by the Earth, whose center of mass is assumed to be located at the origin, would be

$$\varphi(\mathbf{r}) = -\frac{MG}{r}, \tag{11.56}$$

where $r = |\mathbf{r}|$, M is the mass of the Earth and $G = 6.67 \times 10^{-11}\,\mathrm{m^3\,kg^{-1}\,s^{-2}}$ is the universal gravitational constant. Since $\varphi(\mathbf{r})$ is the same for all chemical species, Eq. (11.13) becomes simply

$$\mu_i^m + \varphi(\mathbf{r}) = \lambda_i^m; \quad i = 1, 2, \ldots, \kappa. \tag{11.57}$$

Thus for the potential given by Eq. (11.56), Eq. (11.31) for the partial pressure of a multicomponent gas would be replaced by

$$p_i = p_{i0} \exp\left[-\frac{m_i MG}{RT}\left(\frac{1}{r_0} - \frac{1}{r}\right)\right], \tag{11.58}$$

where r is the distance to the center of gravity of the Earth and r_0 is a reference distance where the partial pressure is p_{i0}. This result is not significantly different from Eq. (11.31) for a constant gravitational acceleration unless r varies by large distances compared to r_0. Indeed, for $|(r - r_0)/r_0| \ll 1$, the effective gravitational acceleration would be $g = MG/r_0^2$.

11.4 Rotating Systems

A system undergoing uniform rotation at an angular velocity ω about some axis behaves as if it were in a non-uniform gravitational field with potential (potential energy per unit mass) $\varphi = -r_\perp^2\omega^2/2$, where $r_\perp$ is the distance from the axis of rotation. This result follows because the corresponding force per unit mass would be $-\mathrm{d}\varphi/\mathrm{d}r_\perp = r_\perp\omega^2$ and would be directed radially outward. This is just the centrifugal acceleration (centrifugal force per unit mass) that is experienced in a rotating coordinate system. The work done by this external force when a mass m moves from $r_\perp = 0$ to $r_\perp$ is just

$$\int_0^{r_\perp} m r'_\perp \omega^2 \, \mathrm{d}r'_\perp = m r_\perp^2 \omega^2/2 = -m\varphi. \tag{11.59}$$

Thus Eq. (11.57) becomes

$$\mu_i^m - r_\perp^2 \omega^2/2 = \lambda_i^m; \quad i = 1, 2, \ldots, \kappa. \tag{11.60}$$

For the partial pressure of a multicomponent ideal gas, one therefore obtains

$$p_i = p_{i0} \exp(m_i r_\perp^2 \omega^2/2RT), \tag{11.61}$$

which differs from Eq. (11.31) in two important ways: the exponential depends on the square of the distance $r_\perp$ and ω^2 can be quite large in the sense that $r_\perp \omega^2 \gg g$. Thus, in a fast centrifuge, the components of such a gas with sufficiently different m_i can undergo significant segregation of components. To achieve significant segregation of components with only slightly different m_i, such as isotopes, one would have to make $r_\perp^2 \omega^2$ as large as practical and employ a multi-stage process wherein the enriched portion of each stage is used as the starting sample for the next stage.

Example Problem 11.2. A circular cylinder with axis of symmetry along the z axis contains a monocomponent liquid that is practically incompressible and therefore has constant density ρ. The cylinder is rotated at constant angular velocity ω. The liquid is also in a constant gravitational field g directed downward, antiparallel to z. Find the pressure of the liquid as a function of position in the cylinder. Determine the shape of the isobars and comment on the shape of the upper free surface of the liquid if it is open to the atmosphere and evaporation is negligible.

Solution 11.2. The governing equation for the chemical potential is

$$\mu^m - (x^2 + y^2)\omega^2/2 + gz = \lambda^m. \tag{11.62}$$

By taking $x = y = z = 0$, we identify λ^m as the chemical potential at the origin where we take the pressure to be p_0. Then by integration of Eq. (11.17) for a single component and constant ρ, we obtain $\mu^m - \lambda^m = (p - p_0)/\rho$. Thus, Eq. (11.62) becomes

$$p - p_0 = \rho\left[(x^2 + y^2)\omega^2/2 - gz\right]. \tag{11.63}$$

The isobars satisfy

$$z = \frac{\omega^2}{2g}(x^2 + y^2) + \frac{p_0 - p}{\rho} \tag{11.64}$$

and each has the shape of a parabola of revolution whose lowest point is along the axis of rotation. The upper free surface of the liquid will also have this parabolic shape with $p = 1$ atmosphere to the extent that capillary effects (see Chapter 13) are negligible.

11.5 Electric Fields

We consider a single phase multicomponent fluid in the presence of an electric field $\mathbf{E} = -\nabla\phi$, where $\phi(\mathbf{r})$ is the electrical potential. If species i carries an electric charge $z_i|e|$, the work done by the field in moving that charge from a reference position $\mathbf{r}_0$ to position $\mathbf{r}$ is

$$\int_{\mathbf{r}_0}^{\mathbf{r}} z_i|e|\mathbf{E}\cdot d\mathbf{r} = z_i|e|[\phi(\mathbf{r}_0) - \phi(\mathbf{r})] = -\mathcal{W}. \tag{11.65}$$

In electrochemistry, z_i is regarded to be the valence of each species. For convenience we take the reference potential $\phi(\mathbf{r}_0) = 0$ in which case the total potential of Eq. (11.6) can be written in the form

$$\Phi = \int_V \sum_i z_i|e|\phi(\mathbf{r})\mathcal{N}_A c_i\,\mathrm{d}^3x = \int_V \sum_i z_i\phi(\mathbf{r})\mathcal{F}c_i\,\mathrm{d}^3x, \tag{11.66}$$

where c_i is the concentration of species i in moles per unit volume, $\mathcal{N}_A$ is Avogadro's number and $\mathcal{F} = |e|\mathcal{N}_A = 96{,}485$ coulomb/mol is the Faraday constant. Then Eq. (11.21) can be replaced by

$$\int_V \sum_i [\mu_i + z_i\phi(\mathbf{r})\mathcal{F} - \lambda_i]\,\delta c_i\,\mathrm{d}^3x \geq 0 \tag{11.67}$$

which leads to the equilibrium equations

$$\mu_i + z_i\phi(\mathbf{r})\mathcal{F} = \lambda_i. \tag{11.68}$$

In this case we see that the electrochemical potentials $\mu_i + z_i\phi(\mathbf{r})\mathcal{F}$ are uniform at equilibrium. It is often convenient to use the chemical potential per atom (or molecule) instead of per mole. If we designate this quantity by μ_i^a, the equilibrium equations can be written in the form

$$\mu_i^a + q_i\phi(\mathbf{r}) = \lambda_i^a, \tag{11.69}$$

where q_i is the charge carried by species i.

When dealing with heterogeneous equilibrium among phases of various composition, a word of caution is in order because relative electrical potentials can become ill-defined due to surface potentials and different chemical environments that a test charge encounters when entering a material from infinity, where the potential is taken to be zero. For a discussion of the equilibrium for transfer between phases, see Denbigh [18, p. 86].

12 Chemical Reactions

We regard chemical reactions to be the formation or dissociation of chemical molecules or compounds in which no chemical elements are created or destroyed. In other words, *we exclude nuclear reactions* in which new nuclei can form and during which there is a change Δm_0 in the rest mass m_0, resulting in a change of energy given by the Einstein relation $\Delta E = \Delta m_0 c^2$ where c is the speed of light. By excluding nuclear reactions, both mass and energy are separately conserved during chemical reactions and the first law of thermodynamics, which embodies the conservation of energy, applies in the form presented in Chapter 2 for a chemically closed system. Therefore, if a chemical reaction occurs in an *isolated* system, the change in internal energy from initial to final state $\Delta_{\rm if}U := U_{\rm f} - U_{\rm i} = 0$.[1] Microscopically this makes sense because the internal energy consists of kinetic energy and potential energy associated with chemical bonds or intermolecular forces. The making or breaking of chemical bonds during chemical reactions in an isolated system involves a redistribution of kinetic and potential energy, but no net change of energy.

As in Section 5.7 where we briefly introduced chemical reactions, we break $\mathrm{d}N_i$ into two parts, $\mathrm{d}N_i = \mathrm{d}^{\rm int}N_i + \mathrm{d}^{\rm ext}N_i$, where $\mathrm{d}^{\rm int}N_i$ is due to chemical reactions and $\mathrm{d}^{\rm ext}N_i$ is due to exchanges with the environment. We write a chemical reaction in the symbolic form

$$\sum_i \nu_i A_i = 0, \tag{12.1}$$

where the A_i are chemical symbols and the ν_i are stoichiometric coefficients that are positive for products (on the right-hand side of the reaction equation) and negative for reactants (on the left-hand side of the reaction equation); see Eq. (5.121) and the related example. Then if $\tilde{N}$ is the progress variable for the reaction, we will have[2]

$$\mathrm{d}N_i^{\rm int} = \nu_i\,\mathrm{d}\tilde{N}. \tag{12.2}$$

Here, $\tilde{N}$ has the dimensions of moles; it is zero when the reaction begins and $\tilde{N}_{\rm final}$ when the reaction ends. In this chapter, we consider only chemically closed systems, so $\mathrm{d}^{\rm ext}N_i = 0$ and $\mathrm{d}N_i = \mathrm{d}N_i^{\rm int}$.

[1]In this section we add subscripts to quantities such as $\Delta_{\rm if}U$ and $\Delta_{\rm if}H$ to emphasize that these symbols denote the change from initial to final states. This is to avoid confusion with the standard notation for such quantities as ΔH in Eq. (12.13), which is actually a derivative of H with respect to the progress variable $\tilde{N}$.

[2]The generalization to multiple chemical reactions is straightforward. One needs only to add a superscript s to both quantities on the right hand-side and sum to obtain $\mathrm{d}^{\rm int}N_i = \sum_s \nu_i^s\,\mathrm{d}\tilde{N}^s$.

Thermal Physics. http://dx.doi.org/10.1016/B978-0-12-803304-3.00012-0

12.1 Reactions at Constant Volume or Pressure

Chemical reactions are typically carried out either at constant volume or at constant pressure. Those involving gases can usually be carried out easily at constant volume because the gases can be contained in a strong and nearly inert solid container. Then the work $\mathcal{W} = 0$, so the change in internal energy of the gases is

$$\Delta_{\mathrm{if}} U = Q, \quad \text{constant volume, chemically closed,} \tag{12.3}$$

where the heat Q is positive if added to the gases and negative if extracted from the gases. If the reaction vessel is thermally insulated, the reaction will result in a change of temperature that can be measured. For example, a **bomb calorimeter** is a rigid vessel with a known heat capacity C_{cal} that is large compared to the heat capacity of the gases undergoing reaction. Typically it is filled with oxygen at high pressure and some fuel that is burned to completion during the reaction. If the calorimeter is well insulated from its surroundings and its temperature changes by $\Delta_{\mathrm{if}} T$, then $Q = -C\Delta_{\mathrm{if}} T$, where C is the heat capacity of the calorimeter and the gases. To extent that the heat capacity of the gases can be neglected, $C_{\mathrm{cal}}\Delta_{\mathrm{if}} T$ represents the energy that is converted from chemical bond energy as a result of the reaction.

Of great practical importance, however, are chemical reactions that are carried out such that the only work done is against a constant external pressure p_{ext}. In such reactions, there is a volume change $\Delta_{\mathrm{if}} V$ and there is no attempt to impose the constraint of constant volume, which might be very difficult if only condensed phases are involved. Moreover, the atmosphere might provide the constant external pressure in industrial reactions. The work done by the system will then be $\mathcal{W} = p_{\mathrm{ext}}\Delta_{\mathrm{if}} V$ and from the first law we will have

$$\Delta_{\mathrm{if}} U + p_{\mathrm{ext}}\Delta_{\mathrm{if}} V = Q. \tag{12.4}$$

If the pressure $p = p_{\mathrm{ext}}$ in the initial and final states of the system, we can introduce the enthalpy $H = U + pV$ in which case Eq. (12.4) takes the form

$$\Delta_{\mathrm{if}} H = Q, \quad \text{constant pressure, chemically closed,} \tag{12.5}$$

where Q is the heat added to the reacting system. Thus the enthalpy H plays the same role at constant pressure as the internal energy U plays at constant volume. In general, one can regard the enthalpy to be a function of its natural variables, in which case

$$\mathrm{d}H = T\,\mathrm{d}S + V\,\mathrm{d}p + \sum_i \mu_i\,\mathrm{d}N_i. \tag{12.6}$$

However, for practical purposes it is more convenient to use the temperature instead of the entropy, which results in

$$\mathrm{d}H = \left(\frac{\partial H}{\partial T}\right)_{p,N_i} \mathrm{d}T + \left(\frac{\partial H}{\partial p}\right)_{T,N_i} \mathrm{d}p + \sum_i \bar{H}_i\,\mathrm{d}N_i, \tag{12.7}$$

where the quantities $\bar{H}_i$ are the partial molar enthalpies. We recognize $(\partial H/\partial T)_{p,N_i} = T(\partial S/\partial T)_{p,N_i} = C_p$ as the heat capacity at constant pressure. Furthermore, regarding S to depend on T, p, N_i, we readily establish that $\bar{H}_i = \mu_i - T\bar{S}_i$ and $(\partial H/\partial p)_{T,N_i} = V + T(\partial S/\partial p)_{T,N_i}$. A Maxwell relation based on the differential $\mathrm{d}G = -S\,\mathrm{d}T + V\,\mathrm{d}p + \sum_i \mu_i dN_i$ readily yields $(\partial S/\partial p)_{T,N_i} = -(\partial V/\partial T)_{p,N_i} = -V\alpha$, where α is the coefficient of thermal expansion. Thus Eq. (12.7) can be written

$$\mathrm{d}H = C_p\,\mathrm{d}T + V(1-\alpha T)\,\mathrm{d}p + \sum_i \bar{H}_i\,\mathrm{d}N_i. \tag{12.8}$$

For the chemically closed systems we are considering, Eq. (12.8) takes the form

$$\mathrm{d}H = C_p\,\mathrm{d}T + V(1-\alpha T)\,\mathrm{d}p + (\sum_i \nu_i \bar{H}_i)\,\mathrm{d}\tilde{N}. \tag{12.9}$$

Since T and p are intensive variables, the Euler equation for the enthalpy (see Eq. (5.101)) is just

$$H = \sum_i N_i \bar{H}_i. \tag{12.10}$$

We emphasize that Eq. (12.10) holds as a function of p, T, and N_i, provided that the $\bar{H}_i$ are evaluated at p, T and the corresponding composition. At any stage of the reaction, $N_i = N_i^0 + \nu_i\tilde{N}$, where N_i^0 is the initial value of N_i. The Euler equation (12.10) becomes

$$H(T,p,N_i) = \sum_i (N_i^0 + \nu_i\tilde{N})\bar{H}_i, \tag{12.11}$$

where it is understood that the $\bar{H}_i$ are to be evaluated at the corresponding composition, temperature, and pressure.

Example Problem 12.1. For the chemical reaction given by Eq. (5.122), namely $C+(1/2)O_2 \rightarrow CO$, assume initially that the mole numbers are $N_C^0 = 3$, $N_{O_2}^0 = 1$, and $N_{CO}^0 = 2$. If conditions are such that the reaction goes to the right until one of the reactants is completely used, what is the value of $\tilde{N}_{\text{final}}$ and how many moles of each component will there be? Answer the same question under different conditions for which the reaction goes to the left until all of the CO is used.

Solution 12.1. The stoichiometric coefficients ν_i for C, O_2, and CO are -1, $-1/2$, and 1, respectively. For either the forward or backward reaction we have $N_C = 3-\tilde{N}$, $N_{O_2} = 1-(1/2)\tilde{N}$, and $N_{CO} = 2+\tilde{N}$. The reaction can go to the right until $\tilde{N}_{\text{final}} = 2 = \tilde{N}_{\text{max}}$ in which case $N_C = 1$, $N_{O_2} = 0$, and $N_{CO} = 3$. The reaction can go to the left until $\tilde{N}_{\text{final}} = -1 = \tilde{N}_{\text{min}}$ in which case $N_C = 4$, $N_{O_2} = 3/2$, and $N_{CO} = 0$. The actual direction of the reaction and the extent of reaction will depend on the conditions under which the reaction is carried out, particularly the temperature. For conditions to be discussed below, the reaction may reach equilibrium at some value $\tilde{N}_{\text{min}} \le \tilde{N}_{\text{final}} \le \tilde{N}_{\text{max}}$.

12.1.1 Heat of Reaction

According to Eq. (12.5), the heat $Q_p = -Q$ *liberated* to the environment by the reacting system at constant pressure p is given by

$$-Q_p = H(T_{\text{final}}, p, N_i^0 + \nu_i \tilde{N}_{\text{final}}) - H(T_{\text{initial}}, p, N_i^0). \tag{12.12}$$

But Q_p is not a very useful way to characterize a reaction because it depends specifically on the initial conditions. A much more useful quantity is the derivative of H with respect to the progress variable $\tilde{N}$ at constant temperature and pressure, namely

$$\Delta H \equiv -Q_{\tilde{N}} := \sum_i \nu_i \bar{H}_i = \left(\frac{\partial H}{\partial \tilde{N}}\right)_{T,p}. \tag{12.13}$$

This quantity is commonly called "the ΔH of the reaction" but that is somewhat of a misnomer because it is a derivative. In particular, ΔH should not be confused with $-Q_p$ for a specific reaction, which is the difference in enthalpy between final and initial states given by Eq. (12.12). $Q_{\tilde{N}} = -\Delta H$ is the heat *liberated* by the reaction per unit change of the progress variable at constant p and T. Callen [2, p. 170] refers to ΔH as the **heat of reaction** and suggests that it be evaluated near the equilibrium state; however, depending on conditions, a specific reaction might go to completion before the equilibrium state is reached. For $Q_{\tilde{N}} = -\Delta H > 0$, the reaction is said to be *exothermic* whereas for $Q_{\tilde{N}} = -\Delta H < 0$, the reaction is said to be *endothermic*.[3] In Section 12.3 we will relate ΔH to the ΔG of the reaction.

For the special but often treated case for which the reactants and the products are not in solution, or if gaseous they form an ideal solution, one has $\bar{H}_i = H_i(p, T)$, where $H_i(p, T)$ is the enthalpy per mole of the respective pure component. This follows for a solution of ideal gases because the chemical potentials

$$\mu_i(T, p, X_i) = \mu_i(T, p) + RT \ln X_i, \tag{12.14}$$

where $\mu_i(T, p)$ corresponds to the pure component and X_i is the mole fraction. Note that the total pressure is p and the partial pressure is $p_i = pX_i$. Thus,

$$\bar{H}_i = \frac{\partial(\mu_i(T, p, X_i)/T)}{\partial(1/T)} = \frac{\partial(\mu_i(T, p)/T)}{\partial(1/T)} = H_i(T, p), \tag{12.15}$$

so there is no heat of mixing for an ideal solution. Under these conditions, the initial and final states can be expressed in terms of heterogeneous components and Eq. (12.13) becomes simply

$$\Delta H = \sum_i \nu_i H_i(T, p), \quad \text{heterogeneous components}, \tag{12.16}$$

for which there is extensive tabulation of data as discussed in the next section.

[3] Unfortunately, various authors use different terminology. Kondepudi and Prigogine [14, p. 53] associate the quantity $\left(\partial U/\partial \tilde{N}\right)_{T,V} = \sum_i \nu_i \bar{U}_i$ with endothermic and exothermic reactions. Lupis [5, p. 10] and Kondepudi and Prigogine [14, p. 52] treat ΔH for the case in which the constituents are not in solution.

12.2 Standard States

We shall define the **standard state** of an element or compound to be its *most stable state* at a pressure $p_0 = 101{,}325\,\text{Pa} = 1$ standard atmosphere and at the temperature T of relevance.[4] The enthalpy of *one mole* of an element or compound in its standard state is denoted by $H^0(T, p_0)$. However, we must remember that enthalpy, like energy, is undefined up to an additive constant. Thus for enthalpy, it is customary to tabulate $H^0(T, p_0) - H^0(T_0, p_0)$ for elements and compounds as a function of temperature, where $T_0 = 298.15\,\text{K} = 25\,°\text{C}$. Here, the superscript 0 reminds us that the element or compound is in its standard state at pressure p_0 at *both* T and T_0.

It follows that the ΔH given by Eq. (12.16) for a *heterogeneous reaction* can be related to

$$\Delta H^0(T, p_0) = \sum_i \nu_i H_i^0(T, p_0). \tag{12.17}$$

Note that

$$\Delta H^0(T, p_0) = \left(\frac{\partial H}{\partial \tilde{N}}\right)_{T,p_0}, \quad \text{all constituents in their standard states.} \tag{12.18}$$

If two reactions are added to form a third reaction, $\Delta H^0(T, p_0)$ is additive since it is a state function. This was discovered empirically and is known as Hess's law. A quantity that is tabulated extensively is

$$\Delta H^0(T_0, p_0) = \sum_i \nu_i H_i^0(T_0, p_0), \tag{12.19}$$

which is the value of ΔH^0 at *both* standard temperature T_0 and pressure p_0. It follows that

$$\Delta H^0(T, p_0) = \Delta H^0(T_0, p_0) + \sum_i \nu_i [H_i^0(T, p_0) - H_i^0(T_0, p_0)]. \tag{12.20}$$

The quantity $\Delta H^0(T_0, p_0)$ is especially valuable because for many reactions, the gases in the reaction behave approximately as ideal gases in an ideal solution (even though they react occasionally due to collisions) so the partial molar quantities $\bar{H}_i$ are very nearly equal to the molar values $H_i(T, p)$ for pure constituents (see Eq. (12.15)). Second, for ideal gases, $\alpha = 1/T$, so the term $V(1 - \alpha T)$ in Eq. (12.8) vanishes, and would be expected to be small even for real gases. For heterogeneous solids and liquids that are not in solution, we again have $\bar{H}_i = H_i(T, p)$ and the dependence on pressure is weak because now the molar volume V_i in the term $V_i(1 - \alpha_i T)$ is small. Therefore, for such reactions,

$$\Delta H^0(T_0, p) = \sum_i \nu_i H_i(T_0, p) \approx \sum_i \nu_i H_i(T_0, p_0) = \Delta H^0(T_0, p_0). \tag{12.21}$$

[4]For gases, the standard state is usually defined as the state in which the fugacity is equal to the pressure p, as $p \to 0$, which would make a small difference if the gas did not behave like an ideal gas at p_0. Other definitions of standard states, such as solutions of a specified concentration, are sometimes used. We could also define a standard state to be the most stable state of the pure constituent at temperature T and pressure p.

Of course one could correct for the small difference due to pressure if compressibility data were available. If heat capacity data are known, the difference in heat capacity between reactants and products is

$$\Delta C_p = \sum_i \nu_i C_{pi}, \tag{12.22}$$

where the C_{pi} are heat capacities at constant pressure of the pure reactants and products. Then

$$\Delta H^0(T,p) \approx \Delta H^0(T_0,p_0) + \int_{T_0}^{T} \Delta C_p \, dT. \tag{12.23}$$

12.2.1 Heat of Formation

The enthalpy required to produce one mole of a compound from its elements at temperature T, everything in its standard state at pressure p_0, is called the **heat of formation** and is designated by $H_f^0(T,p_0)$. Since elements cannot be created by chemical reaction,[5] it follows that the heat of formation of an element is zero. Moreover,

$$\Delta H^0(T,p_0) = \sum_i \nu_i H_f^0(T,p_0). \tag{12.24}$$

Note in the sum that the only contribution comes from compounds.

Example Problem 12.2. The heats of formation H_f^0 at $T_0 = 298.15$ K of H_2O, CO_2, and CO are −285.8 kJ/mol, −393.5 kJ/mol, and −110 kJ/mol, respectively. Discuss the relevant chemical reactions. Then compute $\Delta H^0(T_0,p_0)$ for the reaction

$$H_2(\text{gas}) + CO_2(\text{gas}) \to CO(\text{gas}) + H_2O(\text{liquid}). \tag{12.25}$$

Solution 12.2. The relevant reactions for compound formation are:

$$\begin{aligned} H_2(\text{gas}) + (1/2)O_2(\text{gas}) &\to H_2O(\text{liquid}),\\ C(\text{graphite}) + O_2(\text{gas}) &\to CO_2(\text{gas}),\\ C(\text{graphite}) + (1/2)O_2(\text{gas}) &\to CO(\text{gas}). \end{aligned}$$

For the reaction given by Eq. (12.25), we have

$$\Delta H^0(T_0,p_0) = (-285.8 - 110 + 393.5)\,\text{kJ/mol} = -2\,\text{kJ/mol}. \tag{12.26}$$

Note the sign change for CO_2 because it is a reactant in Eq. (12.25).

[5] Recall that nuclear reactions are excluded. If 2 moles of deuterium react to form one mole of ^{3}He and a neutron, about 3×10^8 kJ/mol are released. Heats of formation of most chemical compounds are typically only several hundred kJ/mol.

12.3 Equilibrium and Affinity

We now examine the conditions under which a chemical reaction is in equilibrium and the direction that the reaction will proceed if it is not in equilibrium. For a multicomponent system, the differential of the Gibbs free energy (see Eq. (5.90)) is

$$dG = -S\,dT + V\,dp + \sum_i \mu_i\,dN_i. \tag{12.27}$$

As before, we assume that $dN_i = d^{int}N_i + d^{ext}N_i$, that the system is chemically closed so that $d^{ext}N_i = 0$, and that $dN_i = d^{int}N_i = \nu_i d\tilde{N}$ due to chemical reaction, as in Eq. (12.2). Then[6]

$$dG = -S\,dT + V\,dp + (\sum_i \nu_i\mu_i)\,d\tilde{N}. \tag{12.28}$$

For a chemically closed system at constant p and T, we know that G is a minimum at equilibrium. Therefore, the criterion for equilibrium of a chemical reaction is

$$\left(\frac{\partial G}{\partial \tilde{N}}\right)_{p,T} = \sum_i \nu_i\mu_i = 0. \tag{12.29}$$

The notations

$$\Delta G \equiv -\mathcal{A} := \sum_i \nu_i\mu_i = \left(\frac{\partial G}{\partial \tilde{N}}\right)_{p,T} \tag{12.30}$$

are common. $\mathcal{A}$ is called the **affinity**[7] of the reaction and is usually used in irreversible thermodynamics. The other notation, "the ΔG of the reaction" is somewhat of a misnomer because it is really a derivative of G with respect to the progress variable $\tilde{N}$ and should not be confused with the actual change in G from beginning to end of the reaction, which depends on the initial values N_i^0 and the extent of reaction, possibly limited because of the depletion of some component. The change in the Gibbs free energy for a small change in $\tilde{N}$ at constant p and T is therefore

$$(dG)_{T,p} = \Delta G\,d\tilde{N} = -\mathcal{A}\,d\tilde{N} \le 0, \tag{12.31}$$

where the inequality holds for a natural irreversible process. Thus if $\Delta G < 0$ ($\mathcal{A} > 0$) the reaction will proceed to the right and for $\Delta G > 0$ ($\mathcal{A} < 0$) the reaction will proceed to the left. For $\Delta G = -\mathcal{A} = 0$, which corresponds to a minimum of G, the reaction will be in equilibrium. See Figure 12–1 for a sketch of G and $\mathcal{A}$ near equilibrium. Equilibrium can

[6]Eq. (12.27) implies that G is a function of T, p and the N_i in the field of equilibrium states. If a chemical reaction can occur, the system will not be in an equilibrium state but we can imagine, for thermodynamic purposes, that the reaction proceeds slowly through a set of constrained equilibrium states. It is generally assumed that Eq. (12.28) is valid for small deviations from equilibrium. The same assumption was implicit in Eq. (12.9).

[7]We use a calligraphic symbol $\mathcal{A}$ to avoid confusion with A which in some books is used to denote the Helmholtz free energy, which we denote by F. The name "affinity" is due to T. De Donder, who founded the Belgian school of thermodynamics [14, p. 104].

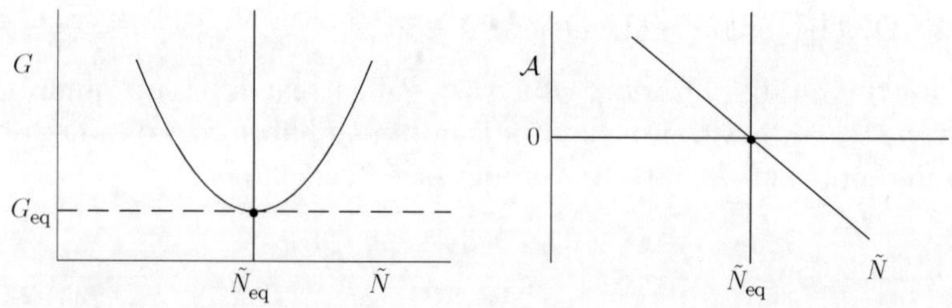

FIGURE 12–1 Sketches of the Gibbs free energy G and the affinity $\mathcal{A}$ as a function of the progress variable $\tilde{N}$ near its equilibrium value $\tilde{N}_{eq}$. At a given value of $\tilde{N}$, the affinity is the negative of the slope of G. If G is nearly parabolic near its minimum value G^{eq}, the affinity $\mathcal{A}$ will be nearly linear. Equilibrium will occur at $\tilde{N}_{eq}$ provided that the initial values N_i^0 of the constituents of the reaction are such that $\tilde{N}_{min} \le \tilde{N}_{eq} \le \tilde{N}_{max}$. Otherwise, the reaction will proceed in the direction of $\tilde{N}_{eq}$ but will stop when $\tilde{N}_{min}$ or $\tilde{N}_{max}$ is reached.

occur at the value of $\tilde{N} = \tilde{N}^{eq}$ that satisfies Eq. (12.29). In an actual situation, equilibrium at this minimum value of G will be achieved provided that the values of the initial mole numbers N_i^0 are such that $\tilde{N}_{min} \le \tilde{N}_{eq} \le \tilde{N}_{max}$. Otherwise the reaction will come to equilibrium at the lowest value of G subject to the constraint that no mole numbers can be negative. Note that the positive quantity $\tilde{N}_{max}$ occurs if one of the reactants goes to zero and the negative quantity $\tilde{N}_{min}$ occurs if one of the products becomes zero. See Example Problem 12.2 for a specific reaction.

The role of the affinity $\mathcal{A}$ can be better understood in terms of entropy production in a chemically closed system. For the case of reversible work, $\delta\mathcal{W} = p\,dV$ and reversible heat flow, $T_r = T$, Eq. (5.127) applies, so

$$dS = \frac{\delta Q}{T} - \sum_i \nu_i \mu_i \, d\tilde{N} = \frac{\delta Q}{T} + \frac{\mathcal{A}}{T} d\tilde{N} > 0 \tag{12.32}$$

for a natural irreversible process. Then according to irreversible thermodynamics, one writes $dS = d^{ext}S + d^{int}S$, where $d^{ext}S = \delta Q/T$ is the entropy exchanged reversibly with the environment and $d^{int}S > 0$ applies to an internal irreversible process. This leads to an entropy production due to chemical reaction given by

$$d^{int}S = \frac{\mathcal{A}}{T} d\tilde{N} > 0, \quad \text{natural irreversible process.} \tag{12.33}$$

Again, $\mathcal{A} > 0$ leads to reaction to the right ($d\tilde{N} > 0$) whereas $\mathcal{A} < 0$ leads to reaction to the left ($d\tilde{N} < 0$). Equilibrium requires prevention of a natural irreversible process for both possible signs of $d\tilde{N}$, which requires $\mathcal{A} = 0$.

We can relate $\Delta G = \left(\partial G/\partial \tilde{N}\right)_{T,p}$ of a reaction to $\Delta H = \left(\partial H/\partial \tilde{N}\right)_{T,p}$ that pertains to the heat of reaction by regarding G, H, and S to be functions of T, p, and $\tilde{N}$. Then by taking $\left(\partial/\partial \tilde{N}\right)_{T,p}$ of $G = H - TS$, we verify that $\Delta G = \Delta H - T\Delta S$, where $\Delta S = \left(\partial S/\partial \tilde{N}\right)_{T,p}$. Hence we have the relations

$$\Delta H = \left(\frac{\partial(\Delta G/T)}{\partial(1/T)}\right)_{p,\tilde{N}} = \Delta G - T\left(\frac{\partial \Delta G}{\partial T}\right)_{p,\tilde{N}}; \quad \Delta S = -\left(\frac{\partial \Delta G}{\partial T}\right)_{p,\tilde{N}}. \tag{12.34}$$

Note that $\Delta S \mathrm{d}\tilde{N} = (\partial \mathcal{A}/\partial T)\,\mathrm{d}\tilde{N}$ is not the same as $\mathrm{d}^{\mathrm{int}}S = \mathcal{A}/T\,\mathrm{d}\tilde{N}$. This arises because

$$\mathrm{d}S = C_p\,\mathrm{d}T - V\alpha\,\mathrm{d}p + \Delta S\,\mathrm{d}\tilde{N}, \tag{12.35}$$

whereas (see Eq. (12.32))

$$\mathrm{d}S = \mathrm{d}U/T + (p/T)\,\mathrm{d}V + (\mathcal{A}/T)\,\mathrm{d}\tilde{N}, \tag{12.36}$$

so different variables are held constant when $\tilde{N}$ changes in these expressions.

12.4 Explicit Equilibrium Conditions

Explicit conditions for equilibrium can be obtained by referring the chemical potentials to **standard states** for pure elements or compounds. For the standard state at p_0, T discussed in Section 12.2, one may express a chemical potential in the form

$$\mu_i = \mu_i^0(T,p) + RT\ln a_i = \mu_i^0(T,p_0) + [\mu_i^0(T,p) - \mu_i^0(T,p_0)] + RT\ln a_i, \tag{12.37}$$

where a_i is a dimensionless quantity called the **activity**. In general, $\mu_i(T,p,\boldsymbol{X})$ and $a_i(T,p,\boldsymbol{X})$ depend on temperature, pressure, and composition which we symbolize by the vector $\boldsymbol{X}$. The quantities $\mu_i^0(T,p)$ and $\mu_i^0(T,p_0)$ are the chemical potentials of the pure component i for the most stable phase at the given temperature and pressures.[8] The activity $a_i = 1$ in the standard state $\mu_i^0(T,p)$.[9] Thus, a_i accounts primarily for the dependence of the chemical potential on composition.

For the pure component i, whether solid, liquid, or gas, we define a dimensionless quantity that we call the **fugacity ratio**[10]

$$\tilde{f}_i(T,p,p_0) := \exp\{[\mu_i^0(T,p) - \mu_i^0(T,p_0)]/RT\}. \tag{12.38}$$

Thus Eq. (12.37) can be written in the form

$$\mu_i(T,p,\boldsymbol{X}) = \mu_i^0(T,p_0) + RT\ln[\tilde{f}_i(T,p,p_0)a_i(T,p,\boldsymbol{X})]. \tag{12.39}$$

In general,

$$\left(\frac{\partial \mu_i^0(T,p)}{\partial p}\right)_T = V_i^0(T,p), \tag{12.40}$$

[8]There could possibly be a phase change between p_0 and p in which case the chemical potential will be a continuous function of pressure but its pressure derivative will be discontinuous at the pressure at which the phase change takes place.

[9]Some authors, such as Kondepudi and Prigogine [14, p. 235], refer the activity to the state $\mu^0(T,p_0)$. Here we follow Lupis [5, p. 108] and refer the activity to the state $\mu^0(T,p)$. One can also employ standard states in which some constituents are in solution.

[10]Note that Eq. (12.38) does not define a fugacity itself. We take this approach because the standard state for the fugacity of a gas is defined to be a state for which the pressure goes to zero; however, for a condensed phase (solid or liquid) it is a state at pressure p_0. In terms of individual fugacities f_i, defined in Section 5.4, one would have $\tilde{f}_i(T,p,p_0) = f_i(T,p)/f_i(T,p_0)$.

where $V_i^0(T,p)$ is the molar volume in the standard state. Thus

$$\mu_i^0(T,p) - \mu_i^0(T,p_0) = \int_{p_0}^{p} V_i^0(T,p')\,\mathrm{d}p'. \tag{12.41}$$

For condensed phases (solids and liquids), usually $pV_i^0/RT \ll 1$ in the range of integration so $\tilde{f}_i(T,p,p_0) \approx 1$ and the dependence on pressure is unimportant. For an ideal gas, one has $V_i^0(T,p)/RT = 1/p$ so

$$\tilde{f}_i(T,p,p_0) = p/p_0 \tag{12.42}$$

and there is considerable dependence on pressure. See Section 5.4 for a more complete discussion of fugacities for real gases and condensed phases. Unless one is dealing with very large pressure differences $|p - p_0|$, the quantity $|\mu_i^0(T,p) - \mu_i^0(T,p_0)|/RT$ is small and usually negligible for solids and liquids but is important and varies considerably with pressure for gases. For condensed phases, usually $\tilde{f}_i a_i \approx a_i$ and the dependence on pressure is unimportant. If these condensed phases are not in solution, $a_i = 1$ so $\tilde{f}_i a_i \approx 1$. For an ideal solution of ideal gases, $a_i = X_i$, the mole fraction, so $\tilde{f}_i a_i = X_i p/p_0 = p_i/p_0$, where $p_i := X_i p$ is the partial pressure of gas i.

Substitution of Eq. (12.39) into the equilibrium condition Eq. (12.29) gives

$$\Delta G = \sum_i \nu_i \mu_i^0(T,p_0) + RT \sum_i \nu_i \ln[\tilde{f}_i(T,p,p_0) a_i(T,p,\boldsymbol{X})] = 0. \tag{12.43}$$

The first term

$$\Delta G^0(T,p_0) \equiv \sum_i \nu_i \mu_i^0(T,p_0) \equiv -RT \ln K(T,p_0) \tag{12.44}$$

refers to the standard states at pressure p_0 and the dimensionless quantity $K(T,p_0)$ is called the **equilibrium constant**.[11] The second term in Eq. (12.43) can be rewritten in the form

$$RT \sum_i \nu_i \ln[\tilde{f}_i(T,p,p_0) a_i(T,p,\boldsymbol{X})] = RT \ln \prod_i [\tilde{f}_i(T,p,p_0) a_i(T,p,\boldsymbol{X})]^{\nu_i}. \tag{12.45}$$

The condition Eq. (12.43) for equilibrium becomes

$$\prod_i [\tilde{f}_i(T,p,p_0) a_i(T,p,\boldsymbol{X})]^{\nu_i} = K(T,p_0). \tag{12.46}$$

The quantity on the left-hand side, often called the **reaction product**, can also be written[12]

$$\prod_i [\tilde{f}_i(T,p,p_0) a_i(T,p,\boldsymbol{X})]^{\nu_i} = \frac{\prod_i^{\text{products}} [\tilde{f}_i(T,p,p_0) a_i(T,p,\boldsymbol{X})]^{\nu_i}}{\prod_i^{\text{reactants}} [\tilde{f}_i(T,p,p_0) a_i(T,p,\boldsymbol{X})]^{|\nu_i|}} \tag{12.47}$$

[11]Most authors would write just $K(T)$ instead of $K(T,p_0)$ because p_0 is fixed at one atmosphere. We carry the extra symbol p_0 to remind ourselves of the standard state that has been used. Our equilibrium constant $K(T,p_0)$ is dimensionless.

[12]Here, a product is on the right-hand side of the chemical equation and has a positive ν_i whereas a reactant is on the left-hand side and has a negative ν_i.

and is sometimes called the **reaction quotient**.

A parallel development can be made in terms of an equilibrium constant that depends on p instead of p_0. In that case, the equilibrium condition Eq. (12.43) is replaced by

$$\Delta G = \sum_i \nu_i \mu_i^0(T,p) + RT \sum_i \nu_i \ln a_i = 0. \tag{12.48}$$

Then one can define

$$\Delta G^0(T,p) \equiv \sum_i \nu_i \mu_i^0(T,p) \equiv -RT \ln K(T,p) \tag{12.49}$$

and the condition for equilibrium becomes[13]

$$\prod_i a_i^{\nu_i} = K(T,p). \tag{12.50}$$

12.4.1 Reactions among Gases

The molecules of ideal gases do not react chemically; however, if all reactants and products are gases whose fugacities and activities can be approximated as if they were ideal gases, we have $\tilde{f}_i a_i \approx p_i/p_0$, where p_i is the partial pressure of gas i, and the equilibrium condition Eq. (12.46) becomes

$$\prod_i (p_i/p_0)^{\nu_i} = \frac{\prod_i^{\text{products}} (p_i/p_0)^{\nu_i}}{\prod_i^{\text{reactants}} (p_i/p_0)^{|\nu_i|}} = K(T,p_0). \tag{12.51}$$

In terms of mole fractions X_i, which are a measure of composition, Eq. (12.51) becomes

$$\prod_i X_i^{\nu_i} = \frac{\prod_i^{\text{products}} X_i^{\nu_i}}{\prod_i^{\text{reactants}} X_i^{|\nu_i|}} = \left(\frac{p}{p_0}\right)^{-\sum_i \nu_i} K(T,p_0), \tag{12.52}$$

which exhibits the role of overall pressure on the reaction. This result also follows from Eq. (12.50) by substitution of $a_i = X_i$. We see in this case that

$$K(T,p) = \left(\frac{p}{p_0}\right)^{-\sum_i \nu_i} K(T,p_0), \quad \text{ideal gases.} \tag{12.53}$$

For a given reaction, it follows that an increase in the overall pressure p will favor the reaction if $-\sum_i \nu_i > 0$, meaning that the number of moles of reactant gases exceeds the number of moles of product gases. If $-\sum_i \nu_i < 0$, an increase of pressure will favor the reverse reaction. The reaction will be independent of pressure if $\sum_i \nu_i = 0$.

[13]This equation illustrates clearly that the equilibrium condition does not depend on the value of p_0. Other conditions that appear to contain p_0 are also independent of p_0 but their individual parts depend on p_0 because data are tabulated at that pressure.

Example Problem 12.3. Discuss the dependence on overall pressure of the reactions H_2(gas) + CO_2(gas) → CO(gas) + H_2O(gas), CO(gas) + (1/2)O_2(gas) → CO_2(gas) and A_2(gas) → 2A(gas).

Solution 12.3. For the first reaction, $\nu_{H_2} = -1$, $\nu_{CO_2} = -1$, $\nu_{CO} = 1$, and $\nu_{H_2O} = 1$, so $-\sum_i \nu_i = 0$ and that reaction is independent of pressure. For the second reaction, $\nu_{CO} = -1$, $\nu_{O_2} = -1/2$, and $\nu_{CO_2} = 1$, so $-\sum_i \nu_i = 1/2$ and that reaction is favored by an increase in pressure because $K(T,p) = (p/p_0)^{1/2} K(T,p_0)$. For the third, $\nu_{A_2} = -1$ and $\nu_A = 2$ so $K(T,p) = (p/p_0)^{-1} K(T,p_0)$. In this last case, we see that the dissociation of argon is impeded by a high pressure, which can be thought of heuristically as a force tending to hold the argon molecule together.

Example Problem 12.4. For the reaction H_2(gas) + CO_2(gas) → CO(gas) + H_2O(gas), some values of the equilibrium constant are $K(1130\,\text{K}, 1\,\text{atm}) = 1.0$ and $K(1500\,\text{K}, 1\,\text{atm}) = 2.16$. Suppose initially that there is one mole of each gas. What will be the composition at equilibrium at 1130 K and 1500 K?

Solution 12.4. After the reaction has progressed by an amount $\tilde{N}$, the numbers of moles of each component will be $N_{H_2} = 1 - \tilde{N}$, $N_{CO_2} = 1 - \tilde{N}$, $N_{CO} = 1 + \tilde{N}$, and $N_{H_2O} = 1 + \tilde{N}$ so Eq. (12.52) becomes

$$\frac{[(1+\tilde{N})/4]^2}{[(1-\tilde{N})/4]^2} = K(T,p_0). \tag{12.54}$$

For $K(1130\,\text{K}, 1\,\text{atm}) = 1.0$, the solution is $\tilde{N} = 0$; the mole fraction of each gas is 0.25 and the reaction was in equilibrium initially. For $K(1500\,\text{K}, 1\,\text{atm}) = 2.16$, the solutions to Eq. (12.54) are $\tilde{N} = 0.19$ and $\tilde{N} = 5.26$ but the latter is unacceptable because it would lead to a negative value of $1 - \tilde{N}$, which would correspond to a negative value of N_{H_2}. Therefore, the composition at equilibrium is $X_{H_2} = X_{CO_2} = 0.20$ and $X_{CO} = X_{H_2O} = 0.30$.

In terms of concentrations $[i] \equiv N_i/V$, one has $p_i = [i]RT$ and Eq. (12.51) becomes

$$\prod_i [i]^{\nu_i} = \frac{\prod_i^{\text{products}} [i]^{\nu_i}}{\prod_i^{\text{reactants}} [i]^{|\nu_i|}} = \left(\frac{RT}{p_0}\right)^{-\sum_i \nu_i} K(T,p_0) \equiv K_c(T,p_0), \tag{12.55}$$

where $K_c(T,p_0)$ is a new equilibrium constant that will not be dimensionless unless $\sum_i \nu_i = 0$.

Any of these forms for ideal gases, especially Eq. (12.55), are referred to as the **law of mass action**. This follows because the rate of gaseous reactions depends on collisions and the collision rate would be expected to depend on a product of concentrations. The rate of forward reaction would be

$$R_f = k_f(T,p_0) \prod_i^{\text{reactants}} [i]^{|\nu_i|}, \tag{12.56}$$

where $k_f(T, p_0)$ is a constant of proportionality. Similarly, the rate of backward reaction would be

$$R_b = k_b(T, p_0) \prod_i^{\text{products}} [i]^{\nu_i}. \tag{12.57}$$

At equilibrium, $R_f = R_b$ leads to Eq. (12.55) with $K_c(T, p_0) = k_f(T, p_0)/k_b(T, p_0)$. This relationship of thermodynamics to kinetics holds provided that the given reaction actually proceeds by an *elementary step* involving the collisions embodied in Eqs. (12.56) and (12.57). On the other hand, if the reaction actually takes place by means of a combination of elementary steps, $K_c(T, p_0)$ can be related to the rate constants of all of these elementary steps. Nevertheless, the value of $K_c(T, p_0)$, since it is a thermodynamic quantity, is independent of the details of the kinetics of the reaction. See Kondepudi and Prigogine [14, p. 241] for further discussion of this point in terms of the principle of detailed balance.

12.4.2 Heterogeneous Solids and Liquids with Gases

Heterogeneous reactions constitute an important special case in which gases react with immiscible solids and liquids. In this case, the activities of the liquids and solids are equal to one. Moreover, as stated in connection with Eq. (12.41), we can neglect the dependence of the chemical potentials of solids and liquids on the overall pressure p. Furthermore, if the gases can be treated as ideal, one arrives at an equation similar to Eq. (12.52) except the reaction product is only over the gases, that is,

$$\prod_i^{\text{gases}} X_i^{\nu_i} = \left(\frac{p}{p_0}\right)^{-\sum_i \nu_i} K(T, p_0). \tag{12.58}$$

This can also be written in terms of partial pressures in a form similar to Eq. (12.52), namely

$$\prod_i^{\text{gases}} (p_i/p_0)^{\nu_i} = K(T, p_0). \tag{12.59}$$

Example Problem 12.5. For the reaction C(graphite) + O_2(gas) → CO_2(gas) one has $\Delta G^0 = -394.4$ kJ/mol, practically independent of temperature. The gas constant $R = 8.314$ J/mol. What is the equilibrium constant $K(T, p_0)$ of this reaction? What is the composition of the gas at equilibrium? What is the fraction of O_2 at 1000 K?

Solution 12.5. We have

$$\frac{X_{CO_2}}{X_{O_2}} = \frac{p_{CO_2}}{p_{O_2}} = K(T, p_0) = \exp(-\Delta G^0/RT) \tag{12.60}$$

with $\Delta G^0/R = -47{,}440$ K. At 1000 K we would have

$$\frac{X_{O_2}}{1 - X_{O_2}} = \exp(-47.44) = 2.5 \times 10^{-21}, \tag{12.61}$$

so this is also practically the value of X_{O_2}. If only graphite and O_2 were present initially, practically all of the oxygen would react to form CO_2 if enough graphite were present. Otherwise, the reaction would stop when all of the graphite is consumed.

12.4.3 Dependence of $K(T, p_0)$ on Temperature

We begin with Eq. (10.30), generalized to a multicomponent system, namely

$$\left(\frac{\partial(G/T)}{\partial(1/T)}\right)_{p,N_i} = H \tag{12.62}$$

in which the Gibbs free energy G and the enthalpy H are expressed in terms of the variable set T, p, N_i. This equation also holds for any chemical component in its standard state, and therefore holds if G is replaced by the sum

$$\Delta G^0(T, p_0) \equiv \sum_i \nu_i \mu_i^0(T, p_0) \tag{12.63}$$

and H is replaced by the sum

$$\Delta H^0(T, p_0) \equiv \sum_i \nu_i H_i^0(T, p_0). \tag{12.64}$$

We therefore obtain

$$\frac{\partial(\Delta G^0(T, p_0)/T)}{\partial(1/T)} = \Delta H^0(T, p_0). \tag{12.65}$$

Since $\partial/\partial(1/T) = -T^2\partial/\partial T$, Eq. (12.65) can also be written

$$\frac{\partial(\Delta G^0(T, p_0)/T)}{\partial T} = -\frac{\Delta H^0(T, p_0)}{T^2}. \tag{12.66}$$

Recalling the definition of $K(T, p_0)$ from Eq. (12.44), we obtain

$$\frac{\partial \ln K(T, p_0)}{\partial T} = \frac{\Delta H^0(T, p_0)}{RT^2}, \tag{12.67}$$

which is known as the **van't Hoff equation.**

Example Problem 12.6. For many chemical reactions, the quantity $\Delta H^0(T, p_0)$ does not depend strongly on T over a significant temperature range and may be treated as a constant, say ΔH_0^0. Determine the dependence of $K(T, p_0)$ on T under these circumstances. Discuss the dependence on temperature for endothermic and exothermic reactions. What is the dependence on temperature of $\Delta G^0(T, p_0)$ in this case?

Solution 12.6. We integrate Eq. (12.67) from T_0 to T to obtain

$$\ln K(T, p_0) = \ln K(T_0, p_0) - \frac{\Delta H_0^0}{R}\left(\frac{1}{T} - \frac{1}{T_0}\right). \tag{12.68}$$

Exponentiation gives

$$K(T,p_0) = K(T_0,p_0)\mathrm{e}^{\Delta H_0^0/RT_0}\,\mathrm{e}^{-\Delta H_0^0/RT}, \tag{12.69}$$

so K increases strongly with temperature for an endothermic reaction $\Delta H_0^0 > 0$ and decreases strongly with temperature for an exothermic reaction $\Delta H_0^0 < 0$. This type of exponential dependence is said to be of **Arrhenius form** and ΔH_0^0 plays the role of an activation energy.

Integration of Eq. (12.65) for constant ΔH^0 gives

$$\frac{\Delta G^0(T,p_0)}{T} = \frac{\Delta G^0(T_0,p_0)}{T_0} + \Delta H_0^0\left(\frac{1}{T} - \frac{1}{T_0}\right). \tag{12.70}$$

Multiplication by T and rearrangement gives

$$\Delta G^0(T,p_0) = \Delta H_0^0 + \frac{T}{T_0}[\Delta G^0(T_0,p_0) - \Delta H_0^0(T_0,p_0)] = \Delta H_0^0 - T\Delta S^0(T_0,p_0). \tag{12.71}$$

We see in this case that $\Delta G^0(T,p_0)$ is linear in T. In general, $\partial\Delta G^0(T,p_0)/\partial T = \Delta S^0(T,p_0)$ so we see from differentiation of Eq. (12.71) with respect to T that $S^0(T,p_0) = S^0(T_0,p_0)$. In other words, the standard entropy difference is also independent of T in this case.

Example Problem 12.7. For the formation of Cu_2O, $\Delta H_f^0(T_0,p_0) = -168.6\,\mathrm{kJ/mol}$ and $\Delta G_f^0(T_0,p_0) = -146.0\,\mathrm{kJ/mol}$. For the formation of Al_2O_3, $\Delta H_f^0(T_0,p_0) = -1675.7\,\mathrm{kJ/mol}$ and $\Delta G_f^0(T_0,p_0) = -1582.3\,\mathrm{kJ/mol}$. In both cases, the metal and its oxide are solids, not in solution, and the oxygen can be treated as an ideal gas. Write chemical equations for the two reactions. Assume that $\Delta H^0(T,p_0)$ can be treated as a constant equal to $\Delta H_f^0(T_0,p_0)$. Determine the equilibrium constants $K(T,p_0)$ as a function of temperature and then determine the equilibrium pressures of oxygen for each reaction at 1000 K.

Solution 12.7. The relevant reactions are $2\mathrm{Cu(solid)} + (1/2)\mathrm{O_2(gas)} \rightarrow \mathrm{Cu_2O(solid)}$ and $2\mathrm{Al(solid)} + (3/2)\mathrm{O_2(gas)} \rightarrow \mathrm{Al_2O_3(solid)}$. From Eq. (12.71),

$$K(T,p_0) = \exp[(\Delta H_f^0(T_0,p_0) - \Delta G_f^0(T_0,p_0))/RT_0]\exp(-\Delta H_f^0/RT). \tag{12.72}$$

For the oxidation of copper,

$$K(T,p_0) = 1.098\times 10^{-4}\exp(20{,}280\,\mathrm{K}/T), \tag{12.73}$$

so $K(1000\,\mathrm{K},p_0) = 2.018\times 10^{10}$. From Eq. (12.59), we see that $(p_{O_2}/p_0)^{-1/2} = K$, so we obtain $p_{O_2} = 2002\times 10^{-10}$ atm.

For the oxidation of aluminum,

$$K(T,p_0) = 4.326\times 10^{-17}\exp(201{,}550\,\mathrm{K}/T), \tag{12.74}$$

so $K(1000\,\mathrm{K},p_0) = 1.475\times 10^{71}$. For this reaction, $(p_{O_2}/p_0)^{-3/2} = K$, so we obtain $p_{O_2} = 3.58\times 10^{-48}$ atm.

Both of these oxygen pressures are very small but their relative values indicate that aluminum oxide is much more stable than copper oxide; an extremely low oxygen pressure would be needed to reduce aluminum oxide to the metallic state.

12.4.4 Dependence of $K(T,p)$ on Pressure

Since $\partial G/\partial p = V$ we have $\partial \mu_i^0(T,p)/\partial p = V_i^0(T,p)$, which is the molar volume of each constituent in its standard state.[14] Therefore,

$$\frac{\partial \Delta G^0(T,p)}{\partial p} = \sum_i \nu_i \frac{\partial \mu_i^0(T,p)}{\partial p} = \sum_i \nu_i V_i^0(T,p) \equiv \Delta V^0(T,p). \tag{12.75}$$

Thus,

$$\frac{\partial \ln K(T,p)}{\partial p} = -\frac{\Delta V^0(T,p)}{RT} = -\sum_i \nu_i \frac{V_i^0(T,p)}{RT}. \tag{12.76}$$

As remarked previously, $V_i^0(T,p)/RT = 1/p$ for ideal gases and $V_i^0(T,p)/RT \ll 1/p$ for condensed phases, so the only really important correction is for gases. If all gases can be treated as ideal,

$$\frac{\partial \ln K(T,p)}{\partial p} = -\sum_i^{\text{gases}} \frac{\nu_i}{p}, \tag{12.77}$$

which can be integrated to give

$$\ln \frac{K(T,p)}{K(T,p_0)} = -\sum_i^{\text{gases}} \nu_i \ln \frac{p}{p_0} = \ln \left[\left(\frac{p}{p_0} \right)^{-\sum_i^{\text{gases}} \nu_i} \right]. \tag{12.78}$$

Exponentiation of this expression gives agreement with Eq. (12.53). If better information is available for non-ideal gases, one could integrate Eq. (12.76).

12.5 Simultaneous Reactions

As mentioned in connection with Eq. (12.2), it is possible to have simultaneous reactions. In that case, there is a progress variable $\tilde{N}^s$ for each reaction and we will have

$$\mathrm{d}^{\text{int}} N_i = \sum_s \nu_i^s \, \mathrm{d}\tilde{N}^s. \tag{12.79}$$

Then for a chemically closed system,

$$\mathrm{d}G_{T,p} = \sum_i \mu_i \, \mathrm{d}^{\text{int}} N_i = \sum_i \mu_i \sum_s \nu_i^s \, \mathrm{d}\tilde{N}^s = -\sum_s \mathcal{A}^s \, \mathrm{d}\tilde{N}^s \le 0, \tag{12.80}$$

where $\mathcal{A}^s = \sum_i \nu_i^s \mu_i$ is the affinity of the reaction s. The corresponding entropy production is

$$\mathrm{d}^{\text{int}} S = \sum_s \frac{\mathcal{A}^s}{T} \, \mathrm{d}\tilde{N}^s \ge 0. \tag{12.81}$$

[14] Here as above, we extend the meaning of standard state to mean the most stable phase of the pure constituent at T and p. A stricter definition (see Lupis [5, p. 120]) that restricts the standard state to T and p_0 for chemical reactions would lead to $\partial \Delta G^0(T,p_0)/\partial p = 0$, in which case $\partial \Delta K(T,p_0)/\partial p = 0$.

Since these chemical reactions can take place at the same spatial position, they may be coupled and it is only necessary that the sum be positive during the reaction. See Lupis [5, p. 122] for examples of uncoupled simultaneous reactions and Kondepudi and Prigogine [14, p. 369] for a discussion of coupled simultaneous reactions in the context of entropy production.

13

Thermodynamics of Fluid-Fluid Interfaces

Until now we have dealt primarily with homogeneous phases or with composite systems consisting of several homogeneous phases. In all of these cases, we have ignored the thermodynamics of the surfaces of these phases or the interfaces that separate them. This was done on the basis that we were interested in the bulk properties of sufficiently large phases that the contributions of surfaces and interfaces could be neglected. Nevertheless, there are many familiar instances where the properties of surfaces cannot be ignored. For example, a glass of water in a gravitational field can be filled somewhat beyond its capacity without spilling; the surface of the water bulges above the top of the glass but is held in place by a force due to "surface tension" that supports the water above the glass. Roughly speaking, the surface of the water has a free energy per unit area in excess of the free energy of the bulk. Thus, the creation of more area requires an increase in free energy and hence to a force per unit length around the perimeter of the area that tries to keep the area from increasing. Another example is the substantial rise (typically a few centimeters) of water in a capillary tube immersed vertically in a large vessel.

In this chapter, we explore in more detail the thermodynamic properties of the thin transition regions that exit in actual systems near these idealized surfaces of discontinuity. We do this under conditions for which the change from one homogeneous phase to another takes place over a region that is thin compared to the extent of the homogeneous phases.[1] We begin by considering a model developed by Gibbs [3, p. 223] that is based on the concept of a dividing surface. Such a surface has zero thickness, so it is a mathematical abstraction. By means of a clever formalism, Gibbs was able to account for the thermodynamic properties of the actual transition region by associating it with the dividing surface. We first consider the Gibbs dividing surface model of planar interfaces in fluid systems, for which the surface tension can be defined unambiguously. Then for planar interfaces, we present Cahn's layer model which allows one to express physically meaningful surface quantities in terms of determinants whose properties illustrate clearly their invariance with respect to the thickness or location of the layer that contains the region of discontinuity. The Gibbs model is seen to be a special case of Cahn's model.

Next, we discuss curved interfaces for fluids, for which the location of the dividing surface must be fixed by some convention, in particular the Gibbs "surface of tension" that we define later. We illustrate surface tension phenomena by examples such as rise or

[1] For curved surfaces, the region of discontinuity must also be thin relative to its radii of curvature.

Thermal Physics. http://dx.doi.org/10.1016/B978-0-12-803304-3.00013-2

depression in a capillary tube, the meniscus that forms at the edge of a submerged plate, a variety of interface shapes for two-dimensional problems, and the shapes of pendant and sessile drops in three dimensions.

13.1 Planar Interfaces in Fluids

We begin by considering a planar interface, such as depicted in Figure 13–1, that separates two essentially homogeneous fluid phases that we denote by superscripts α and β.

The transition region between the phases is assumed to be thin compared to the extent of the phases themselves. The entire system, which is chemically closed, has an internal energy U, entropy S, volume V, and mole numbers N_i of its chemical components. It is assumed to be in equilibrium and to have a temperature T and chemical potentials μ_i that are uniform throughout. Gibbs discusses the need for this uniformity in great detail by imagining the system to be divided into three subsystems by means of imaginary parallel walls that are similarly situated with respect to the transition region. These walls are on opposite sides of a layer L that contains the transition region; they are assumed to be near that region but sufficiently far from it that they are in practically homogeneous regions. They separate the actual thin layer containing the transition region from two homogeneous phases, α_H and β_H. For *immobile walls*, Gibbs assumes that an infinitesimal variation of the energy of the layer L is given by

$$\delta U^L = T^L \delta S^L + \sum_i \mu_i^L \delta N_i^L, \tag{13.1}$$

where S^L is the entropy of the layer and N_i^L is the number of moles of component i in the layer. Gibbs *defines* T^L to be its temperature and the μ_i^L to be its chemical potentials.[2] He then proceeds to show that they must be equal to the temperature and chemical potentials of the bulk phases.

For immobile walls, the energies of the homogeneous subsystems α_H and β_H can vary according to $\delta U^\alpha = T^\alpha \delta S^\alpha + \sum_i \mu_i^\alpha \delta N_i^\alpha$ and $\delta U^\beta = T^\beta \delta S^\beta + \sum_i \mu_i^\beta \delta N_i^\beta$. Then by studying

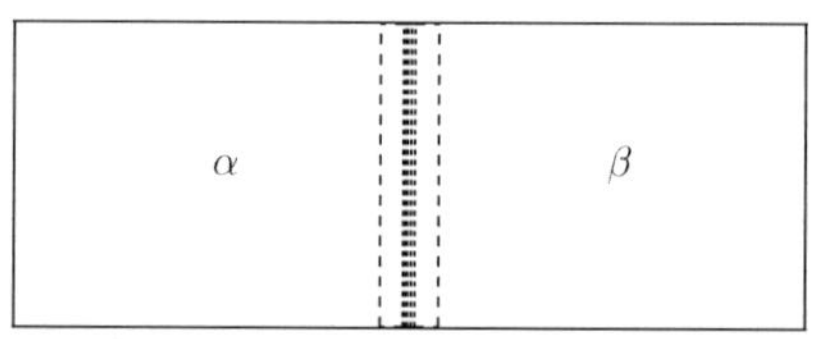

FIGURE 13–1 Schematic diagram showing a planar interface, which is a region of discontinuity (located near the innermost dotted lines) between bulk α and β phases. The dashed lines, which are located near the region of discontinuity but practically in homogeneous regions, are imaginary walls that define a layer that contains the region of discontinuity. This is the same layer L used by Gibbs to define T^L and μ_i^L in Eq. (13.1) and used in Cahn's layer model in Section 13.1.3 to establish Eq. (13.11). The Gibbs dividing surface is any plane parallel to the walls of the layer and can be inside or outside the layer.

[2]Gibbs deals with the masses of the components rather than the number of moles, so his chemical potentials are per unit mass and differ from ours by factors of the molecular weights.

special variations of entropy and mole numbers of each component among the layer L and the homogeneous subsystems α_H and β_H and requiring the total energy $U = U^\alpha + U^L + U^\beta$ to be a minimum at constant total entropy and total mole numbers, Gibbs reasons for variations that can have either sign that the temperature and chemical potentials must be uniform, just as they would be for three bulk systems in heterogeneous equilibrium.

For example, for a special variation in which component j is exchanged between α and L but there is no other change, one has $\delta U = \mu_j^\alpha \delta N_j^\alpha + \mu_j^L \delta N_j^L = (\mu_j^\alpha - \mu_j^L)\delta N_j^\alpha$. Then for variations δN_j^α of either sign, one must have $\mu_j^\alpha - \mu_j^L = 0$ to prevent δU from being negative, which would violate the fact that U must be a minimum at equilibrium. Gibbs also deals carefully with the manner in which extensive quantities can be defined during such variations, which involve infinitesimal discontinuities at the walls [3, p. 224]. Ultimately, $T^\alpha = T^L = T^\beta$ and $\mu_i^\alpha = \mu_i^L = \mu_i^\beta$ for each component i.

The pressures in the bulk phases are also uniform and equal to one another. This must be true for mechanical equilibrium and can be established by means of a variation in which the *entire* layer L is *translated* by an infinitesimal distance in a direction perpendicular to its walls without any change in the layer L itself. This gives rise to a change in volume δV of one homogeneous system and $-\delta V$ of the other, just as if the layer L were absent. Thus the variation of the internal energy of the whole system will be $\delta U = \delta V(p^\alpha - p^\beta)$, so $(p^\alpha - p^\beta)$ must vanish for arbitrary δV of either sign. Therefore, the pressures of the bulk systems must be equal (we shall hereafter denote them both by p) as they would be for bulk systems in heterogeneous equilibrium.[3]

13.1.1 Gibbs Dividing Surface Model

Following Gibbs, we replace the actual system by a model system consisting of two strictly homogeneous phases separated by a single mathematical plane, known as the **Gibbs dividing surface**. In this model system, the homogeneous phases extend uniformly *until they meet* at the dividing surface. This plane is similarly situated with respect to the transition region. For the moment, we assume that it is located *anywhere* in the system, not necessarily in the transition layer, and discuss later the implications of its actual location. One then defines **surface excess quantities** by subtracting the extensive properties of the homogeneous parts of the model system from the corresponding actual parts:

$$U^{\text{xs}} := U - U^\alpha - U^\beta; \tag{13.2}$$

$$S^{\text{xs}} := S - S^\alpha - S^\beta; \tag{13.3}$$

$$N_i^{\text{xs}} := N_i - N_i^\alpha - N_i^\beta; \tag{13.4}$$

$$0 := V - V^\alpha - V^\beta. \tag{13.5}$$

[3]Note that this treatment avoids discussion of the pressure of the subsystem L. In fact, that subsystem is inhomogeneous, so on a microscopic scale it could be characterized by a pressure tensor p_{ij}. If the z direction is perpendicular to the walls of the layer, then p_{zz} must be uniform and equal to the common pressure p of the homogeneous phases. The components $p_{xx} = p_{yy}$ will vary from p near the homogenous phases to negative values within the discontinuity itself, giving rise to a surface tension $\sigma = \int (p - p_{xx})\mathrm{d}z > 0$, where the integration includes the region of discontinuity. See [27, p. 44] for a derivation.

Equation (13.5) is different from the previous three equations because there is no excess volume, due to the fact that the homogeneous phases of the model system meet at the dividing surface, which has no thickness. Since the temperature is uniform, one can also define excesses of the thermodynamic potentials, such as the Helmholtz free energy $F = U - TS$, for which

$$F^{\text{xs}} := F - F^{\alpha} - F^{\beta}. \tag{13.6}$$

It follows that all excess quantities follow the same algebra as their bulk counterparts.

Although these excess quantities can be defined, they usually do *not* have physical significance because they depend on the location of the dividing surface. This is easily illustrated for the case of a single component material in which one bulk phase is a liquid having molar density n^{ℓ} and the other is a gas having molar density n^{g}. Then if the dividing surface is located such that the gas has volume V^{g} and the liquid has volume $V - V^{g}$, where V is the total volume, it follows that

$$N^{\text{xs}} = N - n^{\ell}V + (n^{\ell} - n^{g})V^{g}. \tag{13.7}$$

For $n^{\ell} - n^{g} > 0$, the sum of the first two terms on the right is negative and independent of the location of the dividing surface whereas the last term on the right is positive and depends linearly on V^{g} and hence linearly on the position of the dividing surface. Thus, N^{xs} varies with the position of the dividing surface and can be positive, negative, or zero. Therefore, N^{xs} has no physical significance. One could fix the position of the dividing surface by convention by choosing its location so that $N^{\text{xs}} = 0$; this is known as the **equimolar surface**. Nevertheless, this choice is still artificial. Moreover, in a multi-component system one could only choose the dividing surface to be equimolar relative to one of the components. Similarly, it follows that U^{xs}, S^{xs}, and F^{xs} depend on the location of the dividing surface.

On the other hand, the excess of the Kramers potential[4] $K = F - \sum_i \mu_i N_i$, namely

$$K^{\text{xs}} := K - K^{\alpha} - K^{\beta} = F^{\text{xs}} - \sum_i \mu_i N_i^{\text{xs}} \tag{13.8}$$

turns out to be independent of the location of the dividing surface. This can be seen by noting for a bulk phase that $F - \sum_i \mu_i N_i = F - G = -pV$, so $K^{\alpha} = -pV^{\alpha}$ and $K^{\beta} = -pV^{\beta}$. Therefore

$$K^{\text{xs}} = K + p(V^{\alpha} + V^{\beta}) = K + pV, \tag{13.9}$$

where Eq. (13.5) has been used. The right-hand side of Eq. (13.9) is *independent* of the location of the dividing surface, so K^{xs} is also independent of that location and has

[4]This is also called the grand potential and is often denoted by Ω.

physical meaning. We can therefore divide by the area A of the dividing surface to define the surface free energy[5] (per unit area of interface)

$$\gamma := \frac{K^{xs}}{A} = \frac{F^{xs} - \sum_i \mu_i N_i^{xs}}{A} = \frac{K + pV}{A}, \tag{13.10}$$

which will be independent of the choice of the location of the dividing surface for a planar region of discontinuity.

We now approach the same problem from a different vantage point by considering small reversible changes of the same planar system in contact with a thermal reservoir at temperature T, a pressure reservoir at pressure p, and chemical reservoirs at potentials μ_i. In particular, we allow the system to undergo an infinitesimal change in which its length is unchanged but its cross-sectional area changes by an amount dA. In order to account for work done "by the surface" we write the reversible work done by the system in the form $\delta W = p\,\mathrm{d}V - \sigma\,\mathrm{d}A$, where $p\,\mathrm{d}V$ is the usual quasistatic work done by the pressure and $\sigma\,\mathrm{d}A$ is the extra work done *on* the system because of the surface of discontinuity. The quantity σ is the surface (interfacial) tension, which is a force per unit length that must be applied by an external agent to extend the surface. Thus

$$\mathrm{d}U = T\,\mathrm{d}S - p\,\mathrm{d}V + \sigma\,\mathrm{d}A + \sum_i \mu_i\,\mathrm{d}N_i. \tag{13.11}$$

We shall proceed to show that $\sigma = \gamma$. Indeed, for the bulk systems we have

$$\mathrm{d}U^\alpha = T\,\mathrm{d}S^\alpha - p\,\mathrm{d}V^\alpha + \sum_i \mu_i\,\mathrm{d}N_i^\alpha; \tag{13.12}$$

$$\mathrm{d}U^\beta = T\,\mathrm{d}S^\beta - p\,\mathrm{d}V^\beta + \sum_i \mu_i\,\mathrm{d}N_i^\beta. \tag{13.13}$$

We subtract both of these equations from Eq. (13.11) to obtain

$$\mathrm{d}U^{xs} = T\,\mathrm{d}S^{xs} + \sum_i \mu_i\,\mathrm{d}N_i^{xs} + \sigma\,\mathrm{d}A. \tag{13.14}$$

Equation (13.14) illustrates that U^{xs} can be regarded as a function of S^{xs}, N_i^{xs}, and A. Moreover, by considering systems that have the same values of T, μ_i, and σ but simply different cross-sectional areas, we deduce that

$$U^{xs}(\lambda S^{xs}, \lambda N_i^{xs}, \lambda A) = \lambda U^{xs}(S^{xs}, N_i^{xs}, A) \tag{13.15}$$

[5]The name surface free energy is commonly used, but it is important to remember that the relevant free energy is the Kramers potential. For the case of planar interfaces that is treated here, the pressure is the same in both bulk phases so one can define a Gibbs free energy $G = U - TS + pV$ and note that $K + pV = G - \sum_i \mu_i N_i$. Then γA can be thought of as an excess Gibbs free energy relative to a homogeneous system. For curved surfaces, the pressures in the bulk phases are not equal, so one must resort to the Kramers potential. For the very special case of a single component material with the dividing surface chosen to be the equimolar surface, one has $\gamma = F^{xs}/A$ which is the surface excess of the Helmholtz free energy. The name surface tension is perfectly applicable for γ for a surface of discontinuity between fluids because it can be shown to be the force per unit length needed to extend the surface. For solids, this would be a misnomer because surface can be created but also stretched elastically.

for any positive λ. Thus by the Euler theorem of homogeneous functions of degree one (see Eq. (5.39)), we deduce that

$$U^{\rm xs} = TS^{\rm xs} + \sum_i \mu_i N_i^{\rm xs} + \sigma A. \tag{13.16}$$

Equation (13.15) can be solved for σ to deduce

$$\sigma = \frac{U^{\rm xs} - TS^{\rm xs} - \sum_i \mu_i N_i^{\rm xs}}{A} = \gamma. \tag{13.17}$$

Equation (13.17) together with Eq. (13.11) show that the reversible work associated with an increase in surface area is just $\gamma\, dA$, where γ is the surface excess of the Kramers potential.

13.1.2 Gibbs Adsorption Equation

Differentiation of Eq. (13.16) with σ replaced by γ gives

$$dU^{\rm xs} = T\, dS^{\rm xs} + S^{\rm xs}\, dT + \sum_i \mu_i\, dN_i^{\rm xs} + \sum_i N_i^{\rm xs}\, d\mu_i + \gamma\, dA + A\, d\gamma. \tag{13.18}$$

Comparison with Eq. (13.14), again with $\sigma = \gamma$, gives

$$A\, d\gamma = -S^{\rm xs}\, dT - \sum_i N_i^{\rm xs}\, d\mu_i. \tag{13.19}$$

We divide Eq. (13.19) by A and denote the excess entropy per unit area by $s_A := S^{\rm xs}/A$ and the excess mole numbers per unit area by $\Gamma_i := N_i^{\rm xs}/A$ to obtain the **Gibbs adsorption equation**

$$d\gamma = -s_A\, dT - \sum_i \Gamma_i\, d\mu_i. \tag{13.20}$$

If we define $u_A = U^{\rm xs}/A$ and combine Eq. (13.20) with the differential of γ from Eq. (13.17), we obtain

$$du_A = T\, ds_A + \sum_i \mu_i\, d\Gamma_i, \tag{13.21}$$

which resembles Eq. (13.1) because the special variation considered there was for fixed A and immobile walls.

Equation (13.20) must be handled with great care because the quantities s_A and Γ_i depend on the location of the dividing surface and are therefore not of physical significance. For example, one might be tempted to try to calculate s_A as a derivative $(\partial\gamma/\partial T)_{\mu_i}$ but such a derivative does not exist in this case of planar surfaces. The reason is that the variable set $T, \{\mu_i\}$ is not independent because the bulk phases are each governed by Gibbs-Duhem equations

$$S^\alpha\, dT - V^\alpha\, dp + \sum_i N_i^\alpha\, d\mu_i = 0; \tag{13.22}$$

$$S^\beta\, dT - V^\beta\, dp + \sum_i N_i^\beta\, d\mu_i = 0. \tag{13.23}$$

Introducing the entropy density s_V and the concentrations $c_i = N_i/V$ for bulk α and β and then eliminating $\mathrm{d}p$ gives

$$(s_V^\alpha - s_V^\beta)\,\mathrm{d}T + \sum_{i=1}^{\kappa}(c_i^\alpha - c_i^\beta)\,\mathrm{d}\mu_i = 0. \tag{13.24}$$

Therefore, only κ of the $\kappa + 1$ variables $T, \{\mu_i\}$ are independent. Elimination of one of these variables enables $\mathrm{d}\gamma$ to be expressed in terms of independent variables, and then the corresponding derivatives have physical meaning.

Example Problem 13.1. For the case of a single component, evaluate $\mathrm{d}\gamma/\mathrm{d}T$ and interpret the result physically.

Solution 13.1. Equation (13.20) becomes

$$\mathrm{d}\gamma = -s_A\,\mathrm{d}T - \Gamma^{\mathrm{xs}}\,\mathrm{d}\mu \tag{13.25}$$

and Eq. (13.24) yields

$$\mathrm{d}\mu = -\frac{(s_V^\alpha - s_V^\beta)}{(c^\alpha - c^\beta)}\,\mathrm{d}T. \tag{13.26}$$

Elimination of $\mathrm{d}\mu$ from Eq. (13.25) gives

$$\mathrm{d}\gamma = -\left[s_A - \Gamma^{\mathrm{xs}}\frac{s_V^\alpha - s_V^\beta}{c^\alpha - c^\beta}\right]\mathrm{d}T. \tag{13.27}$$

The required derivative $\mathrm{d}\gamma/\mathrm{d}T$ is the negative of the expression in square brackets. The bulk phases are only in equilibrium along a coexistence curve, say $p = \tilde{f}(T)$, which is a solution of $\mu^\alpha(T,p) = \mu^\beta(T,p)$, so μ depends only on T. The quantity in brackets is an effective surface entropy that governs the dependence of γ on T. It is therefore independent of the choice of the dividing surface. If one adopts the convention of the equimolar surface, then $\Gamma^{\mathrm{xs}} = 0$. In that case, the effective surface entropy reduces to s_A, the equimolar surface entropy.

It is now obvious that we could have used Eq. (13.26) to eliminate $\mathrm{d}T$ instead of $\mathrm{d}\mu$ in Eq. (13.25). In that case

$$\mathrm{d}\gamma = -\left[\Gamma^{\mathrm{xs}} - s_A\frac{c^\alpha - c^\beta}{s_V^\alpha - s_V^\beta}\right]\mathrm{d}\mu. \tag{13.28}$$

Now μ is the only independent variable and the quantity in brackets is an effective surface adsorption, which has physical significance independent of the choice of dividing surface.

In the general case, we can solve Eq. (13.24) for $\mathrm{d}\mu_1$ and then substitute into Eq. (13.20) to obtain

$$\mathrm{d}\gamma = -\left[s_A - \Gamma_1\frac{s_V^\alpha - s_V^\beta}{c_1^\alpha - c_1^\beta}\right]\mathrm{d}T - \sum_{i=2}^{\kappa}\left[\Gamma_i - \Gamma_1\frac{c_i^\alpha - c_i^\beta}{c_1^\alpha - c_1^\beta}\right]\mathrm{d}\mu_i. \tag{13.29}$$

Now the variables $T, \mu_2, \mu_3, \ldots, \mu_\kappa$ are independent, so one may take partial derivatives of γ with respect to them and obtain the quantities in square brackets, which must be independent of the location of the dividing surface.[6]

Gibbs [3, p. 234] discusses Eq. (13.20) by locating the dividing surface to be the equimolar surface for component 1, so that $\Gamma_1 = 0$. He then writes

$$\mathrm{d}\gamma = -s_{A(1)}\,\mathrm{d}T - \sum_{i=2}^{\kappa} \Gamma_{i(1)}\,\mathrm{d}\mu_i, \tag{13.30}$$

where the extra subscript reminds us of that choice. By comparison of partial derivatives of Eqs. (13.29) and (13.30), it is evident for *any* dividing surface that

$$s_{A(1)} = \left[s_A - \Gamma_1 \frac{s_V^\alpha - s_V^\beta}{c_1^\alpha - c_1^\beta} \right]; \tag{13.31}$$

$$\Gamma_{i(1)} = \left[\Gamma_i - \Gamma_1 \frac{c_i^\alpha - c_i^\beta}{c_1^\alpha - c_1^\beta} \right]. \tag{13.32}$$

It should be clear at this stage that other choices of the set of κ independent variables are possible. This freedom of choice is obvious from the generalization developed in the next section.

13.1.3 Cahn's Layer Model

For planar interfaces, Cahn [28] or [29, pp. 379-399] developed a layer model that treats an interfacial region of finite thickness and for which physically meaningful surface quantities can be represented by determinants which are manifestly invariant with respect to the thickness and location of the layer. The layer in Cahn's theory can be taken to be the layer we called L in Section 13.1 and employed by Gibbs (see Figure 13–1). It is only necessary that the planes that bound the layer be sufficiently far from the transition region that they lie in regions that are essentially homogeneous. Outside the layer, one has homogeneous phases α_H and β_H as in Section 13.1. These homogeneous phases are characterized by the same uniform intensive variables T, p, μ_i as in the Gibbs theory but their amounts can differ because they do not occupy the entire volume.[7] Then one can define the content of the various extensive quantities in the layer by

$$U^L := U - U^{\alpha_H} - U^{\beta_H}, \tag{13.33}$$

$$S^L := S - S^{\alpha_H} - S^{\beta_H}, \tag{13.34}$$

$$N_i^L := N_i - N_i^{\alpha_H} - N_i^{\beta_H}, \tag{13.35}$$

$$V^L := V - V^{\alpha_H} - V^{\beta_H}. \tag{13.36}$$

[6]The fact that they are independent of the location of the dividing surface is not obvious from the given expressions, but they can be rewritten in terms of determinants, as shown in Section 13.1.3, in which case their independence is obvious.

[7]This is consistent with the Cahn theory but not essential; one could equally well extend the uniform phases until they met each other at a dividing surface, just as in the Gibbs case, but then the quantity V^L would be zero. In the Cahn theory, $[V] := V^L/A$ depends on the layer thickness, so it is not a fundamental physical quantity.

Note the similarity to Eqs. (13.2)–(13.5) for the excess quantities of Gibbs, with the main exception being that the layer now has a non-vanishing volume V^L and the homogeneous regions do not meet. Following Cahn, we denote the extensive quantities of the layer per unit area by symbols in square brackets, explicitly $[U] := U^L/A$, $[S] := S^L/A$, $[N_i] := N_i^L/A$, and $[V] := V^L/A$, with similar notations for other extensive quantities. Since $[U]$, $[S]$, $[N_i]$, and $[V]$ depend on the position of the walls that bound the layer, they are not fundamental physical quantities.

On the other hand, we know that the quantity $\gamma = (K + pV)/A$ does not depend on any division of the system into L, α_H, and β_H. Moreover, by substitution of Eqs. (13.33)–(13.36) we have[8]

$$\gamma = \frac{U - TS - \sum_i \mu_i N_i + pV}{A} = \frac{U^L - TS^L - \sum_i \mu_i N_i^L + pV^L}{A}, \tag{13.37}$$

where we have used the Euler equations for the bulk phases:

$$U^{\alpha H} - TS^{\alpha H} + pV^{\alpha H} - \sum_i \mu_i N_i^{\alpha H} = 0; \tag{13.38}$$

$$U^{\beta H} - TS^{\beta H} + pV^{\beta H} - \sum_i \mu_i N_i^{\beta H} = 0. \tag{13.39}$$

Therefore

$$\gamma = [U] - T[S] + p[V] - \sum_i \mu_i [N_i]. \tag{13.40}$$

Note especially that the p that multiplies $[V]$ is the pressure of the *bulk* phases. The layer L is inhomogeneous and in a non-hydrostatic state of stress.

For the layer model, one also has $\gamma = \sigma$, the surface tension that enters Eq. (13.11). We can see this by writing Eqs. (13.12) and (13.13) for α_H and β_H and subtracting both from Eq. (13.11) to get

$$\mathrm{d}U^L = T\,\mathrm{d}S^L - p\,\mathrm{d}V^L + \sum_i \mu_i\,\mathrm{d}N_i^L + \sigma\,\mathrm{d}A. \tag{13.41}$$

Note that Eq. (13.14) has no counterpart to the term $p\,\mathrm{d}V^L$ because the Gibbs dividing surface has no volume. Equation (13.40) can be integrated in the same manner as used to obtain Eq. (13.17) and then divided by A to obtain

$$\sigma = \frac{U^L - TS^L + pV^L - \sum_i \mu_i N_i^L}{A} = [U] - T[S] + p[V] - \sum_i \mu_i [N_i] = \gamma. \tag{13.42}$$

In Eq. (13.41), one can employ the definitions of the layer quantities per unit area and use Eq. (13.42) to eliminate the coefficient of $\mathrm{d}A$, resulting in

$$\mathrm{d}[U] = T\,\mathrm{d}[S] - p\,\mathrm{d}[V] + \sum_i \mu_i\,\mathrm{d}[N_i]. \tag{13.43}$$

[8]Note that this is precisely the same definition of γ as for the Gibbs dividing surface, Eq. (13.10). In the present case, however, $K^L = K - K^{\alpha H} - K^{\beta H} = K + p(V^{\alpha H} + V^{\alpha H}) = K + pV - pV^L = A\gamma - pV^L \neq A\gamma$ because $V^L \neq 0$.

Combining Eq. (13.43) with the differential of Eq. (13.40) yields

$$d\gamma = -[S]\,dT + [V]\,dp - \sum_i [N_i]\,d\mu_i, \tag{13.44}$$

which is the counterpart to Eq. (13.20), the Gibbs adsorption equation.

As was the case with the Gibbs dividing surface, the $\kappa + 2$ quantities $T, p, \{\mu_i, i = 1 \cdots \kappa\}$ cannot be varied independently due to the bulk Gibbs-Duhem equations, Eqs. (13.22) and (13.23). Cahn handles this in an elegant and flexible way by solving the set Eq. (13.44), Eq. (13.22), and Eq. (13.23) simultaneously for $d\gamma$ and the differentials of two distinct members of the set $T, p, \{\mu_i, i = 1 \cdots \kappa\}$ which are regarded as dependent variables. This results in an expression for $d\gamma$ in terms of the differentials of the remaining κ independent variables of the set. By means of straightforward application of Cramer's rule, the result can be expressed in terms of determinants of the form[9]

$$[Z/XY] \equiv \frac{\begin{vmatrix} [Z] & [X] & [Y] \\ Z^\alpha & X^\alpha & Y^\alpha \\ Z^\beta & X^\beta & Y^\beta \end{vmatrix}}{\begin{vmatrix} X^\alpha & Y^\alpha \\ X^\beta & Y^\beta \end{vmatrix}}, \tag{13.45}$$

where X, Y, and Z are members of the set $S, V, \{N_i\}$ and X and Y are not the same. In particular, X and Y are the extensive conjugates to the two intensive variables that are chosen to be dependent. The result is

$$d\gamma = -[S/XY]\,dT + [V/XY]\,dp - \sum_{i=1}^{\kappa} [N_i/XY]\,d\mu_i \tag{13.46}$$

in which differentials of the entire variable set $T, p, \{\mu_i, i = 1 \cdots \kappa\}$ appear but in which two of the coefficients are zero because of the structure of the determinant Eq. (13.45). For example, if $X = V$ and $Y = N_1$, one sees that $[V/VN_1] = 0$ and $[N_1/VN_1] = 0$ because two columns in the determinants of their numerators are the same. With that choice, the coefficients of dp and $d\mu_1$ drop out and one is left with

$$d\gamma = -[S/VN_1]\,dT - \sum_{i=2}^{\kappa} [N_i/VN_1]\,d\mu_i. \tag{13.47}$$

Equation (13.47) is the same as Eq. (13.29) or Eq. (13.30) of the Gibbs theory, but here we see from the determinant structure of Eq. (13.45) that the coefficients of these independently variable differentials do not depend on the location of the planes that bound Cahn's

[9] Equations (13.22) and (13.23) were written in terms of amounts of homogeneous α and homogeneous β that meet at the Gibbs dividing surface whereas Cahn's layer theory pertains to homogenous phases α_H and β_H that lie outside the layer. But Eqs. (13.22) and (13.23) are homogeneous so they can be multiplied by any numbers n_α and n_β to increase or reduce the amount of each phase. This will leave the ratio of determinants in Eq. (13.45) unchanged because both numerator and denominator will contain a factor $n_\alpha n_\beta$ which will cancel.

layer or the location of the dividing surface of Gibbs. For a one component system, we can choose the dependent variables to be p and μ, in which case

$$d\gamma = -[S/VN]\,dT, \tag{13.48}$$

so γ depends only on the temperature, as in Eq. (13.27).

At a liquid-vapor interface, γ must go to zero at the critical temperature T_c, at which liquid and vapor become indistinguishable. According to an empirical equation (see [22, p. 474]),

$$\gamma(T) \approx \gamma_0(1 - T/T_c)^{11/9}, \tag{13.49}$$

where γ_0 is a constant. This would correspond to an effective surface entropy,

$$[S/VN] \approx \frac{11}{9}\frac{\gamma_0}{T_c}(1 - T/T_c)^{2/9}, \tag{13.50}$$

which is nearly constant for $T \ll T_c$ but finally becomes zero at $T = T_c$.

We can see the independence of the layer bounds even more clearly by noting that

$$\begin{vmatrix} [Z] & [X] & [Y] \\ Z^\alpha & X^\alpha & Y^\alpha \\ Z^\beta & X^\beta & Y^\beta \end{vmatrix} = \frac{1}{A}\begin{vmatrix} Z^L & X^L & Y^L \\ Z^\alpha & X^\alpha & Y^\alpha \\ Z^\beta & X^\beta & Y^\beta \end{vmatrix} = \frac{1}{A}\begin{vmatrix} Z & X & Y \\ Z^\alpha & X^\alpha & Y^\alpha \\ Z^\beta & X^\beta & Y^\beta \end{vmatrix}, \tag{13.51}$$

where Z, X, Y pertain to the entire system. The last step is true because we can add to the first row whatever multiples of the second and third rows that are needed without changing the value of the determinant. Therefore, we may write

$$[Z/XY] = \frac{1}{A}\frac{\begin{vmatrix} Z & X & Y \\ Z^\alpha & X^\alpha & Y^\alpha \\ Z^\beta & X^\beta & Y^\beta \end{vmatrix}}{\begin{vmatrix} X^\alpha & Y^\alpha \\ X^\beta & Y^\beta \end{vmatrix}}, \tag{13.52}$$

which clearly relates to the *entire system* and has nothing whatsoever to do with the location of any bounding planes of a layer or of a dividing surface. As pointed out in the previous footnote, this expression is independent of the amounts of the homogeneous phases, which must be the case for a physically meaningful interfacial quantity.

As Cahn points out from the structure of Eq. (13.45), the quantity $[Z/XY]$ is the difference, per unit area, in the amount of Z in the layer and portions of homogeneous α and β that, in combination, would have the same values of X and Y as the layer. In other words, if k_α and k_β are chosen so that $k_\alpha X^\alpha + k_\beta X^\beta = A[X]$ and $k_\alpha Y^\alpha + k_\beta Y^\beta = A[Y]$, then $[Z/XY] = [Z] - (k_\alpha Z^\alpha + k_\beta Z^\beta)/A$. It follows from Eq. (13.52) that the same interpretation is true if one considers the entire system instead of the layer.

The foregoing theory is easily extended to the case of planar systems in which multiple homogeneous phases are separated by interfaces. For example, suppose that three phases α, β, and η are in equilibrium with one another. These phases could be separated by two interfaces, one separating α from β and a second separating β from η. Somewhere in the β phase, but very far from both interfaces, one could place an imaginary plane that would divide the system into two parts, one that we will refer to with superscripts $\alpha\beta$ and the other with superscripts $\beta\eta$. Then the quantities

$$\gamma^{\alpha\beta} := \frac{K^{\alpha\beta} + pV^{\alpha\beta}}{A} = \frac{U^{\alpha\beta} - TS^{\alpha\beta} + pV^{\alpha\beta} - \mu_i N_i^{\alpha\beta}}{A}; \tag{13.53}$$

$$\gamma^{\beta\eta} := \frac{K^{\beta\eta} + pV^{\beta\eta}}{A} = \frac{U^{\beta\eta} - TS^{\beta\eta} + pV^{\beta\eta} - \mu_i N_i^{\beta\eta}}{A} \tag{13.54}$$

will be well defined. Both $\gamma^{\alpha\beta}$ and $\gamma^{\beta\eta}$ will depend on the set of intensive variables $T, p, \{\mu_i\}$ which will be uniform throughout the system. But in addition to the Gibbs-Duhem equations (13.22) and (13.23) for the bulk phases α and β, there will be a similar Gibbs-Duhem equation for the bulk η phase. These will constrain three of the intensive variables to be dependent on any others. For a single component material, there are only three variables, T, p, μ, so all would be determined and incapable of change; an equation like Eq. (13.43) would lead to the trivial conclusion $\mathrm{d}\gamma^{\alpha\beta} = 0$ and $\mathrm{d}\gamma^{\beta\eta} = 0$. So we consider at least a binary system, in which case there will be one free variable. Then we will have

$$\mathrm{d}\gamma^{\alpha\beta} = -[S^{\alpha\beta}/XYZ]\,\mathrm{d}T + [V^{\alpha\beta}/XYZ]\,\mathrm{d}p - \sum_{i=1}^{\kappa}[N_i^{\alpha\beta}/XYZ]\,\mathrm{d}\mu_i; \tag{13.55}$$

$$\mathrm{d}\gamma^{\beta\eta} = -[S^{\beta\eta}/XYZ]\,\mathrm{d}T + [V^{\beta\eta}/XYZ]\,\mathrm{d}p - \sum_{i=1}^{\kappa}[N_i^{\beta\eta}/XYZ]\,\mathrm{d}\mu_i, \tag{13.56}$$

where now

$$[W^{\alpha\beta}/XYZ] = \frac{1}{A}\frac{\begin{vmatrix} W^{\alpha\beta} & X^{\alpha\beta} & Y^{\alpha\beta} & Z^{\alpha\beta} \\ W^{\alpha} & X^{\alpha} & Y^{\alpha} & Z^{\alpha} \\ W^{\beta} & X^{\beta} & Y^{\beta} & Z^{\beta} \\ W^{\gamma} & X^{\gamma} & Y^{\gamma} & Z^{\gamma} \end{vmatrix}}{\begin{vmatrix} X^{\alpha} & Y^{\alpha} & Z^{\alpha} \\ X^{\beta} & Y^{\beta} & Z^{\beta} \\ X^{\gamma} & Y^{\gamma} & Z^{\gamma} \end{vmatrix}} \tag{13.57}$$

with a similar expression for $[W^{\beta\eta}/XYZ]$. Here, X, Y, Z must be distinct members of the set $S, V, \{N_i\}$ and W can be any member of the set. Now, three of the coefficients in each of Eqs. (13.55) and (13.56) will be zero. Thus for a binary system there would be only one free variable, say T.

The special case of a single phase interface could occur if the α and β phases are the same but can be distinguished in some other way. Such boundaries can occur in solids, examples being grain boundaries and antiphase boundaries. Although we have not yet discussed the case of solid phases, which involve considerations of surface strain and stress as well as anisotropy, the consequences of having only one Gibbs-Duhem equation will lead to a result of the form

$$\mathrm{d}\gamma = -[S/X]\,\mathrm{d}T + [V/X]\,\mathrm{d}p - \sum_{i=1}^{\kappa}[N_i/X]\,\mathrm{d}\mu_i \tag{13.58}$$

in which X is any member of the set $S, V, \{N_i\}$. We note that

$$[Z/X] = \frac{1}{A}\frac{\begin{vmatrix} Z & X \\ Z^\alpha & X^\alpha \end{vmatrix}}{X^\alpha} = \frac{Z - XZ^\alpha/X^\alpha}{A} \tag{13.59}$$

which obviously does not depend on the total amount of α phase.

Example Problem 13.2. Prove that Cahn's interpretation of the quantity $[Z/XY]$ in the paragraph following Eq. (13.52) is correct.

Solution 13.2. First we choose k_α and k_β so that $k_\alpha X^\alpha + k_\beta X^\beta = X^L$ and $k_\alpha Y^\alpha + k_\beta Y^\beta = Y^L$. Then we substitute for X^L and Y^L into the middle form of Eq. (13.51) to obtain

$$\begin{vmatrix} Z^L & X^L & Y^L \\ Z^\alpha & X^\alpha & Y^\alpha \\ Z^\beta & X^\beta & Y^\beta \end{vmatrix} = \begin{vmatrix} Z^L & k_\alpha X^\alpha + k_\beta X^\beta & k_\alpha Y^\alpha + k_\beta Y^\beta \\ Z^\alpha & X^\alpha & Y^\alpha \\ Z^\beta & X^\beta & Y^\beta \end{vmatrix}. \tag{13.60}$$

We multiply the second row by k_α and the third row by k_β and subtract the resulting rows from the first row to obtain

$$\begin{vmatrix} Z^L & X^L & Y^L \\ Z^\alpha & X^\alpha & Y^\alpha \\ Z^\beta & X^\beta & Y^\beta \end{vmatrix} = \begin{vmatrix} Z^L - k_\alpha Z^\alpha - k_\beta Z^\beta & 0 & 0 \\ Z^\alpha & X^\alpha & Y^\alpha \\ Z^\beta & X^\beta & Y^\beta \end{vmatrix} = (Z^L - k_\alpha Z^\alpha - k_\beta Z^\beta)\begin{vmatrix} X^\alpha & Y^\alpha \\ X^\beta & Y^\beta \end{vmatrix}. \tag{13.61}$$

When we insert this result into the definition of $[Z/XY]$, the 2×2 determinant cancels and we are left with

$$[Z/XY] = (Z^L - k_\alpha Z^\alpha - k_\beta Z^\beta)/A, \tag{13.62}$$

which was to be proven.

13.2 Curved Interfaces in Fluids

To treat curved interfaces in fluids, we return to the comparison system of Gibbs based on a dividing surface. Our main reason for doing this is that the interfacial area can be unambiguously defined; however, for a layer model, one would have to decide what area to use. One side of the dividing surface in the comparison system is assumed to be filled by a homogeneous phase α and the other side by a homogeneous phase β. As in the case of a planar interface, the temperature T and the chemical potentials μ_i are uniform throughout the system. This can be established by considering a layer that extends into the homogeneous phases, similar to the planar case, and studying variations in which there are no changes in the position of the layer. Then one invokes Eq. (13.1) to *define* T and μ_i within this *fixed* layer and then follows the same procedure as for a planar interface to establish uniformity of T and the μ_i throughout the system. Unlike the case of a planar interface, however, the homogeneous phases can have different pressures, p^α and p^β.

Relative to the dividing surface, one can define excess extensive quantities by the same equations, Eqs. (13.2)–(13.6), as in the planar case. Equation (13.8) for the Kramers potential holds as well, but now $K^\alpha = -p^\alpha V^\alpha$ and $K^\beta = -p^\beta V^\beta$ so Eq. (13.9) becomes

$$K^{\rm xs} = K + p^\alpha V^\alpha + p^\beta V^\beta = K + p^\alpha V + (p^\beta - p^\alpha)V^\beta. \tag{13.63}$$

We now define

$$\gamma := \frac{K^{\rm xs}}{A} = \frac{K + p^\alpha V^\alpha + p^\beta V^\beta}{A} = \frac{K + p^\alpha V + (p^\beta - p^\alpha)V^\beta}{A}. \tag{13.64}$$

Unlike the case for a planar interface, γ for a curved interface *depends* on the choice of the location of the dividing surface.

We can illustrate this dependence quite simply for the case in which the β phase is a sphere of radius r surrounded by the α phase. Then $A = 4\pi r^2$ and $V^\beta = (4/3)\pi r^3$ so

$$\gamma = \frac{K + p^\alpha V}{4\pi}\frac{1}{r^2} + \frac{(p^\beta - p^\alpha)}{3}r, \tag{13.65}$$

where the coefficients of $1/r^2$ and r are constants for a given *physical* system. Figure 13–2 shows a sketch of γ as a function of r. We note that γ has a minimum value[10] γ_t at some value r_t. We multiply Eq. (13.65) by r^2 and take its variation with respect to r for a fixed physical system, so $K + p^\alpha V$ and $p^\beta - p^\alpha$ have no variation with r. This results in

$$(p^\beta - p^\alpha) = \frac{2\gamma}{r} + \frac{\partial \gamma}{\partial r}. \tag{13.66}$$

We can now choose $r = r_t$ at which $\partial\gamma/\partial r = 0$ and $\gamma = \gamma_t$, its minimum value. Then Eq. (13.66) reduces to

$$(p^\beta - p^\alpha) = \frac{2\gamma_t}{r_t}. \tag{13.67}$$

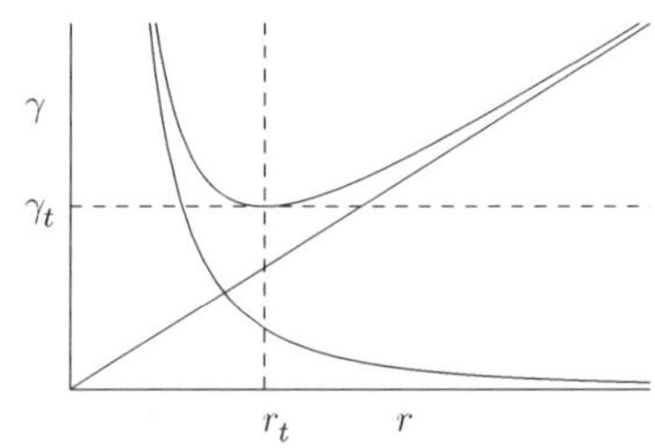

FIGURE 13–2 Plot of γ versus r according to Eq. (13.65) in arbitrary units (top curve). The lower curve and the straight line represent the individual terms. The minimum occurs at the surface of tension where $r = r_t$ and $\gamma = \gamma_t$ and leads to Eq. (13.67) instead of Eq. (13.66).

[10]From stability considerations, γ must be positive. Otherwise, the system could lower its free energy indefinitely by creating an infinite amount of area. For an extensive discussion of stability, see Gibbs [3, pp. 237-252].

This special choice is the so-called **surface of tension** which was introduced by Gibbs following another course of reasoning that we describe below.

Before doing so, however, we derive the counterpart to Eq. (13.67) for a more general surface having principal curvatures $c_1 = 1/R_1$ and $c_2 = 1/R_2$, where R_1 and R_2 are principal radii of curvature. We return to Eq. (13.64) and identify a set of dividing surfaces by means of a parameter λ. One such surface is chosen to be very near the physical region of discontinuity and similarly situated with respect to it, while the others are obtained by shifting a constant distance $\delta\lambda$ along the normals to that chosen surface. Thus, A, V^β, and γ become functions of λ. Multiplying Eq. (13.64) by A and taking its variation with respect to λ, we obtain

$$A\frac{\partial \gamma}{\partial \lambda}\delta\lambda + \gamma\delta A = (p^\beta - p^\alpha)\delta V^\beta, \tag{13.68}$$

where from differential geometry[11]

$$\delta V^\beta = A\delta\lambda; \quad \delta A = (c_1 + c_2)A\delta\lambda. \tag{13.69}$$

We therefore obtain

$$p^\beta - p^\alpha = \gamma(c_1 + c_2) + \frac{\partial \gamma}{\partial \lambda}. \tag{13.70}$$

If we choose the dividing surface to correspond to the generalized surface of tension where $\partial\gamma/\partial\lambda = 0$, Eq. (13.70) becomes[12]

$$p^\beta - p^\alpha = \gamma_t\left(\frac{1}{R_{1t}} + \frac{1}{R_{2t}}\right). \tag{13.71}$$

In the special case of a spherical surface, $R_{1t} = R_{2t} = r_t$ and we recover Eq. (13.67).

An equation of the form of Eq. (13.71) (without the subscripts t) is attributed to Laplace and pertains to a membrane of zero thickness that has the following property: If $\mathrm{d}\boldsymbol{\tau}$ is any infinitesimal vector in that membrane, the membrane on one side of it exerts an attractive force per unit length γ on other side that is perpendicular to $\mathrm{d}\boldsymbol{\tau}$ and tangential to the surface. Equation (13.71) shows that such a relation results from thermodynamic considerations, provided we evaluate the excess surface free energy γ at the surface of tension. As we shall see below, Gibbs arrived at the same equation as an approximation based on the idea that the explicit dependence of γ on the curvature of the dividing surface can be ignored for a dividing surface that is very close to the region of transition, provided that the thickness of the region of transition is small relative to either of its principal radii of curvature. In practice, γ is measured experimentally by assuming it to be equal to γ_t in Eq. (13.71) and by assuming that the principal radii of curvature measured by some technique, usually optical, are essentially the same as R_{1t} and R_{2t}.

[11]These are special cases of the integral formulae $\delta V^\beta = \int \delta\lambda \,\mathrm{d}A$ and $\delta A = \int (c_1 + c_2)\delta\lambda \,\mathrm{d}A$ that hold for a normal shift $\delta\lambda$ that is a function of position on the surface.

[12]The quantity that multiplies γ_t in Eq. (13.71) is called the mean curvature, with the sign convention that the radius is positive for a sphere of the β phase, or for a more general surface if the β phase is on the side of the interface that has net concavity.

Example Problem 13.3. Estimate the capillary rise of water and the capillary depression of mercury in a vertical glass capillary tube of inner diameter $2r_0 = 1.0\,\mathrm{mm}$ at a temperature of 20 °C. Take $\gamma = 0.073\,\mathrm{J/m^2}$ for water and $0.47\,\mathrm{J/m^2}$ for mercury. Assume that the density is $1\,\mathrm{g/cm^3}$ for water and $13.55\,\mathrm{g/cm^3}$ for mercury. See Figure 13–3 for an illustration of the geometry and the contact angle θ where the water meets the glass. For now, assume that the contact angle is an empirical parameter; later in Section 13.3 we provide some theoretical basis for it. The water wets the glass with a contact angle of nearly zero degrees. Mercury does not wet the glass and has an obtuse contact angle of 140°. Assume that the shape of the liquid-gas interface is approximately a portion of a sphere (undistorted by gravity) and that the system is open to the atmosphere at a pressure of p_A. Also assume that these liquids are locally in equilibrium with their vapors but that the rate of evaporation, the solubility of air in either liquid and the density of air are negligible. Take the gravitational acceleration to be $9.8\,\mathrm{m/s^2}$. What would be the corresponding results if the liquids were between large parallel vertical plates separated by a small distance $2x_0$?

Solution 13.3. From the figure, we see that the radius of curvature of the spherical interface is given by $r_t = r_0/\cos\theta$, so $1/R_1 + 1/R_2 = 2/r_0$ for water and $1/R_1 + 1/R_2 = 2\cos(140°)/r_0 = -1.45/r_0$ for mercury. For either substance, the pressure p_h in the liquid at capillary rise height h is given by $p_A - p_h = \rho g h = 2\gamma\cos\theta/r_0$. Here, $\rho g h$ is due to hydrostatic pressure, as explained in Section 11.2. Thus the capillary rise is

$$h = \frac{2\gamma}{\rho g}\frac{\cos\theta}{r_0} = a^2\frac{\cos\theta}{r_0}, \tag{13.72}$$

where

$$a := \sqrt{2\gamma/\rho g} \tag{13.73}$$

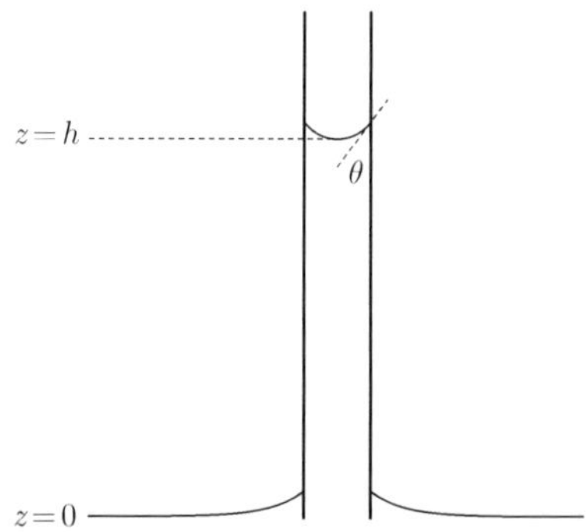

FIGURE 13–3 Sketch of the rise of a liquid in a capillary tube of internal diameter $2r_0$. The rise h is defined to be the distance from the horizontal surface of the bulk liquid to the bottom of the meniscus. The contact angle θ is the angle between the tangent to the meniscus at the tube wall and the tube itself (see Section 13.3). For an obtuse angle θ, the liquid would be depressed below the surface of the bulk liquid.

has dimensions of length and is known as the **capillary length.**[13] At 20 °C, $a \approx 5.5$ mm for water and $a \approx 3.8$ mm for mercury. The quantity a sets the length scale of capillary phenomena. For the example of capillary rise just given, $h = (a^2/r_0)\cos\theta$. For $r_0 = 0.5$ mm, $h = 3$ cm for water and $h = -1.4$ cm for mercury.

For large parallel plates separated by a small distance $2x_0$, one radius of curvature would be infinite and the other would be $\cos\theta/x_0$ so $h = (1/2)(a^2/x_0)\cos\theta$. For $x_0 = r_0$, the magnitude of h would be half as large for parallel plates as for a tube.

13.2.1 Gibbs Coefficients of Curvatures

Gibbs [3, p. 225] proceeded somewhat differently by assuming that the variation of the excess internal energy from a state of equilibrium is given by

$$\delta U^{xs} = T\delta S^{xs} + \mu_i \delta N_i^{xs} + \sigma\delta A + C_1\delta c_1 + C_2\delta c_2, \tag{13.74}$$

where C_1 and C_2 are coefficients of the curvatures. He then employed the identity

$$C_1\delta c_1 + C_2\delta c_2 = (1/2)(C_1 + C_2)\delta(c_1 + c_2) + (1/2)(C_1 - C_2)\delta(c_1 - c_2). \tag{13.75}$$

By considering a spherical surface (so that $C_1 - C_2 = 0$) and two different choices of the dividing surface that are essentially parallel to one another but separated by a small distance, he proceeded to show [3, p. 227] that the coefficient $C_1 + C_2$ can be made to change sign by means of a small shift of the dividing surface. Therefore, one can choose a dividing surface such that $C_1 + C_2 = 0$. He then argued that since $C_1 - C_2 = 0$ for a planar interface, it should not differ much from zero for an interface that does not differ too much from planarity, as would be the case if both principal radii of curvature are large compared to the thickness of the inhomogeneous region. On this basis, he neglected the term $(1/2)(C_1 - C_2)\delta(c_1 - c_2)$, ultimately resulting in

$$\delta U^{xs} = T\delta S^{xs} + \sum_i \mu_i \delta N_i^{xs} + \sigma\delta A, \tag{13.76}$$

where σ is now dependent on the choice of the dividing surface. Provided that both principal radii of curvature are large compared to the thickness of the inhomogeneous region, this dividing surface will be quite close to the inhomogeneous region as measured optically in an experiment.

Essentially, the Gibbs argument boils down to the following: The position of the dividing surface can be chosen such that the dependence of U^{xs} on warping due to curvatures that are not too large can be neglected. Curvatures only affect U^{xs} indirectly through their influence on δA.

We can compare our approach to that of Gibbs as follows. We return to Eq. (13.74) and integrate at constant T, μ_i, c_1, and c_2, by just making the system larger,[14] to obtain the Euler equation

[13] More accurately one should write $a^2 = 2\gamma/[(\rho^\ell - \rho^{air})g]$ but the density ρ^{air} is usually negligible.

[14] For example, for a system consisting of a spherical surface and bulk systems inside a cone with its apex at the center of the sphere, all extensive quantities will be proportional to the size of the cone.

$$U^{\mathrm{xs}} = TS^{\mathrm{xs}} + \sum_i \mu_i N_i^{\mathrm{xs}} + \sigma A. \tag{13.77}$$

Combining this with the Euler equations for the homogeneous phases, we deduce that

$$\sigma = \frac{U^{\mathrm{xs}} - TS^{\mathrm{xs}} - \sum_i \mu_i N_i^{\mathrm{xs}}}{A} = \frac{U - TS - \sum_i \mu_i N_i + p^\alpha V^\alpha + p^\beta V^\beta}{A}, \tag{13.78}$$

so $\sigma = \gamma$ as given by Eq. (13.64).

Next, we take the total variation of $U = U^{\mathrm{xs}} + U^\alpha + U^\beta$ with σ replaced by γ to obtain

$$\delta U = T\delta S - p^\alpha \delta V_\alpha - p^\beta \delta V^\beta + \sum_i \mu_i \delta N_i + \gamma \delta A + C_1 \delta c_1 + C_2 \delta c_2. \tag{13.79}$$

If we now consider a variation at constant external volume $V = V^\alpha + V^\beta$, constant S, and constant N_i, the condition for equilibrium becomes $\delta U = 0$ so

$$(p^\alpha - p^\beta)\delta V^\beta + \gamma \delta A + C_1 \delta c_1 + C_2 \delta c_2 = 0. \tag{13.80}$$

For a normal shift of the interface by an amount $\delta\lambda$, as considered in the derivation of Eq. (13.70), we have Eq. (13.69) and also $\delta c_1 = -c_1^2 \delta\lambda$ and $\delta c_2 = -c_2^2 \delta\lambda$. Thus for arbitrary $\delta\lambda$ Eq. (13.80) becomes

$$(p^\beta - p^\alpha) = (c_1 + c_2)\gamma - \frac{C_1}{A} c_1^2 - \frac{C_2}{A} c_2^2. \tag{13.81}$$

Substitution of Eq. (13.81) into Eq. (13.70) gives

$$\frac{\partial \gamma}{\partial \lambda} = -\frac{(C_1 + C_2)}{2A}(c_1^2 + c_2^2) - \frac{(C_1 - C_2)}{2A}(c_1^2 - c_2^2). \tag{13.82}$$

From Eq. (13.82) we see that the Gibbs choice of location of the dividing surface to make $C_1 + C_2 = 0$ and his neglect of the term in $C_1 - C_2$ is equivalent to choosing the dividing surface to satisfy $\partial\gamma/\partial\lambda = 0$, which leads to Eq. (13.71).

13.3 Interface Junctions and Contact Angles

In this section we investigate briefly the mechanical conditions that must be satisfied at the junctions where several fluid phases meet. We begin by considering the two-dimensional problem of a triple junction where three phases, α, β, and η meet along a line, as illustrated in Figure 13–4. The line where the phases meet is known as a **triple line** and is perpendicular to the plane of the figure. Our objective is to determine the dihedral angles θ_α, θ_β, and θ_η where these phases meet. By studying a simple variation of the position of the triple junction, we shall see that the three tensions satisfy a simple force balance law of the form[15]

$$\gamma^{\alpha\beta}\hat{\boldsymbol{\tau}}^{\alpha\beta} + \gamma^{\beta\eta}\hat{\boldsymbol{\tau}}^{\beta\eta} + \gamma^{\eta\alpha}\hat{\boldsymbol{\tau}}^{\eta\alpha} = 0, \tag{13.83}$$

[15]This equation also follows from a more general result of Gibbs [3, equation 615, p. 281] that holds for curved contact lines and was obtained as part of a complete variation of the system.

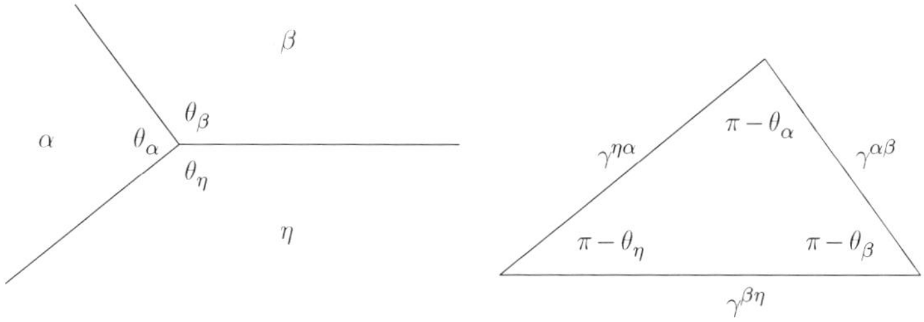

FIGURE 13–4 Three phases α, β, and η that meet at a triple line (left) where they make dihedral angles θ_α, θ_β, and θ_η. On the right is the corresponding force triangle with forces of magnitudes $\gamma^{\alpha\beta}$, $\gamma^{\beta\eta}$, and $\gamma^{\eta\alpha}$ that are directed away from the triple junction, so their vector sum is zero.

where $\hat{\boldsymbol{\tau}}^{\alpha\beta}$ is a unit vector perpendicular to the line of intersection of the three phases, locally tangent to the $\alpha\beta$ interface at the line of intersection, and pointing away from the η phase. The other unit vectors $\hat{\boldsymbol{\tau}}^{\beta\eta}$ and $\hat{\boldsymbol{\tau}}^{\eta\alpha}$ are similarly defined with respect to their phases. One can interpret Eq. (13.83) by regarding the quantity $\gamma^{\alpha\beta}\hat{\boldsymbol{\tau}}^{\alpha\beta}$ to be a force per unit length that acts on the triple line along the $\alpha - \beta$ interface, and similarly $\gamma^{\beta\eta}\hat{\boldsymbol{\tau}}^{\beta\eta}$, and $\gamma^{\eta\alpha}\hat{\boldsymbol{\tau}}^{\eta\alpha}$ are forces per unit length that act on the triple line along their respective interfaces. Thus, Eq. (13.83) is the condition for zero force acting on the triple line.

It also follows that the vectors $\gamma^{\alpha\beta}\hat{\boldsymbol{\tau}}^{\alpha\beta}$, $\gamma^{\beta\eta}\hat{\boldsymbol{\tau}}^{\beta\eta}$, and $\gamma^{\eta\alpha}\hat{\boldsymbol{\tau}}^{\eta\alpha}$ form a triangle, as shown in Figure 13–4, whose internal angles are $\pi - \theta_\alpha$, $\pi - \theta_\beta$, and $\pi - \theta_\eta$. This triangle is known as a **Neumann triangle** and is often drawn in an orientation such that each of its sides is perpendicular to the respective interface. This can be seen by taking the cross product of Eq. (13.83) with a unit vector along the triple line. From the law of sines, and the fact that $\sin(\pi - \theta) = \sin\theta$, it follows that

$$\frac{\sin\theta_\alpha}{\gamma^{\beta\eta}} = \frac{\sin\theta_\beta}{\gamma^{\eta\alpha}} = \frac{\sin\theta_\eta}{\gamma^{\alpha\beta}}. \tag{13.84}$$

For such a Neumann triangle to exist, it is necessary for each of its sides to be less than the sum of the other two sides, for example $\gamma^{\beta\eta} < \gamma^{\eta\alpha} + \gamma^{\alpha\beta}$.

Equation (13.83) can also be generalized to junctions where more than three phases meet, but such configurations might not be stable [3, p. 287]. If crystalline solids are involved, we shall see that γ is anisotropic so Eq. (13.83) must be modified to account for torque terms.

To derive Eq. (13.83) from a variational principle, we suppose that all interfaces are pinned at distances that are far from the triple line and vary the position of the triple line by moving it parallel to itself in the direction of a small vector $\boldsymbol{\epsilon}$. If $\ell^{\alpha\beta}$ is the pinning distance from the $\alpha - \beta$ interface to the original triple line, the distance from the varied triple line is

$$|\ell^{\alpha\beta}\tau^{\alpha\beta} - \boldsymbol{\epsilon}| = \sqrt{(\ell^{\alpha\beta})^2 - 2\boldsymbol{\epsilon}\cdot\hat{\boldsymbol{\tau}}^{\alpha\beta}\ell^{\alpha\beta} + \epsilon^2} = \ell^{\alpha\beta} - \boldsymbol{\epsilon}\cdot\hat{\boldsymbol{\tau}}^{\alpha\beta} \tag{13.85}$$

to first order in $\epsilon/\ell^{\alpha\beta}$. The corresponding change in distance is therefore $-\boldsymbol{\epsilon}\cdot\hat{\boldsymbol{\tau}}^{\alpha\beta}$. By treating the other interfaces in a similar way, we see that the total change in energy per unit length for such a variation is

$$-\boldsymbol{\epsilon}\cdot(\hat{\boldsymbol{\tau}}^{\alpha\beta}\gamma^{\alpha\beta}+\hat{\boldsymbol{\tau}}^{\beta\eta}\gamma^{\beta\eta}+\hat{\boldsymbol{\tau}}^{\eta\alpha}\gamma^{\eta\alpha})=0, \tag{13.86}$$

which has been equated to zero as a condition for equilibrium. For arbitrary $\boldsymbol{\epsilon}$, the quantity in parentheses must vanish, resulting in Eq. (13.83).

It is important to recognize that knowledge of the angles θ_α, θ_β, and θ_η will allow one to determine only the ratios of the quantities $\gamma^{\alpha\beta}$, $\gamma^{\beta\eta}$, and $\gamma^{\eta\alpha}$. This can be seen by noting that multiplication of each of these interfacial energies by some positive number would result in a triangle similar, but different in size, to that depicted in Figure 13–4, so the angles would be unchanged. However, if the ratios of $\gamma^{\alpha\beta}$, $\gamma^{\beta\eta}$, and $\gamma^{\eta\alpha}$ are specified, all three angles are determined uniquely. This can be seen analytically by applying the law of cosines to the triangle in Figure 13–4 to obtain

$$\cos\theta_\alpha=\frac{(\gamma^{\beta\eta})^2-(\gamma^{\eta\alpha})^2-(\gamma^{\alpha\beta})^2}{\gamma^{\eta\alpha}\gamma^{\alpha\beta}} \tag{13.87}$$

and similar expressions for $\cos\beta$ and $\cos\eta$.

13.3.1 Contact Angle

The variational derivation that underlies the force balances represented by Eq. (13.83) must be modified for anisotropic interfaces because the orientation of interfaces can change locally when the position of the triple line is varied. This results in additional torque terms. Nevertheless, the concept of force balances can be used to understand contact angles made by fluids with a rigid amorphous solid. Figure 13–5 shows a triple junction between a liquid L and a gas g on a solid substrate s under conditions for which we assume that $\gamma^{\ell g}$, γ^{sg}, and $\gamma^{s\ell}$ can be defined.[16] Then a variation that involves sliding the triple line along the solid results in the equilibrium condition

$$\gamma^{sg}=\gamma^{\ell g}\cos\theta+\gamma^{s\ell}, \tag{13.88}$$

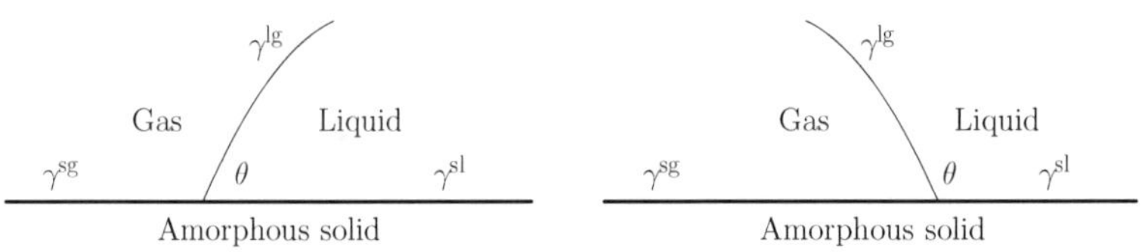

FIGURE 13–5 Contact angle θ for two fluids in contact with a rigid inert amorphous solid. On the left, θ is acute and the liquid is said to wet the solid. On the right, θ is obtuse and the liquid does not wet the solid.

[16]These conditions could deviate considerably from the global equilibrium conditions discussed previously. The solid should behave as if chemically inert, with no solubility of the substances of the liquid or the gas. The gas could contain a substance insoluble in the liquid, and the vapor of the liquid can be in local equilibrium at the solid-liquid interface provided there is negligible evaporation during some period of observation. See Gibbs [3, p. 326] for further discussion.

which may be solved to yield

$$\cos\theta = \frac{\gamma^{sg} - \gamma^{s\ell}}{\gamma^{\ell g}}. \tag{13.89}$$

Equation (13.89) is known as **Young's equation** for the contact angle θ and represents a balance of horizontal forces. Real values of θ only exist when the right-hand side has a magnitude less than or equal to one, which requires $|\gamma^{sg} - \gamma^{s\ell}| \le \gamma^{\ell g}$. If θ exists and $\gamma^{sg} - \gamma^{s\ell} > 0$, $\theta \le \pi/2$ and the liquid is said to wet the solid. For kerosene on glass, one has $\theta \approx 26°$. Complete wetting occurs for $\theta = 0$, which is approximately the case for water on clean glass. If θ exists and $\gamma^{sg} - \gamma^{s\ell} < 0$, $\theta > \pi/2$ and the liquid does not wet the solid. For mercury on glass, one has $\theta \approx 140°$. As long as the solid remains rigid and inert, no vertical variation of the contact line is possible, although it is generally supposed that the solid provides a force of adhesion equal to $\gamma^{\ell g} \sin\theta$ to prevent the ℓg interface from pulling away.

Although Young's equation helps us understand the origin of the contact angle, its derivation suffers from a lack of rigor. Moreover, experimentally measured contact angles are difficult to reproduce and can depend sensitively on impurities as well as surface conditions of the solid. Nevertheless, the use of an empirically measured contact angle can enable one to model liquid shapes in situations of practical importance.

13.4 Liquid Surface Shape in Gravity

The shape of a liquid surface in a uniform gravitational field provides more insight regarding the role of surface tension as well as methods of measuring surface tension experimentally. We shall explore surface shapes for some two-dimensional problems and for some three dimensional problems with axial symmetry.

In all cases we assume that Eq. (13.71) applies and drop the subscripts t with the understanding that, strictly speaking, we are dealing with the surface of tension. For a Cartesian coordinate system with a z axis antiparallel to gravity, we represent the liquid-gas interface by the function $z = z(x, y)$, in which case differential geometry (see Section C.4 of Appendix C for a derivation) leads to a total curvature

$$\mathcal{K} \equiv \frac{1}{R_1} + \frac{1}{R_2} = -\frac{\left[z_{xx}(1+z_y^2) - 2z_x z_y z_{xy} + z_{yy}(1+z_x^2)\right]}{\left(1+z_x^2+z_y^2\right)^{3/2}}, \tag{13.90}$$

where subscripts on z denote partial differentiation. We treat an isothermal case and single component fluids of densities ρ_g and ρ_ℓ and neglect any dependence of γ on the pressure difference[17] $p^\ell - p^g$. Moreover, we have $\mathrm{d}(p^\ell - p^g) = -(\rho^\ell - \rho^g)g\,\mathrm{d}z$. Over the small distances that are important in capillary phenomena (see the capillary length defined by

[17]This is equivalent to eliminating the terms $C_1\delta c_1 + C_2\delta c_2$ in Eq. (13.74) by choice of the surface of tension or an equivalent approximation.

Eq. (13.73)) the densities ρ^ℓ and ρ^g can be taken to be constants.[18] Thus, to an excellent approximation,

$$p^\ell - p^g = -(\rho^\ell - \rho^g)gz + C, \tag{13.91}$$

where C is a constant equal to the value of $p^\ell - p^g$ in the plane $z = 0$. Equation (13.71) therefore becomes[19]

$$\pm \frac{\left[z_{xx}(1+z_y^2) - 2z_x z_y z_{xy} + z_{yy}(1+z_x^2)\right]}{\left(1+z_x^2+z_y^2\right)^{3/2}} = -\frac{(\rho^\ell - \rho^g)gz}{\gamma} + C. \tag{13.92}$$

The $\pm$ sign in Eq. (13.92) must be chosen in accordance with the sign convention inherent in Eq. (13.71), namely that for net positive $\mathcal{K}$, the fluid with the greater pressure is on the side of the interface with the greater concavity. Moreover, z could be positive or negative or even change sign in the domain of interest. One can treat shapes for which the liquid is above the gas or below the gas. In many cases, $z(x, y)$ will be a multiple valued function of x and y so one must be careful to treat each portion of the surface separately. Explicit choices of the correct sign are best left to examples.

13.4.1 Examples in Two Dimensions

For two-dimensional problems, there is only one finite radius of curvature and Eq. (13.92) can be simplified to the form

$$\frac{z_{xx}}{\left(1+z_x^2\right)^{3/2}} = \frac{2}{a^2}z, \tag{13.93}$$

where $a^2 = 2\gamma/[(\rho^\ell - \rho^g)g]$. In Eq. (13.93), the sign of the curvature term has been chosen so that for $z > 0$ one must have $z_{xx} > 0$, which will be the case for a gas at essentially constant pressure p^g on the upper concave side of an interface with a liquid on the lower side whose pressure $p^\ell \le p^g$ is decreasing with increasing z. Alternatively one could have $z < 0$ and $z_{xx} < 0$, which will be the case for a gas at essentially constant pressure p^g on the upper convex side of an interface with a liquid on the lower side whose pressure $p^\ell \ge p^g$ is increasing with decreasing z. We substitute $p = z_x$ and note that $z_{xx} = p\,\mathrm{d}p/\mathrm{d}z$ to obtain

$$\frac{p\,\mathrm{d}p}{\left(1+p^2\right)^{3/2}} = \frac{2}{a^2}z\,\mathrm{d}z, \tag{13.94}$$

which may be integrated to yield

$$-\frac{1}{\left(1+p^2\right)^{1/2}} = \frac{z^2}{a^2} - 1; \quad z^2 \le a^2. \tag{13.95}$$

Here, the constant of integration was evaluated by setting $p = 0$ for $z = 0$.

[18] For liquids, the compressibility is very small. For an ideal gas, $\rho \propto \exp(-mgz/k_B T)$ and $k_B T/mg$ is the order of magnitude of 10 m. Usually $\rho^\ell \gg \rho^g$ so we could neglect ρ^g but we retain it in the formulae which would still be valid if the gas were replaced by a nearly incompressible liquid.

[19] In a gravitational field, γ can depend implicitly on z (see Eqs. (613-614) of Gibbs [3, p. 281]) but this weak dependence is usually negligible.

Example Problem 13.4. Calculate the height $z_{\rm max}$ of the meniscus at the contact line for a large plate of glass immersed vertically in a pool of liquid of density ρ for a contact angle $0 \le \theta \le \pi/2$. This situation is depicted in Figure 13–6 (vertical plate case). Calculate the depression if $\pi/2 \le \theta \le \pi$.

Solution 13.4. Since the plate is large, we ignore end effects and treat the problem as two dimensional. First we treat the case $0 \le \theta \le \pi/2$. At the line of contact with the glass, $p = \cot\theta \ge 0$ so $(1+p^2)^{-1/2} = \sin\theta \ge 0$. Thus, Eq. (13.95) yields

$$z_{\rm max} = a\sqrt{1-\sin\theta} = \sqrt{\frac{2\gamma}{(\rho^\ell - \rho^g)g}}\sqrt{1-\sin\theta}. \tag{13.96}$$

For $\pi/2 \le \theta \le \pi$, $p = \cot\theta \le 0$ at the line of contact. For this contact angle, $\sin\theta \ge 0$ but $\cos\theta \le 0$. Equation (13.95) still applies and can be solved for z^2 but we must now take the negative square root to obtain

$$z_{\rm min} = -a\sqrt{1-\sin\theta} = -\sqrt{\frac{2\gamma}{(\rho^\ell - \rho^g)g}}\sqrt{1-\sin\theta}. \tag{13.97}$$

For the same value of $\sin\theta$, the interface shapes for acute and obtuse contact angles are mirror images of one another in the plane $z = 0$.

Unfortunately, the right-hand side of Eq. (13.95) changes sign at $z = a$ so it cannot hold for both $z^2 > a^2$ and $z^2 < a^2$. This can be handled by introducing the angle ψ where $\sin\psi = {\rm d}z/{\rm d}s$ and $\cos\psi = {\rm d}x/{\rm d}s$ where s is arc length. Then Eq. (13.93) takes the form

$$\frac{{\rm d}\psi}{{\rm d}s} = \frac{2}{a^2}z. \tag{13.98}$$

In this parametric representation, ψ will continue to increase as z and s increase, so the curvature ${\rm d}\psi/{\rm d}s$ will remain positive as z increases from zero, or remain negative as z decreases from zero. We can now scale all lengths with a by defining $X = x/a$, $Z = z/a$, and $S = s/a$ to get the following set of parametric equations:

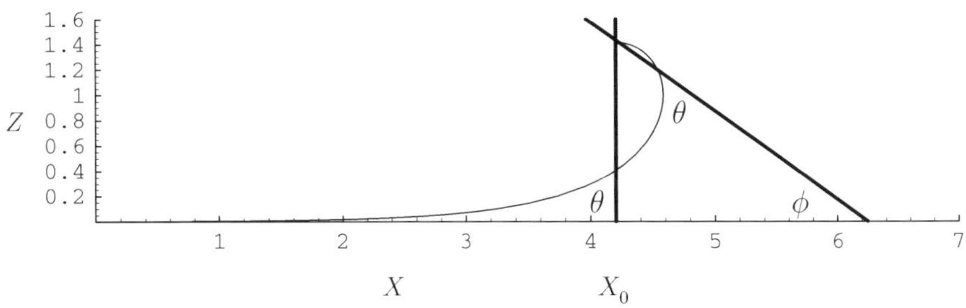

FIGURE 13–6 Shape of the meniscus of a fluid drawn up by a vertical plate at some distance X_0 along the X axis. Of course the actual curve stops at X_0 where it makes the appropriate contact angle θ with the plate. The maximum height, $Z_{\rm max} = \sqrt{1-\sin\theta}$, occurs at the line of contact. By removing the vertical plate, the same curve can be used to treat a plate that makes an angle ϕ with the X axis. If the contact angle is obtuse, one uses a downward sloping curve that is the mirror reflection $Z \to -Z$. ψ is the angle made by a tangent to the curve and the X axis.

$$\frac{d\psi}{dS} = 2Z; \tag{13.99}$$

$$\frac{dX}{dS} = \cos\psi; \tag{13.100}$$

$$\frac{dZ}{dS} = \sin\psi. \tag{13.101}$$

One could integrate the set Eqs. (13.99)–(13.101) by starting from some contact angle where $\cos\psi = \sin\theta$, $\sin\psi = \pm\cos\theta$, and $Z = \pm(1-\sin\theta)^{1/2}$, but a more useful approach is to begin at a very small value $Z = \epsilon$ at $X = Z = S = 0$ and to integrate numerically. This gives a universal curve that can be stopped at any value of X that corresponds to the correct contact angle. To get started, however, one needs a compatible starting value of ψ. This can be obtained by combining Eqs. (13.99) and (13.101) to obtain $d\psi/dZ = 2Z/\sin\psi$ which can be readily integrated to give $\cos\psi = 1 - Z^2$. Thus, the starting values for ψ satisfy $\cos\psi = 1-\epsilon^2$ and $\sin\psi = \epsilon(2-\epsilon^2)^{1/2}$. The result of numerical integration is shown in Figure 13–6.

Equations (13.96) and (13.97) can be generalized for a large plate that makes an angle ϕ with the surface of the bulk fluid, as shown in Figure 13–6. We still have the solution $Z^2 = 1 - \cos\psi$. For an acute contact angle θ, we have $\psi = \pi - (\phi + \theta) \geq 0$ so instead of Eq. (13.96) we have

$$z_{\max} = a\sqrt{1 + \cos(\phi + \theta)}. \tag{13.102}$$

In Eq. (13.102), the values of ϕ are limited because we need $\psi \geq 0$ which requires $\phi \leq \pi - \theta$. In fact, when $\phi = \pi - \theta$, $z_{\max} = 0$ and the interface remains flat. The opposite limit is $\phi = 0$, which yields $z_{\max} = a\sqrt{1+\cos\theta}$. As is evident from Figure 13–6, a value of ϕ very near to zero corresponds to a case where the X coordinate of the intersection of the plate with $Z = 0$ tends to $X \to \infty$. In other words, moving the plate toward $\phi = 0$ “squeezes” the fluid above $z > 0$ in the negative X direction.

For an obtuse contact angle $\theta' > \pi/2$, the corresponding relation at the contact line is $\psi = \pi - (\phi + \theta') \leq 0$. Thus $z_{\min} = -a\sqrt{1+\cos(\phi+\theta')}$. For $\theta = \pi - \theta'$ and $\phi \to \pi - \phi$, we have $z_{\min} = -z_{\max}$ where the latter is given by Eq. (13.102).

The detailed shape of the meniscus for two-dimensional problems can be expressed analytically in terms of incomplete elliptic integrals but such a solution is not very enlightening because those integrals must ultimately be evaluated numerically. With the availability of fast computers and software such as NDSolve in Mathematica^{TM}, it is a simple matter to integrate a first order system such as Eqs. (13.99)–(13.101) numerically and then use a parametric plotting routine to display the result. A number of interesting solutions can be obtained by choosing $\psi = \alpha$ at the origin $X = Z = S = 0$, where $0 \leq \alpha \leq \pi$. In view of Eq. (13.99), the curvature is zero at $Z = 0$ and changes sign along any curve that passes through $Z = 0$, where the curve has an inflection. Such an inflection point might not appear on a curve of physical interest, but portions of the remainder of the curve may be relevant. Since the equations are nonlinear, the behavior as α changes is quire diverse as illustrated in the next few figures.

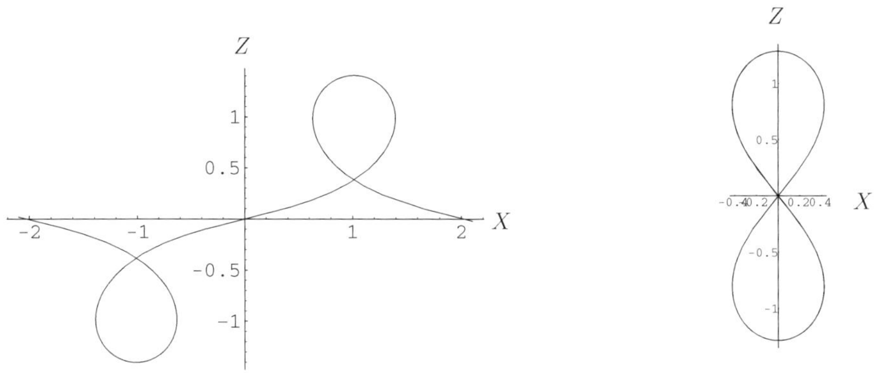

FIGURE 13–7 Two-dimensional interfaces obtained by numerical integration of the system Eqs. (13.99)–(13.101) for $\alpha = 0.25$ (left) and $\alpha = 0.86$ (right). At about $\alpha = 0.86$, the upper and lower loops meet at the origin. For either value of α, the upper loop could represent a two-dimensional bubble and the lower loop could represent a two-dimensional drop. Either of these could be cut off by horizontal surfaces at appropriate contact angles. A portion of the curve for $\alpha = 0.25$ could represent the interface between two vertical plates, the right plate wetted (acute contact angle) and the left plate not wetted (obtuse contact angle) and the inflection somewhere between the plates.

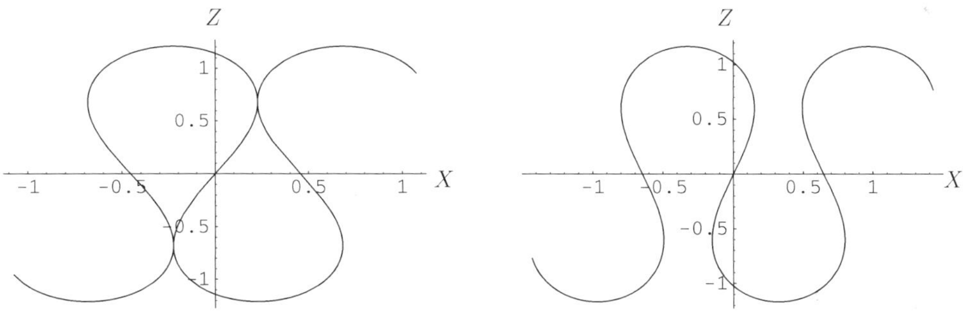

FIGURE 13–8 Two-dimensional interfaces obtained by numerical integration of the system Eqs. (13.99)–(13.101) for $\alpha = 1.09$ (left) and $\alpha = 1.20$ (right). The loops that nearly coalesced for $\alpha = 0.86$ separate. For $\alpha = 1.20$, portions of the curve can be used to represent a two-dimensional bubble or a two-dimensional drop with a neck. For $\alpha = 1.09$, the neck nearly vanishes.

Figure 13–7 shows some results for a small and intermediate value of α. The nature of the curve changes for about $\alpha = 0.86$ where the loops coalesce. Only a portion of any curve will be needed to satisfy wetting conditions at bounding surfaces. Part of an upper loop could represent a two-dimensional bubble and part of a lower loop could represent a two-dimensional drop. As α increases, the character of the solution changes, as illustrated in Figure 13–8. Portions of curves could represent a two-dimensional bubble or a two-dimensional drop with a neck. Curves for still larger values of α are depicted in Figure 13–9. By using an obtuse angle α, one can rename variables to get equations of the same form as Eqs. (13.99)–(13.101) except for a relative minus sign in Eq. (13.99). Portions of these curves could represent an inverted meniscus (liquid on top) between wetted and non-wetted vertical plates with the inflection occurring between the plates.

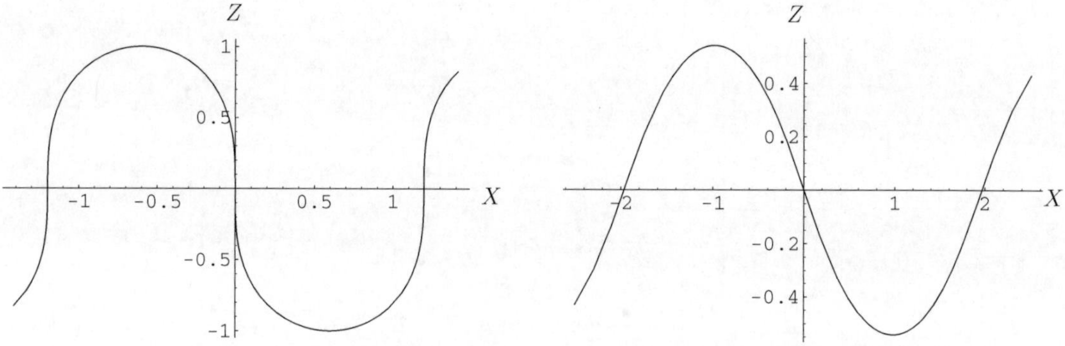

FIGURE 13–9 Two-dimensional interfaces obtained by numerical integration of the system Eqs. (13.99)–(13.101) for $\alpha = \pi/2 = 1.5708$ (left) and $\alpha = 3\pi/4 = 2.3562$ (right). The neck seen for $\alpha = 1.20$ vanishes at $\alpha = \pi/2$ where the curve has vertical points of inflection. For $\alpha = 3\pi/4$ the inflection is no longer vertical and the amplitude has decreased. The amplitude would decrease to zero at $\alpha = \pi$. Portions of curves that contain the inflection could represent an inverted meniscus (liquid on top) between wetted and non-wetted vertical plates with the inflection occurring between the plates. Other portions that do not contain the inflection could represent two-dimensional bubbles or drops.

13.4.2 Examples in Three Dimensions

For three-dimensional interfaces, both principal curvatures, $1/R_2$ and $1/R_2$, must be taken into account. In the general case, one must return to Eq. (13.92) which presents considerable difficulty. We therefore explore briefly the special case where there is axial symmetry about the z axis. In that case, z depends only on $r = \sqrt{x^2 + y^2}$, so

$$\frac{\left[z_{xx}(1 + z_y^2) - 2z_x z_y z_{xy} + z_{yy}(1 + z_x^2)\right]}{\left(1 + z_x^2 + z_y^2\right)^{3/2}} = \frac{z_{rr}}{\left(1 + z_r^2\right)^{3/2}} + \frac{1}{r}\frac{z_r}{\left(1 + z_r^2\right)^{1/2}}. \tag{13.103}$$

The individual terms on the right-hand side of Eq. (13.103) are principal curvatures, the first being the curvature in the r, z plane and the second being in a perpendicular plane. The center of curvature of this second curvature lies on the z axis, as is well known. The problem associated with z being a multiple-valued function on different parts of the surface can be alleviated by introducing the angle ψ between the tangent to the surface and the r axis, specifically by $\mathrm{d}r/\mathrm{d}s = \cos\psi$ and $\mathrm{d}z/\mathrm{d}s = \sin\psi$, where s is arc length measured from the origin $r = z = 0$. With this parametric representation, the right-hand side of Eq. (13.103) takes the form

$$\frac{\mathrm{d}\psi}{\mathrm{d}s} + \frac{\sin\psi}{r}, \tag{13.104}$$

which is valid even when z is a multiple-valued function of r.

For a pendant drop, which is a liquid drop hanging from a syringe and surrounded by a gas, we take $z = 0$ at the bottom of the drop so z is positive on the actual interface.

In Eq. (13.92), the constant $C = 2/R_0 > 0$ where R_0 is the radius of curvature at the drop tip,[20] so we obtain

$$\frac{d\psi}{ds} + \frac{\sin\psi}{r} = -2\frac{z}{a^2} + \frac{2}{R_0} \tag{13.105}$$

with $0 \le \psi \le \pi$ and $a^2 = 2\gamma/[(\rho^\ell - \rho^g)g]$. For a sessile drop, which is a liquid drop resting on a solid surface and surrounded by a gas, we take $z = 0$ to correspond to the maximum height of the drop, so z will be negative on the actual interface and $C = -2/R_0 < 0$. Then Eq. (13.92) becomes

$$\frac{d\psi}{ds} + \frac{\sin\psi}{r} = -2\frac{z}{a^2} - \frac{2}{R_0} \tag{13.106}$$

with $0 \ge \psi \ge -\pi$. Rather than work with negative z and ψ in Eq. (13.106), one can make the transformation $z \to -z$ and $\psi \to -\psi$ and then multiply the equation through by -1 to get an equation for the shape of a "sessile bubble." Then one can combine this equation with Eq. (13.105) to obtain

$$\frac{d\psi}{ds} + \frac{\sin\psi}{r} = \pm 2\frac{z}{a^2} + \frac{2}{R_0}; \quad \begin{array}{l} - \text{ for a pendant drop} \\ + \text{ for a sessile bubble} \end{array} \tag{13.107}$$

with $z \ge 0$ and $0 \le \psi \le \pi$.

Equation (13.107) is a nonlinear differential equation that must be integrated numerically to determine the detailed shape of the drop. To do this, we introduce dimensionless length variables $R := r/R_0$, $Z := z/R_0$, and $S := s/R_0$ to obtain the following set of parametric equations:

$$\frac{d\psi}{dS} = 2 - \frac{\sin\psi}{R} \pm 2\frac{R_0^2}{a^2}Z; \tag{13.108}$$

$$\frac{dR}{dS} = \cos\psi; \tag{13.109}$$

$$\frac{dZ}{dS} = \sin\psi. \tag{13.110}$$

This set of first order equations can be integrated numerically from starting values $\psi = 0$, $Z = S = 0$ and the limiting value $(\sin\psi)/R \to 1$ as $R \to 0$ to produce a shape that depends on the shape parameter R_0^2/a^2. By fitting to an experimentally measured shape and measuring the value of R_0, one can determine the capillary length a, and hence γ. For a very small drop or bubble, $R_0 2/a^2 \ll 1$, the last term in Eq. (13.108) becomes very small and the shape becomes nearly spherical. For a very large drop or bubble, $R_0 2/a^2 \gg 1$ and the shapes will be nearly flat at their tips.

Several sessile shapes computed by numerical integration of Eqs. (13.108)–(13.110) with the plus sign are shown in Figure 13–10. Of course the actual shapes stop when they

[20]At the drop tip, the second term in Eq. (13.104) is not easy to evaluate because $r \to 0$ but also $\sin\psi \to 0$. By symmetry, both radii of curvature are equal at the drop tip, so the total curvature there is just twice the curvature in the r, z plane. Alternatively, one can see this from the Cartesian representation on the left-hand side of Eq. (13.103).

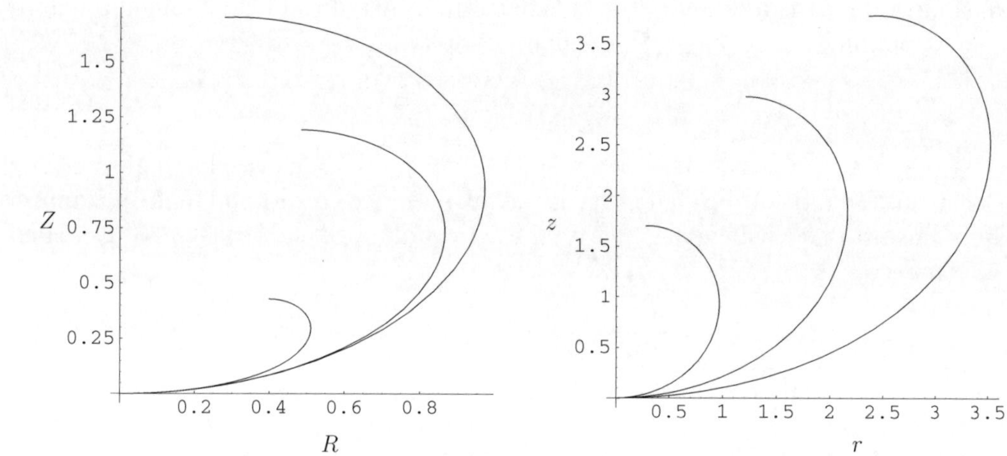

FIGURE 13–10 Curves on the left represent sessile shapes in dimensionless coordinates R, Z scaled with the radius of curvature R_0 at the drop tip. The top curve is for $R_0^2/a^2 = 0.1$ and the middle and bottom curves are for R_0 larger by factors of 2.5 and 5, respectively. Note that the shape becomes less spherical for larger R_0. The curves on the right (which occur in inverted order) are rescaled to reflect true distance in units of $a/\sqrt{10} = a/3.16$, about 1.74 mm for water. The orientation shown is for sessile bubbles; if they are turned upside down they represent sessile drops.

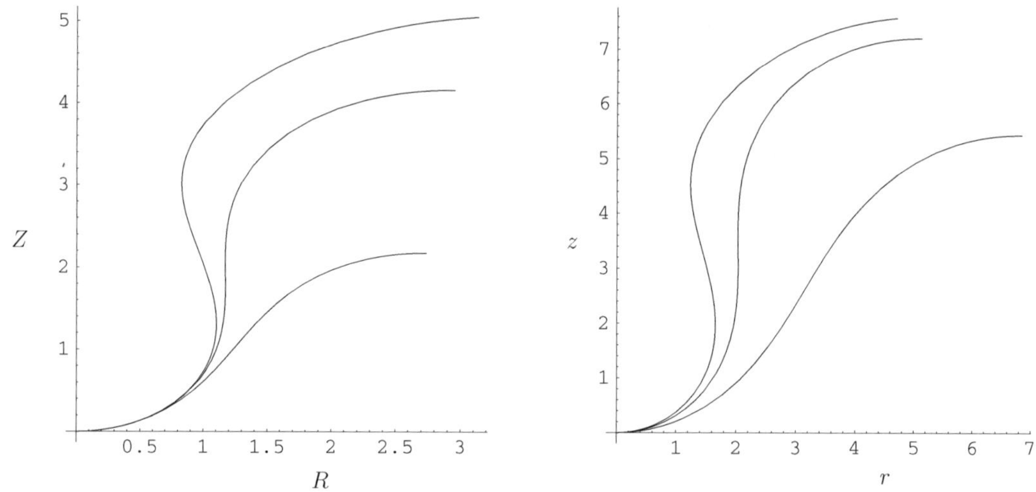

FIGURE 13–11 Curves on the left represent pendant shapes in dimensionless coordinates R, Z scaled with the radius of curvature R_0 at the drop tip. The top curve is for $R_0^2/a^2 = 0.225$, the middle curve for $R_0^2/a^2 = 0.300$ and bottom curve for $R_0^2/a^2 = 0.625$. Note that the shape develops a neck for approximately $R_0^2/a^2 < 0.300$. The curves on the right (that occur in the same order) are rescaled to reflect true distance in units of $a/\sqrt{10} = a/3.16$, about 1.74 mm for water. The orientation shown is for pendant drops; if they are turned upside down they represent pendant bubbles.

reach some bounding surface. For the orientation shown, the shapes represent sessile bubbles attached to the top of some container. If they are turned upside down, they represent sessile drops that rest on a bottom surface.

Several pendant shapes computed by numerical integration of Eqs. (13.108)–(13.110) with the minus sign are shown in Figure 13–11. In this case, a sufficiently small value of approximately $R_0^2/a^2 < 0.300$ gives rise to a drop shape with a neck. As shown, the curves pertain to pendant drops hanging from a syringe; if they are turned upside down they represent "pendant bubbles" that could be produced by means of a syringe that injects gas at the bottom of a liquid.

For a discussion of many other possible shapes of three-dimensional interfaces, including shapes that do not intersect the z axis, which is the axis of revolution, see Princen [30, chapter 1].

14

Thermodynamics of Solid-Fluid Interfaces

In this chapter, we take up solid-fluid interfaces which are regions of discontinuity between an amorphous solid or a crystal and a fluid. This is an advanced topic whose detailed treatment requires some knowledge of elasticity tensors and surface differential geometry. Those not familiar with elasticity tensors can skip Sections 14.1.1 and 14.1.2, the results of which are not used in the remainder of the chapter. Needed aspects of surface differential geometry are covered in Appendix C.

Many aspects of solid-fluid interfaces are the same as those for fluid-fluid interfaces treated in Chapter 13. Nevertheless, some aspects are very different because solids are quite rigid and can support states of shear over considerable periods of time. In other words, they can behave elastically, except at very high temperatures where they can deform plastically by creep. Consequently, the mechanical properties of solids must be described in terms of stress and strain tensors. Moreover, crystalline solids are anisotropic, which results in anisotropy of the interfacial free energy, γ. Solid surfaces can change their areas in two distinct ways, by stretching and by accretion at their boundaries. Therefore, they must be characterized by strain variables absent for a fluid.

We begin by considering planar solid-fluid interfaces, essentially parallel to our treatment of fluid-fluid interfaces but with new complications, including the fact that the interfacial free energy can be referenced to the area of either a stretched or an unstretched interface. The corresponding Gibbs adsorption equation contains the surface stress tensor that must be defined carefully with respect to the state of strain of the solid. This surface stress tensor is anisotropic for a crystal. Anisotropy of γ is treated by means of an auxiliary vector field, the $\boldsymbol{\xi}$-vector field, introduced by Gibbs, whose properties were developed by Hoffman and Cahn. Anisotropy gives rise to torques that arise when the orientation of a surface element is changed. These torques affect the equilibrium conditions at triple junctions where phases meet. The $\boldsymbol{\xi}$-vector formalism can be extended to orientations for which the derivatives of γ are not well defined by identifying $\boldsymbol{\xi}$ with the diameter of a Herring sphere used to determine the Gibbs-Wulff equilibrium shape of a small crystal. Large planar crystal surfaces whose surface orientations are not present on the equilibrium shape can lower their surface free energy by faceting. General conditions for equilibrium at curved surfaces of crystals, described parametrically, are derived by using a variational technique. The equilibrium shape is shown to be similar to that of a polar plot of the $\boldsymbol{\xi}$-vector, suitably truncated to form a convex body. By using a Monge representation of a crystal surface, the Herring formula for local equilibrium is derived. It is shown that

Thermal Physics. http://dx.doi.org/10.1016/B978-0-12-803304-3.00014-4

the surface free energy, per unit area of a reference plane from which the surface height is measured, is the Legendre transform of the equilibrium shape, and *vice versa*.

Finally, we make a few remarks about solid-solid interfaces.

14.1 Planar Solid-Fluid Interfaces

We now treat planar interfaces, such as depicted in Figure 13–1, except that one phase is a solid (superscript s) and the other is a fluid (superscript F). The bulk solid is assumed to be homogeneous; in particular, it is in a state of homogeneous stress and strain. If the solid is a crystal, we treat a constrained equilibrium such that the planar interface has a definite direction with respect to the crystallographic axes. Such an interface might not be stable with respect to breakup into a hill and valley structure (made up of facets) but we will examine this possibility later in Section 14.4. For amorphous solids, stability of a planar interface with respect to faceting is not an issue.

When a bulk solid is in equilibrium at temperature T with a bulk fluid across a hypothetical surface element with normal $\hat{\mathbf{n}}$ pointing from solid to fluid, the following boundary conditions are valid at that surface element [31–33]:

$$u_v^s - Ts_v^s - \sum_i \mu_i^F \rho_i^s = -p^F; \tag{14.1}$$

$$\sum_\alpha n_\beta \sigma_{\alpha\beta} = -p^F n_\beta, \tag{14.2}$$

where u_v^s is the internal energy density of the solid, with α and β representing Cartesian coordinates, s_v^s is the entropy density of the solid, ρ_i^s is the partial density of chemical component i in the solid, $\sigma_{\alpha\beta}$ is the symmetric Cauchy stress tensor in the solid, p^F is the pressure of the fluid, and μ_i^F is the chemical potential of component i in the fluid. Equation (14.2) is just a balance of forces at the surface element. If we take the surface element to be perpendicular to the z-axis, it becomes $\sigma_{zz} = -p^F$, $\sigma_{xz} = \sigma_{yz} = 0$, consistent with the fact that a fluid in equilibrium cannot support shear. Equation (14.1) is a thermodynamic condition. If the mobility of the chemical components of the solid was unrestricted and the solid was in chemical equilibrium,[1] its chemical potentials $\mu_i^s = \mu_i^F$. If the solid behaved like a fluid, it would be in a state of hydrostatic stress, so $\sigma_{\alpha\beta} = -p^s\delta_{\alpha\beta}$ and the left-hand side of Eq. (14.1), *via* its Euler equation, would be the negative of its

[1]Generally speaking, solids are quite rigid and mobility of chemical components within them is quite slow, although not zero. On practical time scales, mobility of such components can sometimes be ignored. This leads to the concept of a Gibbs solid in which the "substance of the solid" is fixed and immobile. Alternatively, movement of solid components can be allowed to occur but restricted to obey certain rules. For example, in a Larché-Cahn (LC) solid [31, 32], components that can only reside on a lattice are allowed to move only by virtue of exchange with point defects, namely lattice vacancies. For the LC solid, and with vacancies assumed to be a conserved species within a single crystal, LC define diffusion potentials $\mathcal{M}_i$ that are formally equal to the differences of chemical potentials of chemical components and chemical potentials of vacancies, calculated in that extended description. So for an LC solid, their $\mathcal{M}_i$ would be equal to our μ_i^s.

pressure p^s. In that case, Eqs. (14.1) and (14.2) would coalesce and become simply $p^s = p^F$. But a solid in a general state of stress has no such simple Euler equation.

For a homogeneous solid, however, the left-hand side of Eq. (14.1) is uniform (independent of position) so one can multiply by the volume of the solid V^s to obtain

$$U^s - TS^s - \mu_i^F N_i^s = -p^F V^s, \quad \text{pseudo-Euler, homogeneous cylindrical solid,} \tag{14.3}$$

which resembles an Euler equation except that μ_i^F and p^F pertain to the fluid. If such a homogeneous solid were surrounded by a fluid, Eq. (14.2) would compel the solid to be in a state of hydrostatic stress. On the other hand, for a *constrained equilibrium* in which a homogeneous cylindrical solid that is only in contact with the fluid across a planar interface perpendicular to the generators of its cylindrical surface, Eq. (14.3) applies and the solid can be in a state of nonhydrostatic stress. The fact that Eq. (14.3) also applies to a homogeneous solid and a homogeneous liquid separated by an actual planar region of discontinuity can be seen by considering a layer bounded by imaginary planes located in homogeneous phases on opposite sides of the region of discontinuity, just as was done for fluid-fluid interfaces. Then one can study variations in which the layer is unchanged but translates intact in either direction perpendicular to the planes that bound it. For such variations, changes of the homogeneous phases are the same as if the layer did not exist and were replaced by a mathematical plane.

Armed with the pseudo-Euler Eq. (14.3), we can *define* an excess pseudo-Kramers potential for a system having a homogeneous solid and a homogeneous liquid and a planar solid-fluid planar Gibbs dividing surface by means of the equation

$$\begin{aligned} K^{xs} &= U - TS - \sum_i \mu_i^F N_i - (U^s - TS^s - \mu_i^F N_i^s) - (U^F - TS^F - \mu_i^F N_i^F) \\ &= U^{xs} - TS^{xs} - \sum_i \mu_i^F N_i^{xs} = U - TS - \sum_i \mu_i^F N_i + p^F V, \end{aligned} \tag{14.4}$$

where the last expression is clearly independent of the location of the dividing surface that separates the homogeneous solid from the homogeneous fluid. Here, $U^{xs} = U - U^s - U^F$, $S^{xs} = S - S^s - S^F$, and $N_i^{xs} = N_i - N_i^s - N_i^F$ but $V^{xs} = V - V^s - V^F = 0$, since the bulk phases meet at the dividing surface. Then we can define an excess potential per unit area by dividing by a suitable area. Following Cahn [28] or [29, pp. 379-399], we distinguish two cases, γ obtained by dividing by the area A of the actual strained state and γ_0 obtained by dividing by the area A_0 of a homogeneous reference state of the solid, by definition the state of zero strain. Specifically,

$$\gamma A = \gamma_0 A_0 = U - TS - \sum_i \mu_i^F N_i + p^F V = U^{xs} - TS^{xs} - \sum_i \mu_i^F N_i^{xs}. \tag{14.5}$$

We could also use these same definitions of γ and γ_0 for a *layer model*, similar to that for the fluid-fluid case, Eqs. (13.33)–(13.36), which gives

$$\gamma A = \gamma_0 A_0 = U - TS - \sum_i \mu_i^F N_i + p^F V = U^L - TS^L - \sum_i \mu_i^F N_i^L + p^F V^L, \tag{14.6}$$

where the bulk phases only extend to the imaginary planes that bound the layer, so $V^L = V - V^s - V^F \neq 0$. As in the fluid case, most of these excess quantities and layer quantities have no physical meaning because they depend on the location of the dividing surface or the bounding planes, but their combinations used to form γ or γ_0 are independent of these locations and do have physical meaning.

We treat the more general layer model first and then indicate the slight modification needed to treat the dividing surface model. To do this, we adopt an equation for $\mathrm{d}U^L$ of the form

$$\mathrm{d}U^L = T\mathrm{d}S^L - p^F\mathrm{d}V^L + \sum_i \mu_i^F \mathrm{d}N_i^L + A\sum_{\alpha\beta} f_{\alpha\beta}^L \mathrm{d}\varepsilon_{\alpha\beta}, \tag{14.7}$$

which is similar to Eq. (13.41) except for the last term. This last term accounts for *stretching* of the interface that accompanies straining of the *bulk solid* and replaces $\gamma\mathrm{d}A$ for a fluid-fluid interface. Here, the Cartesian indices α and β take on the values x and y for an interface perpendicular to the z-direction, as above. The 2×2 tensor $\varepsilon_{\alpha\beta}$ is a symmetric strain tensor (see Eq. (14.17)) *measured in the bulk homogeneous solid* and $f_{\alpha\beta}^L$ is a symmetric stress tensor. This stress tensor must be consistent with the symmetry of the underlying solid, anisotropic if crystalline and isotropic if amorphous.

14.1.1 Adsorption Equation in the Reference State

By combining Eq. (14.7) with the differential of Eq. (14.6), and recognizing that $\mathrm{d}A_0 = 0$, we obtain

$$\mathrm{d}\gamma_0 = -\frac{S^L}{A_0}\mathrm{d}T + \frac{V^L}{A_0}\mathrm{d}p^F - \sum_i \frac{N_i^L}{A_0}\mathrm{d}\mu_i^F + \frac{A}{A_0}\sum_{\alpha,\beta} f_{\alpha\beta}^L\,\mathrm{d}\varepsilon_{\alpha\beta}, \tag{14.8}$$

which is the counterpart to the Gibbs adsorption equation for fluids, Eq. (13.44). Similar to the fluid case, the variables T, μ_i^F, p^F, and $\varepsilon_{\alpha\beta}$ are not independent because of the equations of bulk equilibrium of the solid and fluid phases. Two of these can be chosen as dependent variables and their differentials expressed in terms of the differentials of the others, most elegantly by using the determinant formalism discussed in terms of Cahn's layer model of fluids in Section 13.1.3. To do this, we need a Gibbs-Duhem equation for the fluid, which is just

$$S^F\,\mathrm{d}T - V^F\,\mathrm{d}p^F + \sum_i N_i^F\,\mathrm{d}\mu_i^F = 0, \tag{14.9}$$

but also an equivalent Gibbs-Duhem equation for a cylinder of homogeneous bulk solid in equilibrium with that fluid across a plane perpendicular to the z-axis. This equation can be written in the form

$$S^s\,\mathrm{d}T - V^s\,\mathrm{d}p^F + \sum_i N_i^s\,\mathrm{d}\mu_i^F - V^s\sum_{\alpha\beta}\sigma_{\alpha\beta}^{\mathrm{lat}}\,\mathrm{d}\varepsilon_{\alpha\beta} = 0, \tag{14.10}$$

where

$$\sigma_{\alpha,\beta}^{\mathrm{lat}} \equiv \sum_{\kappa,\lambda} \frac{\partial x'_\alpha}{\partial x_\kappa}(\sigma_{\kappa\lambda} + p^F\delta_{\kappa\lambda})\frac{\partial x'_\beta}{\partial x_\lambda}. \tag{14.11}$$

The last term in Eq. (14.10), in which all sums are only over x and y coordinates, is present because for a cylindrical solid in a nonhydrostatic homogeneous state of stress, the forces needed to stretch it laterally are different from those needed to stretch it longitudinally. Here, $\sigma_{\kappa\lambda}$ is the Cauchy stress tensor of the homogeneous solid, the coordinates $\mathbf{x}'$ are those of a hydrostatic reference state, and the coordinates $\mathbf{x}$ are those of the actual state. If the solid were in a state of hydrostatic stress in its actual state, $\sigma_{\kappa\lambda} = -p^F\delta_{\kappa\lambda}$ and the last term would vanish. We can therefore write (with a notation similar to Eq. (13.46))

$$d\gamma_0 = -[(S^L/A_0)/XY]\,dT + [(V^L/A_0)/XY]\,dp^F - \sum_{i=1}^{\kappa}[(N_i^L/A_0)/XY]\,d\mu_i^F + \sum_{\alpha,\beta} f_{\alpha\beta}^C\,d\varepsilon_{\alpha\beta}, \tag{14.12}$$

where X and Y are the extensive conjugates to two distinct intensive variables of the set T, μ_i^F, p^F that are chosen to be dependent variables.[2] As with fluids, the two coefficients of the dependent intensive variables will vanish. Here,

$$f_{\alpha\beta}^C \equiv \frac{1}{A_0}\frac{\begin{vmatrix} Af_{\alpha\beta}^L & X^L & Y^L \\ V^s\sigma_{\alpha\beta}^{\mathrm{lat}} & X^s & Y^s \\ 0 & X^F & Y^F \end{vmatrix}}{\begin{vmatrix} X^s & Y^s \\ X^F & Y^F \end{vmatrix}} \tag{14.13}$$

and is independent of the choice of the planes that bound the layer, although it does depend on the choice of independent variables. Consequently,

$$\left(\frac{\partial\gamma_0}{\partial\varepsilon_{\alpha\beta}}\right)_{A_0 \text{ and } \kappa \text{ independent intensive variables}} = f_{\alpha\beta}^C. \tag{14.14}$$

The 2×2 tensor $f_{\alpha\beta}^C$ is the surface stress defined by Cahn [28] or [29, pp. 379-399]. As he points out, the application of tractions to the bulk solid usually produces only a small shift in the other intensive variables and can frequently be ignored. If the actual state of the solid is hydrostatic, $\sigma_{\alpha\beta}^{\mathrm{lat}} = 0$ and we have simply $f_{\alpha\beta}^C = (A/A_0)f_{\alpha\beta}^L$. If the actual state is taken to be a state of zero strain (coincident with a hydrostatic reference state), then $f_{\alpha\beta}^C = f_{\alpha\beta}^L$.

Note that Eq. (14.12) also holds formally for the Gibbs excess quantities with the understanding that $V^L = 0$, which does not require the coefficient of dp^F to be zero unless either X or Y is chosen to be V.

[2] We retain $\varepsilon_{\alpha\beta}$ as independent variables because the application of tractions to the solid makes only a small second-order change to the relationships among the set T, μ_i^F, p^F. An estimate by Sekerka and Cahn [34] for a single component Gibbs solid shows that the equilibrium temperature would be lowered by about 10^{-3} K for a shear stress of the order of 10 MPa.

14.1.2 Adsorption Equation in the Actual State

We now examine the parallel development when γ is defined with reference to the area A of the actual state. For the layer model, we combine the differential of Eq. (14.6) with Eq. (14.7) to obtain

$$\mathrm{d}\gamma = -\frac{S^L}{A}\,\mathrm{d}T + \frac{V^L}{A}\,\mathrm{d}p^F - \sum_i \frac{N_i^L}{A}\,\mathrm{d}\mu_i^F + \sum_{\alpha,\beta} f_{\alpha\beta}^L \mathrm{d}\varepsilon_{\alpha\beta} - \gamma\frac{\mathrm{d}A}{A}. \tag{14.15}$$

From a well-known relation [35, p. 16] in elasticity theory with coordinates $\mathbf{x}$ in the actual state and $\mathbf{x}'$ in the reference state, one has (in a 2×2 space)

$$\frac{\mathrm{d}A}{A} = \mathrm{d}\ln A = \mathrm{d}\ln\det\left[\frac{\partial x_\alpha}{\partial x'_\beta}\right] = \sum_{\kappa\lambda}\frac{\partial x'_\lambda}{\partial x_\kappa}\mathrm{d}\frac{\partial x_\kappa}{\partial x'_\lambda} = \sum_{\kappa\lambda\nu}\frac{\partial x'_\lambda}{\partial x_\kappa}\frac{\partial x'_\nu}{\partial x_\kappa}\mathrm{d}\varepsilon_{\lambda\nu}, \tag{14.16}$$

where

$$\varepsilon_{\lambda\nu} = \frac{1}{2}\left[\frac{\partial x_\rho}{\partial x'_\lambda}\frac{\partial x_\rho}{\partial x'_\nu} - \delta_{\lambda\nu}\right] = \frac{1}{2}\left[\frac{\partial u_\nu}{\partial x'_\lambda} + \frac{\partial u_\lambda}{\partial x'_\nu} + \sum_\rho \frac{\partial u_\rho}{\partial x'_\lambda}\frac{\partial u_\rho}{\partial x'_\nu}\right] \tag{14.17}$$

is the full nonlinear strain tensor and $\mathbf{u} = \mathbf{x} - \mathbf{x}'$ is the displacement. Thus, the last two terms in Eq. (14.15) can be combined to yield

$$\sum_{\alpha,\beta} f_{\alpha\beta}^L \mathrm{d}\varepsilon_{\alpha\beta} - \gamma\frac{\mathrm{d}A}{A} = \sum_{\alpha,\beta}\left[f_{\alpha\beta}^L - \gamma\sum_\kappa \frac{\partial x'_\alpha}{\partial x_\kappa}\frac{\partial x'_\beta}{\partial x_\kappa}\right]\mathrm{d}\varepsilon_{\alpha\beta}. \tag{14.18}$$

Then the counterpart to Eq. (14.12) is

$$\mathrm{d}\gamma = -[(S^L/A)/XY]\,\mathrm{d}T + [(V^L/A)/XY]\,\mathrm{d}p^F - \sum_{i=1}^{\kappa}[(N_i^L/A)/XY]\,\mathrm{d}\mu_i^F + \sum_{\alpha,\beta} f_{\alpha\beta}^A\,\mathrm{d}\varepsilon_{\alpha\beta}, \tag{14.19}$$

where

$$f_{\alpha\beta}^A \equiv \frac{1}{A}\frac{\begin{vmatrix} A[f_{\alpha\beta}^L - \gamma\sum_\kappa(\partial x'_\alpha/\partial x_\kappa)(\partial x'_\beta/\partial x_\kappa)] & X^L & Y^L \\ V^s\sigma_{\alpha\beta}^{\mathrm{lat}} & X^s & Y^s \\ 0 & X^F & Y^F \end{vmatrix}}{\begin{vmatrix} X^s & Y^s \\ X^F & Y^F \end{vmatrix}}. \tag{14.20}$$

Consequently

$$\left(\frac{\partial\gamma}{\partial\varepsilon_{\alpha\beta}}\right)_{A_0 \text{ and } \kappa \text{ independent intensive variables}} = f_{\alpha\beta}^A. \tag{14.21}$$

If the actual state of the solid is chosen to be hydrostatic, we have noted previously that $\sigma_{\alpha\beta}^{\mathrm{lat}} = 0$, in which case $f_{\alpha\beta}^A = f_{\alpha\beta}^L - \gamma\sum_\kappa(\partial x'_\alpha/\partial x_\kappa)(\partial x'_\beta/\partial x_\kappa)$. If the actual state of the bulk solid is coincident with a hydrostatic reference state, then simply $f_{\alpha\beta}^A = f_{\alpha\beta}^L - \gamma\delta_{\alpha\beta}$.

Returning to the general case, we can expand the determinant in the numerator of Eq. (14.20) to obtain

$$Af^A_{\alpha\beta} = \frac{\begin{vmatrix} Af^L_{\alpha\beta} & X^L & Y^L \\ V^s\sigma^{\text{lat}}_{\alpha\beta} & X^s & Y^s \\ 0 & X^F & Y^F \end{vmatrix}}{\begin{vmatrix} X^s & Y^s \\ X^F & Y^F \end{vmatrix}} - A\gamma \sum_\kappa \frac{\partial x'_\alpha}{\partial x_\kappa}\frac{\partial x'_\beta}{\partial x_\kappa}. \tag{14.22}$$

Then comparison with Eq. (14.13) shows that

$$f^C_{\alpha\beta} = \frac{A}{A_0}\left[\gamma \sum_\kappa \frac{\partial x'_\alpha}{\partial x_\kappa}\frac{\partial x'_\beta}{\partial x_\kappa} + f^A_{\alpha\beta}\right], \tag{14.23}$$

which can also be written

$$\frac{\partial \gamma_0}{\partial \varepsilon_{\alpha\beta}} = \frac{A}{A_0}\left[\gamma \sum_\kappa \frac{\partial x'_\alpha}{\partial x_\kappa}\frac{\partial x'_\beta}{\partial x_\kappa} + \frac{\partial \gamma}{\partial \varepsilon_{\alpha\beta}}\right]. \tag{14.24}$$

In the case that the actual state of the bulk solid is coincident with a hydrostatic reference state, Eq. (14.24) becomes

$$\frac{\partial \gamma_0}{\partial \varepsilon_{\alpha\beta}} = \gamma\delta_{\alpha\beta} + \frac{\partial \gamma}{\partial \varepsilon_{\alpha\beta}}. \tag{14.25}$$

If the solid behaved like a fluid, then $\partial\gamma/\partial\varepsilon_{\alpha\beta} = 0$ and the surface stress would be isotropic and equal to γ, as we found previously for a fluid-fluid interface.

As Cahn points out, the relationship Eq. (14.25) can be based on the fact that $\gamma_0 A_0 = \gamma A$ and the geometrical relationship of A to A_0 because of strain. Then by using $A_0 \mathrm{d}\gamma_0 = \gamma\, \mathrm{d}A +$ $\mathrm{d}\gamma$ and Eq. (14.16), one would obtain the full nonlinear result Eq. (14.24). For small strain, one has simply $A/A_0 = 1 + \sum_\nu \varepsilon_{\nu\nu}$ so

$$\frac{\partial \gamma_0}{\partial \varepsilon_{\alpha\beta}} \approx \left(1 + \sum_\nu \varepsilon_{\nu\nu}\right)\left[\gamma(\delta_{\alpha\beta} - 2\varepsilon_{\alpha\beta}) + \frac{\partial \gamma}{\partial \varepsilon_{\alpha\beta}}\right] \approx \gamma\delta_{\alpha\beta} + \frac{\partial \gamma}{\partial \varepsilon_{\alpha\beta}} \tag{14.26}$$

to lowest order. This is a linearized version of Eq. (14.24) and happens to agree with the exact Eq. (14.25) for the special states chosen in that case. So Eq. (14.23) and equivalently Eq. (14.24) are always true for geometrical reasons, even in the nonlinear case. Note, however, that these derivatives of γ_0 and of γ are only simply related to $f^L_{\alpha\beta}$ when $\sigma^{\text{lat}}_{\alpha\beta}$ is zero unless the actual bulk solid is hydrostatic or the small effect of shear stress on the bulk equilibrium (embodied by $V^s\sigma^{\text{lat}}_{\alpha\beta}$) is negligible.

14.2 Anisotropy of γ

In Section 14.1, we defined the excess free energy γ for a constrained equilibrium such that the planar interface of a homogeneous bulk crystal in equilibrium with a fluid has a definite direction with respect to the crystallographic axes. In this section, we treat the explicit dependence of γ on the interface orientation, which we characterize by its unit normal vector $\hat{\mathbf{n}}$. Thus we write $\gamma(\hat{\mathbf{n}})$, where $\hat{\mathbf{n}}$ points from crystal[3] to fluid, and in which all other variables that γ depends on have been suppressed.[4] Since $n_x^2 + n_y^2 + n_z^2 = 1$, the components of $\hat{\mathbf{n}}$ are not independent, so one cannot take partial derivatives with respect to one of them while holding the other two constant. Therefore, to treat the angular derivatives of $\gamma(\mathbf{n})$ in a manner that is independent of any coordinate system, we resort to an auxiliary vector field $\boldsymbol{\xi}$ that was introduced by Gibbs [3] and whose properties were developed in detail by Hoffman and Cahn [36, 37].

In order to facilitate the definition of $\boldsymbol{\xi}$, we introduce a three-dimensional vector field

$$\mathbf{P} := P\hat{\mathbf{n}}, \tag{14.27}$$

where the magnitude P of $\mathbf{P}$ can vary. Then one defines a function

$$\tilde{\gamma}(\mathbf{P}) := P\gamma(\hat{\mathbf{n}}) = P\gamma(\mathbf{P}/P), \tag{14.28}$$

which is a homogeneous function of degree 1 in the components P_α of $\mathbf{P}$. Thus, by means of the Euler theorem of homogeneous functions, one can take partial derivatives of $\tilde{\gamma}(\mathbf{P})$ to obtain[5]

$$\sum_\alpha P_\alpha \frac{\partial \tilde{\gamma}(\mathbf{P})}{\partial P_\alpha} = \tilde{\gamma}(\mathbf{P}). \tag{14.29}$$

We now define the vector field

$$\xi_\alpha(\hat{\mathbf{n}}) := \frac{\partial \tilde{\gamma}(\mathbf{P})}{\partial P_\alpha} \tag{14.30}$$

or more succinctly

$$\boldsymbol{\xi}(\hat{\mathbf{n}}) := \nabla_P \tilde{\gamma}(\mathbf{P}) = \nabla_P[P\gamma(\mathbf{P}/P)], \tag{14.31}$$

where the operator ∇_P is the gradient in $\mathbf{P}$ space. The fact that $\boldsymbol{\xi}$ depends only on $\hat{\mathbf{n}}$ follows because it is a homogenous function of degree 0 in the P_α. Then combining Eqs. (14.29) and (14.30) gives $\tilde{\gamma} = \mathbf{P} \cdot \boldsymbol{\xi}$ from which

$$\mathbf{n} \cdot \boldsymbol{\xi} = \gamma(\hat{\mathbf{n}}). \tag{14.32}$$

Moreover,

$$\mathrm{d}\tilde{\gamma} = \boldsymbol{\xi} \cdot \mathrm{d}\mathbf{P} + \mathbf{P} \cdot \mathrm{d}\boldsymbol{\xi}, \tag{14.33}$$

but also

$$\mathrm{d}\tilde{\gamma} = \sum_\alpha \frac{\partial \tilde{\gamma}(\mathbf{P})}{\partial P_\alpha} \mathrm{d}P_\alpha = \boldsymbol{\xi} \cdot \mathrm{d}\mathbf{P}, \tag{14.34}$$

[3]For some crystals, $\gamma(\hat{\mathbf{n}}) \neq \gamma(-\hat{\mathbf{n}})$ in which case one needs to distinguish which interface is being considered.

[4]These are the independent variables that appear in Eq. (14.19). In this section, we write $\mathrm{d}\gamma$ for brevity but it should be understood that the only variable that is allowed to change is the orientation $\hat{\mathbf{n}}$.

[5]For now we assume that $\gamma(\hat{\mathbf{n}})$ is differentiable with continuous derivatives and later discuss the implications of discontinuities in its derivatives.

by Eq. (14.30), so the second term on the right of Eq. (14.33) must vanish. Thus

$$\hat{\mathbf{n}} \cdot \mathrm{d}\boldsymbol{\xi} = 0. \tag{14.35}$$

By combining Eq. (14.35) with the differential of Eq. (14.32), we deduce that

$$\mathrm{d}\gamma(\hat{\mathbf{n}}) = \boldsymbol{\xi} \cdot \mathrm{d}\hat{\mathbf{n}}. \tag{14.36}$$

Equation (14.32) plus either Eq. (14.35) or Eq. (14.36) completely defines the vector field $\boldsymbol{\xi}(\hat{\mathbf{n}})$, although Eq. (14.31) is convenient for its actual calculation. We observe that the normal component of $\boldsymbol{\xi}$ is $\gamma(\hat{\mathbf{n}})$ itself and since $\mathrm{d}\hat{\mathbf{n}}$ is perpendicular to $\hat{\mathbf{n}}$, we see that the tangential component of $\boldsymbol{\xi}$ shows how $\gamma(\hat{\mathbf{n}})$ varies with orientation.

Example Problem 14.1. For a crystal having cubic symmetry,

$$\gamma(\hat{\mathbf{n}}) = \gamma_0 + \gamma_4(n_x^4 + n_y^4 + n_z^4), \tag{14.37}$$

where γ_0 and γ_4 are constants, represents the lowest order anisotropy in the components of $\hat{\mathbf{n}}$. No anisotropy is possible to second order because $n_x^2 + n_y^2 + n_z^2 = 1$. Calculate $\boldsymbol{\xi}(\hat{\mathbf{n}})$ and demonstrate explicitly that Eqs. (14.32), (14.35), and (14.36) are satisfied.

Solution 14.1. We have

$$\tilde{\gamma}(\mathbf{P}) = \gamma_0 P + \gamma_4 \frac{(P_x^4 + P_y^4 + P_z^4)}{P^3}, \tag{14.38}$$

so

$$\nabla_P \tilde{\gamma}(\mathbf{P}) = \gamma_0 \frac{\mathbf{P}}{P} - 3\gamma_4 \frac{\mathbf{P}}{P} \frac{(P_x^4 + P_y^4 + P_z^4)}{P^4} + 4\gamma_4 \frac{1}{P^3}(P_x^3\,\hat{\mathbf{i}} + P_y^3\,\hat{\mathbf{j}} + P_z^3\,\hat{\mathbf{k}}). \tag{14.39}$$

Thus,

$$\begin{aligned} \boldsymbol{\xi}(\hat{\mathbf{n}}) &= \gamma_0 \hat{\mathbf{n}} - 3\gamma_4 \hat{\mathbf{n}}(n_x^4 + n_y^4 + n_z^4) + 4\gamma_4(n_x^3\,\hat{\mathbf{i}} + n_y^3\,\hat{\mathbf{j}} + n_z^3\,\hat{\mathbf{k}}) \\ &= \hat{\mathbf{n}}\,\gamma(\hat{\mathbf{n}}) + 4\gamma_4\left[(n_x^3\,\hat{\mathbf{i}} + n_y^3\,\hat{\mathbf{j}} + n_z^3\,\hat{\mathbf{k}}) - \hat{\mathbf{n}}(n_x^4 + n_y^4 + n_z^4)\right], \end{aligned} \tag{14.40}$$

where the term with square brackets in the second line is the part of $\boldsymbol{\xi}$ that is perpendicular to $\hat{\mathbf{n}}$. Obviously Eq. (14.32) is satisfied. We calculate

$$\begin{aligned} \mathrm{d}\boldsymbol{\xi} &= \gamma_0\,\mathrm{d}\hat{\mathbf{n}} - 3\gamma_4(n_x^4 + n_y^4 + n_z^4)\,\mathrm{d}\hat{\mathbf{n}} - 12\gamma_4 \hat{\mathbf{n}}(n_x^3\,\mathrm{d}n_x + n_y^3\,\mathrm{d}n_y + n_z^3\,\mathrm{d}n_z) \\ &\quad + 12\gamma_4(n_x^2\,\hat{\mathbf{i}}\,\mathrm{d}n_x + n_y^3\,\hat{\mathbf{j}}\,\mathrm{d}n_y + n_z^3\,\hat{\mathbf{k}}\,\mathrm{d}n_z). \end{aligned} \tag{14.41}$$

When we dot $\hat{\mathbf{n}}$ into $\mathrm{d}\boldsymbol{\xi}$, the first two terms vanish because $\hat{\mathbf{n}}$ is perpendicular to $\mathrm{d}\hat{\mathbf{n}}$ and the remaining terms cancel one another, so Eq. (14.35) is satisfied. Of course $\mathrm{d}\hat{\mathbf{n}} = \hat{\mathbf{i}}\,\mathrm{d}n_x + \hat{\mathbf{j}}\,\mathrm{d}n_y + \hat{\mathbf{k}}\,\mathrm{d}n_z$ so

$$\boldsymbol{\xi} \cdot \mathrm{d}\hat{\mathbf{n}} = 4\gamma_4(n_x^3\,\mathrm{d}n_x + n_y^3\,\mathrm{d}n_y + n_z^3\,\mathrm{d}n_z) = \mathrm{d}\gamma, \tag{14.42}$$

and Eq. (14.36) is satisfied.

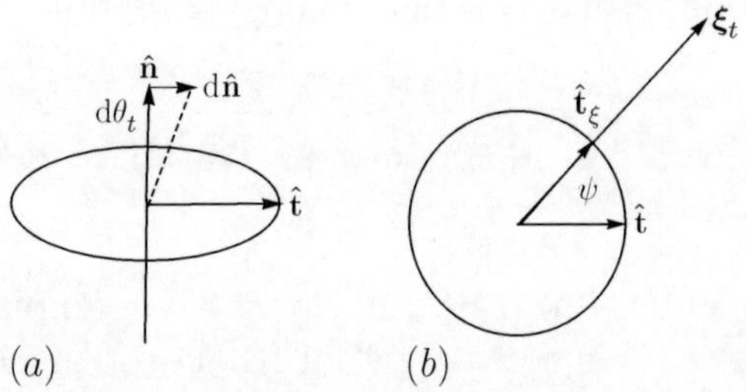

FIGURE 14–1 (a) Illustration of $d\theta_t$, the angle between the surface normal $\hat{\mathbf{n}}$ and $\mathbf{n}+d\hat{\mathbf{n}}$. The unit vector $\hat{\mathbf{t}}$ is in the direction of $d\hat{\mathbf{n}}$. (b) View looking along $-\hat{\mathbf{n}}$ showing the relationship of $\hat{\mathbf{t}}$ to the unit vector $\hat{\mathbf{t}}_\xi$ that lies along the tangential component ξ_t of the ξ-vector.

Let $d\theta_t$ be the angle between $\hat{\mathbf{n}}+d\hat{\mathbf{n}}$ and $\hat{\mathbf{n}}$ and $\hat{\mathbf{t}}$ be a unit vector tangent to $d\hat{\mathbf{n}}$, as illustrated in Figure 14–1. Then

$$d\hat{\mathbf{n}} = \hat{\mathbf{t}}\,d\theta_t \tag{14.43}$$

and Eq. (14.36) can be written

$$\frac{\partial\gamma(\hat{\mathbf{n}})}{\partial\theta_t} = \boldsymbol{\xi}\cdot\hat{\mathbf{t}} = \boldsymbol{\xi}_t\cdot\hat{\mathbf{t}}, \tag{14.44}$$

where the part of $\boldsymbol{\xi}$ that is tangential to the surface is

$$\boldsymbol{\xi}_t = \boldsymbol{\xi} - \hat{\mathbf{n}}(\hat{\mathbf{n}}\cdot\boldsymbol{\xi}) = \boldsymbol{\xi} - \hat{\mathbf{n}}\gamma. \tag{14.45}$$

If we denote the special value of $\hat{\mathbf{t}}$ in the direction of $\boldsymbol{\xi}_t$ by $\hat{\mathbf{t}}_\xi$ and the corresponding value of θ_t by θ_ξ , it follows that

$$\boldsymbol{\xi}_t = |\boldsymbol{\xi}_t|\hat{\mathbf{t}}_\xi = \frac{\partial\gamma(\hat{\mathbf{n}})}{\partial\theta_\xi}\,\hat{\mathbf{t}}_\xi. \tag{14.46}$$

Then Eq. (14.44) can be written

$$\frac{\partial\gamma(\hat{\mathbf{n}})}{\partial\theta_t} = \frac{\partial\gamma(\hat{\mathbf{n}})}{\partial\theta_\xi}\,\hat{\mathbf{t}}_\xi\cdot\hat{\mathbf{t}} = \frac{\partial\gamma(\hat{\mathbf{n}})}{\partial\theta_\xi}\,\cos\psi, \tag{14.47}$$

where ψ is the angle between $\hat{\mathbf{t}}$ and $\hat{\mathbf{t}}_\xi$, as illustrated in Figure 14–1. It follows that $\partial\gamma(\hat{\mathbf{n}})/\partial\theta_\xi$ is the maximum value of $\partial\gamma(\hat{\mathbf{n}})/\partial\theta_t$.

Consider next a small planar element of area A having normal $\hat{\mathbf{n}}$ and define $\mathbf{A}=\hat{\mathbf{n}}A$. Its free energy is $\gamma A = \boldsymbol{\xi}\cdot\mathbf{A}$. The work that must be done by an external agent to change its area and its orientation by infinitesimal amounts is

$$\delta w = d(\gamma A) = d(\boldsymbol{\xi}\cdot\mathbf{A}) = \boldsymbol{\xi}\cdot d\mathbf{A}, \tag{14.48}$$

because $\mathbf{A}\cdot d\boldsymbol{\xi} = 0$. But

$$\boldsymbol{\xi}\cdot d\mathbf{A} = \boldsymbol{\xi}\cdot(\hat{\mathbf{n}}\,dA + A\,d\hat{\mathbf{n}}) = \boldsymbol{\xi}_n\cdot\hat{\mathbf{n}}\,dA + A\,\boldsymbol{\xi}_t\cdot d\hat{\mathbf{n}}, \tag{14.49}$$

where $\boldsymbol{\xi}_n = \gamma\hat{\mathbf{n}}$ is the part of $\boldsymbol{\xi}$ parallel to $\hat{\mathbf{n}}$. The first term

$$\boldsymbol{\xi}_n\cdot\hat{\mathbf{n}}\,dA = \gamma\,dA \tag{14.50}$$

is the work necessary to change[6] the area A. The second term

$$A\,\boldsymbol{\xi}_t \cdot \mathrm{d}\hat{\mathbf{n}} = A\frac{\partial\gamma(\hat{\mathbf{n}})}{\partial\theta_t}\,\mathrm{d}\theta_t = A\frac{\partial\gamma(\hat{\mathbf{n}})}{\partial\theta_\xi}\cos\psi\,\mathrm{d}\theta_t \tag{14.51}$$

is the work needed to rotate the element of area. The quantity $\partial\gamma(\hat{\mathbf{n}})/\partial\theta_t$ given by Eq. (14.47) is the torque per unit area and depends on the axis of rotation $\mathbf{n}\times\hat{\mathbf{t}}$. This torque is largest for rotation about the axis $\mathbf{n}\times\hat{\mathbf{t}}_\xi$.

The forces that give rise to these free energy changes can be understood by considering the case of a planar area $\mathbf{A} = \mathbf{L}_1 \times \mathbf{L}_2$, where $\mathbf{L}_1$ and $\mathbf{L}_2$ are two sides of a parallelogram, as depicted in Figure 14–2. If $\mathbf{L}_2$ is replaced by $\mathbf{L}_2' = \mathbf{L}_2 + \mathbf{dS}$, where $\mathbf{dS}$ is an infinitesimal vector, one can form a rotated and stretched planar area $\mathbf{A}' = \mathbf{L}_1 \times \mathbf{L}_2'$, where $\mathbf{L}_1$ and $\mathbf{L}_2'$ are two sides of a new parallelogram. The work required to do this will be

$$\delta w = \mathrm{d}(\gamma A) = \boldsymbol{\xi}\cdot\mathrm{d}\mathbf{A} = \boldsymbol{\xi}\cdot(\mathbf{A}'-\mathbf{A}) = \boldsymbol{\xi}\cdot\mathbf{L}_1\times\mathbf{dS} = (\boldsymbol{\xi}\times\mathbf{L}_1)\cdot\mathbf{dS}. \tag{14.52}$$

The external force needed to translate one side of the original parallelogram $\mathbf{L}_1$ is therefore $\boldsymbol{\xi}\times\mathbf{L}_1$. Let $\hat{\boldsymbol{\ell}}_1$ be a unit vector along $\mathbf{L}_1$. Then the force per unit length needed to translate $\hat{\boldsymbol{\ell}}_1$ will be

$$\boldsymbol{\sigma} = \boldsymbol{\xi}\times\hat{\boldsymbol{\ell}}_1. \tag{14.53}$$

As pointed out by Hoffman and Cahn [36], the tip of $\boldsymbol{\sigma}$ traces out an ellipse in a plane perpendicular to $\boldsymbol{\xi}$ as the tip of $\hat{\boldsymbol{\ell}}_1$ traces out a circle in the plane of the original surface element. This can be seen by taking $\boldsymbol{\xi}$ along the z-axis and assuming that the plane in which $\hat{\boldsymbol{\ell}}_1$ rotates by an angle θ is tilted by an acute angle ϕ by means of rotation about the y-axis. Then the x component of $\hat{\boldsymbol{\ell}}_1$ can be written $\cos\theta\cos\phi$, its y component can be written $\sin\theta$, and its z component is irrelevant. We therefore compute

$$\boldsymbol{\sigma} = -\xi\sin\theta\,\hat{\mathbf{i}} + \xi\cos\theta\cos\phi\,\hat{\mathbf{j}}, \tag{14.54}$$

which, as θ varies over 2π, describes an ellipse with major axis ξ and minor axis $\xi\cos\phi$.

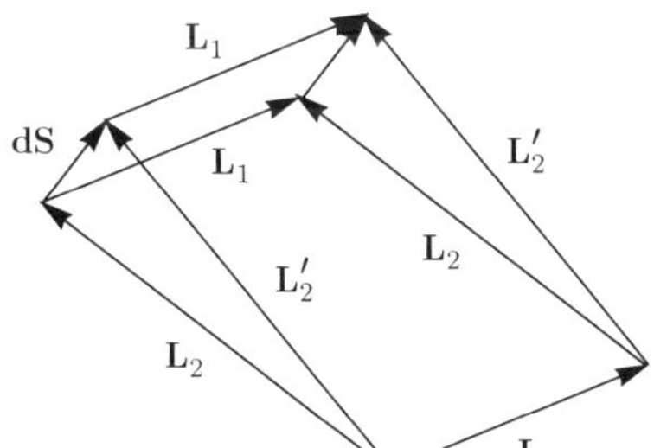

FIGURE 14–2 Planar surface in the shape of a parallelogram bounded by vectors $\mathbf{L}_1$ and $\mathbf{L}_2$ and then rotated and stretched to form another parallelogram bounded by vectors $\mathbf{L}_1$ and $\mathbf{L}_2'$, where $\mathbf{L}_2' = \mathbf{L}_2 + \mathbf{dS}$.

[6]This change must be done without straining the underlying solid; otherwise surface strain terms as in Eq. (14.19) must be taken into account.

We can also decompose $\boldsymbol{\sigma}$ as follows:

$$\boldsymbol{\sigma} = (\boldsymbol{\xi}_n + \boldsymbol{\xi}_t) \times \hat{\boldsymbol{\ell}}_1 = \gamma\, \hat{\mathbf{n}} \times \hat{\boldsymbol{\ell}}_1 + \frac{\partial \gamma(\hat{\mathbf{n}})}{\partial \theta_\xi}\, \hat{\mathbf{t}}_\xi \times \hat{\boldsymbol{\ell}}_1. \tag{14.55}$$

Since $\hat{\mathbf{n}}$ and $\hat{\boldsymbol{\ell}}_1$ are perpendicular, $\hat{\mathbf{t}}_1 := \hat{\mathbf{n}} \times \hat{\boldsymbol{\ell}}_1$ is a unit vector in the plane of the surface element and is perpendicular $\hat{\mathbf{n}}$ and $\hat{\boldsymbol{\ell}}_1$. The force per unit length $\gamma\, \hat{\mathbf{n}} \times \hat{\boldsymbol{\ell}}_1 = \gamma \hat{\mathbf{t}}_1$ therefore has magnitude γ and is the isotropic force per unit length encountered in Eq. (14.50) and needed to enlarge the area of the element. On the other hand, both $\hat{\mathbf{t}}_\xi$ and $\hat{\boldsymbol{\ell}}_1$ lie in the plane of the surface element so $\hat{\mathbf{t}}_\xi \times \hat{\boldsymbol{\ell}}_1$ lies along $\pm\hat{\mathbf{n}}$. Since $\hat{\boldsymbol{\ell}}_1 = \hat{\mathbf{t}}_1 \times \hat{\mathbf{n}}$, we readily compute $\hat{\mathbf{t}}_\xi \times \hat{\boldsymbol{\ell}}_1 = -(\hat{\mathbf{t}}_\xi \cdot \hat{\mathbf{t}}_1)\, \hat{\mathbf{n}}$. Therefore, the force per unit area in Eq. (14.55) related to $\boldsymbol{\xi}_t$ can be expressed as

$$\boldsymbol{\xi}_t \times \hat{\boldsymbol{\ell}}_1 = \frac{\partial \gamma(\hat{\mathbf{n}})}{\partial \theta_\xi}\, \hat{\mathbf{t}}_\xi \times \hat{\boldsymbol{\ell}}_1 = -\frac{\partial \gamma(\hat{\mathbf{n}})}{\partial \theta_\xi}\, (\hat{\mathbf{t}}_\xi \cdot \hat{\mathbf{t}}_1)\, \hat{\mathbf{n}}. \tag{14.56}$$

Its magnitude is similar in form to that of the torque per unit area given by Eq. (14.47).[7] The total force per unit length can therefore be written in the form

$$\boldsymbol{\sigma} = \gamma \hat{\mathbf{t}}_1 - (\boldsymbol{\xi}_t \cdot \hat{\mathbf{t}}_1)\, \hat{\mathbf{n}} = \gamma \hat{\mathbf{t}}_1 - \frac{\partial \gamma(\hat{\mathbf{n}})}{\partial \theta_\xi} (\hat{\mathbf{t}}_\xi \cdot \hat{\mathbf{t}}_1)\, \hat{\mathbf{n}}. \tag{14.57}$$

Note especially that $\hat{\mathbf{t}}_1$ is perpendicular to $\hat{\boldsymbol{\ell}}_1$ and points away from the area being considered. In this respect, the sign convention that applies to Eq. (14.53) should be considered carefully, namely $\hat{\boldsymbol{\ell}}_1 = \hat{\mathbf{t}}_1 \times \hat{\mathbf{n}}$. This is opposite to the sign convention used for the Stokes theorem in which positive circulation on a curve with line elements $\mathrm{d}\boldsymbol{\ell}$ that bounds an area having normal $\hat{\mathbf{n}}$ is in the direction of the right-hand rule. In that case, the unit vector $\hat{\boldsymbol{\ell}} := \mathrm{d}\boldsymbol{\ell}/\mathrm{d}\ell$ satisfies $\hat{\boldsymbol{\ell}} = -\hat{\mathbf{t}}_1 \times \hat{\mathbf{n}}$. Thus, $\hat{\boldsymbol{\ell}}_1 = -\hat{\boldsymbol{\ell}}$ and Eq. (14.53) would become

$$\boldsymbol{\sigma} = -\boldsymbol{\xi} \times \hat{\boldsymbol{\ell}}. \tag{14.58}$$

We point this out because use of the calculus of variations for curved surfaces, which we will consider later, shows that Eq. (14.57) or its equivalent Eq. (14.58) apply generally, not just for the edge of a planar element.

The equilibrium conditions at **triple junctions** are also affected by anisotropy, which requires Eq. (13.83) to be modified. By requiring that no work be done by any small translation of the triple line where three phases meet, one concludes that the net force must be zero, resulting in

$$\boldsymbol{\sigma}^{\alpha\beta} + \boldsymbol{\sigma}^{\beta\eta} + \boldsymbol{\sigma}^{\eta\alpha} = 0. \tag{14.59}$$

From Eq. (14.57), we see that this balance equation accounts for forces due to both tensions and torques, as discussed by Herring [38, p. 157]. By using Eq. (14.58), Eq. (14.59) can be written in the form

[7] Great care must be made in comparing these formulae, even though they look very much alike. In Eq. (14.47), $\hat{\mathbf{t}}$ is fixed for the entire area in the direction of $\mathrm{d}\hat{\mathbf{n}}$, but in Eq. (14.56) $\hat{\mathbf{t}}_1$ is related to the orientation of $\hat{\boldsymbol{\ell}}_1$ by $\hat{\mathbf{t}}_1 = \hat{\mathbf{n}} \times \hat{\boldsymbol{\ell}}_1$.

$$(\boldsymbol{\xi}^{\alpha\beta} + \boldsymbol{\xi}^{\beta\eta} + \boldsymbol{\xi}^{\eta\alpha}) \times \hat{\boldsymbol{\ell}}^{\alpha\beta\eta} = 0, \tag{14.60}$$

where the normals to the surfaces must be chosen consistently to point from the first named phase to the second. Here, $\hat{\boldsymbol{\ell}}^{\alpha\beta\eta}$ is a unit vector along the line of the triple junction. Thus the vector $(\boldsymbol{\xi}^{\alpha\beta} + \boldsymbol{\xi}^{\beta\eta} + \boldsymbol{\xi}^{\eta\alpha})$ has no component perpendicular to the triple line. For a detailed discussion of a quad junction where four phases meet, see Hoffman and Cahn [36, Eq. (28)].

14.3 Curved Solid-Fluid Interfaces

For an infinitesimal area of a curved solid-fluid interface, one can assume that γ is approximately the same as it would be for a planar solid-fluid interface having the same normal $\hat{\mathbf{n}}$, provided that the thickness of the region of discontinuity is small compared to the local radii of curvature. This is similar to what is done for fluid-fluid interfaces except in that case one is able to locate the interface at the surface of tension. No such surface of tension is identified for a curved solid-fluid interface, which is assumed to be located within the physical region of discontinuity with sufficient accuracy. If more rigor is required, the interface could possibly be located at the equimolar surface of some component.

A polar plot

$$\mathbf{r} = \hat{\mathbf{n}}\,\gamma(\hat{\mathbf{n}}) \tag{14.61}$$

is commonly known as a **gamma-plot**, or γ-plot for short. It gives a pictorial representation of γ as a function of the orientation of $\hat{\mathbf{n}}$ and has a unique positive value for each $\hat{\mathbf{n}}$. In spherical coordinates, its equation is $r = \gamma(\theta, \varphi)$, so γ is the distance from the origin to the γ-plot at given θ and φ. Since $\boldsymbol{\xi}$ is a function of $\hat{\mathbf{n}}$, one can also obtain a corresponding **xi-plot**, or $\boldsymbol{\xi}$-plot for short, by allowing $\hat{\mathbf{n}}$ to take on all orientations. In this case, the magnitude ξ of $\boldsymbol{\xi}$ can be a multiple-valued function of the unit vector $\hat{\mathbf{N}} = \boldsymbol{\xi}/\xi$ that points in the direction of $\boldsymbol{\xi}$. An example in two dimensions is shown in Figure 14–3. We shall prove in Section 14.5 that the inner convex hull of the $\boldsymbol{\xi}$-plot has the same shape as the equilibrium shape of a crystal. Note especially that $\boldsymbol{\xi}(\hat{\mathbf{n}})$ is a parametric representation of $\boldsymbol{\xi}$ in terms of the orientation of its surface normal $\hat{\mathbf{n}}$. In particular, it is *not* a representation of $\boldsymbol{\xi}$ in terms of its own orientation $\hat{\mathbf{N}}$.

The **equilibrium shape** of a crystal, also known as the **Gibbs-Wulff shape** or sometimes simply the Wulff shape, is that shape taken on by a crystal by minimizing its total surface free energy subject to the constraint of fixed volume. For kinetic reasons, only small crystals can achieve this shape in a reasonable time. For a given γ-plot, this shape can be found by means of the following construction due to Wulff [39]. At every point $\hat{\mathbf{n}}\gamma(\hat{\mathbf{n}})$ of the γ-plot, erect a plane perpendicular to $\hat{\mathbf{n}}$ and passing through that point. Then the inner convex hull of those **Wulff planes** is the equilibrium shape. This so-called Wulff theorem was stated without proof by Wulff in the context of polyhedral shapes but has since been studied extensively and applies also to curved shapes. In Section 14.5, we

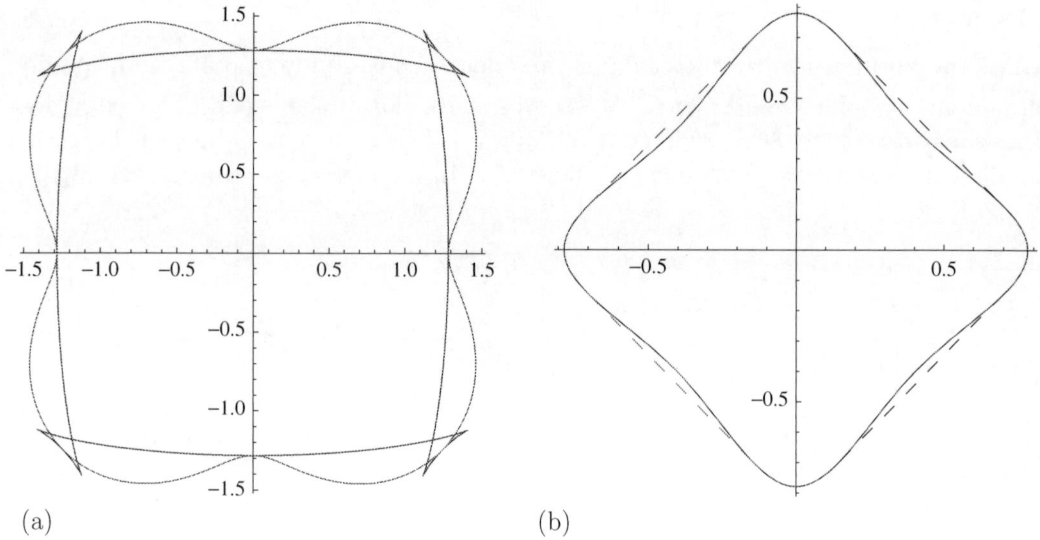

FIGURE 14–3 (a) Illustration of a γ-plot and a ξ-plot in two dimensions for $\gamma = 1 + \sqrt{2n_x^2 n_y^2 + 0.08}$ in arbitrary units. The ξ-plot is not a single-valued function of its polar angle in this case but has "ears" that must be truncated leaving a convex body that would have the equilibrium shape of a two-dimensional crystal. (b) Inverted γ-plot (full curve) whose properties are discussed in Section 14.3.2. The dashed lines are added to "convexify" (to make convex) the plot.

derive an analytical formula for this shape in terms of the ξ-vector for differentiable $\gamma(\hat{\mathbf{n}})$. Immediately below, we show how ξ can be defined for cases for which $\gamma(\hat{\mathbf{n}})$ is not differentiable. Moreover, the analysis of faceting in Section 14.4 can be used to show that a surface whose orientation does not appear on the Gibbs-Wulff shape is unstable with respect to faceting, consistent with the Gibbs-Wulff shape being the equilibrium shape.

14.3.1 Discontinuous Derivatives of γ

The definition of ξ, and hence the ξ-plot, can be extended to cover cases in which the derivatives of γ are discontinuous. In particular, $\gamma(\hat{\mathbf{n}})$ can have sharp grooves (knife edges) or inwardly directed sharp points, including cusps, at special orientations that correspond to low index planes in three dimensions.[8] One way of handling this situation is to consider a slight rounding of the grooves or sharp points and then take the limit as this rounding tends to zero. For example, in the vicinity of $n_x \ll 1$, $n_y \approx 1$, and $n_z \ll 1$, suppose that

$$\gamma(\hat{\mathbf{n}}) = \gamma_0 \left[1 + \alpha \sqrt{n_x^2 + \epsilon^2} \right], \tag{14.62}$$

where γ_0, α, and $\epsilon \ll 1$ are positive constants. Then one readily calculates

[8]In strictly two dimensions, these grooves or sharp points become rounded at finite temperatures due to entropic effects. See Mullins [40, p. 28] and Herring [41, p. 18] for further discussion.

$$\xi_x = \gamma_0 \left[n_x + \alpha n_x \frac{1+\epsilon^2}{\sqrt{n_x^2+\epsilon^2}} \right] \approx \gamma_0 \left[n_x + \alpha \frac{n_x}{|n_x|} \right]; \tag{14.63}$$

$$\xi_y = \gamma_0 \left[n_y + \alpha n_y \frac{\epsilon^2}{\sqrt{n_x^2+\epsilon^2}} \right] \approx \gamma_0 n_y; \tag{14.64}$$

$$\xi_z = \gamma_0 \left[n_z + \alpha n_z \frac{\epsilon^2}{\sqrt{n_x^2+\epsilon^2}} \right] \approx \gamma_0 n_z, \tag{14.65}$$

where the approximate forms are in the limit $\epsilon = 0$, where $\gamma = \gamma_0[1 + \alpha|n_x|]$. In this limit, we see that γ is continuous but ξ_x is discontinuous at $n_x = 0$, where it jumps from $-\gamma_0\alpha$ to $\gamma_0\alpha$. In the plane $n_z = 0$, we see for finite but small ϵ that $\xi_y \approx \gamma_0(1 - n_x^2)^{1/2}$ is very nearly equal to γ_0 for small n_x while ξ_x changes considerably as n_x makes small changes near zero, as shown in Figure 14–4. As ϵ becomes very small, the γ-plot tends toward a V-shaped groove and the tips of the $\boldsymbol{\xi}$-vector lie nearly along a straight line segment corresponding to $\xi_y = \gamma_0$ that extends from $-\gamma_0\alpha$ to $\gamma_0\alpha$. Thus it would be natural for a sharp groove ($\varepsilon = 0$) to *define* $\boldsymbol{\xi}$, for $n_x = 0$, to be multiple-valued, namely the **fan of vectors** having $\xi_z = 0$, $\xi_y = \gamma_0$, and $-\gamma_0\alpha \le \xi_x \le \gamma_0\alpha$. The tails of these vectors are at the origin and their tips lie along a straight line. The corresponding three-dimensional portion of the $\boldsymbol{\xi}$-plot would be a ruled surface. By analogous reasoning, the $\boldsymbol{\xi}$-vectors corresponding to a sharp inwardly directed point of γ would form a cone whose tips lie along a portion of a plane, a facet of the $\boldsymbol{\xi}$-plot.

An elegant way of defining $\boldsymbol{\xi}$ for general γ that is fully consistent with the foregoing limiting arguments can be based on the famous sphere construction of Herring, as illustrated in Figure 14–5. The **Herring sphere** construction is based on the fact that

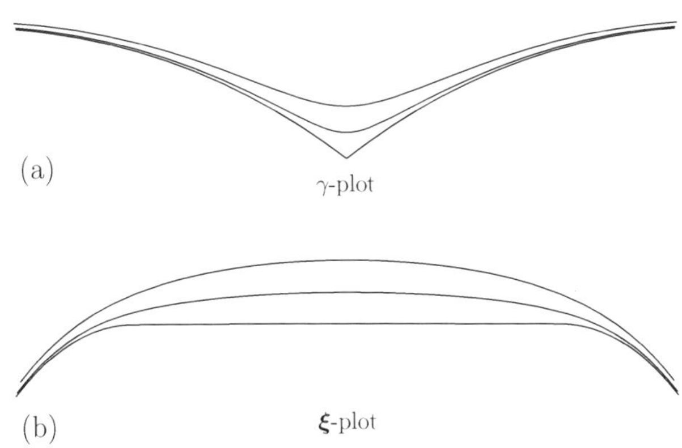

FIGURE 14–4 Portions of a γ-plot and the corresponding ξ-plot near a groove for three values of ϵ in the plane $n_z = 0$ according to Eqs. (14.62)–(14.64). For plotting purposes, $\gamma_0 = 1$, $\alpha = \sqrt{2}/2$, and $\epsilon = 0.1$, 0.05, and 0.01 from top to bottom. (a) The upper curves are plots of γ versus n_x with the origin located at the root of the V-shaped groove. (b) The lower curves are parametric plots of ξ_y versus ξ_x for very small changes of n_x near $n_x = 0$.

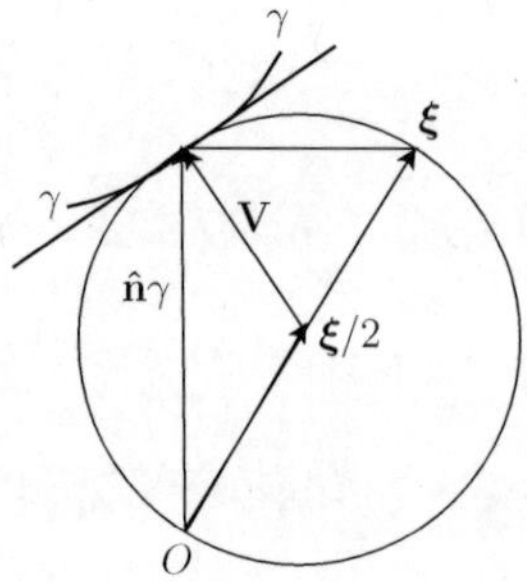

FIGURE 14–5 Herring sphere (actually a circle in two dimension) tangent to a segment of the γ-plot and passing through its origin, O. The ξ-vector lies along a diameter and the center of the sphere is located at $\xi/2$. The vector **V** has magnitude $\xi/2$ and is perpendicular to the local tangent plane.

any angle inscribed in a hemisphere is a right angle. Therefore, according to the Wulff theorem, any point on the surface of an equilibrium shape must be located at end of the diameter $\boldsymbol{\xi}(\hat{\mathbf{n}})$ of a sphere that passes through the origin of the γ-plot and satisfies $\boldsymbol{\xi}\cdot\hat{\mathbf{n}} = \gamma$. Furthermore, this sphere must either be tangent to the γ-plot at a place where its tangent is well defined or else touch the γ-plot at some sharp groove or sharp point where the tangent to the γ-plot is not well defined. For a surface element with definite orientation to actually appear on the equilibrium shape, it is necessary and sufficient that no portion of the γ-plot lies inside the Herring sphere corresponding to that orientation. This is true because a Wulff plane corresponding to a portion of the γ-plot that lies inside the Herring sphere would cut through its diameter and exclude that orientation from the inner convex hull. As shown below, the resulting equilibrium shape can have curved sections and flat sections, as well as edges and sharp corners that correspond to missing orientations.

Therefore, at any point on the γ-plot for some orientation $\hat{\mathbf{n}}$ at which its derivatives are well defined and continuous, one can erect a Herring sphere that passes through the origin and is tangent to the γ-plot at that point. Then the vector $\boldsymbol{\xi}$ is the unique vector from the origin of the γ-plot that passes through the center of that sphere and terminates on its opposite side. For points on the γ-plot for which its derivatives are undefined, $\boldsymbol{\xi}$ is multiple-valued because one can construct a continuum of Herring spheres that pass through that point and the origin of the γ-plot and define a fan or cone of $\boldsymbol{\xi}$-vectors, as illustrated in Figure 14–6. This leads to a $\boldsymbol{\xi}$-plot that can have curved surfaces, ruled surfaces, and planar sections. The resulting $\boldsymbol{\xi}$-plot can be nonconvex and have "ears" that must, however, be truncated to form the equilibrium shape, which is convex.

Since any angle inscribed in a hemisphere is a right angle, this extended definition of $\boldsymbol{\xi}$ will satisfy $\gamma = \boldsymbol{\xi}\cdot\hat{\mathbf{n}}$. Other relations can be established as follows. For any point $\mathbf{r} = \hat{\mathbf{n}}\gamma(\hat{\mathbf{n}})$ on the γ-plot, one can erect a vector **V** that points from the center of a Herring sphere to **r** and passes through the origin of the γ-plot. This construction satisfies

$$\frac{\boldsymbol{\xi}}{2} + \mathbf{V} = \gamma\hat{\mathbf{n}} = \mathbf{r} \tag{14.66}$$

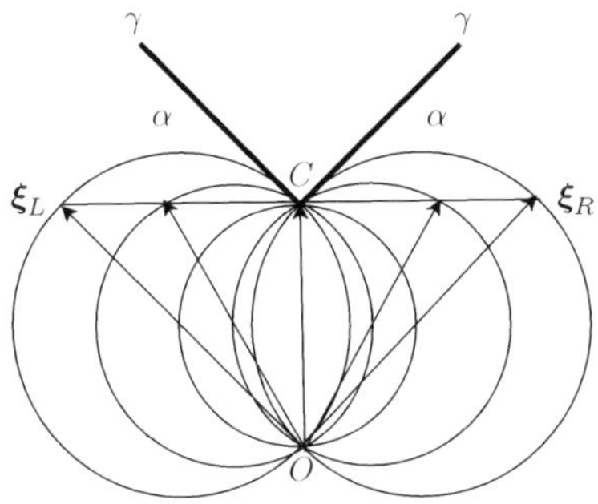

FIGURE 14–6 Fan of ξ-vectors corresponding to a V-shaped portion of a γ-plot. Five Herring spheres (actually circles in two dimensions) that pass through the origin O of the γ-plot and the point C are shown along with the ξ-vectors that lie along their diameters. The two largest circles are tangent to the γ-plot at C. The other three circles pass through C but are not tangent to the γ-plot; there is a continuum of such circles that would have that same property. The tips of the ξ-vectors lie along a line of length $2\gamma_c \cos\alpha$ that stretches from ξ_L to ξ_R, where γ_c is the value of γ at the point C where each segment of the V-shaped plot makes an angle α with the horizontal.

with $|\mathbf{V}| = |\boldsymbol{\xi}|/2$. For a particular point corresponding to $\hat{\mathbf{n}}_0$, suppose that the γ-plot has a well-defined tangent plane whose equation is $(\mathbf{r}-\mathbf{r}_0)\cdot\mathbf{V}_0 = 0$. For a small change $\mathbf{dr} = \mathbf{r}-\mathbf{r}_0$ in this plane, we will have

$$\mathbf{dr} \cdot \mathbf{V}_0 = 0. \tag{14.67}$$

But since this plane is tangent to the γ-plot, we also have $\mathbf{dr} = \mathrm{d}(\gamma\hat{\mathbf{n}}) = \gamma_0 \mathrm{d}\hat{\mathbf{n}} + \hat{\mathbf{n}}_0 \mathrm{d}\gamma$. By substituting for $\mathbf{V}_0$ from Eq. (14.66), Eq. (14.67) can be written

$$(\gamma_0\, \mathrm{d}\hat{\mathbf{n}} + \hat{\mathbf{n}}_0\, \mathrm{d}\gamma) \cdot (\gamma_0 \hat{\mathbf{n}}_0 - \boldsymbol{\xi}_0/2) = 0. \tag{14.68}$$

By computing the dot products and dividing by $\gamma_0/2 \neq 0$ we obtain

$$\mathrm{d}\gamma = \boldsymbol{\xi} \cdot \mathrm{d}\hat{\mathbf{n}}, \quad \gamma\text{-plot has well-defined tangent plane,} \tag{14.69}$$

where we have dropped the subscript 0 on $\boldsymbol{\xi}$ with the understanding that it is to be evaluated at $\mathrm{d}\hat{\mathbf{n}} = 0$. Then by using $\mathrm{d}\gamma = \mathrm{d}(\boldsymbol{\xi} \cdot \hat{\mathbf{n}})$, we can use Eq. (14.69) to obtain $\hat{\mathbf{n}} \cdot \mathrm{d}\boldsymbol{\xi} = 0$. Thus, wherever the γ-plot has a well-defined tangent plane, Eqs. (14.32), (14.35), and (14.36) are all satisfied, as expected.

On the other hand, if a Herring sphere that passes through the origin touches the γ-plot at a point where it does not have a well-defined tangent plane, the position of its center and its size can vary while holding γ and $\hat{\mathbf{n}}$ constant. In other words, $\boldsymbol{\xi}$ can vary while holding γ and $\hat{\mathbf{n}}$ constant. Thus we can take the differential of $\gamma = \boldsymbol{\xi} \cdot \hat{\mathbf{n}}$ to obtain

$$\hat{\mathbf{n}} \cdot \mathrm{d}\boldsymbol{\xi} = 0, \quad \text{always,} \tag{14.70}$$

which is the same as for a well-defined tangent plane. But now the vector $\mathbf{V}$ touches the γ-plot but it is no longer normal to it at the point of touching. Thus Eq. (14.67) must be replaced by

$$\mathbf{dr} \cdot \mathbf{V}_0 \geq 0, \quad \text{where the } \gamma\text{-plot has no well-defined tangent plane.} \tag{14.71}$$

The equality holds only if $\mathbf{dr}$ lies along and is tangent to the curve of a knife edge. Accordingly, Eq. (14.69) is replaced by

$$d\gamma \geq \boldsymbol{\xi} \cdot d\hat{\mathbf{n}}, \quad \text{where the } \gamma\text{-plot has no well-defined tangent plane.} \tag{14.72}$$

If we write $d\hat{\mathbf{n}} = \hat{\mathbf{t}}d\theta_t$ as in Eq. (14.43), Eq. (14.72) becomes

$$\frac{\partial \gamma(\hat{\mathbf{n}})}{\partial \theta_t} \geq \boldsymbol{\xi}_t \cdot \hat{\mathbf{t}}, \quad \text{where the } \gamma\text{-plot has no well-defined tangent plane.} \tag{14.73}$$

In this case, Eq. (14.73) replaces Eq. (14.44). Thus, the multiple values of $\boldsymbol{\xi}_t$ that are associated with a ruled or flat portion of a surface determine the range of torques that can be supported.

14.3.2 Inverted γ-Plot

The Herring sphere criterion can be used to determine which orientations are missing on the equilibrium shape. This is easy to state but hard to apply. An equivalent criterion that is much easier to apply has been discussed by Frank [42]. It can be obtained by considering the **inverted gamma-plot**, namely the polar plot

$$\mathbf{R} = \hat{\mathbf{n}} \frac{1}{\gamma(\hat{\mathbf{n}})}, \tag{14.74}$$

or simply $R = 1/\gamma(\hat{\mathbf{n}})$. On inversion through the origin, a sphere becomes a plane.[9] Thus a Herring sphere that is tangent to the γ-plot becomes a tangent plane to the $1/\gamma$-plot. Any portion of a γ-plot that lies inside a Herring sphere will contribute to planes that cut the $1/\gamma$-plot. It follows that for all orientations to appear on the equilibrium shape, it is necessary and sufficient that the $1/\gamma$-plot be convex. Furthermore, if the $1/\gamma$-plot is not convex, one can form a convex body by means of enveloping it by portions of touching planes that do not cut the plot. See Figure 14–3b for an example in two dimensions, where the dashed lines are added to "convexify" the plot. The orientations on the nonconvex $1/\gamma$-plot that do not appear on the enveloped convex plot are those that are missing from the equilibrium shape. They actually appear on the ears of the $\boldsymbol{\xi}$-plot.

It is easy to show that the normal to the $1/\gamma(\hat{\mathbf{n}})$-plot is in the direction $\boldsymbol{\xi}(\hat{\mathbf{n}})$. Indeed, the equation of the $1/\gamma(\hat{\mathbf{n}})$-plot can be written in the form $R\gamma(\hat{\mathbf{n}}) = 1$ with $\hat{\mathbf{n}} = \mathbf{R}/R$. Then we can let $\mathbf{R}$ play the role of $\mathbf{P}$ in Eq. (14.31) to obtain

$$\boldsymbol{\xi}(\hat{\mathbf{n}}) = \nabla_R [R\gamma(\mathbf{R}/R)], \tag{14.75}$$

which is clearly in the direction of the normal to the $1/\gamma(\hat{\mathbf{n}})$-plot. This expression can be used to compute the Gauss curvature of the $1/\gamma(\hat{\mathbf{n}})$-plot and determine when it changes sign, which defines the limits of its convexity. This gives an analytical criterion for the onset of missing orientations on a three-dimensional equilibrium shape [43]. Herring [44]

[9] In particular, consider the sphere $\mathbf{r} = \hat{\mathbf{n}}(\hat{\mathbf{n}} \cdot \boldsymbol{\xi})$, where $\hat{\mathbf{n}}$ varies for fixed $\boldsymbol{\xi}$. When inverted, this sphere becomes $\mathbf{Q} = \hat{\mathbf{n}}(\hat{\mathbf{n}} \cdot \boldsymbol{\xi})^{-1}$ or $\mathbf{Q} \cdot \boldsymbol{\xi} = 1$, which is linear in the components of $\mathbf{Q}$. Thus it is a plane that passes through the point $\mathbf{Q} = \boldsymbol{\xi}/\xi^2$.

has given an extensive discussion of the qualitative characteristics of the γ-plot and the resulting equilibrium shapes, with particular attention to corners, cylindrical portions, and facets.

14.4 Faceting of a Large Planar Face

Herring [44] has also considered the possibility that a large planar surface of a crystal could break up into a hill-and-valley structure composed of facets. Such a consideration is important for kinetic reasons because a large amount of transport would be required to convert a large crystal to its equilibrium shape. It therefore makes sense to consider a state that could occur on a time scale that is very short compared to the time needed to transform an entire crystal to its equilibrium shape.

To analyze this problem, consider a small area a_0 on the planar face of a large crystal having unit normal $\hat{\mathbf{n}}_0$ and free energy $\gamma(\hat{\mathbf{n}}_0) \equiv \gamma_0$ per unit area. We then investigate the stability of this planar area with respect to being replaced by a pyramid[10] having three noncoplanar orientations $\hat{\mathbf{n}}_1$, $\hat{\mathbf{n}}_2$, and $\hat{\mathbf{n}}_3$, corresponding to facets having respective areas a_1, a_2, and a_3, as illustrated in Figure 14–7. From Gauss's theorem in the form $\int_V \nabla\cdot\mathbf{k}\,\mathrm{d}^3x = \int_A \mathbf{k}\cdot\hat{\mathbf{n}}\,\mathrm{d}^2x$, where $\mathbf{k}$ is an arbitrary but constant vector, we can deduce that $\int_A \hat{\mathbf{n}}\,\mathrm{d}^2x = 0$. By applying this result to the pyramid just described, we obtain

$$\hat{\mathbf{n}}_0 = f_1\hat{\mathbf{n}}_1 + f_2\hat{\mathbf{n}}_2 + f_3\hat{\mathbf{n}}_3, \tag{14.76}$$

where $f_i = a_i/a_0$ are area fractions.[11] By using reciprocal vectors $\boldsymbol{\tau}_i$ defined such that $\boldsymbol{\tau}_i \cdot \hat{\mathbf{n}}_\mathbf{j} = \delta_{ij}$ for $i, j = 1, 2, 3$, we deduce that $f_i = \boldsymbol{\tau}_i \cdot \hat{\mathbf{n}}_0$. Thus the $\boldsymbol{\tau}_i$, but not necessarily the $\hat{\mathbf{n}}_i$ as required by Herring [45], must have positive projections on $\hat{\mathbf{n}}_0$ in order to obtain a real pyramid with positive f_i. The free energy associated with the three faces of the pyramid, measured per unit area of the large planar face, is

$$\gamma_h = f_1\gamma_1 + f_2\gamma_2 + f_3\gamma_3. \tag{14.77}$$

Thus

$$\gamma_h = \mathbf{c}\cdot\hat{\mathbf{n}}_0, \tag{14.78}$$

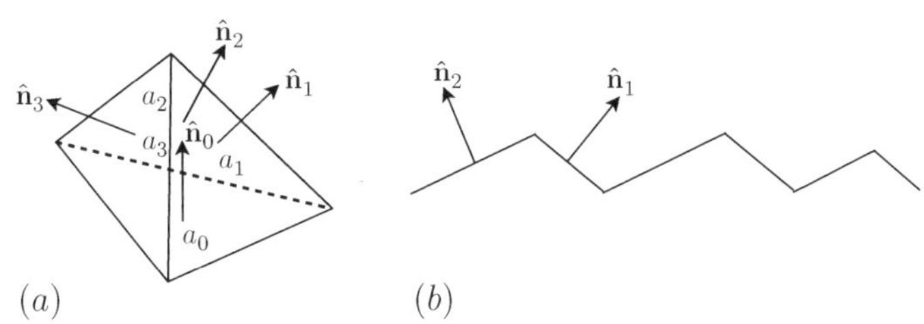

FIGURE 14–7 (a) Typical pyramid for faceting of a surface. The $\hat{\mathbf{n}}_i$ are unit normals and a_i are respective areas of the faces. (b) Faceted surface in two dimensions, showing facets of different sizes but the same orientation.

[10]For the entire planar face, this is to be done by using many pyramids but without a change in volume.
[11]Note that $\hat{\mathbf{n}}_0$ is the outward normal to the large planar face so it is an inner normal to the pyramid.

where $\mathbf{c} := \tau_i\gamma_1 + \tau_i\gamma_2 + \tau_i\gamma_3$ can be interpreted geometrically as the vector from the origin of the γ-plot to a point defined by the intersections of three Wulff planes drawn perpendicular to $\hat{\mathbf{n}}_1$, $\hat{\mathbf{n}}_2$, and $\hat{\mathbf{n}}_3$ at the points where they intersect the γ-plot. This interpretation follows because $\mathbf{c} \cdot \hat{\mathbf{n}}_i = \gamma_i$ for $i = 1, 2, 3$. We observe that $\mathbf{c}$ lies along the diameter of a sphere that passes through four points, $\hat{\mathbf{n}}_1\gamma_1$, $\hat{\mathbf{n}}_2\gamma_2$, $\hat{\mathbf{n}}_3\gamma_3$, and the origin.

From the above considerations, it follows that the large planar face will be stable against faceting if $\gamma_h > \gamma_0$, which means that the point $(\mathbf{c} \cdot \hat{\mathbf{n}}_0)\hat{\mathbf{n}}_0 = \gamma_h\hat{\mathbf{n}}_0$ will lie outside the γ-plot. On the other hand, if the point $\gamma_h\hat{\mathbf{n}}_0$ lies inside the γ-plot, the large planar face will be unstable with respect to this type of faceting. However, if the orientation $\hat{\mathbf{n}}_0$ occurs on the equilibrium shape, it is impossible for point $\gamma_h\hat{\mathbf{n}}_0$ to lie inside the γ-plot because at least one of the Wulff planes corresponding to $\hat{\mathbf{n}}_1$, $\hat{\mathbf{n}}_2$, or $\hat{\mathbf{n}}_3$ would cut it off. This results in **Herring's theorem** [44]:

> *If a given macroscopic surface of a crystal does not coincide in orientation with some portion of the boundary of the equilibrium shape, there will always exist a hill-and-valley structure which has a lower free energy than a flat surface, while if the given surface does occur in the equilibrium shape, no hill-and-valley structure can be more stable.*

With keen geometrical insight, Frank [42] observed that Herring's faceting criterion has a very simple interpretation in terms of the inverted γ-plot. In particular, the tip of the inverted vector $\hat{\mathbf{n}}/\gamma_h$ lies on the plane that passes through the points $\hat{\mathbf{n}}_1/\gamma_1$, $\hat{\mathbf{n}}_2/\gamma_2$, and $\hat{\mathbf{n}}_3/\gamma_3$. To see this, let $\hat{\mathbf{p}}$ be a unit vector perpendicular to that plane and pointing away from the origin. Then the distance from the origin to that plane is given by $d = \hat{\mathbf{p}} \cdot \hat{\mathbf{n}}_1/\gamma_1 = \hat{\mathbf{p}} \cdot \hat{\mathbf{n}}_2/\gamma_2 = \hat{\mathbf{p}} \cdot \hat{\mathbf{n}}_3/\gamma_3$, from which we deduce that $\hat{\mathbf{p}} = d(\tau_i\gamma_1 + \tau_i\gamma_2 + \tau_i\gamma_3) = \mathbf{c}\,d$. Thus $\hat{\mathbf{p}} \cdot \hat{\mathbf{n}}/\gamma_h = d$, confirming Frank's observation. Therefore, we can compare $\hat{\mathbf{n}}/\gamma_h$ with $\hat{\mathbf{n}}/\gamma$ and deduce that the free energy will be lowered by faceting only if $\hat{\mathbf{n}}/\gamma$ lies inside the plane (nearer to the origin) that passes through $\hat{\mathbf{n}}_1/\gamma_1$, $\hat{\mathbf{n}}_2/\gamma_2$, and $\hat{\mathbf{n}}_3/\gamma_3$. This analysis also clarifies that the orientations that are unstable with respect to faceting are those that lie on the ears of the ξ-plot, which result from nonconvex portions of the $1/\gamma$-plot. Indeed, the very notion of a value of γ for unstable orientations requires the concept of a constrained equilibrium state for which faceting is prevented.

Herring's analysis was extended by Mullins and Sekerka [45] by using linear programming theory to analyze faceting into shapes having an arbitrary number of orientations. It was shown that a minimum value of γ_h can always be obtained by using *no more than three* orientations; however, degeneracies can occur such that more than three orientations can lead to the same minimum value of γ_h. Moreover, the minimum value of γ_h that can be achieved by faceting corresponds to the distance $\Gamma(\hat{\mathbf{n}})$ from the origin to a so-called **contact plane** of the Gibbs-Wulff shape, the latter being a plane that is perpendicular to $\hat{\mathbf{n}}$ and touches but does not cut that shape. In fact, $\hat{\mathbf{n}}\Gamma(\hat{\mathbf{n}})$ is the **minimum gamma-plot** (contained in all others) that gives the same Gibbs-Wulff shape as $\gamma(\hat{\mathbf{n}})$. Figure 14–8

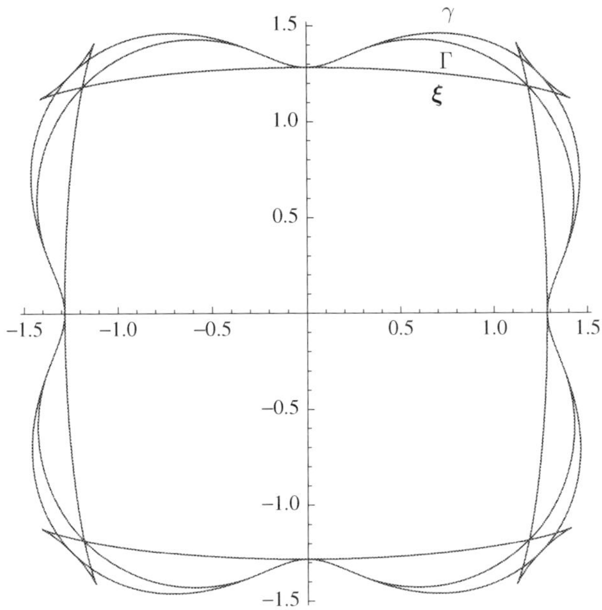

FIGURE 14–8 A γ-plot (outer curve) and a ξ-plot (inner curve) in two dimensions for $\gamma = 1 + \sqrt{2n_x^2 n_y^2 + 0.08}$ in arbitrary units. The equilibrium Gibbs-Wulff shape is the convex shape found by truncating the ears of the ξ-plot. The middle curve is the Γ-plot, which is the smallest that will lead to the same Gibbs-Wulff shape. The distance along any $\hat{\mathbf{n}}$ between $\gamma(\hat{\mathbf{n}})$ and $\Gamma(\hat{\mathbf{n}})$ represents the maximum possible energy reduction by faceting. Orientations for which this difference is zero appear on the Gibbs-Wulff shape.

illustrates $\gamma(\hat{\mathbf{n}})$, $\Gamma(\hat{\mathbf{n}})$, and $\boldsymbol{\xi}(\hat{\mathbf{n}})$ in two dimensions. For orientations such that a contact plane is actually tangent to the Gibbs-Wulff shape, that orientation appears on the shape and the corresponding plane is not unstable with respect to faceting. The inverted plot $\hat{\mathbf{n}}\Gamma(\hat{\mathbf{n}})$ is just the convex plot obtained by enveloping the plot $\hat{\mathbf{n}}/\gamma(\hat{\mathbf{n}})$ by portions of planes, as illustrated in Figure 14–3b. The portions of planes invert to portions of spheres on $\Gamma(\hat{\mathbf{n}})$ that correspond to orientations for which the contact plane is not tangent to the Gibbs-Wulff shape.

It is important to recognize that this analysis of faceting provides no size scale for the facets; it deals only with their orientation. In other words, surfaces with large facets have the same free energy as those with small facets. However, one would expect there to be a mixture of facet sizes on a given surface (e.g., colonies of large facets and small facets, as suggested by Figure 14–7b in two dimensions) and the resulting configurational entropy would further lower the free energy of a faceted surface. Modification of the theory to allow for excess energies at edges and corners would change the invariance to size scale. Of course it would also require modification of our concept of an equilibrium shape, which would only be valid for crystals sufficiently large that excess energies at edges and corners are negligible.

14.5 Equilibrium Shape from the ξ-Vector

Provided that $\gamma(\hat{\mathbf{n}})$ is differentiable, we can proceed to find an analytical formula for the equilibrium shape of a solid in contact with a fluid. Places where it is not differentiable can be handled as limiting cases as explained in Section 14.3.1. We proceed to minimize the grand potential K for the entire solid, assumed to be constrained to have a fixed volume and maintained at fixed temperature T and chemical potentials μ_i. We write this potential in the form

$$K = \int_{V_s} \omega_v^s \, \mathrm{d}V + \int_{V_F} \omega_v^F \, \mathrm{d}V + \int_A \gamma(\hat{\mathbf{n}}) \, \mathrm{d}A, \tag{14.79}$$

where ω_v^s is the grand potential per unit volume in the solid, which may be crystalline or amorphous, ω_v^F is the grand potential per unit volume in the fluid, V_s is the volume of the solid, V_F is the volume of the fluid, A is the area of the interface that separates the solid from the fluid, and $\hat{\mathbf{n}}$ points from solid to fluid. For the moment, we assume that the area A is bounded by some closed curve C. We presume that the interface can be represented in terms of parameters u, v by the equation $\mathbf{r} = \mathbf{r}(u, v)$, as discussed in detail in Appendix C, Section C.2. We write the equilibrium shape in the form $\mathbf{r} = \mathbf{r}_0(u, v)$ and make an infinitesimal normal variation to a new position,

$$\mathbf{r} = \mathbf{r}_0(u, v) + \hat{\mathbf{n}}_0(u, v)\eta(u, v) \equiv \mathbf{r}_0(u, v) + \delta\mathbf{r}(u, v), \tag{14.80}$$

where the infinitesimal quantity $\eta(u, v)$ is arbitrary but differentiable. Then the variation of the total Kramers (grand) potential is

$$\delta K = \int_A (\omega_v^s - \omega_v^F)\eta(u, v) \, \mathrm{d}A + \delta \int_A \boldsymbol{\xi} \cdot \hat{\mathbf{n}} \, \mathrm{d}A, \tag{14.81}$$

where we have replaced γ by $\boldsymbol{\xi} \cdot \hat{\mathbf{n}}$. The second area integral can be written in the form

$$\delta \int_A \boldsymbol{\xi} \cdot \hat{\mathbf{n}} \, \mathrm{d}A = \delta \int_{u,v} \boldsymbol{\xi} \cdot \mathbf{H} \, \mathrm{d}u \, \mathrm{d}v, \tag{14.82}$$

where $\mathbf{H} = \mathbf{r}_u \times \mathbf{r}_v$, with $\mathbf{r}_u = \partial\mathbf{r}/\partial u$ and $\mathbf{r}_v = \partial\mathbf{r}/\partial v$. Note that $\hat{\mathbf{n}} = \mathbf{H}/H$. The second integral in Eq. (14.82) is over a fixed domain in u, v space. Thus we can take δ inside the integral to obtain

$$\delta \int_{u,v} \boldsymbol{\xi} \cdot \mathbf{H} \, \mathrm{d}u \, \mathrm{d}v = \int_{u,v} \boldsymbol{\xi} \cdot \delta\mathbf{H} \, \mathrm{d}u \, \mathrm{d}v, \tag{14.83}$$

where $\mathbf{H} \cdot \delta\boldsymbol{\xi} = 0$ because of Eq. (14.35). Then, as shown by Eq. (C.46) in Appendix C, Eq. (14.80) leads to

$$\delta\mathbf{H} = \hat{\mathbf{n}}_0(u, v)H_0(u, v)\eta(u, v)\,\mathcal{K}_0(u, v) - H_0(u, v)\nabla_s\eta(u, v), \tag{14.84}$$

where $\mathcal{K}_0(u, v)$ is the mean curvature of the surface in the unvaried state (see Eq. (C.28) for a general formula) and ∇_s is the surface gradient operator defined by Eq. (C.35). Thus Eq. (14.81) becomes

$$\delta\mathcal{K} = \int_A (\omega_v^s - \omega_v^F)\eta(u, v) \, \mathrm{d}A + \int_A \left[\gamma\,\mathcal{K}\,\eta(u, v) - \boldsymbol{\xi} \cdot \nabla_s\eta(u, v)\right] \mathrm{d}A, \tag{14.85}$$

where we have dropped the zero subscripts. Next, we prepare to integrate by parts by writing

$$-\boldsymbol{\xi}\cdot\nabla_s\eta(u,v) = -\nabla_s\cdot[\boldsymbol{\xi}\eta(u,v)] + \eta(u,v)\nabla_s\cdot\boldsymbol{\xi}, \tag{14.86}$$

where the terms on the right-hand side contain the surface divergence. According to the surface divergence theorem, Eq. (C.49), we have

$$\int_A \nabla_s\cdot[\boldsymbol{\xi}\eta(u,v)]\,\mathrm{d}A = \oint_C \boldsymbol{\xi}_t\cdot\hat{\mathbf{t}}\eta(u,v)\,\mathrm{d}\ell + \int_A \gamma\,\mathcal{K}\eta(u,v)\,\mathrm{d}A, \tag{14.87}$$

where $\hat{\mathbf{t}}$ is a unit tangent vector pointing out of the area along the curve C, specifically $\hat{\mathbf{t}}\,\mathrm{d}\ell = \mathrm{d}\boldsymbol{\ell}\times\hat{\mathbf{n}}$, where the direction of $\hat{\mathbf{t}}\mathrm{d}\ell$ is determined by the right-hand rule. Thus

$$\delta\mathcal{K} = \int_A\Big[(\omega_v^s-\omega_v^F)+\nabla_s\cdot\boldsymbol{\xi}\Big]\eta(u,v)\,\mathrm{d}A - \oint_C \boldsymbol{\xi}_t\cdot\hat{\mathbf{t}}\,\eta(u,v)\,\mathrm{d}\ell. \tag{14.88}$$

To guarantee that no work is done along the curve C, we can take $\eta(u,v)=0$ along C and the equilibrium criterion becomes

$$0 = \delta\mathcal{K} = \int_A\Big[(\omega_v^s-\omega_v^F)+\nabla_s\cdot\boldsymbol{\xi}\Big]\eta(u,v)\,\mathrm{d}A. \tag{14.89}$$

Then since $\eta(u,v)$ is arbitrary over the area A, the integrand must vanish, and we obtain the equilibrium condition

$$\omega_v^F-\omega_v^s = \nabla_s\cdot\boldsymbol{\xi}. \tag{14.90}$$

If the solid is amorphous and therefore isotropic, $\boldsymbol{\xi}=\gamma\hat{\mathbf{n}}$, $\nabla_s\cdot\boldsymbol{\xi}=\gamma\mathcal{K}$ by Eq. (C.38), and $\omega_v^F-\omega_v^s = p_s-p_F$, so the Laplace equation (Eq. (13.71)) for fluids would apply.

Equation (14.90) is a nonlinear partial differential equation for the solid-fluid interface shape, so one would have to find a solution that attached to the bounding curve C, a difficult task. On the other hand, for a *closed* surface, the curve C closes back on itself and the line integral in Eq. (14.88) vanishes without restriction on $\eta(u,v)$. Then, since $\nabla_s\cdot\mathbf{r}=2$ (see Eq. (C.41)) an obvious solution to Eq. (14.90) is

$$\mathbf{r} = \frac{2}{(\omega_v^F-\omega_v^s)}\boldsymbol{\xi}, \tag{14.91}$$

which is the equation for the equilibrium shape of a crystal.[12] Note that a result of the same form would be obtained if one varied only the shape of the body while holding its volume constant. This could be done by using a Lagrange multiplier to put in the volume constraint. The present method identifies that Lagrange multiplier in terms of physical quantities so we obtain the size of the crystal in addition to its shape.

Let us return to Eq. (14.88) for a bounding curve C that can move in a manner described by Eq. (14.80). Then the work done by an external force work $\mathbf{f}_L$ per unit length is given by

[12]As explained in connection with the Wulff theorem, one must truncate the ears to get a convex body if there are missing orientations. Note in two dimensions that $\nabla_s\cdot\mathbf{r}=1$, so the factor of 2 would be missing.

$$\delta W = \oint_C \mathbf{f}_L \cdot \hat{\mathbf{n}}\,\eta(u,v)\,\mathrm{d}\ell. \tag{14.92}$$

The equilibrium criterion now becomes $\delta\mathcal{K} - \delta W = 0$. Equation (14.90) still holds over the area, but along the curve C we would need

$$-\oint_C \boldsymbol{\xi}_t \cdot \hat{\mathbf{t}}\,\eta(u,v)\,\mathrm{d}\ell - \oint_C \mathbf{f}_L \cdot \hat{\mathbf{n}}\,\eta(u,v)\,\mathrm{d}\ell = 0. \tag{14.93}$$

By using $\hat{\mathbf{t}}\,\mathrm{d}\ell = \mathrm{d}\boldsymbol{\ell} \times \hat{\mathbf{n}}$ in the first integral, Eq. (14.93) becomes

$$\oint_C \left[\boldsymbol{\xi}_t \times (\mathrm{d}\boldsymbol{\ell}/\mathrm{d}\ell) + \mathbf{f}_L\right] \cdot \hat{\mathbf{n}}\,\eta(u,v)\,\mathrm{d}\ell = 0. \tag{14.94}$$

Since $\eta(u,v)$ is arbitrary along C, we conclude that

$$\left[\mathbf{f}_L + \boldsymbol{\xi}_t \times (\mathrm{d}\boldsymbol{\ell}/\mathrm{d}\ell)\right] \cdot \hat{\mathbf{n}} = 0, \tag{14.95}$$

which gives a normal force for curved surfaces that is the same as given by Eq. (14.58) for a planar surface. By considering a variation of the curve C in the tangential direction $\hat{\mathbf{t}}$ instead of Eq. (14.80) one can obtain the tangential component of Eq. (14.58). It must be borne in mind, however, that this is only valid if the state of strain of the solid is not affected by the variation; otherwise, one would obtain the surface stress instead of γ for a tangential force per unit length, as discussed in Section 14.1.2.

To evaluate the quantity $\omega_v^F - \omega_v^s$ for a single component, one usually uses

$$\mathrm{d}(\omega_v^F - \omega_v^s) = -(s^F n^F - s^s n^s)\,\mathrm{d}T - (n^F - n^s)\,\mathrm{d}\mu, \tag{14.96}$$

which is only valid if the effect of shear strain in the solid can be ignored. Here, s^F is the entropy per mole of the fluid, s^s is the entropy per mole of the solid, n^F is the molar density of the fluid, n^s is the molar density of the solid, T is temperature, and μ is chemical potential. For the fluid, one also has

$$\mathrm{d}\mu = -s^F\,\mathrm{d}T + (1/n^F)\,\mathrm{d}p^F, \tag{14.97}$$

where p^F is the pressure of the fluid. Then

$$\mathrm{d}(\omega_v^F - \omega_v^s) = -n^s(s^F - s^s)\,\mathrm{d}T - \frac{(n^F - n^s)}{n^F}\,\mathrm{d}p^F. \tag{14.98}$$

We can examine two states, both at the same value of p^F. One such state corresponds to a planar interface for an infinite crystal, so $\omega_v^F - \omega_v^s = 0$ and T becomes the nominal melting point T_{M} for that chosen pressure, where we have chosen the fluid to be a liquid for the sake of illustration. The other state corresponds to the equilibrium state of a small crystal in equilibrium with its liquid melt at temperature T. Then integration of Eq. (14.98) with $L_{\mathrm{V}} := n^s(s^F - s^s)T_{\mathrm{M}}$, the latent heat per unit volume of solid, assumed to be constant, gives

$$(\omega_v^F - \omega_v^s) = L_V \frac{(T_M - T)}{T_M}. \tag{14.99}$$

Thus Eq. (14.90) becomes

$$T = T_M - (T_M/L_V)\nabla_s \cdot \boldsymbol{\xi}, \tag{14.100}$$

which is a form of the **Gibbs-Thomson equation** for anisotropic γ. For isotropic γ it becomes

$$T = T_M - (T_M\gamma/L_V)\mathcal{K}, \tag{14.101}$$

which is well known.

Another important option is to keep T fixed in Eq. (14.96) but allow μ to vary from its value μ_∞ for an infinite crystal with $\omega_v^F - \omega_v^s = 0$ to its value μ for a finite crystal. Then by treating $\Omega_0 =: (n_s - n_F)^{-1}$ as a constant, Eq. (14.98) can be integrated to obtain

$$\mu = \mu_\infty + \Omega_0 \nabla_s \cdot \boldsymbol{\xi}. \tag{14.102}$$

If $\nabla_s \cdot \boldsymbol{\xi}$ is evaluated at a point on the surface, Eq. (14.100) is equivalent to the Herring equation, which usually pertains to the case in which the fluid is a gas with negligible density, so that $\Omega_0 \approx (n_s)^{-1}$. In the next section, we will develop that equation in detail.

The derivation of Eq. (14.90) was carried out in the context of global equilibrium between a crystal and a fluid, so that Eq. (14.91) is an equation for the equilibrium shape of the crystal. Under those conditions, the fluid is homogeneous, so its temperature and chemical potentials are uniform. On the other hand, Eqs. (14.100) and (14.102) are frequently regarded as **local equilibrium** conditions that apply at the surface of a crystal having any shape. In that case, for example, Eq. (14.102) would lead to a chemical potential that varied along the surface of the crystal. Such a nonuniform chemical potential would provide a driving force for diffusion processes that would lead to shape changes of the crystal and eventually to an equilibrium shape and a uniform chemical potential. For multicomponent systems, an equation similar in form to Eq. (14.100) can be obtained if the chemical potentials μ_i^F of the fluid can be maintained at fixed values. Then only the term in dT in Eq. (14.96) survives and Eq. (14.100) applies with L_V replaced by $L_V' = (s^F n^F - s^s n^s)T_M$, where now T_M is understood to be the local bulk melting point of the multicomponent alloy. Note that it is not so easy to extend Eq. (14.102) to a multicomponent system because $\omega = u_V - Ts_V - \sum \mu_i n_i$, so more than one chemical potential is involved. Such an extension is sometimes made, however, to the case of a Gibbs solid that has a *fixed* composition (Gibbs called this the substance of the solid) and does not contain other chemical components (if any) that are contained in the fluid. For such a solid, the chemical potential μ^A of the substance of the solid (regarded as a supercomponent A that is made up of appropriate components of the fluid in fixed proportions) would obey Eq. (14.102) at its surface. The chemical potentials of any other components of the fluid would be unconstrained.

14.6 Herring Formula

By explicitly evaluating the quantity $\nabla_s \cdot \boldsymbol{\xi}$ that appears in Eq. (14.90) at some point on a surface, one can obtain a formula due to Herring that is often used to calculate local equilibrium at a crystal-fluid interface. This can be accomplished by going to a Monge representation of the surface which requires one to adopt a specific parameterization of the surface of the form

$$x = u; \quad y = v; \quad z = w(u, v), \tag{14.103}$$

where x, y, and z are Cartesian coordinates, as in Section C.3 of Appendix C. This amounts to writing $z = z(x, y)$, where z on the right represents the function w of x and y whereas z on the left represents the value of that function, a common shorthand notation. Then with $p = z_x$ and $q = z_y$, where the subscripts denote partial derivatives (see Eq. (C.70)),

$$\omega_v^F - \omega_v^s = \nabla_s \cdot \boldsymbol{\xi} = -(\Phi_{pp} z_{xx} + 2\Phi_{pq} z_{xy} + \Phi_{qq} z_{yy}), \tag{14.104}$$

where $\Phi(p, q) = \gamma(p, q)\sqrt{1 + p^2 + q^2}$ is the value of the surface free energy γ per unit area of the x, y plane and subscripts indicate partial derivatives. Explicit values of the derivatives of Φ are given by Eq. (C.71). Equation (14.104) is a rather complicated nonlinear partial differential equation for the shape of the surface $z = z(x, y)$.

A formula that applies at a given point x_0, y_0 on the surface can be obtained by choosing the z-axis to lie along the normal $\hat{\mathbf{n}}_0$ at that point with the x, y plane locally tangent to the shape. In that case, $p = q = 0$ when evaluated at the chosen point, which gives

$$\omega_v^F - \omega_v^s = -(\gamma + \gamma_{pp}) z_{xx} - 2\gamma_{pq} z_{xy} - (\gamma + \gamma_{qq}) z_{yy}, \quad \text{at a point } x_0, y_0,\ z \text{ along } \hat{\mathbf{n}}_0, \tag{14.105}$$

where the derivatives are to be evaluated at $p = q = 0$, $x = x_0$, and $y = y_0$. This is a more general than the Herring formula because it does not require location of the principal axes of the shape under consideration.

If x and y are chosen along principal axes, Eq. (14.105) can be simplified further because $z_{xy} = 0$. Then $-z_{xx} = K_1 = 1/R_1$ and $-z_{yy} = K_2 = 1/R_2$ are principal curvatures. Thus

$$\omega_v^F - \omega_v^s = \frac{\gamma + \gamma_{pp}}{R_1} + \frac{\gamma + \gamma_{qq}}{R_2}, \quad \text{at a point } x_0, y_0,\ z \text{ along } \hat{\mathbf{n}}_0, \text{ principal axes.} \tag{14.106}$$

The **Herring formula** can be obtained by rewriting Eq. (14.106) in terms of the angles θ_1 and θ_2 made between the normal $\mathbf{n}$ and $\mathbf{n}_0$ near x_0, y_0 and measured in principal planes. Specifically, $\tan\theta_1 = \pm p$ and $\tan\theta_2 = \pm q$. This results in

$$\omega_v^F - \omega_v^s = \frac{\gamma + \gamma_{\theta_1\theta_1}}{R_1} + \frac{\gamma + \gamma_{\theta_2\theta_2}}{R_2}, \quad \text{at a point } x_0, y_0, \text{ principal planes,} \tag{14.107}$$

which is a somewhat more general version of Herring's result. (See Section C.4 for details of this variable change.) The original Herring formula [38, 41] pertained to the case of a solid-vapor interface for a single component for which Eq. (14.102) becomes[13]

$$\mu = \mu_\infty + \Omega_0 \left[\frac{\gamma + \gamma_{\theta_1\theta_1}}{R_1} + \frac{\gamma + \gamma_{\theta_2\theta_2}}{R_2} \right], \quad \text{at a point } x_0, y_0, \text{ principal planes.} \tag{14.108}$$

It should be emphasized that the Herring formula applies only at a point on the surface. In particular, it is not a partial differential equation for the surface shape, such as Eq. (14.104). It can, however, serve as a local equilibrium condition for a nonequilibrium shape, as discussed at the end of Section 14.5.

14.7 Legendre Transform of the Equilibrium Shape

In a Monge representation, $z = z(x, y)$ as introduced in Eq. (14.103), an interesting reciprocal relationship exists between equilibrium shape and the interfacial free energy expressed per unit area of the x, y plane, namely the quantity[14]

$$\Phi(p, q) = \frac{\gamma}{n_z} = \frac{\gamma}{\cos\theta} = \gamma(p, q)\sqrt{1 + p^2 + q^2}, \tag{14.109}$$

where $p = z_x$ and $q = z_y$ as introduced previously in connection with Eq. (14.104). We shall see that $\Phi(p, q)$ and $z = z(x, y)$ are essentially Legendre transforms of one another, with an appropriate constant scale factor λ. According to Eq. (14.91), which we write in the form $\boldsymbol{\xi} = \lambda\mathbf{r}$ with $\lambda = (\omega_v^F - \omega_v^s)/2$, we could equally well represent the equilibrium shape in a Monge representation of the form $Z = Z(X, Y)$, where $X \equiv \xi_x$, $Y \equiv \xi_y$, and $Z \equiv \xi_z$. With this notation we shall show that

$$\Phi = Z - pX - qY = Z - X\frac{\partial Z}{\partial X} - Y\frac{\partial Z}{\partial Y}, \tag{14.110}$$

whose inverse is

$$Z = \Phi + pX + qY = \Phi - p\frac{\partial \Phi}{\partial p} - q\frac{\partial \Phi}{\partial q}. \tag{14.111}$$

To relate directly to the equilibrium shape, just substitute $X = \lambda x$, $Y = \lambda y$, and $Z = \lambda z$.

We begin with Eq. (14.35) which we write in the form

$$n_x\,\mathrm{d}X + n_y\,\mathrm{d}Y + n_z\,\mathrm{d}Z = 0. \tag{14.112}$$

Then we calculate

$$p = \left(\frac{\partial Z}{\partial X}\right)_Y = -\frac{n_x}{n_z}; \quad q = \left(\frac{\partial Z}{\partial Y}\right)_X = -\frac{n_y}{n_z}, \tag{14.113}$$

[13]Herring actually treated a substitutional crystal with atoms and vacancies located on lattice sites, so μ here is actually equal to the difference between his chemical potential of atoms and his chemical potential of vacancies in an extended variable set. μ is also equal to the chemical potential of atoms in the vapor.

[14]With a Monge representation, it is necessary to use single-valued functions to represent various parts of a body, which amounts to choosing the sign of $n_z = \cos\theta$. Here we choose $\cos\theta > 0$ to obtain a positive value of Φ, with the understanding that the square root is positive.

which allows Eq. (14.112) to be written in the form

$$dZ = p\,dX + q\,dY. \tag{14.114}$$

By using Eq. (14.32), we deduce

$$\Phi = \frac{n_x X + n_y Y + n_z Z}{n_z} = Z - pX - qY, \tag{14.115}$$

which establishes Eq. (14.110). Note that $n_z^2 = 1 - n_x^2 - n_y^2 = 1 - n_z^2(p^2 + q^2)$ which can be solved for n_z, resulting in

$$1/n_z = \sqrt{1 + p^2 + q^2}, \tag{14.116}$$

consistent with Eq. (14.109). To establish the inverse transform, we calculate

$$d\Phi = dZ - p\,dX - X\,dp - q\,dY - Y\,dq = -X\,dp - Y\,dq, \tag{14.117}$$

where Eq. (14.114) has been used. From this differential we calculate

$$X = -\left(\frac{\partial \Phi}{\partial p}\right)_q; \quad Y = -\left(\frac{\partial \Phi}{\partial q}\right)_p, \tag{14.118}$$

which justifies the second part of Eq. (14.111).

These Legendre transforms can also be established without using the $\boldsymbol{\xi}$-vector by using the calculus of variations with the surface represented by a Monge representation, as shown in Appendix C, Section C.4.1.

14.8 Remarks About Solid-Solid Interfaces

Solid-solid interfaces are quite complicated if both phases are crystals, which are anisotropic. For example, specification of the interface between two crystals has five degrees of freedom (geometrical parameters), three to specify the relative orientations of the crystals (say, three Euler angles) and two to specify the orientation of the interface (grain boundary) that separates them. Structure and properties vary considerably with these five parameters because certain angles give rise to special lattice matchings. Moreover, most such interfaces are characterized by rather intricate dislocation arrays. The situation would be much simpler if one or both solids were amorphous.

For a detailed treatment of crystal-crystal interfaces, the reader is referred to *Interfaces in Crystalline Materials* by Sutton and Ballufi [46]. The first four chapters are devoted to interface structure and are well beyond the scope of the present book. Chapter 5 is devoted to thermodynamics of interfaces. Much of the formalism resembles that for fluid-fluid and solid-fluid interfaces but the variable set T, p, $\{\mu_i\}$ must be augmented by interfacial strain variables $\varepsilon_{\alpha\beta}$ and the geometrical parameters mentioned above. Considerations of excess free energies and forces that are used for fluid-fluid and solid-fluid interfaces can sometimes be used for solid-solid interfaces; however, they should be used with great care and might be completely inapplicable in some cases. For reasons of stability against cleavage, the size of γ for a solid-solid interface cannot exceed the sum of the free energies per unit area of the individual free surfaces. Heterophase interfaces typically have values of γ that are several times larger than those for homophase interfaces. Grain boundary free energies per unit area are usually less than those of a free surface because the number of nearest neighbors of an atom in a grain boundary is comparable to that of an atom in the bulk.

PART II

Statistical Mechanics

15

Entropy and Information Theory

The entropy S, which was introduced in Chapter 3 as a state function in connection with the second law of thermodynamics, plays a special role in statistical mechanics. Unlike the internal energy U, whose existence is an extension, although not a trivial one, of the concept of energy in mechanics, the entropy is intrinsically statistical and has no counterpart in mechanics. In thermodynamics, it is the conjugate variable to the absolute temperature, which also has no counterpart in mechanics. Nevertheless, the entropy, known since the time of Rudolf Clausius circa 1854, has its roots in information theory. This connection had been appreciated for a long time but not quantified. In a letter to Irving Langmuir, August 5, 1930, Gilbert Norton Lewis wrote [47, p. 400]:

> *It is not easy for a person brought up in the ways of classical thermodynamics to come around to the idea that the gain of entropy eventually is nothing more nor less than loss of information.*

The quantification of information in the context of communication theory was developed somewhat later (1948) by Claude Shannon [48, 49]. Subsequently, Shannon's communication theory was given a firm basis in probability theory by A.I. Khinchin [50].

15.1 Entropy as a Measure of Disorder

In order to understand the physical basis of entropy, it is often stated that entropy is a measure of disorder in a system, although this concept is sometimes objected to on the basis that common notions of disorder can disagree. Nevertheless, Shannon's information function, which we shall represent by the symbol D and refer to as the **disorder function**, gives a mathematically precise definition of information, the opposite of which is disorder. This disorder function is in complete agreement with the entropy of statistical mechanics, for all of its ensembles, except for the value of a multiplicative constant that simply accounts for units that are compatible with those chosen for energy and temperature. In addition, the function D plays the same role (except for opposite sign) as the dynamical function $\mathrm{H}(t)$ (also denoted by $\mathrm{E}(t)$ in Boltzmann's writings) that enters Boltzmann's Eta theorem [51].

15.1.1 The Disorder Function

We consider a set of mutually exclusive events A_i for $i = 1, 2, \ldots, N$ that have respective probabilities p_i. Mutual exclusivity means that only one of them can occur for a given trial. Khinchin calls this a "finite scheme." We introduce a disorder function

Thermal Physics. http://dx.doi.org/10.1016/B978-0-12-803304-3.00015-6

$$D\{p_i\} \equiv D(p_1, p_2, \ldots, p_N) \tag{15.1}$$

that depends only on the set of probabilities $\{p_i\}$ and has the following additional properties:

1. $D\{p_i\}$ takes on its minimum value, zero, when one of the p_i is equal to unity and all of the others are zero. This makes sense because the outcome of a trial is then certain, so there is complete information and therefore no disorder.
2. $D\{p_i\}$ takes on its maximum value $J(N)$ when the probabilities are all equal, that is, $p_i = 1/N$ for all i. This is reasonable because the outcome of any trial could equally well result in any of the possible events, so as little as possible is known about the outcome. Specifically,

$$J(N) := D(1/N, 1/N, \ldots, 1/N). \tag{15.2}$$

3. $J(N)$ should be a monotonically increasing function of N because a measure of disorder (lack of information) should increase if there is a larger possible number of outcomes.
4. The measure D should be independent of any manner in which the events are batched and the disorder of the batching is added to the weighted respective disorders of the batches. Thus if there are B batches (necessarily $B \leq N$) labeled by an index j with each batch having a probability q_j, then

$$D\{p_i\} = D\{q_j\} + \sum_{j=1}^{B} q_j D\{p_i/q_j\}_{p_i \subset q_j}. \tag{15.3}$$

Here, the notation $p_i \subset q_j$ means that p_i belongs to the batch q_j in which case $q_j = \sum_i^{p_i \subset q_j} p_i$. It therefore follows that $\sum_i^{p_i \subset q_j} p_i/q_j = 1$, so p_i/q_j is the probability of A_i within the batch j. For example, for $N = 5$ such a batching might be $q_1 = p_1 + p_2$ and $q_2 = p_3 + p_4 + p_5$, so $p_1/q_1 = p_1/(p_1 + p_2)$, $p_2/q_1 = p_2/(p_1 + p_2)$, etc.

Under these conditions, we shall proceed to demonstrate that the disorder function is

$$D\{p_i\} = -k \sum_i p_i \ln p_i, \tag{15.4}$$

where k is a positive constant. For a rigorous proof that this result is unique provided that $D\{p_i\}$ is continuous with respect to all of its arguments, see Khinchin [50, p. 9].

We first consider a special case of Eq. (15.3) for which all $p_i = 1/(BN)$ where B and N are integers, so $D\{p_i\} = J(BN)$. We divide into B batches, each with N elements, so $q_j = (1/B)$ and $p_i/q_j = 1/N$. Thus Eq. (15.3) becomes

$$J(BN) = J(B) + \sum_{j=1}^{B} \frac{1}{B} J(N), \tag{15.5}$$

so

$$J(BN) = J(B) + J(N). \tag{15.6}$$

We differentiate Eq. (15.6) partially with respect to B and then set $B = 1$ in the result to obtain

$$NJ'(N) = J'(1) \equiv k = \text{constant}, \tag{15.7}$$

where the prime indicates the derivative of a function with respect to its argument. The solution to this differential equation, subject to $J(1) = 0$ which follows from condition (1), is[1]

$$J(N) = k \ln N. \tag{15.8}$$

Armed with the functional form Eq. (15.8), we return to Eq. (15.3) with $p_i = 1/N$ for all i but with arbitrary batches having probabilities q_j, in which case it becomes

$$J(N) = D\{q_j\} + \sum_j q_j J(Nq_j) \tag{15.9}$$

or

$$k \ln N = D\{q_j\} + \sum_j q_j (k \ln q_j + k \ln N). \tag{15.10}$$

After cancellation of the term $k \ln N$, Eq. (15.10) becomes

$$D\{q_j\} = -k \sum_j q_j \ln q_j, \tag{15.11}$$

in agreement with Eq. (15.4).

From Eq. (15.11), condition (2) can be demonstrated by setting the result of partial differentiation with respect to q_i equal to zero, subject to the constraint $\sum_j q_j = 1$, which is easily handled by means of a Lagrange multiplier λ, that is, $\partial/\partial q_i [D\{q_j\} - \lambda \sum_j q_j] = 0$. Details are left to the reader to show that all q_i will be equal.

Since logarithms to any base are simply proportional to those for any other base, it is customary in information theory to use logarithms to the base two and then to set the overall constant equal to unity, resulting in the function

$$H_2\{p_j\} = -\sum_j p_j \log_2 p_j, \tag{15.12}$$

in which case the units of H_2 are known as bits. For the 128 ASCII characters, the maximum value of H_2 would be $J(128) = \log_2 128 = 7$ bits.

To obtain the entropy function of statistical mechanics, one retains the natural logarithm but chooses k to be Boltzmann's constant $k_B = 1.38065 \times 10^{-23}$ J K^{-1}. Thus the entropy

$$S\{p_j\} = -k_B \sum_j p_j \ln p_j, \tag{15.13}$$

[1]Since we seek only to discover a functional form, we have treated B as a continuous variable but the result also satisfies Eq. (15.6) when B is an integer.

where the p_j are the probabilities of the quantum microstates of the system subject to the constraints of the ensemble under consideration. For example, for the microcanonical ensemble to be considered in Chapter 16, one considers an isolated system having known energy E and makes the assumption that all compatible stationary quantum microstates, which number Ω, are equally probable. Then $p_j = 1/\Omega$ for all j, so

$$S\{p_j\} = -k_B \sum_{j=1}^{\Omega} (1/\Omega) \ln(1/\Omega) = k_B \ln \Omega. \tag{15.14}$$

For the canonical ensemble (see Chapter 19), one finds that each p_j is proportional to its Boltzmann factor, which results in $S = (U - F)/T$, a valid thermodynamic equation. In Chapter 22, we examine in detail the relationship between the disorder function D and the entropy S for any ensemble, with due respect to the constraints of that ensemble.

Example Problem 15.1. Suppose[2] that Eq. (15.13) gives the entropy of two systems, (1) and (2), so

$$S^{(1)} = -k_B \sum_i p_i^{(1)} \ln p_i^{(1)}; \quad S^{(2)} = -k_B \sum_j p_j^{(2)} \ln p_j^{(2)}, \tag{15.15}$$

where $p_i^{(1)}$ is the probability of state i for system (1) and $p_j^{(2)}$ is the probability of state j for system (2). If these two systems are combined to form a composite system having states ij with probabilities P_{ij}, the entropy of the composite system will be

$$S = -k_B \sum_{ij} P_{ij} \ln P_{ij}. \tag{15.16}$$

If the subsystems (1) and (2) are very weakly interacting, show that the entropy is additive, $S = S^{(1)} + S^{(2)}$.

Solution 15.1. If the systems interact sufficiently weakly that they are statistically independent, then $P_{ij} = p_i^{(1)} p_j^{(2)}$ so

$$S = -k_B \sum_{ij} p_i^{(1)} p_j^{(2)} \left[\ln p_i^{(1)} + \ln p_j^{(2)}\right] = -k_B \left[\sum_i p_i^{(1)} \ln p_i^{(1)} + \sum_j p_j^{(2)} \ln p_j^{(2)}\right], \tag{15.17}$$

where we have used the normalizations $\sum_i p_i^{(1)} = 1$ and $\sum_j p_j^{(2)} = 1$.

Example Problem 15.2. Suppose Eqs. (15.15) and (15.16) apply but the systems interact such that they are no longer statistically independent, so there are correlations and $P_{ij} \neq p_i^{(1)} p_j^{(2)}$. Show that $S < S^{(1)} + S^{(2)}$.

[2]This example and the one immediately following are based on problems in the book by Reif [52, p. 236].

Solution 15.2. We can use the general relations

$$p_i^{(1)} = \sum_j P_{ij}; \quad p_j^{(2)} = \sum_i P_{ij}; \quad \sum_{ij} P_{ij} = 1. \tag{15.18}$$

We have

$$\frac{S - S^{(1)} - S^{(2)}}{k_B} = -\sum_{ij} P_{ij} \ln P_{ij} + \sum_i p_i^{(1)} \ln p_i^{(1)} + \sum_j p_j^{(2)} \ln p_j^{(2)}. \tag{15.19}$$

The second two terms on the right can be converted to double sums, so

$$\sum_i p_i^{(1)} \ln p_i^{(1)} + \sum_j p_j^{(2)} \ln p_j^{(2)} = \sum_{ij} \left[P_{ij} \ln p_i^{(1)} + P_{ij} \ln p_j^{(2)} \right]. \tag{15.20}$$

Thus,

$$\frac{S - S^{(1)} - S^{(2)}}{k_B} = \sum_{ij} P_{ij} \ln \left(\frac{p_i^{(1)} p_j^{(2)}}{P_{ij}} \right). \tag{15.21}$$

However, the inequality $\ln x \le (x - 1)$ holds for all positive x with equality only for $x = 1$. This is true because $x = 1$ is also a point of tangency with slope 1 of the line $y = 1 - x$ and the curve $y = \ln x$, but elsewhere the slope of $\ln x$, namely $1/x$, is less than 1 for $x > 0$ and greater than 1 for $x < 0$. Therefore

$$\frac{S - S^{(1)} - S^{(2)}}{k_B} \le \sum_{ij} P_{ij} \left(\frac{p_i^{(1)} p_j^{(2)}}{P_{ij}} - 1 \right) = \sum_{ij} \left(p_i^{(1)} p_j^{(2)} - P_{ij} \right) = 0, \tag{15.22}$$

with equality holding only for the uncorrelated case $P_{ij} = p_i^{(1)} p_j^{(2)}$. We see that correlations reduce the disorder, and therefore lead to a smaller entropy S of the composite system.

15.2 Boltzmann Eta Theorem

In 1872, Ludwig Boltzmann (1838-1906), one of the pioneers of statistical mechanics, proved an important theorem, known as the Eta theorem, relating to the dynamics of an ideal gas of hard spheres as it approaches equilibrium by means of elastic collisions [51]. The treatment was, of course, classical since quantum mechanics did not emerge until about 1925. Boltzmann therefore described a homogenous gas in terms of a distribution function $f(\mathbf{v}, t)$ such that[3]

[3]A more general Boltzmann equation for a nonhomogeneous gas can be based on a distribution function $f(\mathbf{r}, \mathbf{v}, t)$ such that $f(\mathbf{r}, \mathbf{v}, t)\, \mathrm{d}^3 r\, \mathrm{d}^3 v$ is the probability that a sphere will have a position located in a volume element $\mathrm{d}^3 r$ centered about position $\mathbf{r}$ and a velocity located in a volume element $\mathrm{d}^3 v$ centered about velocity $\mathbf{v}$. See Reif [52, p. 523] for such a treatment.

$$f(\mathbf{v},t)\,\mathrm{d}^3 v \tag{15.23}$$

is the number of hard spheres per unit volume of actual space that have a velocity located in a volume element $\mathrm{d}^3 v$ in velocity space centered about velocity $\mathbf{v}$.

15.2.1 Boltzmann Equation

Boltzmann proceeded to derive a differential equation, known today as the **Boltzmann equation**, to describe the time evolution of the gas as it approached equilibrium. He assumed that only binary collisions occur and that the probability distribution function for pairs is given by $f(\mathbf{v}_1,\mathbf{v}_2,t) = f(\mathbf{v}_1,t)f(\mathbf{v}_2,t)$ which he based on an assumption of "molecular chaos," which we discuss later. He then set up a balance equation according to which

$$\frac{\partial f(\mathbf{v},t)}{\partial t} = r^{\mathrm{si}} - r^{\mathrm{so}}, \tag{15.24}$$

where $r^{\mathrm{si}}\mathrm{d}^3 v$ is the rate per unit volume of actual space at which spheres are scattered into $\mathrm{d}^3 v$ centered about $\mathbf{v}$ and $r^{\mathrm{so}}\mathrm{d}^3 v$ is the rate per unit volume of actual space at which spheres are scattered out of the volume element $\mathrm{d}^3 v$ in velocity space, all by means of binary elastic collisions.

To treat a hard sphere gas, we first consider only two hard spheres, subscripts 1 and 2, having velocities $\mathbf{v}_1$ and $\mathbf{v}_2$ before a collision and $\mathbf{v}_1'$ and $\mathbf{v}_2'$ after that collision. Each sphere is assumed to have the same diameter a. Later we will integrate over all possible collisions.

Since energy and momentum are conserved for elastic collision of hard spheres, we have

$$v_1^2 + v_2^2 = v_1'^2 + v_2'^2 \tag{15.25}$$

and

$$\mathbf{v}_1 + \mathbf{v}_2 = \mathbf{v}_1' + \mathbf{v}_2'. \tag{15.26}$$

Collisions can be understood better, however, in terms of the relative velocities before and after collision, namely

$$\mathbf{g} = \mathbf{v}_2 - \mathbf{v}_1; \quad \mathbf{g}' = \mathbf{v}_2' - \mathbf{v}_1'. \tag{15.27}$$

By squaring $\mathbf{g}$ and $\mathbf{g}'$ as well as squaring Eq. (15.26) and using Eq. (15.25), one sees readily that $g^2 - g'^2 = 0$, so $\mathbf{g}$ and $\mathbf{g}'$ have the same magnitude but differ in direction, as expected for an elastic collision. Let $\hat{\boldsymbol{\ell}}$ be a unit vector along the line of centers of the spheres at the time of collision. Then an elastic collision satisfies

$$\mathbf{g}\cdot\hat{\boldsymbol{\ell}} = -\mathbf{g}'\cdot\hat{\boldsymbol{\ell}}; \quad \mathbf{g}\times\hat{\boldsymbol{\ell}} = \mathbf{g}'\times\hat{\boldsymbol{\ell}}. \tag{15.28}$$

By crossing $\hat{\boldsymbol{\ell}}$ into the second member of Eq. (15.28), one readily obtains $\mathbf{g}' = \mathbf{g} - 2(\mathbf{g}\cdot\hat{\boldsymbol{\ell}})\hat{\boldsymbol{\ell}}$, so the second member of Eq. (15.27) becomes

$$\mathbf{v}_2' - \mathbf{v}_1' = \mathbf{g} - 2(\mathbf{g}\cdot\hat{\boldsymbol{\ell}})\hat{\boldsymbol{\ell}}. \tag{15.29}$$

Equation (15.29) can now be solved simultaneously with Eq. (15.26) to obtain

$$\mathbf{v}_1' = \mathbf{v}_1 + [(\mathbf{v}_2 - \mathbf{v}_1)\cdot\hat{\boldsymbol{\ell}}]\hat{\boldsymbol{\ell}} \tag{15.30}$$

$$\mathbf{v}_2' = \mathbf{v}_2 - [(\mathbf{v}_2 - \mathbf{v}_1)\cdot\hat{\boldsymbol{\ell}}]\hat{\boldsymbol{\ell}}, \tag{15.31}$$

which completely determines $\mathbf{v}_1'$ and $\mathbf{v}_2'$ as functions of $\mathbf{v}_1$ and $\mathbf{v}_2$ and two spherical angles, θ and ϕ, that determine the unit vector $\hat{\boldsymbol{\ell}}$.

Next we integrate over all possible collisions. In an infinitesimal increment $\mathrm{d}t$ of time, all spheres 2 in a cylinder of real space of length $|g|\,\mathrm{d}t$ and cross sectional area $(a^2/4)\,\mathrm{d}\Omega$ will collide with sphere 1 and scatter into solid angle $\mathrm{d}\Omega = \sin\theta\,\mathrm{d}\theta\,\mathrm{d}\phi$, so[4]

$$r^{\mathrm{so}}\mathrm{d}^3 v_1 = \frac{a^2}{4}\int \mathrm{d}\Omega \int \mathrm{d}^3 v_2 \left\{|\mathbf{v}_2 - \mathbf{v}_1| f(\mathbf{v}_2, t)\right\} f(\mathbf{v}_1, t)\,\mathrm{d}^3 v_1 \tag{15.32}$$

$$r^{\mathrm{si}}\mathrm{d}^3 v_1 = \frac{a^2}{4}\int \mathrm{d}\Omega \int \mathrm{d}^3 v_2' \left\{|\mathbf{v}_2' - \mathbf{v}_1'| f(\mathbf{v}_2', t) f(\mathbf{v}_1', t)\right\} \mathrm{d}^3 v_1'. \tag{15.33}$$

In Eq. (15.32), the integrations are only over Ω and $\mathbf{v}_2$ with $\mathbf{v}_1$ fixed. In Eq. (15.33), the integrations are only over Ω and $\mathbf{v}_2'$ with $\mathbf{v}_1$ fixed. Thus in Eq. (15.33), we need to think of $\mathbf{v}_1'$ as a function of $\mathbf{v}_2'$, $\mathbf{v}_1$, and $\hat{\boldsymbol{\ell}}$ through the relation $\mathbf{v}_1' = \mathbf{v}_1 - [(\mathbf{v}_2' - \mathbf{v}_1')\cdot\hat{\boldsymbol{\ell}}]\hat{\boldsymbol{\ell}}$.

By differentiation of Eqs. (15.30) and (15.31), however, one can show that

$$\mathrm{d}v_1'^2 + \mathrm{d}v_2'^2 = \mathrm{d}v_1^2 + \mathrm{d}v_2^2, \tag{15.34}$$

which means that the transformation is orthogonal and has Jacobian unity. Thus, the volume element $\mathrm{d}^3 v_1'\,\mathrm{d}^3 v_2' = \mathrm{d}^3 v_1\,\mathrm{d}^3 v_2$. This is to be expected because it is a linear transformation that is length preserving (see Eq. (15.25)), actually a rotation in a six-dimensional space. We therefore obtain (after cancellation of $\mathrm{d}^3 v_1$)

$$r^{\mathrm{so}} = \frac{a^2}{4}\int \mathrm{d}\Omega \int \mathrm{d}^3 v_2 \left\{|\mathbf{v}_2 - \mathbf{v}_1| f(\mathbf{v}_2, t) f(\mathbf{v}_1, t)\right\} \tag{15.35}$$

$$r^{\mathrm{si}} = \frac{a^2}{4}\int \mathrm{d}\Omega \int \mathrm{d}^3 v_2 \left\{|\mathbf{v}_2 - \mathbf{v}_1| f(\mathbf{v}_2', t) f(\mathbf{v}_1', t)\right\}. \tag{15.36}$$

In Eq. (15.36), $\mathbf{v}_1'$ and $\mathbf{v}_2'$ are to be regarded as functions of $\mathbf{v}_1$, $\mathbf{v}_2$, and $\hat{\boldsymbol{\ell}}$. Substitution into Eq. (15.24) then leads to the Boltzmann equation

$$\frac{\partial f(\mathbf{v}_1, t)}{\partial t} = \frac{a^2}{4}\int \mathrm{d}\Omega \int \mathrm{d}^3 v_2 \left\{|\mathbf{v}_2 - \mathbf{v}_1| [f(\mathbf{v}_2', t) f(\mathbf{v}_1', t) - f(\mathbf{v}_2, t) f(\mathbf{v}_1, t)]\right\}. \tag{15.37}$$

This is a complicated equation and its solution is a field of study unto itself. See Reif [52, chapters 13-14] for more detail on its derivation and approximate solution. It can be generalized to treat inhomogeneous systems as well as particles other than hard spheres. We shall follow Boltzmann by using the simplified Boltzman equation (Eq. (15.37)) to prove a very important theorem.

[4]See [53] for a more detailed discussion of the collision geometry that leads to the factor $a^2/4$. For collision of particles other than hard spheres, this factor can be replaced by the collision cross section $A(g,\theta)$ under the integral sign.

15.2.2 Eta Theorem

We follow Boltzmann in assuming that $f(\mathbf{v}, t)$ is a solution to the Boltzmann equation and defining the function of time

$$\mathrm{Eta}(t) \equiv \mathrm{E}(t) \equiv \mathrm{H}(t) = \int \mathrm{d}^3 v\, f(\mathbf{v}, t) \ln f(\mathbf{v}, t). \tag{15.38}$$

Then by clever manipulation, Boltzmann showed that

$$\frac{\mathrm{dH}(t)}{\mathrm{d}t} \le 0. \tag{15.39}$$

Thus the function $\mathrm{H}(t)$ decays to a minimum value, which corresponds to equilibrium. We recognize that $\mathrm{H}(t)$ is a dynamical analog of the negative of the disorder function $D\{p_j\}$ of the previous section. In other words, $\mathrm{H}(t)$ is expected to relate to the negative of the entropy.

To prove Eq. (15.39), we first differentiate Eq. (15.38) to obtain

$$\frac{\mathrm{dH}(t)}{\mathrm{d}t} = \int \mathrm{d}^3 v_1 \frac{\partial f(\mathbf{v}_1, t)}{\partial t} \left[1 + \ln f(\mathbf{v}_1, t)\right]. \tag{15.40}$$

Then we substitute for the time derivative of f from the Boltzmann equation to obtain

$$\begin{aligned}\frac{\mathrm{dH}(t)}{\mathrm{d}t} = \frac{a^2}{4} \int \mathrm{d}^3 v_1 \int \mathrm{d}^3 v_2 \int \mathrm{d}\Omega \{|\mathbf{v}_2 - \mathbf{v}_1| [f(\mathbf{v}_2', t) f(\mathbf{v}_1', t) \\ - f(\mathbf{v}_2, t) f(\mathbf{v}_1, t)]\} \left[1 + \ln f(\mathbf{v}_1, t)\right].\end{aligned} \tag{15.41}$$

Because of the symmetry of the portion of the integrand in brackets $\{\}$, we can get the same result by interchanging $\mathbf{v}_1$ and $\mathbf{v}_2$. Thus the integrand can also be written

$$\{|\mathbf{v}_2 - \mathbf{v}_1| [f(\mathbf{v}_2', t) f(\mathbf{v}_1', t) - f(\mathbf{v}_2, t) f(\mathbf{v}_1, t)]\} \left[1 + \ln f(\mathbf{v}_2, t)\right]. \tag{15.42}$$

Adding these results and dividing by 2 gives

$$\begin{aligned}\frac{\mathrm{dH}(t)}{\mathrm{d}t} = \frac{a^2}{4} \int \mathrm{d}^3 v_1 \int \mathrm{d}^3 v_2 \int \mathrm{d}\Omega \frac{1}{2} \{|\mathbf{v}_2 - \mathbf{v}_1| [f(\mathbf{v}_2', t) f(\mathbf{v}_1', t) \\ - f(\mathbf{v}_2, t) f(\mathbf{v}_1, t)]\} \left[2 + \ln f(\mathbf{v}_1, t) f(\mathbf{v}_2, t)\right].\end{aligned} \tag{15.43}$$

Next we exchange the primed with the unprimed velocities but recall that $|\mathbf{v}_2' - \mathbf{v}_1'| = |\mathbf{v}_2 - \mathbf{v}_1|$ and $\mathrm{d}^3 v_1' \mathrm{d}^3 v_2' = \mathrm{d}^3 v_1 \mathrm{d}^3 v_2$. Thus, the integrand in Eq. (15.43) can be written

$$\frac{1}{2} \{|\mathbf{v}_2 - \mathbf{v}_1| [f(\mathbf{v}_2, t) f(\mathbf{v}_1, t) - f(\mathbf{v}_2', t) f(\mathbf{v}_1', t)]\} \left[2 + \ln f(\mathbf{v}_1', t) f(\mathbf{v}_2', t)\right]. \tag{15.44}$$

By adding the forms given by Eqs. (15.43) and (15.44) and again dividing by 2, we obtain

$$\begin{aligned}\frac{\mathrm{dH}(t)}{\mathrm{d}t} = \frac{a^2}{4} \int \mathrm{d}^3 v_1 \int \mathrm{d}^3 v_2 \int \mathrm{d}\Omega \frac{1}{4} \{|\mathbf{v}_2 - \mathbf{v}_1| [f(\mathbf{v}_2', t) f(\mathbf{v}_1', t) \\ - f(\mathbf{v}_2, t) f(\mathbf{v}_1, t)]\} \left[\ln f(\mathbf{v}_1, t) f(\mathbf{v}_2, t) - \ln f(\mathbf{v}_1', t) f(\mathbf{v}_2', t)\right].\end{aligned} \tag{15.45}$$

The integrand of Eq. (15.45) has the form $(x - y)(\ln y - \ln x) \le 0$ for any positive x and y, with equality only holding for $x = y$. Therefore, Eq. (15.39) is proven.

At equilibrium, $\mathrm{dH}/dt = 0$ and the distribution functions $f(\mathbf{v}, t)$ become independent of time, so we designate them by $f_0(\mathbf{v})$, resulting in the equilibrium condition

$$f_0(\mathbf{v}_2')f_0(\mathbf{v}_1') = f_0(\mathbf{v}_2)f_0(\mathbf{v}_1). \tag{15.46}$$

By taking the logarithm of Eq. (15.46), we obtain

$$\ln f_0(\mathbf{v}_2') + \ln f_0(\mathbf{v}_1') = \ln f_0(\mathbf{v}_2) + \ln f_0(\mathbf{v}_1), \tag{15.47}$$

which looks like a conservation condition for some property of spheres 1 and 2 before and after collision. It must be related to the conservation of energy and momentum during a collision.

Example Problem 15.3. Show that Eq. (15.47) is satisfied by the Maxwell-Boltzmann distribution function

$$M(\mathbf{v} - \mathbf{v}_r) = A\exp[-(m/2k_BT)(\mathbf{v} - \mathbf{v}_r)^2]; \quad A = [m/(2\pi k_BT)]^{3/2}, \tag{15.48}$$

where $\mathbf{v}_r$ is some reference velocity. This is a slight generalization of Eq. (19.70) that we derive later for the case $\mathbf{v}_r = 0$.

Solution 15.3. We substitute Eq. (15.48) into Eq. (15.47). The quantity $2\ln A$ appears on both sides and can be canceled. Then we divide both sides by $-(m/2k_BT)$ to obtain $(\mathbf{v}_2' - \mathbf{v}_r)^2 + (\mathbf{v}_1' - \mathbf{v}_r)^2 = (\mathbf{v}_2 - \mathbf{v}_r)^2 + (\mathbf{v}_1 - \mathbf{v}_r)^2$. Multiplying out the squares and canceling v_r^2 gives

$$(v_1')^2 + (v_2')^2 + 2\mathbf{v_r} \cdot (\mathbf{v}_2' + \mathbf{v}_1') = (v_1)^2 + (v_2)^2 + 2\mathbf{v_r} \cdot (\mathbf{v}_2 + \mathbf{v}_1). \tag{15.49}$$

In Eq. (15.49), the squared terms cancel by conservation of energy, Eq. (15.25), and the terms dotted into $\mathbf{v_r}$ cancel by conservation of momentum, Eq. (15.26).

In Boltzmann's day, there were many objections to his Eta theorem on the grounds that the equations of classical mechanics are invariant under time reversal. What was not recognized, however, was that Boltzmann's assumption of molecular chaos, $f(\mathbf{v}_1, \mathbf{v}_2, t) = f(\mathbf{v}_1, t)f(\mathbf{v}_2, t)$, (Stosszahlansatz in German, which literally means collision frequency assumption) was a statistical assumption, not based on mechanics. For a detailed discussion, see [53, p. 20]. The assumption of molecular chaos, not deterministic mechanics, causes the system to evolve to a more probable state.

The Eta theorem as presented here should be regarded as a demonstration that for a simple system there is a function $-\mathrm{H}(t)$, given by the negative of Eq. (15.38), that can only increase in time for an isolated system, which of course would have constant energy. This function has the same characteristics that we ascribe to the thermodynamic function S, the entropy, that can only increase for an isolated system that changes from one state to another. Note also that $-\mathrm{H}(t)$ has the same structure as the disorder function $D\{q_j\}$ given by Eq. (15.11). The Eta theorem therefore demonstrates that the assumption of molecular chaos leads to evolution to a more probable state. It is not a substitute for the second law of thermodynamics, for which the entropy of any isolated thermodynamic system is postulated to increase, subject to any internal constraints, until it reaches its maximum possible value at equilibrium. The validity of the second law can be bolstered by statistical analysis of more complex systems, but ultimately rests on agreement with experiments.

16

Microcanonical Ensemble

The general approach to statistical mechanics is based on the idea of an **ensemble**. An ensemble is an imaginary collection of microstates that are compatible with a specified macrostate of a thermodynamic system. A **macrostate** is specified by giving a complete set of *macro*scopic state variables. The number of state variables necessary to constitute a complete set depends on the complexity of the system; usually only a few are necessary. The members (microstates) of the ensemble differ from one another *micro*scopically and are usually extremely large in number, approaching infinity in the thermodynamic limit.

Each microstate that makes up an ensemble occurs with a probability that depends on which set of macroscopic state variables are used to specify the macrostate. This probability is chosen so that the ensemble represents, in a statistical sense, the macrostate. Specifically, the ensemble is chosen such that averages of measurable quantities computed by using it will lead to values representative of the macrostate. Such an approach is necessary in statistical mechanics because specification of a macrostate constitutes incomplete information.

We believe that quantum mechanics ultimately governs all systems, even if classical mechanics is a good approximation for some purposes. Therefore, we identify the microstates of a system in equilibrium with a set of **stationary quantum states** of that system. Since we specify a macroscopic system by a small number of state variables, we cannot know the total wave function of the system, which would constitute a specific knowledge of a linear combination (time-dependent) of its stationary states; such a state is called a **pure state**. All we will know for a system in equilibrium is a set of probabilities of its stationary states. We will delay the formal quantum mechanical description of such quantum systems until Chapter 26 where we introduce density operators and use them to describe pure states and **statistical states**, also known as mixed states. Until then, it will suffice to know only the set of probabilities of the stationary states of an ensemble.

A classical system is described by specification of the coordinates and momenta of every particle in the system at some given time. As time evolves, the particles will move and the system can be imagined to progress through other microstates that are compatible with the given set of macroscopic state variables. Clearly there is a continuum of such microstates, which are infinite in number. Therefore, for a classical system it will only be possible to specify a probability density function for a given hypervolume of **phase space** (the space of all coordinates and momenta) since the probability of having a specific classical microstate would be 0.

In the present chapter, we present the microcanonical ensemble which is applicable to an **isolated system** whose macrostate is specified by its total energy and additional extensive macrovariables, such as volume and number of particles, that altogether constitute

Thermal Physics. http://dx.doi.org/10.1016/B978-0-12-803304-3.00016-8

a complete set. This ensemble is of fundamental theoretical importance but its practical usefulness is limited because of the complexity of computations required to enumerate the microstates. In later chapters, we will introduce other more useful ensembles such as the canonical ensemble, where temperature instead of energy is specified, and the grand canonical ensemble where temperature instead of energy and chemical potential instead of particle number are specified.

16.1 Fundamental Hypothesis of Statistical Mechanics

For an isolated thermodynamical system, we consider all **microstates** compatible with a **macrostate**. For the sake of illustration, we will specify this macrostate by its total energy[1] E, its volume V, and its number of particles $\mathcal{N}$; more complex systems can be treated by adding additional extensive state variables, specifying subsystems, etc. We consider our system to be governed by quantum mechanics and therefore associate each microstate with a stationary quantum state for the given quantities $E, V, \mathcal{N}$. Since our system is macroscopic, the difference between E and the energy levels of one of its particles is very large compared to the differences among the energy levels of a particle. Thus, there is massive degeneracy and hence a huge number of microstates for each macrostate. This ensemble is usually referred to as the **microcanonical ensemble**.

The **fundamental hypothesis** is that every microstate of an isolated thermodynamic system that is compatible with a given macrostate of such a system is *equally probable*. If Ω is the total number of compatible microstates for a given macrostate, then the probability of each microstate is $1/\Omega$.

It follows from this hypothesis that the expected value $\langle y \rangle$ of any property y of the system is given by

$$\langle y \rangle = (1/\Omega) \sum_{\nu=1}^{\Omega} y_\nu, \tag{16.1}$$

where y_ν is the value of y in the microstate with label ν. Furthermore, the entropy of the system is defined to be

$$S = k_{\rm B} \ln \Omega, \tag{16.2}$$

where $k_{\rm B}$ is Boltzmann's constant. Equation (16.2) is exactly the function given by Eq. (15.14) that was based on the disorder function $D\{p_i\}$ for the case $p_i = 1/\Omega$. The classical counterpart to Eq. (16.2) was proposed by Boltzmann and will be discussed in the next chapter.

Since S is a monotonically increasing function of Ω, maximizing Ω is consistent with the second law of thermodynamics, according to which S for an isolated system will be a

[1]This is the total internal energy of the system which excludes macroscopic kinetic energy associated with motion of the center of mass. Here, we use the symbol E instead of U to make a subtle distinction because we *specify* the energy rather than calculating its average value from a knowledge of other state variables, for example, the temperature.

maximum at equilibrium, with respect to variations of its internal extensive parameters and subject to any internal constraints. To better illustrate what this means, we consider a composite system made up of two subsystems with energies E_1 and E_2, volumes V_1 and V_2, and particle numbers $\mathcal{N}_1$ and $\mathcal{N}_2$. Furthermore, we assume that the subsystems are very weakly interacting and statistically independent so that $\Omega = \Omega_1(E_1, V_1, \mathcal{N}_1)\Omega_2(E_2, V_2, \mathcal{N}_2)$, where the total quantities $E = E_1 + E_2$, $V = V_1 + V_2$, and $\mathcal{N} = \mathcal{N}_1 + \mathcal{N}_2$. So even with E, V, and $\mathcal{N}$ held constant, $\Omega(E, V, \mathcal{N}, E_1, V_1, \mathcal{N}_1)$ still depends on the partitioning of energy, volume, and particle numbers between the two subsystems. Since $\Omega = \Omega_1\Omega_2$, the total entropy is simply additive, resulting in

$$S = k_B \ln \Omega = k_B \ln(\Omega_1 \Omega_2) = k_B \ln(\Omega_1) + k_B \ln(\Omega_2) = S_1 + S_2. \tag{16.3}$$

With E, V, $\mathcal{N}$, V_1, and $\mathcal{N}_1$ held constant, maximizing $\ln \Omega$ with respect to E_1 gives

$$\frac{\partial \ln \Omega}{\partial E_1} = \frac{1}{\Omega_1}\frac{\partial \Omega_1}{\partial E_1} + \frac{1}{\Omega_2}\frac{\partial \Omega_2}{\partial E_2}\frac{\partial E_2}{\partial E_1} = \frac{1}{\Omega_1}\frac{\partial \Omega_1}{\partial E_1} - \frac{1}{\Omega_2}\frac{\partial \Omega_2}{\partial E_2} = 0. \tag{16.4}$$

But $(1/\Omega_1)\partial\Omega_1/\partial E_1 = (1/k_B)\partial S_1/\partial E_1 = T_1/k_B$ and similarly $(1/\Omega_2)\partial\Omega_2/\partial E_2 = T_2/k_B$, so Eq. (16.4) is equivalent to equality of absolute temperature, $T_1 = T_2$. Similarly, maximizing with respect to V_1 gives equality of pressure, and maximizing with respect to $\mathcal{N}_1$ gives equality of chemical potential. If a system cannot be decomposed into statistically independent subsystems, Ω will not factor and it will be difficult to enumerate the microstates, but the maximization of Ω and hence S will still be a valid criterion for equilibrium.

The considerations of the preceding paragraph are still valid if the two subsystems are actually portions of the same system that are initially isolated from one another by constraints that forbid exchange of energy, volume, and mole numbers. Then $\Omega = \Omega'(E', V', \mathcal{N}')\Omega''(E'', V'', \mathcal{N}'')$. As these constraints are gradually relaxed, each subsystem can proceed through a series of equilibrium states until $\Omega = \Omega'\Omega''$ is maximized. We can therefore say that Ω is proportional to the probability of a macrostate. The quantity $1/\Omega$ is the probability of each microstate of the ensemble that represents that macrostate. For further support that Ω is proportional to the probability of a macrostate, the reader is referred to Section 19.1.3 in which the number of ways W_{ens} of constructing an ensemble from members, each of which is in a single eigenstate i, is related to the probability P_i of occurrence of that eigenstate in the ensemble. If every member of the ensemble has the *same* energy, as it would for the microcanonical ensemble, W_{ens} will be a maximum when all probabilities are equal. See also Chapter 22 where the entropy of any ensemble is discussed in terms of maximizing a probability. Moreover, since S given by Eq. (16.2) is proportional to the maximum value of the disorder function $D\{q_i\}$ given by Eq. (15.11), a macrostate of maximum entropy, and hence maximum Ω, is a state of maximum disorder, equivalent to a state with minimum information content, compatible with constraints.

In some books on statistical mechanics, there is an attempt to justify the fundamental assumption of equal probability of each compatible microstate from other considerations.

The basic idea for classical systems, see Landau and Lifshitz [7], is that a macroscopic system in equilibrium is assumed to progress in time through phase space so that it visits every allowed volume of phase space with equal probability. This is sometimes called the **ergodic hypothesis**. The quantum analog would be to assume that every microstate is visited with equal probability over a time τ that is long compared to some characteristic relaxation time. The measurement of some physical property of a thermodynamic system is given by a time average of the form

$$\bar{y} = \frac{1}{\tau}\int_0^{\tau} y(t)\,\mathrm{d}t. \tag{16.5}$$

Thus, the time average would be equal to the ensemble average, that is,

$$\langle y \rangle = \bar{y} \tag{16.6}$$

for a system in equilibrium. In this book, we shall assume that Eq. (16.6) is true. In the last analysis, Eqs. (16.1), (16.2), and (16.6) are hypotheses that have borne up under the test of experiment. In Chapter 26, we introduce the statistical density operator and give a more detailed discussion of ensemble averages and time averages in the context of quantum mechanics.

Under conditions for which a macroscopic system can be described approximately by a set of $\mathcal{N}$ particles that obey classical mechanics, we do not have access to the concept of stationary quantum states. As discussed in the next chapter, we take Ω to be proportional to the volume of phase space (volume of momentum space times volume of actual space) in a thin shell near a hypersurface that corresponds to the total energy, E. It turns out that the volume of the shell itself is not important. Indeed, isolation of a system is only an idealization and therefore an approximation, so there will always be a small uncertainty in its energy. Landau and Lifshitz [7] refer to such systems as being quasi-isolated. Nevertheless, to agree ultimately with quantum mechanics, it is necessary to assume that the number of microstates in a volume element $(\mathrm{d}p\,\mathrm{d}q)^{3\mathcal{N}}$ is given by $(\mathrm{d}p\,\mathrm{d}q/h)^{3\mathcal{N}}$ where h is Planck's constant.[2] This is an artificial prescription that has no real basis in classical mechanics, for which Planck's constant is effectively 0.

The microcanonical ensemble is easy to define but very hard to use because of the difficulty in cataloging and enumerating the microstates that are compatible with specification of the macrostate. For simple systems this is possible, as we shall illustrate for two-state subsystems, simple harmonic oscillators and ideal gases. The main value of the microcanonical ensemble is its theoretical importance, and we shall use it later to derive the canonical ensemble and the grand canonical ensemble which are much more tractable.

[2]This result holds for **identical but distinguishable particles**. For an ideal gas, the number of microstates is approximately $(\mathrm{d}p\,\mathrm{d}q/h)^{3\mathcal{N}}(1/\mathcal{N}!)$ at high temperature and low density. The extra Gibbs factor of $1/\mathcal{N}!$ is needed to make the entropy an extensive function and follows from quantum mechanics for identical indistinguishable particles. See Section 16.4 for further details.

16.2 Two-State Subsystems

We consider a system consisting of $\mathcal{N}$ subsystems (particles) fixed in a solid, each having two quantum states that correspond to nondegenerate energy levels, 0 and ε. The particles are assumed to be identical except for their locations (their positions in the solid) which makes them distinguishable. We might think of each particle as a quantum system having spin 1/2 under conditions for which the two states "spin up" and "spin down" have different energies, perhaps because of an externally applied magnetic field. The particles are assumed to interact very weakly, so that their interaction energy is negligible but still sufficient to enable them to come to equilibrium with one another.

If we consider a quantum state of the system in which $\mathcal{M}$ particles are in the state with energy ε, the total energy of that quantum state is $E = \mathcal{M}\varepsilon$ and the number of microstates (its degeneracy) is

$$\Omega(\mathcal{N}, E) = \frac{\mathcal{N}!}{(\mathcal{N} - \mathcal{M})!\mathcal{M}!} = g(\mathcal{N}, \mathcal{M}). \tag{16.7}$$

The quantity $g(\mathcal{N}, \mathcal{M})$ is sometimes called the **multiplicity function**. Thus

$$S = k_B \ln \Omega = k_B[\ln \mathcal{N}! - \ln(\mathcal{N} - \mathcal{M})! - \ln \mathcal{M}!]. \tag{16.8}$$

By using the first two terms in Stirling's approximation in the form of Eq. (A.1), we obtain

$$S \sim k_B[\mathcal{N} \ln \mathcal{N} - (\mathcal{N} - \mathcal{M}) \ln(\mathcal{N} - \mathcal{M}) - \mathcal{M} \ln \mathcal{M}]. \tag{16.9}$$

Note that the second term in Stirling's approximation (see Appendix A) makes no contribution to this result because of an exact cancellation. Other terms in Stirling's approximation have been dropped because they would lead to sub-extensive results of order $\ln \mathcal{N}$ or smaller.

To see that the entropy given by Eq. (16.9) is an extensive function, we can write it in the form

$$S = -\mathcal{N} k_B[(1 - \mathcal{M}/\mathcal{N}) \ln(1 - \mathcal{M}/\mathcal{N}) + (\mathcal{M}/\mathcal{N}) \ln(\mathcal{M}/\mathcal{N})], \tag{16.10}$$

in which the substitution $\mathcal{M} = E/\varepsilon$ yields the fundamental equation, $S(E, \mathcal{N})$, of the system. The ratios $\mathcal{M}/\mathcal{N} = E/(\varepsilon\mathcal{N})$ are intensive[3] variables, so the form of Eq. (16.10) shows it to be the product of an extensive quantity $\mathcal{N}$ and an intensive quantity in square brackets, which shows explicitly that S is extensive. This is not apparent from Eq. (16.9) because expressions such as $\mathcal{N} \ln \mathcal{N}$ are not extensive.

Figure 16–1 shows a plot of $S/(\mathcal{N} k_B)$ versus $\mathcal{M}/\mathcal{N}$. For convenience we show a continuous curve, although only integer values of $\mathcal{M}$ are allowed. This is justified because $\mathcal{N}$ is very large, perhaps of order 10^{20}, so changes of $\mathcal{M}/\mathcal{N}$ are of order 10^{-20} as $\mathcal{M}$ changes by unity. We observe that the curve is symmetric about its maximum, which occurs at $\mathcal{M}/\mathcal{N} = 1/2$. It has a positive slope for $\mathcal{M}/\mathcal{N} \leq 1/2$, zero slope for $\mathcal{M} = 1/2$, and a negative

[3]We know that Stirling's approximation is only valid for large numbers, so one might worry about small values of $\mathcal{M}$. However, only values of $\mathcal{M}$ that are comparable to $\mathcal{N} \gg 1$ will lead to significant results. Therefore, $\mathcal{M} = E/\varepsilon$ must be regarded as an extensive quantity.

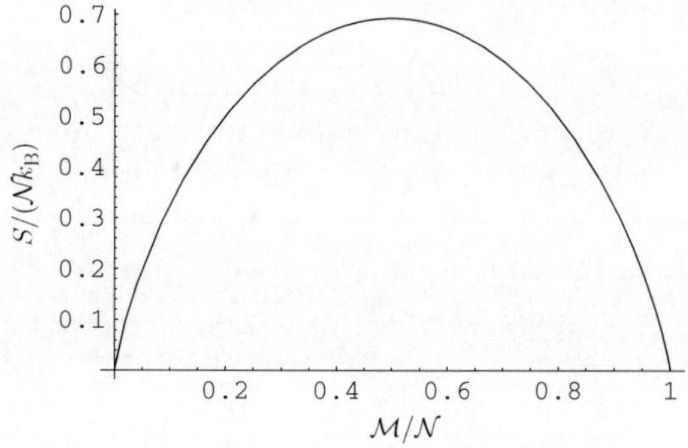

FIGURE 16–1 Dimensionless entropy $S/(\mathcal{N}k_B)$ versus dimensionless energy $\mathcal{M}/\mathcal{N} = E/(\mathcal{N}\varepsilon)$ for a two-state system according to Eq. (16.10). The portion of the curve for $\mathcal{M}/\mathcal{N} > 1/2$ corresponds to negative temperatures, which do not represent equilibrium states in thermodynamics.

slope for $\mathcal{M}/\mathcal{N} > 1/2$. As we shall see subsequently, this slope is proportional to $1/T$ so only $0 < \mathcal{M}/\mathcal{N} < 1/2$ corresponds to positive finite temperatures. Indeed, $\mathcal{M}/\mathcal{N} = 1/2$ corresponds to $T = \infty$. The range of values $1/2 < \mathcal{M}/\mathcal{N} < 1$ corresponds to negative temperatures, which are not allowed in thermodynamics.[4]

Since $\mathcal{M}$ is a nearly continuous variable, we can calculate the absolute temperature as a derivative,

$$\frac{1}{T} = \left(\frac{\partial S}{\partial E}\right)_{\mathcal{N}} = k_B \frac{d\mathcal{M}}{dE}\left(\frac{\partial \ln g}{\partial \mathcal{M}}\right)_{\mathcal{N}} = -\frac{k_B}{\varepsilon}\ln\left(\frac{\mathcal{M}/\mathcal{N}}{1-\mathcal{M}/\mathcal{N}}\right). \tag{16.11}$$

Figure 16–2 shows a plot of k_BT/ε as a function of $\mathcal{M}/\mathcal{N}$. We see that only the range $\mathcal{M}/\mathcal{N} < 1/2$ leads to positive values of T, as anticipated above. Equation (16.11) can be solved to yield

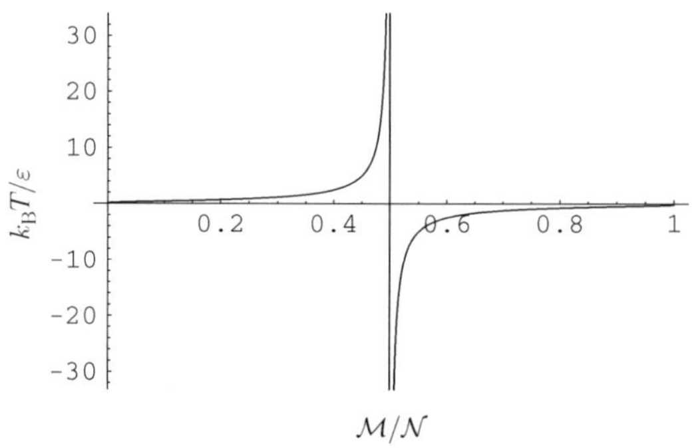

FIGURE 16–2 Dimensionless temperature k_BT/ε versus dimensionless energy $\mathcal{M}/\mathcal{N} = E/(\mathcal{N}\varepsilon)$ for a two-state system according to Eq. (16.11). Only positive values of T are thermodynamically significant and these correspond to $\mathcal{M}/\mathcal{N} < 1/2$. See Figure 16–3 for better resolution of the shape near $T = 0$.

[4]Fictitious negative temperatures are sometimes used to characterize nonequilibrium states in which the populations of states having high energies are greater than those having lower energies.

$$p_1 := \frac{\mathcal{M}}{\mathcal{N}} = \frac{\exp(-\varepsilon/k_B T)}{1+\exp(-\varepsilon/k_B T)} \tag{16.12}$$

for the probability p_1 of occupation of the state ε. The probability of occupation of the ground state is

$$p_0 := \frac{\mathcal{N}-\mathcal{M}}{\mathcal{N}} = 1 - p_1 = \frac{1}{1+\exp(-\varepsilon/k_B T)}. \tag{16.13}$$

The quantity $\exp(-\varepsilon/k_B T)$ is called a **Boltzmann factor**; it is very small for low temperatures and nearly equal to 1 at very high temperatures. Thus, at very high temperatures, $p_0 = p_1 = 1/2$ and the levels have equal population, a condition known as **equipartition**. Negative values of T correspond to nonequilibrium states in which the population is inverted, that is, $p_1 > p_0$. Such states do not occur naturally at equilibrium but can be brought about by "pumping" of the upper state from some higher state that decays, as might take place in a laser.

Substitution of $\mathcal{M} = E/\varepsilon$ into Eq. (16.12) gives the energy[5] of the system as a function of temperature, namely

$$E = \mathcal{N}\varepsilon \frac{\exp(-\varepsilon/k_B T)}{1+\exp(-\varepsilon/k_B T)}. \tag{16.14}$$

As T increases from 0, the energy increases from 0 to $\mathcal{N}\varepsilon/2$, as shown in Figure 16–3. An expression for the entropy as a function of temperature can be obtained by substitution of Eqs. (16.12) and (16.13) into Eq. (16.10) to obtain

$$S = -\mathcal{N}k_B(p_0 \ln p_0 + p_1 \ln p_1). \tag{16.15}$$

A plot of S versus T is shown in Figure 16–4. As T increases from 0, we see that S increases from 0, in agreement with the third law of thermodynamics, and becomes equal to $\mathcal{N}k_B \ln 2$ at very high temperatures.

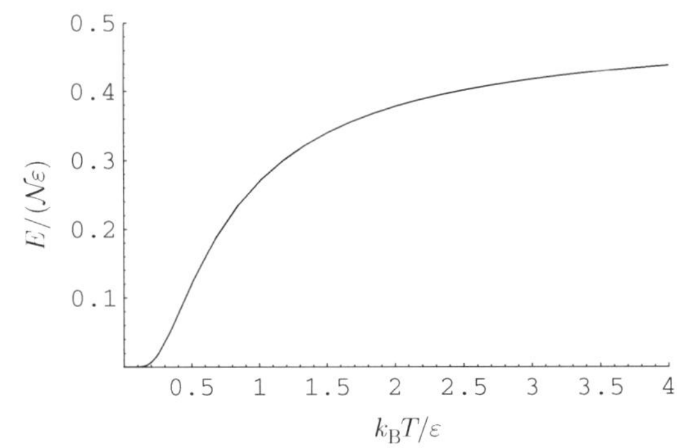

FIGURE 16–3 Dimensionless energy $E/(\mathcal{N}\varepsilon)$ versus dimensionless temperature k_BT/ε for a two-state system according to Eq. (16.14). At $T = 0$, all particles are in the ground state. As $T \to \infty$, the states become equally populated so that the average energy per particle is $\varepsilon/2$.

[5]If the ground state of the particle had been at energy ε_0, its excited state would have energy $\varepsilon_0 + \varepsilon$ and the total energy E would contain an additional term $\mathcal{N}\varepsilon_0$.

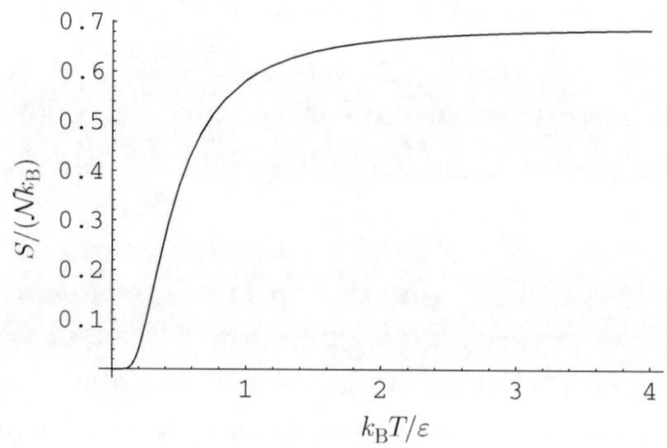

FIGURE 16–4 Dimensionless entropy $S/(\mathcal{N}k_B)$ versus dimensionless temperature k_BT/ε for a two-state system according to Eq. (16.15). At $T=0$, all particles are in the ground state and $S=0$. As $T\to\infty$, the states become equally populated so $S/(\mathcal{N}k_B)\to\ln 2=0.693$.

The case of two-state subsystems is sufficiently simple that one can calculate explicitly the relationship between the multiplicity functions of two subsystems and the multiplicity function of the combined system that results when these subsystems are combined and come to thermal equilibrium. In their book *Thermal Physics* [6], Kittel and Kroemer show that the multiplicity function for a spin system is highly peaked about the value $s=0$ where $s=\mathcal{N}/2-\mathcal{M}$ is called the **spin excess** of the ground state. Thus, $2s$ is the number of spins in the ground state ("spin up") minus the number of spins in the state having higher energy ("spin down"). When two spin subsystems are combined and come to thermal equilibrium, the combined system will have a multiplicity function that depends on the narrow region of overlap between the high peaks of the subsystems. We follow Kittel and Kroemer and carry out this calculation in more detail in Appendix D where we also demonstrate explicitly the additivity of entropy when two such systems are combined.

Example Problem 16.1. Consider an isolated system having energy E and consisting of $\mathcal{N}$ identical but distinguishable subsystems, each having two energy levels that are degenerate. One level has energy 0 and degeneracy d_0 and the other level has energy ε and degeneracy d_1. Use the microcanonical ensemble to compute the entropy of this system and then compute its temperature and the probability of occupation of the level with energy ε.

Solution 16.1. If $\mathcal{M}=E/\varepsilon$ is the number of particles in the level having energy ε, the multiplicity function is

$$\Omega_d(\mathcal{N},E)=\frac{\mathcal{N}!}{(\mathcal{N}-\mathcal{M})!\mathcal{M}!}(\mathcal{N}-\mathcal{M})^{d_0}(\mathcal{M})^{d_1}. \tag{16.16}$$

Thus the entropy is given by

$$S/k_B=\mathcal{N}\ln\mathcal{N}-(\mathcal{N}-\mathcal{M})\ln(\mathcal{N}-\mathcal{M})-\mathcal{M}\ln\mathcal{M}+(\mathcal{N}-\mathcal{M})\ln d_0+\mathcal{M}\ln d_1. \tag{16.17}$$

The temperature is given by

$$\frac{1}{k_{\mathrm{B}}T} = \frac{\partial \Omega_d}{\partial E} = \frac{1}{\varepsilon}\ln\left[\frac{(\mathcal{N}-\mathcal{M})d_0}{\mathcal{M}d_1}\right], \tag{16.18}$$

which can be solved for $\mathcal{M}$ to give the required probability

$$\frac{\mathcal{M}}{\mathcal{N}} = \frac{d_1 \exp(-\varepsilon/k_{\mathrm{B}}T)}{d_0 + d_1 \exp(-\varepsilon/k_{\mathrm{B}}T)}. \tag{16.19}$$

The form of this result can be understood by using the canonical ensemble, Chapter 19.

16.3 Harmonic Oscillators

We consider a system of $\mathcal{N}$ harmonic oscillators, each fixed in a solid and having an infinite number of nondegenerate energy levels[6] $\varepsilon_n = n\varepsilon = n\hbar\omega$ where the integer $n = 0, 1, 2, \ldots$. The symbol $\hbar = h/2\pi$, where $h = 6.626069 \times 10^{-34}$ J s is Planck's constant, is pronounced "h-bar" and appears frequently in the equations of quantum mechanics. It has the value $\hbar = 1.054572 \times 10^{-34}$ J s = 1.054572×10^{-27} erg s. A useful related constant is $\hbar/k_B = 7.638234$ K s. See http://physics.nist.gov/cuu/constants for the latest internationally recommended values of physical constants. We consider a macrostate of the system having total energy $E = \mathcal{M}\varepsilon$ and denote the multiplicity function for this state by $g_H(\mathcal{N}, \mathcal{M})$.

We can calculate g_H by means of the following device illustrated in Figure 16–5. Between fixed ends of a box, we place $\mathcal{N} - 1$ movable partitions. These partitions divide the box into $\mathcal{N}$ intervals, each interval representing a particle. The number of X symbols in an interval designates the energy of that particle in units of ε. The total energy is $E = \mathcal{M}\varepsilon$, where $\mathcal{M}$ is the total number of X symbols. The multiplicity function $g_H(\mathcal{N}, \mathcal{M})$ is the number of distinct arrangements of the X symbols and the movable partitions, namely

$$\Omega(\mathcal{N}, E) = \frac{(\mathcal{N} - 1 + \mathcal{M})!}{(\mathcal{N} - 1)!\mathcal{M}!} = g_H(\mathcal{N}, \mathcal{M}). \tag{16.20}$$

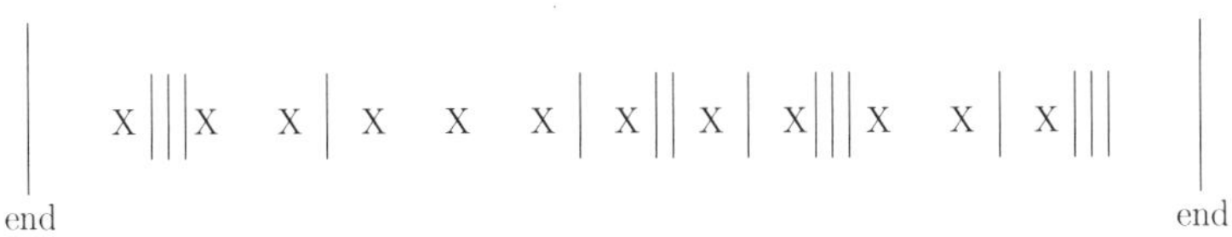

FIGURE 16–5 Diagram illustrating an algorithm for calculating the multiplicity function for the harmonic oscillator. Between fixed ends of a box (long lines) we place $\mathcal{N} - 1 = 15$ movable partitions (short lines). These partitions divide the box into $\mathcal{N} = 16$ intervals, each representing a particle. The number of X symbols in an interval designates the energy of that particle in units of ε. The total energy is $E = \mathcal{M}\varepsilon = 12\varepsilon$, where $\mathcal{M} = 12$ is the number of X symbols. As shown, there are eight particles with energy 0, five particles with energy ε, two particles with energy 2ε, and one particle with energy 3ε. The multiplicity function $g_H(\mathcal{N}, \mathcal{M}) = g_H(16, 12)$ is the number of distinct arrangements of the X symbols and the movable partitions, namely 27!/(15! 12!) = 17,383,860.

[6]We omit the zero point energy $\hbar\omega/2$ for simplicity since it only shifts the overall energy by $\mathcal{N}\hbar\omega/2$.

This same result can be derived by using a generating function, as shown below in Section 16.3.1.

For a thermodynamic system consisting of $\mathcal{N}$ harmonic oscillators, the entropy will be

$$S = k_{\rm B} \ln g_H(\mathcal{N}, \mathcal{M}); \quad \mathcal{M} = E/\varepsilon. \tag{16.21}$$

Therefore,

$$S = k_{\rm B}[\ln(\mathcal{N} - 1 + \mathcal{M})! - \ln(\mathcal{N} - 1)! - \mathcal{M} \ln \mathcal{M}!]. \tag{16.22}$$

Again we use Stirling's approximation and consistent with this we approximate $\mathcal{N} - 1 \approx \mathcal{N}$ to obtain[7]

$$\begin{aligned} S &\sim k_{\rm B}[(\mathcal{N} + \mathcal{M}) \ln(\mathcal{N} + \mathcal{M}) - \mathcal{N} \ln \mathcal{N} - \mathcal{M} \ln \mathcal{M}] \\ &= -\mathcal{N} k_{\rm B} \left[\frac{\mathcal{M}}{\mathcal{N}} \ln \frac{\mathcal{M}}{\mathcal{N} + \mathcal{M}} + \ln \frac{\mathcal{N}}{\mathcal{N} + \mathcal{M}} \right], \end{aligned} \tag{16.23}$$

where the second form demonstrates that S is extensive.

Thus the absolute temperature can be calculated from

$$\frac{1}{T} = \left(\frac{\partial S}{\partial E} \right)_{\mathcal{N}} = \frac{\partial \mathcal{M}}{\partial E} \left(\frac{\partial S}{\partial \mathcal{M}} \right)_{\mathcal{N}} = \frac{k_{\rm B}}{\varepsilon} \ln \frac{\mathcal{N} + \mathcal{M}}{\mathcal{M}}, \tag{16.24}$$

where we have used the first form of Eq. (16.23) to ease calculation of the derivative with respect to $\mathcal{M}$. Hence, Eq. (16.24) yields

$$\frac{\mathcal{M}}{\mathcal{N}} = \frac{1}{\exp(\varepsilon/k_{\rm B}T) - 1}, \tag{16.25}$$

or equivalently

$$E = \mathcal{N} \varepsilon \frac{1}{\exp(\varepsilon/k_{\rm B}T) - 1}. \tag{16.26}$$

Since $\mathcal{M}$ is equal to the sum of the actual number of quantum numbers n of a single oscillator, the average quantum number of an oscillator is given by

$$\langle n(T) \rangle := \frac{\mathcal{M}}{\mathcal{N}} = \frac{1}{\exp(\varepsilon/k_{\rm B}T) - 1}. \tag{16.27}$$

As T increases from 0, $\langle n \rangle$ increases very slowly and tends to the value $k_{\rm B}T/\varepsilon$ as T becomes large. Since a given oscillator has an infinite number of states, both $\langle n \rangle$ and E continue to increase linearly with T at large T. In fact, $E \approx \mathcal{N} k_{\rm B} T$ independent of ε for large T. This is consistent with the fact that the average thermal energy of a classical harmonic oscillator is independent of its oscillation frequency, as will be seen later.

By returning to Eq. (16.23), we can write the entropy in the form

$$S = -\mathcal{N} k_{\rm B} \left[\ln \left(\frac{1}{1 + \langle n \rangle} \right) + \langle n \rangle \ln \left(\frac{\langle n \rangle}{1 + \langle n \rangle} \right) \right]. \tag{16.28}$$

[7]By using Stirling's approximation, we are already assuming that $\mathcal{N} \gg 1$ and dropping terms of order $\ln \mathcal{N}$ and smaller relative to $\mathcal{N}$. It would be inconsistent to keep 1 relative to $\mathcal{N}$.

This expression can also be written in the form

$$S = -\mathcal{N} k_{\rm B} \left[\ln(1 - {\rm e}^{-x}) - \frac{x\,{\rm e}^{-x}}{(1 - {\rm e}^{-x})} \right], \tag{16.29}$$

where $x = \varepsilon / k_{\rm B} T$. As T increases from 0, S increases slowly from 0. For large T, x becomes very small and we can expand Eq. (16.29) to obtain

$$S \approx \mathcal{N} k_{\rm B} \left[-\ln x + 1 \right] = \mathcal{N} k_{\rm B} \left[\ln(k_{\rm B} T/\varepsilon) + 1 \right]. \tag{16.30}$$

Thus, due to the availability of an infinite number of energy levels of a given oscillator, the entropy continues to increase with T. See Section 18.3 for an alternative derivation and relevant graphs.

16.3.1 Generating Function

As mentioned above, we can derive the multiplicity function $g_H(\mathcal{N}, \mathcal{M})$, given by Eq. (16.20) for the simple harmonic oscillator, by another method involving a generating function. We observe that $g_H(\mathcal{N}, \mathcal{M})$ is the coefficient of $t^{\mathcal{M}}$ in the series expansion

$$\left(\sum_{j=0}^{\infty} t^j \right)^{\mathcal{N}} = (1 + t + t^2 + \cdots)^{\mathcal{N}} = \sum_{\mathcal{M}} g_H(\mathcal{N}, \mathcal{M}) t^{\mathcal{M}}. \tag{16.31}$$

But

$$\sum_{j=0}^{\infty} t^j = \frac{1}{1 - t}, \tag{16.32}$$

so we have

$$g_H(\mathcal{N}, \mathcal{M}) = \frac{1}{\mathcal{M}!} \left(\frac{\rm d}{{\rm d}t} \right)^{\mathcal{M}} \left(\frac{1}{1 - t} \right)^{\mathcal{N}} \Bigg|_{t=0}. \tag{16.33}$$

Carrying out the required differentiation and evaluation at $t = 0$ readily yields Eq. (16.20).

In principle, this method can also be used for subsystems having any finite number of states equally spaced in energy, but the results can be cumbersome. Thus, for two-state subsystems one could find the coefficient of $t^{\mathcal{M}}$ in the expansion of $(1 + t)^{\mathcal{N}}$ which would lead immediately to Eq. (16.7). For subsystems with three equally spaced states, one would require the coefficient of $t^{\mathcal{M}}$ in the expansion of $(1 + t + t^2)^{\mathcal{N}}$.

16.4 Ideal Gas

The monatomic ideal gas is a tractable example of a system of **identical indistinguishable particles** that can be treated by means of the microcanonical ensemble. It is especially important because it serves as a link between quantum statistical mechanics and classical statistical mechanics, as we shall see in Chapter 17. Atoms of the gas are confined to a box of volume V that they *share*. Therefore, our macroscopic view of the system precludes knowledge of which atoms occupy any particular sub-volume of the box. This is quite different from the case of identical particles in a solid that are essentially immobile and may be distinguished by virtue of their position.

An important advance in proper counting of such microstates was made by Gibbs in the context of classical statistical mechanics. For a monatomic gas of $\mathcal{N}$ particles, Gibbs reasoned that one should divide the volume of phase space occupied by the particles by $\mathcal{N}!$ to correct for the fact that there are $\mathcal{N}!$ permutations of the particles that do not lead to states of the system that can be distinguished macroscopically. Without this division by $\mathcal{N}!$, the calculated entropy is not an extensive function, so it is surely incorrect and leads to an inconsistency known as the **Gibbs paradox.**[8] As we shall see, we can readily count the states of $\mathcal{N}$ free quantum particles in a box as if the particles were distinguishable. If we use the Gibbs correction factor of $1/\mathcal{N}!$ to correct this count, we obtain an entropy that is extensive. This is satisfying and gives a good result for an ideal gas at high temperatures and low density. It is, however, not a correct quantum mechanical result under other conditions. To get a correct quantum mechanical result, one must construct a total wave function of the gas that is either an antisymmetric or a symmetric function on interchange of any two particles, depending on whether they are fermions or bosons, respectively. This will result in corrections that are important at low temperatures and high number densities. We defer detailed treatment of such quantum statistical effects until Section 26.7. Here, we give only an approximate treatment based on the Gibbs correction factor.

16.4.1 Monatomic Ideal Gas with Gibbs Correction Factor

We give a pseudo-quantum mechanical treatment of a monatomic ideal gas, including the Gibbs correction factor. We use periodic boundary conditions and wave functions that are also eigenfunctions of the momentum operator. For a single free particle in a cubical box of volume V, the wave function is

$$\psi_{\mathbf{k}}(\mathbf{r}) = V^{-1/2}\exp(i\mathbf{k}\cdot\mathbf{r}), \tag{16.34}$$

which satisfies

$$\hat{\mathcal{H}}\psi_{\mathbf{k}} = \varepsilon_{\mathbf{k}}\psi_{\mathbf{k}} \tag{16.35}$$

with $\varepsilon_{\mathbf{k}} = \hbar^2k^2/(2m)$. Here, $\hat{\mathcal{H}} = \hat{p}^2/2m$, where $\hat{p} = (\hbar/i)\nabla$ is the momentum operator. For periodic boundary conditions, the allowed values of $\mathbf{k}$ are

$$\mathbf{k} = \frac{2\pi}{V^{1/3}}[n_x\,\hat{\mathbf{i}} + n_y\,\hat{\mathbf{j}} + n_z\,\hat{\mathbf{k}}], \tag{16.36}$$

[8]According to the Gibbs paradox, the additional entropy from mixing identical gases, each having the same temperature and pressure as the final mixture, turns out to be positive rather than zero, as it should be for a process that is clearly reversible. See Pathria [8, p. 22] for details.

where $\hat{\mathbf{i}}$, $\hat{\mathbf{j}}$, $\hat{\mathbf{k}}$ are the Cartesian unit vectors and n_x, n_y, n_z are positive and negative integers and zero. The states of a single particle of energy ε lie on the surface of the sphere

$$n_x^2 + n_y^2 + n_z^2 = 2m\varepsilon\, V^{2/3}/h^2. \tag{16.37}$$

The total energy E will be the sum of the energies of the individual particles because particles of an ideal gas do not, by definition, have an interaction energy. The number of quantum states for $\mathcal{N}$ *distinguishable* particles is therefore equal to the number of allowed solutions to[9]

$$\sum_{r=1}^{3\mathcal{N}} n_r^2 = 2mE\, V^{2/3}/h^2 = R^2, \tag{16.38}$$

where $R := (2mE\, V^{2/3}/h^2)^{1/2}$ is the dimensionless radius of a hypersphere in $3\mathcal{N}$-dimensional space. Allowed solutions are those for which each n_r is a positive or negative integer or zero. Of course no such solutions exist unless R^2 is an integer, but to circumvent this technicality we shall actually count solutions in a thin shell corresponding to energies between $E - \Delta E$ and E, or equivalently between $R - \Delta R$ and R, where $\Delta R/R \approx \Delta E/(2E)$.

The number of solutions to Eq. (16.38) for any radius less than or equal to R will be equal to the volume V_R of the hypersphere given by Eq. (16.38). Obviously this volume will be proportional to $R^{3\mathcal{N}}$ but the proportionality constant will depend on the dimensionality $3\mathcal{N}$ of the space. A simple derivation is given by Pathria [8, p. 504] and results in

$$V_R = \frac{\pi^{3\mathcal{N}/2}}{(3\mathcal{N}/2)!} R^{3\mathcal{N}} = V^{\mathcal{N}} \frac{(mE/2\pi\hbar^2)^{3\mathcal{N}/2}}{(3\mathcal{N}/2)!}. \tag{16.39}$$

Here, $(3\mathcal{N}/2)!$ should be interpreted as the gamma function $\Gamma(3\mathcal{N}/2 + 1)$ in case $\mathcal{N}$ is an odd integer. Since, however, $3\mathcal{N}$ is extremely large, almost all of these solutions lie near the surface of the hypersphere. In fact, the fraction of the volume of the hypersphere that lies within ΔR of the surface is just

$$\mathcal{F} := 1 - \left(1 - \frac{\Delta R}{R}\right)^{3\mathcal{N}} \approx 1 - \exp(-3\mathcal{N}\Delta R/R) \approx 1 - \exp(-3\mathcal{N}\Delta E/2E). \tag{16.40}$$

The number of solutions to Eq. (16.38) in the thin shell near E is therefore

$$\Omega_0 = \mathcal{F} V_R = [1 - \exp(-3\mathcal{N}\Delta E/2E)] V_R \approx V_R. \tag{16.41}$$

[9]To simplify the notation, for particle number 1 we let $n_x = n_1$, $n_y = n_2$, $n_z = n_3$ and for particle number 2 we let $n_x = n_4$, $n_y = n_5$, $n_z = n_6$, etc. The wave function of the whole system can be made up of products of the wave functions of the individual particles, consistent with the additivity of particle energies. Nevertheless, true quantum mechanical considerations also restrict the symmetry of the wave function under an interchange of identical particles, which is discussed in detail in Section 26.7. Here, in the spirit of treating a classical ideal gas, we omit that complication but make up for it by using the Gibbs factor $\mathcal{N}!$ in Eq. (16.44) to correct approximately the count of the number of microstates.

Thus,[10]

$$\ln \Omega_0 = \ln V_R + \ln[1 - \exp(-3\mathcal{N}\Delta E/2E)] \approx \ln V_R - \exp(-3\mathcal{N}\Delta E/2E). \tag{16.42}$$

The second term is clearly negligible, so substitution of Eq. (16.39) and use of Stirling's approximation gives

$$\ln \Omega_0 \approx \ln V_R \sim \mathcal{N} \ln \left\{ V \left(\frac{mE}{3\pi\hbar^2 \mathcal{N}} \right)^{3/2} \right\} + \frac{3}{2}\mathcal{N}, \tag{16.43}$$

essentially independent of any reasonable choice for ΔE. As shown in the following example and elsewhere [8, p. 17], this result is the same as would be obtained for wave functions that vanish on the walls of the box.

One might be tempted to equate the entropy to $k_B \ln \Omega_0$ but that would be incorrect because $\ln \Omega_0$ is not an extensive function. The argument of the logarithm contains the ratio $E/\mathcal{N}$ which is intensive, but it also contains V without $\mathcal{N}$. To get a corrected value for the number of states, we must account for the fact in observing such a system we have no way of distinguishing the particles. We follow Gibbs and divide Ω_0 by the number of indistinguishable states $\mathcal{N}!$ to get the corrected number of microstates

$$\Omega \approx \frac{\Omega_0}{\mathcal{N}!} = V^{\mathcal{N}} \frac{(mE/2\pi\hbar^2)^{3\mathcal{N}/2}}{\mathcal{N}!(3\mathcal{N}/2)!}. \tag{16.44}$$

Thus

$$S = k_B \ln \Omega \approx k_B[\ln \Omega_0 - \mathcal{N} \ln \mathcal{N} + \mathcal{N}] \tag{16.45}$$

(where Stirling's approximation has been used), which results in

$$S = \mathcal{N} k_B \ln \left\{ \frac{V}{\mathcal{N}} \left(\frac{mE}{3\pi\hbar^2 \mathcal{N}} \right)^{3/2} \right\} + \frac{5}{2}\mathcal{N} k_B. \tag{16.46}$$

The entropy given by Eq. (16.46) is clearly an extensive function. The temperature is given by $1/T = (\partial S/\partial E)_{V,\mathcal{N}} = (3/2)\mathcal{N} k_B/E$, so in terms of the temperature we can write the entropy in the form

$$S = \mathcal{N} k_B \ln (n_Q/n) + \frac{5}{2}\mathcal{N} k_B, \tag{16.47}$$

where

[10]Many treatments take the volume of the spherical shell to be $(dV_R/dR)\Delta R = V_R 3\mathcal{N}\Delta R/R = V_R 3\mathcal{N}\Delta E/2E$ and then argue that $\ln(V_R 3\mathcal{N}\Delta E/2E) \approx \ln V_R$. That result would be obtained if the exponential in the expression for $\mathcal{F}$ were expanded, which procedure is incorrect for huge $\mathcal{N}$. A more accurate expression can be obtained by setting $y = (1 - \Delta R/R)^{3\mathcal{N}}$ so $\ln y = 3\mathcal{N} \ln(1 - \Delta R/R) \approx -3\mathcal{N}\Delta R/R$, and then exponentiating. In fact, isolation of a system is only an idealization which is the rationale for some finite ΔE, notwithstanding implications of the uncertainty principle. Thus if atoms near the surface of a body were to interact weakly with its environment, one might expect $\Delta E/E \sim \mathcal{N}^{2/3}/\mathcal{N} = \mathcal{N}^{-1/3}$ so $3\mathcal{N}\Delta E/2E \sim \mathcal{N}^{2/3}$ which is huge. After taking $\ln \Omega_0$, the additive term $\ln \mathcal{F}$ is negligible with respect to $\ln V_R$, so ultimately one arrives at the same result as usually quoted. Our more precise analysis shows that the neglected term is much smaller than usually claimed.

$$n_Q(T) := \left(\frac{mk_BT}{2\pi\hbar^2}\right)^{3/2} \tag{16.48}$$

is known as the quantum concentration and $n := \mathcal{N}/V$ is the actual concentration. Division of Ω_0 by $\mathcal{N}!$ is a good approximation to Ω if the probability of multiple occupation of single particle states is negligible. Thus, Eq. (16.47) is valid provided that the actual concentration n is small compared to the quantum concentration n_Q. This will be the case at sufficiently high temperatures and low densities and will be borne out by a complete quantum mechanical analysis. From Eq. (16.46), we can calculate the chemical potential

$$\mu = -T\left(\frac{\partial S}{\partial \mathcal{N}}\right)_{E,V} = k_BT\ln(n/n_Q) = k_BT\ln[p/(n_Qk_BT)]. \tag{16.49}$$

The quantity n_Qk_BT can be thought of as a quantum pressure. Note that $(\partial S/\partial\mathcal{N})_{E,V} \neq (\partial S/\partial\mathcal{N})_{T,V}$ because $E = (3/2)\mathcal{N}k_BT$. On the other hand, the relationship between E and T does not involve V for an ideal gas. Thus the pressure of an ideal gas can be computed by holding either E or T constant, resulting in

$$p/T = \left(\frac{\partial S}{\partial V}\right)_{\mathcal{N},E} = \left(\frac{\partial S}{\partial V}\right)_{\mathcal{N},T} = \mathcal{N}k_B/V, \tag{16.50}$$

which is the familiar ideal gas equation of state.

Example Problem 16.2. Show for an ideal gas in a box having the shape of a rectangular parallelepiped with dimensions H, K, L that one obtains the same result for Ω_0 as given by Eq. (16.41) for periodic boundary conditions and for boundary conditions for which the wave function $\psi = 0$ on the walls of the box.

Solution 16.2. We still have $\varepsilon = \hbar^2k^2/2m$ and for periodic boundary conditions,

$$\mathbf{k} = 2\pi\left[\frac{n_x}{H}\hat{\mathbf{i}} + \frac{n_y}{K}\hat{\mathbf{j}} + \frac{n_z}{L}\hat{\mathbf{k}}\right], \tag{16.51}$$

where n_x, n_y, n_z are positive and negative integers and zero. For $\psi = 0$ on the walls, the solutions are of the form $\psi \propto \sin(k_xx)\sin(k_yy)\sin(k_zz)$ with

$$\mathbf{k} = \pi\left[\frac{n_x}{H}\hat{\mathbf{i}} + \frac{n_y}{K}\hat{\mathbf{j}} + \frac{n_z}{L}\hat{\mathbf{k}}\right], \tag{16.52}$$

but now n_x, n_y, n_z are only positive integers (because negative integers would only result in a change of phase, not a linearly independent eigenfunction). In the case of Eq. (16.51), $\Delta n_x\Delta n_y\Delta n_z = HKL/(2\pi)^3\Delta k_x\Delta k_y\Delta k_z$ so the density of states in $\mathbf{k}$ space for a single particle is $V/(2\pi)^3$ where the volume $V = HKL$. For $\mathcal{N}$ particles, the density of states is therefore $\left[V/(2\pi)^3\right]^{\mathcal{N}}$. In the case of Eq. (16.52), $\Delta n_x\Delta n_y\Delta n_z = HKL/(\pi)^3\Delta k_x\Delta k_y\Delta k_z$ so the density of states in $\mathbf{k}$ space for a single particle is $V/(\pi)^3$, and for $\mathcal{N}$ particles it is $[V/(\pi)^3]^{\mathcal{N}}$. The volume of an entire hypersphere in $3\mathcal{N}$-dimensional $\mathbf{k}$ space with radius $(2mE/\hbar^2)^{1/2}$ is

$$V_k = \frac{\pi^{3\mathcal{N}/2}}{(3\mathcal{N}/2)!}(2mE/\hbar^2)^{3\mathcal{N}/2}. \tag{16.53}$$

To get the corresponding number of states in the case for periodic boundary conditions, we multiply V_k by $\left[V/(2\pi)^3\right]^{\mathcal{N}}$. But for the case of $\psi = 0$ on the walls, only positive values of k_i are allowed, so we must first multiply V_k by $(1/2^3)^{\mathcal{N}}$ and then by $[V/(\pi)^3]^{\mathcal{N}}$, resulting in the same net factor $[V/(2\pi)^3]^{\mathcal{N}}$ of V_k. So in either case, the number of states (not yet corrected by the Gibbs factor) is

$$\left(\frac{V}{(2\pi)^3}\right)^{\mathcal{N}} V_k = V^{\mathcal{N}}\frac{(mE/2\pi\hbar^2)^{3\mathcal{N}/2}}{(3\mathcal{N}/2)!}, \tag{16.54}$$

the same as V_R given by Eq. (16.39).

16.4.2 Scaling Analysis

As noted by Pathria [8, p. 16], many important results for the ideal gas can be ascertained from a simple scaling analysis without actually calculating Ω in detail. For E and $\mathcal{N}$ fixed, it can be argued that the number of states for a single particle is proportional to V, so for $\mathcal{N}$ particles we expect Ω to be proportional to $V^{\mathcal{N}}$. Moreover, the form of Eq. (16.38) shows that Ω will depend on E and V only in the combination $EV^{2/3}$, so we can immediately express Ω in the functional form

$$\Omega = \tilde{\Omega}(\mathcal{N})V^{\mathcal{N}}E^{(3\mathcal{N}/2)}, \tag{16.55}$$

where $\tilde{\Omega}(\mathcal{N})$ is some unknown function of $\mathcal{N}$. Then from

$$S = k_{\rm B}[\ln\tilde{\Omega}(\mathcal{N}) + \mathcal{N}\ln V + (3\mathcal{N}/2)\ln E], \tag{16.56}$$

we readily deduce the following:

1. From $1/T = (\partial S/\partial E)_{V,\mathcal{N}} = (3\mathcal{N}k_{\rm B}/2E)$, we see that E is independent of V and has the form

$$E = (3\mathcal{N}/2)k_{\rm B}T. \tag{16.57}$$

2. From $p/T = (\partial S/\partial V)_{E,\mathcal{N}} = \mathcal{N}k_{\rm B}/V$, we deduce the familiar ideal gas law

$$pV = \mathcal{N}k_{\rm B}T. \tag{16.58}$$

Combining Eqs. (16.57) and (16.58) gives $p = (2/3)(E/V)$, which relates pressure to energy density.

3. For an isentropic transformation at constant $\mathcal{N}$, the constancy of S requires $VE^{3/2}$ = constant. If we differentiate this equation, we deduce

$$dE = -(2/3)(E/V)\,dV = -p\,dV, \tag{16.59}$$

so the only change in energy for this reversible adiabatic transformation comes from reversible work $\delta W = p\,dV$ done by the system. By eliminating E from $VE^{3/2}$ =

constant, we also deduce the scaling laws $VT^{3/2}$ = constant, $T^{5/2}/p$ = constant, and $pV^{5/3}$ = constant for an isentropic transformation of a monatomic ideal gas.

4. The enthalpy $H = E + pV = (5/3)E = (5/2)\mathcal{N} k_B T$. Therefore, the heat capacities are

$$C_V = (\partial E/\partial T)_{V,\mathcal{N}} = (3/2)\mathcal{N} k_B; \quad C_p = (\partial H/\partial T)_{p,\mathcal{N}} = (5/2)\mathcal{N} k_B, \tag{16.60}$$

so $C_p/C_V = 5/3$.

16.5 Multicomponent Ideal Gas

We next treat a multicomponent ideal gas in the same approximation used above for a monocomponent gas. It will suffice to treat only a gas having A and B atoms because generalization to a larger number of chemical components is straightforward.

We consider $\mathcal{N}_A$ atoms of A, each with mass m_A, giving rise to a total energy E_A for all A atoms, and similarly for B atoms. Applying Eq. (16.44) to each gas we obtain

$$\Omega_A = \left(\frac{V}{h^3}\right)^{\mathcal{N}_A} \frac{(2\pi m_A E_A)^{3\mathcal{N}_A/2}}{\mathcal{N}_A!(3\mathcal{N}_A/2)!}; \quad \Omega_B = \left(\frac{V}{h^3}\right)^{\mathcal{N}_B} \frac{(2\pi m_B E_B)^{3\mathcal{N}_B/2}}{\mathcal{N}_B!(3\mathcal{N}_B/2)!}. \tag{16.61}$$

What we would like to calculate is $\Omega(E)$ for the whole system where $E = E_A + E_B$ is the total energy; however, we do not yet know how the energies of A and B are partitioned. Hence, we will have to accept all possible partitions of energy and sum over them to obtain

$$\Omega(E) = \sum_{E_B} \Omega_A(E - E_B)\Omega_B(E_B) \tag{16.62}$$

in an abbreviated notation where symbols other than the energy are suppressed. This would appear to be a very difficult calculation were it not for the fact that all we need is a sufficient approximation to $\ln \Omega$ which is given by the largest term in the sum. McQuarrie [54, p. 25] refers to this approximation as the **maximum term method**. Following McQuarrie, we let $\mathcal{T}_{\max}$ be the largest term in a sum $\mathcal{S}$ of $\mathcal{M}$ positive terms. Then $\mathcal{T}_{\max} \le \mathcal{S} \le \mathcal{M}\mathcal{T}_{\max}$. Thus

$$\ln \mathcal{T}_{\max} \le \ln \mathcal{S} \le \ln \mathcal{T}_{\max} + \ln \mathcal{M}. \tag{16.63}$$

If, in order of magnitude, $\mathcal{T}_{\max} \sim A^{\mathcal{M}}$ where $A = \mathcal{O}(1)$ and $\mathcal{M} \gg 1$, we have $\ln \mathcal{T}_{\max} \sim \mathcal{M} \ln A$. Therefore, for sufficiently large $\mathcal{M}$, the term $\ln \mathcal{M}$ is negligible with respect to $\mathcal{M} \ln A$ and

$$\ln \mathcal{S} \approx \ln \mathcal{T}_{\max}. \tag{16.64}$$

In our case, each term in the sum is proportional to

$$E_A^{3\mathcal{N}_A/2} E_B^{3\mathcal{N}_B/2} = (E - E_B)^{3\mathcal{N}_A/2} E_B^{3\mathcal{N}_B/2} \tag{16.65}$$

and $\mathcal{N}_A$ and $\mathcal{N}_B$ are huge numbers for all cases of interest. To find a maximum, we differentiate partially with respect to E_B holding E constant to obtain

$$-(3\mathcal{N}_A/2)\, E_A^{(3\mathcal{N}_A/2)-1} E_B^{3\mathcal{N}_B/2} + (3\mathcal{N}_B/2)\, E_A^{3\mathcal{N}_A/2} E_B^{(3\mathcal{N}_B/2)-1} = 0, \tag{16.66}$$

which simplifies to

$$\mathcal{N}_A/E_A = \mathcal{N}_B/E_B = \mathcal{N}/E, \tag{16.67}$$

where the last equality follows from the properties of proportions. Thus we have $E_A = E\mathcal{N}_A/\mathcal{N}$ and $E_B = E\mathcal{N}_B/\mathcal{N}$ where $\mathcal{N} = \mathcal{N}_A + \mathcal{N}_B$, in which case[11]

$$S = k\ln\left[\Omega_A\left(E\frac{\mathcal{N}_A}{\mathcal{N}}\right)\Omega_B\left(E\frac{\mathcal{N}_B}{\mathcal{N}}\right)\right] = k\ln\Omega_A\left(E\frac{\mathcal{N}_A}{\mathcal{N}}\right) + k\ln\Omega_B\left(E\frac{\mathcal{N}_B}{\mathcal{N}}\right)$$

$$= \mathcal{N}_A k\ln\left\{\frac{V}{\mathcal{N}_A}\left(\frac{4\pi m_A E}{3h^2\mathcal{N}}\right)^{3/2}\right\} + \frac{5}{2}\mathcal{N}_A k + \mathcal{N}_B k\ln\left\{\frac{V}{\mathcal{N}_B}\left(\frac{4\pi m_B E}{3h^2\mathcal{N}}\right)^{3/2}\right\} + \frac{5}{2}\mathcal{N}_B k. \tag{16.68}$$

From Eq. (16.68) we compute the temperature:

$$\frac{1}{T} = \left(\frac{\partial S}{\partial E}\right)_{V,\mathcal{N}_A,\mathcal{N}_B} = \frac{3}{2}k\frac{\mathcal{N}}{E}, \tag{16.69}$$

which leads immediately to

$$S(T,V,\mathcal{N}_A,\mathcal{N}_B) = \mathcal{N}_A k\ln\left(\frac{V}{\mathcal{N}_A}n_{QA}\right) + \mathcal{N}_B k\ln\left(\frac{V}{\mathcal{N}_B}n_{QB}\right) + \frac{5}{2}\mathcal{N}k, \tag{16.70}$$

where $n_{QA} := [m_A kT/(2\pi\hbar^2)]^{3/2} = [2\pi m_A kT/h^2]^{3/2}$ is the quantum concentration of A and n_{QB} is defined similarly.

Examination of Eq. (16.70) in view of Eq. (16.47) allows for an immediate physical interpretation, namely that the entropy of the combined ideal gases of A and B atoms at temperature T in a volume V is the sum of the entropies they would have if each were at temperature T' and occupied the volume V separately. According to Callen [2, p. 69], this is often referred to as **Gibbs's theorem**.[12] In fact, Eq. (16.67) is precisely the condition that gases A and B have the same temperature.[13] This treatment clearly generalizes to multicomponent ideal gases.

The pressure p of the gas mixture may be computed from Eq. (16.68), resulting in

$$\frac{p}{T} = \left(\frac{\partial S}{\partial V}\right)_{E,\mathcal{N}_A,\mathcal{N}_B} = k\frac{\mathcal{N}}{V}, \tag{16.71}$$

so $pV = \mathcal{N}kT$, the same as for a monocomponent gas. From Eq. (16.50), we see that the pressures of A and B separately in volume V would be $p_A = \mathcal{N}_A kT/V = (\mathcal{N}_A/\mathcal{N})p$ and $p_B = \mathcal{N}_B kT/V = (\mathcal{N}_B/\mathcal{N})p$. These are called **partial pressures** of A and B. Such an additivity of partial pressures is unique to ideal gases because they do not interact.

[11]In this and the next section, we deal with A and B gases so we drop the subscript B on the Boltzmann constant k_B to avoid confusion.

[12]An equivalent statement is that the Helmholtz free energy of a mixture of ideal gases at temperature T in a volume V is additive, that is, $F(T,V,\mathcal{N}_A,\mathcal{N}_B) = F_A(T,V,\mathcal{N}_A) + F_B(T,V,\mathcal{N}_B)$. It would be incorrect to apply this formula to a pure gas by assuming that A and B atoms are identical. The correct procedure would be to let $\mathcal{N}_B = 0$ to get a pure gas of A atoms.

[13]From Eq. (16.67), it follows that $\mathcal{N}_A/E_A = \mathcal{N}_B/E_B = \mathcal{N}/E = 2/(3k_B T)$.

Similarly, one can compute the chemical potential of A and B in the mixture or, alternatively, as if each gas occupied the volume V separately. In either case, the result turns out to be the same and one obtains

$$\mu_{\mathrm{A}} = -T\left(\frac{\partial S}{\partial \mathcal{N}_{\mathrm{A}}}\right)_{E,V,\mathcal{N}_{\mathrm{B}}} = -T\left(\frac{\partial S_{\mathrm{A}}(E_{\mathrm{A}},V,\mathcal{N}_{\mathrm{A}})}{\partial \mathcal{N}_{\mathrm{A}}}\right)_{E_{\mathrm{A}},V} = kT\ln\left[\frac{p_{\mathrm{A}}}{n_{\mathrm{QA}}kT}\right]. \tag{16.72}$$

Thus, from the standpoint of chemical potential, the presence of another species in a mixture of ideal gases in a volume V at temperature T is irrelevant.[14]

16.5.1 Entropy of Mixing

The entropy of mixing of ideal gases is defined to be the entropy of the mixture of gases at temperature T and volume V minus the entropies of the separate gases (unmixed state) each at temperature T but confined to separate sub-volumes of V such that each has the same pressure p as the mixture. Equal pressure is guaranteed by equal number density. Thus, in the case of our mixture of A and B atoms, the A atoms would need to be confined to a sub-volume $V_{\mathrm{A}} = V\mathcal{N}_{\mathrm{A}}/\mathcal{N}$ and they would have entropy

$$S_{\mathrm{A}}(E_{\mathrm{A}},V_{\mathrm{A}},\mathcal{N}_{\mathrm{A}}) = \mathcal{N}_{\mathrm{A}}k\ln\left\{\frac{V_{\mathrm{A}}}{\mathcal{N}_{\mathrm{A}}}\left(\frac{4\pi m_{\mathrm{A}}E_{\mathrm{A}}}{3h^2\mathcal{N}_{\mathrm{A}}}\right)^{3/2}\right\} + \frac{5}{2}\mathcal{N}_{\mathrm{A}}k. \tag{16.73}$$

The reciprocal temperature of these A atoms would be

$$\left(\frac{\partial S_{\mathrm{A}}}{\partial E_{\mathrm{A}}}\right)_{V_{\mathrm{A}},\mathcal{N}_{\mathrm{A}}} = \frac{3}{2}k\frac{\mathcal{N}_{\mathrm{A}}}{E_{\mathrm{A}}} = \frac{3}{2}k\frac{\mathcal{N}}{E} = \frac{1}{T} \tag{16.74}$$

and the ratio of their pressure to their temperature would be

$$\left(\frac{\partial S_{\mathrm{A}}}{\partial V_{\mathrm{A}}}\right)_{E_{\mathrm{A}},\mathcal{N}_{\mathrm{A}}} = k\frac{\mathcal{N}_{\mathrm{A}}}{V_{\mathrm{A}}} = k\frac{\mathcal{N}}{V} = \frac{p}{T}. \tag{16.75}$$

They therefore have the same temperature and pressure as the mixture. We write

$$S_{\mathrm{A}}(T,V_{\mathrm{A}},\mathcal{N}_{\mathrm{A}}) = \mathcal{N}_{\mathrm{A}}k\ln\left(\frac{V_{\mathrm{A}}}{\mathcal{N}_{\mathrm{A}}}n_{\mathrm{QA}}\right) + \frac{5}{2}\mathcal{N}_{\mathrm{A}}k \tag{16.76}$$

and similarly for $S_{\mathrm{B}}(T,V_{\mathrm{B}},\mathcal{N}_{\mathrm{B}})$ with $V_{\mathrm{B}} = V\mathcal{N}_{\mathrm{B}}/\mathcal{N}$. The entropy of mixing is therefore given by

$$\begin{aligned}\Delta S^{\mathrm{mix}} &:= S(T,V,\mathcal{N}_{\mathrm{A}},\mathcal{N}_{\mathrm{B}}) - S_{\mathrm{A}}(T,V_{\mathrm{A}},\mathcal{N}_{\mathrm{A}}) - S_{\mathrm{B}}(T,V_{\mathrm{B}},\mathcal{N}_{\mathrm{B}})\\ &= \mathcal{N}_{\mathrm{A}}k\ln(V/V_{\mathrm{A}}) + \mathcal{N}_{\mathrm{B}}k\ln(V/V_{\mathrm{B}})\\ &= -k\left[\mathcal{N}_{\mathrm{A}}\ln(\mathcal{N}_{\mathrm{A}}/\mathcal{N}) + \mathcal{N}_{\mathrm{B}}\ln(\mathcal{N}_{\mathrm{B}}/\mathcal{N})\right] > 0.\end{aligned} \tag{16.77}$$

[14]This statement may seem counterintuitive to those familiar with solution chemistry but a different scenario is used in that case. In solution chemistry, one generally considers the difference between the chemical potential of a pure gas A at some temperature T and pressure p (a "standard state") and a mixture of gases at temperature T and *total* pressure p. In a mixture of ideal gases, the gas of atoms A has only a partial pressure $p_{\mathrm{A}} = p\mathcal{N}_{\mathrm{A}}/\mathcal{N}$ so the difference in chemical potential per atom, compared to the standard state, is $kT\ln(p_{\mathrm{A}}/p) = kT\ln(\mathcal{N}_{\mathrm{A}}/\mathcal{N})$. In our case above, the gas of $\mathcal{N}_{\mathrm{A}}$ atoms in volume V has the same pressure p_{A} whether it is alone or in the presence of other gases.

This result should be compared to the case when a monocomponent gas occupies the entire volume V, in which case its entropy is given by Eq. (16.47) which we symbolize in the form

$$S(T, V, \mathcal{N}) = \mathcal{N}k \ln\left(\frac{V}{\mathcal{N}} n_Q\right) + \frac{5}{2}\mathcal{N}k. \tag{16.78}$$

If such a gas is partitioned such that $\mathcal{N}' + \mathcal{N}'' = \mathcal{N}$ with $\mathcal{N}'$ atoms occupying volume $V' = V\mathcal{N}'/\mathcal{N}$ and $\mathcal{N}''$ atoms occupying $V'' = V\mathcal{N}''/\mathcal{N}$, both the unpartitioned gas and the partitioned gases will have the same pressure because $V'/N' = V''/N'' = V/N$. The entropies of the partitioned gases will be

$$S'(T, V', \mathcal{N}') = \mathcal{N}'k \ln\left(\frac{V'}{\mathcal{N}'} n_Q\right) + \frac{5}{2}\mathcal{N}'k \tag{16.79}$$

and

$$S''(T, V'', \mathcal{N}'') = \mathcal{N}''k \ln\left(\frac{V''}{\mathcal{N}''} n_Q l\right) + \frac{5}{2}\mathcal{N}''k. \tag{16.80}$$

The entropy change when the partitioned gases are mixed will then be

$$\Delta S = S(T, V, \mathcal{N}) - S'(T, V', \mathcal{N}') - S''(T, V'', \mathcal{N}'') = 0, \tag{16.81}$$

as expected.

More insight about the entropy of mixing of ideal gases can be gained by the following. If we start with unmixed gases of A and B atoms, each at temperature T and pressure p and form from them a mixed gas having the same T and p, the number of configurations of $\mathcal{N}_A$ atoms of A and $\mathcal{N}_B$ atoms of B that can be obtained by arranging particles is

$$\Omega^{\text{mix}} = \frac{\mathcal{N}!}{\mathcal{N}_A!\mathcal{N}_B!}, \tag{16.82}$$

where $\mathcal{N} = \mathcal{N}_A + \mathcal{N}_B$. Then with Stirling's approximation,

$$\Delta S^{\text{mix}} = k \ln \Omega^{\text{mix}} = -k\left[\mathcal{N}_A \ln(\mathcal{N}_A/\mathcal{N}) + \mathcal{N}_B \ln(\mathcal{N}_B/\mathcal{N})\right], \tag{16.83}$$

which is the same as given by Eq. (16.77). In other words, the ideal entropy of mixing results simply from the number of distinct configurations of A and B atoms at temperature T and pressure p. We note that ΔS^{mix} is exactly the same quantity that we called ΔS^{ideal} in Section 10.2 where we treated so-called ideal solutions thermodynamically.

17 Classical Microcanonical Ensemble

In Chapter 16, we explored the microcanonical ensemble in the context of quantum statistical mechanics. First of all, we believe that quantum mechanics is correct whereas classical mechanics is just an asymptotic (but very useful) approximation. Second, however, quantum statistical mechanics is easier to understand because implementing the fundamental hypothesis is, in principle, a matter of counting quantum states and deciding on their statistical weight (e.g., equally probable for the microcanonical ensemble). In classical mechanics, however, we deal with *continuous* variables so we have to replace counting with integration over a continuous weighting function. On the other hand, classical statistical mechanics was the first to be developed and its study allows us to gain some physical intuition about statistical mechanics without dealing with the abstractions and statistical nature of quantum mechanics itself. Moreover, there are systems and situations for which quantum effects are not important and for which a treatment by classical statistical mechanics is more tractable. We shall therefore discuss briefly the foundations of classical statistical mechanics and explore briefly the classical version of the microcanonical ensemble.

We consider a three-dimensional classical system consisting of $\mathcal{N}$ identical particles and characterized by generalized coordinates $\mathbf{q} = q_1, q_2, \ldots, q_i, \ldots, q_{3\mathcal{N}}$ and generalized conjugate momenta $\mathbf{p} = p_1, p_2, \ldots, p_i, \ldots, p_{3\mathcal{N}}$. These variables span a $6\mathcal{N}$-dimensional space known as **phase space**. We denote them collectively by a $6\mathcal{N}$-dimensional vector $\boldsymbol{\omega}$. We write the volume element in this space in the form $\mathrm{d}^{3\mathcal{N}}p\,\mathrm{d}^{3\mathcal{N}}q \equiv \mathrm{d}p\,\mathrm{d}q \equiv \mathrm{d}\omega$. We denote the Hamiltonian for this system by $\mathcal{H}(\mathbf{p}, \mathbf{q}; t)$. Then the system evolves in time according to Hamilton's equations

$$\dot{q}_i = \frac{\partial \mathcal{H}}{\partial p_i}; \quad \dot{p}_i = -\frac{\partial \mathcal{H}}{\partial q_i}, \tag{17.1}$$

where a dot above a variable denotes differentiation with respect to time. As time evolves, the point $\mathbf{p}, \mathbf{q}$ traces out a trajectory in phase space. For the case $\mathcal{H}(\mathbf{p}, \mathbf{q}; t) = \mathcal{H}(\mathbf{p}, \mathbf{q})$, explicitly independent of time, the total energy E is conserved and this trajectory lies on the **hypersurface** $\mathcal{H}(\mathbf{p}, \mathbf{q}) = E$. For an isolated system, the energy will be constant and a fundamental assumption of classical statistical mechanics is that all points on that hypersurface are equally probable.[1] This leads to the classical microcanonical ensemble.

[1]In fact, one usually considers a thin hypershell such that $E - \Delta E \leq \mathcal{H}(\mathbf{p}, \mathbf{q}) \leq E$ and then assumes that every volume element in that *hypershell* is equally probable. The entropy is assumed to be proportional to the logarithm of the volume of that hypershell.

Thermal Physics. http://dx.doi.org/10.1016/B978-0-12-803304-3.00017-X

17.1 Liouville's Theorem

To gain more insight into the basis for classical statistical mechanics, we digress to discuss Liouville's theorem. We consider an ensemble of identical classical systems governed by Hamilton's equations. Each member of the ensemble corresponds to the same macrostate of some macroscopic system under consideration, but the members of the ensemble differ from one another microscopically, that is, they represent different microstates. Each member of the ensemble is represented by a point in phase space that moves in time from its initial point, which will differ for each member of the ensemble. We assume that there is an enormous number of such points that form a virtual continuum in all accessible parts of phase space. To quantify this swarm of points, we denote by $\rho(\mathbf{p},\mathbf{q};t) \equiv \rho(\boldsymbol{\omega};t)$ a distribution function such that $\rho(\mathbf{p},\mathbf{q};t)\,\mathrm{d}p\,\mathrm{d}q \equiv \rho(\boldsymbol{\omega};t)\,\mathrm{d}\omega$ is the number of members of the ensemble in the phase space volume element $\mathrm{d}\omega$. For a macroscopic system of interest, we take the point of view that observed quantities can be calculated by means of an ensemble average. Thus if $y(\boldsymbol{\omega})$ is some property that depends on the coordinates and momenta of the particles, its ensemble average would be

$$\langle y\rangle = \frac{\int y(\boldsymbol{\omega})\rho(\boldsymbol{\omega};t)\,\mathrm{d}\omega}{\int \rho(\boldsymbol{\omega};t)\,\mathrm{d}\omega}. \tag{17.2}$$

In this case, $\int \rho(\boldsymbol{\omega};t)\,\mathrm{d}\omega = \mathcal{N}_{\mathrm{ens}}$, the total number of members of the ensemble. We could equally well regard ρ to be a probability density function, in which case it would be normalized such that $\int \rho(\boldsymbol{\omega};t)\,\mathrm{d}\omega = 1$. In that case, the denominator in Eq. (17.2) would not be needed. Interpretation as a probability density is necessary in the limit $\mathcal{N}_{\mathrm{ens}} \to \infty$

Liouville's theorem deals with the evolution of ρ in phase space. We consider some fixed sub-volume ω' of phase space and equate the time rate of change of microsystems in that volume to the net rate at which microsystems enter that volume. Thus

$$\frac{\mathrm{d}}{\mathrm{d}t}\int_{\omega'} \rho(\boldsymbol{\omega};t)\,\mathrm{d}\omega = -\int_{a'} \rho(\boldsymbol{\omega};t)\,\dot{\boldsymbol{\omega}}\cdot\hat{\mathbf{n}}\,\mathrm{d}a', \tag{17.3}$$

where a' is the area of the sub-volume ω' and $\hat{\mathbf{n}}$ is its unit outward normal. Here, $\dot{\boldsymbol{\omega}}$ is the time rate of change of the $6\mathcal{N}$-dimensional vector $\boldsymbol{\omega}$, so that $\rho(\boldsymbol{\omega};t)\,\dot{\boldsymbol{\omega}}$ represents the flux of systems in phase space. We apply Gauss's theorem to the right-hand member of Eq. (17.3) and take the time derivative of the left-hand member inside the integral to obtain

$$\int_{\omega'}\left[\frac{\partial\rho}{\partial t} + \nabla_{\omega}\cdot(\rho\dot{\boldsymbol{\omega}})\right]\mathrm{d}\omega = 0, \tag{17.4}$$

where ∇_{ω} acts on the components of the vector $\boldsymbol{\omega}$. We assert that Eq. (17.4) is true for any arbitrary sub-volume of phase space, so the integrand itself must vanish, which gives

$$\frac{\partial\rho}{\partial t} + \nabla_{\omega}\cdot(\rho\dot{\boldsymbol{\omega}}) = 0. \tag{17.5}$$

We note that Eq. (17.5) is analogous to the continuity equation for conservation of mass of a classical fluid; in that case, ρ represents the density of the fluid and $\dot{\boldsymbol{\omega}}$ represents the barycentric fluid velocity $\mathbf{v}$.

The second term in Eq. (17.5) can be expanded to obtain

$$\nabla_\omega \cdot (\rho\dot{\boldsymbol{\omega}}) = \dot{\boldsymbol{\omega}} \cdot \nabla_\omega \rho + \rho \nabla_\omega \cdot \dot{\boldsymbol{\omega}}. \tag{17.6}$$

We shall proceed to show that the second term on the right in Eq. (17.6) vanishes. Indeed,

$$\nabla_\omega \cdot \dot{\boldsymbol{\omega}} = \sum_{i=1}^{3\mathcal{N}} \left[\frac{\partial}{\partial p_i} \dot{p}_i + \frac{\partial}{\partial q_i} \dot{q}_i \right] = \sum_{i=1}^{3\mathcal{N}} \left[-\frac{\partial^2 \mathcal{H}}{\partial p_i \partial q_i} + \frac{\partial^2 \mathcal{H}}{\partial q_i \partial p_i} \right] = 0, \tag{17.7}$$

where Hamilton's equations (Eq. (17.1)) have been used. Equation (17.7) is analogous to the equation of classical fluid dynamics, $\nabla \cdot \mathbf{v} = 0$, which is often referred to as incompressible flow. Its interpretation is that the fluid flows in closed loops, which is known as **solenoidal flow**. The first term on the right in Eq. (17.6) can also be written in terms of $\mathbf{p}$ and $\mathbf{q}$ in the form

$$\dot{\boldsymbol{\omega}} \cdot \nabla_\omega \rho = \sum_{i=1}^{3\mathcal{N}} \left[\frac{\partial \rho}{\partial q_i} \frac{\partial \mathcal{H}}{\partial p_i} - \frac{\partial \rho}{\partial p_i} \frac{\partial \mathcal{H}}{\partial q_i} \right] \equiv \{\rho, \mathcal{H}\} \tag{17.8}$$

and is known in classical mechanics as a **Poisson bracket**. It is analogous to a commutator in quantum mechanics. Equation (17.5) can therefore be written in the form

$$\frac{\mathcal{D}\rho}{\mathcal{D}t} \equiv \frac{\partial \rho}{\partial t} + \dot{\boldsymbol{\omega}} \cdot \nabla_\omega \rho = \frac{\partial \rho}{\partial t} + \{\rho, \mathcal{H}\} = 0. \tag{17.9}$$

The quantity $\mathcal{D}\rho/\mathcal{D}t$ is a total time derivative of ρ as one follows members of the ensemble through phase space; it is analogous to the substantial derivative of classical fluid dynamics, which is a total time derivative as one follows mass through real space. Equation (17.9), which is essentially Liouville's theorem, states that the density of members of the ensemble, as they move through phase space, does not change. An equivalent interpretation is that the volume of phase space occupied by a dense set of points representing members of the ensemble does not change with time, although it can change position and shape.

Further information can be obtained from Eq. (17.9) if one requires ρ to correspond to a state of equilibrium, in which case it should not depend on time explicitly. Then

$$\frac{\partial \rho}{\partial t} = 0, \quad \text{equilibrium ensemble}, \tag{17.10}$$

and Eq. (17.9) yields

$$\{\rho, \mathcal{H}\} = 0. \tag{17.11}$$

For a system in equilibrium, we shall require $\rho(\boldsymbol{\omega}; t) = \rho(\boldsymbol{\omega})$, explicitly independent of t. Physical measurements of such a system, which will disturb the system slightly, are really time averages over times that are large compared to the time it takes a system to relax to equilibrium. The system therefore passes through an enormous number of "equilibrium" states during a physical measurement, and its initial state is irrelevant. The time average of an ensemble average is therefore the same as the ensemble average of a time average

[8, p. 37]. In statistical mechanics, one adopts the hypothesis that the observed value of $y(\omega)$ in some macroscopic equilibrium state is its ensemble average. For further discussion, see [7, chapter 1].

17.2 Classical Microcanonical Ensemble

Equation (17.11) is a requirement for an acceptable distribution function and shows the close relationship of ρ to $\mathcal{H}$, and hence to the energy E. One way to satisfy Eq. (17.11) is to take ρ to be a constant. If the energy is precisely fixed at the value E, the members of the ensemble move in phase space on a subspace of phase space that we can regard as an energy hypersurface. We could represent ρ as a delta function, with some constant strength, that vanishes except on that energy hypersurface. Alternatively, and what is usually done, is to consider a thin shell of width Δ near the energy surface, that is, $E - \Delta E \le \mathcal{H} \le E$ and then take ρ to be a constant within that shell and zero elsewhere. This choice actually corresponds to the classical microcanonical ensemble. The constant value of ρ, which depends on the normalization of ρ, cancels in Eq. (17.2) which becomes

$$\langle y \rangle = \frac{1}{\Delta\omega} \int_{\Delta\omega} y(\omega)\, \mathrm{d}\omega. \tag{17.12}$$

Here, $\Delta\omega$ corresponds to the volume of phase space within the energy shell where ρ is not equal to zero. Equation (17.12) is the classical analog of the quantum mechanical formula (see Eq. (16.1))

$$\langle y \rangle = \frac{1}{\Omega} \sum_{\nu=1}^{\Omega} y_\nu, \tag{17.13}$$

where Ω is the degeneracy (multiplicity function) for a fixed energy and ν labels the compatible quantum states of the system, for which y has values y_ν.

It remains to establish a relationship between $\Delta\omega$, which is some measure of the volume of phase space available to the system, and the entropy S. From the quantum mechanical point of view, we need to relate $\Delta\omega$ to the number of allowed microstates of the system. In other words, we need to know what volume ω_0 of phase space corresponds to one microstate. Classical mechanics provides no answer to this question. We could write

$$S = k_{\mathrm{B}} \ln(\Delta\omega/\omega_0), \tag{17.14}$$

where k_{B} is Boltzmann's constant, but the entropy would still remain undetermined up to an additive "constant," although we would expect ω_0 to depend on $\mathcal{N}$. We can, however, appeal to quantum mechanics and choose ω_0 so that classical mechanics and quantum mechanics will agree in the asymptotic limit where classical mechanics is valid. This can only be done for simple systems, for which the problem is tractable, but presumably ω_0 will be the same for all systems, so we can determine it in a simple case. For an ideal gas, one can work out both the classical and quantum mechanical cases and make a comparison (see also Pathria [8, p. 39] and Chandler [12, p. 191]), as we do in the next section.

17.2.1 Classical Ideal Gas

To treat a classical ideal gas confined to a volume V in the microcanonical ensemble, we calculate the volume of phase space in a thin energy shell between energies $E - \Delta E$ and E. This volume is

$$\Delta\omega = \int \mathrm{d}^{3\mathcal{N}}q\,\mathrm{d}^{3\mathcal{N}}p = V^{\mathcal{N}}\int \mathrm{d}^{3\mathcal{N}}p, \tag{17.15}$$

where the momentum integral is over the hyperspherical shell

$$2m(E-\Delta E) \le \sum_{r=1}^{3\mathcal{N}} p_r^2 \le 2mE \tag{17.16}$$

of outer radius $(2mE)^{1/2}$. Proceeding as in the pseudo-quantum mechanical case, we know that the volume of this hyperspherical shell is just the factor $\mathcal{F}$ in Eq. (16.40) times the volume of the entire hypersphere, so

$$\Delta\omega = V^{\mathcal{N}}\mathcal{F}\frac{(2\pi mE)^{3\mathcal{N}/2}}{(3\mathcal{N}/2)!} \approx V^{\mathcal{N}}\frac{(2\pi mE)^{3\mathcal{N}/2}}{(3\mathcal{N}/2)!}. \tag{17.17}$$

The entropy is given by Eq. (17.14) with

$$\frac{\Delta\omega}{\omega_0} = \frac{V^{\mathcal{N}}}{\omega_0}\frac{(2\pi mE)^{3\mathcal{N}/2}}{(3\mathcal{N}/2)!}. \tag{17.18}$$

To agree with our pseudo-quantum mechanical treatment, specifically Eq. (16.44) for Ω, we deduce that

$$\omega_0 = h^{3\mathcal{N}}\mathcal{N}!, \quad \text{identical and indistinguishable particles.} \tag{17.19}$$

The factor $h^{3\mathcal{N}}$ in Eq. (17.19) has the same dimensions as the volume of phase space and can be thought of as dividing phase space into cells. The volume of each cell would be h^3 per particle, consistent with the Heisenberg uncertainty principle. The factor $h^{3\mathcal{N}}$ will make the ratio $(\Delta\omega/\omega_0)$ dimensionless. The factor of $\mathcal{N}!$ is the Gibbs correction factor that corrects for indistinguishable particles and makes the entropy an extensive function. For a dilute gas at high temperatures, it would occur automatically from quantum mechanical considerations that are designed from the start to deal systematically with indistinguishable particles.

Although we have derived this factor for an ideal gas, it is presumed to be a universal factor for all classical statistical systems consisting of indistinguishable particles. Of course the $\mathcal{N}!$ factor is to be omitted for classical identical but distinguishable particles such as identical classical harmonic oscillators imbedded in a solid and distinguished by their positions.

Since ω_0 depends only on $\mathcal{N}$, it will make no contribution to the calculation of $1/T = (\partial S/\partial E)_{V,\mathcal{N}}$ or to $p/T = (\partial S/\partial V)_{E,\mathcal{N}}$, but a knowledge of $\omega_0(\mathcal{N})$ is necessary to get the quantum mechanically correct entropy or any of the thermodynamic potentials such as F and G that depend on S.

For future reference, we remark that this same factor appears in the corrected expression for the classical partition function in the canonical ensemble, as given by Eq. (20.7), which can also be written as

$$Z_C^* = \frac{1}{h^{3\mathcal{N}}\mathcal{N}!}\int \exp[-\beta\,\mathcal{H}(\boldsymbol{\omega})]\,\mathrm{d}\omega = \int \exp[-\beta\,\mathcal{H}(\boldsymbol{\omega})]\,\mathrm{d}(\omega/\omega_0). \tag{17.20}$$

This dimensionless partition function is the analog of the quantum partition function

$$Z = \sum_\nu \exp[-\beta\,E_\nu]. \tag{17.21}$$

Example Problem 17.1. Consider a classical harmonic oscillator with spring constant k in one dimension x for a particle of mass m having linear momentum p. Its energy E is given by

$$E = \frac{p^2}{2m} + \frac{1}{2}kx^2 = \frac{p^2}{2m} + \frac{m\omega^2}{2}x^2, \tag{17.22}$$

where $\omega = \sqrt{k/m}$ is its angular frequency. The well-known solution to this equation has the form $x = A\sin(\omega t + \varphi)$, where t is the time and A and φ are constants. Show that the trajectory of the particle orbit in phase space is an ellipse and determine the sizes of its semiaxes. Then compute the area of the shell in phase space that lies between energies E and $E - \Delta E$ and compare with the corresponding energy ΔE for a quantum harmonic oscillator having quantized energies $E = (n + 1/2)\hbar\omega$. From your result, determine the number of quantum states per area of phase space.

Solution 17.1. Equation (17.22) can be written in the form

$$\frac{p^2}{a^2} + \frac{x^2}{b^2} = 1, \tag{17.23}$$

which is the equation of an ellipse with semiaxes $a = \sqrt{2mE}$ and $b = \sqrt{2E/m\omega^2}$. The momentum $p = m\mathrm{d}x/\mathrm{d}t = mA\omega\cos(\omega t + \varphi)$, with $A = \sqrt{2E/m\omega^2}$, so an elliptical trajectory is traversed periodically as time increases. The area of the ellipse is $\pi ab = 2\pi E/\omega$. The phase space area in a shell between E and $E - \Delta E$ is therefore

$$2\pi E/\omega - 2\pi(E - \Delta E)/\omega = 2\pi\,\Delta E/\omega. \tag{17.24}$$

For a quantum oscillator, the corresponding energy increment is

$$\Delta E = \Delta(n + 1/2)\hbar\omega = \hbar\omega\Delta n. \tag{17.25}$$

The number of quantum states per area of phase space is therefore

$$\frac{\Delta n}{2\pi\,\Delta E/\omega} = \frac{1}{2\pi\hbar} = \frac{1}{h}. \tag{17.26}$$

17.2.2 Classical Harmonic Oscillators in Three Dimensions

For $\mathcal{N}$ classical harmonic oscillators at fixed locations in three-dimensional space, the Hamiltonian is

$$\mathcal{H} = \sum_{i=1}^{3\mathcal{N}} \frac{p_i^2}{2m} + \frac{m\omega^2}{2} \sum_{i=1}^{3\mathcal{N}} x_i^2. \tag{17.27}$$

The volume of phase space for $\mathcal{H} \le E$ can be computed by mapping the hyperellipsoid described by Eq. (17.27) into a unit hypersphere S_1 given by

$$\sum_{i=1}^{6\mathcal{N}} X_i^2 = 1, \tag{17.28}$$

by means of the transformation

$$X_i = \frac{p_i}{\sqrt{2mE}}; \quad X_{i+3\mathcal{N}} = x_i \sqrt{m\omega^2/2E}; \quad i = 1, 2, \ldots, 3\mathcal{N}. \tag{17.29}$$

The corresponding volume of phase space within the entire hyperellipsoid is therefore

$$\int_E \mathrm{d}^{3\mathcal{N}} p \, \mathrm{d}^{3\mathcal{N}} x = \int_{S_1} J \mathrm{d}^{6\mathcal{N}} X = J \frac{\pi^{6\mathcal{N}/2}}{(6\mathcal{N}/2)!}, \tag{17.30}$$

where the Jacobian $J = (\sqrt{2mE})^{3\mathcal{N}} (\sqrt{2E/m\omega^2})^{3\mathcal{N}} = (2E/\omega)^{3\mathcal{N}}$. As was the case for a hypersphere (see Eq. (17.17)), the volume $\Delta\omega$ of a hyperellipsoidal shell between energies $E \ge \mathcal{H} \ge E - \Delta E$ is practically the same as the volume of the entire hyperellipsoid, so

$$\Delta\omega \approx \frac{(2\pi E/\omega)^{3\mathcal{N}}}{(3\mathcal{N})!}. \tag{17.31}$$

The entropy is therefore

$$\frac{S}{k_\mathrm{B}} = \ln \frac{\Delta\omega}{\omega_0} = \ln \frac{(2\pi E/\omega)^{3\mathcal{N}}}{h^{3\mathcal{N}} (3\mathcal{N})!} = 3\mathcal{N} \left[\ln \left(\frac{E}{3\mathcal{N}\hbar\omega} \right) + 1 \right], \tag{17.32}$$

where Stirling's approximation has been used in the last step. Here, we have used

$$\omega_0 = h^{3\mathcal{N}}, \quad \text{identical but distinguishable particles}, \tag{17.33}$$

because the oscillators are distinguishable due to their fixed locations. The temperature is therefore given by $1/T = (\partial S/\partial E)_{\mathcal{N}}$, resulting in

$$E = 3\mathcal{N} k_\mathrm{B} T. \tag{17.34}$$

In terms of temperature, the entropy is

$$S = 3\mathcal{N}k_B\left[\ln\left(\frac{k_B T}{\hbar\omega}\right) + 1\right]. \tag{17.35}$$

Comparison with Eqs. (16.26) and (16.29) shows that Eqs. (17.34) and (17.35) are only valid at high temperatures, quantum effects having been lost in the classical limit.[2]

[2]The dependence of Eq. (17.35) on $\hbar$ results from identification of $\omega_0 = h^{3\mathcal{N}}$ from quantum mechanical considerations. From a strictly classical point of view, ω_0 would be unknown so the entropy would only be determined up to an additive constant, namely $S = 3\mathcal{N}k_B \ln T + \text{constant}$. The factor of 3 would be absent for one-dimensional oscillators.

18

Distinguishable Particles with Negligible Interaction Energies

In Chapters 16 and 17, we introduced the microcanonical ensemble. This ensemble was useful for stating the fundamental postulates on which statistical mechanics is based, but not useful for practical calculations. From the microcanonical ensemble, we can derive other ensembles, such as the canonical ensemble (Chapter 19) and the grand canonical ensemble (Chapter 21) that are more tractable. Before doing so, however, we pause to develop the special case of the statistical mechanics at constant temperature T of **identical but distinguishable particles** having negligible interaction energies. This is a *special case* of the canonical ensemble and allows us to quickly and easily obtain a number of useful results of practical importance without complication. In Chapter 19, we will derive the canonical ensemble from the microcanonical ensemble for a large system in contact with a heat reservoir at temperature T. In Section 19.2.1, we will show that the results in the current chapter can be deduced by means of the factorization theorem. *A more sophisticated reader can skip this chapter temporarily and go directly to Chapter 19.*

We consider a system consisting of $\mathcal{N}$ identical but distinguishable quantum subsystems that we shall refer to as "particles." Each particle has stationary states having energies $\varepsilon_1, \varepsilon_2, \ldots, \varepsilon_i, \ldots$. The particles are assumed to be distinguishable because of their *fixed location* (e.g., in a solid) but are otherwise the same. The states of each particle may be finite or infinite in number, and some of them may be degenerate.[1] Moreover, the energies ε_i could possibly depend on the volume V of the system. In the derivation that follows, we shall suppress any dependence of ε_i on V until needed. The particles are assumed to interact sufficiently weakly that their interaction energy is negligible, but to a degree that will allow them eventually to come to equilibrium.

18.1 Derivation of the Boltzmann Distribution

We examine a **configuration** $\{\mathcal{N}_i\} = \mathcal{N}_1, \mathcal{N}_2, \ldots, \mathcal{N}_\kappa$ of the system such that $\mathcal{N}_1$ particles are in a quantum state[2] with energy ε_1, $\mathcal{N}_2$ particles are in a quantum state with energy ε_2, etc. Such a configuration is subject to the constraints

[1]Degeneracy arises when there are stationary states of the subsystems having the same energy but a different set of quantum numbers.

[2]For brevity we use a single index to denote a quantum state but in fact many quantum numbers may be necessary. Moreover, there can be degeneracy if different quantum states have the same energy.

Thermal Physics. http://dx.doi.org/10.1016/B978-0-12-803304-3.00018-1

$$\sum_i \mathcal{N}_i = \mathcal{N} \tag{18.1}$$

and

$$\sum_i \mathcal{N}_i \varepsilon_i = E, \tag{18.2}$$

where E is the total energy of the system. Since the particles are distinguishable, the number of ways of making a given configuration is[3]

$$W\{\mathcal{N}_i\} := \frac{\mathcal{N}!}{\mathcal{N}_1!\mathcal{N}_2!\cdots\mathcal{N}_\kappa!}. \tag{18.3}$$

We proceed to maximize $W\{\mathcal{N}_i\}$, considered to be a function of the $\mathcal{N}_i$, subject to the constraints expressed by Eqs. (18.1) and (18.2). Since $\ln x$ is a monotonically increasing function of x, we actually maximize $\ln W$ subject to these same constraints. To handle the constraints, we introduce Lagrange multipliers β and α and solve the problem

$$\frac{\partial}{\partial \mathcal{N}_j}[\ln W\{\mathcal{N}_i\} - \beta E - \alpha \mathcal{N}] = 0. \tag{18.4}$$

By virtue of the Lagrange multipliers, all $\mathcal{N}_j$ in Eq. (18.4) can be regarded as independent variables, which will turn out to be functions of β and α. We can then choose β and α to satisfy the constraints.

In order to differentiate $\ln W\{\mathcal{N}_i\}$, we use Stirling's approximation (see Appendix A) and obtain

$$\frac{\partial}{\partial \mathcal{N}_j} \ln W\{\mathcal{N}_i\} \sim \frac{\partial}{\partial \mathcal{N}_j}\left[\mathcal{N}\ln\mathcal{N} - \sum_i \mathcal{N}_i \ln \mathcal{N}_i\right] = -\ln\frac{\mathcal{N}_j}{\mathcal{N}}. \tag{18.5}$$

Thus Eq. (18.4) becomes

$$-\ln\frac{\mathcal{N}_j}{\mathcal{N}} - \beta\varepsilon_j - \alpha = 0. \tag{18.6}$$

The solution to Eq. (18.6) is

$$\frac{\mathcal{N}_j}{\mathcal{N}} = \mathrm{e}^{-\alpha}\mathrm{e}^{-\beta\varepsilon_j}. \tag{18.7}$$

Applying the constraint Eq. (18.1) we obtain

$$1 = \sum_j \frac{\mathcal{N}_j}{\mathcal{N}} = \mathrm{e}^{-\alpha}\sum_j \mathrm{e}^{-\beta\varepsilon_j}, \tag{18.8}$$

which results in

$$\mathrm{e}^{-\alpha} = 1/z, \tag{18.9}$$

[3]Here, W plays the same role as Ω for the microcanonical ensemble, but we use a different notation because Ω corresponds to constrained values of E and $\mathcal{N}$. In the present case, these constraints are replaced by Eqs. (18.1) and (18.2). Ultimately we will specify the temperature T and then determine E from the probabilities p_i of occupation of the quantum states.

where

$$z = \sum_j \mathrm{e}^{-\beta\varepsilon_j} \tag{18.10}$$

is known as the **partition function**.[4] Thus Eq. (18.7) becomes

$$p_i := \frac{\mathcal{N}_i}{\mathcal{N}} = \frac{\mathrm{e}^{-\beta\varepsilon_i}}{z}, \tag{18.11}$$

where we have also introduced the symbol p_i, the probability of occupation of the ith state of a particle.

The internal energy can now be determined from Eq. (18.2) to be[5]

$$U := \langle E\rangle = \mathcal{N}\sum_i p_i\varepsilon_i = \mathcal{N}\sum_i \frac{\varepsilon_i \mathrm{e}^{-\beta\varepsilon_i}}{z} = -\frac{\mathcal{N}}{z}\frac{\partial z}{\partial\beta} = -\mathcal{N}\frac{\partial \ln z}{\partial\beta}. \tag{18.12}$$

To obtain the entropy, we use[6]

$$S = k_\mathrm{B} \ln W\{\mathcal{N}_i\} \tag{18.13}$$

with $\mathcal{N}_i$ given by Eq. (18.11). With the aid of Stirling's approximation, Eq. (18.13) becomes

$$S = k_\mathrm{B}\left[\mathcal{N}\ln\mathcal{N} - \sum_i \mathcal{N}_i \ln \mathcal{N}_i\right] = -k_\mathrm{B}\sum_i \mathcal{N}_i \ln(\mathcal{N}_i/\mathcal{N}) = -\mathcal{N}k_\mathrm{B}\sum_i p_i \ln p_i. \tag{18.14}$$

We now proceed to identify the remaining Lagrange multiplier β. In principle, we could do this by specifying the total energy and solving Eq. (18.2) for β, with $\mathcal{N}_i$ given by Eq. (18.11), but this would necessitate solving a complicated transcendental equation. Instead, we suppose that our system is in equilibrium at fixed temperature T and appeal to thermodynamics to identify β. We do this by relating the above expressions for U and S by means of the thermodynamic equation $\mathrm{d}U = T\,\mathrm{d}S - p\,\mathrm{d}V$ which holds at constant $\mathcal{N}$ for a system that can do reversible work $p\,\mathrm{d}V$.[7] From Eq. (18.12), the differential of the internal energy is

$$\mathrm{d}U = \mathcal{N}\sum_i \varepsilon_i\,\mathrm{d}p_i + \mathcal{N}\sum_i p_i \mathrm{d}\varepsilon_i = \mathcal{N}\sum_i \varepsilon_i\,\mathrm{d}p_i + \mathcal{N}\sum_i p_i \frac{\partial\varepsilon_i}{\partial V}\,\mathrm{d}V, \tag{18.15}$$

[4]In Eq. (18.10), z is the partition function for an individual particle. We reserve the symbol Z for the partition function of the whole system that we shall later relate to z.

[5]Since the p_i are probabilities, Eq. (18.12) actually gives the most probable value $\langle E\rangle$ of energy which we identify with the internal energy U that we will ultimately compute from a knowledge of the temperature.

[6]To get the entropy, we should really compute the logarithm of the total number of microstates by summing all values of $\ln W\{\mathcal{N}_i\}$ that are compatible with the constraints. Instead, we approximate this sum by its overwhelmingly largest term.

[7]See Section 19.1.3 for a similar treatment for a more general system.

where we have assumed that the energies of the states depend on the volume of the system. From Eq. (18.14), the differential of the entropy is

$$dS = -\mathcal{N}k_B \sum_i (\ln p_i + 1)\, dp_i = -\mathcal{N}k_B \sum_i \ln p_i\, dp_i$$

$$= -\mathcal{N}k_B \sum_i (-\beta\varepsilon_i - \ln z)\, dp_i = \mathcal{N}k_B\beta \sum_i \varepsilon_i\, dp_i, \tag{18.16}$$

where we have used $\sum_i dp_i = 0$ because $\sum_i p_i = 1$. By combining Eq. (18.15) with Eq. (18.16) we obtain

$$dU = \frac{1}{k_B\beta}\, dS + \mathcal{N} \sum_i p_i \frac{\partial \varepsilon_i}{\partial V}\, dV. \tag{18.17}$$

Comparison with $dU = T\, dS - p\, dV$ shows that

$$\beta = \frac{1}{k_B T}. \tag{18.18}$$

We also obtain a useful equation for the pressure, namely

$$p = -\mathcal{N} \sum_i p_i \frac{\partial \varepsilon_i}{\partial V}, \tag{18.19}$$

which by means of Eq. (18.17) with $dS = 0$ is seen to be equivalent to $p = -(\partial U/\partial V)_{S,\mathcal{N}}$ with $U = \mathcal{N}\sum_i p_i\varepsilon_i$. By using Eq. (18.11) to rewrite $\ln p_i$, the entropy given by Eq. (18.14) can be written in the form

$$S = \frac{U}{T} + \mathcal{N}k_B \ln z. \tag{18.20}$$

Equation (18.20) can then be combined with the equation $F = U - TS$ to deduce a useful formula for the Helmholtz free energy

$$F = -\mathcal{N}k_B T \ln z = -\frac{\mathcal{N}}{\beta} \ln z. \tag{18.21}$$

In Section 19.1.3, this derivation will be generalized to an **ensemble** of complicated systems (instead of a collection of weakly interacting particles). Such an ensemble is known as a **canonical ensemble** and allows for each complicated system to consist of many interacting particles. For such complicated systems, determination of the quantum states and the resulting partition functions can be quite difficult.

18.1.1 Summary of Results

The probability p_i of occupation of the i state is

$$p_i := \frac{\mathcal{N}_i}{\mathcal{N}} = \frac{e^{-\beta\varepsilon_i}}{z}, \tag{18.22}$$

where $\beta = 1/(k_B T)$, k_B is Boltzmann's constant, T is the absolute temperature, and

$$z = \sum_j e^{-\beta \varepsilon_j} \tag{18.23}$$

is the partition function. The entropy is

$$S = -\mathcal{N} k_B \sum_{i=1} p_i \ln p_i = -\mathcal{N} k_B \beta^2 \frac{\partial}{\partial \beta}\left[\frac{\ln z}{\beta}\right]. \tag{18.24}$$

The internal energy is

$$U = \mathcal{N} \sum_{i=1} p_i \varepsilon_i = -\mathcal{N} \frac{\partial}{\partial \beta} \ln z \tag{18.25}$$

and the Helmholtz free energy is

$$F = -\frac{\mathcal{N}}{\beta} \ln z. \tag{18.26}$$

In solving problems, one usually proceeds as follows:

- Determine the subsystem states i having energies ε_i from a model or from experimental data.
- Calculate the partition function z and deduce the Helmholtz free energy F by using Eq. (18.26).
- Obtain the entropy from $S = -(\partial F/\partial T)_{V,\mathcal{N}}$ or from Eq. (18.24).
- Obtain the internal energy from $U = F + TS$ or from Eq. (18.25).
- Obtain the chemical potential per particle from $\mu = (\partial F/\partial \mathcal{N})_{T,V}$.
- If the dependence of the ε_i on volume V is known, determine the pressure from $p = -(\partial F/\partial V)_{T,\mathcal{N}}$.

In following this procedure, it should be recognized that Eq. (18.26) yields F as a function of its natural variables T, V, and N, where $N = \mathcal{N}/\mathcal{N}_A$ is the mole number of particles, $\mathcal{N}_A$ being Avogadro's number. We therefore recover the usual thermodynamic description of a monocomponent system. The volume V might enter because each particle occupies a volume $V/\mathcal{N}$ on which the energy levels ε_i might depend. These particles, although identical, are supposed to be distinguishable by virtue of their position. If all particles were to *share* the same volume and were identical, they would not be distinguishable. Such would be the case for a monatomic ideal gas, so to treat such a system the above equations would have to be modified.

18.2 Two-State Subsystems

We apply the results of the previous section to a number $\mathcal{N}$ of identical but distinguishable two-state subsystems, each having nondegenerate energy levels ε_1 and ε_2. These subsystems are distinguishable because each is assumed to have a fixed location. In order to focus ideas, we consider the case in which each of our two-state systems is a particle having

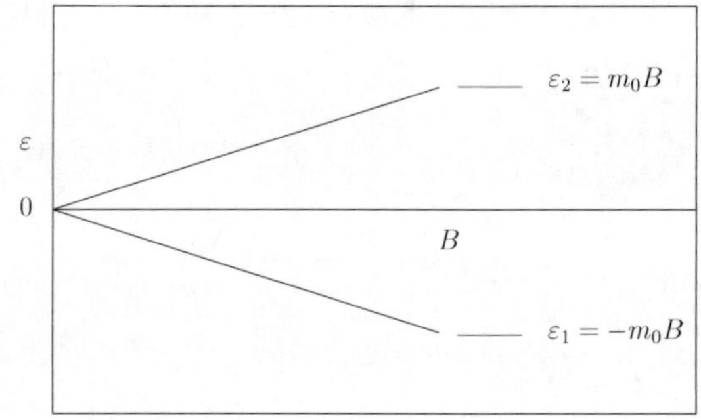

FIGURE 18–1 Energy levels $\varepsilon_1 = -m_0 B$ and $\varepsilon_2 = m_0 B$ due to splitting by a magnetic field B for a spin 1/2 particle having magnetic moment $m_0 > 0$ for "spin up."

spin $1/2$ in a magnetic field of strength B. Each particle can exist in two states, a state with "spin up" having energy $\varepsilon_1 = -m_0 B$ and a state with "spin down" having energy $\varepsilon_2 = m_0 B$, where the magnetic moment $m_0 > 0$.[8] See Figure 18–1 for an energy level diagram.

From Eq. (18.23), the partition function is

$$z = \mathrm{e}^{-\beta\varepsilon_1} + \mathrm{e}^{-\beta\varepsilon_2} = \mathrm{e}^{m_0 B\beta} + \mathrm{e}^{-m_0 B\beta}. \tag{18.27}$$

From Eq. (18.22), the probabilities of occupation of each state are

$$p_1 = \frac{\mathrm{e}^{-\beta\varepsilon_1}}{z} = \frac{\mathrm{e}^{m_0 B\beta}}{\mathrm{e}^{m_0 B\beta} + \mathrm{e}^{-m_0 B\beta}} = \frac{1}{1 + \mathrm{e}^{-2m_0 B\beta}} \tag{18.28}$$

and

$$p_2 = \frac{\mathrm{e}^{-\beta\varepsilon_2}}{z} = \frac{\mathrm{e}^{-m_0 B\beta}}{\mathrm{e}^{m_0 B\beta} + \mathrm{e}^{-m_0 B\beta}} = \frac{\mathrm{e}^{-2m_0 B\beta}}{1 + \mathrm{e}^{-2m_0 B\beta}}. \tag{18.29}$$

The latter expressions in Eqs. (18.28) and (18.29) involve only the energy gap $\varepsilon := 2m_0 B$ between the states and could be obtained directly by shifting the zero of energy so that the ground state would have zero energy[9] and the excited state would have energy ε. The ratio of these populations is

$$\frac{p_2}{p_1} = \mathrm{e}^{-2m_0 B\beta} = \mathrm{e}^{-2m_0 B/k_\mathrm{B} T} = \mathrm{e}^{-\beta\varepsilon} \to \begin{cases} 0 \text{ as } T \to 0 \\ 1 \text{ as } T \to \infty. \end{cases} \tag{18.30}$$

Thus at high temperatures, $p_1 = p_2 = 1/2$ and the states are equally probable.[10]

[8]We regard the state with "spin up" as having its magnetic moment in the same direction as the magnetic field, and hence the lower energy. This unambiguous sign convention avoids the question of the connection between direction of the spin and the sign of the charge of a particle having spin.

[9]From the forms of Eqs. (18.22) and (18.23), it is clear for any system that the probabilities p_i are invariant if $\varepsilon_i \to \varepsilon_i + \varepsilon$ for all energies, so the p_i are independent of the zero of energy.

[10]A common misconception by new students of statistical mechanics is that all of the subsystems will be in their highest energy state as $T \to \infty$. Nothing could be further from the truth! At sufficiently high temperatures, the entropy dominates the free energy F, and the internal energy becomes irrelevant. A state in which $p_2 > p_1$ would correspond formally to a negative temperature. Negative temperatures have been used to represent nonequilibrium states in which the population of the excited state has been "pumped" to some high level by means of some external stimulation, but such negative temperatures are outside the scope of conventional thermodynamics.

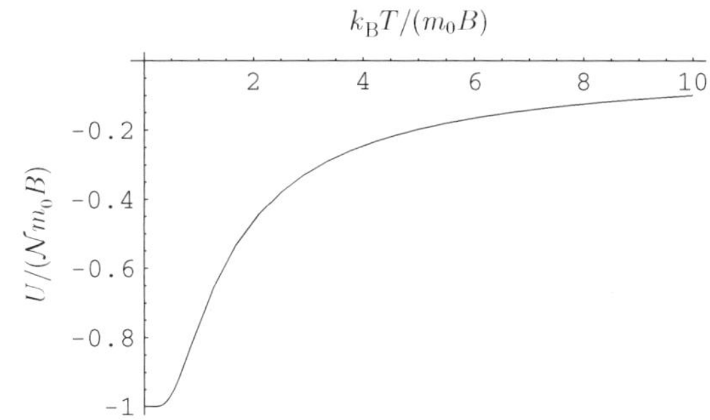

FIGURE 18–2 Dimensionless internal energy $U/(\mathcal{N} m_0 B)$ versus dimensionless temperature $k_B T/(m_0 B)$ for a two-state magnetic system according to Eq. (18.31). At $T = 0$, all spins are aligned with the magnetic field. As $T \to \infty$, half of the spins are aligned with the field and half are aligned opposite to the field, so their energies cancel.

The energy can be calculated directly from Eq. (18.2), resulting in

$$U = \mathcal{N} \frac{(-m_0 B\, e^{m_0 B\beta} + m_0 B\, e^{-m_0 B\beta})}{e^{m_0 B\beta} + e^{-m_0 B\beta}} = -\mathcal{N} m_0 B \tanh(x), \tag{18.31}$$

where $x = m_0 B\beta = m_0 B/k_B T$ and $\tanh(x) = \sinh(x)/\cosh(x)$ is the hyperbolic tangent function, $\sinh x := (e^x - e^{-x})/2$ is the hyperbolic sine function, and $\cosh x := (e^x + e^{-x})/2$ is the hyperbolic cosine function. Figure 18–2 shows a plot of U versus temperature in dimensionless units. We observe that

$$U \to \begin{cases} -\mathcal{N} m_0 B & \text{as } T \to 0 \\ 0 & \text{as } T \to \infty. \end{cases} \tag{18.32}$$

For this simple system, the **magnetic moment** $\mathcal{M}$ is given by[11]

$$\mathcal{M} = -\frac{U}{B} = \mathcal{M}_0 \tanh(x), \tag{18.33}$$

where $\mathcal{M}_0 := \mathcal{N} m_0$ is called the *saturation* magnetic moment. $\mathcal{M}$ decreases from $\mathcal{M}_0$ at $T = 0$ to zero as $T \to \infty$, as shown in Figure 18–3. This type of magnetism, for which the interaction energy between subsystems having a magnetic moment is negligible, in known as **paramagnetism**. For $B = 0$, there is no splitting of the states, and no net magnetic moment. Ferromagnetic systems, in which there are strong interactions between magnetic subsystems, can have a magnetic moment without an applied magnetic field B.

The entropy can be calculated from Eq. (18.24), resulting in

$$S = \mathcal{N} k_B \left[x + \ln(1 + e^{-2x}) - x \tanh(x) \right]. \tag{18.34}$$

Figure 18–4 shows a plot of S versus T in dimensionless units. We observe that

[11] See Section 19.6 for a general definition of the magnetic moment, $\mathcal{M} = -(\partial F/\partial B)_T = -(\partial U/\partial B)_S$.

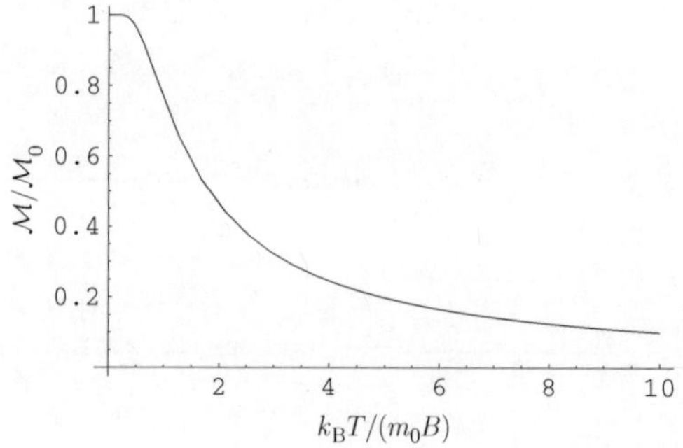

FIGURE 18–3 Dimensionless magnetic moment $\mathcal{M}/\mathcal{M}_0$ versus dimensionless temperature $k_BT/(m_0B) = 1/x$ for a two-state magnetic system according to Eq. (18.33). At $T = 0$, all spins are aligned with the magnetic field, so $\mathcal{M} = \mathcal{M}_0$. As T increases, more spins are promoted to the upper state so the magnetic moment weakens. For large T, $\mathcal{M}$ is approximately proportional to $1/T$, which is known as Curie's law.

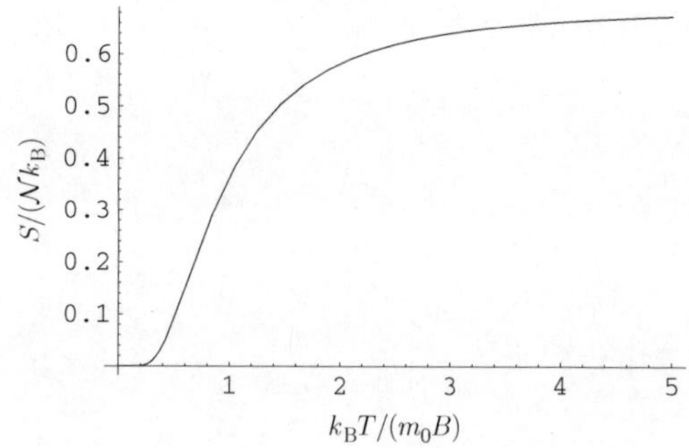

FIGURE 18–4 Dimensionless entropy $S/(\mathcal{N}k_B)$ versus dimensionless temperature $k_BT/(m_0B) = 1/x$ for a two-state magnetic system according to Eq. (18.34). At $T = 0$, all spins are aligned with the magnetic field, so $S = 0$. As $T \to \infty$, half of the spins are aligned with the field and half are aligned opposite to the field, so $S/(\mathcal{N}k_B) \to \ln 2 = 0.693$.

$$S \to \begin{cases} 0 & \text{as } T \to 0 \\ \mathcal{N}k_B \ln 2 & \text{as } T \to \infty. \end{cases} \tag{18.35}$$

This last result can be obtained easily by substituting $p_1 = p_2 = 1/2$ into the middle member of Eq. (18.24).[12]

Alternatively, we can use Eq. (18.26) to obtain the Helmholtz free energy

$$F = -\mathcal{N}k_BT\ln(e^x + e^{-x}) = -\mathcal{N}k_BT[x + \ln(1 + e^{-2x})]. \tag{18.36}$$

Note that Eq. (18.36) also results from $F = U - TS$ with U given by Eq. (18.31) and S given by Eq. (18.34). We can also differentiate Eq. (18.36) with respect to T to obtain $-S$, and then obtain the internal energy from $U = F + TS$. A plot of F versus T in dimensionless units[13] is shown in Figure 18–5.

We note that

$$F \to \begin{cases} -\mathcal{N}m_0B & \text{as } T \to 0 \\ -\mathcal{N}k_BT\ln 2 & \text{as } T \to \infty. \end{cases} \tag{18.37}$$

At low temperatures, F behaves like the internal energy; however, at high temperatures, it behaves like $-TS$ and becomes linear in T as S saturates to a constant value.

[12]For a system having q states, $p_1 \to 1/q$ as $T \to \infty$. Then the middle member of Eq. (18.24) yields $S = \mathcal{N}k_B \ln q$.

[13]Note from Eq. (18.36) that $F/(\mathcal{N}m_0B) = -(1/x)[x + \ln(1 + e^{-2x})]$.

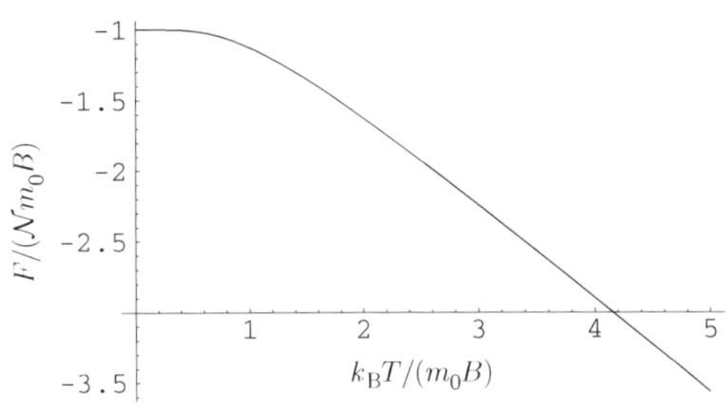

FIGURE 18–5 Dimensionless Helmholtz free energy $F/(\mathcal{N}m_0B)$ versus dimensionless temperature $k_BT/(m_0B) = 1/x$ for a two-state magnetic system according to Eq. (18.36). At $T = 0$, F is equal to U. For large T, F is nearly equal to $-TS$ with S nearly constant.

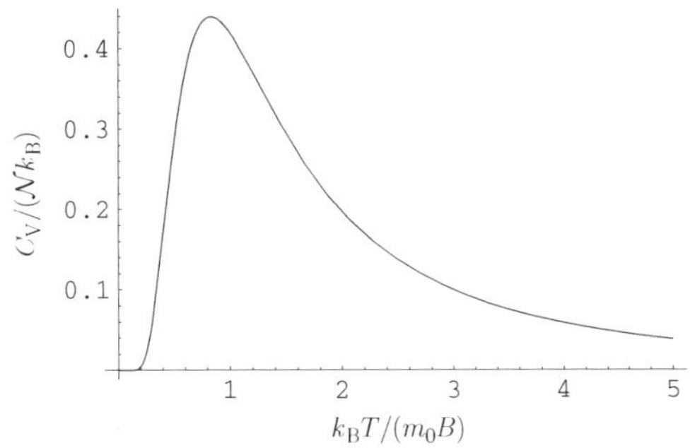

FIGURE 18–6 Dimensionless heat capacity $C_V/(\mathcal{N}k_B)$ versus dimensionless temperature $k_BT/(m_0B) = 1/x$ for a two-state magnetic system according to Eq. (18.38). A Schottky peak occurs at $k_BT/(m_0B) \approx 0.834$ as spins are promoted to the upper state with increasing T. As T becomes very large, the population of the upper state becomes nearly equal to that of the lower state and can increase very little as T increases, resulting in C_V decreasing to zero.

Finally, we can differentiate U with respect to T to obtain the heat capacity C_V at constant volume, resulting in

$$C_V = \mathcal{N}k_B \frac{4x^2}{(e^x + e^{-x})^2} = \mathcal{N}k_B x^2 \operatorname{sech}^2(x), \tag{18.38}$$

where $\operatorname{sech} x = 1/\cosh x$ is the hyperbolic secant function. A plot of C_V versus T in dimensionless units is shown in Figure 18–6.

The peak[14] near $m_0B = k_BT$ is called a **Schottky peak** and occurs when the population of the upper level increases at maximum rate with increasing T. At high T, $C_V \to 0$ because the populations of the states become equal and no more increase in energy is possible as T increases.

18.3 Harmonic Oscillators

We consider the case in which each of our particles is a one-dimensional harmonic oscillator, fixed in location, with Hamiltonian

$$\mathcal{H} = \frac{p^2}{2m} + \frac{1}{2}kx^2, \tag{18.39}$$

where p is the momentum, x is the coordinate, m is the mass, and k is the spring constant. The quantum energy levels of such an oscillator can be obtained in the Schrödinger

[14]The actual position of the peak occurs at the positive root of $\tanh x = 1/x$, which we estimate to be $x = 1.19968$.

picture by using the momentum operator $p = -i\hbar\partial/\partial x$ and solving the time-independent Schrödinger equation

$$\mathcal{H}\psi_n = \left(-\frac{\hbar^2}{2m}\frac{\partial^2}{\partial x^2} + \frac{1}{2}kx^2\right)\psi_n = \epsilon_n\psi_n \tag{18.40}$$

to determine the wave functions ψ_n and the energies ϵ_n of the stationary states. The fact that the wave functions ψ_n have to go to zero far outside the potential well $(k/2)x^2$ leads to a set of allowable wave functions having parity $(-1)^n$, where $n = 0, 1, 2, \ldots$ is zero or a positive integer and nondegenerate energy levels[15]

$$\epsilon_n = \hbar\omega(n + 1/2), \tag{18.41}$$

where $\omega := \sqrt{k/m}$ is the classical angular frequency of the oscillator. The quantity $\hbar\omega/2$ is known as the zero-point energy. Since energies in thermodynamics have an arbitrary zero, we will calculate the thermodynamic functions by using the shifted set of energy levels

$$\varepsilon_n = \hbar\omega n. \tag{18.42}$$

Using Eq. (18.42) rather than Eq. (18.41) will affect the partition function but will not affect the probabilities p_n or the entropy S. The internal energy and all other thermodynamic potentials will be lowered by the constant amount $\mathcal{N}\hbar\omega/2$.

The partition function

$$z = \sum_{n=0}^{\infty} \exp(-\beta\varepsilon_n) = \sum_{n=0}^{\infty} \exp(-\beta\hbar\omega n) = \sum_{n=0}^{\infty} y^n, \tag{18.43}$$

where $\beta = 1/(k_B T)$ and $y := e^{-x}$, with $x := \beta\hbar\omega$. The geometric series in Eq. (18.43) can be summed by noting[16] that $yz = z - 1$ which leads to $z = 1/(1-y)$ or

$$z = \frac{1}{1 - e^{-x}}. \tag{18.44}$$

From Eq. (18.26), we determine the Helmholtz free energy to be

$$F = \mathcal{N}k_B T \ln(1 - e^{-x}). \tag{18.45}$$

The entropy is therefore

$$S = -\frac{\partial F}{\partial T} = -\mathcal{N}k_B \left[\ln(1 - e^{-x}) - \frac{x\,e^{-x}}{1 - e^{-x}}\right], \tag{18.46}$$

where we have used $\partial x/\partial T = -x/T$. The internal energy is

$$U = F + TS = \mathcal{N}k_B T \frac{x\,e^{-x}}{1 - e^{-x}} = \mathcal{N}\hbar\omega\frac{1}{e^x - 1}. \tag{18.47}$$

[15]See practically any book on quantum mechanics for details. See Appendix I for an algebraic solution by means of creation and annihilation operators.

[16]For any finite temperature, $e^{-x} < 1$, so the series converges.

Equation (18.47) was derived originally by Planck [55]. In view of Eq. (18.42), the quantity

$$\langle n(T) \rangle := \frac{1}{e^x - 1} = \frac{1}{\exp(\hbar\omega/k_B T) - 1} \tag{18.48}$$

can be thought of as the thermal average $\langle n(T) \rangle$ of the quantum number n. At low temperatures, $\langle n(T) \rangle \approx \exp(-\hbar\omega/k_B T)$, so

$$U \approx \mathcal{N}\hbar\omega \exp(-\hbar\omega/k_B T), \quad \text{low } T. \tag{18.49}$$

At high T, we have

$$\langle n(T) \rangle = \frac{1}{1 + x + x^2/2 + \cdots - 1} = \frac{1}{x + x^2/2 + \cdots} \approx \frac{1}{x} = \frac{k_B T}{\hbar\omega}. \tag{18.50}$$

Thus

$$U \approx \mathcal{N} k_B T, \quad \text{high } T. \tag{18.51}$$

Note that Eq. (18.51) is independent of ω and so would be true for *any* one-dimensional harmonic oscillator, irrespective of mass or force constant. We shall see later that the result given by Eq. (18.51) is the same as would be given by classical statistical mechanics (continuum of energies, no quantum states) at all temperatures. Indeed, as $\omega \to 0$ we have $x \to 0$ so the expansion in Eq. (18.50) would be valid for any $T > 0$. Planck recognized that the result at low temperatures would be significantly different if the energy levels were quantized.

Figure 18–7 shows a plot of the internal energy versus temperature. At low temperatures, hardly any oscillators can be excited to the first excited state, so $\langle n(T) \rangle \ll 1$. Therefore, U remains very small until $x \approx 1$, or $k_B T \approx \hbar\omega$, at which temperature U begins to rise significantly, ultimately becoming linear in T as more and more quantum states become significantly occupied.

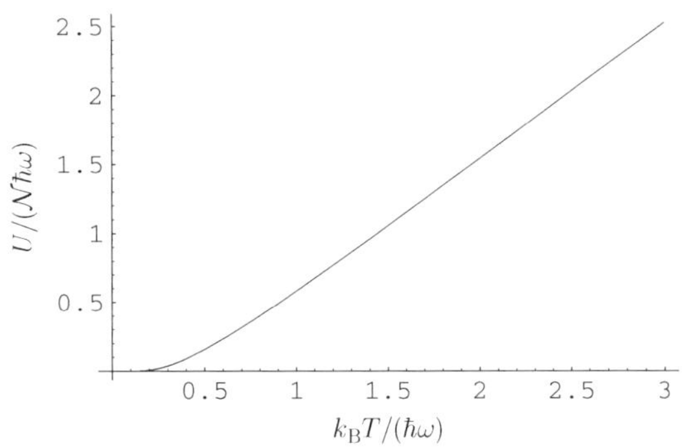

FIGURE 18–7 Dimensionless internal energy $U/(\mathcal{N}\hbar\omega) = \langle n(T) \rangle$ versus dimensionless temperature $k_B T/(\hbar\omega) = 1/x$ for one-dimensional harmonic oscillators according to Eq. (18.47). As T increases from zero, Eq. (18.49) shows that U increases very little. For large T, Eq. (18.51) shows that U increases nearly linearly with T.

Example Problem 18.1. Calculate the probability p_n that a single oscillator is the quantum state n. Then calculate directly the average value of n and hence verify directly Eq. (18.48).

Solution 18.1. From Eqs. (18.11), (18.42), and (18.44) we have

$$p_n = \exp[n\beta\hbar\omega]/z = \mathrm{e}^{-nx}/z = \mathrm{e}^{-nx}(1 - \mathrm{e}^{-x}), \tag{18.52}$$

where $x = \beta\hbar\omega$. Thus

$$\langle n \rangle = \sum_{n=0}^{\infty} n p_n = z^{-1} \sum_{n=0}^{\infty} n\,\mathrm{e}^{-nx} = -z^{-1}\frac{\partial}{\partial x} z = -\frac{\partial}{\partial x} \ln z = 1/(\mathrm{e}^x - 1). \tag{18.53}$$

Equation (18.53) is equivalent to calculating the average energy of a single oscillator from $-\partial/\partial\beta \ln z$ and then dividing by $\hbar\omega$. Indeed, Eq. (18.12) could have been used to calculate U in Eq. (18.47) directly from $\ln z$ rather than from F and S.

The heat capacity of $\mathcal{N}$ one-dimensional harmonic oscillators is

$$C_V = \mathcal{N} k_\mathrm{B} \frac{x^2\,\mathrm{e}^x}{(\mathrm{e}^x - 1)^2}, \tag{18.54}$$

which is plotted in Figure 18–8. Note at high temperatures that C_V approaches the constant value $\mathcal{N} k_\mathrm{B}$. Unlike the two-state system, the harmonic oscillator has an infinite number of states, so U continues to increase with T as described by Eq. (18.51).

Similarly, the entropy does not saturate as T increases, as it would for subsystems having a finite number of states. Figure 18–9 shows a plot of entropy versus temperature.

As T increases from zero, S remains practically zero until the first excited state becomes significantly occupied. Equation (18.46) shows that

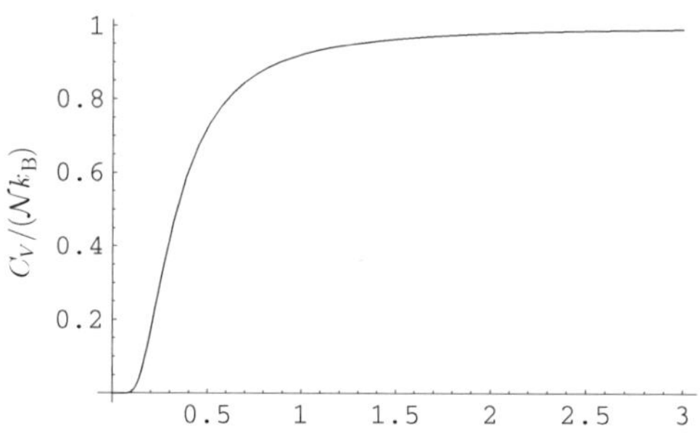

FIGURE 18–8 Dimensionless heat capacity $C_V/(\mathcal{N}k_\mathrm{B})$ versus dimensionless temperature $k_\mathrm{B}T/(\hbar\omega) = 1/x$ for one-dimensional harmonic oscillators according to Eq. (18.54). At high temperatures, C_V tends to a constant, $\mathcal{N}k_\mathrm{B}$, a behavior very different from that of two-state subsystems, Figure 18–6.

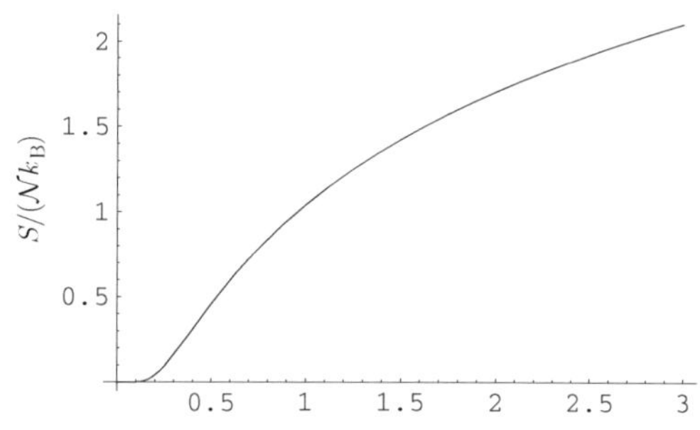

FIGURE 18–9 Dimensionless entropy $S/(\mathcal{N}k_\mathrm{B})$ versus dimensionless temperature $k_\mathrm{B}T/(\hbar\omega)$ for one-dimensional harmonic oscillators according to Eq. (18.46). At low temperatures, S remains near zero until the first excited state is populated. At high temperatures, S continues to increase with T because there is an infinite number of states to occupy.

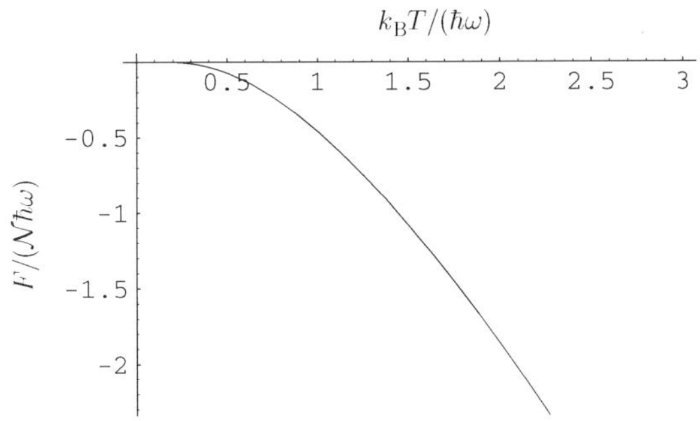

FIGURE 18–10 Dimensionless Helmholtz free energy $F/(\mathcal{N}\hbar\omega)$ versus dimensionless temperature $k_BT/(\hbar\omega)$ for one-dimensional harmonic oscillators according to Eq. (18.45). F decreases with increasing T at an ever increasing rate, as described by Eq. (18.56).

$$S \approx \mathcal{N}k_B\left[1+\ln(k_BT/\hbar\omega)\right], \quad \text{high } T. \tag{18.55}$$

Figure 18–10 shows a plot of the Helmholtz free energy versus temperature. Since $\partial F/\partial T = -S < 0$, F decreases with increasing T. From Eq. (18.45) we see that it diverges logarithmically, that is,

$$F \approx -\mathcal{N}k_BT\ln(k_BT/\hbar\omega), \quad \text{high } T. \tag{18.56}$$

18.3.1 Application: Heat Capacity of a Crystal

The heat capacity of a one-dimensional crystal can be modeled by considering a system made up of one-dimensional harmonic oscillators. Atoms in a crystal vibrate about their equilibrium positions with increasing amplitudes as the temperature increases. The Einstein model is a simple model based on the idea that the solid can be characterized by a harmonic oscillator of a single effective frequency, ω_E, the Einstein frequency. Thus the heat capacity C_E is given by Eq. (18.54) which we rewrite in the form

$$C_E = \mathcal{N}k_B\frac{x_E^2\exp(x_E)}{[\exp(x_E)-1]^2}; \quad x_E := \frac{T_E}{T}, \tag{18.57}$$

where the Einstein temperature $T_E := \hbar\omega_E/k_B$. Of course a graph of $C_E/(\mathcal{N}k_B)$ versus T/T_E looks just like Figure 18–8. The point of inflection is located at about $T = 0.4261T_E$, so T_E is roughly at the knee of the curve, after which C_E is practically constant. The Einstein model yields a curve with about the right shape, but it is wrong in detail at low temperatures. For a three-dimensional solid, the corresponding heat capacity would be larger by a factor of 3 because oscillations in different directions are decoupled.

A better model can be based on a treatment that allows for vibrating atoms to be coupled to one another. In solid state physics courses, it is shown that oscillations of the atoms can be described in terms of a set of spatially delocalized waves, each with its own frequency. Furthermore, each of these waves has the same nondegenerate energy levels as a one-dimensional harmonic oscillator at some appropriate frequency. For nearest neighbor interactions only, it can be shown that the allowed angular frequencies are distributed according to a distribution function

$$\mathcal{D}(\omega) = \frac{2\mathcal{N}}{\pi} \frac{1}{\sqrt{1-(\omega/\omega_0)^2}} \frac{1}{\omega_0} \text{ for } \omega \le \omega_0; \quad \mathcal{D}(\omega) = 0 \text{ for } \omega > \omega_0. \tag{18.58}$$

Thus, the number of oscillators that have frequencies between ω and $\omega + \mathrm{d}\omega$ is $\mathcal{D}(\omega)\,\mathrm{d}\omega$. Here, ω_0 represents the maximum frequency of any oscillator, and can be related to the "spring constant" and mass of a vibrating atom. The function $\mathcal{D}(\omega)$ is normalized so that

$$\int_0^{\omega_0} \mathcal{D}(\omega)\,\mathrm{d}\omega = \mathcal{N}. \tag{18.59}$$

To get the total internal energy, we form the integral

$$U = \int_0^{\omega_0} \mathcal{D}(\omega)\langle n\rangle\hbar\omega\,\mathrm{d}\omega = \int_0^{\omega_0} \mathcal{D}(\omega)\frac{\hbar\omega}{\exp(\hbar\omega/k_B T)-1}\,\mathrm{d}\omega, \tag{18.60}$$

where Eq. (18.48) has been used. At high temperatures, we can expand the exponential in Eq. (18.60) to get

$$U \approx \int_0^{\omega_0} \mathcal{D}(\omega)\,k_B T\,\mathrm{d}\omega = \mathcal{N}k_B T, \tag{18.61}$$

independent of the details of $\mathcal{D}(\omega)$, the same result as Eq. (18.51).

At any temperature, we can calculate the heat capacity

$$C = \frac{\partial U}{\partial T} = k_B \int_0^{\omega_0} \mathcal{D}(\omega)\frac{\exp(\hbar\omega/k_B T)}{(\exp(\hbar\omega/k_B T)-1)^2}\left(\frac{\hbar\omega}{k_B T}\right)^2 \mathrm{d}\omega. \tag{18.62}$$

By substituting $y = \hbar\omega/k_B T$ into this integral, we obtain

$$C = \mathcal{N}k_B \frac{2}{\pi}\left(\frac{k_B T}{\hbar\omega_0}\right)\int_0^{y_0} \frac{y^2\,\mathrm{e}^y}{(\mathrm{e}^y-1)^2}\frac{1}{\sqrt{1-(y/y_0)^2}}\,\mathrm{d}y, \tag{18.63}$$

where $y_0 = \hbar\omega_0/k_B T$. At very low temperatures, we have $y_0 \approx \infty$ and the integral can be evaluated to yield $\pi^2/3$. Therefore, we obtain

$$C = \mathcal{N}k_B \frac{2\pi}{3}\left(\frac{k_B T}{\hbar\omega_0}\right), \quad \text{low } T. \tag{18.64}$$

Equation (18.64) shows that C is linear in T at low T and not exponentially small, as it would be for the Einstein model (see Eq. (18.54) for large x). In three dimensions, calculations along similar lines show that $C = 3\mathcal{N}k_B T$ at high T and $C \propto T^3$ at low T.

18.3.2 Application: Blackbody Radiation

Planck [55, 56] reasoned that radiation from a very small hole in a cavity, which is known as **blackbody radiation**, could be treated by assuming that the radiation was in equilibrium with harmonic oscillators that make up the vibrating atoms of the cavity. In particular, Planck assumed that the energy of that radiation at frequency ν could only be emitted in amounts $h\nu$ where $h = 6.626 \times 10^{-34}\,\mathrm{m^2\,kg\,s^{-1}}$ is what we now call Planck's constant. This turned out to be an inspired guess that can now be fully justified by quantum mechanics. The name "blackbody" stems from the fact that any radiation that enters the very small hole will reflect many times from the cavity walls and is very unlikely to exit, so the body behaves like a nearly perfect absorber.[17]

Radiation is made up of electromagnetic waves having electric and magnetic vectors perpendicular to their direction of propagation. The electric field for such a wave can be represented in complex notation by

$$\mathbf{E} = \mathbf{E}_0\, \mathrm{e}^{\mathrm{i}(\mathbf{k}\cdot\mathbf{r}-\omega t)}, \tag{18.65}$$

where it is understood that we will take the real part to get the actual field. Here, $\mathbf{E}_0$ is a complex amplitude, $\mathbf{r} = (x, y, z)$ is the position vector in Cartesian coordinates, $\mathbf{k} = (k_x, k_y, k_z)$ is a wave vector that points in the direction of propagation, ω is an angular frequency, and t is time. This field must satisfy the wave equation

$$\nabla^2\mathbf{E} = \frac{1}{c^2}\frac{\partial^2\mathbf{E}}{\partial t^2}, \tag{18.66}$$

which results in[18]

$$\omega = ck, \tag{18.67}$$

where $k = (k_x^2 + k_y^2 + k_z^2)^{1/2}$ is the magnitude of the wave vector. The electric field must also satisfy $\nabla \cdot \mathbf{E} = 0$ which requires $\mathbf{k} \cdot \mathbf{E}_0 = 0$, so $\mathbf{E}$ is perpendicular to the direction of propagation. Thus, there are two independent modes, known as polarizations, corresponding to two orthogonal orientations of the electric field in the plane perpendicular to the direction of propagation. Accompanying the electric field given by Eq. (18.65) is a magnetic field $\mathbf{B}$ that can be written in the same form. Then from the Maxwell equation $\nabla \times \mathbf{E} = -(1/c)\partial\mathbf{B}/\partial t$, we can deduce $\mathbf{B} = \hat{\mathbf{k}} \times \mathbf{E}$, where $\hat{\mathbf{k}} = \mathbf{k}/k$ is a unit vector in the direction of propagation. This shows that the corresponding magnetic field is at right angles to the electric field and in phase.

We must also apply boundary conditions to account for the walls of the cavity. This can be done simply by assuming the cavity to be a cubical box of edge length L whose edges are parallel to Cartesian axes. This idealization is meaningful because the radiation emitted by two different blackbodies at the same temperature must be the same; otherwise, radiant energy could be transmitted from one body to another in the absence of a temperature difference, a violation of the second law of thermodynamics. Moreover, this must be true in each frequency range by means of the same argument with the addition of a filter to eliminate other frequencies. We could use real functions for the fields and make them vanish on the walls of the box, but for traveling waves of the form of Eq. (18.65) it easier

[17]In German, it is known as hohlraum, literally hollow space, or cavity.

[18]By working in Cartesian coordinates, it is easy to show that $\nabla \cdot \mathbf{E} = i\mathbf{k} \cdot \mathbf{E}$, $\nabla^2\mathbf{E} = -k^2\mathbf{E}$ and $\nabla \times \mathbf{E} = i\mathbf{k} \times \mathbf{E}$.

to use periodic boundary conditions. Thus, we require $\mathbf{E}(x, y, z, t) = \mathbf{E}(x + L, y, z, t)$ and similarly for the y- and z-directions to deduce

$$k_x = 2\pi n_x/L; \quad k_y = 2\pi n_y/L; \quad k_z = 2\pi n_z/L, \tag{18.68}$$

where n_x, n_y, and n_z are integers, both positive and negative.[19] Thus, the frequencies $\omega(k) = \omega(|\mathbf{k}|)$ of the modes are known and form a discrete set. However, it is important to recognize that this is all based on classical electromagnetic theory and has nothing to do with quantum mechanics. The quantization actually enters from the quantum theory of fields which is beyond the scope of this book.[20] According to that theory, the allowed energies of each mode are given by $nh\nu = n\hbar\omega$, where $n = 0, 1, 2, \ldots$ is an integer, just as for a harmonic oscillator of that same angular frequency, in agreement with Planck's inspired hypothesis.

We can therefore use Eq. (18.48) for the thermal average, $\langle n(T)\rangle$, of the quantum number n. Thus, the thermal energy of the radiation is given by

$$U = 2\sum_{\mathbf{k}} \frac{\hbar\omega(|\mathbf{k}|)}{\exp(\hbar\omega(|\mathbf{k}|)/k_B T) - 1}, \tag{18.69}$$

where $\omega(|\mathbf{k}|)$ is the frequency of an electromagnetic wave having wave vector $\mathbf{k}$. This sum can be converted to an integral by recognizing that for sufficiently large L the allowed values of $\mathbf{k}$ are closely spaced and therefore virtually continuous. According to Eq. (18.67), for each polarization there will be one mode for each volume of $\mathbf{k}$ space of size $\Delta k_x \Delta k_y \Delta k_z = (2\pi/L)^3 = (2\pi)^3/V$, where V is the volume of the box. Thus, for some function $\mathcal{F}(\mathbf{k})$,

$$\sum_{\mathbf{k}} \mathcal{F}(\mathbf{k}) = \sum_{k_x}\sum_{k_y}\sum_{k_z} \frac{V}{(2\pi)^3} \Delta k_x \Delta k_y \Delta k_z\, \mathcal{F}(\mathbf{k}) \to \frac{V}{(2\pi)^3} \int_{\text{all } \mathbf{k}} \mathrm{d}^3k\, \mathcal{F}(\mathbf{k}). \tag{18.70}$$

But for the special case in which $\mathcal{F}$ depends only on the magnitude of $\mathbf{k}$, we can use spherical coordinates in $\mathbf{k}$ space with volume element $\mathrm{d}^3k = 4\pi k^2\, \mathrm{d}k$ to obtain the well-known result

$$\sum_{\mathbf{k}} \mathcal{F}(|\mathbf{k}|) \to \frac{V}{(2\pi)^3} \int_0^\infty 4\pi k^2\, \mathrm{d}k\, \mathcal{F}(|\mathbf{k}|). \tag{18.71}$$

Finally, in the case that $\mathcal{F}(|\mathbf{k}|) = f(\omega(|\mathbf{k}|))$, we can convert to an integral over $\omega = ck$ by using $\mathrm{d}k = (\mathrm{d}k/\mathrm{d}\omega)\, \mathrm{d}\omega = \mathrm{d}\omega/c$ to obtain

$$\sum_{\mathbf{k}} \mathcal{F}(|\mathbf{k}|) \to \frac{V}{2\pi^2} \frac{1}{c^3} \int_0^\infty \omega^2\, \mathrm{d}\omega f(\omega). \tag{18.72}$$

[19] If real functions that vanish at the walls of the box are used, these wave vectors are reduced by a factor of 2, but the integers are only positive. This leads to a different density of modes in $\mathbf{k}$ space that is eight times larger, so the final outcome of calculations will be the same. See Example Problem 16.2 and Section 23.1 for treatment of a rectangular box and more detail in a related context.

[20] See, for example, Schiff [57, p. 517].

Thus, Eq. (18.69) becomes

$$U = \frac{V}{\pi^2}\frac{1}{c^3}\int_0^\infty \omega^2\,\mathrm{d}\omega\frac{\hbar\omega}{\exp(\hbar\omega/k_BT)-1}. \tag{18.73}$$

We substitute $x = \hbar\omega/k_BT$ in this integral to obtain

$$U = \frac{V(k_BT)^4}{\pi^2c^3\hbar^3}\int_0^\infty \mathrm{d}x\frac{x^3}{\mathrm{e}^x-1}. \tag{18.74}$$

The value of the integral turns out to be $\pi^4/15$, so the energy density

$$u_V := \frac{U}{V} = \frac{(k_BT)^4\pi^2}{15c^3\hbar^3}. \tag{18.75}$$

Equation (18.74) can be used to calculate the flux, J, of blackbody radiation from a small hole in the cavity by recognizing that the radiation propagates at speed c and falls onto a given area from all directions that point from a hemisphere to its center. The flux results from the normal component of that radiation, so $J = cu_Vf_{\text{geo}}$, where the geometrical factor

$$f_{\text{geo}} := \frac{1}{4\pi}\int_0^{2\pi}\mathrm{d}\varphi\int_0^{\pi/2}\sin\theta\cos\theta\,\mathrm{d}\theta = \frac{1}{4}. \tag{18.76}$$

Thus

$$J = \frac{1}{4}cu_V = \sigma_{\text{SB}}T^4, \tag{18.77}$$

where

$$\sigma_{\text{SB}} := \frac{\pi^2k_B^4}{60\hbar^3c^2} = 4.67\times10^{-8}\text{ watt m}^2\text{ K}^{-4} \tag{18.78}$$

is known as the **Stefan-Boltzmann constant**. This T^4 law for blackbody radiation has been confirmed experimentally.

We can also deduce the **spectral distribution** of blackbody radiation by multiplying Eq. (18.73) by $c/4V$ and extracting the integrand to obtain

$$j_\omega = \frac{\hbar}{4\pi^2c^2}\frac{\omega^3}{\exp(\hbar\omega/k_BT)-1};\quad \int_0^\infty j_\omega\,\mathrm{d}\omega = J. \tag{18.79}$$

The quantity $j_\omega\,\mathrm{d}\omega$ is therefore the power per unit area of radiation emitted in the angular frequency interval $\mathrm{d}\omega$ centered about ω.

We can investigate the shape of this spectral distribution as a function of temperature by introducing an arbitrary reference temperature T_0, a dimensionless temperature $t := T/T_0$, and a dimensionless angular frequency $W := \hbar\omega/k_BT_0$. Then

$$j_\omega\,\mathrm{d}\omega = \mathcal{J}_W\,\mathrm{d}W = \frac{15}{\pi^4}\sigma_{\text{SB}}T_0^4\frac{W^3}{\exp(W/t)-1}\,\mathrm{d}W. \tag{18.80}$$

Figure 18–11 shows a plot of the dimensionless spectral distribution of radiation according to Eq. (18.80) as a function of dimensionless frequency, W, for three dimensionless

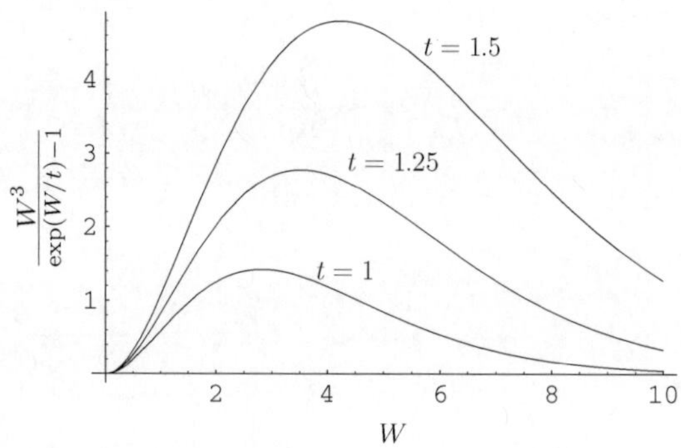

FIGURE 18–11 Plot of the dimensionless spectral distribution according to Eq. (18.80) as a function of dimensionless frequency, W, for three dimensionless temperatures, $t = 1$, $t = 1.25$, and $t = 1.5$. The peaks of these curves increase in height as $1.42\,t^3$, broaden in proportion to t, and move to higher frequencies $W = 2.82\,t$ with increasing t. The area under each curve is $(\pi^4/15)t^4$.

temperatures, t. As t increases, the peaks of these curves increase in height and move to higher frequencies. The peaks occur for $\hbar\omega/k_B T = W/t = 2.82$, which is evident from the lower curve for which $T = T_0$. This is sometimes referred to as **Wien's displacement law** and can be used to estimate the temperature of stars from their dominant frequency of radiation. The peak heights of the curves in Figure 18–11 are $1.42\,t^3$ and the area under each curve is $(\pi^4/15)t^4$, so $\int_0^\infty \mathcal{J}_W\,\mathrm{d}W = \sigma_{\mathrm{SB}} T^4 = J$, in agreement with Eq. (18.77).

Note that the spectral distribution of radiation given by Eq. (18.80) is in agreement with the following familiar observation: As a body is heated to higher and higher temperatures, it begins to glow, first a dull cherry red, then a somewhat brighter orange, then yellow, then white, then bluish white, with ever increasing intensity.

Returning to Eq. (18.79), we note in the classical limit $\hbar \to 0$ that $\hbar\omega/[\exp(\hbar\omega/k_B T) - 1] \to k_B T$, so

$$\int_0^\infty j_\omega\,\mathrm{d}\omega = \frac{1}{4\pi^2 c^2}\int_0^\infty k_B T\omega^2\,\mathrm{d}\omega = 2\pi c\int_0^\infty k_B T\lambda^{-4}\,\mathrm{d}\lambda, \tag{18.81}$$

where the wavelength $\lambda = 2\pi c/\omega$. These integrals do not converge, the former at $\omega \to \infty$ and the latter at $\lambda \to 0$. Prior to the advent of quantum mechanics, this was known as the ultraviolet catastrophe. Quantum mechanics resolves this problem at large ω because $j_\omega \sim \omega^3 \exp(-\hbar\omega/k_B T)$ which is strongly damped because of the extremely low probability of exciting the high energy quanta $\hbar\omega$. Planck's energy quantization hypothesis [55, 56] was the key to removing this singularity and stimulated the development of quantum theory.[21]

[21]The citation of Planck's 1918 Nobel Prize in Physics reads: "In recognition of the services he rendered to the advancement of Physics by his discovery of energy quanta."

18.4 Rigid Linear Rotator

We consider a system for which each particle is a rigid linear rotator, such as a diatomic molecule with only two degrees of rotational freedom. See Section 21.3.2 and Appendix F for context and more detail. The quantized energy levels are

$$\varepsilon(j) = j(j+1)\varepsilon_0, \tag{18.82}$$

where j is zero or a positive integer, that is, $j = 0, 1, 2, \ldots$ and the constant $\varepsilon_0 = \hbar^2/2\mathcal{I}$. Here, $\mathcal{I}$ is the moment of inertia[22] of the rotator with two degrees of freedom having principal moments of inertia $(\mathcal{I}, \mathcal{I}, 0)$. In this case, the energy levels are degenerate, each corresponding to $2j+1$ states. The partition function for each particle is therefore

$$z = \sum_{j=0}^{\infty}(2j+1)\exp[-j(j+1)x], \tag{18.83}$$

where $x := \varepsilon_0/k_\mathrm{B}T$. In Eq. (18.83), it is important to note that the degeneracy factor $2j+1$ appears because the partition function is a sum over quantum *states*, not just energy levels.

For high temperatures, x is small and the energy levels are practically continuous. We can therefore replace the sum over j by an integral. Thus

$$z \approx \int_0^{\infty}(2j+1)\exp[-j(j+1)x]\,\mathrm{d}j. \tag{18.84}$$

We set $y = j(j+1)x$ in which case $\mathrm{d}y = (2j+1)x\,\mathrm{d}j$ and Eq. (18.84) becomes

$$z = \frac{1}{x}\int_0^{\infty}\exp[-y]\,\mathrm{d}y = \frac{1}{x} = \frac{k_\mathrm{B}T}{\varepsilon_0} = \frac{1}{\beta\varepsilon_0}. \tag{18.85}$$

This result can also be derived from classical statistical mechanics (see Eq. (20.123)). From Eq. (18.25) we readily obtain

$$U = \mathcal{N}\frac{\partial}{\partial\beta}(\ln\beta + \ln\varepsilon_0) = \frac{\mathcal{N}}{\beta} = \mathcal{N}k_\mathrm{B}T \tag{18.86}$$

independent of ε_0. This turns out to be the same result as would be obtained for a classical rotator. The corresponding heat capacity is $C = \mathcal{N}k_\mathrm{B}$, which explains why diatomic gases have a correspondingly higher heat capacity (by R per mole) than monatomic gases.

For low temperatures, x is very large and the exponential series cuts off very quickly, which leads to

$$z \approx 1 + 3\mathrm{e}^{-2x} + 5\mathrm{e}^{-6x} + \cdots. \tag{18.87}$$

From Eq. (18.25), we obtain

$$\frac{U}{\mathcal{N}\varepsilon_0} = -\frac{\partial}{\partial x}\ln z = \frac{6\mathrm{e}^{-2x} + 30\mathrm{e}^{-6x}}{1 + 3\mathrm{e}^{-2x} + 5\mathrm{e}^{-6x}} + \cdots. \tag{18.88}$$

[22]For a diatomic molecule made up of point masses m_1 and m_2 separated by a distance ℓ_0, $\mathcal{I} = \ell_0^2 m_1 m_2/(m_1 + m_2)$.

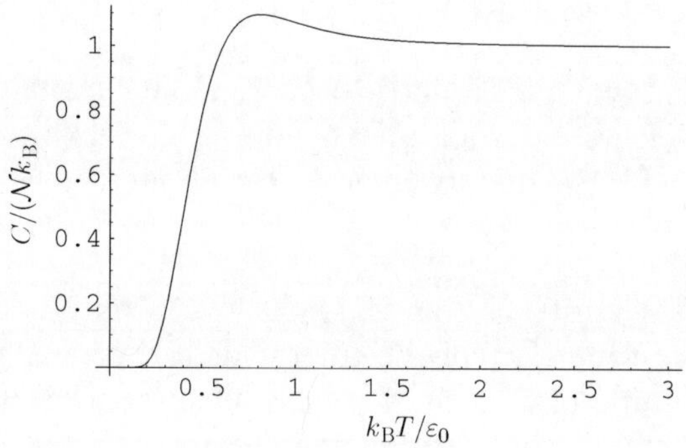

FIGURE 18–12 Plot of the dimensionless heat capacity $C/(\mathcal{N}k_B)$ versus dimensionless temperature k_BT/ε_0 for a linear rigid rotator. Note especially the overshoot of the asymptotic value, which is quite different from the monotonic increase of C for the harmonic oscillator.

To leading order, the low-temperature heat capacity is

$$C = 12\mathcal{N}k_B x^2 e^{-2x}. \tag{18.89}$$

We observe that C vanishes exponentially as $T \to 0$ and therefore rises very slowly as T first increases.

For intermediate values of the temperature, one must resort to series expansions or numerical computations. A plot of the dimensionless heat capacity versus dimensionless temperature is shown in Figure 18–12. Unlike the heat capacity of the harmonic oscillator, which increases monotonically with T, C for the rigid rotator passes through a maximum before becoming asymptotic to its value at high temperatures.

An approximate solution at high temperatures can be obtained by using the Euler-Maclaurin sum formula discussed in Appendix H and results in

$$\frac{C}{Nk_B} = \left(1 + \frac{x^2}{45} + \frac{16x^3}{945} + O(x^4)\right). \tag{18.90}$$

Eq. (18.90) shows clearly that C asymptotes $\mathcal{N}k_B$ from larger values as $T \to \infty$.

19

Canonical Ensemble

In Chapter 16, we introduced the microcanonical ensemble which is based on the fundamental hypothesis that all microstates of an isolated system, compatible with a given macrostate and having a fixed energy and other specified macrovariables, are equally probable. This ensemble is of great theoretical importance but difficult to use because of the formidable problem of counting the number of microstates. We shall therefore use it to derive a more useful ensemble, known as the **canonical ensemble**, that is much more tractable. To do this, we give up a precise knowledge of the energy of our system of interest and specify instead its temperature. Nevertheless, its average energy will still be known to high precision and will play the role of the internal energy of thermodynamics. The temperature of our system can be imposed by contact with a heat reservoir, in which case our system is not isolated. The classical version of this ensemble, discussed in the next chapter, was developed by Gibbs who named it "the distribution of phase called canonical" [4, p. 32].

19.1 Three Derivations

The canonical ensemble can be derived in a number of ways, all of which lead to the same final result in the thermodynamic limit. Because of the importance of this ensemble, we present three derivations, each of which emphasizes an aspect of the ensemble that is not transparent from the others. The methodology of the second derivation will be used in Chapter 21 to derive the grand canonical ensemble and the methodology of the third derivation will be used in Chapter 22 to derive a number of ensembles from a general expression for the entropy.

19.1.1 Derivation from Microcanonical Ensemble I

We derive the canonical ensemble from the microcanonical ensemble by applying the fundamental hypothesis to an isolated total system with fixed energy $E_{\rm T}$, consisting of a reservoir R and a system $\mathcal{I}$ of interest. The system $\mathcal{I}$ may, itself, be very large and consist of a number of subsystems, or particles, that may interact with one another. We assume that the system $\mathcal{I}$ has quantum states $\mathcal{E}_i$ and that its extensive macrovariables, other than energy, are fixed. The index i indicates a specific quantum state, so it actually represents a complete set of quantum numbers.

Suppose that the system $\mathcal{I}$ is in a *definite* quantum state i having energy $\mathcal{E}_i$. Then the reservoir has energy $E_{\rm T} - \mathcal{E}_i$. For the total system, the number of microstates can be expressed as a product of the number of microstates of the reservoir, $\Omega_{\rm R}$, and the number of microstates of the system of interest, Ω, in the form

Thermal Physics. http://dx.doi.org/10.1016/B978-0-12-803304-3.00019-3

$$\Omega_{\rm T}^{i} = \Omega_{\rm R}(E_{\rm T} - \mathcal{E}_i)\Omega(\mathcal{E}_i) = \Omega_{\rm R}(E_{\rm T} - \mathcal{E}_i) \times 1 = \Omega_{\rm R}(E_{\rm T} - \mathcal{E}_i). \tag{19.1}$$

In other words, the system $\mathcal{I}$ is in a single definite microstate, so for the system of interest, $\Omega(\mathcal{E}_i) = 1$; therefore, only the number of microstates of the reservoir must be counted. An equation similar to Eq. (19.1) holds if the system $\mathcal{I}$ is in the quantum state j. As explained in Section 16.1, the probability of a system being in a given macrostate with energy E, volume V, and number of particles $\mathcal{N}$ is proportional to $\Omega(E, V, \mathcal{N})$, which is the sum of its number of equally probable microstates. Therefore, the ratio of the probability P_i of system $\mathcal{I}$ being in the eigenstate i to the probability P_j of system $\mathcal{I}$ being in the eigenstate j is

$$\frac{P_i}{P_j} = \frac{\Omega_{\rm R}(E_{\rm T} - \mathcal{E}_i)}{\Omega_{\rm R}(E_{\rm T} - \mathcal{E}_j)} = \frac{\exp[S_{\rm R}(E_{\rm T} - \mathcal{E}_i)/k_{\rm B}]}{\exp[S_{\rm R}(E_{\rm T} - \mathcal{E}_j)/k_{\rm B}]}, \tag{19.2}$$

where $S_{\rm R}(E_{\rm R})$ is the entropy of the reservoir in a state having energy $E_{\rm R}$. We now assume that the reservoir R is very large so that $|\mathcal{E}_j - \mathcal{E}_i| \ll |E_{\rm T} - \mathcal{E}_j|$ for any states of $\mathcal{I}$. Then by expanding in a Taylor series we obtain[1]

$$\begin{aligned} S_{\rm R}(E_{\rm T} - \mathcal{E}_i) &= S_{\rm R}[(E_{\rm T} - \mathcal{E}_j) + (\mathcal{E}_j - \mathcal{E}_i)] = S_{\rm R}(E_{\rm T} - \mathcal{E}_j) + (\mathcal{E}_j - \mathcal{E}_i)\frac{\partial S_{\rm R}}{\partial E_{\rm R}} + \cdots \\ &= S_{\rm R}(E_{\rm T} - \mathcal{E}_j) + \frac{\mathcal{E}_j - \mathcal{E}_i}{T_{\rm R}} + \cdots, \end{aligned} \tag{19.3}$$

where $T_{\rm R}$ is the temperature of the reservoir. Substitution into Eq. (19.2) and cancellation of the factor $\exp[S_{\rm R}(E_{\rm T} - \mathcal{E}_j)/k_{\rm B}]$ gives

$$\frac{P_i}{P_j} = \frac{\exp(-\mathcal{E}_i/k_{\rm B}T_{\rm R})}{\exp(-\mathcal{E}_j/k_{\rm B}T_{\rm R})}. \tag{19.4}$$

Equation (19.4) states that the probability P_i of system $\mathcal{I}$ being in eigenstate i is proportional to its Boltzmann factor $\exp(-\mathcal{E}_i/k_{\rm B}T)$ where we have dropped the subscript R on T for simplicity.[2] We can obtain a normalized probability by dividing by the total partition function

$$Z = \sum_j \exp(-\beta\mathcal{E}_j) \tag{19.5}$$

to obtain

$$P_i = \frac{\exp(-\beta\mathcal{E}_i)}{Z}, \tag{19.6}$$

where $\beta = 1/(k_{\rm B}T)$. In Eq. (19.5), the sum is over all of the quantum states of the system of interest. Equation (19.5) resembles our former equation for the occupation probabilities $p_i = \exp(-\beta\varepsilon_i)/z$ of weakly interacting distinguishable subsystems except that we are now

[1]The ratio of the second-order term to the first-order term is $-(\mathcal{E}_j - \mathcal{E}_i)/(2C_{\rm R}T_{\rm R})$, where $C_{\rm R}$ is the heat capacity of the reservoir. We assume that $C_{\rm R}$ is so large that this term and higher order terms are negligible. This is essentially the definition of a heat reservoir.

[2]We must still bear in mind, however, that the canonical ensemble applies to a system in contact with a heat reservoir of constant temperature T. Given that other extensive variables of the system are held constant in this derivation, the canonical ensemble will relate thermodynamically to the Helmholtz free energy.

dealing with the states and energy levels of a *whole system*. The internal energy of our system is

$$U = \sum_i P_i \mathcal{E}_i = -\frac{\partial \ln Z}{\partial \beta}, \tag{19.7}$$

which resembles $U = -\mathcal{N}\partial \ln z/\partial\beta$ except that the factor of $\mathcal{N}$ is now missing because we are dealing with Z for the whole system.

Finally, we obtain the Helmholtz free energy F of system $\mathcal{I}$. Since $F = U - TS$ and $S = -\partial F/\partial T$, we see that F satisfies the differential equation

$$F - T\frac{\partial F}{\partial T} = U, \tag{19.8}$$

which, in terms of β, can be rewritten in the form

$$F + \beta\frac{\partial F}{\partial \beta} = -\frac{\partial \ln Z}{\partial \beta}. \tag{19.9}$$

The left-hand side of Eq. (19.9) is recognized immediately to be $\partial(\beta F)/\partial\beta$, so it may be integrated to obtain

$$F = -\frac{1}{\beta}\ln Z + \frac{a}{\beta}, \tag{19.10}$$

where a is a function of integration (independent of β). The entropy is therefore

$$S = \frac{U - F}{T} = \frac{U}{T} + k_B \ln Z - k_B a. \tag{19.11}$$

But when $T \to 0$, only the ground state with degeneracy g_0 and energy $\mathcal{E}_0$ is occupied, so $Z \to g_0 \exp(-\beta\mathcal{E}_0)$ and $\ln Z \to \ln g_0 - \beta\mathcal{E}_0$. Similarly, as $T \to 0$ we have $U \to \mathcal{E}_0$, so Eq. (19.11) becomes

$$S(T \to 0) = k_B \ln g_0 - k_B a. \tag{19.12}$$

Consistent with Eq. (16.2), however, we require[3]

$$S(T \to 0) = k_B \ln g_0, \tag{19.13}$$

which means that the function of integration $a = 0$. Thus Eq. (19.10) becomes

$$F = -\frac{1}{\beta}\ln Z \tag{19.14}$$

[3]Note that the value of $S(T \to 0)$ according to Eq. (19.13) is not zero due to the possibility of degeneracy of the ground state. It would be strictly zero for a nondegenerate ground state for which $g_0 = 1$. If this degeneracy were massive, say of order $g_0 = q^{\mathcal{N}}$, where q is some integer and $\mathcal{N}$ is the number of subsystems or "particles" in the system, then $S(T \to 0)$ would be $\mathcal{N} k_B \ln q$ which would be extensive and significant. Otherwise, $S(T \to 0)$ is practically zero.

which resembles our former result for weakly interacting identical but distinguishable subsystems with $\mathcal{N}$ missing and z replaced by Z. Equation (19.14) can also be written in the form

$$\sum_j \exp(-\beta\mathcal{E}_j) = \exp(-\beta F), \tag{19.15}$$

which shows the relationship between the microscopic picture (on the left) and the macroscopic picture (on the right). Moreover,

$$S = \frac{U}{T} - \frac{F}{T} = k_B\beta U + k_B \ln Z = -k_B \sum_{i=1}^{\kappa} P_i \ln P_i = -k_B\beta^2 \frac{\partial}{\partial\beta}\left[\frac{\ln Z}{\beta}\right]. \tag{19.16}$$

Note that the quantity $-\sum_{i=1}^{\kappa} P_i \ln P_i = D\{P_i\}$ is the disorder function of information theory discussed in Section 15.1.

19.1.2 Derivation from Microcanonical Ensemble II

We give an alternative derivation of the canonical ensemble from the microcanonical ensemble by following the procedure of the preceding section but calculating directly the probability P_i of a given microstate of the system of interest. This probability is the ratio of Ω_T^i given by Eq. (19.1) to the total number of microstates $\Omega_T(E_T)$ when the system of interest is *not restricted* to a specific microstate. Thus

$$P_i = \frac{\Omega_T^i}{\Omega_T(E_T)} = \frac{\Omega_R(E_T - \mathcal{E}_i)}{\Omega_T(E_T)} = \frac{\exp[S_R(E_T - \mathcal{E}_i)/k_B]}{\exp[S_T(E_T)/k_B]}. \tag{19.17}$$

Since the entropy of a composite system is additive, we have

$$S_T(E_T) = S_R(E_T - U) + S(U), \tag{19.18}$$

where U is the (average) internal energy of the system of interest at equilibrium in its unrestricted state. We can therefore recast Eq. (19.17) in the form

$$P_i = \frac{\exp[-S(U)/k_B]\exp[S_R(E_T - \mathcal{E}_i)/k_B]}{\exp[S_R(E_T - U)/k_B]}. \tag{19.19}$$

But

$$S_R\,[E_T - \mathcal{E}_i] = S_R[(E_T - U) + (U - \mathcal{E}_i)] = S_R(E_T - U) + \frac{U - \mathcal{E}_i}{T_R} + \cdots, \tag{19.20}$$

where we have expanded on the basis that $|U - \mathcal{E}_i|/|E_T - U| \ll 1$. Substitution into Eq. (19.19) yields

$$P_i = \exp[-S(U)/k_B]\exp[U/k_B T_R]\exp[-\mathcal{E}_i/k_B T_R]. \tag{19.21}$$

Dropping the subscript on T_R and using $\beta = 1/k_B T$, Eq. (19.21) can be written in the succinct form

$$P_i = \exp(\beta F)\exp(-\beta\mathcal{E}_i), \tag{19.22}$$

where $F = U - TS$ is the Helmholtz free energy. Since $\sum_i P_i = 1$, Eq. (19.22) yields

$$\exp(-\beta F) = \sum_i \exp(-\beta \mathcal{E}_i) = Z \tag{19.23}$$

in agreement with our previous results, Eqs. (19.14) and (19.15).

19.1.3 Derivation III: Most Probable Distribution

In this section, we give yet another derivation of the canonical ensemble but from the point of view of the most probable distribution. We consider a large number $\mathcal{N}_{\text{ens}}$ of identical systems, each having the same volume V and the same number of particles $\mathcal{N}$, and each in a stationary quantum state.[4] These systems constitute the ensemble and they *share* a constant total energy $\mathcal{N}_{\text{ens}}\bar{E}$, where $\bar{E}$ is the average energy per system. $\mathcal{N}_i$ members of the ensemble are in an eigenstate having energy $\mathcal{E}_i$ such that the probability of occurrence of that eigenstate is $P_i = \mathcal{N}_i/\mathcal{N}_{\text{ens}}$. The set $\{\mathcal{N}_i\} = \mathcal{N}_1, \mathcal{N}_2, \ldots, \mathcal{N}_r$ is such that $\sum_{i=1}^{r} \mathcal{N}_i = \mathcal{N}_{\text{ens}}$ which is equivalent to[5]

$$\sum_{i=1}^{r} P_i = 1. \tag{19.24}$$

Since $\sum_{i=1}^{r} \mathcal{N}_i \mathcal{E}_i = \mathcal{N}_{\text{ens}}\bar{E}$, we also have

$$\sum_{i=1}^{r} P_i \mathcal{E}_i = \bar{E}. \tag{19.25}$$

The number of ways of constructing such an ensemble is

$$W_{\text{ens}}\{\mathcal{N}_i\} := \frac{\mathcal{N}_{\text{ens}}!}{\mathcal{N}_1!\mathcal{N}_2!\cdots\mathcal{N}_r!}. \tag{19.26}$$

We would like to choose the set $\{\mathcal{N}_i\}$ to maximize $W_{\text{ens}}\{\mathcal{N}_i\}$ subject to the constraints above to give the most probable distribution. Since $\mathrm{d}\ln x = (1/x)\mathrm{d}x$, the maximum of $\ln x$ occurs at the same value of x as the maximum of x. Therefore, for convenience, we maximize $\ln W_{\text{ens}}$. With the aid of Stirling's approximation we have

$$\ln W_{\text{ens}} = \mathcal{N}_{\text{ens}} \ln \mathcal{N}_{\text{ens}} - \sum_{i=1}^{r} \mathcal{N}_i \ln \mathcal{N}_i = -\mathcal{N}_{\text{ens}} \sum_{i=1}^{r} P_i \ln P_i. \tag{19.27}$$

Since $\mathcal{N}_{\text{ens}}$ is a constant, we can just maximize the function

$$D\{P_i\} = -\sum_{i=1}^{r} P_i \ln P_i, \tag{19.28}$$

[4]If other extensive variables are necessary to specify our system of interest, they are also the same for all members of the ensemble.

[5]If a system has a number of eigenstates r, we certainly need $\mathcal{N}_i > r$ to represent the ensemble. But ultimately we can take the limit $\mathcal{N}_{\text{ens}} \to \infty$ in such a way that $\mathcal{N}_i \to \infty$ but the ratio $P_i = \mathcal{N}_i \mathcal{N}_{\text{ens}}$ remains finite. Thus, there is essentially no problem even if $r \to \infty$.

subject to the constraints Eqs. (19.24) and (19.25). Once the set $\{\mathcal{N}_i\}$ is determined, $W_{\rm ens}$ is the number of microstates of the whole ensemble, so $k_{\rm B} \ln W_{\rm ens}$ represents the entropy of the whole ensemble. Thus, $S = (1/\mathcal{N}_{\rm ens}) k_{\rm B} \ln W_{\rm ens}$ is the entropy per system of the ensemble. It therefore plays the role of the thermodynamic entropy of the system that the ensemble represents. We therefore have

$$S = -k_{\rm B} \sum_{i=1}^{r} P_i \ln P_i = k_{\rm B} D\{P_i\}, \tag{19.29}$$

where $D\{P_i\}$ is seen to be a dimensionless measure of the entropy. We note that $D\{P_i\}$ is the same as the disorder function of Section 15.1, where we have shown (see the first example problem) from its form that it is additive for a composite system. But here we are maximizing $D\{P_i\}$ subject to the additional constraint Eq. (19.25) on the *average* energy of members of an ensemble that have different energies. For the microcanonical ensemble, all members of the ensemble have the same energy.

In the maximization process, we handle the constraints by means of Lagrange multipliers β and α and solve the problem

$$\frac{\partial}{\partial P_j}\left(-\sum_{i=1}^{r} P_i \ln P_i - \beta \sum_{i=1}^{r} P_i \mathcal{E}_i - \alpha \sum_{i=1}^{r} P_i\right) = 0 \tag{19.30}$$

with each P_j now (temporarily) considered to be an independent variable. We obtain

$$-\ln P_j - 1 - \beta \mathcal{E}_j - \alpha = 0, \tag{19.31}$$

which may be exponentiated to give

$$P_j = {\rm e}^{-\alpha-1}\,{\rm e}^{-\beta \mathcal{E}_j}. \tag{19.32}$$

Summing Eq. (19.32) over all values of j and applying the constraint Eq. (19.24) allows us to determine that $\exp(-\alpha - 1) = 1/Z$, where $Z = \sum_j \exp(-\beta \mathcal{E}_j)$ is the partition function as given by Eq. (19.5). Therefore, Eq. (19.32) becomes

$$P_j = \frac{\exp(-\beta \mathcal{E}_j)}{Z}. \tag{19.33}$$

It remains to determine the Lagrange multiplier β. Formally, this could be done in terms of $\bar{E}$ by satisfying the constraint Eq. (19.25) but this would lead to a difficult transcendental equation for β. Therefore, one takes instead an alternative approach by appealing to thermodynamics which allows β to be identified as a physical quantity. To strengthen this identification, we recognize that the energies $\mathcal{E}_i$ of the eigenstates depend on the volume V of the system[6] and its number of particles $\mathcal{N}$. Then

[6]This is convenient but not essential to the identification of β. It simply allows the system to do reversible work $\delta W = p\,{\rm d}V$. If $\mathcal{E}_i$ were to depend on a set of extensive mechanical parameters Y_j instead of just V, one could write the reversible work in the form $\sum_j f_j\,{\rm d}Y_j$, where the f_j are generalized forces. Then $f_j = -\sum_i P_i \partial \mathcal{E}_i / \partial Y_j$.

$$\mathrm{d}\bar{E} = \sum_{i=1}^{r} \mathcal{E}_i \,\mathrm{d}P_i + \sum_{i=1}^{r} P_i \,\mathrm{d}\mathcal{E}_i = \sum_{i} \mathcal{E}_i \,\mathrm{d}P_i + \sum_{i=1}^{r} P_i \frac{\partial \mathcal{E}_i}{\partial V} \,\mathrm{d}V + \sum_{i=1}^{r} P_i \frac{\partial \mathcal{E}_i}{\partial \mathcal{N}} \,\mathrm{d}\mathcal{N}. \tag{19.34}$$

From Eq. (19.29), the differential of the entropy is

$$\mathrm{d}S = -k_\mathrm{B} \sum_{i=1}^{r} (1 + \ln P_i) \,\mathrm{d}P_i = -k_\mathrm{B} \sum_{i=1}^{r} \ln P_i \,\mathrm{d}P_i = k_\mathrm{B}\beta \sum_{i=1}^{r} \mathcal{E}_i \,\mathrm{d}P_i, \tag{19.35}$$

where $\sum_i \mathrm{d}P_i = 0$ has been used in the second and third steps. Substitution of Eq. (19.35) into Eq. (19.34) then gives

$$\mathrm{d}\bar{E} = (k_\mathrm{B}\beta)^{-1} \,\mathrm{d}S + \sum_{i=1}^{r} P_i \frac{\partial \mathcal{E}_i}{\partial V} \,\mathrm{d}V + \sum_{i=1}^{r} P_i \frac{\partial \mathcal{E}_i}{\partial \mathcal{N}} \,\mathrm{d}\mathcal{N}. \tag{19.36}$$

Comparison of Eq. (19.36) with $\mathrm{d}U = T\,\mathrm{d}S - p\,\mathrm{d}V + \mu\,\mathrm{d}\mathcal{N}$ and the identification $\bar{E} = U$ shows that $\beta = 1/(k_\mathrm{B}T)$ as expected. We also deduce

$$p = -\sum_{i=1}^{r} P_i \frac{\partial \mathcal{E}_i}{\partial V}; \quad \mu = \sum_{i=1}^{r} P_i \frac{\partial \mathcal{E}_i}{\partial \mathcal{N}}. \tag{19.37}$$

According to Eq. (19.37), the pressure p can be interpreted heuristically as if $\partial \mathcal{E}_i/\partial V$ were a force per unit area associated with each state and $\partial \mathcal{E}_i/\partial \mathcal{N}$ were an energy per particle associated with each state. From the forms of Eqs. (19.35) and (19.36), we see that a change $\mathrm{d}S$ in entropy results from a change in populations P_i at fixed $\mathcal{E}_i$; however, reversible work results from a change $\mathrm{d}\bar{E}$ of energy at constant S, and therefore from a change of $\mathcal{E}_i$ at constant population P_i and fixed particle number $\mathcal{N}$. Similarly, the chemical potential μ results from a change in $\mathcal{E}_i$ with $\mathcal{N}$ at constant population P_i and fixed V.

Recognizing that the philosophy of this ensemble is to specify T and take whatever $\bar{E}$ corresponds, we return to the notation of thermodynamics and write Eq. (19.25) in the form

$$U = \sum_{i=1}^{r} P_i \mathcal{E}_i, \tag{19.38}$$

where the summation is over all states. From Eqs. (19.29) and (19.33) we deduce that

$$TS = -k_\mathrm{B}T \sum_{i=1}^{r} P_i(-\beta\mathcal{E}_i - \ln Z) = U + k_\mathrm{B}T \ln Z. \tag{19.39}$$

Thus the Helmholtz free energy

$$F = U - TS = -k_\mathrm{B}T \ln Z \tag{19.40}$$

in agreement with Eq. (19.14) or (19.23).

As an alternative procedure, we could have identified β by comparing $\mathrm{d}U$ with $\mathrm{d}S$ at constant V and $\mathcal{N}$, in which case $\mathrm{d}U = T\,\mathrm{d}S$. Then we could calculate S from Eq. (19.39) and the results in Eq. (19.37) could be obtained from $p = -\partial F/\partial V$, $\mu = -\partial F/\partial \mathcal{N}$ and Eq. (19.33).

Before leaving this section, we remark that instead of the most probable values of P_j or, equivalently $\mathcal{N}_j = \mathcal{N}_{\text{ens}} P_j$, we could deal with the *mean* values $\langle \mathcal{N}_j \rangle$ with respect to the quantities $W_{\text{ens}}\{\mathcal{N}_i\}$ given by Eq. (19.26). Specifically,

$$\langle \mathcal{N}_j \rangle = \frac{\sum_{\{\mathcal{N}_i\}} \mathcal{N}_j W_{\text{ens}}\{\mathcal{N}_i\}}{\sum_{\{\mathcal{N}_i\}} W_{\text{ens}}\{\mathcal{N}_i\}}, \tag{19.41}$$

where the sums are to be taken over *all* values of the set $\{\mathcal{N}_i\}$ that are compatible with the constraint Eqs. (19.24) and (19.25), written in terms of the $\mathcal{N}_i$. By means of a somewhat technical and lengthy calculation (e.g., see Schrödinger [99, p. 27] or Pathria [8, p. 46]), it can be shown that $\langle \mathcal{N}_j \rangle$ and $\mathcal{N}_j$ calculated for the most probable distribution are the same in the limit $\mathcal{N}_{\text{ens}} \to \infty$.

19.2 Factorization Theorem

If our system of interest can be decomposed into a number M of *distinguishable elements* that have *negligible interaction energy* and whose *quantum states can be occupied independently* of the occupation of the quantum states of the other elements, then the partition function of the system factors into the product of partition functions of the elements. Thus

$$Z = \prod_{\ell=1}^{M} Z^{(\ell)}, \tag{19.42}$$

where $Z^{(\ell)}$ is the partition function of the element (ℓ).

We shall prove this theorem for two elements but the result can clearly be extended to any number of elements by further decomposition. We replace the single quantum number i by the composite quantum numbers jk and write

$$\mathcal{E}_{jk} = \mathcal{E}_j^{(1)} + \mathcal{E}_k^{(2)}, \tag{19.43}$$

where the superscripts pertain to the two elements. The partition function becomes

$$Z = \sum_{jk} \exp[-\beta \mathcal{E}_{jk}] = \sum_{jk} \exp[-\beta \mathcal{E}_j^{(1)}] \exp[-\beta \mathcal{E}_k^{(2)}] = Z^{(1)} Z^{(2)}, \tag{19.44}$$

where

$$Z^{(1)} = \sum_j \exp[-\beta \mathcal{E}_j^{(1)}]; \quad Z^{(2)} = \sum_k \exp[-\beta \mathcal{E}_k^{(2)}]. \tag{19.45}$$

19.2.1 Distinguishable Particles with Negligible Interaction

We can recover our former results (Chapter 18) for $\mathcal{N}$ very weakly interacting (meaning negligible energy of interaction) identical but distinguishable particles (subsystems) by noting that the energy $\mathcal{E}_i$ of each state is a sum of energies $\varepsilon_n^{(m)}$ of individual particles m in quantum states n. Thus

$$Z = \sum_i \exp(-\beta\mathcal{E}_i) = \sum_{jk\ell\cdots} \exp[-\beta(\varepsilon_j^{(1)} + \varepsilon_k^{(2)} + \varepsilon_\ell^{(3)} + \cdots)] = \prod_m^{\mathcal{N}} \sum_n \exp(-\beta\varepsilon_n^{(m)}) = z^{\mathcal{N}}. \tag{19.46}$$

From Eq. (19.46), we obtain

$$\ln Z = \mathcal{N} \ln z, \quad \text{identical distinguishable particles,} \tag{19.47}$$

and our former equations (see Section 18.1.1 for a summary) for identical but distinguishable particles with negligible interaction energies are recovered. *The reader is encouraged to study the numerous examples in Chapter 18.*

If the particles are identical but not distinguishable, for example particles of an ideal gas that share the same volume, then occupation of individual particle states does not constitute an independent state and the factorization theorem requires modification, as illustrated in the next section for a classical ideal gas. If the particles are identical fermions or identical bosons, their wave functions must obey quantum statistics so the occupation of their quantum states is correlated and factorization of the canonical partition function is not possible. In Chapter 21, we introduce the grand canonical ensemble which enables factorization of the grand partition function for ideal Fermi and Bose gases.

19.3 Classical Ideal Gas

For a classical ideal gas, the identical particles do not interact, but since they *share* the same volume they are not distinguishable. In this case, the simple decomposition that led to Eq. (19.46) is not applicable. This is because an interchange of particles does not constitute a new quantum state. Nevertheless, if the gas is very dilute, in the sense that the number of particles is much smaller than the number of accessible single particle quantum states, the probability of multiply-occupied states will be very small. By **accessible quantum state**, we mean a state whose Boltzmann factor makes a significant contribution to the single-particle partition function at the temperature under consideration. We can therefore correct Eq. (19.46) by dividing by $\mathcal{N}!$ which is the number of permutations of $\mathcal{N}$ particles among $\mathcal{N}$ distinct single particle states. Then approximately

$$Z \approx \frac{z^{\mathcal{N}}}{\mathcal{N}!}, \quad \text{dilute indistinguishable particles,} \tag{19.48}$$

so that

$$\ln Z \approx \mathcal{N} \ln z - \ln \mathcal{N}! \approx \mathcal{N} \ln(z/\mathcal{N}) + \mathcal{N}, \quad \text{dilute indistinguishable particles,} \tag{19.49}$$

where Stirling's approximation for $\ln \mathcal{N}!$ has been used.

Note that Eq. (19.48) is based on

$$\mathcal{E}_i \to \varepsilon_j^{(1)} + \varepsilon_k^{(2)} + \varepsilon_\ell^{(3)} + \cdots \tag{19.50}$$

and the fact that it no longer matters *which* particles (subsystems) are in a given state. If all of the terms on the right-hand side of Eq. (19.50) correspond to *different* single particle states, the result in Eq. (19.46) would be too large by exactly a factor of $\mathcal{N}!$. If some of the single particle states are the same, then $\mathcal{N}!$ would be an overestimate. If, however, the system is dilute in the sense that the probability of multiple occupation of a single particle state is negligible, then $\mathcal{N}!$ is a good estimate of the overcount and Eq. (19.48) holds approximately.[7] The factor $1/\mathcal{N}!$ is the same Gibbs correction factor that we discussed in Section 16.4.1 in connection with the microcanonical ensemble, but now we are in a position to better understand the conditions for its applicability.

19.3.1 Free Particle in a Box

A structureless free particle of mass m in a rectangular box of dimensions H, K, L has eigenstates with energies $\varepsilon = \hbar^2 k^2/2m$ as discussed in Section 16.4.1. For periodic boundary conditions $\psi(x+H,y,z) = \psi(x,y,z)$, $\psi(x,y+K,z) = \psi(x,y,z)$, and $\psi(x,y,z+L) = \psi(x,y,z)$, the wave vector $\mathbf{k}$ is given by Eq. (16.51). We expand our notation and write

$$\varepsilon_{n_x,n_y,n_z} = \frac{\hbar^2}{2m}(2\pi)^2\left[\left(\frac{n_x}{H}\right)^2 + \left(\frac{n_y}{K}\right)^2 + \left(\frac{n_z}{L}\right)^2\right]. \tag{19.51}$$

Then the single particle partition function is

$$\begin{aligned} z &= \sum_{n_x,n_y,n_z} \exp(-\beta\varepsilon_{n_x,n_y,n_z}) \\ &= \sum_{n_x} \exp\left[-\beta\frac{\hbar^2}{2m}\left(\frac{2\pi n_x}{H}\right)^2\right] \sum_{n_y} \exp\left[-\beta\frac{\hbar^2}{2m}\left(\frac{2\pi n_y}{K}\right)^2\right] \sum_{n_z} \exp\left[-\beta\frac{\hbar^2}{2m}\left(\frac{2\pi n_z}{L}\right)^2\right] \\ &= \sum_{k_x} \exp\left[-\beta\frac{\hbar^2}{2m}k_x^2\right] \sum_{k_y} \exp\left[-\beta\frac{\hbar^2}{2m}k_y^2\right] \sum_{k_z} \exp\left[-\beta\frac{\hbar^2}{2m}k_z^2\right]. \end{aligned} \tag{19.52}$$

Equation (19.52) shows that the single particle partition function factors, one factor for each direction in three-dimensional space. Moreover, if $k_\mathrm{B}T$ is large compared to the splittings between states, the sums in Eq. (19.52) can be approximated by integrals, *viz.*,

$$\sum_{n_x} \exp\left[-\beta\frac{\hbar^2}{2m}\left(\frac{2\pi n_x}{H}\right)^2\right] = \sum_{k_x} \exp\left[-\beta\frac{\hbar^2}{2m}k_x^2\right] \approx \frac{H}{2\pi}\int_{-\infty}^{\infty} \mathrm{d}k_x \exp\left[-\beta\frac{\hbar^2}{2m}k_x^2\right]. \tag{19.53}$$

The factor of $H/(2\pi)$ on the right-hand side of Eq. (19.53) arises because k_x changes by $2\pi/H$ as n_x changes by one. Applying Eq. (19.53) to each of the products in Eq. (19.52) gives

[7] An alternative derivation of this result can be based on use of the grand canonical ensemble, which allows the number of particles to be indefinite but specifies the chemical potential. See Section 21.2.4 and Chandler [12, pp. 100-103] for further discussion.

$$z = \frac{HKL}{(2\pi)^3} \int_{-\infty}^{\infty} \mathrm{d}k_x \exp\left[-\beta \frac{\hbar^2}{2m} k_x^2\right] \int_{-\infty}^{\infty} \mathrm{d}k_y \exp\left[-\beta \frac{\hbar^2}{2m} k_y^2\right] \int_{-\infty}^{\infty} \mathrm{d}k_z \exp\left[-\beta \frac{\hbar^2}{2m} k_z^2\right]$$

$$= \frac{V}{(2\pi)^3} \int_{\mathbf{k}} \mathrm{d}^3 k \exp\left[-\beta \frac{\hbar^2}{2m} k^2\right] = V \left(\frac{m k_\mathrm{B} T}{2\pi \hbar^2}\right)^{3/2}. \tag{19.54}$$

In Eq. (19.54), the integral on the second line is over all of **k** space. To get the final result, one can either do the individual Cartesian integrals, each of which has the same value $(m k_\mathrm{B} T/2\pi\hbar^2)^{1/2}$, or do the three-dimensional integral in polar coordinates. The prescription

$$\frac{1}{V} \sum_{\mathbf{k}} \to \frac{1}{(2\pi)^3} \int_{\mathbf{k}} \mathrm{d}^3 k, \tag{19.55}$$

which is valid when the state energies are closely spaced compared to $k_\mathrm{B} T$, becomes exact in the limit $V \to \infty$, in which case the sum on the left-hand side is over an infinite number of states whose separations tend to zero.

Our result in Eq. (19.54) can be written in the form

$$z = V n_\mathrm{Q}, \tag{19.56}$$

where

$$n_\mathrm{Q} := \left(\frac{m k_\mathrm{B} T}{2\pi \hbar^2}\right)^{3/2} \tag{19.57}$$

is known as the **quantum concentration**. The de Broglie wavelength is $\lambda_\mathrm{B} = 2\pi\hbar/p$, where p is the momentum of a particle. If we estimate $p^2/(2m) \sim \pi k_\mathrm{B} T$, we obtain the **thermal wavelength**

$$\lambda_\mathrm{B} \sim \lambda_\mathrm{T} := \left(\frac{2\pi \hbar^2}{m k_\mathrm{B} T}\right)^{1/2}, \tag{19.58}$$

which leads to

$$n_\mathrm{Q} = \frac{1}{\lambda_\mathrm{T}^3}. \tag{19.59}$$

To see when our approximation of a dilute gas is valid, we note that the magnitude of the partition function z is a rough measure of the number of single particle quantum states accessible to a particle at a given temperature. We therefore want z to be much greater than the number of particles, that is, $z/\mathcal{N} \gg 1$. By substituting z from Eq. (19.56), we obtain $n_\mathrm{Q} \gg \mathcal{N}/V =: n$ or alternatively $(\lambda_\mathrm{T}^3) \ll V/\mathcal{N} = 1/n$. In other words, the concentration of the gas must be sufficiently low that the volume per particle, $1/n$ is very large compared to the cube of the thermal wavelength. For situations in which this inequality is not satisfied, quantum effects become important and the particles must be treated either as fermions or bosons, depending on whether their spin is half integral or integral.

Substitution of Eq. (19.56) into Eq. (19.49) yields

$$\ln Z = \mathcal{N}\ln(V/\mathcal{N}) + \mathcal{N}\ln n_Q + \mathcal{N}, \quad \text{ideal gas.} \tag{19.60}$$

The Helmholtz free energy is therefore

$$F = -\mathcal{N}k_B T\ln(V/\mathcal{N}) - \mathcal{N}k_B T\ln n_Q - \mathcal{N}k_B T, \quad \text{ideal gas,} \tag{19.61}$$

which is an extensive function, as required. By applying Eq. (19.7) to $\ln Z$ expressed by Eq. (19.60), we obtain

$$U = -\frac{\partial}{\partial\beta}[\mathcal{N}\ln(V/\mathcal{N}) + \mathcal{N}\ln n_Q + \mathcal{N}] = \frac{3}{2}\mathcal{N}\frac{\partial\ln\beta}{\partial\beta} = \frac{3}{2}\mathcal{N}k_B T, \quad \text{ideal gas,} \tag{19.62}$$

consistent with a constant heat capacity of $C_V = (3/2)\mathcal{N}k_B$ as expected. Note that only the term in n_Q contributes to Eq. (19.62) so this same result would have been obtained if we had used the (incorrect) partition function $Z = z^{\mathcal{N}}$. The pressure can be determined by differentiation of Eq. (19.61) to obtain

$$p = -\frac{\partial F}{\partial V} = \mathcal{N}k_B T\frac{\partial\ln V}{\partial V} = \frac{\mathcal{N}k_B T}{V} = \frac{NRT}{V}, \quad \text{ideal gas,} \tag{19.63}$$

which is recognized as the ideal gas law. Again, this same result would have been obtained if we had used the (incorrect) partition function $Z = z^{\mathcal{N}}$. On the other hand, the entropy can be obtained by differentiation of Eq. (19.61) to obtain

$$S = -\frac{\partial F}{\partial T} = \mathcal{N}k_B[\ln(n_Q V/\mathcal{N}) + (5/2)], \quad \text{ideal gas.} \tag{19.64}$$

Equation (19.64) is known as the **Sackur-Tetrode equation** and requires use of the correct partition function $Z = z^{\mathcal{N}}/\mathcal{N}!$. The entropy constant in Eq. (19.64) has been verified by experiment.[8] Note that this constant depends on $\hbar$ (through n_Q) so its origin is quantum mechanical. Classical thermodynamics alone would yield

$$S = \mathcal{N}k_B[\ln(V/\mathcal{N}) + (3/2)\ln T] + \mathcal{N}s_0, \quad \text{ideal gas,} \tag{19.65}$$

and it would not be possible to determine the constant s_0. Similarly, the chemical potential (per particle) is

$$\mu = \frac{\partial F}{\partial\mathcal{N}} = k_B T\ln[\mathcal{N}/(Vn_Q)] = k_B T\ln[p/(n_Q k_B T)], \quad \text{ideal gas,} \tag{19.66}$$

and requires use of the correct partition function. In the second form of this expression, the quantity $p_Q := n_Q k_B T \propto T^{5/2}$ plays the role of a quantum pressure.

[8] See Fermi [1, chapter VIII] for an excellent discussion of the entropy of mercury vapor.

19.4 Maxwell-Boltzmann Distribution

We can obtain the well-known Maxwell-Boltzmann (MB) distribution function for the velocities of ideal gas molecules by using Eq. (19.51) in the classical limit

$$\frac{\hbar^2 k^2}{2m} \to \frac{1}{2} m v^2, \tag{19.67}$$

where $v^2 = v_x^2 + v_y^2 + v_z^2$. By the argument following Eq. (19.50), we note that the approximate correction factor $1/\mathcal{N}!$ gives equal weighting to every single particle state, so it can be ignored in calculating the probability density function $M(\mathbf{v})$ for the velocity $\mathbf{v} = v_x \hat{\mathbf{i}} + v_y \hat{\mathbf{j}} + v_z \hat{\mathbf{k}}$ of a single particle, which takes the form

$$M(\mathbf{v}) = A \exp\left(-\frac{mv^2}{2k_B T}\right). \tag{19.68}$$

The constant A is to be chosen by normalization. Specifically, $M(\mathbf{v})\,\mathrm{d}^3 v$ is the probability of a gas molecule having a velocity in the infinitesimal volume element $\mathrm{d}^3 v$ centered about $\mathbf{v}$. Therefore, the normalization is

$$1 = \int M(\mathbf{v})\,\mathrm{d}^3 v = 4\pi A \int_0^\infty v^2 \exp\left(-\frac{mv^2}{2k_B T}\right) \mathrm{d}v = A\left(\frac{2\pi k_B T}{m}\right)^{3/2}, \tag{19.69}$$

which leads to

$$M(\mathbf{v}) = \left(\frac{m}{2\pi k_B T}\right)^{3/2} \exp\left(-\frac{mv^2}{2k_B T}\right). \tag{19.70}$$

As shown in Section 20.1, Eq. (19.70) can also be obtained by using the classical canonical ensemble rather than from the classical limit of the quantum mechanical result as presented here.

Equation (19.70) is sketched as a function of $|\mathbf{v}|$ in Figure 19–1a. Note that $M(\mathbf{v})$ is isotropic, and hence depends only on v, the magnitude of $\mathbf{v}$, which we refer to as the

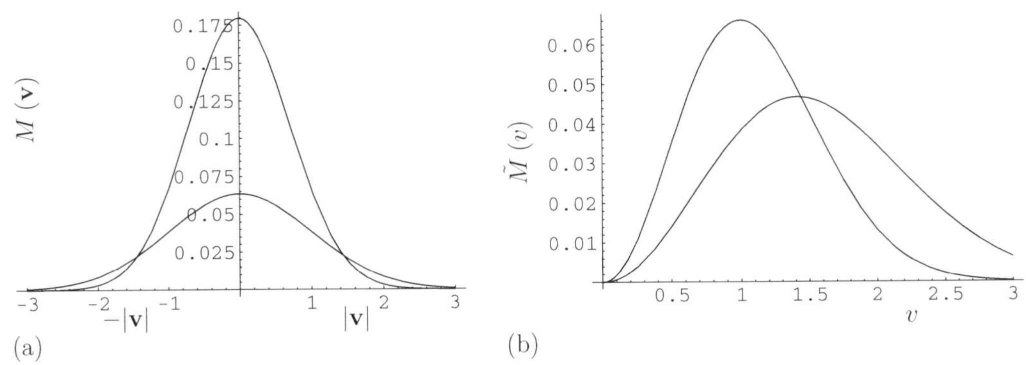

FIGURE 19–1 Maxwell-Boltzmann distributions for an ideal gas. (a) The velocity distribution, $M(\mathbf{v})$, according to Eq. (19.70). (b) The speed distribution $\tilde{M}(v)$ according to Eq. (19.75). In both (a) and (b), for the sake of illustration, the curves with the higher peaks correspond to $2k_B T/m = 1$ and those with the lower peaks to $2k_B T/m = 2$, in arbitrary units.

speed. The form on the right-hand side of Eq. (19.70) is known as a **normalized Gaussian distribution**. The mean *velocity* is

$$\langle \mathbf{v} \rangle := \int \mathbf{v} M(\mathbf{v}) \, \mathrm{d}^3 v = \left(\frac{m}{2\pi k_{\mathrm{B}} T} \right)^{3/2} \int \mathbf{v} \exp\left(-\frac{m v^2}{2 k_{\mathrm{B}} T} \right) \mathrm{d}^3 v = 0. \tag{19.71}$$

This can be seen by writing $\mathbf{v} = v_x \hat{\mathbf{i}} + v_y \hat{\mathbf{j}} + v_z \hat{\mathbf{k}}$, $v^2 = v_x^2 + v_y^2 + v_z^2$, $\mathrm{d}^3 v = \mathrm{d}v_x \, \mathrm{d}v_y \, \mathrm{d}v_z$ and doing the integrals in Cartesian coordinates, *viz.*,

$$\int_{-\infty}^{\infty} v_x \exp\left(-\frac{m v_x^2}{2 k_{\mathrm{B}} T} \right) \mathrm{d}v_x = 0. \tag{19.72}$$

The integral vanishes because the integrand is a product of an odd function and an even function of v_x. In fact, the velocity distribution factors into normalized distribution functions for each Cartesian velocity component:[9]

$$M(\mathbf{v}) \, \mathrm{d}^3 v = \prod_{i=x,y,z} \left(\frac{m}{2\pi k_{\mathrm{B}} T} \right)^{1/2} \exp\left(-\frac{m v_i^2}{2 k_{\mathrm{B}} T} \right) \mathrm{d}v_i. \tag{19.73}$$

Therefore, the average value of any odd power of a velocity component will vanish. The mean squared velocity is

$$\langle v^2 \rangle := \int v^2 M(\mathbf{v}) \, \mathrm{d}^3 v \tag{19.74}$$

and does not vanish. Since v^2 is independent of direction, we can do the integral in spherical polar coordinates, as we did in Eq. (19.69). To facilitate this approach, we write the volume element in the form $v^2 \sin\Theta \, \mathrm{d}\Theta \, \mathrm{d}\Phi \, \mathrm{d}v$ and integrate over angles to define the *speed* distribution function

$$\tilde{M}(v) := \int_0^{2\pi} \mathrm{d}\Phi \int_0^{\pi} \sin\Theta \, \mathrm{d}\Theta \, v^2 M(\mathbf{v}) = 4\pi \left(\frac{m}{2\pi k_{\mathrm{B}} T} \right)^{3/2} v^2 \exp\left(-\frac{m v^2}{2 k_{\mathrm{B}} T} \right). \tag{19.75}$$

$\tilde{M}(v)$ is normalized such that

$$\int_0^{\infty} \tilde{M}(v) \, \mathrm{d}v = 1. \tag{19.76}$$

The speed distribution function $\tilde{M}(v)$ is sketched in Figure 19–1b. Note that this function peaks at a positive value of v because of the v^2 that comes from the volume element $\mathrm{d}^3 v$. $\tilde{M}(v) \, \mathrm{d}v$ is therefore the probability of finding a particle with speed between v and $v + \mathrm{d}v$, or alternatively the probability of finding a particle with velocity in a spherical shell of inner radius v and outer radius $v + \mathrm{d}v$. Equation (19.74) may therefore be written in the form

[9]Given N numbers $a_1, \ldots, a_N$, their product is denoted by $\Pi_{j=1}^{N} a_j = a_1 \times a_2 \times \cdots \times a_N$. Note that if α is constant, this implies that $\Pi_{j=1}^{N} \alpha a_j = \alpha^N \Pi_{j=1}^{N} a_j$.

$$\langle v^2 \rangle = \int_0^\infty v^2\,\tilde{M}(v)\,\mathrm{d}v = 4\pi\left(\frac{m}{2\pi k_\mathrm{B}T}\right)^{3/2}\int_0^\infty v^4 \exp\left(-\frac{mv^2}{2k_\mathrm{B}T}\right)\mathrm{d}v = \frac{3k_\mathrm{B}T}{m}. \tag{19.77}$$

In view of Eq. (19.73), we have $\langle v_x^2\rangle = \langle v_y^2\rangle = \langle v_z^2\rangle = (1/3)\langle v^2\rangle = k_\mathrm{B}T/m$. According to Eq. (19.77), the average kinetic energy is

$$\frac{1}{2}m\langle v^2\rangle = \frac{3}{2}k_\mathrm{B}T. \tag{19.78}$$

Equation (19.78) can be interpreted to mean that there is $(1/2)k_\mathrm{B}T$ of average kinetic energy associated with each of the translational degrees of freedom in the three Cartesian (x, y, z) directions, which is consistent with the principle of equipartition of energy which is valid in the classical limit of high temperatures (see Sections 20.2 and 20.3). The heat capacity of one mole of an ideal gas would therefore be $3R/2$, or about 3 cal/mol. Recall that the average energy of a one-dimensional harmonic oscillator at high temperature is $k_\mathrm{B}T$; in this case, there is also equipartition of energy, but $(1/2)k_\mathrm{B}T$ comes from kinetic energy and $(1/2)k_\mathrm{B}T$ comes from potential energy. Thus the heat capacity of one mole of a solid, which behaves as if each atom were a three-dimensional harmonic oscillator, would be $3R$, or about 6 cal/mol at high temperatures.

Example Problem 19.1. Find the average speed of a particle according to the MB distribution.

Solution 19.1. We use the speed distribution function given by Eq. (19.75) to obtain

$$\langle v \rangle = \int_0^\infty v\tilde{M}(v)\,\mathrm{d}v = 4\pi\left(\frac{m}{2\pi k_\mathrm{B}T}\right)^{3/2}\int_0^\infty v^3 \exp\left(-\frac{mv^2}{2k_\mathrm{B}T}\right)\mathrm{d}v = \left(\frac{8k_\mathrm{B}T}{\pi m}\right)^{1/2}. \tag{19.79}$$

Although the average velocity vanishes, the average of the always-positive speed does not.

Example Problem 19.2. Find the average speed of a particle that moves only in the x-direction according to the MB distribution.

Solution 19.2. We integrate over v_y and v_z the velocity distribution function given by Eq. (19.73) to obtain

$$M(v_x)\,\mathrm{d}v_x = \left(\frac{m}{2\pi k_\mathrm{B}T}\right)^{1/2}\exp\left(-\frac{mv_x^2}{2k_\mathrm{B}T}\right)\mathrm{d}v_x. \tag{19.80}$$

Here, $M(v_x)$ is a velocity distribution function for velocity in the x-direction, normalized on the interval $-\infty$ to ∞. The speed in the x-direction is $|v_x|$ so its average is

$$\langle |v_x| \rangle = \int_{-\infty}^\infty |v_x| M(v_x)\,\mathrm{d}v_x = 2\int_0^\infty v_x\left(\frac{m}{2\pi k_\mathrm{B}T}\right)^{1/2}\exp\left(-\frac{mv_x^2}{2k_\mathrm{B}T}\right)\mathrm{d}v_x. \tag{19.81}$$

This is equivalent to using only positive v_x and renormalizing. The result for the average speed in the x-direction is $\langle |v_x|\rangle = (2k_\mathrm{B}T/\pi m)^{1/2}$, which is not simply related to the average speed in three dimensions because $v = \sqrt{v_x^2 + v_y^2 + v_z^2}$.

19.5 Energy Dispersion

The canonical ensemble is applicable to a system in equilibrium with a heat reservoir such that its temperature T is the same as that of the reservoir. Therefore, the energy of such a system is not precisely fixed, even though its average energy $\langle E\rangle$, which we identify as the internal energy of thermodynamics U, is known and given by Eq. (19.7). In other words, the energy of a system held at constant T has some dispersion and can deviate from its average value, $\langle E\rangle = U$. Dynamically speaking, we can think of the energy of such a system as fluctuating in time. This dispersion can be quantified by calculating higher moments of the energy with respect to the probabilities given by Eq. (19.6). We proceed to calculate its second moment relative to its average value, namely

$$\langle(\Delta E)^2\rangle := \langle(E-U)^2\rangle = \langle E^2 - 2EU + U^2\rangle = \langle E^2\rangle - U^2, \tag{19.82}$$

where

$$\langle E^2\rangle = \sum_i \mathcal{E}_i^2 P_i. \tag{19.83}$$

By differentiation of Eq. (19.5), we note that

$$\frac{\partial^2 Z}{\partial \beta^2} = \sum_i \mathcal{E}_i^2 \exp(-\beta \mathcal{E}_i), \tag{19.84}$$

which yields

$$\langle E^2\rangle = \frac{1}{Z}\left(\frac{\partial^2 Z}{\partial \beta^2}\right)_{\mathcal{N},V}. \tag{19.85}$$

Therefore,

$$\langle(\Delta E)^2\rangle = \frac{1}{Z}\frac{\partial^2 Z}{\partial \beta^2} - \frac{1}{Z^2}\left(\frac{\partial Z}{\partial \beta}\right)^2 = \frac{\partial^2 \ln Z}{\partial \beta^2} = -\left(\frac{\partial U}{\partial \beta}\right)_{\mathcal{N},V}. \tag{19.86}$$

Since $\mathrm{d}T/\mathrm{d}\beta = -k_\mathrm{B}T^2$, this result can also be written

$$\langle(\Delta E)^2\rangle = k_\mathrm{B}T^2 C_V, \tag{19.87}$$

where the heat capacity $C_V = \partial U/\partial T$.

For a system having a large number $\mathcal{N}$ of particles, we can see that this dispersion is quite small in the following sense. We define the heat capacity per particle as $c_V := C_V/\mathcal{N}$ and take the square root of Eq. (19.87) to obtain

$$\frac{\sqrt{\langle(\Delta E)^2\rangle}}{\mathcal{N}} = \frac{\sqrt{k_\mathrm{B}T^2 c_V}}{\sqrt{\mathcal{N}}}, \tag{19.88}$$

where the expression on the left is a measure of the dispersion of energy per particle. Typically, c_V is of the order of k_B, so the right-hand side of Eq. (19.88) is of the order of $k_\mathrm{B}T/\sqrt{\mathcal{N}} = 10^{-11}k_\mathrm{B}T$ for $\mathcal{N} = 10^{22}$. For example, for a monatomic ideal gas, $c_V = (3/2)k_\mathrm{B}$ and Eq. (19.88) becomes

$$\frac{\sqrt{\langle(\Delta E)^2\rangle}}{\mathcal{N}} = \sqrt{3/2}\,\frac{k_B T}{\sqrt{\mathcal{N}}}. \tag{19.89}$$

Alternatively for a monatomic ideal gas, we have $U = (3/2)\mathcal{N}k_B T$ relative to a zero of energy such that $U = 0$ when $T = 0$. In that case, Eq. (19.87) leads to

$$\frac{\sqrt{\langle(\Delta E)^2\rangle}}{U} = \sqrt{2/3}\,\frac{1}{\sqrt{\mathcal{N}}}. \tag{19.90}$$

In any case, as $\mathcal{N} \to \infty$, there is no dispersion of energy, which is the limit in which thermodynamics becomes precise. For the microcanonical ensemble, we regard the energy to be fixed precisely, so the temperature is not precisely defined. Later we shall consider the grand canonical ensemble, for which even the number of particles of a system has dispersion about its average value. In the thermodynamic limit, however, this dispersion also tends to zero.

19.6 Paramagnetism

The phenomenon of paramagnetism pertains to systems that have no net magnetic moment in the absence of an applied magnetic field but acquire a net magnetic moment in the direction of an applied magnetic field. Roughly speaking, it can be thought of as resulting from the alignment of magnetic dipoles when a magnetic field is applied. As the temperature increases at fixed magnetic field strength, entropic effects become more important, the degree of alignment decreases and the net magnetic moment decreases.

We consider a system having a number of particles $\mathcal{N}$ for which the internal energy $U(S, V, B, \mathcal{N})$ can be expressed as a function of the entropy S, the volume V, and the magnetic field strength[10] B. Thus,

$$\mathrm{d}U = T\,\mathrm{d}S - p\,\mathrm{d}V + \left(\frac{\partial U}{\partial B}\right)_{S,V,\mathcal{N}} \mathrm{d}B + \mu\,\mathrm{d}\mathcal{N}. \tag{19.91}$$

The differential of the Helmholtz free energy $F = U - TS$ is therefore

$$\mathrm{d}F = -S\,\mathrm{d}T - p\,\mathrm{d}V + \left(\frac{\partial U}{\partial B}\right)_{S,V,\mathcal{N}} \mathrm{d}B + \mu\,\mathrm{d}\mathcal{N}, \tag{19.92}$$

from which we see that

$$\left(\frac{\partial F}{\partial B}\right)_{T,V,\mathcal{N}} = \left(\frac{\partial U}{\partial B}\right)_{S,V,\mathcal{N}}. \tag{19.93}$$

Accordingly, we define the net magnetic moment

$$\mathcal{M} := -\left(\frac{\partial F}{\partial B}\right)_{T,V,\mathcal{N}} = -\left(\frac{\partial U}{\partial B}\right)_{S,V,\mathcal{N}}, \tag{19.94}$$

[10]The magnetic field is a vector but for simplicity we consider a magnetically isotropic system and represent the z component of the magnetic field by B.

which is an extensive thermodynamic quantity. The above differentials therefore become

$$dU = T\,dS - p\,dV - \mathcal{M}\,dB + \mu\,d\mathcal{N}; \tag{19.95}$$

$$dF = -S\,dT - p\,dV - \mathcal{M}\,dB + \mu\,d\mathcal{N}. \tag{19.96}$$

B is an intensive variable, so the corresponding Euler equations are $U = TS - pV + \mu\mathcal{N}$ and $F = -pV + \mu\mathcal{N}$, which have the same form as in the absence of a magnetic field.

One can also employ potentials $\tilde{U} := U + B\mathcal{M}$ and $\tilde{F} := \tilde{U} - TS = U - TS + B\mathcal{M}$ which are Legendre transforms of U and F. Then

$$d\tilde{U} = T\,dS - p\,dV + B\,d\mathcal{M} + \mu\,d\mathcal{N}; \tag{19.97}$$

$$d\tilde{F} = -S\,dT - p\,dV + B\,d\mathcal{M} + \mu\,d\mathcal{N}, \tag{19.98}$$

with corresponding Euler equations $\tilde{U} = TS - pV + B\mathcal{M} + \mu\mathcal{N}$ and $\tilde{F} = -pV + B\mathcal{M} + \mu\mathcal{N}$.

One often treats the special case in which the partition function depends on β and B only as a product βB, so $Z = Q(\beta B)$, where Q is some differentiable function. Then

$$F = -(1/\beta)\ln Q(\beta B) \tag{19.99}$$

from which we readily compute

$$\mathcal{M} = \frac{1}{\beta}\left(\frac{\partial \ln Z}{\partial B}\right)_{T,V,\mathcal{N}} = \frac{Q'(\beta B)}{Q(\beta B)}; \quad U = -\left(\frac{\partial \ln Z}{\partial \beta}\right)_{B,V,\mathcal{N}} = -\frac{Q'(\beta B)}{Q(\beta B)}B. \tag{19.100}$$

Here, Q' is just the derivative of Q with respect to its argument. Thus, in this special case, we have

$$\mathcal{M} = -\frac{U}{B}, \tag{19.101}$$

which is a ratio rather than a derivative. For more general systems, however, Eq. (19.101) does not hold and one must compute $\mathcal{M}$ by differentiation, according to Eq. (19.94). Note in this special case that the Legendre transformed potentials $\tilde{U} = 0$ and $\tilde{F} = -TS$. This occurs because the functional form $Z = Q(\beta B)$ is valid whenever the only relevant energy levels have energies that are proportional to B. For a more detailed discussion of energy in magnetic systems, see Callen [2, appendix B], but note that his U is the same as our $\tilde{U}$.

19.6.1 Classical Treatment

For historical reasons, we first calculate the magnetization by means of classical statistical mechanics.[11] The classical energy of a dipole of magnetic moment $\boldsymbol{\mu}_\mathbf{c}$ that makes an angle θ with a magnetic field $\mathbf{B}$ is

$$\varepsilon_\theta = -\boldsymbol{\mu}_\mathbf{c} \cdot \mathbf{B} = -\mu_c B \cos\theta. \tag{19.102}$$

[11]See Eq. (20.3) and Chapter 20 for details of the classical partition function. For present purposes, we only need to integrate the relevant Boltzmann factor over angles in phase space and the overall constant is irrelevant.

Thus, the classical partition function of a single dipole is

$$z_{\mathrm{C}} = \mathrm{const} \int_0^{2\pi} \mathrm{d}\varphi \int_0^{\pi} \sin\theta \, \mathrm{d}\theta \exp(\beta \mu_{\mathrm{C}} B \cos\theta) = \mathrm{const}\, 4\pi \frac{\sinh(\beta\mu_{\mathrm{C}} B)}{\beta \mu_{\mathrm{C}} B}. \tag{19.103}$$

Accordingly, for $\mathcal{N}$ identical but independent distinguishable dipoles, the total partition function $Z_{\mathrm{C}} = z_{\mathrm{C}}^{\mathcal{N}}$ so $F = -(\mathcal{N}/\beta) \ln z_{\mathrm{C}}$. Thus

$$\mathcal{M} = \frac{\mathcal{N}}{\beta} \frac{\partial}{\partial B} \ln \frac{\sinh(\beta\mu_{\mathrm{C}} B)}{\beta\mu_{\mathrm{C}} B} = \mathcal{N} \mu_{\mathrm{C}} L(x_{\mathrm{C}}), \tag{19.104}$$

where $x_{\mathrm{C}} := \beta\mu_{\mathrm{C}} B$ and the Langevin function

$$L(x) := \coth x - 1/x. \tag{19.105}$$

The Langevin function has the properties

$$L(x) = \begin{cases} \dfrac{x}{3} - \dfrac{x^3}{45} + \dfrac{2x^5}{945} + O(x^7) & x \ll 1 \\ 1 - \dfrac{1}{x} + 2\,\mathrm{e}^{-2x} & x \gg 1 \end{cases} \tag{19.106}$$

and is depicted in Figure 19–2. At very low temperatures or very high magnetic fields, $x_{\mathrm{C}} \gg 1$ and the magnetic moment saturates at a value $\mathcal{M} = \mathcal{N}\mu_{\mathrm{C}}$. For high temperatures or very very weak fields, $x_{\mathrm{C}} \ll 1$ and

$$\mathcal{M} \approx \frac{\mathcal{N}\mu_{\mathrm{C}}^2}{3k_{\mathrm{B}}T} B. \tag{19.107}$$

The magnetic susceptibility is then given by **Curie's law**,

$$\chi := \frac{\partial \mathcal{M}}{\partial B} = \frac{\mathcal{N}\mu_{\mathrm{C}}^2}{3k_{\mathrm{B}}T} = \frac{C}{T}, \tag{19.108}$$

where $C := \mathcal{N}\mu_{\mathrm{C}}^2/(3k_{\mathrm{B}}T)$ is known as the Curie constant. The fact that χ varies inversely with T at high temperatures is well known experimentally and enables μ_{C} to be deter-

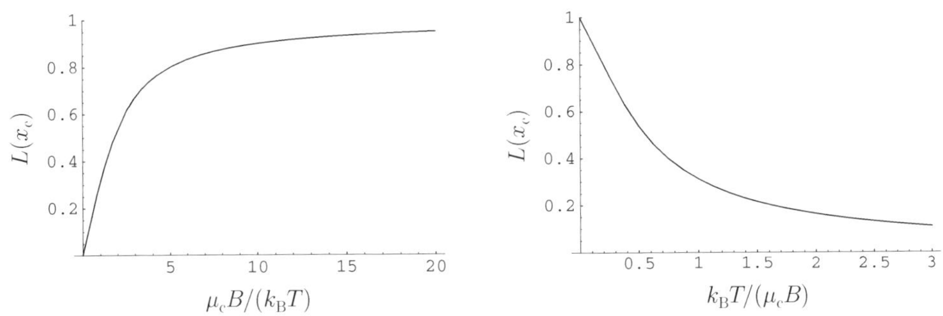

FIGURE 19–2 The Langevin function $L(x_{\mathrm{C}})$ given by Eq. (19.105) with $x_{\mathrm{C}} = \mu_{\mathrm{C}} B/(k_{\mathrm{B}}T)$. The plot on the left can be interpreted as the dimensionless magnetic moment as a function of dimensionless magnetic field strength at constant T. The plot on the right is against $1/x_{\mathrm{C}}$ and can be interpreted as the dimensionless magnetic moment versus dimensionless temperature at fixed B; it gives incorrect results at small T, including a nonzero slope, because it does not account properly for quantum effects.

mined. From the quantum mechanical treatment to follow, we shall see that the Langevin function gives incorrect answers at low temperatures, so the saturation magnetic moment $\mathcal{N}\mu_c$ and the shape of the curve at low temperatures are incorrect.

19.6.2 Quantum Treatment

For an atom[12] in a uniform magnetic field **B** along the z-axis, the part of the Hamiltonian that depends on **B** can be written in Gaussian units in the form[13]

$$\mathcal{H}'_{\rm B} = \frac{e\hbar}{2mc}(\mathbf{L} + 2\mathbf{S}) \cdot \mathbf{B} + \frac{e^2 B^2}{8mc^2} \sum_i (x_i^2 + y_i^2), \tag{19.109}$$

where e is the magnitude of the charge on the electron, c is the speed of light, **L** is the total orbital angular momentum, and **S** is the total spin angular momentum. Both angular momenta are measured in units of $\hbar$ and are therefore dimensionless. The sum on i is over all electrons. The term in B^2 contributes to diamagnetism, but here we deal only the linear term in **B**, which is usually written in the form

$$\mathcal{H}_{\rm B} = \mu_{\rm B}(\mathbf{L} + 2\mathbf{S}) \cdot \mathbf{B}, \tag{19.110}$$

where the quantity $\mu_{\rm B} := e\hbar/(2mc)$ is known as the **Bohr magneton.**[14] Furthermore, we shall confine our treatment to cases for which the only important states of the atom are its ground states that are degenerate in the absence of a magnetic field and are eigenstates of the operators $\hat{L}^2$, $\hat{S}^2$, $\hat{J}^2$, and $\hat{J}_z$, where $\hat{\mathbf{J}} = \hat{\mathbf{L}} + \hat{\mathbf{S}}$. Such states $|LSJM\rangle$ satisfy the relations

$$\begin{aligned} \hat{L}^2|LSJM\rangle &= L(L+1)|LSJM\rangle; \qquad \hat{J}^2|LSJM\rangle = J(J+1)|LSJM\rangle \\ \hat{S}^2|LSJM\rangle &= S(S+1)|LSJM\rangle; \qquad \hat{J}_z|LSJM\rangle = M|LSJM\rangle \end{aligned} \tag{19.111}$$

and have a degeneracy of $2J + 1$ because $M = -J, -J+1, \ldots, J-1, J$. Based on addition theorems for angular momenta,[15] one can show that

$$\langle LSJM'|\hat{L}_z + 2\hat{S}_z|LSJM\rangle = gM\,\delta_{MM'}, \tag{19.112}$$

where

$$g := \frac{3}{2} + \frac{1}{2}\left[\frac{S(S+1) - L(L+1)}{J(J+1)}\right] \tag{19.113}$$

[12] We use the word atom but we will frequently actually treat an ion in some crystal. For example, the rare earth elements (atomic numbers 58-71) have similar chemistry governed by a pair of $6s$ valence electrons. They form salts that contain rare earth ions, each having from 1 to 14 electrons in their inner f-shells. These ions have net magnetic moments that can be aligned by a magnetic field. For a table summarizing details, see Ashcroft and Mermin [58, p. 652]. For an extensive discussion, see van Vleck [100, p. 228].

[13] For a derivation, see [58, p. 646]. To covert to SI units, replace eB/c by eB. The g-factor for spin, which is approximately 2.0023, has been taken to be exactly 2 for simplicity.

[14] $\mu_{\rm B} = 9.274 \times 10^{-21}$ erg/gauss. In SI units, $\mu_{\rm B} = e\hbar/2m = 9.274 \times 10^{-24}$ joule/tesla.

[15] The proof is based on the Wigner-Eckart theorem which leads to operator equivalents [59, p. 707]. For a thorough discussion of the allowable ground states and examples of ions having partially filled d- or f-shells that can be treated by Hund's rules, see Ashcroft and Mermin [58, p. 650].

is known as the **Lande g-factor**. In fact, within the subspace of such states having the same values of L, S, J, and M, one has the operator equivalence

$$\hat{\mathbf{L}} + 2\hat{\mathbf{S}} = g\hat{\mathbf{J}} \tag{19.114}$$

for all vector components. Therefore, one can define a magnetic moment operator

$$\hat{\boldsymbol{\mu}} := -\mu_{\mathrm{B}} g\hat{\mathbf{J}} \tag{19.115}$$

in terms of which the Hamiltonian

$$\hat{\mathcal{H}}_{\mathrm{B}} = -\hat{\boldsymbol{\mu}} \cdot \mathbf{B}, \tag{19.116}$$

which resembles the classical expression for the energy of a dipole of magnetic moment $\boldsymbol{\mu}$ in a magnetic field $\mathbf{B}$. For a magnetic field along the z-axis, we therefore have

$$\hat{\mathcal{H}}_{\mathrm{B}}|LSJM\rangle = \mu_{\mathrm{B}} gMB|LSJM\rangle, \tag{19.117}$$

so the $J(J+1)$ degenerate[16] states for zero magnetic field are split into states having energies $\mu_{\mathrm{B}} gBM$ that are equally spaced.

The canonical partition function for a single atom is therefore

$$z = \sum_{M=-J}^{J} \exp(\beta\mu_{\mathrm{B}} gBM) = \sum_{M=-J}^{J} [\mathrm{e}^{x/J}]^{M}, \tag{19.118}$$

where $x = \beta\mu_{\mathrm{B}} gBJ$. The variable x is equal to βB times the maximum eigenvalue $\mu_{\mathrm{B}} gJ$ of the magnetic moment operator $\hat{\mu}_z$. We shall see that x plays almost the same role as x_{c} in the classical treatment, but they are somewhat different. The geometric series in Eq. (19.118) can be readily summed to yield

$$z = \sinh\left[x\left(1 + \frac{1}{2J}\right)\right] \Big/ \sinh\left(\frac{x}{2J}\right). \tag{19.119}$$

From the total partition function $Z = z^{\mathcal{N}}$ and Eq. (19.100), we readily compute

$$\mathcal{M} = \mathcal{N}\mu_{\mathrm{B}} gJB_J(x), \tag{19.120}$$

where

$$B_J(x) = \left(1 + \frac{1}{2J}\right)\coth\left[x\left(1 + \frac{1}{2J}\right)\right] - \left(\frac{1}{2J}\right)\coth\left(\frac{x}{2J}\right); \quad J \neq 0, \tag{19.121}$$

is called the **Brillouin function**. It is depicted in Figure 19–3 and has the following properties:

[16]For the special case $J = 0$, one has no degeneracy, $M = 0$ and there is no first-order effect of a magnetic field. In that case, the ground state has no magnetic moment and one must consider interaction with excited states as well as the second-order term in Eq. (19.109).

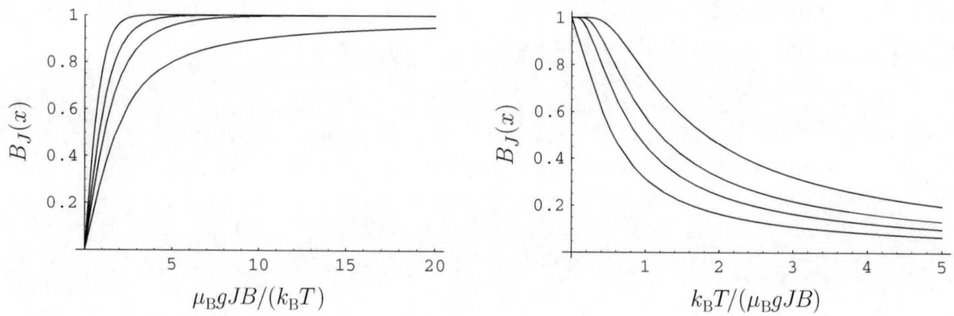

FIGURE 19–3 The Brillouin function $B_J(x)$ given by Eq. (19.121) with $x = \mu_B gJB/(k_B T)$. From the top down, the curves are for $J = 1/2$, $J = 1$, and $J=2$. The bottom curve is the Langevin function $L(x)$. The plot on the left can be interpreted as the dimensionless magnetic moment as a function of dimensionless magnetic field strength at constant T. The plot on the right is against $1/x$ and can be thought of as the dimensionless magnetic moment versus dimensionless temperature at fixed B. Note that the Brillouin function versus T has zero slope at $T = 0$ because quantum effects result in a very small population of the first excited state at low temperatures.

$$B_J(x) = \frac{1}{3}\left(1+\frac{1}{J}\right)x - \frac{1}{45}\left(1+\frac{2}{J}+\frac{3}{2J^2}+\frac{1}{2J^3}\right)x^3 + O(x^5); \quad x \ll 1, \tag{19.122}$$

$$B_J(x) = 1 - \frac{1}{J}\exp(-x/J); \quad x \gg 1 \text{ with } J \text{ finite}, \tag{19.123}$$

$$B_J(x) = L(x); \quad J \to \infty \text{ with } x \text{ finite}. \tag{19.124}$$

For high temperatures, Eq. (19.122) is valid and the first term gives

$$\mathcal{M} = \frac{\mathcal{N}\mu_B^2 g^2 J(J+1)}{3k_B T}B. \tag{19.125}$$

Comparison with Eq. (19.107) for the classical treatment gives the correspondence

$$\mu_c = \mu_B g\sqrt{J(J+1)}. \tag{19.126}$$

Equation (19.126) is the correct relationship between the quantum mechanical treatment and the classical treatment because the latter is only valid at high temperatures. It leads to the correspondence

$$x_c = x\sqrt{(J+1)/J}. \tag{19.127}$$

It would be incorrect to make a comparison by matching the saturation magnetic moments at low temperatures and high magnetic field strengths, in which case both x and x_c become very large, because the classical treatment is not valid under those conditions. The saturation magnetic moment for the quantum treatment is $\mathcal{N}\mu_B gJ$ whereas for the classical treatment it is $\mathcal{N}\mu_c$. By using Eq. (19.126), we see that $\mathcal{N}\mu_c$ is a factor of $\sqrt{(J+1)/J}$ larger than the quantum mechanical value $\mathcal{N}\mu_B gJ$.

We can make a comparison between quantum results and classical results as follows. We fix the value of μ_c and choose the product $g\sqrt{J(J+1)}$ so that Eq. (19.126) is satisfied.

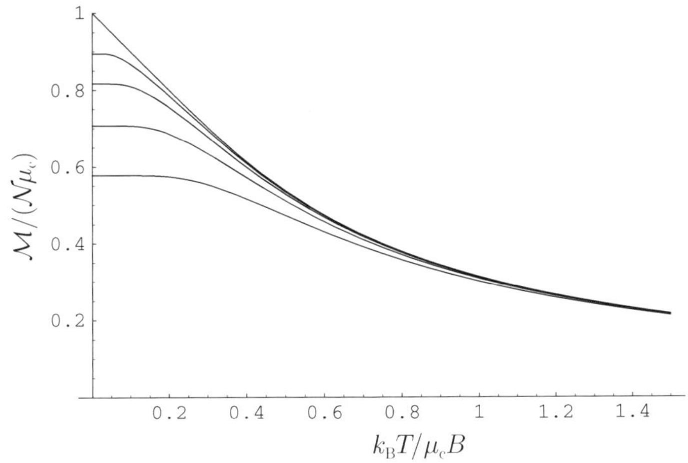

FIGURE 19–4 Comparison of quantum and classical results under the constraint that both agree at high temperatures. The top curve is the Langevin function $L(x_c)$. The other curves are calculated from the Brillouin function in the form of Eq. (19.128). From the bottom up, they correspond to $J = 1/2, 1, 2, 4$. The respective saturation values for the quantum results are $\sqrt{J/(J+1)} = 0.57735, 0.707107, 0.816497, 0.894427$.

This will make quantum and classical results agree for high temperatures. Then x will be related to x_c by Eq. (19.127), which allows Eq. (19.120) to be written

$$\mathcal{M} = \mathcal{N}\mu_c\sqrt{J/(J+1)}B_J\left(x_c\sqrt{J/(J+1)}\right). \tag{19.128}$$

Figure 19–4 shows a plot of $\mathcal{M}/\mathcal{N}\mu_c$ versus $1/x_c = k_B T/(\mu_c B)$ for the classical result (Langevin function) and for quantum results for several values of J. For fixed high temperature moment μ_c, we see that the quantum results saturate at smaller values than the classical result, the smallest occurring for $J = 1/2$. Of course the actual quantum saturation values are

$$\mathcal{M}^* = \mathcal{N}\mu_B gJ = \mathcal{N}\mu_B\left\{\frac{3J}{2} + \frac{1}{2}\left[\frac{S(S+1) - L(L+1)}{J+1}\right]\right\}, \tag{19.129}$$

where Eq. (19.113) has been used, so one should take great care in discussing general trends with J.

19.6.3 Properties of Paramagnetic Systems

We digress here to explore some useful properties of the paramagnetic system treated in Section 19.6.2 that are not necessarily obvious. The first concerns the sign of the magnetic moment $\mathcal{M}$. From Eq. (19.101), we see for this particular model that $\mathcal{M} = -U/B$ so $\mathcal{M}$ has the opposite sign of the internal energy U. In general, the internal energy is undefined up to an additive constant, so it can be positive or negative, but Eq. (19.101) is only true because the partition function depends on B and β only in the combination $y = B\beta$, as in Eq. (19.99). This arises because the energies of the states given by Eq. (19.117) are of the form $\varepsilon_i = a_i B$ such that for every value of i there is also a state with energy $\varepsilon_i = -a_i B$. Therefore, the partition function for a single particle can be written the form

$$z = \sum_i \mathrm{e}^{\beta a_i B} = \frac{1}{2}\sum_i (\mathrm{e}^{\beta a_i B} + \mathrm{e}^{-\beta a_i B}) = \sum_i \cosh(a_i y). \tag{19.130}$$

Thus,

$$U = -\mathcal{N}\frac{\partial}{\partial \beta}\ln z = -\frac{\mathcal{N}}{z}\sum_i a_i \tanh(a_i y) B \leq 0 \tag{19.131}$$

and it follows that $\mathcal{M} \geq 0$ with the equal sign corresponding to $B = 0$ or $T = \infty$.

Next, we turn to the magnetic susceptibility $\chi = \partial \mathcal{M}/\partial B$ and show that $\chi \geq 0$. We could do this for the specific model of Section 19.6.2 but instead we proceed to derive a more general relation that is even more interesting. We consider a many-particle system with Hamiltonian $\mathcal{H}$ and define a *total* magnetic moment operator $\hat{M} = -\partial \mathcal{H}/\partial B = \mathcal{N}\mu_z$. For clarity, we now denote the magnetic moment itself by $\langle \hat{M} \rangle$, which is the thermal average of $\hat{M}$ in the canonical ensemble. Then it follows that

$$\frac{\mathrm{tr}[\mathrm{e}^{-\beta\mathcal{H}}(\hat{M} - \langle \hat{M} \rangle)]}{\mathrm{tr}[\mathrm{e}^{-\beta\mathcal{H}}]} = \langle \hat{M} - \langle \hat{M} \rangle \rangle = 0. \tag{19.132}$$

Here, to achieve more generality, we have used the trace, denoted by tr, to write thermal averages in an invariant form whereas until now we have used only the energy representation (see Chapter 26 for more detail). We now differentiate the numerator of the first term in Eq. (19.132) with respect to B to obtain

$$\frac{\partial}{\partial B}\mathrm{tr}[\mathrm{e}^{-\beta\mathcal{H}}(\hat{M} - \langle \hat{M} \rangle)] = \mathrm{tr}[\mathrm{e}^{-\beta\mathcal{H}}\beta\hat{M}(\hat{M} - \langle \hat{M} \rangle) - \mathrm{e}^{-\beta\mathcal{H}}\partial\langle \hat{M} \rangle/\partial B] = 0. \tag{19.133}$$

We then divide by $\mathrm{tr}[\mathrm{e}^{-\beta\mathcal{H}}]$ and recognize $\partial\langle \hat{M} \rangle/\partial B = \chi$, the susceptibility, to obtain

$$\langle \hat{M}^2 \rangle - \langle \hat{M} \rangle^2 = \chi/\beta, \tag{19.134}$$

where $\langle \hat{M}^2 \rangle$ denotes the thermal average of $\hat{M}^2$. But we know that $\langle \hat{M}^2 \rangle - \langle \hat{M} \rangle^2 = \langle (\hat{M} - \langle \hat{M} \rangle)^2 \rangle$, which leads to

$$\chi/\beta = \langle (\hat{M} - \langle \hat{M} \rangle)^2 \rangle > 0. \tag{19.135}$$

Thus the susceptibility χ is positive at any finite temperature. We note the similarity of Eqs. (19.87)–(19.135) for the heat capacity in terms of the dispersion of energy, which could have been derived in the same way. One subtle difference, however, is that the Hamiltonian always commutes with itself but there could be cases for which parts of the Hamiltonian do not commute with the magnetic moment operator, in which case the above derivation would not hold. We remark that for the special case we have been treating for which $\mathcal{M} \equiv \langle \hat{M} \rangle$ depends only on $y = B\beta$, Eq. (19.135) becomes

$$\frac{\mathrm{d}\mathcal{M}}{\mathrm{d}y} = \langle (\hat{M} - \mathcal{M})^2 \rangle > 0. \tag{19.136}$$

Finally, we shall show for the model of Section 19.6.2 that the entropy S is a monotonically increasing function of $1/y = k_B T/B$. To do this, we substitute $U = -\mathcal{M}B$ into Eq. (19.95) at constant V and $\mathcal{N}$ to obtain

$$-\,\mathrm{d}(\mathcal{M}B) = T\,\mathrm{d}S - \mathcal{M}\,\mathrm{d}B, \tag{19.137}$$

which yields

$$\mathrm{d}S/k_B = -\,y\,\mathrm{d}\mathcal{M} = -\,y\frac{\mathrm{d}\mathcal{M}}{\mathrm{d}y}\,\mathrm{d}y = y^3\frac{\mathrm{d}\mathcal{M}}{\mathrm{d}y}\,\mathrm{d}(1/y). \tag{19.138}$$

From Eq. (19.136) we see that the coefficient of $\mathrm{d}(1/y)$ in Eq. (19.138) is positive, so S is a monotonically increasing function of $1/y = k_B T/B$. This result will be used in the next section.

19.6.4 Adiabatic Demagnetization

Adiabatic demagnetization is an experimental technique that can be used to cool magnetic samples to extremely low temperatures. A sample is first cooled and maintained at a very low temperature T_0, for example, by contact with liquid helium, while an extremely strong magnetic field B_0 is applied. Then the sample is thermally insulated and the magnetic field is slowly and carefully lowered to as small a value as possible,[17] say B_E. As we shall show subsequently, the temperature of the sample will be lowered to

$$T_E = T_0\, B_E/B_0. \tag{19.139}$$

This simple result can be understood by examining the entropy S of the sample. Since $S = (U - F)/T$, we can use Eqs. (19.99) and (19.100) to obtain

$$S/k_B = \ln Q(\beta B) - (\beta B)Q'(\beta B)/Q(\beta B). \tag{19.140}$$

The entropy is therefore only a function of the product βB, or for our purposes the ratio T/B. The stage of the process in which the sample is thermally insulated and the magnetic field is slowly and carefully lowered is adiabatic and practically reversible, so it is approximately isentropic, that is, $S = \text{constant}$. If T/B is constant, then surely S will be constant. In Section 19.6.3, however, we showed that S is a monotonically increasing function of $1/y = k_B T/B$. Therefore, if S is constant, T/B will also be constant and Eq. (19.139) follows.

We can gain more insight by examining the details of a simple case. For example, for the case $J = 1/2$, Eq. (19.119) simplifies to $z = 2\cosh x$ and Eq. (19.140) yields

$$S/(\mathcal{N}k_B) = \ln(2\cosh x) - x\tanh x \tag{19.141}$$

as illustrated in Figure 19–5. Results for other values of J are qualitatively similar. During reversible adiabatic demagnetization from the point 0, the dimensionless entropy would remain at the value 0.6 and the temperature would drop in proportion to the field strength.

[17]The lowest possible field B_E will probably be the order of the magnetic field of the Earth, about 0.5 gauss $= 5 \times 10^{-5}$ tesla.

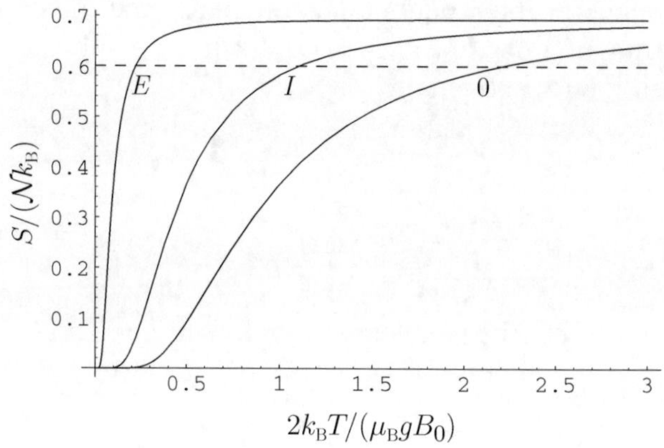

FIGURE 19–5 Entropy as a function of temperature for $J = 1/2$. From right to left, the curves are for $B = B_0$, $B_0/2$, and $B_0/10$. For sufficiently high T, all curves would saturate at $\ln 2 = 0.693$. In a hypothetical process of adiabatic demagnetization, suppose that the sample were magnetized in a strong field B_0 at temperature T_0, so that it is represented by the point 0 which has dimensionless entropy 0.6. If the sample were now insulated and reversibly demagnetized isentropically to the point I, its temperature would become $T_0/2$. If the isentropic demagnetization were continued to the point E, its temperature would become $T_0/10$.

One might wonder how the temperature of the system could drop without extracting heat. The answer to this mystery lies in the initial stage of the process wherein the system is magnetized by applying the high field B_0. If the cooling fluid is able to maintain the system at temperature T_0 throughout this process, and if the process is reversible, an amount of heat $|Q| = -T_0\Delta S$ would be extracted from the system. We know that $\Delta S < 0$ because S increases with T at fixed B and therefore S decreases with B at fixed T. If the initial magnetization process is not reversible, even more heat would have to be extracted. Similarly, if the demagnetization process is not quite reversible, the entropy of the system will go up slightly and one will achieve a final temperature slightly higher than that calculated for the reversible process.

19.7 Partition Function and Density of States

Under suitable circumstances, the energy levels of the quantum states of a system can be treated as quasi-continuous. Specifically, the spacing between levels must be small compared to k_BT, which is often possible for large systems if the temperature is not too low. Under those circumstances, the sum over *states* that is used to calculate the partition function, namely[18]

$$Z(\beta) = \sum_j \exp(-\beta\mathcal{E}_j), \tag{19.142}$$

[18] Z will generally depend on other parameters such as the volume V but we suppress these variables for simplicity.

can be approximated by an integral of the form

$$Z(\beta) = \int_0^\infty e^{-\beta E}\, \mathcal{D}(E)\, dE, \tag{19.143}$$

where $\mathcal{D}(E)$ is known as the **density of states** and accounts for the spacing and degeneracy of the quantum states. Specifically, $\mathcal{D}(E)$ is a distribution function such that $\mathcal{D}(E)\,dE$ is the number of quantum states in the energy interval between E and $E+dE$. Equation (19.143) has the same form as a Laplace transform with transform variable β. Therefore, one can use the Laplace inversion formula

$$\mathcal{D}(E) = \frac{1}{2\pi i}\int_{Br} e^{\beta E} Z(\beta)\, d\beta \tag{19.144}$$

to compute $\mathcal{D}(E)$ from a knowledge of $Z(\beta)$. In Eq. (19.144), β is regarded as a complex variable and the integration is over a contour Br in the complex plane known as the Bromwich contour. This contour starts out at $\beta = -i\infty$, goes to the right of all singularities[19] of $Z(\beta)$ and ends up at $\beta = i\infty$. One can use Cauchy's theorem to deform the contour and thus calculate $\mathcal{D}(E)$ by standard methods of contour integration.

Example Problem 19.3. Calculate the Laplace transform $Z(\beta)$ of the partition function for $\mathcal{N}$ atoms of a monotonic ideal gas to determine its density of states $\mathcal{D}(E)$ and relate $\mathcal{D}(E)$ to the corresponding function $\Omega(E)$ of the microcanonical ensemble.

Solution 19.3. By combining Eq. (19.48) with Eq. (19.56), we see that the partition function for $\mathcal{N}$ atoms of a monatomic ideal gas is given by

$$Z(\beta) = \frac{(V n_Q)^{\mathcal{N}}}{\mathcal{N}!} = \frac{V^{\mathcal{N}}}{\mathcal{N}!}\left(\frac{m}{2\pi\hbar^2\beta}\right)^{3\mathcal{N}/2}. \tag{19.145}$$

Thus,

$$\mathcal{D}(E) = \frac{V^{\mathcal{N}}}{\mathcal{N}!}\left(\frac{m}{2\pi\hbar^2}\right)^{3\mathcal{N}/2}\frac{1}{2\pi i}\int_{Br}\frac{e^{\beta E}}{\beta^{3\mathcal{N}/2}}\, d\beta. \tag{19.146}$$

The integrand certainly has a singularity at $\beta = 0$ but if $\mathcal{N}$ is an odd integer, one also needs a branch cut, usually taken from $\beta = 0$ to $\beta = -\infty$ along the real axis to make it analytic. But $\mathcal{N}$ is large so we do not really care if it is odd or even. Therefore, we temporarily pretend that it is even, in which case the integrand has a pole of order $3\mathcal{N}/2$ at the origin. We can therefore close the contour in the left half plane and apply Cauchy's theorem to shrink the contour to a small circle around $\beta = 0$. The result of integration is then well known to be

$$\int_{Br}\frac{e^{\beta E}}{\beta^{3\mathcal{N}/2}}\, d\beta = 2\pi i\,\text{Residue}\left[\frac{e^{\beta E}}{\beta^{3\mathcal{N}/2}}\right] = 2\pi i\frac{E^{3\mathcal{N}/2-1}}{(3\mathcal{N}/2-1)!}, \tag{19.147}$$

where Residue means to extract the coefficient of $1/\beta$. Thus,

$$\mathcal{D}(E) = \frac{V^{\mathcal{N}}}{\mathcal{N}!(3\mathcal{N}/2-1)!}\left(\frac{mE}{2\pi\hbar^2}\right)^{3\mathcal{N}/2}\frac{1}{E}. \tag{19.148}$$

[19]Such singularities are poles where $Z(\beta)$ becomes infinite or branch cuts needed to make it single-valued.

We note that $\mathcal{D}(E)$ has dimensions of $1/E$ so that $\mathcal{D}(E)\,\mathrm{d}E$ is dimensionless, as a probability should be. In the present case, we can easily check our result because Eq. (16.44) gives an expression for Ω, which is the total number of microstates having energies less than E. Differentiation with respect to E shows that $(\partial\Omega/\partial E)_{\mathcal{N},V} = \mathcal{D}(E)$ as it should (see Eq. (19.154) for more detail).

Note that Eq. (19.148) can be written in terms of the gamma function in the form

$$\mathcal{D}(E) = \frac{V^{\mathcal{N}}}{\mathcal{N}!\Gamma(3\mathcal{N}/2)}\left(\frac{mE}{2\pi\hbar^2}\right)^{3\mathcal{N}/2}\frac{1}{E}. \tag{19.149}$$

Of course $\Gamma(3\mathcal{N}/2)$ makes sense even when $3\mathcal{N}/2$ is a half integer, so we suspect that Eq. (19.149) might hold in general. Substitution into Eq. (19.143) shows that this conjecture is true.

We remark that this same Laplace transform relationship holds between the density of states $\mathcal{D}_1(\varepsilon)$ of a single particle and its partition function $z(\beta)$. Thus for an ideal gas,

$$z(\beta) = Vn_Q = V\left(\frac{m}{2\pi\hbar^2\beta}\right)^{3/2}, \tag{19.150}$$

so

$$\mathcal{D}_1(\varepsilon) = \frac{V}{\Gamma(3/2)}\left(\frac{m\varepsilon}{2\pi\hbar^2}\right)^{3/2}\frac{1}{\varepsilon} = \frac{V}{(1/2)\pi^{1/2}}\left(\frac{m\varepsilon}{2\pi\hbar^2}\right)^{3/2}\frac{1}{\varepsilon}, \quad \text{no spin degeneracy,} \tag{19.151}$$

where we have used $\Gamma(x+1) = x\Gamma(x)$ and $\Gamma(1/2) = \pi^{1/2}$. This result is the same as the density of states $G(\varepsilon)/2$ given by Eq. (25.13), where the division by 2 is necessary because $G(\varepsilon)$ contains a factor of 2 due to spin degeneracy. Since $\mathcal{D}_1(\varepsilon)$ is proportional to V, one often deals with the intensive quantity

$$\frac{\mathcal{D}_1(\varepsilon)}{V} = \frac{1}{(1/2)\pi^{1/2}}\left(\frac{m\varepsilon}{2\pi\hbar^2}\right)^{3/2}\frac{1}{\varepsilon}, \tag{19.152}$$

which is also called the density of states and has units of (volume energy)$^{-1}$. One must therefore be careful to ascertain from the context just what density of states is being used!

Strictly speaking, one should have $\mathcal{D}(E) = (1/\mathcal{N}!)\,\partial V_R/\partial E$, where V_R given by Eq. (16.39) is the total number of microstates for all energies $\leq E$. But $(1/\mathcal{N}!)\,V_R = \Omega(E)/\mathcal{F}$, where $\mathcal{F}$ is given by Eq. (16.40). Therefore,

$$\mathcal{D}(E) = \frac{\partial(\Omega/\mathcal{F})}{\partial E} = \frac{1}{\mathcal{F}}\frac{\partial\Omega}{\partial E} + \frac{\Omega}{\mathcal{F}^2}\frac{3\mathcal{N}\Delta E}{2E^2}\exp\left(-\frac{3\mathcal{N}\Delta E}{2E}\right). \tag{19.153}$$

The second term in Eq. (19.153) is negligible compared to the first (because of the exponential) and $\mathcal{F} \approx 1$, so

$$\mathcal{D}(E) \approx \frac{\partial\Omega(E)}{\partial E} \tag{19.154}$$

to an excellent approximation.

Finally, we make one more connection between the microcanonical ensemble and the canonical ensemble as follows. For the microcanonical ensemble, we have $S = k_B \ln \Omega(E)$; however, for the canonical ensemble

$$S = \frac{U - F}{T} = k_B \left(\ln Z + \frac{U}{k_B T} \right) = k_B \ln \left(Z \, e^{U/k_B T} \right). \tag{19.155}$$

If E and U are nearly the same, we should have

$$\ln(\Omega(E)) \sim \ln(\tilde{\Omega}(U)) \equiv \ln(Z \, e^{U/k_B T}). \tag{19.156}$$

However, Eq. (19.156) must be interpreted very carefully because the systems being compared are not quite the same. $\Omega(E)$ relates to the microcanonical ensemble for which the energy E of each microstate is specified, whereas $\tilde{\Omega}(U)$ relates to the canonical ensemble for which the temperature is specified, so only the average energy $U(T)$ is specified. Therefore, if we exponentiate both sides of Eq. (19.156) we obtain

$$\Omega(E) \sim \tilde{\Omega}(U) \equiv Z \, e^{U/k_B T}, \tag{19.157}$$

which only holds to the extent that $\ln \Omega(E) \sim \ln \tilde{\Omega}(U)$ when sub-extensive terms are neglected.

For example, for an ideal gas, for which $U = (3/2)\mathcal{N} k_B T$, we have

$$\tilde{\Omega}(U) = \frac{1}{\mathcal{N}!} V^{\mathcal{N}} \left(\frac{mU}{3\pi\hbar^2 \mathcal{N}} \right)^{3\mathcal{N}/2} e^{3\mathcal{N}/2}. \tag{19.158}$$

According to Eq. (16.44), we have

$$\Omega(E) = V^{\mathcal{N}} \frac{(mE/2\pi\hbar^2)^{3\mathcal{N}/2}}{\mathcal{N}!(3\mathcal{N}/2)!}. \tag{19.159}$$

We observe that the factors that multiply $U^{3\mathcal{N}/2}$ and $E^{3\mathcal{N}/2}$ are not quite the same, but since $\mathcal{N}$ is large we can use $(3\mathcal{N}/2)! \sim \mathcal{N}^{3\mathcal{N}/2} e^{-3\mathcal{N}/2} \sqrt{3\pi\mathcal{N}}$ to write Eq. (19.158) in the form

$$\tilde{\Omega}(U) \sim \sqrt{3\pi\mathcal{N}} \, V^{\mathcal{N}} \frac{(mU/2\pi\hbar^2)^{3\mathcal{N}/2}}{\mathcal{N}!(3\mathcal{N}/2)!}. \tag{19.160}$$

Thus, in the thermodynamic limit of extremely large $\mathcal{N}$, we have

$$\ln \tilde{\Omega}(U) = \ln \Omega(E) + (1/2) \ln(3\pi\mathcal{N}) \tag{19.161}$$

in which the last term is sub-extensive, and therefore negligible. It is also illuminating to use Eq. (19.148) with $E \to U$ to express $\tilde{\Omega}(U)$ in terms of the density of states evaluated at energy U, which results in

$$\tilde{\Omega}(U) \sim \sqrt{2\pi} \frac{U}{\sqrt{3\mathcal{N}/2}} \mathcal{D}(U) = \sqrt{2\pi} \sqrt{3\mathcal{N}/2} \, k_B T \, \mathcal{D}(U). \tag{19.162}$$

In view of Eq. (19.89), we recognize the factor $\sqrt{3\mathcal{N}/2}\,k_B T = \sqrt{\langle(\Delta E)^2\rangle}$ to be a measure of the spread of energy at temperature T. Thus Eq. (19.162) can be written

$$\tilde{\Omega}(U) \sim \sqrt{2\pi}\sqrt{\langle(\Delta E)^2\rangle}\,\mathcal{D}(U), \tag{19.163}$$

which demonstrates clearly that the density of states $\mathcal{D}(U)$ must be multiplied by the spread of energy to approximate the number of microstates $\tilde{\Omega}(U)$.

For a single ideal gas particle, the correspondence implied by Eq. (19.157) would give

$$\Omega_1(\varepsilon) \sim z\exp(\langle\varepsilon\rangle/k_B T) = z\,\mathrm{e}^{3/2}, \tag{19.164}$$

which illustrates that z is essentially a measure of the number of states available to an individual particle at temperature T.

Another way of evaluating $\tilde{\Omega}$ in Eq. (19.157) is to evaluate approximately the partition function

$$Z = \int_0^\infty \mathcal{D}(E)\,\mathrm{e}^{-\beta E}\,\mathrm{d}E = \int_0^\infty \mathrm{e}^{[-\beta E + \ln D(E)]}\,\mathrm{d}E \tag{19.165}$$

by expanding about the most probable state. To do this, we recognize that $\mathcal{D}(E)$ is a rapidly increasing function of E and $\mathrm{e}^{-\beta E}$ is a rapidly decreasing function of E. Thus the integrand has a sharp maximum at the most probable value E^* that satisfies

$$0 = \frac{\partial}{\partial E}[-\beta E + \ln D(E)]_{E^*} = -\beta + [\ln D(E^*)]', \tag{19.166}$$

where the prime indicates a derivative. We can therefore expand the exponent in the right-hand integrand in Eq. (19.165) to second order to obtain

$$-\beta E + \ln D(E) = -\beta E^* + \ln D(E^*) - (1/2)\alpha(E - E^*)^2 + \cdots, \tag{19.167}$$

where

$$\alpha := -[\ln D(E^*)]'' > 0 \tag{19.168}$$

is positive because E^* corresponds to a sharp maximum. Thus with $\xi = E - E^*$,

$$Z \approx \mathcal{D}(E^*)\,\mathrm{e}^{-\beta E^*}\int_{-E^*}^\infty \mathrm{e}^{-(\alpha/2)\xi^2}\,\mathrm{d}\xi \approx \sqrt{2\pi}\frac{1}{\sqrt{\alpha}}\mathcal{D}(E^*)\,\mathrm{e}^{-\beta E^*}, \tag{19.169}$$

where the lower limit in the second integral has been approximated by $-\infty$ because the peak is so sharp. See Widom [17, Eq. 1.25] for an equivalent result with his $\delta E = \sqrt{2/\alpha}$. Thus

$$\tilde{\Omega} \sim \sqrt{2\pi}\frac{1}{\sqrt{\alpha}}\mathcal{D}(E^*)\,\mathrm{e}^{-\beta(E^* - U)}. \tag{19.170}$$

But the difference[20] between E^* and U is of order $k_B T$, so the exponential in Eq. (19.170) gives a numerical factor of order 1 and the result greatly resembles Eq. (19.162).

[20]In this Gaussian approximation, there is negligible difference between E^* and U if the lower limit of the integral in Eq. (19.169) can be approximated by $-\infty$.

Specifically for a monatomic ideal gas, Eq. (19.148) shows that $\mathcal{D}(E) = AE^{3\mathcal{N}/2-1}$ so $E^* = (3\mathcal{N}/2 - 1)k_B T$ and $\alpha = (3\mathcal{N}/2 - 1)/(E^*)^2$. Since $U = (3\mathcal{N}/2)k_B T$, Eq. (19.170) becomes

$$\tilde{\Omega} \sim \sqrt{2\pi}\frac{E^*}{\sqrt{3\mathcal{N}/2 - 1}}\mathcal{D}(E^*)\,\mathrm{e} = \sqrt{2\pi}\frac{U}{\sqrt{3\mathcal{N}/2 - 1}}\mathcal{D}(U)\,\mathrm{e}\,(E^*/U)^{3\mathcal{N}/2}. \tag{19.171}$$

But

$$(E^*/U)^{3\mathcal{N}/2} = \left(\frac{3\mathcal{N}/2 - 1}{3\mathcal{N}/2}\right)^{3\mathcal{N}/2} = \left(1 - \frac{2}{3\mathcal{N}}\right)^{3\mathcal{N}/2} \sim \left(\mathrm{e}^{-2/3\mathcal{N}}\right)^{3\mathcal{N}/2} = \mathrm{e}^{-1}. \tag{19.172}$$

Therefore, Eq. (19.170) reduces to

$$\tilde{\Omega} \sim \sqrt{2\pi}\frac{U}{\sqrt{3\mathcal{N}/2 - 1}}\mathcal{D}(U) = \sqrt{2\pi}\sqrt{\langle(\Delta E)^2\rangle}\,\mathcal{D}(U), \tag{19.173}$$

in excellent agreement with Eq. (19.163) because the 1 in the square root is negligible.

20 Classical Canonical Ensemble

For the canonical ensemble,[1] the temperature rather than the energy is fixed. Therefore, the members of the ensemble have various energies. Members of the ensemble having a given energy must still obey the Liouville theorem and hence Eq. (17.11). This possibility can be accommodated by choosing the density ρ in phase space to be some function of the classical Hamiltonian $\mathcal{H}$, in which case Eq. (17.11) becomes

$$\{\rho(\mathcal{H}), \mathcal{H}\} = \frac{\mathrm{d}\rho}{\mathrm{d}\mathcal{H}}\{\mathcal{H}, \mathcal{H}\} = 0. \tag{20.1}$$

Proceeding with the same arguments as in the quantum mechanical case, it can be inferred for a system in contact with a heat reservoir at temperature T that the probability of a system having energy E is proportional to the Boltzmann factor $\exp(-\beta E)$, where $\beta = 1/(k_B T)$ as usual. Since $\mathcal{H} = E$ for such a system, the appropriate probability distribution function is

$$\mathcal{P}(\mathbf{p}, \mathbf{q}) := \frac{\exp[-\beta\, \mathcal{H}(\mathbf{p}, \mathbf{q})]}{Z_C}, \tag{20.2}$$

where where $\mathbf{p}$ and $\mathbf{q}$ are $3\mathcal{N}$-dimensional vectors representing the canonical momenta and coordinates, respectively. The function

$$Z_C := \int \exp[-\beta\, \mathcal{H}(\mathbf{p}, \mathbf{q})]\, \mathrm{d}\omega, \tag{20.3}$$

where $\mathrm{d}\omega \equiv \mathrm{d}^{3\mathcal{N}}p\, \mathrm{d}^{3\mathcal{N}}q$ and the integration is over all phase space. $\mathcal{P}(\mathbf{p}, \mathbf{q})\, \mathrm{d}\omega$ is the probability that the system will be in the volume element $\mathrm{d}\omega$ of phase space centered about the point $\mathbf{p}, \mathbf{q}$. The factor Z_C in the denominator of Eq. (20.2) is needed to insure normalization, that is,

$$\int \mathcal{P}(\mathbf{p}, \mathbf{q})\, \mathrm{d}\omega = 1. \tag{20.4}$$

If $Y(\mathbf{p}, \mathbf{q})$ is some function of $\mathbf{p}$ and $\mathbf{q}$, then the average value of Y is

$$\langle Y \rangle := \int Y(\mathbf{p}, \mathbf{q})\, \mathcal{P}(\mathbf{p}, \mathbf{q})\, \mathrm{d}\omega. \tag{20.5}$$

[1]Those interested in the historical development of classical statistical mechanics are encouraged to read the original work of J.W. Gibbs [4]. Based on Hamilton's classical dynamical equations that we discussed in Chapter 17, Gibbs developed the classical canonical ensemble in Chapter IV, the microcanonical ensemble in Chapter X, and the grand canonical ensemble in Chapter XV. The integral form of Liouville's theorem that we presented in Section 17.1 is what Gibbs called the "conservation of probability of phase." If $\rho = e^{\eta}$ is the probability density function in phase space, Gibbs called η the "index of probability." Then he referred to a "canonical distribution" as one in which the index of probability is a linear function of the energy.

Thermal Physics. http://dx.doi.org/10.1016/B978-0-12-803304-3.00020-X

Comparison of Eqs. (20.2) and (20.3) with Eqs. (19.5) and (19.6) shows that the function Z_C plays the role of a classical partition function. In fact, the formula Eq. (19.7) for the average internal energy has exactly the same form in the classical case. Thus,

$$U := \langle \mathcal{H} \rangle = \int \mathcal{H}(\mathbf{p}, \mathbf{q})\, \mathcal{P}(\mathbf{p}, \mathbf{q})\, \mathrm{d}\omega = -\frac{1}{Z_C}\frac{\partial Z_C}{\partial \beta} = -\frac{\partial \ln Z_C}{\partial \beta}. \tag{20.6}$$

But in some other respects, the correspondence of Z_C with the quantum mechanical partition function is incorrect. Unlike the quantum partition function, Z_C is not dimensionless and does not account for the number of quantum states that need to be associated with a volume of phase space. For $\mathcal{N}$ identical particles that occupy the same volume, one can define a dimensionless classical partition function

$$Z_C^* := \frac{Z_C}{\omega_0} = \frac{1}{\omega_0}\int \exp[-\beta\, \mathcal{H}(\mathbf{p}, \mathbf{q})]\, \mathrm{d}\omega. \tag{20.7}$$

The factor ω_0 is the same factor as in Eq. (17.14) that allows us to convert from volume of phase space to microscopic states. For identical distinguishable particles we have $\omega_0 = h^{3\mathcal{N}}$ and for identical indistinguishable classical particles we have approximately $\omega_0 = h^{3\mathcal{N}}\mathcal{N}!$. In other words, $\mathrm{d}\omega$ has been replaced by the dimensionless quantity $\mathrm{d}\omega/\omega_0$, which is the differential of the number of microscopic states. Doing this gives rise to the correct entropy constant at high temperatures, where classical statistics are valid approximately. In this respect, we could view Eq. (20.2) in the form

$$\mathcal{P}(\mathbf{p}, \mathbf{q})\, \mathrm{d}\omega = \frac{\exp[-\beta\, \mathcal{H}(\mathbf{p}, \mathbf{q})]}{Z_C^*}\left(\frac{\mathrm{d}\omega}{\omega_0}\right). \tag{20.8}$$

In this manner, we can also relate properly to the Helmholtz free energy, namely

$$F = -k_B T \log Z_C^*, \tag{20.9}$$

and the entropy will be correctly given by

$$S = -\frac{\partial F}{\partial T}. \tag{20.10}$$

20.1 Classical Ideal Gas

To illustrate the classical canonical ensemble, we shall treat a classical ideal gas and rederive the Maxwell-Boltzmann distribution function. The Hamiltonian is

$$\mathcal{H} = \sum_{i=1}^{3\mathcal{N}} \frac{p_i^2}{2m} \tag{20.11}$$

for $\mathcal{N}$ particles of mass m. The p_i are just the Cartesian momenta of the particles. We can use Eq. (20.2) to calculate the average value of some function $f(\mathbf{v}_1) = f(\mathbf{p}_1/m)$ of the velocity of particle number 1, resulting in

$$\langle f(\mathbf{v}_1)\rangle = \int f(\mathbf{p}_1/m)\, \mathcal{P}(\mathbf{p}, \mathbf{q})\, \mathrm{d}\omega = \int f(\mathbf{p}_1/m)\, \frac{\exp[-\beta\, \mathcal{H}(\mathbf{p}, \mathbf{q})]}{Z_C}\, \mathrm{d}^{3\mathcal{N}}p\, \mathrm{d}^{3\mathcal{N}}q. \tag{20.12}$$

In Eq. (20.12), all integrals relating to q and to particles other than particle number 1 cancel with the corresponding factors in Z_C. We therefore obtain

$$\langle f(\mathbf{v}_1)\rangle = \frac{\int f(\mathbf{p}_1/m)\exp[-\beta\, p_1^2/2m]\,\mathrm{d}^3 p_1}{\int \exp[-\beta\, p_1^2/2m]\,\mathrm{d}^3 p_1} = \frac{\int f(\mathbf{v}_1)\exp[-\beta\, m v_1^2/2]\,\mathrm{d}^3 v_1}{\int \exp[-\beta\, m v_1^2/2]\,\mathrm{d}^3 v_1}. \tag{20.13}$$

The denominator on the right-hand side can be evaluated to give

$$\int \exp[-\beta\, m v_1^2/2]\,\mathrm{d}^3 v_1 = \left(\frac{2\pi k_B T}{m}\right)^{3/2}. \tag{20.14}$$

Thus

$$\langle f(\mathbf{v}_1)\rangle = \left(\frac{m}{2\pi k_B T}\right)^{3/2} \int_{-\infty}^{\infty} f(\mathbf{v}_1)\exp[-\beta\, m v_1^2/2]\,\mathrm{d}^3 v_1. \tag{20.15}$$

A similar result would be obtained for any other particle since they are all equivalent. The normalized distribution function for the velocity **v** of any particle is therefore

$$M(\mathbf{v}) = \left(\frac{m}{2\pi k_B T}\right)^{3/2} \exp[-mv^2/2k_B T], \tag{20.16}$$

in agreement with Eq. (19.70) and known as the **Maxwell-Boltzmann distribution** function.

Note that Eq. (20.16) can be factored into normalized distributions for each Cartesian component by writing $v^2 = v_x^2 + v_y^2 + v_z^2$ and apportioning the normalization factor, resulting in

$$M(\mathbf{v}) = \prod_{i=x,y,z} \left(\frac{m}{2\pi k_B T}\right)^{1/2} \exp[-mv_i^2/2k_B T], \tag{20.17}$$

which is the same as Eq. (19.73).

Example Problem 20.1. Find the distribution function for the speed $v_\perp$ for motion perpendicular to the z-axis.

Solution 20.1. First, we calculate the distribution function for velocity $\mathbf{v}_\perp$ perpendicular to the z-axis by integrating $M(\mathbf{v})$ over v_z, which we no longer care about. This gives

$$M(\mathbf{v}_\perp) = \int_{-\infty}^{\infty} M(\mathbf{v})\,\mathrm{d}v_z = \prod_{i=x,y} \left(\frac{m}{2\pi k_B T}\right)^{1/2} \exp[-mv_i^2/2k_B T]. \tag{20.18}$$

Then we go to polar coordinates so that $v_x = v_\perp \cos\phi$ and $v_y = v_\perp \sin\phi$ to obtain

$$M(\mathbf{v}_\perp)\,\mathrm{d}v_x\,\mathrm{d}v_y = \left(\frac{m}{2\pi k_B T}\right) \exp[-mv_\perp^2/2k_B T]\,\mathrm{d}\phi\, v_\perp\,\mathrm{d}v_\perp. \tag{20.19}$$

Next we integrate on ϕ from 0 to 2π to get the speed distribution function

$$\check{M}(v_\perp) = \left(\frac{m}{k_B T}\right) \exp[-mv_\perp^2/2k_B T]\, v_\perp. \tag{20.20}$$

Thus the average speed for motion perpendicular to the z-axis is

$$\langle v_\perp \rangle = \int_0^\infty \check{M}(v_\perp) v_\perp \, dv_\perp = \int_0^\infty \left(\frac{m}{k_B T}\right) \exp[-m v_\perp^2 / 2k_B T] \, v_\perp^2 \, dv_\perp = \left(\frac{\pi k_B T}{2m}\right)^{1/2}. \tag{20.21}$$

We can also evaluate the partition function Z_C^* by substitution of Eq. (20.11) into Eq. (20.7). The integral over $\mathbf{q}$ just gives a factor of $V^{\mathcal{N}}$ and the integral over $\mathbf{p}$ can be performed in Cartesian coordinates, resulting in

$$Z_C^* = \frac{V^{\mathcal{N}}}{h^{3\mathcal{N}} \mathcal{N}!} \left[\int \exp[-\beta p^2/(2m)] \, dp \right]^{3\mathcal{N}} = \frac{V^{\mathcal{N}}}{h^{3\mathcal{N}} \mathcal{N}!} (2\pi m k_B T)^{3\mathcal{N}/2}. \tag{20.22}$$

To evaluate $\ln Z_C^*$, we use Stirling's approximation for $\ln \mathcal{N}!$ to obtain

$$\ln Z_C^* = \mathcal{N} \ln(V/\mathcal{N}) + \mathcal{N} \ln n_Q + \mathcal{N}, \tag{20.23}$$

where the quantum concentration

$$n_Q = \left(\frac{2\pi m k_B T}{h^2}\right)^{3/2} = \left(\frac{m k_B T}{2\pi \hbar^2}\right)^{3/2}. \tag{20.24}$$

Equations (20.23) and (20.24) are the same as Eqs. (19.57) and (19.60) derived from quantum statistical mechanics in the high temperature limit. For high temperatures, the resulting thermodynamic functions are therefore the same as derived in Chapter 19.

20.1.1 Effusion of an Ideal Classical Gas

The slow leaking of a gas through a small hole in a box containing the gas is a process known as **effusion**. The hole is assumed to be so small that the gas inside the box can be assumed to be practically in equilibrium at each instant of time, as described by the Maxwell-Boltzmann velocity distribution function $M(\mathbf{v})$ given by Eq. (20.16). For convenience, we treat a monatomic gas and assume that the hole has an area a in a plane perpendicular to the z-axis. Let J be the flux of gas atoms that exit the hole; J has units of atoms/(area time) so that $Ja\,dt$ is the number of atoms that effuse (exit the box) in an infinitesimal time dt. These atoms can have any values of v_x and v_y but they must have $v_z > 0$. In an infinitesimal time dt, atoms in a rectangular parallelepiped of volume $av_z\,dt$ will exit. Thus if n is the number density of the gas, the flux will be given by

$$J = n \int_{-\infty}^{\infty} dv_x \int_{-\infty}^{\infty} dv_y \int_0^\infty dv_z \, v_z M(\mathbf{v}). \tag{20.25}$$

We now transform to spherical coordinates for which the volume element is $v^2 \sin\theta \, d\varphi \, d\theta \, dv$ and also write $v_z = v\cos\theta$. This gives

$$J = n \int_0^{2\pi} d\varphi \int_0^\infty dv \, v^2 \int_0^{\pi/2} d\theta \, v \cos\theta \sin\theta M(\mathbf{v}). \tag{20.26}$$

Since $M(\mathbf{v})$ depends only on v^2 and is therefore independent of θ and φ, the trigonometric integrals give factors of 2π and $1/2$. We therefore obtain

$$J = n\pi \int_0^\infty \mathrm{d}v\, v^3 M(\mathbf{v}) = \frac{n}{4}\int_0^\infty \mathrm{d}v\, v\, 4\pi v^2 M(\mathbf{v}) = \frac{n}{4}\int_0^\infty \mathrm{d}v\, v\tilde{M}(v), \tag{20.27}$$

where $\tilde{M}(v) = 4\pi v^2 M(\mathbf{v})$ is the speed distribution function given by Eq. (19.75). We readily compute $J = n(k_B T/2\pi m)^{1/2}$. This result could have been obtained more easily by just evaluating Eq. (20.25) in Cartesian coordinates, but the weighting of the speed in Eq. (20.27) would hold for any gas (e.g., an ultra-relativistic gas) for which $M(\mathbf{v})$ depends only on $|\mathbf{v}|$, as pointed out by Pathria [8, p. 139].

A similar calculation can be used to obtain an expression for the pressure of the gas. Instead of effusing, each gas atom that strikes an area a of a closed box will rebound and have the z-component of its momentum reversed.[2] This requires the wall of the box to exert a force $2mv_z$. Therefore, the pressure is given by

$$p = n\int_{-\infty}^\infty \mathrm{d}v_x \int_{-\infty}^\infty \mathrm{d}v_y \int_0^\infty \mathrm{d}v_z\, v_z 2mv_z M(\mathbf{v}). \tag{20.28}$$

Converting to polar coordinates as above gives

$$p = n\int_0^{2\pi} \mathrm{d}\varphi \int_0^\infty \mathrm{d}v\, v^2 \int_0^{\pi/2} \mathrm{d}\theta\, 2mv^2 \cos^2\theta \sin\theta M(\mathbf{v}). \tag{20.29}$$

Therefore,

$$p = \frac{4\pi n}{3}\int_0^\infty \mathrm{d}v\, mv^4 M(\mathbf{v}) = \frac{n}{3}\int_0^\infty \mathrm{d}v\, mv^2 \tilde{M}(v) = nk_B T. \tag{20.30}$$

Example Problem 20.2. Compute the energy flux J_E associated with effusion.

Solution 20.2. Each atom will carry an energy $(1/2)mv^2$ so

$$J_E = n\int_{-\infty}^\infty \mathrm{d}v_x \int_{-\infty}^\infty \mathrm{d}v_y \int_0^\infty \mathrm{d}v_z\, v_z (1/2)mv^2 M(\mathbf{v}). \tag{20.31}$$

By using spherical polar coordinates as above, we readily obtain

$$J_E = \frac{n}{4}\int_0^\infty \mathrm{d}v (1/2)mv^2 v\tilde{M}(v) = 2k_B T\, n(k_B T/2\pi m)^{1/2}. \tag{20.32}$$

We note that the average energy per effused atom is $J_E/J = 2k_B T$, which is greater than the average energy $(3/2)k_B T$ per atom of gas in the box. This arises because there is a preference for the faster atoms to effuse.

Example Problem 20.3. How would the fluxes J and J_E for effusion be modified if instead of a monatomic gas we have a molecular gas?

Solution 20.3. As long as the partition function for a molecule can be factored into a translational partition function and an internal partition function (see Section 21.3 for details), the

[2]This assumption appears to attribute special properties to the walls of the box, such as specular reflection, but it must be true on average in order to maintain equilibrium.

Maxwell-Boltzmann distribution for the velocity will still apply. Therefore, J given by Eq. (20.27) is still valid except the mass in the distribution function $\tilde{M}(v)$ must be replaced by the mass of the molecule. On the other hand, Eqs. (20.31) and (20.32) must be modified by replacing $(1/2)mv^2$ with $(1/2)mv^2 + u_{\rm int}$, where $u_{\rm int}$ is the energy per molecule due to internal degrees of freedom. Thus,

$$J_{\rm E} = \frac{n}{4}\int_0^\infty {\rm d}v\, v[(1/2)mv^2 + u_{\rm int}]\tilde{M}(v). \tag{20.33}$$

The crucial difference is that $u_{\rm int}$ is independent of v, so we obtain simply

$$J_{\rm E} = n(2\pi)^{-1/2}2m(k_{\rm B}T/m)^{3/2} + Ju_{\rm int}. \tag{20.34}$$

In this case, the average energy per effused atom is $J_{\rm E}/J = 2k_{\rm B}T + u_{\rm int}$ and we see that there is no enhancement of the internal energy per molecule, as there is for the kinetic energy.

Example Problem 20.4. For the case of a monatomic gas without internal structure, how would the number of atoms and the temperature of the gas in the box decay with time due to effusion?

Solution 20.4. We have

$$\frac{{\rm d}\mathcal{N}}{{\rm d}t} = aJ; \quad \frac{{\rm d}U}{{\rm d}t} = \frac{3}{2}\mathcal{N}k_{\rm B}\frac{{\rm d}T}{{\rm d}t} + \frac{3}{2}k_{\rm B}T\frac{{\rm d}\mathcal{N}}{{\rm d}t} = aJ_{\rm E}. \tag{20.35}$$

One can eliminate ${\rm d}\mathcal{N}/{\rm d}t$ from the second equation to get an ordinary differential equation for T that can be integrated subject to the initial condition $T = T_0$. This can be used to obtain an ordinary differential equation for $\mathcal{N}$ that can be integrated subject to the initial condition $\mathcal{N} = \mathcal{N}_0$. The results are

$$\frac{T}{T_0} = \frac{1}{(1+rt)^2}; \quad \frac{\mathcal{N}}{\mathcal{N}_0} = \frac{1}{(1+rt)^6}, \tag{20.36}$$

where $r = (a/6V)(2\pi)^{-1/2}(k_{\rm B}T_0/m)^{1/2}$ and V is the volume of the box. Of course we need $rt \ll 1$ for the effusion to be slow enough for the quasi-equilibrium assumption to hold.

20.2 Law of Dulong and Petit

An important application of classical statistical mechanics pertains to the heat capacity of a system for which the Hamiltonian is a quadratic function of both the p_i and the q_i:

$$\mathcal{H} = \sum_i \frac{p_i^2}{2m} + \sum_{i,j} q_i L_{ij} q_j. \tag{20.37}$$

Here, m is the particle mass and the L_{ij} are coupling constants. To a first approximation, the energy of a solid, in excess of its equilibrium potential (binding) energy, can be approximated by Eq. (20.37), which is known as a **harmonic Hamiltonian**.

We proceed to evaluate the classical partition function Z_C by substitution of Eq. (20.37) into Eq. (20.3). To do this, we use scaled variables $P_i := p_i\sqrt{\beta}$ and $Q_i = q_i\sqrt{\beta}$. Then

$$Z_C = \beta^{-3\mathcal{N}} \int \exp\left\{-\left[\sum_i \frac{P_i^2}{2m} + \sum_{i,j} Q_i L_{ij} Q_j\right]\right\} d^{3\mathcal{N}}P\, d^{3\mathcal{N}}Q. \tag{20.38}$$

With respect to β, the integral in Eq. (20.38) is just a constant which we do not need to evaluate! Hence Eq. (20.6) becomes

$$U = -\frac{\partial}{\partial\beta}[-3\mathcal{N}\ln\beta + \text{constant}] = 3\mathcal{N}k_B T. \tag{20.39}$$

This amazingly simple result is independent of the mass and the coupling constants!

The corresponding heat capacity is therefore

$$C_V = \left(\frac{\partial U}{\partial T}\right)_{V,\mathcal{N}} = 3\mathcal{N}k_B, \tag{20.40}$$

which is called the **law of Dulong and Petit**. Of course it is only valid at high temperatures and for a Hamiltonian that is strictly a quadratic function of the momenta and coordinates (i.e., strictly harmonic). It is exactly twice the heat capacity of an ideal gas. It is in agreement with the **equipartition principle**, according to which each translational degree of freedom contributes $(1/2)k_B$ per particle and each vibrational degree of freedom also contributes $(1/2)k_B$ per particle. The factor of 3 comes from the dimensionality of space. For one mole of such a solid, $C_V = 3R = 5.96$ cal/(mol K) ≈ 6 cal/(mol K), a good number to remember. Experimental values of C_V approach but generally lie a bit below the value given by the law of Dulong and Petit, even at very high temperatures, presumably due to anharmonic effects. Moreover, quantum effects, which lower C_V to zero as $T \to 0$, may still persist at apparently high temperatures as one approaches the melting point of a solid. See [58, p. 428] for some experimental curves for noble-gas solids.

20.3 Averaging Theorem and Equipartition

The law of Dulong and Petit is actually a special case of a more general theorem that concerns classical thermal averages. We shall proceed to show that under suitable conditions

$$\left\langle \omega_i \frac{\partial \mathcal{H}}{\partial \omega_j}\right\rangle = \delta_{ij}k_B T, \tag{20.41}$$

where ω_k is a component of the $6\mathcal{N}$-dimensional vector $\boldsymbol{\omega}$ composed of the $3\mathcal{N}$ coordinates $\boldsymbol{q}$ and the $3\mathcal{N}$ momenta $\boldsymbol{p}$. Explicit versions of Eq. (20.41) are therefore

$$\left\langle q_i \frac{\partial \mathcal{H}}{\partial q_j}\right\rangle = -\langle q_i \dot{p}_j\rangle = \delta_{ij}k_B T, \tag{20.42}$$

$$\left\langle q_i \frac{\partial \mathcal{H}}{\partial p_j}\right\rangle = \langle q_i \dot{q}_j\rangle = 0, \tag{20.43}$$

$$\left\langle p_i \frac{\partial \mathcal{H}}{\partial q_j} \right\rangle = -\langle p_i \dot{p}_j \rangle = 0, \tag{20.44}$$

$$\left\langle p_i \frac{\partial \mathcal{H}}{\partial p_j} \right\rangle = \langle p_i \dot{q}_j \rangle = \delta_{ij} k_B T, \tag{20.45}$$

where Hamilton's equations (Eq. (17.1)) have been used.

To prove Eq. (20.41), we note that

$$\begin{aligned}\left\langle \omega_i \frac{\partial \mathcal{H}}{\partial \omega_j} \right\rangle &= \frac{1}{Z_C} \int \omega_i \frac{\partial \mathcal{H}}{\partial \omega_j} \mathrm{e}^{-\beta \mathcal{H}}\, \mathrm{d}\omega = -\frac{1}{\beta Z_C} \int \omega_i \frac{\partial \mathrm{e}^{-\beta \mathcal{H}}}{\partial \omega_j}\, \mathrm{d}\omega \\ &= -\frac{1}{\beta Z_C} \int \left[\frac{\partial}{\partial \omega_j} \left(\omega_i \mathrm{e}^{-\beta \mathcal{H}} \right) - \delta_{ij} \mathrm{e}^{-\beta \mathcal{H}} \right] \mathrm{d}\omega \\ &= -\frac{1}{\beta Z_C} \int \omega_i\, \mathrm{e}^{-\beta \mathcal{H}} \Big|_{\omega_j^*}^{\omega_j^{**}} \mathrm{d}\omega^{(j)} + \delta_{ij} k_B T,\end{aligned} \tag{20.46}$$

where ω_j^* and ω_j^{**} represent the limits of integration of ω_j and $\mathrm{d}\omega^{(j)}$ denotes the phase space volume element $\mathrm{d}\omega$ with $\mathrm{d}\omega_j$ missing. Under suitable conditions, the first term on the last line will vanish. This could occur if $\mathcal{H} \to \infty$ at ω_j^* and ω_j^{**}. In such cases, Eq. (20.41) will hold. In other cases, however, the integrated term will not vanish or else the integration by parts makes no sense. For example, in the case of a free particle, $\mathcal{H}$ is independent[3] of all coordinates q_j so $\partial \mathcal{H}/\partial q_j = 0$ and Eq. (20.42) would not hold.

Example Problem 20.5. Calculate the average kinetic energy for a Hamiltonian of the form

$$\mathcal{H} = \sum_{i=1}^{3\mathcal{N}} \frac{p_i^2}{2m} + \mathcal{V}(q_1, q_2, \ldots, q_{3\mathcal{N}}), \tag{20.47}$$

where the first term is the kinetic energy $\mathcal{T}$ and the second term is the potential energy due to interactions among $\mathcal{N}$ particles.

Solution 20.5. We note that $\partial \mathcal{H}/\partial p_j = p_j/m$ so from Eq. (20.45) we obtain $\langle p_j^2/m \rangle = k_B T$. Therefore,

$$\langle \mathcal{T} \rangle = \left\langle \sum_{i=1}^{3\mathcal{N}} \frac{p_i^2}{2m} \right\rangle = \frac{3}{2} \mathcal{N} k_B T. \tag{20.48}$$

Note especially that this result holds not only for an ideal gas but for any system governed by classical statistical mechanics, even when there are interactions among the particles, provided that the potential energy depends only on the coordinates q_i.

---

[3]For a free particle confined to a box, one could modify $\mathcal{H}$ to account for forces due to the walls of the box, but this is better handled by the virial theorem discussed in Section 20.4.

A Hamiltonian of the form of Eq. (20.37) is a homogeneous function of degree 2 in the variables ω. In other words, $\mathcal{H}(\lambda\boldsymbol{\omega}) = \lambda^2\mathcal{H}(\boldsymbol{\omega})$. Applying Euler's theorem (which amounts to differentiating with respect to λ and then setting $\lambda = 1$) gives

$$\sum_i \omega_i \frac{\partial \mathcal{H}}{\partial \omega_i} = 2\mathcal{H}. \tag{20.49}$$

Thus,

$$\langle \mathcal{H} \rangle = \frac{1}{2} k_B T f_s, \tag{20.50}$$

where f_s is the number of active degrees of freedom, which is equal to the number of nonvanishing terms in the sum in Eq. (20.49). If only the kinetic energy terms contribute, as would be the case if the potential energy were zero, we would have $f_s = 3\mathcal{N}$ and the result would be $\langle \mathcal{H} \rangle = (3/2)\mathcal{N} k_B T$, as for the classical ideal gas. If all coordinates and momenta contributed, we would have $f_s = 6\mathcal{N}$, and $\langle \mathcal{H} \rangle = 3\mathcal{N} k_B T$, in agreement with the law of Dulong and Petit in Section 20.2. In the general case, one would have to use a canonical transformation to transform Eq. (20.37) into diagonal form for all generalized coordinates and momenta to see if any terms are missing [8, p. 64]. Moreover, we should eliminate any degrees of freedom that are not activated for quantum mechanical reasons, namely when the corresponding energy levels are so far apart that no excited states are appreciably occupied. For example, consider the degrees of freedom of a diatomic molecule consisting of two point particles. Two point particles would have six degrees of freedom, three translational degrees for each, so one might expect to have $f_s = 6\mathcal{N}$. However, if the particles of the molecule are strongly bound together at some fixed separation ℓ_0, the molecule will behave like a rigid rotator. It has three translational degrees of freedom and it can rotate. Therefore, $f_s = 5\mathcal{N}$ and $\langle \mathcal{H} \rangle = (5/2)\mathcal{N} k_B T$, leading to the well-known heat capacity $C_V = (5/2)\mathcal{N} k_B$. See Section 21.3 for a more detailed discussion, including the possibility of a vibrational degree of freedom[4] if the distance between the particles varies from the constant ℓ_0.

If the particles are not point particles, one might think of including a rotational degree of freedom that amounts to spinning about the axis connecting the particles. However, the moment of inertia for spinning of actual atoms about the axis that connects their centers is so small that the associated quantum energy levels, which are proportional to its inverse, are very high above the ground state. Therefore, only two rotational degrees of freedom are activated at any reasonable temperature. For a detailed analysis, see Section F.8 in Appendix F.

[4]It is worth noting that the vibrational zero point energy $\hbar\omega/2$ per molecule cannot be avoided, even if $\hbar\omega \gg k_B T$ so that excited vibrational states are negligible. This, however, just adds a constant $\mathcal{N}\hbar\omega/2$ to the total energy and does not contribute to the heat capacity, so it is seldom mentioned.

20.4 Virial Theorem

A topic that is closely related to the averaging results in Section 20.3 is the **virial theorem**. The results of Section 20.3, however, are based on *ensemble* averages computed from the classical canonical ensemble. The virial theorem, on the other hand, is based on *time averages* in a classical system. Comparison of these results helps to substantiate that ensemble averages are equivalent to time averages for systems in equilibrium.

We begin by considering the quantity

$$G := \sum_{i=1}^{3\mathcal{N}} q_i p_i, \tag{20.51}$$

where q_i are the canonical coordinates and p_i are the canonical momenta for a classical system of $\mathcal{N}$ particles in three dimensions. Then differentiation with respect to time yields

$$\frac{\mathrm{d}G}{\mathrm{d}t} = \sum_{i=1}^{3\mathcal{N}} \dot{q}_i p_i + \sum_{i=1}^{3\mathcal{N}} q_i \dot{p}_i. \tag{20.52}$$

We define the time average of any function $Q(t)$ of time by the equation

$$\overline{Q} := \frac{1}{\tau} \int_0^\tau Q(t)\,\mathrm{d}t. \tag{20.53}$$

Accordingly, the time average of $\mathrm{d}G/\mathrm{d}t$ is

$$\overline{\frac{\mathrm{d}G}{\mathrm{d}t}} = \frac{G(\tau) - G(0)}{\tau}. \tag{20.54}$$

We now assume that G is bounded, which it certainly will be if the coordinates and momenta themselves are bounded. We also take τ to be arbitrarily large. Since the quantity $G(\tau) - G(0)$ will also be bounded, we obtain

$$\lim_{\tau\to\infty} \overline{\frac{\mathrm{d}G}{\mathrm{d}t}} = \frac{G(\tau) - G(0)}{\tau} = 0. \tag{20.55}$$

Under these circumstances, Eq. (20.52) becomes[5]

$$\overline{\sum_{i=1}^{3\mathcal{N}} \dot{q}_i p_i} = -\overline{\sum_{i=1}^{3\mathcal{N}} q_i \dot{p}_i}. \tag{20.56}$$

According to Eq. (20.45), we would have $\langle \sum_{i=1}^{3\mathcal{N}} \dot{q}_i p_i \rangle = 3\mathcal{N} k_\mathrm{B} T$ and from Eq. (20.42) we would have $\langle \sum_{i=1}^{3\mathcal{N}} q_i \dot{p}_i \rangle = -3\mathcal{N} k_\mathrm{B} T$. Therefore, Eq. (20.56) is consistent with the results of Section 20.3 for systems in equilibrium if the time averages are replaced by ensemble averages.

[5]Textbooks and other references are quite inconsistent on which quantity is called the virial. Some consider the quantity G to be the virial; others consider the right-hand side of Eq. (20.56) or half that quantity or the negative of that quantity to be the virial. We avoid these inconsistencies by not defining any quantity to be the virial and allowing the equations to speak for themselves.

By means of Hamilton's equations, we note that Eq. (20.56) can also be written in the form

$$\overline{\sum_{i=1}^{3\mathcal{N}} p_i \frac{\partial \mathcal{H}}{\partial p_i}} = \overline{\sum_{i=1}^{3\mathcal{N}} q_i \frac{\partial \mathcal{H}}{\partial q_i}}. \tag{20.57}$$

For example, for a Hamiltonian that is homogeneous of degree 2 in all of its coordinates and momenta, Eq. (20.49) applies and its time average gives

$$\overline{\mathcal{H}} = \frac{1}{2}\left[\overline{\sum_{i=1}^{3\mathcal{N}} p_i \frac{\partial \mathcal{H}}{\partial p_i}} + \overline{\sum_{i=1}^{3\mathcal{N}} q_i \frac{\partial \mathcal{H}}{\partial q_i}}\right]. \tag{20.58}$$

According to Eq. (20.57), each sum contributes equally to $\overline{\mathcal{H}}$. If time averages were replaced by ensemble averages, each sum would contribute $3\mathcal{N} k_{\mathrm{B}} T$.

Equation (20.56) can be interpreted readily if we use Cartesian coordinates and a vector notation, in which case it takes the form

$$\overline{\sum_{i=1}^{\mathcal{N}} m_i \left(\frac{\mathrm{d}\mathbf{r}_i}{\mathrm{d}t}\right)^2} = -\overline{\sum_{i=1}^{\mathcal{N}} m_i \, \mathbf{r}_i \cdot \frac{\mathrm{d}^2\mathbf{r}_i}{\mathrm{d}t^2}} = -\overline{\sum_{i=1}^{\mathcal{N}} \mathbf{r}_i \cdot \mathbf{f}_i}, \tag{20.59}$$

where $\mathbf{f}_i$ is the force on particle i having mass m_i. We recognize the left-hand side as twice the time average of the kinetic energy, namely $2\overline{\mathcal{T}}$, which yields

$$\overline{\mathcal{T}} = -\frac{1}{2}\overline{\sum_{i=1}^{\mathcal{N}} \mathbf{r}_i \cdot \mathbf{f}_i}. \tag{20.60}$$

In the form of Eq. (20.59), the virial theorem is often used to relate the average kinetic energy to a total potential $\mathcal{V}(\mathbf{r}_1, \mathbf{r}_2, \ldots, \mathbf{r}_{\mathcal{N}})$ from which the force can be derived. If

$$\mathbf{f}_i = -\nabla_i \mathcal{V} \equiv -\frac{\partial \mathcal{V}}{\partial \mathbf{r}_i}, \tag{20.61}$$

Equation (20.60) takes the form

$$\overline{\mathcal{T}} = \frac{1}{2}\overline{\sum_{i=1}^{\mathcal{N}} \mathbf{r}_i \cdot \frac{\partial \mathcal{V}}{\partial \mathbf{r}_i}}. \tag{20.62}$$

In the case that $\mathcal{V}$ is a homogeneous function of degree α in the coordinates, Eq. (20.62) takes the simple form

$$\overline{\mathcal{T}} = \frac{\alpha}{2}\overline{\mathcal{V}}, \tag{20.63}$$

which relates the average total kinetic energy to the average total potential energy. For the case of a harmonic potential, such as given by Eq. (20.37), we would have $\alpha = 2$ which would lead to $\overline{\mathcal{T}} = \overline{\mathcal{V}} = E/2$, where E is the constant total energy. This is a well-known result for a simple harmonic oscillator but we see here that it is also true for coupled harmonic motion of a number of oscillators.

For the case of gravitational forces, we can take $\alpha = -1$, as shown below, and obtain the often quoted result

$$\overline{\mathcal{T}} = -\frac{1}{2}\overline{\mathcal{V}}. \tag{20.64}$$

Then the total energy, which is a constant, would be

$$E = \overline{\mathcal{T}} + \overline{\mathcal{V}} = \overline{\mathcal{T}} - 2\overline{\mathcal{T}} = -\overline{\mathcal{T}} < 0. \tag{20.65}$$

In this latter case of gravitation, one often sees derivations of Eq. (20.64) in which the interparticle forces are written out explicitly in terms of the relative coordinates of particles, but the result also follows directly from a slightly modified version of the Euler theorem applied to the gravitational potential. Indeed, for gravitational interaction among particles, one can take

$$\mathcal{V}(\mathbf{r}_1, \mathbf{r}_2, \ldots, \mathbf{r}_\mathcal{N}) = -\frac{1}{2}\sum_{j\neq k}\frac{\mathcal{G}m_j m_k}{|\mathbf{r}_i - \mathbf{r}_k|}, \tag{20.66}$$

where $\mathcal{G}$ is the gravitational constant. For $\lambda \neq 0$ it follows that

$$\mathcal{V}(\lambda\mathbf{r}_1, \lambda\mathbf{r}_2, \ldots, \lambda\mathbf{r}_\mathcal{N}) = -\frac{1}{2}\sum_{j\neq k}\frac{\mathcal{G}m_j m_k}{|\lambda\mathbf{r}_i - \lambda\mathbf{r}_k|} = \frac{1}{|\lambda|}\mathcal{V}(\mathbf{r}_1, \mathbf{r}_2, \ldots, \mathbf{r}_\mathcal{N}). \tag{20.67}$$

Therefore,[6]

$$\frac{\partial}{\partial\lambda}\sum_{i=1}^{\mathcal{N}}\mathcal{V}(\lambda\mathbf{r}_1, \lambda\mathbf{r}_2, \ldots, \lambda\mathbf{r}_\mathcal{N}) = \sum_{i=1}^{\mathcal{N}}\mathbf{r}_i \cdot \frac{\partial\mathcal{V}(\lambda\mathbf{r}_1, \lambda\mathbf{r}_2, \ldots, \lambda\mathbf{r}_\mathcal{N})}{\partial(\lambda\mathbf{r}_i)} = -\frac{\lambda}{|\lambda|^3}\mathcal{V}(\mathbf{r}_1, \mathbf{r}_2, \ldots, \mathbf{r}_\mathcal{N}). \tag{20.68}$$

Setting $\lambda = 1$ then gives

$$\sum_{i=1}^{\mathcal{N}}\mathbf{r}_i \cdot \frac{\partial\mathcal{V}(\mathbf{r}_1, \mathbf{r}_2, \ldots, \mathbf{r}_\mathcal{N})}{\partial\mathbf{r}_i} = -\mathcal{V}(\mathbf{r}_1, \mathbf{r}_2, \ldots, \mathbf{r}_\mathcal{N}), \tag{20.69}$$

which is needed to obtain Eq. (20.64) from Eq. (20.62).

20.5 Virial Coefficients

We can use the virial theorem to compute the lowest order correction to the pressure of a classical monatomic gas that accounts approximately for the effect of pairwise forces between atoms. This amounts to calculating what is known as the second virial coefficient $B_2(T)$ in a **virial expansion** of the form

$$\frac{p}{nk_\mathrm{B}T} = 1 + \sum_{\nu=2}^{\infty}B_\nu(T)n^{\nu-1}. \tag{20.70}$$

The first virial coefficient is $B_1 = 1$ and is seldom mentioned.

[6]We use $|\lambda| = \lim_{\epsilon\to 0}\sqrt{\lambda^2 + \epsilon^2}$ to take the derivative of $1/|\lambda|$ on the right-hand side of Eq. (20.67).

We begin by considering a gas in equilibrium in a box of volume V at temperature T to which Eqs. (20.48) and (20.60) apply. Combining these equations by equating the time average and the ensemble average gives

$$\frac{3}{2}\mathcal{N}k_{\mathrm{B}}T = -\overline{\frac{1}{2}\sum_{i=1}^{\mathcal{N}}\mathbf{r}_i\cdot\mathbf{f}_i}. \tag{20.71}$$

Next, we recognize that the forces $\mathbf{f}_i$ come from the walls of the box and from internal forces due to interparticle interactions. The time average forces due to the walls can be accounted for by means of an average pressure p so that

$$-\overline{\frac{1}{2}\sum_{i=1}^{\mathcal{N}}\mathbf{r}_i\cdot\mathbf{f}_i^{\mathrm{walls}}} = p\frac{1}{2}\int_A \mathbf{r}\cdot\hat{\mathbf{n}}\,\mathrm{d}A = p\frac{1}{2}\int_V \nabla\cdot\mathbf{r}\,\mathrm{d}V = \frac{3}{2}pV. \tag{20.72}$$

Thus, Eq. (20.71) becomes

$$p = nk_{\mathrm{B}}T + \overline{\frac{1}{3V}\sum_{i=1}^{\mathcal{N}}\mathbf{r}_i\cdot\mathbf{f}_i^{\mathrm{int}}}, \tag{20.73}$$

where $n = \mathcal{N}/V$ is the number density and $\mathbf{f}_i^{\mathrm{int}}$ are the internal forces due to interparticle interactions.

We proceed to compute the effect of these internal forces for pairwise interactions and central forces between particles i and j that may be calculated from a potential $u(|\mathbf{r}_i - \mathbf{r}_j|)$. Each pair i, j of particles contributes

$$-\mathbf{r}_i\cdot\nabla_i u - \mathbf{r}_j\cdot\nabla_j u = -\frac{\partial u}{\partial r_{ij}}\left[\mathbf{r}_i\cdot\frac{(\mathbf{r}_i-\mathbf{r}_j)}{r_{ij}} + \mathbf{r}_j\cdot\frac{(\mathbf{r}_j-\mathbf{r}_i)}{r_{ij}}\right] = -r_{ij}\frac{\partial u(r_{ij})}{\partial r_{ij}}, \tag{20.74}$$

where $r_{ij} = |\mathbf{r}_i - \mathbf{r}_j|$. Therefore, the interaction term may be written in the form

$$\overline{\frac{1}{3V}\sum_{i=1}^{\mathcal{N}}\mathbf{r}_i\cdot\mathbf{f}_i^{\mathrm{int}}} = \frac{\mathcal{N}}{3V}\overline{\sum_{j\neq 0}\mathbf{r}_0\cdot\mathbf{f}_{j\to 0}^{\mathrm{int}}} = -\frac{\mathcal{N}}{6V}\overline{\sum_{j\neq 0} r_{0j}\frac{\partial u(r_{0j})}{\partial r_{0j}}}, \tag{20.75}$$

where $\mathbf{r}_0$ designates a specific particle and $\mathbf{f}_{j\to 0}^{\mathrm{int}}$ designates the forces on it due to all other particles j. To compute this average, we introduce the **pair distribution function** $g(r)$ defined such that the average number density of particles at a distance r from the center of a particle located at $r = 0$ is $ng(r)$, where $n = \mathcal{N}/V$ is the overall number density. In other words, the average number of particles in a small cube of volume d^3r located at a distance r from the center of a given particle is $ng(r)\,\mathrm{d}^3r$. Thus

$$\overline{\sum_{j\neq 0} r_{0j}\frac{\partial u(r_{0j})}{\partial r_{0j}}} = \int_V r\frac{\partial u(r)}{\partial r}\,ng(r)\,\mathrm{d}^3r = n\int_0^\infty r\frac{\partial u(r)}{\partial r}\,4\pi r^2 g(r)\,\mathrm{d}r. \tag{20.76}$$

We therefore find

$$\frac{p}{nk_{\mathrm{B}}T} = 1 - \frac{n}{6k_{\mathrm{B}}T}\int_0^\infty r\frac{\partial u(r)}{\partial r}\,4\pi r^2 g(r)\,\mathrm{d}r. \tag{20.77}$$

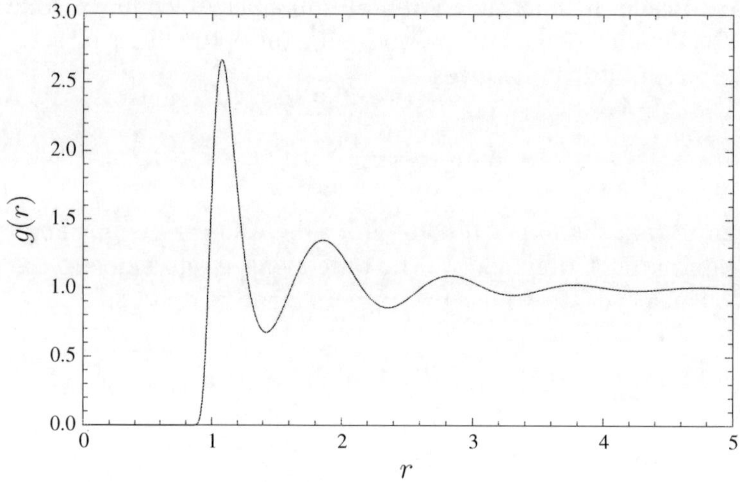

FIGURE 20–1 Sketch of the pair distribution function $g(r)$ versus r for r measured in units of the atomic diameter. For a gas of hard spheres, the rise at $r = 1$ would be vertical and the first peak would be sharp.

The quantity $n4\pi r^2 g(r)$ is the number of particles in a spherical shell between r and $r + \mathrm{d}r$. It is worth noting that the dependence of the second term in Eq. (20.77) on T is more complicated than is apparent because the distribution function $g(r)$ depends on T. At lowest order for a dilute gas, correlations among particles are negligible so $g(r) = \exp(-\beta u)$ is given by the Boltzmann distribution for the pair potential $u(r)$, with the convention $u(\infty) = 0$. Thus, $g(\infty) = 1$ and $g(r) \approx 1$ for r greater than the range of the potential where $u(r) \approx 0$. More generally, $g(r) = \exp(-\beta u) + ng_1(r) + n^2 g_2(r) + \cdots$. Figure 20–1 shows a sketch of $g(r)$ versus r. Typically, $g(r)$ is zero at $r = 0$ and remains at that value until r approaches the atomic diameter; then it rises rapidly to a maximum and undergoes decaying oscillations about the value 1 for a few more atomic diameters, finally decaying to the value 1 for larger r. These oscillations are due to short-range order as rings of neighbors, second nearest neighbors, etc. are reached. For a general definition of the pair distribution function as well as graphs for a hard sphere gas and for argon, see Pathria and Beale [9, p. 332]. For its connection to the direct correlation function and the Ornstein-Zernike equation, see McQuarrie [54, p. 268].

Example Problem 20.6. If the pair distribution function is given to lowest order in n by $g(r) = \mathrm{e}^{-\beta u}$, show that p can be expressed in terms of a volume integral of the Mayer function $f(r) = \mathrm{e}^{-\beta u} - 1$.

Solution 20.6. We note that $\partial f/\partial r = -\beta g(r)\partial u/\partial r$ and substitute into the second term in Eq. (20.77) to obtain

$$\int_0^\infty \frac{\partial u(r)}{\partial r} g(r)\, r^3\, \mathrm{d}r = -\frac{1}{\beta}\int_0^\infty \frac{\partial f}{\partial r} r^3\, \mathrm{d}r = \frac{3}{\beta}\int_0^\infty f\, r^2\, \mathrm{d}r, \tag{20.78}$$

where we have integrated by parts in the last step and noted that $r^3 f(r)$ vanishes at both $r = 0$ and $r = \infty$. Therefore,

$$\frac{p}{nk_B T} = 1 - \frac{n}{2} \int_0^\infty f(r)\, 4\pi r^2 \, dr. \tag{20.79}$$

Example Problem 20.7. Calculate $B_2(r)$ for a gas of hard spheres of diameter σ. When two such spheres just touch, their centers are at a distance σ from each other, which can be accounted for by assuming that there is an infinite potential within a radius $r = \sigma$ from the center of a given sphere.

Solution 20.7. The relevant functions are shown in the following table:

Region	$r < \sigma$	$\sigma < r$
$u(r)$	∞	0
$g(r)$	0	1
$f(r)$	-1	0

Thus,

$$B_2(T) = -\frac{1}{2} \int_0^\infty f(r) 4\pi r^2 \, dr = 2\pi \int_0^\sigma r^2 \, dr = \frac{2\pi}{3} \sigma^3 = 4 v_0, \tag{20.80}$$

where v_0 is the volume of a single hard sphere. In this case, $p = nk_B T(1 + 4 v_0 n) \approx \mathcal{N} k_B T/(V - 4\mathcal{N} v_0)$, which has the form of an ideal gas with an excluded volume equal to four times the volume of all the hard spheres.

Example Problem 20.8. Calculate $B_2(r)$ for a gas having a potential that is infinite for $r < \sigma$, has a square well of depth ε for r between σ and $\sigma + a$, and is zero for $r > \sigma + a$.

Solution 20.8. The relevant functions are shown in the following table:

Region	$r < \sigma$	$\sigma < r < \sigma + a$	$\sigma + a < r$
$u(r)$	∞	$-\varepsilon$	0
$g(r)$	0	$\exp(\beta\varepsilon)$	1
$f(r)$	-1	$\exp(\beta\varepsilon) - 1$	0

The integrations are straightforward and result in

$$B_2(T) = \frac{2\pi}{3} \sigma^3 \left[1 - (e^{\beta\varepsilon} - 1) \frac{(\sigma + a)^3 - \sigma^3}{\sigma^3} \right], \tag{20.81}$$

which agrees with the hard sphere gas for $a = 0$. For $\beta\varepsilon \ll 1$, we can expand the exponential to get the high-temperature result

$$B_2(T) = \frac{2\pi}{3} \sigma^3 \left[1 - \frac{\varepsilon}{k_B T} \frac{(\sigma + a)^3 - \sigma^3}{\sigma^3} \right]. \tag{20.82}$$

Then the equation for p can be written approximately in the van der Waals form

$$p + a_0/v^2 = k_B T/(v - b_0), \tag{20.83}$$

where $v = 1/n$, $a_0 = \varepsilon(2\pi/3)[(\sigma + a)^3 - \sigma^3]$, and $b_0 = (2\pi/3)\sigma^3$.

■ ■ ■

An alternative definition of the pair distribution function is that

$$g(|\mathbf{r}_2 - \mathbf{r}_1|)\frac{d^3 r_1}{V}\frac{d^3 r_2}{V} \tag{20.84}$$

is the probability of finding a pair of particles (any particles) in volumes $d^3 r_1$ and $d^3 r_2$, respectively, that are separated by a distance $|\mathbf{r}_2 - \mathbf{r}_1|$. It follows that $ng(r)\,d^3 r$ is the average number of particles in $d^3 r$ located at $\mathbf{r}$, given that there is a particle at the origin. Thus an alternative way of evaluating the left-hand term in Eq. (20.75) is

$$\frac{1}{3V}\overline{\sum_{i=1}^{\mathcal{N}} \mathbf{r}_i \cdot \mathbf{f}_i^{\text{int}}} = -\frac{1}{3V}\overline{\sum_{\text{pairs}} r_{ij}\frac{\partial u(r_{ij})}{\partial r_{ij}}}. \tag{20.85}$$

The number of pairs is $\mathcal{N}(\mathcal{N} - 1)/2 \approx \mathcal{N}^2/2$ so

$$-\frac{1}{3V}\overline{\sum_{\text{pairs}} r_{ij}\frac{\partial u(r_{ij})}{\partial r_{ij}}} = -\frac{\mathcal{N}^2}{6V}\int\frac{d^3 r_1}{V}\int\frac{d^3 r_2}{V}g(|\mathbf{r}_2 - \mathbf{r}_1|)r_{12}\frac{\partial u(r_{12})}{\partial r_{12}}. \tag{20.86}$$

To do the integrals, one uses the relative coordinate $\mathbf{r} = \mathbf{r}_2 - \mathbf{r}_1$ and the coordinate $\mathbf{r}_1$. Then $\int d^3 r_1/V = 1$ and we are left with

$$-\frac{1}{3V}\overline{\sum_{\text{pairs}} r_{ij}\frac{\partial u(r_{ij})}{\partial r_{ij}}} = -\frac{\mathcal{N}^2}{6V^2}\int d^3 r\, g(r) r\frac{\partial u(r)}{\partial r} = -\frac{n^2}{6}\int_0^\infty r\frac{\partial u(r)}{\partial r}4\pi r^2 g(r)\,dr, \tag{20.87}$$

in agreement with the second term of Eq. (20.77) multiplied by $nk_B T$.

With the use of Eq. (20.84), we can write an expression for the internal energy per particle due to particle-particle interactions. This energy $U_{\text{int}}/\mathcal{N}$ is the difference between the total internal energy $U/\mathcal{N}$ per particle and the energy per particle $(3/2)k_B T$ of an ideal gas and is given by

$$\frac{U_{\text{int}}}{\mathcal{N}} = \frac{\mathcal{N}}{2}\int\frac{d^3 r_1}{V}\int\frac{d^3 r_2}{V}u(r_{12})g(|\mathbf{r}_2 - \mathbf{r}_1|) = \frac{n}{2}\int_0^\infty u(r)4\pi r^2 g(r)\,dr. \tag{20.88}$$

■ ■ ■

Example Problem 20.9. Beginning with the virial expansion Eq. (20.70) and the fact that one obtains an ideal gas if all $B_r(T) = 0$ for $r \geq 2$, determine series expansions for the following: the Helmholtz free energy per particle, f; the entropy per particle, s; the internal energy per particle, u; the heat capacity (at constant volume) per particle, c; and the chemical potential, μ.

Solution 20.9. We begin with

$$df = -s\,dT - p\,dv = -s\,dT + (p/n^2)\,dn \tag{20.89}$$

and integrate p/n^2 over n at constant T. This yields

$$f = k_B T \ln n + w(T) + k_B T \sum_{r=2}^{\infty} B_r(T) \frac{n^{r-1}}{r-1}, \tag{20.90}$$

where $w(T)$ is a function of integration that depends only on T. We determine $w(T)$ by recognizing that $k_B T \ln n + w(T)$ must be the value $f^{\text{ideal}} = (\mu - pv)^{\text{ideal}}$ for an ideal gas. Thus

$$f = k_B T[\ln(n/n_Q) - 1] + k_B T \sum_{r=2}^{\infty} B_r(T) \frac{n^{r-1}}{r-1}, \tag{20.91}$$

where $n_Q = (mk_B T/2\pi\hbar^2)^{3/2}$ is the quantum concentration. Thus,

$$s = -(\partial f/\partial T)_n = k_B[\ln(n_Q/n) + 5/2] - k_B \sum_{r=2}^{\infty} \frac{\mathrm{d}[TB_r(T)]}{\mathrm{d}T} \frac{n^{r-1}}{r-1}, \tag{20.92}$$

$$u = f + Ts = (3/2)k_B T - k_B T^2 \sum_{r=2}^{\infty} \frac{\mathrm{d}B_r(T)}{\mathrm{d}T} \frac{n^{r-1}}{r-1}, \tag{20.93}$$

$$c = (\partial u/\partial T)_n = (3/2)k_B - k_B T \sum_{r=2}^{\infty} \frac{\mathrm{d}^2[TB_r(T)]}{\mathrm{d}T^2} \frac{n^{r-1}}{r-1}, \tag{20.94}$$

$$\mu = f + p/n = k_B T \ln(n/n_Q) + k_B T \sum_{r=2}^{\infty} B_r(T) \frac{r\, n^{r-1}}{r-1}. \tag{20.95}$$

Example Problem 20.10. Show that the pressure given by Eq. (20.79) and the particle-particle interaction energy given by Eq. (20.88) are compatible with the $r = 2$ terms in the general expansions Eqs. (20.70) and (20.93).

Solution 20.10. Agreement of the interaction energies given by Eqs. (20.88) and (20.93) requires

$$-k_B T^2 \frac{\mathrm{d}B_2(T)}{\mathrm{d}T} = \frac{1}{2} \int_0^{\infty} u(r) 4\pi r^2 g(r)\, \mathrm{d}r. \tag{20.96}$$

Agreement of the pressures given by Eqs. (20.70) and (20.79) requires

$$B_2(T) = -\frac{1}{2} \int_0^{\infty} f(r)\, 4\pi r^2\, \mathrm{d}r. \tag{20.97}$$

By differentiation of Eq. (20.97) we obtain

$$\frac{\mathrm{d}B_2(T)}{\mathrm{d}T} = -\frac{1}{2} \int_0^{\infty} \frac{\partial f(r)}{\partial T} 4\pi r^2\, \mathrm{d}r = -\frac{1}{2k_B T^2} \int_0^{\infty} u(r)\, 4\pi r^2 g(r)\, \mathrm{d}r, \tag{20.98}$$

in agreement with Eq. (20.96). Note that we had to use the explicit form $g(r) = \mathrm{e}^{\beta u}$, which gives the correct second-order virial coefficient. More generally, this is only the leading term in an expansion of $g(r)$ in powers of n. Including such terms would lead to virial coefficients of higher order.

20.6 Use of Canonical Transformations

Evaluation of the classical partition function can be facilitated greatly by using canonical transformations to perform the required integrals. Such transformations leave the form of Hamilton's equations unchanged. Simple examples of canonical transformations are coordinate transformations, such as from Cartesian to cylindrical or spherical coordinates. General canonical transformations are discussed in Appendix E. In particular, one transforms from one set of generalized coordinates $q = q_1, q_2, \ldots, q_N$ and their conjugate momenta $p = p_1, p_2, \ldots, p_N$ to another independent set $Q = Q_1, Q_2, \ldots, Q_N$ and $P = P_1, P_2, \ldots, P_N$ according to relations of the form[7]

$$q_i = q_i(Q, P); \quad p_i = p_i(Q, P). \tag{20.99}$$

The corresponding transformation of the partition function would be

$$Z_C = \int \exp[-\beta\,\mathcal{H}(\mathbf{p}, \mathbf{q})]\,\mathrm{d}p\,\mathrm{d}q = \int \exp[-\beta\,\mathcal{K}(\mathbf{P}, \mathbf{Q})]|J|\,\mathrm{d}P\,\mathrm{d}Q, \tag{20.100}$$

where $\mathcal{K}(\mathbf{P}, \mathbf{Q})$ is the new Hamiltonian and $|J|$ is the absolute value of the Jacobian of the transformation. See Eq. (E.4) for an explicit representation of J. As shown in Appendix E, canonical transformations are members of the **symplectic group** for which it is shown (see Eq. (E.32)) that $J = \pm 1$. Thus $|J| = 1$ and the transformation of the partition function integral becomes simply

$$Z_C = \int \exp[-\beta\,\mathcal{H}(\mathbf{p}, \mathbf{q})]\,\mathrm{d}p\,\mathrm{d}q = \int \exp[-\beta\,\mathcal{K}(\mathbf{P}, \mathbf{Q})]\,\mathrm{d}P\,\mathrm{d}Q. \tag{20.101}$$

The simple form of Eq. (20.101) might seem strange to those accustomed to seeing scale factors such as $r^2 \sin\theta$ appearing in Jacobians that arise in transformation of volume integrals from Cartesian to spherical polar coordinates. But for canonical transformations, *both* the coordinates and the momenta transform in just such a way that the corresponding volumes in phase space are the same. Nevertheless, it is frequently the case that familiar scale factors for the coordinate integrals will appear after performing the momentum integrals. Finally, it is sometimes expedient to use transformations that are not canonical to do the necessary integrals. See the next section for explicit examples.

[7] In Appendix E, we allow these transformations to depend explicitly on time as well, but here we are interested in equilibrium ensembles so we omit explicit dependence on time. Such transformations are called restricted canonical transformations and are treated explicitly in Section E.2. For restricted canonical transformations, the old and new Hamiltonians are numerically equal at corresponding points.

Example Problem 20.11. Compute the classical partition function Z_C^* for a single diatomic molecule consisting of point particles having masses m_1 and m_2 separated by a fixed distance ℓ (no vibrational mode) and having a magnetic moment μ_B that makes an angle θ with an applied magnetic field B. The molecule is confined to a box of volume V and its Hamiltonian has the form (which we need to write in terms of canonical momenta)

$$\mathcal{H} = \frac{m_1}{2}\dot{\mathbf{r}}_1^2 + \frac{m_2}{2}\dot{\mathbf{r}}_2^2 - \mu_B B\cos\theta. \tag{20.102}$$

Solution 20.11. We make a canonical transformation to a center of mass coordinate $\mathbf{R} = (m_1\mathbf{r}_1 + m_2\mathbf{r}_2)/(m_1 + m_2)$ and a relative coordinate $\mathbf{r} = \mathbf{r}_1 - \mathbf{r}_2$ but then an additional transformation of the relative coordinate into spherical polar coordinates with fixed radius $r = \ell$, azimuthal angle φ and polar angle θ. The kinetic energy in $\mathcal{H}$ takes the well-known form

$$\frac{m_1}{2}\dot{\mathbf{r}}_1^2 + \frac{m_2}{2}\dot{\mathbf{r}}_2^2 = \frac{M}{2}\dot{\mathbf{R}}^2 + \frac{\mu_{\rm red}}{2}[\ell^2\dot{\theta}^2 + \ell^2\sin^2\theta\dot{\varphi}^2], \tag{20.103}$$

where $M = m_1 + m_2$ is the total mass and $\mu_{\rm red} = m_1 m_2/M$ is the reduced mass. The canonical momenta are $P_i = \partial\mathcal{H}/\partial\dot{R}_i = M\dot{R}_i$, $p_\theta = \partial\mathcal{H}/\partial\dot{\theta} = \mu\ell^2\dot{\theta}$, and $p_\varphi = \partial\mathcal{H}/\partial\dot{\varphi} = \mu\ell^2\sin^2\theta\,\dot{\varphi}$, so the Hamiltonian becomes

$$\mathcal{H} = \sum_{i=1}^{3}\frac{P_i^2}{2M} + \frac{p_\theta^2}{2I} + \frac{p_\varphi^2}{2I\sin^2\theta} - \mu_B B\cos\theta, \tag{20.104}$$

where $I = \mu_{\rm red}\ell^2$ is the moment of inertia about the center of mass of the molecule. See Section F.1.1 in Appendix F for an explicit evaluation of I. Since the transformation is canonical, the partition function is

$$Z_C^* = \frac{1}{h^5}\int \exp[-\beta\mathcal{H}]\,dP_1\,dP_2\,dP_3\,dR_1\,dR_2\,dR_3\,dp_\theta\,dp_\varphi\,d\theta\,d\varphi. \tag{20.105}$$

The factor of h^{-5}, rather than h^{-6}, arises because the magnitude of ℓ has been assumed to be constant (no vibrational mode). Since $\exp[-\beta\mathcal{H}]$ factors, we can perform the R and P integrals immediately to obtain

$$\frac{1}{h^3}\int \exp\left[-\beta\sum_{i=1}^{3}\frac{P_i^2}{2M}\right]dP_1\,dP_2\,dP_3\,dR_1\,dR_2\,dR_3 = Vn_Q, \tag{20.106}$$

where $n_Q = (Mk_BT/2\pi\hbar^2)^{3/2}$. The integrals over dp_θ and dp_φ are well-known Gaussian integrals, resulting in

$$\frac{1}{h}\int \exp\left[-\beta\frac{p_\theta^2}{2I}\right]dp_\theta = \left(\frac{2\pi I}{h^2\beta}\right)^{1/2};\quad \frac{1}{h}\int \exp\left[-\beta\frac{p_\varphi^2}{2I\sin^2\theta}\right]dp_\varphi = \sin\theta\left(\frac{2\pi I}{h^2\beta}\right)^{1/2}. \tag{20.107}$$

The crucial point to note is that the scale factor $\sin\theta$ gets trapped inside the remaining θ integral, resulting in

$$Z_C^* = Vn_Q\left(\frac{2\pi I}{h^2\beta}\right)\int \exp[\beta\mu_B B\cos\theta]\sin\theta\,d\theta\,d\varphi = Vn_Q\left(\frac{2I}{\hbar^2\beta}\right)\frac{\sinh(\beta\mu_B B)}{\beta\mu_B B}. \tag{20.108}$$

We see that the partition function is the product of three factors, Vn_Q which is the partition function for translation of a structureless ideal gas of molecules having molecular mass M, a factor $(2I/\hbar^2\beta)$, which is the partition function for a rigid diatomic molecule rotating about its center of mass, and a factor $\sinh(\beta\mu_B B)/\beta\mu_B B$ which is the partition function for a magnetic dipole.

20.7 Rotating Rigid Polyatomic Molecules

In the approximation of classical statistical mechanics, we can calculate the partition function by integrating over canonical coordinates and momenta in phase space and dividing by appropriate powers of h. Such a partition function should agree with a quantum mechanical result at high temperatures. Rigid rotation of a polyatomic molecule is a three-dimensional problem for a body that can have three different principal moments of inertia, $\mathcal{I}_1, \mathcal{I}_2$, and $\mathcal{I}_3$. See Appendix F for details. As shown in Section F.6, the orientation of the molecule can be expressed in terms of three Euler angles, ϕ, θ, and ψ, where we have adopted the notation and convention of Goldstein [60, p. 107]. As shown in Section F.7, the Hamiltonian can be written in the forms

$$\mathcal{H} = \frac{1}{2}(\mathcal{I}_1\omega_1^2 + \mathcal{I}_2\omega_2^2 + \mathcal{I}_3\omega_3^2) = \frac{L_1^2}{2\mathcal{I}_1} + \frac{L_2^2}{2\mathcal{I}_2} + \frac{L_3^2}{2\mathcal{I}_3}. \tag{20.109}$$

Here, ω_1, ω_2, and ω_3 are principal angular velocities and $L_i = \mathcal{I}_i\omega_i$ are the corresponding principal angular momenta. The ω_i can be expressed in terms of the Euler angles and their time derivatives (see Eq. (F.59)). Then the canonical momenta, p_ϕ, p_θ, and p_ψ, can be calculated by differentiation and are given explicitly by Eqs. (F.61)–(F.63). Thus

$$z = \frac{1}{h^3}\int \exp(-\beta\mathcal{H})\,\mathrm{d}p_\phi\,\mathrm{d}p_\theta\,\mathrm{d}p_\psi\,\mathrm{d}\phi\,\mathrm{d}\theta\,\mathrm{d}\psi. \tag{20.110}$$

One could proceed by solving Eqs. (F.61)–(F.63) for L_1, L_2, L_3 and using the results to eliminate these quantities from Eq. (20.109). This results in a very cumbersome expression for $\mathcal{H}$ as a function of the canonical momenta and the Euler angles and poses a rather unwieldy integration. An alternative procedure is to transform the integration variables to $L_1, L_2, L_3, \phi, \theta, \psi$ by means of a Jacobian $J^{\rm poly}$ so that

$$z = \frac{1}{h^3}\int \exp(-\beta\mathcal{H})\,|J^{\rm poly}|\,\mathrm{d}L_1\,\mathrm{d}L_2\,\mathrm{d}L_3\,\mathrm{d}\phi\,\mathrm{d}\theta\,\mathrm{d}\psi, \tag{20.111}$$

where

$$|J^{\rm poly}| = \left|\frac{\partial(p_\phi, p_\theta, p_\psi, \phi, \theta, \psi)}{\partial(L_1, L_2, L_3, \phi, \theta, \psi)}\right| = \left|\frac{\partial(p_\phi, p_\theta, p_\psi)}{\partial(L_1, L_2, L_3)}\right|. \tag{20.112}$$

This is *not* a canonical transformation, so the magnitude of the Jacobian is

$$|J^{\rm poly}| = \left|\det\begin{pmatrix} \sin\theta\sin\psi & \sin\theta\cos\psi & \cos\theta \\ \cos\psi & -\sin\psi & 0 \\ 0 & 0 & 1 \end{pmatrix}\right| = |-\sin\theta| = \sin\theta. \tag{20.113}$$

The partition function therefore becomes[8]

$$
\begin{aligned}
z &= \frac{1}{h^3}\int \exp(-\beta\mathcal{H})\sin\theta\, \mathrm{d}L_1\,\mathrm{d}L_2\,\mathrm{d}L_3\,\mathrm{d}\phi\,\mathrm{d}\theta\,\mathrm{d}\psi \\
&= \frac{8\pi^2}{h^3}\int \exp\left[-\beta\left(\frac{L_1^2}{2\mathcal{I}_1}+\frac{L_2^2}{2\mathcal{I}_2}+\frac{L_3^2}{2\mathcal{I}_3}\right)\right]\mathrm{d}L_1\,\mathrm{d}L_2\,\mathrm{d}L_3.
\end{aligned} \tag{20.114}
$$

We are left with the product of three Gaussian integrals of the form

$$
\int_{-\infty}^{\infty} \exp[-\beta L_1^2/(2\mathcal{I}_1)]\,\mathrm{d}L_1 = (2\pi\mathcal{I}_1 k_{\mathrm{B}}T)^{1/2}. \tag{20.115}
$$

We therefore obtain

$$
z = \pi^{1/2}\left(\frac{2\mathcal{I}_1 k_{\mathrm{B}}T}{\hbar^2}\right)^{1/2}\left(\frac{2\mathcal{I}_2 k_{\mathrm{B}}T}{\hbar^2}\right)^{1/2}\left(\frac{2\mathcal{I}_3 k_{\mathrm{B}}T}{\hbar^2}\right)^{1/2}. \tag{20.116}
$$

This result will be used in Section 21.3.3 in the context of a gas of polyatomic molecules that can also vibrate.

For a diatomic molecule, only two degrees of freedom are considered because $\mathcal{I}_3$ is essentially zero[9] and the two remaining moments of inertia are equal, say to $\mathcal{I}$. Thus

$$
\mathcal{H} = \frac{\mathcal{I}}{2}(\omega_1^2+\omega_2^2) = \frac{\mathcal{I}}{2}(\sin^2\theta\dot{\phi}^2+\dot{\theta}^2) = \frac{1}{2\mathcal{I}}(L_1^2+L_2^2). \tag{20.117}
$$

Now the only canonical momenta are[10]

$$
p_\phi = \mathcal{I}\sin^2\theta\dot{\phi} = L_1\sin\theta\sin\psi + L_2\sin\theta\cos\psi \tag{20.118}
$$

and

$$
p_\theta = \mathcal{I}\dot{\theta} = L_1\cos\psi - L_2\sin\psi. \tag{20.119}
$$

Therefore,

$$
z^{\mathrm{dia}} = \frac{1}{h^2}\int \exp(-\beta\mathcal{H})\,\mathrm{d}p_\phi\,\mathrm{d}p_\theta\,\mathrm{d}\phi\,\mathrm{d}\theta = \frac{1}{h^2}\int \exp(-\beta\mathcal{H})|J^{\mathrm{dia}}|\,\mathrm{d}L_1\,\mathrm{d}L_2\,\mathrm{d}\phi\,\mathrm{d}\theta, \tag{20.120}
$$

where the magnitude of the Jacobian

$$
|J^{\mathrm{dia}}| = \left|\frac{\partial(p_\phi, p_\theta)}{\partial(L_1, L_2)}\right| = \sin\theta. \tag{20.121}
$$

The integrals over ϕ and θ give a factor of 4π and we obtain

$$
z^{\mathrm{dia}} = \frac{4\pi}{h^2}\int \exp\left[-\beta\left(\frac{L_1^2}{2\mathcal{I}}+\frac{L_2^2}{2\mathcal{I}}\right)\right]\mathrm{d}L_1\,\mathrm{d}L_2. \tag{20.122}
$$

[8]The ranges of the Euler angles are $0 \le \phi \le 2\pi$, $0 \le \theta \le \pi$, and $0 \le \psi \le 2\pi$. Landau and Lifshitz [7, p. 149] give a verbal argument that an integral over three unspecified angles gives a factor of $8\pi^2$ and then proceed to integrate over only L_1, L_2, L_3, but no justification in terms of canonical momenta is given.

[9]As shown in Section F.8 of Appendix F, the quantum states associated with $\mathcal{I}_3$ have such high energies that they are not excited at any reasonable temperature.

[10]Here, we continue to use the same Euler angles as for the polyatomic molecule for the sake of a parallel treatment.

Integration results in two equal factors having the form of Eq. (20.115) which yields the result

$$z^{\mathrm{dia}} = \frac{2\mathcal{I}k_B T}{\hbar^2}, \tag{20.123}$$

in agreement with the high-temperature quantum mechanical result given by Eq. (18.85).

In this simple case, the Hamiltonian can be written in terms of the canonical momenta in the form

$$\mathcal{H} = \frac{1}{2\mathcal{I}}\left(\frac{p_\phi^2}{\sin^2\theta} + p_\theta^2\right), \tag{20.124}$$

so there is not much advantage in transforming to an integral over L_1 and L_2.

On the other hand, when $\mathcal{I}_1$, $\mathcal{I}_2$, and $\mathcal{I}_3$ are all different, there is great simplification in transforming to L_1, L_2, L_3. For example, a normalized probability distribution function for the angular momenta L_i would be

$$M(\mathbf{L}) = \left(\frac{\beta}{2\pi\mathcal{I}_1}\right)^{1/2}\left(\frac{\beta}{2\pi\mathcal{I}_2}\right)^{1/2}\left(\frac{\beta}{2\pi\mathcal{I}_3}\right)^{1/2}\exp\left[-\beta\left(\frac{L_1^2}{2\mathcal{I}_1} + \frac{L_2^2}{2\mathcal{I}_2} + \frac{L_3^2}{2\mathcal{I}_3}\right)\right]. \tag{20.125}$$

The quantity $M(\mathbf{L})\,\mathrm{d}L_1\,\mathrm{d}L_2\,\mathrm{d}L_3$ is the probability of finding an angular momentum in a cube of infinitesimal volume $\mathrm{d}L_1\,\mathrm{d}L_2\,\mathrm{d}L_3$ centered at $\mathbf{L}$. The average square of the angular momentum is

$$\langle L^2\rangle = \int M(\mathbf{L})(L_1^2 + L_2^2 + L_3^2)\,\mathrm{d}L_1\,\mathrm{d}L_2\,\mathrm{d}L_3 = k_B T(\mathcal{I}_1 + \mathcal{I}_2 + \mathcal{I}_3). \tag{20.126}$$

Alternatively, we can transform the integration variables to $\omega_1, \omega_2, \omega_3, \phi, \theta, \psi$ in the partition function Eq. (20.110) by means of the Jacobian

$$|J_\omega| = \left|\frac{\partial\,(p_\phi, p_\theta, p_\psi)}{\partial\,(\omega_1, \omega_2, \omega_3)}\right| = \mathcal{I}_1\mathcal{I}_2\mathcal{I}_3\sin\theta. \tag{20.127}$$

This leads to the same partition function as in Eq. (20.116) but we can also deduce that the normalized distribution function for the ω_i is

$$M^*(\boldsymbol{\omega}) = \left(\frac{\beta\mathcal{I}_1}{2\pi}\right)^{1/2}\left(\frac{\beta\mathcal{I}_2}{2\pi}\right)^{1/2}\left(\frac{\beta\mathcal{I}_3}{2\pi}\right)^{1/2}\exp\left[-\beta\left(\frac{\mathcal{I}_1\omega_1^2}{2} + \frac{\mathcal{I}_2\omega_2^2}{2} + \frac{\mathcal{I}_3\omega_3^2}{2}\right)\right]. \tag{20.128}$$

This leads to an average value

$$\langle\omega^2\rangle = \int M^*(\boldsymbol{\omega})(\omega_1^2 + \omega_2^2 + \omega_3^2)\,\mathrm{d}\omega_1\,\mathrm{d}\omega_2\,\mathrm{d}\omega_3 = k_B T\left(\frac{1}{\mathcal{I}_1} + \frac{1}{\mathcal{I}_2} + \frac{1}{\mathcal{I}_3}\right). \tag{20.129}$$

It is interesting and physically reasonable that average values of the squares of the principal angular velocities are inversely proportional to their respective moments of inertia.

21

Grand Canonical Ensemble

In Chapter 19, we derived the canonical ensemble by starting with the microcanonical ensemble. The microcanonical ensemble applies to an isolated system which therefore has a fixed energy; on the other hand, the canonical ensemble applies to a system that has a fixed temperature. The derivation is accomplished by considering the system of interest to be a subsystem of a total system that is isolated and to which the microcanonical ensemble applies. The remainder of the total system, exclusive of the system of interest, acts as a heat reservoir whose temperature is imposed on the system of interest.

In the present chapter, we introduce the grand canonical ensemble (GCE) which applies to a system having a fixed temperature and a fixed chemical potential, but not a fixed energy or a fixed number of particles. Other extensive parameters of the system, which we take to be only the volume V in the development that follows, are fixed.[1] Our system of interest is again considered to be a subsystem of a total system that is isolated and therefore has a fixed energy and a fixed number of particles. In this case, the remainder of the total system, exclusive of the system of interest, acts as both a heat reservoir and a particle reservoir for the system of interest. Thus, it imposes its temperature and its chemical potential on the system of interest. But the system of interest will have an average energy, U, and an average number of particles, $\langle \mathcal{N} \rangle$, which together with its volume V will be sufficient for its thermodynamic description.

By using the GCE for which the number of particles is not specified, we gain the flexibility to treat systems that have quantum mechanical constraints on the number of particles that can occupy a quantum state. We shall use it to treat ideal Fermi and Bose gases whose wave functions must be antisymmetric or symmetric, respectively, when its identical particles are interchanged. For such quantum ideal gases, the grand canonical partition function factors, which is not the case for the canonical partition function. The classical ideal gas will be shown to be a limiting case of either a Fermi gas or a Bose gas. Thus the approximations used to treat the classical ideal gas by means of the canonical ensemble with the Gibbs correction factor $\mathcal{N}!$ can be clarified. Accordingly we treat a classical ideal gas of molecules having internal structure. Dilute systems for which the constituents can be regarded as independent subsystems can also be treated by a grand canonical partition function that factors. We shall illustrate its use to treat adsorption from a gas that imposes its chemical potential on a surface having dilute adsorption

[1] These are usually the parameters that allow the system to do work. A system without a volume might have an area or a length that is relevant. A system could also have a fixed number of sites that can be occupied by particles.

Thermal Physics. http://dx.doi.org/10.1016/B978-0-12-803304-3.00021-1

sites. Finally, we use the same methodology as used to derive the GCE to develop a pressure ensemble that we illustrate by treating point defects in crystals under conditions of constant temperature and pressure.

21.1 Derivation from Microcanonical Ensemble

We derive the GCE from the microcanonical ensemble by generalization of the second derivation of the canonical ensemble given in Section 19.1.2. For an alternative treatment based on the evaluation of an integral by the method of steepest descent, see Schrödinger [99, p. 41]. We consider a total isolated system (subscript "T") having a fixed energy $E_{\rm T}$ and a fixed number of particles $\mathcal{N}_{\rm T}$. We regard the total system to consist of a reservoir R and a system $\mathcal{I}$ of interest. The system $\mathcal{I}$ may, itself, be very large. We consider a situation in which the system $\mathcal{I}$ is in a quantum state having a specific number of particles $\mathcal{N}_s$ and quantum states $\mathcal{E}_{rs}(V) \equiv \mathcal{E}_r(\mathcal{N}_s, V)$, where its volume V is fixed. For simplicity of notation, we will suppress the argument V in the development that follows.

When the system of interest is in a specific quantum state, the energy of the reservoir will be $E_{\rm T} - \mathcal{E}_{rs}$ and its number of particles will be $N_{\rm T} - \mathcal{N}_s$. Then we represent the multiplicity function (number of microstates) of the reservoir by the symbol $\Omega_{\rm R}(E_{\rm T} - \mathcal{E}_{rs}, \mathcal{N}_{\rm T} - \mathcal{N}_s)$. This is also equal to the multiplicity function of the total system because the system of interest is in a definite quantum state, so its multiplicity function is $\Omega(\mathcal{E}_{rs}, \mathcal{N}_s) = 1$. Symbolically,

$$\Omega_{\rm T}^{rs} = \Omega_{\rm R}(E_{\rm T} - \mathcal{E}_{rs}, \mathcal{N}_{\rm T} - \mathcal{N}_s)\Omega(\mathcal{E}_{rs}, \mathcal{N}_s) = \Omega_{\rm R}(E_{\rm T} - \mathcal{E}_{rs}, \mathcal{N}_{\rm T} - \mathcal{N}_s), \tag{21.1}$$

which is a generalization of Eq. (19.1). The probability of the system of interest being in this definite quantum state is therefore

$$P_{rs} = \frac{\Omega_{\rm T}^{rs}}{\Omega_{\rm T}(E_{\rm T}, \mathcal{N}_{\rm T})} = \frac{\Omega_{\rm R}(E_{\rm T} - \mathcal{E}_{rs}, N_{\rm T} - \mathcal{N}_s)}{\Omega_{\rm T}(E_{\rm T}, \mathcal{N}_{\rm T})} = \frac{\exp[S_{\rm R}(E_{\rm T} - \mathcal{E}_{rs}, N_{\rm T} - \mathcal{N}_s)/k_{\rm B}]}{\exp[S_{\rm T}(E_{\rm T}, \mathcal{N}_{\rm T})/k_{\rm B}]}, \tag{21.2}$$

which is a generalization of Eq. (19.17). Since the entropy of a composite system is additive, we have

$$S_{\rm T}(E_{\rm T}, \mathcal{N}_{\rm T}) = S_{\rm R}(E_{\rm T} - U, \mathcal{N}_{\rm T} - \langle\mathcal{N}\rangle) + S(U, \langle\mathcal{N}\rangle), \tag{21.3}$$

where $U = \langle E\rangle$ is the average internal energy of the system of interest and $\langle\mathcal{N}\rangle$ is its average number of particles in its unrestricted state in equilibrium with the reservoir. We can therefore recast Eq. (21.2) in the form

$$P_{rs} = \frac{\exp[-S(U, \langle\mathcal{N}\rangle)/k_{\rm B}]\exp[S_{\rm R}(E_{\rm T} - \mathcal{E}_{rs}, N_{\rm T} - \mathcal{N}_s)/k_{\rm B}]}{\exp[S_{\rm R}(E_{\rm T} - U, \mathcal{N}_{\rm T} - \langle\mathcal{N}\rangle)/k_{\rm B}]}. \tag{21.4}$$

We write

$$S_{\rm R}\left[E_{\rm T} - \mathcal{E}_{rs}, N_{\rm T} - \mathcal{N}_s\right] = S_{\rm R}\left[(E_{\rm T} - U) + (U - \mathcal{E}_{rs}), (N_{\rm T} - \langle\mathcal{N}\rangle) + (\langle\mathcal{N}\rangle - \mathcal{N}_s)\right] \tag{21.5}$$

and then expand on the basis that $|U - \mathcal{E}_{rs}|/|E_{\rm T} - U| \ll 1$ and $|\langle\mathcal{N}\rangle - \mathcal{N}_s|/|N_{\rm T} - \langle\mathcal{N}\rangle| \ll 1$ to obtain

$$S_R(E_T - \mathcal{E}_{rs}, N_T - \mathcal{N}_s) = S_R(E_T - U, N_T - \langle \mathcal{N} \rangle) + \frac{U - \mathcal{E}_{rs}}{T_R} - \frac{\mu_R}{T_R}(\langle \mathcal{N} \rangle - \mathcal{N}_s) + \cdots . \tag{21.6}$$

Substitution into Eq. (21.4) yields

$$P_{rs} = \exp[(U - T_R S - \mu_R \langle \mathcal{N} \rangle)/(k_B T_R)] \exp[-\mathcal{E}_{rs}/(k_B T_R)] \exp[\mu_R \mathcal{N}_s/(k_B T_R)]. \tag{21.7}$$

Dropping the subscripts on T_R and μ_R and defining the Kramers potential (also known as the grand potential),

$$K := U - TS - \mu \langle \mathcal{N} \rangle = F - \mu \langle \mathcal{N} \rangle, \tag{21.8}$$

Equation (21.7) can be written in the form

$$P_{rs} = \exp(\beta K) \exp(-\beta \mathcal{E}_{rs}) \exp(\beta \mu \mathcal{N}_s), \tag{21.9}$$

where $\beta = 1/(k_B T)$ as usual. The quantity $\exp(-\beta \mathcal{E}_{rs}) \exp(\beta \mu \mathcal{N}_s)$ is referred to as a **Gibbs factor** by Kittel and Kroemer [6], by analogy to a Boltzmann factor.

At constant T and μ, we see that the ratio of the probabilities of any two states is equal to the ratio of their Gibbs factors. We recall for the canonical ensemble that at constant T and $\mathcal{N}$, the ratio of the probabilities of any two states is equal to the ratio of their Boltzmann factors.

If we sum the probabilities P_{rs} over all values of $\mathcal{E}_{rs}$ and $\mathcal{N}_s$ (which Kittel and Kroemer [6] call "all states and numbers," abbreviated by "ASN"), Eq. (21.9) yields

$$1 = \exp(\beta K) \sum_{rs} \exp(-\beta \mathcal{E}_{rs}) \exp(\beta \mu \mathcal{N}_s). \tag{21.10}$$

This allows us to define a grand partition function (also known as the Gibbs sum [6] over ASN),

$$\mathcal{Z} := \sum_{rs} \exp(-\beta \mathcal{E}_{rs}) \exp(\beta \mu \mathcal{N}_s) = \exp(-\beta K). \tag{21.11}$$

For the GCE, Eq. (21.11) defines a grand partition function $\mathcal{Z}$ that plays the same role as the canonical partition function Z. The probabilities can be written in the form

$$P_{rs} = \frac{\exp(-\beta \mathcal{E}_{rs}) \exp(\beta \mu \mathcal{N}_s)}{\mathcal{Z}}, \tag{21.12}$$

which is similar in form to Eq. (19.6).

In order to recover the thermodynamic functions, we write Eq. (21.11) in the form

$$K = -k_B T \ln \mathcal{Z} \tag{21.13}$$

and note[2] from the Legendre transformation $K = F - \mu \langle \mathcal{N} \rangle$ and the differential $\mathrm{d}F$ that

$$\mathrm{d}K = -S\,\mathrm{d}T - p\,\mathrm{d}V - \langle \mathcal{N} \rangle\,\mathrm{d}\mu. \tag{21.14}$$

[2]In dealing with the canonical ensemble, to which F is related, we regard the number of particles $\mathcal{N}$ to be specified; however, for the GCE, we relate K to the average number of particles $\langle \mathcal{N} \rangle$. In thermodynamics, we can ignore the distinction between $\mathcal{N}$ and $\langle \mathcal{N} \rangle$, but for the GCE, this distinction must be made, so we write $\langle \mathcal{N} \rangle$ in Eq. (21.14).

Thus we have[3]

$$S = -\left(\frac{\partial K}{\partial T}\right)_{V,\mu} \quad ; \quad p = -\left(\frac{\partial K}{\partial V}\right)_{T,\mu} \quad ; \quad \langle \mathcal{N} \rangle = -\left(\frac{\partial K}{\partial \mu}\right)_{T,V}. \tag{21.15}$$

The last of these equations can be used to find $\langle \mathcal{N} \rangle$ if μ is known, but for a thermodynamic system, one usually regards $\langle \mathcal{N} \rangle$ to be known. In principle, one can take this point of view, specify $\langle \mathcal{N} \rangle$ and solve for μ, but since μ is contained in a transcendental equation, this can only be done approximately.

To get an expression for the internal energy U, we return to Eq. (21.8) and use Eq. (21.15) to obtain

$$U = K - T\left(\frac{\partial K}{\partial T}\right)_{V,\mu} - \mu\left(\frac{\partial K}{\partial \mu}\right)_{T,V}. \tag{21.16}$$

Substitution of Eq. (21.13) into Eq. (21.16) then leads to

$$U = k_B T^2\left(\frac{\partial \ln \mathcal{Z}}{\partial T}\right)_{V,\mu} + k_B T\mu\left(\frac{\partial \ln \mathcal{Z}}{\partial \mu}\right)_{T,V} = -\left(\frac{\partial \ln \mathcal{Z}}{\partial \beta}\right)_{V,\mu} + \frac{\mu}{\beta}\left(\frac{\partial \ln \mathcal{Z}}{\partial \mu}\right)_{\beta,V}. \tag{21.17}$$

We will obtain a more convenient expression for U in terms of other variables.

Several remarks about the interpretation and structure of the grand partition function are in order:

1. Since $\mathcal{E}_{rs} \equiv \mathcal{E}_r(\mathcal{N}_s)$, the double sum in Eq. (21.11) can first be performed on r to yield

$$\mathcal{Z} = \sum_s \exp(\beta\mu\mathcal{N}_s) \sum_r \exp(-\beta\mathcal{E}_{rs}) = \sum_s \exp(\beta\mu\mathcal{N}_s) Z_{\mathcal{N}_s}, \tag{21.18}$$

   where, according to Eq. (19.5),

$$Z_{\mathcal{N}_s} := \sum_r \exp(-\beta\mathcal{E}_{rs}) \tag{21.19}$$

   is the canonical partition function for a system having exactly $\mathcal{N}_s$ particles.
2. We need to specify clearly the variable set on which $\mathcal{Z}$ and K depend. Until now, as reflected by Eq. (21.14), we have considered the variable set to be T, μ, and V or, since k_B is a constant, the equivalent set β, μ, and V. But one can also introduce the **absolute activity**

$$\lambda := \exp(\beta\mu) \tag{21.20}$$

   and consider instead the variable set T, λ, and V or equivalently β, λ, and V. Then Eq. (21.18) can be written in the form of a power series

$$\mathcal{Z} = \sum_s \lambda^{\mathcal{N}_s} \sum_r \exp(-\beta\mathcal{E}_{rs}) = \sum_{\mathcal{N}_s} \lambda^{\mathcal{N}_s} Z_{\mathcal{N}_s} \tag{21.21}$$

[3]Strictly speaking, S is an average entropy and p is an average pressure, but we have omitted the averaging brackets because they were absent in Eq. (21.14) which is of thermodynamic origin. In the thermodynamic limit, the distinction is irrelevant.

whose coefficients are just the canonical partition functions that can be extracted by means of the formula

$$Z_{\mathcal{N}_s} = \frac{1}{\mathcal{N}_s!}\left(\frac{\partial}{\partial \lambda}\right)^{\mathcal{N}_s} \mathcal{Z}\Bigg|_{\lambda=0}. \tag{21.22}$$

Then the probabilities can be written in the form

$$P_{rs} = \frac{\lambda^{\mathcal{N}_s}\exp(-\beta\mathcal{E}_{rs})}{\mathcal{Z}}. \tag{21.23}$$

3. In the expression for $\mathcal{Z}$, it is often convenient to run the sum formally from $\mathcal{N}_s = 0$ to $\mathcal{N}_s = \infty$ which would require a reservoir of infinite size. This does not give rise to a problem because we are interested in systems with finite $\langle \mathcal{N} \rangle$ and we shall see that the important values of $\mathcal{N}_s$ are those near $\langle \mathcal{N} \rangle$ because

$$\frac{\sqrt{\langle(\mathcal{N}_s - \langle\mathcal{N}\rangle)^2\rangle}}{\langle\mathcal{N}\rangle} \sim \frac{1}{\sqrt{\langle\mathcal{N}\rangle}} \tag{21.24}$$

which turns out to be exact for an ideal gas (see Eqs. (21.51) and (21.52)). For a system having a finite number of absorption sites (see Section 21.2.1), the sum is finite.

4. The state with $\mathcal{N}_s = 0$ is known as the **vacuum state**. We regard it to be a nondegenerate state having zero energy, $Z(\mathcal{N}_s = 0) = 1$. Then Eq. (21.18) can be written in the form

$$\mathcal{Z} = 1 + \sum_{\mathcal{N}_s} \lambda^{\mathcal{N}_s} Z_{\mathcal{N}_s}. \tag{21.25}$$

All other energies are to be measured relative to the vacuum state. This convention is consistent with representation of many particle states by means of occupation numbers of orbitals, which are quantum states of single particles (see Eq. (21.63)).

21.1.1 Kramers Function

Somewhat more transparent results for the thermodynamic functions can be written in terms of the Kramers dimensionless function[4]

$$q(\beta, V, \lambda) := \ln \mathcal{Z}, \tag{21.26}$$

where $\mathcal{Z}$ is expressed in terms of the variables β, λ , and V according to Eq. (21.21). Since $(\partial\lambda/\partial\beta)_{V,\mu} = \lambda\mu$ and $(\partial\lambda/\partial\mu)_{V,\beta} = \lambda\beta$, we find

$$\left(\frac{\partial \ln \mathcal{Z}}{\partial \beta}\right)_{V,\mu} = \left(\frac{\partial q}{\partial \beta}\right)_{V,\lambda} + \lambda\mu\left(\frac{\partial q}{\partial \lambda}\right)_{V,\beta} \tag{21.27}$$

and

$$\left(\frac{\partial \ln \mathcal{Z}}{\partial \mu}\right)_{V,\beta} = \lambda\beta\left(\frac{\partial q}{\partial \lambda}\right)_{V,\beta}. \tag{21.28}$$

[4]The Kramers potential (grand potential) K is related to the dimensionless Kramers function q by the equation $K = -k_B T q$, but the variables on which they are usually regarded to depend are different.

Thus Eq. (21.17) becomes

$$U = -\left(\frac{\partial q}{\partial \beta}\right)_{V,\lambda}. \tag{21.29}$$

Similar conversion of the derivatives in Eq. (21.15) gives

$$\langle \mathcal{N} \rangle = k_B T \left(\frac{\partial \ln \mathcal{Z}}{\partial \mu}\right)_{V,\beta} = \lambda \left(\frac{\partial q}{\partial \lambda}\right)_{V,\beta} \tag{21.30}$$

and

$$p = k_B T \left(\frac{\partial \ln \mathcal{Z}}{\partial V}\right)_{\beta,\mu} = \frac{1}{\beta}\left(\frac{\partial q}{\partial V}\right)_{\beta,\lambda}. \tag{21.31}$$

These results can be summarized in terms of the differential

$$dq = -U\,d\beta + \beta p\,dV + \frac{\langle \mathcal{N} \rangle}{\lambda}\,d\lambda. \tag{21.32}$$

Note that Eqs. (21.29) and (21.30) could have been obtained directly from definitions of average quantities in terms of the probabilities $P_{rs} = \lambda^{\mathcal{N}_s} \exp(-\beta \mathcal{E}_{rs})/\mathcal{Z}$ with $\mathcal{Z}$ expressed as a function of β, λ, and V. For example, $U = \sum_{rs} P_{rs}\mathcal{E}_{rs} = (\partial \ln \mathcal{Z}/\partial \beta)_{\lambda,V}$ because differentiation with respect to β introduces $\mathcal{E}_{rs}$ inside the sum.

Equations (21.29) and (21.30) can be written as weighted averages by using the explicit form of Eq. (21.21) for q, namely

$$q = \ln\left[\sum_{\mathcal{N}_s} \lambda^{\mathcal{N}_s} Z_{\mathcal{N}_s}\right]. \tag{21.33}$$

We recall that the average energy for a system of exactly $\mathcal{N}_s$ particles in the canonical ensemble is

$$U_{\mathcal{N}_s} = -\frac{1}{Z_{\mathcal{N}_s}}\left(\frac{\partial Z_{\mathcal{N}_s}}{\partial \beta}\right)_{V,\mathcal{N}_s}. \tag{21.34}$$

Thus Eq. (21.29) takes the form

$$U = \frac{\sum_{\mathcal{N}_s} \lambda^{\mathcal{N}_s} Z_{\mathcal{N}_s} U_{\mathcal{N}_s}}{\sum_{\mathcal{N}_s} \lambda^{\mathcal{N}_s} Z_{\mathcal{N}_s}}, \tag{21.35}$$

which is a weighed average of $U_{\mathcal{N}_s}$ with weighting factors $\lambda^{\mathcal{N}_s} Z_{\mathcal{N}_s}$. Equation (21.30) takes a similar form

$$\langle \mathcal{N} \rangle = \frac{\sum_{\mathcal{N}_s} \lambda^{\mathcal{N}_s} Z_{\mathcal{N}_s} \mathcal{N}_s}{\sum_{\mathcal{N}_s} \lambda^{\mathcal{N}_s} Z_{\mathcal{N}_s}}. \tag{21.36}$$

In fact, Eqs. (21.35) and (21.36) follow directly from the probabilities given by Eq. (21.12). The reader is invited to show that the pressure is a similar weighted average of the pressures, calculated from the canonical ensemble, for systems having definite values of $\mathcal{N}_s$.

The pressure can also be related directly to the q function by an algebraic equation. The only extensive variable on which K depends is V, so its Euler equation is just

$$K = -pV, \tag{21.37}$$

which is consistent with $U = TS - pV + \mu\langle\mathcal{N}\rangle$. Therefore,

$$q = \frac{pV}{k_B T}. \tag{21.38}$$

Both K and q are extensive variables that depend on β, λ, and V, but β and λ are intensive, so both K and q must be proportional to V.

Indeed, comparison of Eq. (21.38) with Eq. (21.31) requires

$$\left(\frac{\partial q}{\partial V}\right)_{\beta,\lambda} = \frac{q}{V}, \tag{21.39}$$

which can be integrated at constant β and λ to give

$$\ln q = \ln V + \ln q_0(\beta, \lambda). \tag{21.40}$$

The last term in Eq. (21.40) is a function of integration, independent of V, that could equally well depend on T and μ. Therefore, q is of the form

$$q = V q_0(\beta, \lambda), \tag{21.41}$$

which is proportional to V. Comparison with Eq. (21.38) shows that

$$p = k_B T q_0(\beta, \lambda), \tag{21.42}$$

so the intensive variable p can be expressed in terms of only the intensive variables β and λ, or equivalently T and μ, independent of V as expected.

It sometimes happens that the GCE is used to treat systems that do not contain the volume V as a variable, in which case $\mathcal{Z}$, and therefore q and K, are independent of V. Examples of such systems would be identical spins or harmonic oscillators, fixed in position and distinguishable by virtue of their position, as in a rigid solid, or a set of adsorption sites on a surface. In such cases, equations such as Eqs. (21.31) and (21.37)–(21.42) are not applicable. Formally, such systems have zero pressure or equivalents to pressure that can be defined in spaces of lower dimensionality. As long as these systems are well defined, we can still use equations such as Eqs. (21.35) and (21.36), or their equivalents, because they can be justified on the basis of

$$U = \sum_{r,s} P_{rs}\mathcal{E}_{rs} = -\left(\frac{\partial q}{\partial \beta}\right)_{\lambda} \tag{21.43}$$

and

$$\langle\mathcal{N}\rangle = \sum_{r,s} P_{rs}\mathcal{N}_s = \lambda\left(\frac{\partial q}{\partial \lambda}\right)_{\beta}. \tag{21.44}$$

Example Problem 21.1. Show that the entropy can be expressed in the form

$$S = -k_B \sum_{rs} P_{rs} \ln P_{rs}. \tag{21.45}$$

Solution 21.1. Whether or not there is dependence on V, we have $K = U - TS - \mu \langle \mathcal{N} \rangle$ so $S/k_B = q + \beta U - \langle \mathcal{N} \rangle \ln \lambda$. But

$$-\sum_{rs} P_{rs} \ln P_{rs} = -\sum_{rs} P_{rs}[-q - \beta \mathcal{E}_{rs} + \mathcal{N}_s \ln \lambda] = q + \beta U - \langle \mathcal{N} \rangle \ln \lambda.$$

The entropy takes the same form as Eq. (21.45) in any ensemble.

21.1.2 Particle Number Dispersion

In Section 19.5, we showed for the canonical ensemble that there was dispersion of the internal energy for a system held at constant temperature. This is also true for the GCE; however, for the GCE, the chemical potential is held constant and equal to that of a reservoir. Therefore, for a system described by the GCE, there is also dispersion of the number of particles relative to the average number of particles $\langle \mathcal{N} \rangle = (1/\beta)(\partial \ln \mathcal{Z}/\partial \mu)_{\beta,V}$ given by Eq. (21.15).

We can quantify this dispersion of particle number by calculating its second moment relative to its average value, namely

$$\langle (\Delta \mathcal{N})^2 \rangle := \langle (\mathcal{N} - \langle \mathcal{N} \rangle)^2 \rangle = \langle \mathcal{N}^2 \rangle - \langle \mathcal{N} \rangle^2, \tag{21.46}$$

where

$$\langle \mathcal{N}^2 \rangle = \sum_{rs} \mathcal{N}_s^2 P_{rs}, \tag{21.47}$$

with P_{rs} given by Eq. (21.12). By differentiation of Eq. (21.11), we note that

$$\frac{\partial^2 \mathcal{Z}}{\partial \mu^2} = \beta^2 \sum_{rs} \mathcal{N}_s^2 \exp(-\beta \mathcal{E}_{rs}) \exp(\beta \mu \mathcal{N}_s), \tag{21.48}$$

which yields

$$\langle \mathcal{N}^2 \rangle = \frac{1}{\beta^2} \frac{1}{\mathcal{Z}} \left(\frac{\partial^2 \mathcal{Z}}{\partial \mu^2} \right)_{\beta,V}. \tag{21.49}$$

Therefore,

$$\langle (\Delta \mathcal{N})^2 \rangle = \frac{1}{\beta^2} \frac{1}{\mathcal{Z}} \frac{\partial^2 \mathcal{Z}}{\partial \mu^2} - \frac{1}{\beta^2} \frac{1}{\mathcal{Z}^2} \left(\frac{\partial \mathcal{Z}}{\partial \mu} \right)^2 = \frac{1}{\beta^2} \frac{\partial^2 \ln \mathcal{Z}}{\partial \mu^2} = \frac{1}{\beta} \left(\frac{\partial \langle \mathcal{N} \rangle}{\partial \mu} \right)_{\beta,V}. \tag{21.50}$$

Since μ and β are intensive, the right-hand side of Eq. (21.50) is $O(\langle \mathcal{N} \rangle)$. Therefore, in agreement with Eq. (21.24), we have

$$\frac{\sqrt{\langle (\Delta \mathcal{N})^2 \rangle}}{\langle \mathcal{N} \rangle} = O\left(\frac{1}{\sqrt{\langle \mathcal{N} \rangle}} \right). \tag{21.51}$$

For an ideal gas, Eq. (19.66) applies,[5] so $\partial\langle\mathcal{N}\rangle/\partial\mu=\beta\langle\mathcal{N}\rangle$ and we have exactly

$$\frac{\sqrt{\langle(\Delta\mathcal{N})^2\rangle}}{\langle\mathcal{N}\rangle}=\frac{1}{\sqrt{\langle\mathcal{N}\rangle}}. \tag{21.52}$$

For a system having a large number of particles, we see that this dispersion is quite small.

The result in Eq. (21.50) can be related to the isothermal compressibility, $\kappa_T=(1/v)(\partial v/\partial p)_\beta$, where $v := V/\langle\mathcal{N}\rangle$ is the volume per particle. At constant β, one has $\mathrm{d}\mu = v\,\mathrm{d}p$, where p is the pressure. Thus

$$\left(\frac{\partial\mu}{\partial v}\right)_\beta = v\left(\frac{\partial p}{\partial v}\right)_\beta = -1/\kappa_T. \tag{21.53}$$

Therefore,

$$\left(\frac{\partial\langle\mathcal{N}\rangle}{\partial\mu}\right)_{\beta,V}=\left(\frac{\partial(V/v)}{\partial\mu}\right)_{\beta,V}=-\frac{V}{v^2}\left(\frac{\partial v}{\partial\mu}\right)_\beta=\frac{\langle\mathcal{N}\rangle}{v}\kappa_T. \tag{21.54}$$

Substitution into Eq. (21.50) leads to

$$\frac{\sqrt{\langle(\Delta\mathcal{N})^2\rangle}}{\langle\mathcal{N}\rangle}=\sqrt{\frac{k_BT\kappa_T}{v\langle\mathcal{N}\rangle}} \tag{21.55}$$

in agreement with Eq. (21.51). For an ideal gas, $\kappa_T=1/p$ and $k_BT/(pv)=1$, so Eq. (21.55) becomes Eq. (21.52). Since $v\langle\mathcal{N}\rangle = V$, we observe that fluctuations in particle numbers could be large if very small sub-volume of a large volume of gas is observed.

21.1.3 Energy Dispersion

Energy dispersion in the GCE is different from that calculated for the canonical ensemble in Section 19.5 because in the GCE the number of particles has dispersion. From Eq. (21.29) we have $U=\langle E\rangle = -(1/\mathcal{Z})(\partial\mathcal{Z}/\partial\beta)_{\lambda,V}$. By analogy to Eq. (19.85) we have

$$\langle E^2\rangle=\sum_{rs}\mathcal{E}_{rs}^2P_{rs}=\frac{1}{\mathcal{Z}}\left(\frac{\partial^2\mathcal{Z}}{\partial\beta^2}\right)_{\lambda,V}. \tag{21.56}$$

Then by following a procedure similar to that in Section 19.5, we find

$$\langle(\Delta E)^2\rangle=\langle E^2\rangle-\langle E\rangle^2=\left(\frac{\partial\ln\mathcal{Z}}{\partial\beta^2}\right)_{\lambda,V}=-\left(\frac{\partial U}{\partial\beta}\right)_{\lambda,V}. \tag{21.57}$$

We can relate the derivative in Eq. (21.57) to the heat capacity at constant volume, C_V, by thinking of U as a function of β, V, and $\langle\mathcal{N}\rangle$ and writing

$$\left(\frac{\partial U}{\partial\beta}\right)_{\lambda,V}=\left(\frac{\partial U}{\partial\beta}\right)_{\langle\mathcal{N}\rangle,V}+\left(\frac{\partial U}{\partial\langle\mathcal{N}\rangle}\right)_{\beta,V}\left(\frac{\partial\langle\mathcal{N}\rangle}{\partial\beta}\right)_{\lambda,V}. \tag{21.58}$$

[5]Equation (19.66) applies to the canonical ensemble, for which the system is regarded as having an exact number of particles $\mathcal{N}$, which therefore corresponds to the symbol $\langle\mathcal{N}\rangle$ of the GCE.

The first term in Eq. (21.58) is just $-k_B T^2 C_V$ and if substituted alone into Eq. (21.57) would give Eq. (19.87) for the canonical ensemble. To evaluate the remaining terms in Eq. (21.58), we make use of a Maxwell relation derived from Eq. (21.32), namely

$$\left(\frac{\partial(\langle\mathcal{N}\rangle/\lambda)}{\partial\beta}\right)_{\lambda,V} = -\left(\frac{\partial U}{\partial\lambda}\right)_{\beta,V}, \tag{21.59}$$

which becomes

$$\left(\frac{\partial\langle\mathcal{N}\rangle}{\partial\beta}\right)_{\lambda,V} = -\left(\frac{\partial U}{\partial\ln\lambda}\right)_{\beta,V} = -\frac{1}{\beta}\left(\frac{\partial U}{\partial\mu}\right)_{\beta,V}. \tag{21.60}$$

But

$$\left(\frac{\partial U}{\partial\mu}\right)_{\beta,V} = \left(\frac{\partial U}{\partial\langle\mathcal{N}\rangle}\right)_{\beta,V}\left(\frac{\partial\langle\mathcal{N}\rangle}{\partial\mu}\right)_{\beta,V} = \left(\frac{\partial U}{\partial\langle\mathcal{N}\rangle}\right)_{\beta,V}\beta\langle(\Delta\mathcal{N})^2\rangle, \tag{21.61}$$

where Eq. (21.50) has been used in the last step. We therefore obtain finally

$$\langle(\Delta E)^2\rangle = k_B T^2 C_V + \left\{\left(\frac{\partial U}{\partial\langle\mathcal{N}\rangle}\right)_{\beta,V}\right\}^2\langle(\Delta\mathcal{N})^2\rangle. \tag{21.62}$$

Equation (21.62) shows that the energy dispersion is the sum of two terms, the first term being the same as for the canonical ensemble and the second term arising from dispersion of the number of particles in the system.

21.2 Ideal Systems: Orbitals and Factorization

The GCE can be used to treat the important case of ideal systems of identical particles that can be described in terms of single-particle quantum states called **orbitals**. As defined by Kittel and Kroemer [6, p. 152], an orbital is a term often used by chemists to denote a single-particle quantum state characterized by specification of the quantum numbers of its spatial wave function *and its spin*. A system for which particles interact very weakly can be approximated by an ideal system in which the particles do not interact at all. For a system having $\mathcal{N}_s$ noninteracting particles, the total wave function can be formed as a sum of products of the wave functions of the orbitals, [8, p. 116], which requires specification of the numbers (called **occupation numbers**) of particles that occupy each orbital. Frequently there is an infinite number of possible orbitals. If the particles are fermions (half integral spin), the total wave function must be antisymmetric under interchange of particles, which requires that each orbital be either unoccupied or occupied by only one particle (the **Pauli exclusion principle**). If the particles are bosons, the total wave function must be symmetric under interchange of particles, which allows each orbital to be unoccupied or multiply-occupied. Classical particles are an approximation to fermions or bosons in a dilute limit to be discussed below.

We denote each orbital by the single symbol ε which denotes its energy *but also carries the information about all of its quantum numbers, including spin.* The number of such orbitals in a quantum state r of the whole system having $\mathcal{N}_s$ particles is denoted by n_ε^{rs}.

These occupation numbers, n_ε^{rs}, completely specify the state of the system. Since the particles are identical, we cannot distinguish which particles are in a given orbital. The energy of a quantum state having $\mathcal{N}_s$ particles is then given by[6]

$$\mathcal{E}_{rs} = \sum_\varepsilon n_\varepsilon^{rs} \varepsilon, \tag{21.63}$$

where

$$\sum_\varepsilon n_\varepsilon^{rs} = \mathcal{N}_s, \quad \text{all allowed states } r. \tag{21.64}$$

The grand partition function is therefore

$$\mathcal{Z} = \sum_s \lambda^{\mathcal{N}_s} \sum_r^{*} \exp\left(-\beta \sum_\varepsilon n_\varepsilon^{rs} \varepsilon\right) = \sum_s \sum_r^{*} \prod_\varepsilon [\lambda \exp(-\beta\varepsilon)]^{n_\varepsilon^{rs}}. \tag{21.65}$$

For fixed $\mathcal{N}_s$, the allowed values of n_ε^{rs} are constrained by Eq. (21.64) and also by the constraints for fermions or bosons. The asterisk "$*$" on the r sum is intended to remind us of those constraints. Since, however, *all* values of r and s are summed over in Eq. (21.65), we do not need to *correlate* the values of n_ε^{rs} for a given orbital ε. The double sum over s and the restricted sum over r can therefore be replaced a single sum over quantum-mechanically allowed occupation numbers that are *uncorrelated* for each orbital ε. The remaining sum and the product commute, so we obtain

$$\mathcal{Z} = \prod_\varepsilon \left\{ \sum_n [\lambda \exp(-\beta\varepsilon)]^n \right\}. \tag{21.66}$$

In other words, $\mathcal{Z}$ factors into the product of the grand partition functions for the individual orbitals, each of the form

$$\mathcal{Z}_1(\varepsilon) := \sum_n [\lambda \exp(-\beta\varepsilon)]^n. \tag{21.67}$$

In Eq. (21.67), n is only constrained by the rules for occupation of orbitals by fermions $(n = 0, 1)$ or bosons $(n = 0, 1, 2, 3, \ldots)$, depending on which kind of particles we are considering. Specifically,

$$\mathcal{Z} = \prod_\varepsilon \mathcal{Z}_1(\varepsilon), \tag{21.68}$$

so contributions of the orbitals to $\ln \mathcal{Z}$ and physical properties are simply additive. Any restrictions of orbital occupation were already taken into account in computing $\mathcal{Z}_1(\varepsilon)$.

Alternatively, Eq. (21.68) can be justified on physical grounds, a point of view taken by Kittel and Kroemer [6, p. 154]. They consider all but one orbital to be part of the reservoir

[6]Here, we have used a shorthand notation $\mathcal{E}_{rs} \equiv \mathcal{E}_r(\mathcal{N}_s)$ and $n_\varepsilon^{rs} \equiv n_\varepsilon^r(\mathcal{N}_s)$. For the vacuum state $\mathcal{N}_s = 0$, all occupation numbers $n_\varepsilon^r(\mathcal{N}_s = 0) = 0$, in which case Eq. (21.63) gives $\mathcal{E}_r(\mathcal{N}_s = 0) = 0$. This is consistent with the convention $Z(\mathcal{N}_s = 0) = 1$ used to establish Eq. (21.25).

with which that orbital interacts, and thus calculate its grand partition function separately. For the entire system, they appeal to additivity over ε of $\ln \mathcal{Z}_1(\varepsilon)$, so that

$$\ln \mathcal{Z} = \sum_{\varepsilon} \ln \mathcal{Z}_1(\varepsilon), \tag{21.69}$$

which is equivalent to Eq. (21.68). In a similar spirit, they also present examples [6, pp. 140-146] in which the general formula for $\mathcal{Z}$ is applied to noninteracting subsystems that can be unoccupied or occupied by one or two particles. The resulting partition function is used to calculate the probability of each state. A similar example is presented by Callen [2, p. 389] in which sites on a surface can be empty, singly occupied or doubly occupied by adsorbed gas molecules, and factorization of the grand partition function is assumed because the sites are so sparsely distributed that they do not interact.

We can generalize these examples as follows. If $\mathcal{Z}^{(\nu)}(\beta, \lambda)$ are the grand partition functions for a set of subsystems labeled by ν and such subsystems are independent and have negligible interaction energy, the grand partition function for the entire system is

$$\mathcal{Z} = \prod_{\nu} \mathcal{Z}^{(\nu)}(\beta, \lambda); \quad \ln \mathcal{Z} = \sum_{\nu} \ln \mathcal{Z}^{(\nu)}(\beta, \lambda), \tag{21.70}$$

where

$$\mathcal{Z}^{(\nu)}(\beta, \lambda) = \sum_{n} \lambda^n \sum_{r} \exp(-\beta \varepsilon_{rn}^{(\nu)}). \tag{21.71}$$

Each subsystem is in equilibrium with the reservoir and therefore with each other. Note that Eq. (21.71) is more general than Eq. (21.67) because the energies $\varepsilon_{rn}^{(\nu)}$ do not have to be multiples of the same quantity ε.,

21.2.1 Factorization for Independent Sites

In this section, we present several examples of factorization of the grand partition function for cases in which particles can reside on a number $\mathcal{N}_{\rm tot}$ of noninteracting sites that can be occupied by one or more particles. This would be expected if such sites are sufficiently dilute; they are separated by distances that are large relative to the range of forces applicable to each site. In these examples, we shall suppose for simplicity that the chemical potential μ, and hence the activity $\lambda = \exp(\beta\mu)$ is imposed by a classical ideal monatomic gas.

Example Problem 21.2. Calculate the probability of adsorption of an ideal gas on $\mathcal{N}_{\rm tot}$ independent sites that are either unoccupied, with energy zero, or singly occupied with energy ε_1.

Solution 21.2. The grand partition function for a single site is $\mathcal{Z}^{(1)} = 1 + \lambda\, \mathrm{e}^{-\beta\varepsilon_1}$ so the total grand partition function is $\mathcal{Z} = (\mathcal{Z}^{(1)})^{\mathcal{N}_{\rm tot}}$. The average number of adsorbed atoms, which in this case happens to equal the number of occupied sites, is

$$\langle \mathcal{N} \rangle = \lambda \frac{\partial q}{\partial \lambda} = \mathcal{N}_{\text{tot}} \frac{\lambda\, \mathrm{e}^{-\beta\varepsilon_1}}{1 + \lambda\, \mathrm{e}^{-\beta\varepsilon_1}} \tag{21.72}$$

and the average energy is

$$U = -\frac{\partial q}{\partial \beta} = \mathcal{N}_{\text{tot}} \frac{\varepsilon_1 \lambda\, \mathrm{e}^{-\beta\varepsilon_1}}{1 + \lambda\, \mathrm{e}^{-\beta\varepsilon_1}}. \tag{21.73}$$

Except for the very important factors of λ, Eq. (21.73) resembles the energy for independent two-state systems. In order for the gas to adsorb at low temperatures, we want $\varepsilon_1 < 0$. The fraction of occupied sites is $\theta = \langle \mathcal{N} \rangle / \mathcal{N}_{\text{tot}}$, so

$$\theta = \frac{\lambda\, \mathrm{e}^{-\beta\varepsilon_1}}{1 + \lambda\, \mathrm{e}^{-\beta\varepsilon_1}} = \frac{\lambda\, \mathrm{e}^{-\beta\varepsilon_1}}{\mathcal{Z}^{(1)}}. \tag{21.74}$$

Of course the fraction of unoccupied sites is $1 - \theta = 1/\mathcal{Z}^{(1)}$, so these results could have been deduced entirely from the ratios of the corresponding terms in $\mathcal{Z}^{(1)}$ to $\mathcal{Z}^{(1)}$ itself. From Eq. (19.66), the chemical potential of an ideal gas is $\mu = k_{\mathrm{B}} T \ln(n/n_{\mathrm{Q}}) = k_{\mathrm{B}} T \ln(p/(n_{\mathrm{Q}} k_{\mathrm{B}} T))$, where n is the number density and $n_{\mathrm{Q}}(T) = (m k_{\mathrm{B}} T / 2\pi \hbar^2)^{3/2}$ is the quantum concentration. The absolute activity is therefore

$$\lambda = \frac{n}{n_{\mathrm{Q}}(T)} = \frac{p}{n_{\mathrm{Q}}(T) k_{\mathrm{B}} T}, \tag{21.75}$$

which is the ratio of the actual pressure to a quantum pressure. We can therefore define a temperature-dependent pressure

$$p_0(T) := n_{\mathrm{Q}}(T) k_{\mathrm{B}} T\, \mathrm{e}^{\beta\varepsilon_1} = n_{\mathrm{Q}}(T) k_{\mathrm{B}} T\, \mathrm{e}^{-\beta|\varepsilon_1|}, \tag{21.76}$$

for $\varepsilon_1 < 0$, which increases with temperature. Then Eq. (21.74) takes the simple form

$$\theta = \frac{p}{p_0 + p}. \tag{21.77}$$

Equation (21.77) has the form of a Langmuir adsorption isotherm and is plotted in Figure 21–1. See Kittel and Kroemer [6, p. 142] for a plot of data for adsorption of an oxygen molecule by a heme group of myoglobin, which closely follows such an isotherm.

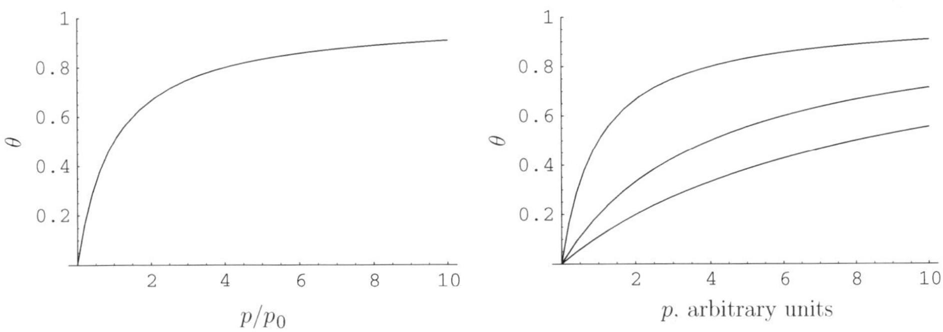

FIGURE 21–1 Langmuir adsorption isotherms for the fractional adsorption of an ideal gas on $\mathcal{N}_{\text{tot}}$ independent sites. The curves on the right correspond to temperatures in the ratios 1:4:8, from left to right.

Example Problem 21.3. Calculate the probability of adsorption of an ideal gas on $\mathcal{N}_{\rm tot}$ independent sites that are either unoccupied, with energy zero, or singly occupied with partition function $z(T)$. What is the canonical partition function $Z_{\mathcal{N}}$ for a system having $\mathcal{N}$ occupied sites?

Solution 21.3. The grand partition function for a single site is

$$\mathcal{Z}^{(1)} = 1 + \lambda z(T), \tag{21.78}$$

so Eq. (21.77) still applies; however, the pressure in Eq. (21.76) is replaced by

$$p_0(T) := n_Q k_B T / z(T). \tag{21.79}$$

The canonical partition function for $\mathcal{N}$ adsorbed atoms is the coefficient of $\lambda^{\mathcal{N}}$ in $\mathcal{Z} = (\mathcal{Z}^{(1)})^{\mathcal{N}_{\rm tot}}$ which is readily found from the binomial theorem to be

$$Z_{\mathcal{N}} = \frac{\mathcal{N}_{\rm tot}!}{\mathcal{N}!(\mathcal{N}_{\rm tot} - \mathcal{N})!}[z(T)]^{\mathcal{N}}. \tag{21.80}$$

The binomial coefficient accounts for the degeneracy that arises because we do not know which of the $\mathcal{N}_{\rm tot}$ sites are occupied, but they are distinguishable by virtue of their position. The reader is invited to verify that the chemical potential for such a system is

$$\mu = -k_B T \frac{\partial \ln Z_{\mathcal{N}}}{\partial \mathcal{N}} = k_B T \ln\left[\frac{\mathcal{N}}{\mathcal{N}_{\rm tot} - \mathcal{N}} \frac{1}{z(T)}\right] = k_B T \ln\left[\frac{\theta}{1-\theta} \frac{1}{z(T)}\right]. \tag{21.81}$$

Equating this μ to that of a classical ideal gas, Eq. (21.75), gives $p/p_0(T) = \theta/(1 - \theta)$ with $p_0(T)$ given by Eq. (21.79). Then solving for θ gives the consistent result Eq. (21.77).

Example Problem 21.4. Calculate the probability of adsorption of a monatomic ideal gas on $\mathcal{N}_{\rm tot}$ independent sites that are either unoccupied with energy zero or singly occupied with energy ε_1 or doubly occupied with energy ε_2. Note that ε_2 is not necessarily equal to $2\varepsilon_1$, so atoms on a doubly-occupied site can interact.

Solution 21.4. The grand partition function for a single site is $\mathcal{Z}^{(1)} = 1 + \lambda\, e^{-\beta\varepsilon_1} + \lambda^2\, e^{-\beta\varepsilon_2}$. The probabilities of a site being unoccupied, singly occupied, or doubly occupied are

$$p_0 = 1/\mathcal{Z}^{(1)}; \quad p_1 = \lambda\, e^{-\beta\varepsilon_1}/\mathcal{Z}^{(1)}; \quad p_2 = \lambda^2\, e^{-\beta\varepsilon_2}/\mathcal{Z}^{(1)}. \tag{21.82}$$

The average number of adsorbed gas atoms is therefore $\langle\mathcal{N}\rangle = \mathcal{N}_{\rm tot}(p_1 + 2p_2)$, where the factor of 2 enters because of the double occupancy. Alternatively, one can use the total grand partition function $\mathcal{Z} = [\mathcal{Z}^{(1)}]^{\mathcal{N}_{\rm tot}}$ to calculate

$$\langle\mathcal{N}\rangle = \lambda \frac{\partial q}{\partial \lambda} = \mathcal{N}_{\rm tot} \frac{\lambda\, e^{-\beta\varepsilon_1} + 2\lambda^2\, e^{-\beta\varepsilon_2}}{1 + \lambda\, e^{-\beta\varepsilon_1} + \lambda^2\, e^{-\beta\varepsilon_2}}, \tag{21.83}$$

where the factor of 2 occurs automatically, or

$$U = -\left(\frac{\partial q}{\partial \beta}\right)_\lambda = \mathcal{N}_{\text{tot}} \frac{\varepsilon_1 \lambda\, \mathrm{e}^{-\beta\varepsilon_1} + \varepsilon_2 \lambda^2\, \mathrm{e}^{-\beta\varepsilon_2}}{1 + \lambda\, \mathrm{e}^{-\beta\varepsilon_1} + \lambda^2\, \mathrm{e}^{-\beta\varepsilon_2}} \tag{21.84}$$

where there is no such factor of 2.

Example Problem 21.5. Calculate the probability of adsorption of either an A atom or a B atom on $\mathcal{N}_{\text{tot}}$ independent sites that are either unoccupied with energy zero or singly occupied with energies ε_A and ε_B, respectively.

Solution 21.5. See Eq. (21.169) for an obvious generalization of the GCE to a binary system. In the present case, we have

$$\mathcal{Z}^{(1)} = 1 + \lambda_A\, \mathrm{e}^{-\beta\varepsilon_A} + \lambda_B\, \mathrm{e}^{-\beta\varepsilon_B}. \tag{21.85}$$

Thus the fractional occupations are

$$\theta_A = \frac{\lambda_A\, \mathrm{e}^{-\beta\varepsilon_A}}{1 + \lambda_A\, \mathrm{e}^{-\beta\varepsilon_A} + \lambda_B\, \mathrm{e}^{-\beta\varepsilon_B}}; \quad \theta_B = \frac{\lambda_B\, \mathrm{e}^{-\beta\varepsilon_B}}{1 + \lambda_A\, \mathrm{e}^{-\beta\varepsilon_A} + \lambda_B\, \mathrm{e}^{-\beta\varepsilon_B}}, \tag{21.86}$$

and the fraction of unoccupied sites is $1 - \theta_A - \theta_B$. We would have to determine λ_A and λ_B from the chemical potentials of the environment, say ideal gases of A and B. We see in this case that the A and B atoms compete for occupancy of the sites. Moreover, a small difference between ε_A and ε_B can make an enormous difference between the relative adsorption of A and B if $|\beta\varepsilon_i| \gg 1$.

For the examples in this section, $\mathcal{Z} = (\mathcal{Z}^{(1)})^{\mathcal{N}_{\text{tot}}}$, so viewed as a series in powers of λ, the series cuts off after a finite number of terms. For the first two examples, the highest power is $(\lambda)^{\mathcal{N}_{\text{tot}}}$ and for the third example it is $(\lambda)^{2\mathcal{N}_{\text{tot}}}$. These cutoffs occur because of the restrictions on maximum occupancy of a site. In terms of the general formula Eq. (21.21), they can be imposed formally by assuming that any state of the entire system having greater occupancy than allowed would have an infinite energy, so its Boltzmann factor would be zero. On the other hand, for ideal Fermi and Bose gases, there are an infinite number of orbitals available for occupation, so the expression for $\mathcal{Z}$ for those gases contains all powers of λ, as shown in the next section.

21.2.2 Fermi-Dirac Distribution

For a single orbital of a gas of noninteracting fermions, Eq. (21.67) becomes

$$\mathcal{Z}_1(\varepsilon) := \sum_{n=0,1} [\lambda \exp(-\beta\varepsilon)]^n = 1 + \lambda \exp(-\beta\varepsilon). \tag{21.87}$$

The average number of particles that occupy that orbital is therefore

$$f_{\text{FD}}(\varepsilon) := \frac{\lambda \exp(-\beta\varepsilon)}{1 + \lambda \exp(-\beta\varepsilon)} = \frac{1}{\lambda^{-1} \exp(\beta\varepsilon) + 1} = \frac{1}{\exp[\beta(\varepsilon - \mu)] + 1}, \tag{21.88}$$

which is known as the **Fermi-Dirac distribution** function. Equation (21.88) can be deduced by inspection or by applying Eq. (21.30) to $\mathcal{Z}_1(\varepsilon)$. If the fermions each have spin s and no magnetic field is present, there will be $2s+1$ orbitals as compared to orbitals with spin degeneracy ignored. Thus Eq. (21.66) will contain a factor of $[\mathcal{Z}_1(\varepsilon)]^{2s+1}$ for each orbital with spin degeneracy ignored, which will contribute $(2s+1)\ln \mathcal{Z}_1(\varepsilon)$ to $\ln \mathcal{Z}$. The total average number of particles in the entire system is therefore

$$\langle \mathcal{N} \rangle = \sum_{\varepsilon} f_{\mathrm{FD}}(\varepsilon) = (2s+1) \sum_{\varepsilon}{}' f_{\mathrm{FD}}(\varepsilon), \tag{21.89}$$

where the primed sum is over orbitals with spin degeneracy ignored. In practice, $\langle \mathcal{N} \rangle$ is usually specified and Eq. (21.89) is used to determine the chemical potential μ, which turns out to be a function of β and $\langle \mathcal{N} \rangle / V$ because the sum will turn out to be proportional to V. By similar reasoning, the total internal energy is given by

$$U = \sum_{\varepsilon} \varepsilon f_{\mathrm{FD}}(\varepsilon) = (2s+1) \sum_{\varepsilon}{}' \varepsilon f_{\mathrm{FD}}(\varepsilon). \tag{21.90}$$

For the important case of a free electron gas, $s = 1/2$ so $2s+1 = 2$.

21.2.3 Bose-Einstein Distribution

For a single orbital of a gas of noninteracting bosons, Eq. (21.67) becomes

$$\mathcal{Z}_1(\varepsilon) := \sum_{n=0}^{\infty} [\lambda \exp(-\beta\varepsilon)]^n = \frac{1}{1 - \lambda \exp(-\beta\varepsilon)}, \tag{21.91}$$

where the sum only converges for $\lambda \exp(-\beta\varepsilon) < 1$. The average number of particles that occupy that orbital can be deduced by applying Eq. (21.30) to $\mathcal{Z}_1(\varepsilon)$ to obtain

$$f_{\mathrm{BE}}(\varepsilon) := \frac{\lambda \exp(-\beta\varepsilon)}{1 - \lambda \exp(-\beta\varepsilon)} = \frac{1}{\lambda^{-1} \exp(\beta\varepsilon) - 1} = \frac{1}{\exp[\beta(\varepsilon - \mu)] - 1}, \tag{21.92}$$

which is known as the **Bose-Einstein distribution** function. We note that $f_{\mathrm{BE}}(\varepsilon)$ differs from $f_{\mathrm{FD}}(\varepsilon)$ only by a sign, but this difference is crucial. For example, $f_{\mathrm{FD}}(\varepsilon) \leq 1$ but $f_{\mathrm{BE}}(\varepsilon)$ can be greater than 1, reflecting the possible multiple occupancy of boson orbitals. Moreover, for $\varepsilon = \mu$, $f_{\mathrm{FD}} = 1/2$, which presents no problem, but $f_{\mathrm{BE}} = \infty$, which cannot be allowed.[7] In the absence of a magnetic field, each energy level has a degeneracy of $2s+1$ due to spin, which can be treated in a manner similar to that for fermions. We therefore have

$$\langle \mathcal{N} \rangle = \sum_{\varepsilon} f_{\mathrm{BE}}(\varepsilon) = (2s+1) \sum_{\varepsilon}{}' f_{\mathrm{BE}}(\varepsilon) \tag{21.93}$$

and

$$U = \sum_{\varepsilon} \varepsilon f_{\mathrm{BE}}(\varepsilon) = (2s+1) \sum_{\varepsilon}{}' \varepsilon f_{\mathrm{BE}}(\varepsilon). \tag{21.94}$$

[7]The minimum value of $\varepsilon - \mu$ can be related to the phenomenon of Bose condensation in the ground state.

21.2.4 Classical Ideal Gas

A classical ideal gas can be thought of as a quantum ideal gas, of either fermions or bosons, in the limit of high temperature and low density. In particular, the temperature must be so high (β so small) and the density so low that the ratio of the number of particles to the number of accessible single particle states is very small. In other words, the average number of particles that occupy a single orbital must be small. This will be true for either $f_{\rm FD}$ or $f_{\rm BE}$ provided that

$$\lambda^{-1}\exp(\beta\varepsilon) = \exp[\beta(\varepsilon-\mu)] \gg 1 \tag{21.95}$$

for all ε at small β. This restriction is most severe for $\varepsilon = 0$, so it requires $\lambda = \exp(\mu\beta) \ll 1$ in the limit of small β and low density. If Eq. (21.95) holds, then either $f_{\rm FD}$ or $f_{\rm BE}$ becomes the classical occupation number

$$f_{\rm CL}(\varepsilon) = \exp(\beta\mu)\exp(-\beta\varepsilon). \tag{21.96}$$

We can evaluate the factor $\exp(\beta\mu)$ by applying Eq. (21.89) or (21.93) but for $f_{\rm CL}$. Thus

$$\langle \mathcal{N}\rangle = \sum_\varepsilon f_{\rm CL}(\varepsilon) = \exp(\beta\mu)\sum_\varepsilon \exp(-\beta\varepsilon), \tag{21.97}$$

which yields

$$\exp(\beta\mu) = \frac{\langle \mathcal{N}\rangle}{z}, \tag{21.98}$$

where

$$z = \sum_\varepsilon \exp(-\beta\varepsilon) = (2s+1)\sum_\varepsilon{}' \exp(-\beta\varepsilon) \tag{21.99}$$

is the canonical partition function for a single particle. Hence,

$$\frac{f_{\rm CL}(\varepsilon)}{\langle \mathcal{N}\rangle} = \frac{\exp(-\beta\varepsilon)}{z}. \tag{21.100}$$

The left-hand side of Eq. (21.100) is the probability of finding a particle in the orbital (quantum state including spin) corresponding to ε and the right-hand side is the familiar Boltzmann distribution, the same as given by Eq. (18.11).

As a further bonus, we can use Eq. (19.56) that applies for particles without spin to give the single free particle partition function $z = (2s+1)Vn_{\rm Q}$, where V is the volume and $n_{\rm Q}(T) = (mk_{\rm B}T/2\pi\hbar^2)^{3/2}$ is the quantum concentration. Substitution into Eq. (21.98) gives

$$\beta\mu = \ln\left[\frac{\langle \mathcal{N}\rangle}{V}\frac{1}{n_{\rm Q}(T)}\right] - \ln(2s+1) \tag{21.101}$$

in agreement with Eq. (19.66) for $s = 0$. The second term in Eq. (21.101) arises because of the spin degeneracy, which has no classical counterpart and which also contributes a term $\mathcal{N}k_{\rm B}\ln(2s+1)$ to the entropy. The above condition $\lambda = \exp(\beta\mu) \ll 1$ is seen to be equivalent to $\langle \mathcal{N}\rangle/(Vn_{\rm Q}) \ll 1$, which is true if the actual concentration $n = \langle \mathcal{N}\rangle/V$ is small compared with the quantum concentration $n_{\rm Q}(T)$. This will be true for low density and high temperature. It serves to quantify the sense in which the gas is dilute, which is the same condition discussed immediately following Eq. (19.59).

In this same approximation, we can evaluate the canonical partition function $Z_{\mathcal{N}}$ for a system having exactly $\mathcal{N}$ particles. For a Fermi gas, we have

$$\ln \mathcal{Z} = \ln \prod_{\varepsilon} \{1 + \exp[\beta(\mu - \varepsilon)]\}, \tag{21.102}$$

whereas for a Bose gas

$$\ln \mathcal{Z} = \ln \prod_{\varepsilon} \frac{1}{\{1 - \exp[\beta(\mu - \varepsilon)]\}}. \tag{21.103}$$

These can be combined and rewritten in the form

$$\ln \mathcal{Z} = \pm \sum_{\varepsilon} \ln \{1 \pm \exp[\beta(\mu - \varepsilon)]\}. \tag{21.104}$$

For a classical ideal gas for which $\lambda \ll 1$ holds, we can expand the logarithm in Eq. (21.104) to obtain

$$\ln \mathcal{Z} = \sum_{\varepsilon} \exp[\beta(\mu - \varepsilon)] = \lambda z = \langle \mathcal{N} \rangle, \tag{21.105}$$

where Eq. (21.98) has been used in the last step. Therefore,

$$\mathcal{Z} = \mathrm{e}^{\lambda z} = \sum_{\mathcal{N}=0}^{\infty} \lambda^{\mathcal{N}} \frac{z^{\mathcal{N}}}{\mathcal{N}!}. \tag{21.106}$$

Comparison with Eq. (21.21) shows that

$$Z_{\mathcal{N}} = \frac{z^{\mathcal{N}}}{\mathcal{N}!} \tag{21.107}$$

in agreement with Eq. (19.48). Since $\ln \mathcal{Z} = pV/(k_B T)$, we observe that Eq. (21.105) is equivalent to the ideal gas law.

21.2.5 Fermi, Bose, and Classical Gases

As shown by Pathria [8, p. 134], the main results for ideal Fermi, Bose, and classical gases can be summarized conveniently as follows. One invents a parameter a that takes on the values $a = 1$ for the Fermi gas, $a = -1$ for the Bose gas, and $a = 0$ for the classical gas. Then the distribution function

$$f(\varepsilon; a) := \frac{1}{\lambda^{-1} \exp(\beta\varepsilon) + a} = \frac{1}{\exp[\beta(\varepsilon - \mu)] + a} \tag{21.108}$$

encompasses all three results. The three distribution functions are plotted as a function of $\beta(\varepsilon - \mu)$ in Figure 21–2. The consolidated formula

$$\ln \mathcal{Z} = \frac{1}{a} \sum_{\varepsilon} \ln \{1 + a\lambda \exp(-\beta\varepsilon)\} = \frac{1}{a} \sum_{\varepsilon} \ln \{1 + a \exp[\beta(\mu - \varepsilon)]\} \tag{21.109}$$

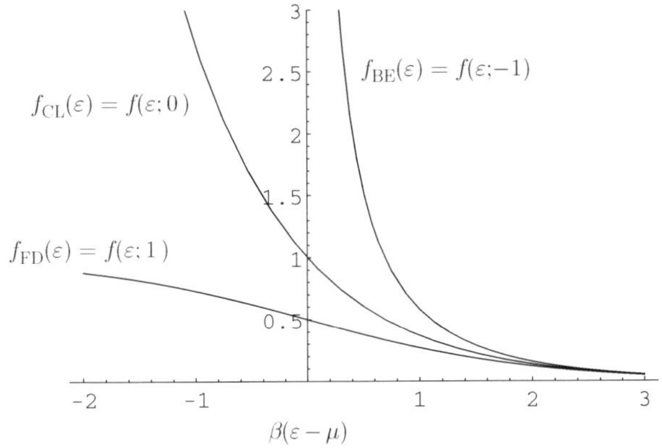

FIGURE 21–2 Plots of the distribution functions for ideal Fermi, classical, and Bose gases as a function of $\beta(\varepsilon - \mu)$. Note that the three distributions merge for large values of $\beta(\varepsilon - \mu)$ which is the same as the limit $\lambda^{-1} \gg 1$.

can be written for the function $q = \ln \mathcal{Z}$. For $a = \pm 1$ we obtain Eq. (21.104) but for $a \to 0$ the formal limit is

$$\ln \mathcal{Z} = \sum_{\varepsilon} \lambda \exp(-\beta\varepsilon) = \lambda z, \tag{21.110}$$

where z is the canonical partition function for a single particle, in agreement with Eq. (21.105). Note that we need $\lambda \ll 1$ to be in the classical limit.

Following Pathria [8, p. 137], we can use these consolidated formulae to get an interesting general equation for the pressure of any of these gases. From Eq. (21.38) we have

$$p = \frac{1}{\beta V} q = \frac{1}{\beta V} \frac{1}{a} \sum_{\varepsilon} \ln\left[1 + a\lambda \exp(-\beta\varepsilon)\right]. \tag{21.111}$$

For a system with very large volume, the energy levels are quasi-continuous and we can replace the summation with integration over k according to Eq. (19.55), although we must add a spin degeneracy factor $g_0 = 2s + 1$. Thus we obtain

$$p = \frac{g_0}{\beta(2\pi)^3} \frac{1}{a} \int_0^\infty \ln\left[1 + a\lambda\, \mathrm{e}^{-\beta\varepsilon(k)}\right] 4\pi k^2\, \mathrm{d}k. \tag{21.112}$$

We write $k^2 = (\mathrm{d}/\mathrm{d}k)k^3/3$ and integrate by parts to obtain

$$p = \frac{g_0}{2a\beta\pi^2} \left\{ \ln\left[1 + a\lambda\, \mathrm{e}^{-\beta\varepsilon(k)}\right] \frac{k^3}{3}\bigg|_0^\infty - \int_0^\infty \frac{k^3}{3} \frac{\mathrm{d}}{\mathrm{d}k} \ln\left[1 + a\lambda\, \mathrm{e}^{-\beta\varepsilon(k)}\right] \mathrm{d}k \right\}. \tag{21.113}$$

The integrated part vanishes and we are left with

$$p = \frac{g_0}{6\pi^2} \int_0^\infty \frac{\lambda\, \mathrm{e}^{-\beta\varepsilon(k)}}{\left[1 + a\lambda\, \mathrm{e}^{-\beta\varepsilon(k)}\right]} k \frac{\mathrm{d}\varepsilon(k)}{\mathrm{d}k} k^2\, \mathrm{d}k = \frac{g_0}{6\pi^2} \int_0^\infty f(\varepsilon; a) k \frac{\mathrm{d}\varepsilon(k)}{\mathrm{d}k} k^2\, \mathrm{d}k. \tag{21.114}$$

Similarly, from Eq. (21.30) we have

$$\langle \mathcal{N} \rangle = \lambda \frac{\partial}{\partial \lambda} \frac{1}{a} \sum_{\varepsilon} \ln\left[1 + a\lambda \exp(-\beta\varepsilon)\right] = \frac{g_0 V}{2\pi^2} \int_0^\infty f(\varepsilon; a)\, k^2\, dk. \tag{21.115}$$

Then Eqs. (21.114) and (21.115) can be combined to give

$$p = \frac{\langle \mathcal{N} \rangle}{3V} \left\langle k \frac{d\varepsilon(k)}{dk} \right\rangle, \tag{21.116}$$

where

$$\left\langle k \frac{d\varepsilon(k)}{dk} \right\rangle \equiv \frac{\int_0^\infty f(\varepsilon; a)[k\,d\varepsilon(k)/dk]k^2\, dk}{\int_0^\infty f(\varepsilon; a) k^2\, dk} \tag{21.117}$$

is an average value of $k\,d\varepsilon(k)/dk$. In simple cases for which $\varepsilon(k) \propto k^s$, where s is a constant, we have $k\,d\varepsilon(k)/dk = s\varepsilon(k)$ which yields

$$p = \frac{\langle \mathcal{N} \rangle}{3V} s \langle \varepsilon(k) \rangle = \frac{s}{3} u, \tag{21.118}$$

where $u = U/V$ is the energy density. For a nonrelativistic particle in a box, $s = 2$ and we have the result $p = 2u/3$. This result is familiar for a classical ideal gas (see Eq. (19.78)), for which $\langle \varepsilon \rangle = (3/2)k_B T$ and $p = \mathcal{N} k_B T/V$, but we see that it is also true for an ideal Fermi gas and an ideal Bose gas. For a particle in a box in the extreme relativistic limit, $s = 1$ and we have the result $p = u/3$.

21.2.6 Orbital Populations for Ideal Gases

We can apply Eq. (21.30) for $\langle \mathcal{N} \rangle$ and Eq. (21.50) for $\langle (\Delta\mathcal{N})^2 \rangle$ to the case of ideal Fermi or Bose gases, for which the grand partition function factors, as represented generally by Eq. (21.68). Since $\mathcal{Z}_1(\varepsilon)$ is a function of only the variable $\beta(\mu - \varepsilon)$ and β is held constant in the derivatives that follow, we note that

$$\left(\frac{\partial \mathcal{Z}_1(\varepsilon)}{\partial \mu} \right)_{V,\beta,\varepsilon} = - \left(\frac{\partial \mathcal{Z}_1(\varepsilon)}{\partial \varepsilon} \right)_{V,\beta,\mu}. \tag{21.119}$$

Thus

$$\langle \mathcal{N} \rangle = \frac{1}{\beta} \frac{\partial \ln \mathcal{Z}}{\partial \mu} = \frac{1}{\beta} \sum_{\varepsilon} \frac{\partial \ln \mathcal{Z}_1(\varepsilon)}{\partial \mu} = -\frac{1}{\beta} \sum_{\varepsilon} \frac{\partial q_\varepsilon}{\partial \varepsilon} = \sum_{\varepsilon} \langle n_\varepsilon \rangle, \tag{21.120}$$

where $q_\varepsilon := \ln \mathcal{Z}_1(\varepsilon)$ and

$$\langle n_\varepsilon \rangle := -\frac{1}{\beta} \left(\frac{\partial q_\varepsilon}{\partial \varepsilon} \right)_{V,\beta,\mu}. \tag{21.121}$$

Similarly,

$$\langle (\Delta\mathcal{N})^2 \rangle = \frac{1}{\beta^2} \frac{\partial^2 \ln \mathcal{Z}}{\partial \mu^2} = \frac{1}{\beta^2} \sum_{\varepsilon} \frac{\partial^2 \ln \mathcal{Z}_1(\varepsilon)}{\partial \mu^2} = \frac{1}{\beta^2} \sum_{\varepsilon} \frac{\partial^2 q_\varepsilon}{\partial \varepsilon^2} = \sum_{\varepsilon} \langle (\Delta n_\varepsilon)^2 \rangle, \tag{21.122}$$

where

$$\langle(\Delta n_\varepsilon)^2\rangle := -\frac{1}{\beta}\left(\frac{\partial\langle n_\varepsilon\rangle}{\partial\varepsilon}\right)_{V,\beta,\mu}. \tag{21.123}$$

Thus the average number of particles and its variance are additive over the orbitals labeled by ε. According to Eq. (21.104), we have explicitly[8]

$$\langle n_\varepsilon\rangle = \frac{1}{\exp[\beta(\varepsilon-\mu)]\pm 1} \tag{21.124}$$

and

$$\frac{\langle(\Delta n_\varepsilon)^2\rangle}{\langle n_\varepsilon\rangle^2} = \exp[\beta(\varepsilon-\mu)] = \frac{1}{\langle n_\varepsilon\rangle} \mp 1. \tag{21.125}$$

For a classical ideal gas, we would have $\langle n_\varepsilon\rangle \ll 1$ so the ∓ 1 in Eq. (21.125) is negligible. This is called a *normal* fluctuation. For fermions, the result is $1/\langle n_\varepsilon\rangle - 1$ which nearly vanishes for temperatures sufficiently low that $k_BT \ll \mu - \varepsilon > 0$, in which case $\langle n_\varepsilon\rangle \approx 1$. Such fluctuations are called *infranormal* or *subnormal.* For bosons, the result is $1/\langle n_\varepsilon\rangle + 1$ which is *above normal* or *extranormal.*

For classical ideal gases, the populations follow a **Poisson distribution**. This can be seen by returning to Eq. (21.105) from which we obtain

$$\mathcal{Z} = \exp\langle\mathcal{N}\rangle = \sum_{\mathcal{N}=0}^{\infty}\frac{\langle\mathcal{N}\rangle^{\mathcal{N}}}{\mathcal{N}!}. \tag{21.126}$$

Since $\langle\mathcal{N}\rangle$ is proportional to λ, the probability of exactly $\mathcal{N}$, for the entire ensemble, is the $\mathcal{N}$th term in this sum divided by $\mathcal{Z}$, namely

$$P_{\mathcal{N}} = \frac{\langle\mathcal{N}\rangle^{\mathcal{N}}}{\mathcal{N}!}\exp(-\langle\mathcal{N}\rangle), \tag{21.127}$$

which is a Poisson distribution. From Eq. (21.110),

$$\ln\mathcal{Z}_1(\varepsilon) = \lambda\exp(-\beta\varepsilon) = \langle n_\varepsilon\rangle \tag{21.128}$$

according to Eq. (21.124). Therefore,

$$\mathcal{Z}_1(\varepsilon) = \mathrm{e}^{\langle n_\varepsilon\rangle} = \sum_{n_\varepsilon}\frac{\langle n_\varepsilon\rangle^{n_\varepsilon}}{n_\varepsilon!}. \tag{21.129}$$

The probability of occupation of the orbital ε is therefore

$$p_{n_\varepsilon} = \frac{\langle n_\varepsilon\rangle^{n_\varepsilon}}{n_\varepsilon!}\exp(-\langle n_\varepsilon\rangle), \tag{21.130}$$

which is also a Poisson distribution.

[8]The upper sign is for fermions and the lower sign is for bosons.

Example Problem 21.6. Compare the occupation probabilities of orbitals for fermions, bosons, and classical particles and discuss the limit where they become essentially the same.

Solution 21.6. For simplicity, we define $\gamma = \lambda \exp(-\beta\varepsilon)$. For fermions, there are only two probabilities, $p_0 = 1/(1+\gamma)$ and $p_1 = \gamma/(1+\gamma)$. For bosons, one has $p_n = \gamma^n(1-\gamma)$. For classical particles $p_n = \gamma^n \exp(-\gamma)/n!$. The result for classical particles is only valid for $\gamma \ll 1$. In that limit, all three distributions become approximately $p_0 = 1 - \gamma$, $p_1 = \gamma$, and $p_n = 0, n \geq 1$. Thus, when conditions for a classical gas are valid, there is essentially no double occupancy of orbitals, which explains why the Gibbs correction factor of $\mathcal{N}!$ leads to the correct partition function.

21.3 Classical Ideal Gas with Internal Structure

In Sections 21.2.4 and 21.2.5, we treated ideal gases without internal structure, except for spin, which was necessary to distinguish between Fermi and Bose gases and which led to a degeneracy factor of $2s + 1$. In the present section, we show how to treat gases whose particles are atoms or molecules having internal structure, not only due to nuclear spin but also due to electronic and molecular structure.

We return to Eq. (21.105) and expand the notation such that $\varepsilon \to \varepsilon_t + \varepsilon_i$, where ε_t is the energy due to translation and ε_i is due to internal structure, including nuclear spin and electronic and molecular structure. We assume that these energies are separable, which means that the internal degrees of freedom are not affected by translation and *vice versa*. Thus we obtain

$$\ln \mathcal{Z} = \sum_{t,i} \exp[\beta(\mu - \varepsilon_t - \varepsilon_i)] = \lambda z_{\text{int}} z_t = \langle \mathcal{N} \rangle, \tag{21.131}$$

where the translational partition function is

$$z_t = \sum_t \exp(-\beta\varepsilon_t) = V n_{\text{Q}} \tag{21.132}$$

and the internal partition function is

$$z_{\text{int}} = \sum_i \exp(-\beta\varepsilon_i). \tag{21.133}$$

Thus Eq. (21.107) holds with z replaced by $z_{\text{int}} z_t$, resulting in

$$Z = \frac{(z_{\text{int}} z_t)^{\mathcal{N}}}{\mathcal{N}!}. \tag{21.134}$$

The corresponding Helmholtz free energy is

$$F = -\mathcal{N} k_{\text{B}} T[\ln(V n_{\text{Q}}/\mathcal{N}) + 1] - \mathcal{N} k_{\text{B}} T \ln z_{\text{int}}. \tag{21.135}$$

We see that the effect of the internal structure is additive. In the case that z_{int} is only due to spin degeneracy, we have $z_{\text{int}} = 2s + 1$ and we can recover Eq. (21.101) by taking $\partial/\partial\mathcal{N}$ of Eq. (21.135).

For a gas of molecules or of atoms having structure, one usually assumes that the electronic, vibrational, rotational, and nuclear degrees of freedom are decoupled from each other. This is based partly on the Born-Oppenheimer approximation, which is supposed to hold because nucleons are more massive and move more slowly than electrons. Therefore, the internal partition function factors to give

$$z_{\text{int}} = z_{\text{elec}} z_{\text{vib}} z_{\text{nuc}} z_{\text{rot}} \tag{21.136}$$

so we have

$$\ln z_{\text{int}} = \ln z_{\text{elec}} + \ln z_{\text{vib}} + \ln z_{\text{nuc}} + \ln z_{\text{rot}}. \tag{21.137}$$

In other words, the contributions of the internal degrees of freedom are additive. However, in the case of homonuclear molecules (e.g., H_2 that we treat later) it is important to correlate the nuclear and rotational partition functions such that the product $z_{\text{nuc}} z_{\text{rot}}$ is replaced by $z_{\text{nuc-rot}}$ which is based on antisymmetric wave functions for fermions and symmetric ones for bosons.

21.3.1 Monatomic Gas

For a monatomic gas, we only have to deal with the nuclear and electronic partition functions. To avoid any ambiguity, we choose the zero of energy to be the nuclear and electronic ground state, as well as zero translational energy.

The hyperfine structure due to the nuclear spin has energy splittings that are very small compared to $k_B T$ in most cases of interest, so $z_{\text{nuc}} = 2I + 1$, where I is the nuclear spin. There is no contribution to the energy and the heat capacity, but the entropy is changed by $\mathcal{N} k_B \ln(2I+1)$. The free energy and chemical potential are changed by $-\mathcal{N} k_B T \ln(2I+1)$ and $-k_B T \ln(2I + 1)$, respectively.

The value of z_{elec} due to the electronic structure depends on the electronic orbital angular momentum L and the electronic spin angular momentum S. If $L = S = 0$, the state is nondegenerate and $z_{\text{elect}} = 1$. If $L = 0$ but $S \neq 0$, which is typical of alkali atoms such as Na, K, and Rb, there is no fine structure and we have $z_{\text{elect}} = 2S + 1$ due to electronic spin degeneracy. In the general case, $L \neq 0$ and $S \neq 0$, we have

$$z_{\text{elect}} = \sum_{\varepsilon_{\text{elect}}} \exp(-\beta\varepsilon_{\text{elect}}), \tag{21.138}$$

where the sum is over all electronic states, having energies $\varepsilon_{\text{elect}}$. Usually, only the ground state of degeneracy g_{0e} and the first excited state of degeneracy g_{1e} and energy Δ_e are important because the rest of the states have such high energies that they are practically unoccupied. It is therefore often sufficient to take

$$z_{\text{elect}} = g_{0e} + g_{1e} \exp(-\beta\Delta_e). \tag{21.139}$$

This leads to contributions to the energy and the heat capacity of the forms

$$U_{\text{elect}} = \mathcal{N}\Delta_e \frac{g_{1e}\exp(-\beta\Delta_e)}{g_{0e} + g_{1e}\exp(-\beta\Delta_e)} \tag{21.140}$$

and

$$C_{\text{elect}} = \mathcal{N}k_{\text{B}}(\beta\Delta_e)^2 \frac{g_{0e}g_{1e}\exp(-\beta\Delta_e)}{[g_{0e} + g_{1e}\exp(-\beta\Delta_e)]^2}. \tag{21.141}$$

This electronic heat capacity is zero at low temperatures, passes through a maximum, and then decays again to zero at high temperatures, assuming that no higher energy levels come into play. Contributions to the entropy and the chemical potential are

$$S_{\text{elect}} = \mathcal{N}k_{\text{B}}\ln[g_{0e} + g_{1e}\exp(-\beta\Delta_e)] + \mathcal{N}k_{\text{B}}\beta\Delta_e \frac{g_{1e}\exp(-\beta\Delta_e)}{g_{0e} + g_{1e}\exp(-\beta\Delta_e)} \tag{21.142}$$

and

$$\mu_{\text{elect}} = -k_{\text{B}}T\ln[g_{0e} + g_{1e}\exp(-\beta\Delta_e)]. \tag{21.143}$$

For future reference, the entire internal partition function is given approximately by

$$z_{\text{int}} = z_{\text{elect}}z_{\text{nuc}} = [g_{0e} + g_{1e}\exp(-\beta\Delta_e)](2I+1). \tag{21.144}$$

21.3.2 Diatomic Molecular Gas

For diatomic molecules, we take the zero of energy to be the nuclei in their ground states and the atoms to be in their electronic ground states for a completely dissociated molecule, that is, infinite separation of the atoms. Homonuclear diatomic molecules are indistinguishable if rotated 180° about their center of mass, thus exchanging identical particles. Therefore, their nuclear and rotational partition functions must be correlated to satisfy requirement of quantum statistics. No such requirements exist for heteronuclear molecules because their nuclei are distinguishable. Therefore, we first treat the simpler case of heteronuclear molecules and then treat homonuclear molecules.

Heteronuclear Molecules

For heteronuclear diatomic molecules *AB*, composed of *A* and *B* atoms, the nuclei remain in their ground states so the nuclear partition function $z_{\text{nuc}} = (2I_{\text{A}} + 1)(2I_{\text{B}} + 1)$ only accounts for degeneracy.

The relevant electronic structure is now that of the molecule. This is usually described in terms of a potential that is strongly repulsive (positively infinite) at short distances of separation, becomes negative reaching the bottom of a potential well at a negative energy $\varepsilon_{0m} = -D$, and then rises to zero at infinite separation. Usually one needs to consider only the electronic ground state and the first excited state, having energy ε_{1m}, because occupation of higher molecular electronic states would lead to dissociation. Therefore, the electronic partition function of the molecule can be represented approximately by

$$z_{\rm elect} = \exp(\beta D)[g_{0m} + g_{1m}\exp(-\beta\Delta_m)], \tag{21.145}$$

where $\Delta_m = \varepsilon_{1m} - \varepsilon_{0m}$ is the *separation* between the first excited electronic state and the electronic ground state, and g_{0m} and g_{1m} are the respective degeneracies. This expression resembles Eq. (21.139) for the monatomic gas except for the prefactor $\exp(\beta D)$ that arises because of the depth of the potential well. Since

$$\ln z_{\rm elect} = \beta D + \ln[g_{0m} + g_{1m}\exp(-\beta\Delta_m)], \tag{21.146}$$

the only contribution of the factor $\exp(\beta D)$ is to add an energy $-D$ per molecule. The remaining term in Eq. (21.146) makes contributions exactly analogous to those made by $z_{\rm elect}$ for the monatomic case.

A new effect, however, comes from vibrations about the equilibrium separation of the atoms, giving rise to quantum states that can be approximated by those of a harmonic oscillator with nondegenerate energy levels $(1/2 + n)\hbar\omega_0$, where n is zero or a positive integer and ω_0 is the angular frequency of vibration. The partition function is

$$z_{\rm vib} = \exp(-\Theta_{\rm v}/2T)\frac{1}{1-\exp(-\Theta_{\rm v}/T)}, \tag{21.147}$$

where $\Theta_{\rm v} := \hbar\omega/k_{\rm B}$ is a characteristic temperature. The contribution to the total energy is therefore

$$U_{\rm vib} = \mathcal{N}k_{\rm B}\Theta_{\rm v}/2 + \mathcal{N}\frac{k_{\rm B}\Theta_{\rm v}}{\exp(\Theta_{\rm v}/T)-1}. \tag{21.148}$$

Here again, the prefactor $\exp(-\Theta_{\rm v}/2T)$ in the partition function just adds a constant energy $k_{\rm B}\Theta_{\rm v}/2 = \hbar\omega_0/2$ per molecule. The cumulative shift in the energy per molecule from the dissociated state, which was taken as the zero of energy, is therefore $-(D-\hbar\omega_0/2)$. Typical measured values are $D - \hbar\omega_0/2 \approx$ 1 to 10 eV per molecule. The corresponding heat capacity is given by Eq. (18.54) which we rewrite in the form

$$C_{\rm vib} = \mathcal{N}k_{\rm B}\frac{(\Theta_{\rm v}/T)^2\exp(\Theta_{\rm v}/T)}{[\exp(\Theta_{\rm v}/T)-1]^2}. \tag{21.149}$$

Figure 18–8 depicts a graph of this heat capacity versus temperature. Typical values of this characteristic vibration temperature are $\Theta_{\rm v} = 1000$ to 4000 K, which corresponds to an energy of about 0.1-0.3 eV. For $T \ll \Theta_{\rm v}$, which is typical, we have $C_{\rm vib} \approx 0$ and the vibrational mode is said to be "frozen out." For high temperatures, $C_{\rm vib} \approx \mathcal{N}k_{\rm B}$ and the heat capacity would be increased by a constant amount. However, the molecule will probably dissociate before one observes the maximum heat capacity due to vibration.

Most interesting are the rotational degrees of freedom, for which we will treat the atoms as point particles. Classically, we can think of such a degree of freedom as a rotation of a rigid diatomic molecule about an axis perpendicular to the line joining the atoms and passing through the center of mass of the molecule. There are two degrees of freedom of this rotation because we must consider rotations about two perpendicular axes, each also perpendicular to the axis separating the atoms. The quantum energy levels are

$$\varepsilon_{\rm rot}(j) = j(j+1)\varepsilon_0, \tag{21.150}$$

where $\varepsilon_0 = \hbar^2/(2\mathcal{I})$ and the moment of inertia $\mathcal{I} = \ell_0^2 m_1 m_2/(m_1+m_2)$ for atoms of masses m_1 and m_2 separated by a distance ℓ_0. Each energy level has degeneracy $2j+1$. This problem was treated in Section 18.4 and the corresponding heat capacity is depicted in Figure 18–12. We define $\Theta_r := \varepsilon_0/k_{\rm B} = \hbar^2/(2\mathcal{I}k_{\rm B})$. For $T \ll \Theta_r$, $C_{\rm rot} \approx 0$ and the rotation does not contribute, whereas for $T \gg \Theta_r$, $C_{\rm rot} \approx \mathcal{N}k_{\rm B}$ and the rotation contributes $(1/2)k_{\rm B}$ per molecule for each of its two rotational degrees of freedom, consistent with equipartition. For most diatomic molecules, Θ_r is only a few Kelvin degrees [61, p. 92], so the rotational mode is fully excited and the total heat capacity due to rotation is

$$C_{\rm rot} = \mathcal{N}k_{\rm B}, \quad \text{diatomic molecules, } T \gg \Theta_r. \tag{21.151}$$

Since typically $\Theta_r \ll \Theta_{\rm V}$, the heat capacity at constant volume for a classical ideal gas composed of diatomic molecules varies with temperature as follows: For temperatures high enough to be treated as a classical gas, the heat capacity has the translational value $(3/2)\mathcal{N}k_{\rm B}$, rises after a slight overshoot[9] to $(5/2)\mathcal{N}k_{\rm B}$ at temperatures above Θ_r, and finally rises to $(7/2)\mathcal{N}k_{\rm B}$ for temperatures above $\Theta_{\rm V}$. This behavior is sketched in Figure 21–3 under conditions for which all temperature ranges are accessible. In a practical temperature range, however, one might only observe the value $(5/2)\mathcal{N}k_{\rm B}$.

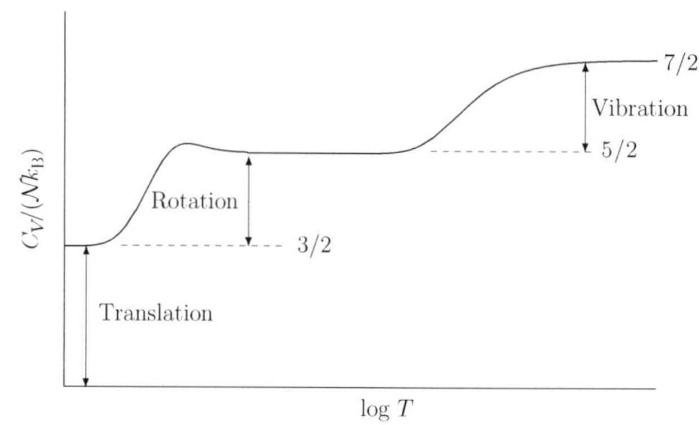

FIGURE 21–3 Sketch of the heat capacity C_V in units of $\mathcal{N}k_{\rm B}$ of a diatomic molecule as a function of $\log T$. The first level at 3/2 at low temperatures $T < \Theta_r$ results from translational degrees of freedom. The second level at 5/2 at intermediate temperatures $\Theta_r < T < \Theta_{\rm V}$ results from translation plus rotation. The final level at 7/2 at high temperatures $\Theta_{\rm V} < T$ results from translation, rotation, and vibration. Since Θ_r is typically a few degrees K and $\Theta_{\rm V}$ is typically a few thousand degrees K, only the middle value 5/2 is usually observed. This simple picture omits corrections due to the electronic degrees of freedom, which are similar in form to those for a monatomic gas, Eq. (21.142).

[9]See Section 18.4 for details and a graph.

For an excellent and more detailed treatment of heteronuclear diatomic molecules, including data for a number of actual molecules, see McQuerrie [54, p. 91].

Homonuclear Molecules

The situation for homonuclear molecules, such as hydrogen H_2 or deuterium D_2 is more complicated because quantum statistics for fermions and bosons comes into play and requires correlation of $z_{\rm nuc}$ and $z_{\rm rot}$ to produce a net result $z_{\rm nuc-rot}$ that corresponds to a wave function that has the correct symmetry under interchange of the nuclei. In case the nuclei are fermions, as for H_2, the Pauli exclusion principle applies so the total wave function must be antisymmetric under interchange of the nuclei. This requires each net spin state of the combined nuclei to be paired with a rotational state that has the correct symmetry.

For H_2, each nucleus has spin 1/2 so the combined nuclear spin states have spin components $|1\rangle$, $|0\rangle$, and $|-1\rangle$. The states corresponding to $|1\rangle$ and $|-1\rangle$ come from $|1/2, 1/2\rangle$ and $|-1/2, -1/2\rangle$ and are symmetric. Of the states corresponding to $|0\rangle$, one is $(|1/2, -1/2\rangle + |-1/2, 1/2\rangle)/\sqrt{2}$ and is symmetric; the other is $(|1/2, -1/2\rangle - |-1/2, 1/2\rangle)/\sqrt{2}$ and is antisymmetric. Rotational states are symmetric and antisymmetric according to whether j is even or odd. So we have to pair the three symmetric spin states with odd-j rotational states and the one antisymmetric spin state with even-j rotational states. These rotational partition functions are

$$z_{\rm rot}(\text{even}) := \sum_{j=0,2,4,\ldots} (2j+1)\exp[-\beta j(j+1)\varepsilon_0] \tag{21.152}$$

and

$$z_{\rm rot}(\text{odd}) := \sum_{j=1,3,5,\ldots} (2j+1)\exp[-\beta j(j+1)\varepsilon_0]. \tag{21.153}$$

For H_2, the combined partition function would be

$$z_{\rm nuc-rot}(\text{hydrogen}) = 3z_{\rm rot}(\text{odd}) + z_{\rm rot}(\text{even}). \tag{21.154}$$

If each nucleus has spin I, then $(I+1)(2I+1)$ of the combined spin states have even symmetry and $I(2I+1)$ have odd symmetry under interchange of the nuclei.[10]

Thus more generally,

$$z_{\rm nuc-rot}(\text{fermions}) = (I+1)(2I+1)z_{\rm rot}(\text{odd}) + I(2I+1)z_{\rm rot}(\text{even}). \tag{21.155}$$

For the case of bosons, the total wave function must be symmetric under interchange of atoms. Deuterium is a boson with spin 1; of the combined spin states, six are symmetric and three are antisymmetric. Thus

$$z_{\rm nuc-rot}(\text{deuterium}) = 6z_{\rm rot}(\text{even}) + 3z_{\rm rot}(\text{odd}). \tag{21.156}$$

[10]The rotational state that goes with the larger weight $(I+1)(2I+1)$ is called *ortho* and the one that goes with the smaller weight $I(2I+1)$ is called *para*.

More generally,

$$z_{\text{nuc-rot}}(\text{bosons}) = (I+1)(2I+1)z_{\text{rot}}(\text{even}) + I(2I+1)z_{\text{rot}}(\text{odd}). \tag{21.157}$$

At high temperatures, $z_{\text{rot}} = T/\Theta_r$ and $z_{\text{rot}}(\text{odd}) = z_{\text{rot}}(\text{even}) = T/(2\Theta_r)$. Thus at high temperatures we have

$$z_{\text{nuc-rot}}(\text{fermions}) = z_{\text{nuc-rot}}(\text{bosons}) \approx (2I+1)^2 T/(2\Theta_r). \tag{21.158}$$

Therefore, at high temperatures, the only difference as compared to the heteronuclear case is division by a factor of $\sigma_2 = 2$, known as the **symmetry number**, which affects the entropy but not the heat capacity.

At lower temperatures, however, the results for fermions and bosons would differ from one another due to the differences in weightings in Eqs. (21.155) and (21.157). For hydrogen, the internal energy per molecule due to rotation would be

$$u_{\text{nuc-rot}} = -\frac{\partial \ln z_{\text{nuc-rot}}(\text{hydrogen})}{\partial \beta} = \frac{3z_{\text{rot}}(\text{odd})u_{\text{rot}}(\text{odd}) + z_{\text{rot}}(\text{even})u_{\text{rot}}(\text{even})}{3z_{\text{rot}}(\text{odd}) + z_{\text{rot}}(\text{even})}, \tag{21.159}$$

where

$$u_{\text{rot}}(\text{odd}) = -\partial \ln z_{\text{rot}}(\text{odd})/\partial\beta; \quad u_{\text{rot}}(\text{even}) = -\partial z_{\text{rot}}(\text{even})/\partial\beta. \tag{21.160}$$

The heat capacity per hydrogen molecule would then be

$$c_{\text{nuc-rot}} = \frac{\partial u_{\text{nuc-rot}}}{\partial T} = -k_{\text{B}}\beta^2 \frac{\partial u_{\text{nuc-rot}}}{\partial \beta}. \tag{21.161}$$

Similar expressions for $u_{\text{nuc-rot}}$ and $c_{\text{nuc-rot}}$ but based on Eq. (21.156) would pertain to deuterium.

Ironically, it turns out that the heat capacity given by Eq. (21.161) *does not* lead to agreement with experiments on the heat capacity of hydrogen. The situation for deuterium is similar. This is apparently due to the fact that samples are prepared at room temperature which is well above Θ_r and they do not re-equilibrate during subsequent experiments at low temperatures [62]. At high temperatures, $z_r(\text{odd}) \approx z_{\text{rot}}(\text{even})$, so the contribution to $z_{\text{nuc-rot}}(\text{hydrogen})$ comes 3/4 from molecules in odd rotational states and 1/4 from molecules in even rotational states. But at low temperatures, $z_{\text{rot}}(\text{odd})/z_{\text{rot}}(\text{even}) \approx 3\exp(-2\Theta_r/T) \ll 1$, so $z_{\text{nuc-rot}}(\text{hydrogen})$ comes almost entirely from molecules in even rotational states. These molecules would be required to have antisymmetric spin states. Thus, when hydrogen is cooled from high to low temperatures, equilibrium would require 3/4 of all molecules to change their spin states from symmetric to antisymmetric. Such a change requires molecules to collide at container walls [61, p. 97] and is an extremely slow process. As a result, the gas behaves like a *nonequilibrium mixture* in which the proportion of rotational states is the same as at high temperature. The observed internal energy per molecule of hydrogen due to rotation would then be

$$u_{\text{nuc-rot}}^{\text{neq}}(\text{hydrogen}) = (3/4)u_{\text{rot}}(\text{odd}) + (1/4)u_{\text{rot}}(\text{even}). \tag{21.162}$$

The corresponding heat capacity would be

$$c_{\text{nuc-rot}}^{\text{neq}}(\text{hydrogen}) = (3/4)c_{\text{rot}}(\text{odd}) + (1/4)c_{\text{rot}}(\text{even}), \tag{21.163}$$

where

$$c_{\text{rot}}(\text{odd}) = -k_{\text{B}}\beta^2 \partial u_{\text{rot}}(\text{odd})/\partial\beta; \quad c_{\text{rot}}(\text{even}) = -k_{\text{B}}\beta^2 \partial z_{\text{rot}}(\text{even})/\partial\beta. \tag{21.164}$$

For deuterium we would have

$$c_{\text{nuc-rot}}^{\text{neq}}(\text{deuterium}) = (2/3)c_{\text{rot}}(\text{even}) + (1/3)c_{\text{rot}}(\text{odd}). \tag{21.165}$$

These nonequilibrium values agree with experiment.

21.3.3 Polyatomic Molecular Gas

Polyatomic gas molecules come in many varieties. Each atom has a nuclear spin and the molecule has an electronic structure. A molecule consisting of n atoms has $3n - 5$ vibrational degrees of freedom if is a linear molecule (such as CO_2) and $3n - 6$ vibrational degrees of freedom if it is not a linear molecule (such as CH_4 which has C at the center of a regular tetrahedron with H atoms at each corner).[11] The vibrational degrees of freedom can sometimes be complex (e.g., torsional modes) but can often be treated as normal modes of vibration, *each* of which leads to contributions to the energy and heat capacity of the forms given by Eqs. (21.148) and (21.149).

Generally speaking, one still has $T \ll \Theta_{\text{V}}$ for all of these vibrational modes; they make small contributions but must be taken into account to explain experimentally measured heat capacities. One might also have to take into account a few molecular electronic states. But the main contribution of the internal structure to the heat capacity usually comes from the rotational modes. For linear polyatomic molecules, the rotational modes can be treated in a similar way to diatomic molecules. For polyatomic molecules that are not linear, there is usually considerable simplification because the three principal moments of inertia (see Appendix F), $\mathcal{I}_i$, of polyatomic molecules are usually sufficiently large that the energy quanta $\varepsilon_i := \hbar^2/(2\mathcal{I}_i)$ are small compared with $k_{\text{B}}T$. We can therefore evaluate the partition function in the classical limit, as in Section 20.7. This results in a partition function of the form (see Eq. (20.116))

$$z_{\text{rot}} = \pi^{1/2}\left(\frac{2\mathcal{I}_1 k_{\text{B}}T}{\hbar^2}\right)^{1/2}\left(\frac{2\mathcal{I}_2 k_{\text{B}}T}{\hbar^2}\right)^{1/2}\left(\frac{2\mathcal{I}_3 k_{\text{B}}T}{\hbar^2}\right)^{1/2}. \tag{21.166}$$

All we really need to know is that $z_{\text{rot}} \propto \beta^{-3/2}$ which leads immediately to

$$c_{\text{rot}} = -k_{\text{B}}\beta^2 \frac{\partial^2 \ln z_{\text{rot}}}{\partial\beta^2} = \frac{3}{2}k_{\text{B}} \tag{21.167}$$

[11] In both cases, the total number of degrees of freedom is $3n$ and there are three translational degrees of freedom. A linear molecule has two rotational degrees of freedom and a nonlinear molecule has three. The remaining degrees of freedom are generally termed vibrational modes.

for each molecule. The total heat capacity of a polyatomic gas, to a good approximation, is therefore

$$C_V = \mathcal{N}\frac{3}{2}k_B + \mathcal{N}\sum_{\text{vib}} c_{\text{vib}} + \mathcal{N}c_{\text{elect}} + \mathcal{N}\frac{3}{2}k_B = 3\mathcal{N}k_B \quad \text{plus small corrections.} \tag{21.168}$$

For molecules that are not heteronuclear, such as CO_2 or CH_4, one must correct the partition function by dividing by a "symmetry number" σ_r that is equal to the number of indistinguishable rotational states of the molecule, [54, p. 101], but this does not affect the heat capacity in the high-temperature approximation used above. It does, however, affect the entropy by an amount $-\mathcal{N}k_B \ln \sigma_r$. For example, CO_2 is a linear molecule with $\sigma_r = 2$ because of two indistinguishable rotations of π in orthogonal planes about the carbon atom. For CH_4, $\sigma_r = 12$ because of three indistinguishable rotations of $2\pi/3$ about each of the four C–H bonds that form a tetrahedron. For a more detailed discussion of polyatomic molecules, see McQuarrie [54, p. 129].

For polyatomic molecules, it is also possible to deduce high-temperature distribution functions for the principal angular momenta, Eq. (20.125), and for the vibrational frequencies about the principal axes, Eq. (20.128).

21.4 Multicomponent Systems

The derivation in Section 21.1 can be generalized in a straightforward way to multicomponent systems, which we illustrate for two components, A and B. The probabilities become

$$P_{r\mathcal{N}_A\mathcal{N}_B} = \lambda_A^{\mathcal{N}_A}\lambda_B^{\mathcal{N}_B} \exp[-\mathcal{E}_{r\mathcal{N}_A\mathcal{N}_B}]/\mathcal{Z}, \tag{21.169}$$

where $\lambda_A = \exp(\beta\mu_A)$, $\lambda_B = \exp(\beta\mu_B)$ and

$$\mathcal{Z} = \sum_{\mathcal{N}_A}\sum_{\mathcal{N}_B}\sum_{r} \lambda_A^{\mathcal{N}_A}\lambda_B^{\mathcal{N}_B} \exp[-\mathcal{E}_{r\mathcal{N}_A\mathcal{N}_B}]. \tag{21.170}$$

If the interaction energy between A and B particles is negligible and their states can be occupied independently, as would be the case for ideal gases, then $\mathcal{E}_{r\mathcal{N}_A\mathcal{N}_B} = \mathcal{E}_{r_A\mathcal{N}_A} + \mathcal{E}_{r_B\mathcal{N}_B}$ and we have factorization which results in

$$\mathcal{Z} = \mathcal{Z}_A\mathcal{Z}_B, \tag{21.171}$$

where

$$\mathcal{Z}_A = \sum_{\mathcal{N}_A}\sum_{r_A} \lambda_A^{\mathcal{N}_A} \exp[-\mathcal{E}_{r_A\mathcal{N}_A}]; \quad \mathcal{Z}_B = \sum_{\mathcal{N}_B}\sum_{r_B} \lambda_B^{\mathcal{N}_B} \exp[-\mathcal{E}_{r_B\mathcal{N}_B}]. \tag{21.172}$$

In this case, the probabilities also factor, so

$$P_{r\mathcal{N}_A\mathcal{N}_B} = P_{r_A\mathcal{N}_A}P_{r_B\mathcal{N}_B}. \tag{21.173}$$

For classical ideal gases, we would have $\mathcal{Z}_A = \exp(\lambda_A z_A)$ and $\mathcal{Z}_B = \exp(\lambda_B z_B)$ so Eq. (21.171) would become

$$\mathcal{Z} = \exp(\lambda_A z_A)\exp(\lambda_B z_B) = \sum_{\mathcal{N}_A,\mathcal{N}_B} \lambda_A^{\mathcal{N}_A}\lambda_B^{\mathcal{N}_B} \frac{z_A^{\mathcal{N}_A}}{\mathcal{N}_A!}\frac{z_B^{\mathcal{N}_B}}{\mathcal{N}_B!}. \tag{21.174}$$

The coefficient of $\lambda_A^{\mathcal{N}_A}\lambda_B^{\mathcal{N}_B}$ is therefore the canonical partition function for exactly $\mathcal{N}_A$ particles of A and $\mathcal{N}_B$ particles of B, namely

$$Z = \frac{z_A^{N_A}}{N_A!}\frac{z_B^{N_B}}{N_B!}. \tag{21.175}$$

21.5 Pressure Ensemble

The pressure ensemble can be obtained by using the same procedure as used to derive the GCE. It applies to a system of interest $\mathcal{I}$ with a definite number of particles $\mathcal{N}$ that is held at constant temperature T and constant pressure p. Thus, the volume V_s of the system $\mathcal{I}$ can vary. This is accomplished by putting $\mathcal{I}$ in contact with a thermal and pressure reservoir R. The total system consisting of $\mathcal{I}$ and R is assumed to be isolated and has a fixed energy E_T and a fixed volume V_T. The quantum states of $\mathcal{I}$ have energies $\mathcal{E}_{rs} \equiv \mathcal{E}_r(V_s, \mathcal{N})$. Not surprisingly, the pressure ensemble will be associated with the thermodynamic function G, the Gibbs free energy. Since $\mathcal{N}$ is fixed, we suppress it in the arguments of the functions below.

If the system of interest has a definite volume and is in a definite quantum state, its multiplicity function will be $\Omega(\mathcal{E}_{rs}, V_s) = 1$, so the probability of that state will be given by

$$P_{rs} = \frac{\Omega_\mathrm{R}(E_\mathrm{T} - \mathcal{E}_{rs}, V_\mathrm{T} - V_s)}{\Omega_\mathrm{T}(E_\mathrm{T}, V_\mathrm{T})} = \frac{\exp[S_\mathrm{R}(E_\mathrm{T} - \mathcal{E}_{rs}, V_\mathrm{T} - V_s)/k_\mathrm{B}]}{\exp[S_\mathrm{T}(E_\mathrm{T}, V_\mathrm{T})/k_\mathrm{B}]}, \tag{21.176}$$

which should be compared to Eq. (21.2). The denominator pertains to an unrestricted equilibrium state, so the entropy of the composite system is additive,

$$S_\mathrm{T}(E_\mathrm{T}, V_\mathrm{T}) = S_\mathrm{R}(E_\mathrm{T} - U, V_\mathrm{T} - V) + S(U, V), \tag{21.177}$$

where $U \equiv \langle E \rangle$ is the average energy and $V \equiv \langle V \rangle$ is the average volume of $\mathcal{I}$. In the numerator of Eq. (21.176), we write

$$S_\mathrm{R}\left[E_\mathrm{T} - \mathcal{E}_{rs}, V_\mathrm{T} - V_s\right] = S_\mathrm{R}\left[(E_\mathrm{T} - U) + (U - \mathcal{E}_{rs}), (V_\mathrm{T} - V) + (V - V_s)\right]. \tag{21.178}$$

Then we expand in a Taylor series on the basis that $|U - \mathcal{E}_{rs}|/|E_\mathrm{T} - U| \ll 1$ and $|V - V_s|/|V_\mathrm{T} - V| \ll 1$ to obtain

$$S_\mathrm{R}(E_\mathrm{T} - \mathcal{E}_{rs}, V_\mathrm{T} - V_s) = S_\mathrm{R}(E_\mathrm{T} - E, V_\mathrm{T} - V) + (U - \mathcal{E}_{rs})/T + p(V - V_s)/T, \tag{21.179}$$

where higher order terms are neglected. Substitution of Eqs. (21.177) and (21.179) into Eq. (21.176) gives

$$P_{rs} = \exp[G/k_\mathrm{B}T]\exp[-(\mathcal{E}_{rs} + pV_s)/k_\mathrm{B}T], \tag{21.180}$$

where the Gibbs free energy

$$G = U - TS + pV. \tag{21.181}$$

Since P_{rs} are probabilities, summation[12] over all r and s gives $\sum_{rs} P_{rs} = 1$, so Eq. (21.180) yields

$$\exp[-G/k_\mathrm{B}T] = \sum_{r,s} \exp[-(\mathcal{E}_{rs} + pV_s)/k_\mathrm{B}T] \equiv \mathcal{Z}_p, \tag{21.182}$$

where $\mathcal{Z}_p(T, p, \mathcal{N})$ is the partition function for the pressure ensemble. Then

$$G = -k_\mathrm{B}T \ln \mathcal{Z}_p \tag{21.183}$$

allows for calculation of the other thermodynamic functions by differentiation, recognizing that

$$\mathrm{d}G(T, p, N) = -S\,\mathrm{d}T + V\,\mathrm{d}p + \mu\,\mathrm{d}\mathcal{N}. \tag{21.184}$$

From the Euler equation for the Gibbs free energy, $G = \mu\mathcal{N}$, so the chemical potential is also given by

$$\mu = -(k_\mathrm{B}T/\mathcal{N}) \ln \mathcal{Z}_p, \tag{21.185}$$

which should be compared with the relationship of p to $\ln \mathcal{Z}$ of the GCE.

Rather than working with G and its derivatives, we can define a new variable

$$\lambda_p := \exp(-\beta p) \tag{21.186}$$

and write the partition function in the functional form

$$\mathcal{Z}_p(\beta, \lambda_p, N) = \sum_s \lambda_p^{V_s} \sum_r \exp(-\beta\mathcal{E}_{rs}) = \sum_s \lambda_p^{V_s} Z(T, V_s, \mathcal{N}), \tag{21.187}$$

where $Z(T, V_s, \mathcal{N})$ is the canonical partition function for a system having volume V_s. Note the similarity of Eqs. (21.187) and (21.21). Then we can define a function

$$q_p(\beta, \lambda_p, N) := \ln \mathcal{Z}_p(\beta, \lambda_p, \mathcal{N}) \tag{21.188}$$

whose differential is given by

$$\mathrm{d}q_p = -U\,\mathrm{d}\beta + (V/\lambda_p)\,\mathrm{d}\lambda_p - \beta\mu\,\mathrm{d}\mathcal{N}. \tag{21.189}$$

Equation (21.189) can be verified by using the chain rule of differentiation to convert partial derivatives with respect to the set $T, p, \mathcal{N}$ to those for the set $\beta, \lambda_p, \mathcal{N}$. This allows us to calculate the internal energy from a single differentiation, namely,

$$U = -\left(\frac{\partial q_p}{\partial \beta}\right)_{\lambda_p, \mathcal{N}}. \tag{21.190}$$

The form of Eq. (21.189) is also obvious if Eq. (21.187) is used to compute averages, for example,

$$\lambda_p \frac{\partial q_p}{\partial \lambda_p} = \frac{\sum_s \lambda_p^{V_s} V_s \sum_r \exp(-\beta\mathcal{E}_{rs})}{\sum_s \lambda_p^{V_s} \sum_r \exp(-\beta\mathcal{E}_{rs})} = \sum_{rs} P_{rs} V_s = \langle V \rangle = V. \tag{21.191}$$

[12]To treat V_s as a continuous variable, we would need to replace summation by integration with a probability density function for V_s. We use summation here to parallel the treatment of the GCE.

21.5.1 Vacancies in Monovalent Crystals

As an application of the pressure ensemble, we can calculate the number of **vacancies** on substitutional sites in monovalent crystals having a Bravais lattice. Such vacancies are **point defects** known as **Schottky defects**. They can affect the properties of crystals such as thermal expansion and diffusion by a vacancy mechanism. For the sake of simplicity, we will confine ourselves to metals in which ions occupy lattice sites in a sea of shared electrons. This treatment is a modification of that given by Girifalco [63, p. 195].

For such a crystal at constant temperature T and constant pressure p and having $\mathcal{N}$ ions and $\mathcal{N}_{\rm v}$ vacancies on specific sites, the probability of being in the quantum state $\mathcal{E}_{rs}$ for a crystal of volume V_s is given by

$$P_{rs}(\mathcal{N},\mathcal{N}_{\rm v}) = \frac{1}{\mathcal{Z}_p(\mathcal{N},\mathcal{N}_{\rm v})}\exp[-\beta(\mathcal{E}_{rs}+pV_s)]. \tag{21.192}$$

By summing over all values of r and s as in Eq. (21.182), we can identify a Gibbs free energy $G_0(\mathcal{N},\mathcal{N}_{\rm v})$ given by

$$\exp[-\beta G_0(\mathcal{N},\mathcal{N}_{\rm v})] = \sum_{r,s}\exp[-\beta(\mathcal{E}_{rs}+pV_s)] = \mathcal{Z}_p(\mathcal{N},\mathcal{N}_{\rm v}). \tag{21.193}$$

At this point, we assume that the vacancies are sufficiently dilute (to be justified later) that they do not interact, so $G_0(\mathcal{N},\mathcal{N}_{\rm v})$ does not depend on which sites are occupied. However, since the $\mathcal{N}$ atoms and $\mathcal{N}_{\rm v}$ vacancies are on *specific* sites, $G_0(\mathcal{N},\mathcal{N}_{\rm v})$ does not account for the configurational entropy

$$S^{\rm conf}(\mathcal{N},\mathcal{N}_{\rm v}) = k_{\rm B}\ln w(\mathcal{N},\mathcal{N}_{\rm v}); \quad w(\mathcal{N},\mathcal{N}_{\rm v}) = \frac{(\mathcal{N}+\mathcal{N}_{\rm v})!}{\mathcal{N}!\mathcal{N}_{\rm v}!}. \tag{21.194}$$

We can therefore construct a total Gibbs free energy of the form

$$G(\mathcal{N},\mathcal{N}_{\rm v}) = G_0(\mathcal{N},\mathcal{N}_{\rm v}) - TS^{\rm conf}(\mathcal{N},\mathcal{N}_{\rm v}) = G_0(\mathcal{N},\mathcal{N}_{\rm v}) - k_{\rm B}T\ln w(\mathcal{N},\mathcal{N}_{\rm v}). \tag{21.195}$$

The equilibrium number of vacancies $\mathcal{N}_{\rm v}^{\rm eq}$ can now be determined by minimizing $G(\mathcal{N},\mathcal{N}_{\rm v})$ with respect to $\mathcal{N}_{\rm v}$. With the aid of Stirling's approximation, we differentiate with respect to $\mathcal{N}_{\rm v}$ to obtain

$$0 = \left.\frac{\partial G(\mathcal{N},\mathcal{N}_{\rm v})}{\partial \mathcal{N}_{\rm v}}\right|_{\mathcal{N}_{\rm v}^{\rm eq}} = g_{\rm v} + k_{\rm B}T\ln\left[\frac{\mathcal{N}_{\rm v}^{\rm eq}}{\mathcal{N}+\mathcal{N}_{\rm v}^{\rm eq}}\right], \tag{21.196}$$

where

$$g_{\rm v} := \left.\frac{\partial G_0(\mathcal{N},\mathcal{N}_{\rm v})}{\partial \mathcal{N}_{\rm v}}\right|_{\mathcal{N}_{\rm v}^{\rm eq}} \approx \left.\frac{\partial G_0(\mathcal{N},\mathcal{N}_{\rm v})}{\partial \mathcal{N}_{\rm v}}\right|_{\mathcal{N}_{\rm v}=0}. \tag{21.197}$$

The last expression follows because $\mathcal{N}_{\rm v}^{\rm eq}/\mathcal{N} \ll 1$ as we shall see. Thus

$$\frac{\mathcal{N}_{\rm v}^{\rm eq}}{\mathcal{N}+\mathcal{N}_{\rm v}^{\rm eq}} = \exp[-g_{\rm v}/k_{\rm B}T]. \tag{21.198}$$

Table 21–1 Vacancy and Divacancy Fractions at the Melting Points of Some FCC Crystals Having $\mathcal{N}$ Atoms ($z = 12$ Nearest Neighbors) According to Eqs. (21.199) and (21.207)

FCC crystal	Cu	Ni	Al
T_M	1358 K	1728 K	933 K
h_v	1.05 eV	1.4 eV	0.65 eV
s_v	0.4 k_B	1.5 k_B	0.8 k_B
$\exp(s_v/k_B)$	1.5	4.5	2.2
$\mathcal{N}_v^{eq}/\mathcal{N}$ at T_M	1.9×10^{-4}	3.6×10^{-4}	6.8×10^{-4}
h_b	0.1 eV	0.3 eV	0.3 eV
s_b	–	2 k_B	1 k_B
$\mathcal{N}_d^{eq}/\mathcal{N}$ at T_M	–	8×10^{-7}	4×10^{-5}
f_d	–	0.1%	10%

Notes: f_d given by Eq. (21.208) is the fraction of vacant lattice sites due to divacancies. Data for h_v, s_v, h_b, and s_b, where $g_b = h_b - Ts_b$ is the binding free energy of a divacancy, are from Girifalco [63, p. 217].

The quantity g_V is nearly independent of $\mathcal{N}_V^{eq}$ and can be thought of as the Gibbs free energy needed to create a vacancy by moving an atom from a definite substitutional site to the crystal surface. In order of magnitude, we expect $g_V \sim 1$ eV. At room temperature, $k_B T \sim 1/40$ eV, so the right-hand side of Eq. (21.198) would be about 4×10^{-18}. At $T = 900$ K, it would increase to about 6×10^{-6}.

Since $\mathcal{N}_V^{eq} \ll \mathcal{N}$, it is usual to omit it in the denominator of Eq. (21.198). We can also write $g_V = h_V - Ts_V$, where h_V is an enthalpy and s_V is an entropy, both approximately constant and associated with a single vacancy at a definite location. Then Eq. (21.198) can be written approximately as

$$\mathcal{N}_V^{eq} = \mathcal{N} \exp(s_V/k_B) \exp(-h_V/k_B T), \tag{21.199}$$

where h_V plays the role of an activation energy. Typically the prefactor $\exp(s_V/k_B) \sim 1$ and $\mathcal{N}_V^{eq}/\mathcal{N} \sim 10^{-4}$ near the melting point of a crystal. Table 21–1 gives experimentally determined values of h_V and s_V for some FCC metals as well as values of $\mathcal{N}_V^{eq}$ at their melting points.

According to Eq. (21.197), g_V is nearly independent of $\mathcal{N}_V$ in the range of interest where vacancies are a dilute species, so $G_0(\mathcal{N}, \mathcal{N}_V) \approx G_0(\mathcal{N}, 0) + g_V \mathcal{N}_V$ is nearly linear in $\mathcal{N}_V$. Thus Eq. (21.195) can be written

$$\Delta G(\mathcal{N}, \mathcal{N}_V) = g_V \mathcal{N}_V - k_B T \ln w(\mathcal{N}, \mathcal{N}_V), \tag{21.200}$$

where $\Delta G(\mathcal{N}, \mathcal{N}_V) \equiv G(\mathcal{N}, \mathcal{N}_V) - G_0(\mathcal{N}, 0)$. Equation (21.200) is usually the starting point of a simplified treatment of vacancies and will be used in the next section to explore other point defects.

Nonequilibrium concentrations of vacancies can be obtained by such means as quenching from a higher temperature or irradiation by neutrons. Such concentrations might last for a long time, depending on the rate of vacancy diffusion and the proximity of vacancy sinks such as dislocations and grain boundaries. This can result in the formation of voids. Line defects such as dislocations and area defects such as grain boundaries are not equilibrium defects because the energy to create them is too large to be offset by configurational entropy. They usually result from material preparation, for example, by crystallization, or by mechanical deformation. Prolonged annealing at sufficiently high temperatures can be used to eliminate some line and surface defects but such a process is very slow.

21.5.2 Vacancies, Divacancies, and Interstitials

In monovalent crystals, vacancies (v), vacancies in adjacent lattice sites called "divacancies" (d) and ions in voids of the substitutional lattice called "interstitials" (i) can be considered as point defects. At equilibrium, all of these are dilute species, so we can proceed with a generalization of Eq. (21.200), resulting in

$$\Delta G(\mathcal{N},\mathcal{N}_{\rm v},\mathcal{N}_{\rm d},\mathcal{N}_{\rm i}) = g_{\rm v}\mathcal{N}_{\rm v} + g_{\rm d}\mathcal{N}_{\rm d} + g_{\rm i}\mathcal{N}_{\rm i} - k_{\rm B}T\ln W, \tag{21.201}$$

where $k_{\rm B}\ln W$ is a suitable configurational entropy. We expect $g_{\rm d} < 2g_{\rm v}$ since fewer broken bonds are needed to form a divacancy than to form two isolated vacancies. Interstitials need to crowd surrounding ions, so we expect $g_{\rm i} > g_{\rm v}$, usually leading to interstitials being the most dilute. The number of substitutional lattice sites is

$$\mathcal{S} = N + \mathcal{N}_{\rm v} + 2\mathcal{N}_{\rm d} - \mathcal{N}_{\rm i}. \tag{21.202}$$

The total number of configurations can be expressed as a product of three factors, $W = w_{\rm d}w_{\rm v}w_{\rm i}$, where $w_{\rm d}$ is the number of ways that divacancies can be distributed on the sites $\mathcal{S}$; $w_{\rm v}$ is the number of ways that isolated vacancies can be distributed on the remaining substitutional sites; and $w_{\rm i}$ is the number of ways that interstitials can be distributed on $\mathcal{I}$ interstitial sites. Since we are treating a crystal with a Bravais lattice, we take $\mathcal{I} = \alpha\mathcal{S}$, where α is an integer, for example, 1 for FCC and 3 for BCC. Expressions for $w_{\rm d}$, $w_{\rm v}$, and $w_{\rm i}$ can be quite complex (see Girifalco [63, p. 214]) if proper counting is done to insure, for example, that vacancies are not adjacent to divacancies. But given that all species are dilute, we can make reasonable approximations immediately. For example, if there are $\mathcal{N}_{\rm d}$ divacancies, the isolated vacancies can reside on $\mathcal{S} - 2\mathcal{N}_{\rm d} = \mathcal{N} + \mathcal{N}_{\rm v} - \mathcal{N}_{\rm i}$ sites, but not next to a divacancy or next to each other. But one makes negligible error by assuming that isolated vacancies are distributed over $\mathcal{N} + \mathcal{N}_{\rm v}$ sites, or even over $\mathcal{N}$ sites. Similarly, if z is the number of nearest neighbors of a lattice site, the number of nearest neighbor pairs where a divacancy might reside is $\mathcal{S}z/2$ and some of these could be adjacent. But with negligible error, we can replace $\mathcal{S}z/2$ by $\mathcal{N}z/2$ and ignore the possibility of adjacent sites. Therefore, we adopt the following approximate quantities:[13]

$$w_{\rm d} \approx \frac{(\mathcal{N}z/2)!}{(\mathcal{N}z/2) - (\mathcal{N}_{\rm d})!\mathcal{N}_{\rm d}!}; \quad w_{\rm v} \approx \frac{(\mathcal{N}+\mathcal{N}_{\rm v})!}{\mathcal{N}!\mathcal{N}_{\rm v}!}; \quad w_{\rm i} \approx \frac{(\alpha\mathcal{N})!}{(\alpha\mathcal{N} - \mathcal{N}_{\rm i})!\mathcal{N}_{\rm i}!}, \tag{21.203}$$

[13]In making these approximations, it is important to be sure each quantity has the form of a binomial coefficient.

which completely decouples different point defects. Then proceeding as with vacancies only, we obtain

$$\mathcal{N}_{\rm d}^{\rm eq} \approx \frac{\mathcal{N}z}{2}\exp(-g_{\rm d}/k_{\rm B}T); \quad \mathcal{N}_{\rm v}^{\rm eq} \approx \mathcal{N}\exp(-g_{\rm i}/k_{\rm B}T); \quad \mathcal{N}_{\rm i}^{\rm eq} \approx \alpha\mathcal{N}\exp(-g_{\rm v}/k_{\rm B}T). \tag{21.204}$$

These same results are obtained if the more accurate values of $w_{\rm d}$, $w_{\rm v}$, and $w_{\rm i}$ are used, provided that defect numbers are neglected in comparison with $\mathcal{N}$ in the final results.

The most interesting result is for the divacancies. If we had simply $g_{\rm d} = 2g_{\rm v}$, we would obtain

$$\mathcal{N}_{\rm d}^{\rm eq} = (1/2)[\mathcal{N}\exp(-g_{\rm v}/k_{\rm B}T)][z\exp(-g_{\rm v}/k_{\rm B}T)], \quad \text{no binding energy}, \tag{21.205}$$

which is simply the equilibrium number of vacancies times the probability that one of the z nearest neighbor sites of a vacancy is also occupied by a vacancy, divided by 2 to avoid double counting of pairs. But this does not account for the reduction in free energy (binding energy) that results from the proximity of adjacent vacancies. Thus one can write

$$g_{\rm d} = 2g_{\rm v} - g_{\rm b}, \tag{21.206}$$

where $g_{\rm b} = h_{\rm v} - Ts_{\rm b} > 0$ is a binding (free) energy between adjacent vacancies. Then

$$\mathcal{N}_{\rm d}^{\rm eq} = (z/2)\mathcal{N}\exp[-(2g_{\rm v} - g_{\rm b})/k_{\rm B}T] = (z/2)\mathcal{N}[\exp(-g_{\rm v}/k_{\rm B}T)]^2\exp(g_{\rm b}/k_{\rm B}T). \tag{21.207}$$

Numbers for $g_{\rm b}$ are not very accurate so we give only some estimates of $h_{\rm b}$ and $s_{\rm b}$ to one significant figure in Table 21–1 for Ni and Al, along with calculations of $\mathcal{N}_{\rm d}^{\rm eq}/\mathcal{N}$ at their melting points. A better comparison of divacancies to vacancies can be made by recognizing that a divacancy results in two vacant sites, so the fraction of vacant sites due to divacancies is

$$f_{\rm d} = 2\mathcal{N}_{\rm d}^{\rm eq}/(\mathcal{N}_{\rm v}^{\rm eq} + 2\mathcal{N}_{\rm d}^{\rm eq}). \tag{21.208}$$

For Ni, $f_{\rm d}$ is essentially negligible but for Al it is about 10% at its melting point.

21.5.3 Vacancies and Interstitials in Ionic Crystals

Vacancies and interstitials can occur in crystals with ionic bonding but their formation is subject to additional constraints to insure charge neutrality. We shall illustrate these considerations by treating alkali halides, such as NaCl, and silver halides with formulae of the form AgX in which Ag has the oxidation state +1 and X is a halogen.[14] We consider the following types of point defects:

Positive ion vacancy $\mathcal{N}_{\rm v+}$ in number, each being a region of negative charge $-e$ and capable of existing on $N^{\rm v+}$ sites.

Negative ion vacancy $\mathcal{N}_{\rm v-}$ in number, each being a region of positive charge e and capable of existing on $N^{\rm v-}$ sites.

[14]We exclude AgF_2 in which Ag has the oxidation state +2. At this stage, we do not treat the possibility of color centers in which localized electrons or holes can exist.

Positive ion interstitial $\mathcal{N}_{\mathrm{i}+}$ in number, each being a region of positive charge e and capable of existing on $N^{\mathrm{i}+}$ sites.

Negative ion interstitial $\mathcal{N}_{\mathrm{i}-}$ in number, each being a region of negative charge $-e$ and capable of existing on $N^{\mathrm{i}-}$ sites.

For the sake of simplicity, we will first treat the case in which only pairs of defects are needed to balance charge because the other two types of defects have values of Gibbs free energy per defect that are much larger. For example, we will only need to consider positive ion vacancies balancing the charge of negative ion vacancies if $g_{\mathrm{i}+}$ and $g_{\mathrm{i}-}$ exceed $g_{\mathrm{v}+}$ and $g_{\mathrm{v}-}$ by amounts that are large compared to $k_{\mathrm{B}}T$. This will give rise to two types of vacancies, also known as Schottky defects. On the other hand, we will only need to consider positive ion vacancies balancing the charge of positive ion interstitials if $g_{\mathrm{v}-}$ and $g_{\mathrm{i}-}$ exceed $g_{\mathrm{v}+}$ and $g_{\mathrm{i}+}$ by amounts that are large compared to $k_{\mathrm{B}}T$. Such vacancy-interstitial pairs are known as **Frenkel defects**. In these cases, the constraints on charge neutrality could be applied by immediately setting $\mathcal{N}_{\mathrm{v}+} = \mathcal{N}_{\mathrm{v}-}$ in the Schottky case and $\mathcal{N}_{\mathrm{v}+} = \mathcal{N}_{\mathrm{i}+}$ in the Frenkel case, but a general methodology that can be used if one needs to consider more than two defect types is to use a Lagrange multiplier λ to apply the constraints.

Thus, for the Schottky case we can minimize the function

$$g_{\mathrm{v}+}\mathcal{N}_{\mathrm{v}+} + g_{\mathrm{v}+}\mathcal{N}_{\mathrm{v}+} - k_{\mathrm{B}}T \ln W_{\mathrm{vv}} - \lambda(\mathcal{N}_{\mathrm{v}+} - \mathcal{N}_{\mathrm{v}-}), \tag{21.209}$$

where

$$W_{\mathrm{vv}} = \frac{N^{\mathrm{v}+}!}{(N^{\mathrm{v}+} - \mathcal{N}_{\mathrm{v}+})!\mathcal{N}_{\mathrm{v}+}!}\frac{N^{\mathrm{v}-}!}{(N^{\mathrm{v}-} - \mathcal{N}_{\mathrm{v}-})!\mathcal{N}_{\mathrm{v}-}!}. \tag{21.210}$$

This results in

$$\mathcal{N}^{\mathrm{eq}}_{\mathrm{v}+} = N^{\mathrm{v}+}\exp(-\beta g_{\mathrm{v}+} + \lambda); \quad \mathcal{N}^{\mathrm{eq}}_{\mathrm{v}-} = N^{\mathrm{v}-}\exp(-\beta g_{\mathrm{v}-} - \lambda). \tag{21.211}$$

These can be multiplied to eliminate λ which yields

$$\mathcal{N}^{\mathrm{eq}}_{\mathrm{v}+}\mathcal{N}^{\mathrm{eq}}_{\mathrm{v}-} = N^{\mathrm{v}+}N^{\mathrm{v}-}\exp\left[-\beta(g_{\mathrm{v}+} + g_{\mathrm{v}-})\right]. \tag{21.212}$$

But since the constraint requires $\mathcal{N}^{\mathrm{eq}}_{\mathrm{v}+} = \mathcal{N}^{\mathrm{eq}}_{\mathrm{v}-}$, we obtain[15]

$$\mathcal{N}^{\mathrm{eq}}_{\mathrm{v}+} = \mathcal{N}^{\mathrm{eq}}_{\mathrm{v}-} = (N^{\mathrm{v}+}N^{\mathrm{v}-})^{1/2}\exp\left[-\beta(g_{\mathrm{v}+} + g_{\mathrm{v}-})/2\right]. \tag{21.213}$$

Equation (21.213) depends only on the average of $g_{\mathrm{v}+}$ and $g_{\mathrm{v}-}$, so the smaller of the two compensates for the larger in establishing the effective activation energy. This case is typical for alkali halides.

For the Frenkel case, we can proceed in a similar manner to obtain

$$\mathcal{N}^{\mathrm{eq}}_{\mathrm{v}+} = \mathcal{N}^{\mathrm{eq}}_{\mathrm{i}+} = (N^{\mathrm{v}+}N^{\mathrm{i}+})^{1/2}\exp\left[-\beta(g_{\mathrm{v}+} + g_{\mathrm{i}+})/2\right]. \tag{21.214}$$

This case typically occurs for silver halides. By replacing + with – in Eq. (21.214), we could get a case in which negative ion vacancies and negative ion interstitials are the dominant point defects. By replacing "v" with "i" in Eq. (21.213), we could get a case in which positive

[15]Alternatively we could have set $\mathcal{N}^{\mathrm{eq}}_{\mathrm{v}+} = \mathcal{N}^{\mathrm{eq}}_{\mathrm{v}-}$ in Eq. (21.211) and then solved for $\exp\lambda$.

ion interstitials and negative ion interstitials are the dominant point defects, but this case is not expected to occur because interstitials typically have higher activation energies than vacancies.

Example Problem 21.7. Investigate the case in which g_{v-} and g_{i+} differ from one another by order $k_B T$ but $g_{v+}, g_{v-}, g_{i+} \ll g_{i-}$. Thus, negative ion interstitials can be ignored, so there must be charge balance among the remaining three types of defects.

Solution 21.7. In this case, we apply the charge balance constraint by adding $\lambda(\mathcal{N}_{v+} - \mathcal{N}_{v-} - \mathcal{N}_{i+})$ to ΔG and minimizing to obtain

$$\mathcal{N}_{v+}^{eq} = N^{v+} \exp(-\beta g_{v+} + \lambda); \quad \mathcal{N}_{v-}^{eq} = N^{v-} \exp(-\beta g_{v-} - \lambda); \quad \mathcal{N}_{i+}^{eq} = N^{i+} \exp(-\beta g_{i+} - \lambda). \tag{21.215}$$

By eliminating λ, we obtain

$$\mathcal{N}_{v+}^{eq}\mathcal{N}_{v-}^{eq} = N^{v+}N^{v-} \exp\left[-\beta(g_{v+} + g_{v-})\right]; \quad \mathcal{N}_{v+}^{eq}\mathcal{N}_{i+}^{eq} = N^{v+}N^{i+} \exp\left[-\beta(g_{v+} + g_{i+})\right]. \tag{21.216}$$

Adding the two equations in Eq. (21.216) and applying the constraint $\mathcal{N}_{v+}^{eq} = \mathcal{N}_{v-}^{eq} + \mathcal{N}_{i+}^{eq}$ allows us to solve for

$$\mathcal{N}_{v+}^{eq} = (N^{v+})^{1/2} \exp(-\beta g_{v+}/2) \left[N^{v-} \exp(-\beta g_{v-}) + \mathcal{N}^{i+} \exp(-\beta g_{i+})\right]^{1/2}. \tag{21.217}$$

Then combining this result with Eq. (21.216) gives

$$\mathcal{N}_{v-}^{eq} = \frac{(N^{v+})^{1/2} \exp(-\beta g_{v+}/2)}{\left[N^{v-} \exp(-\beta g_{v-}) + \mathcal{N}_{i+} \exp(-\beta g_{i+})\right]^{1/2}} N^{v-} \exp(-\beta g_{v-}) \tag{21.218}$$

and

$$\mathcal{N}_{i+}^{eq} = \frac{(N^{v+})^{1/2} \exp(-\beta g_{v+}/2)}{\left[N^{v-} \exp(-\beta g_{v-}) + \mathcal{N}_{i+} \exp(-\beta g_{i+})\right]^{1/2}} N^{i+} \exp(-\beta g_{i+}). \tag{21.219}$$

For ionic crystals, there are many other types of point defects, such as those that arise when a small number of Ca^{++} ions are substituted for Na^{+} ions in NaCl, thus stimulating the production of an equal number of Na^{+} vacancies. Such defects can strongly affect electrical conductivity because of vacancy-assisted diffusion of ions. There is also the possibility of color centers that involve localized electrons and holes that have a large influence on optical adsorption. The reader is referred to the book by Ashcroft and Mermin [58, p. 621] for a discussion of these and other defects.

22

Entropy for Any Ensemble

Until now we have introduced four ensembles that are used in statistical mechanics: the microcanonical ensemble in Chapter 16, the canonical ensemble in Chapter 19, the grand canonical ensemble in Chapter 21, and the pressure ensemble in Section 21.5 of Chapter 21. The canonical ensemble and the grand canonical ensemble were derived from the microcanonical ensemble, although an alternative derivation of the canonical ensemble was presented. Moreover, in Chapter 15, we introduced the disorder function $D\{p_i\}$ that gives a precise measure of information based on a set of probabilities $\{p_i\}$ that can be used to characterize a system. In the present chapter, we give a definition of the entropy of a system represented by any ensemble used to define its thermodynamic state statistically. This definition will be based on the methodology of the most probable distribution used in Section 19.1.3 to derive the canonical ensemble. Our definition of entropy will enable us to relate systematically a specific thermodynamic function with the logarithm of the partition function for that ensemble.

22.1 General Ensemble

A general ensemble consists of a very large number $\mathcal{N}_{\rm ens}$ of immaginary systems, each in some quantum state that we can index by a set of numbers, i, j, k, and a set of probabilities P_{ijk} such that a given state will appear $\mathcal{N}_{ijk} = \mathcal{N}_{\rm ens} P_{ijk}$ times in the ensemble. For the sake of illustration, we have assumed that the states of the ensemble can be characterized by three numbers, but more or less could be used depending on the ensemble. In the case of the ensembles heretofore treated, one number i or two numbers i, j is sufficient. To complete the definition of the ensemble, we must specify the set of constraints that must be satisfied. One such constraint,

$$\sum_{i,j,k} P_{ijk} = 1, \tag{22.1}$$

comes from normalization of the set of probabilities and must always be satisfied. If it is the only constraint, a single state index would suffice. But other constraints might also be relevant. These are best illustrated by example for which we select a grand canonical ensemble with two kinds of particles, say A and B. Then we would characterize the states of the ensemble as having $\mathcal{N}_j^A$ of A particles, $\mathcal{N}_k^B$ of B particles and eigenstates with energies

Thermal Physics. http://dx.doi.org/10.1016/B978-0-12-803304-3.00022-3

$\mathcal{E}_{ijk} = \mathcal{E}_i(\mathcal{N}_j^A, \mathcal{N}_k^B, V)$, where V is the volume of the system on which the energies of the eigenstates could depend.[1] In this case, the additional constraint equations would be[2]

$$\sum_{i,j,k} P_{ijk}\, \mathcal{E}_{ijk} = \text{constant}; \tag{22.2}$$

$$\sum_{i,j,k} P_{ijk}\, \mathcal{N}_j^A = \text{constant}; \tag{22.3}$$

$$\sum_{i,j,k} P_{ijk}\, \mathcal{N}_k^B = \text{constant}. \tag{22.4}$$

Given such a general ensemble, the number of ways that the ensemble can be formed is

$$W = \frac{\mathcal{N}_{\text{ens}}!}{\prod_{ijk} \mathcal{N}_{ijk}!} = \frac{\mathcal{N}_{\text{ens}}!}{\prod_{ijk} (\mathcal{N}_{\text{ens}} P_{ijk})!}. \tag{22.5}$$

Then we maximize $\ln W$ subject to the constraints and assert that the entropy of the system represented by the ensemble is given by

$$S = k_{\text{B}}\, \mathcal{N}_{\text{ens}}^{-1} (\ln W)_{\text{max}}, \quad \text{subject to constraints,} \tag{22.6}$$

provided that all Lagrange multipliers employed to incorporate the constraints are identified. Since $\mathcal{N}_{\text{ens}}$ is large, we can use Stirling's approximation to evaluate $\ln W$, resulting in

$$\ln W = \mathcal{N}_{\text{ens}} \ln \mathcal{N}_{\text{ens}} - \sum_{ijk} \mathcal{N}_{\text{ens}} P_{ijk} \ln(\mathcal{N}_{\text{ens}} P_{ijk}) = -\mathcal{N}_{\text{ens}} \sum_{ijk} P_{ijk} \ln P_{ijk}. \tag{22.7}$$

Thus

$$S = -k_{\text{B}} \left(\sum_{ijk} P_{ijk} \ln P_{ijk} \right)_{\text{max}}, \quad \text{subject to constraints,} \tag{22.8}$$

where Lagrange multipliers associated with the constraints must still be identified.

Referring to Chapter 15, we see that Eq. (22.8) amounts to the maximization of the disorder function, but with the important added information that one must maximize the disorder function subject to the constraints of the ensemble under consideration. Thus, Eq. (22.8) provides a general formula for the entropy of a system represented by any ensemble in terms of maximization of the disorder function. The Lagrange multipliers can be identified by comparison with the fundamental differential for $\mathrm{d}S$ according to thermodynamics. For the example given above, this differential would be

$$\mathrm{d}S = T^{-1}\, \mathrm{d}\langle E\rangle + (p/T)\, \mathrm{d}V - (\mu_A/T)\, \mathrm{d}\langle \mathcal{N}^A\rangle - (\mu_B/T)\, \mathrm{d}\langle \mathcal{N}^B\rangle, \tag{22.9}$$

[1] Instead of the volume, $\mathcal{E}_{ijk}$ could depend on a whole set of mechanical variables Y_ℓ if the system can do reversible work by means of generalized forces $p_\ell = -\sum_{ijk} P_{ijk}\, \partial \mathcal{E}_{ijk}/\partial Y_\ell$.

[2] These constraints could be multiplied by $\mathcal{N}_{\text{ens}}$ in which case they are conservation laws for the entire ensemble, which is how they actually originate.

where T is the temperature, p is the pressure, μ_A is the chemical potential of A, μ_B is the chemical potential of B, and $\langle \cdots \rangle$ denotes ensemble averaging.

22.1.1 Example of the Maximization

We proceed to carry out this maximization for the example given above. Introducing Lagrange multipliers α, β, γ_A, and γ_B, we calculate

$$0 = \frac{\partial}{\partial P_{rst}} \left\{ -\sum_{ijk} P_{ijk} \ln P_{ijk} - \sum_{ijk} P_{ijk}[\alpha + \beta \mathcal{E}_{ijk} + \gamma_A \mathcal{N}_j^A + \gamma_B \mathcal{N}_k^B] \right\}. \quad (22.10)$$

By carrying out the differentiation, we obtain

$$-1 - \ln P_{rst} - \alpha - \beta \mathcal{E}_{rst} - \gamma_A \mathcal{N}_s^A - \gamma_B \mathcal{N}_t^B = 0, \quad (22.11)$$

which yields (after a change of indices $r, s, t \to i, j, k$)

$$P_{ijk} = \exp\left\{ -\alpha - 1 - \beta \mathcal{E}_{ijk} - \gamma_A \mathcal{N}_j^A - \gamma_B \mathcal{N}_k^B \right\}. \quad (22.12)$$

By applying the normalization constraint Eq. (22.1), we obtain

$$P_{ijk} = \mathcal{Z}^{-1} \exp(-\gamma_A \mathcal{N}_j^A) \exp(-\gamma_B \mathcal{N}_k^B) \exp(-\beta \mathcal{E}_{ijk}), \quad (22.13)$$

where the grand partition function

$$\mathcal{Z} = \sum_j \exp(-\gamma_A \mathcal{N}_j^A) \sum_k \exp(-\gamma_B \mathcal{N}_k^B) \sum_i \exp(-\beta \mathcal{E}_{ijk}). \quad (22.14)$$

The differential of our expression Eq. (22.8) for the entropy yields

$$\begin{aligned} \mathrm{d}S &= -k_B \sum_{ijk} \left[1 + \ln P_{ijk}\right] \mathrm{d}P_{ijk} \\ &= k_B \sum_{ijk} \left[\gamma_A \mathcal{N}_j^A + \gamma_B \mathcal{N}_k^B + \beta \mathcal{E}_{ijk}\right] \mathrm{d}P_{ijk}, \end{aligned} \quad (22.15)$$

where we have used $\sum_{ijk} \mathrm{d}P_{ijk} = 0$. We also have

$$\langle E \rangle = \sum_{ijk} P_{ijk} \mathcal{E}_{ijk}; \quad \mathrm{d}\langle E \rangle = \sum_{ijk} \mathcal{E}_{ijk}\, \mathrm{d}P_{ijk} + \sum_{ijk} P_{ijk} \frac{\partial \mathcal{E}_{ijk}}{\partial V} \mathrm{d}V; \quad (22.16)$$

$$\langle \mathcal{N}^A \rangle = \sum_{ijk} P_{ijk} \mathcal{N}_j^A; \quad \mathrm{d}\langle \mathcal{N}^A \rangle = \sum_{ijk} \mathcal{N}_j^A\, \mathrm{d}P_{ijk}; \quad (22.17)$$

$$\langle \mathcal{N}^B \rangle = \sum_{ijk} P_{ijk} \mathcal{N}_k^B; \quad \mathrm{d}\langle \mathcal{N}^B \rangle = \sum_{ijk} \mathcal{N}_k^B\, \mathrm{d}P_{ijk}. \quad (22.18)$$

In writing Eq. (22.16), we have recognized that the eigenstates $\mathcal{E}_{ijk} = \mathcal{E}_i(\mathcal{N}_j^A, \mathcal{N}_k^B, V)$, depend on the volume V of the system for a given set of the integers $\mathcal{N}_j^A$ and $\mathcal{N}_k^B$. Substitution of Eqs. (22.16)–(22.18) into Eq. (22.15) gives

$$dS = k_B\gamma_A \, d\langle N^A\rangle + k_B\gamma_B \, d\langle N^B\rangle + k_B\beta \, d\langle E\rangle - \sum_{ijk} P_{ijk}\frac{\partial \mathcal{E}_{ijk}}{\partial V} \, dV. \tag{22.19}$$

Comparison with Eq. (22.9) allows identification of the Lagrange multipliers

$$\beta = \frac{1}{k_B T}; \quad \gamma_A = -\frac{\mu_A}{k_B T}; \quad \gamma_B = -\frac{\mu_B}{k_B T}, \tag{22.20}$$

as well as giving the relation

$$p = -\sum_{ijk} P_{ijk}\frac{\partial \mathcal{E}_i(\mathcal{N}_j^A, \mathcal{N}_k^B, V)}{\partial V} \tag{22.21}$$

for the pressure.

Having now identified the Lagrange multipliers, we can return to Eq. (22.8) and calculate the entropy, resulting in

$$\begin{aligned} S &= -k_B \sum_{ijk} P_{ijk}\left[-\gamma_A \mathcal{N}_j^A - \gamma_B \mathcal{N}_k^B - \beta\mathcal{E}_{ijk} - \ln \mathcal{Z}\right] \\ &= \frac{1}{T}\left[-\mu_A\langle \mathcal{N}^A\rangle - \mu_B\langle \mathcal{N}^B\rangle + \langle E\rangle + k_B T \ln \mathcal{Z}\right]. \end{aligned} \tag{22.22}$$

Thus

$$-k_B T \ln \mathcal{Z} = \langle E\rangle - TS - \mu_A\langle \mathcal{N}^A\rangle - \mu_B\langle \mathcal{N}^B\rangle \equiv K = -pV, \tag{22.23}$$

where the Euler equation for $\langle E\rangle$ has been used in the last step. The Kramers function K should be regarded as a function of its natural variables T, μ_A, μ_B, and V, on which this ensemble depends. Equation (22.23) is an easy way of calculating the pressure in terms of $\ln \mathcal{Z}$, although Eq. (22.21) reveals its physical origin.

22.1.2 Use of the Entropy Formula

The entropy formula Eq. (22.8) can be used practically by inspection to write down the entropy of any ensemble.

For the microcanonical ensemble, there is only one constraint, the normalization of the probabilities Eq. (22.1), for which a single subscript can be used to label the quantum states, all having the same energy. Maximization of the entropy with that constraint shows immediately that all of the P_i are equal, specifically $P_i = 1/\Omega$, where $\Omega(E, V, \mathcal{N})$ is the number of compatible microstates. Therefore, $S(E, V, \mathcal{N}) = -k_B \sum_\Omega (1/\Omega)\ln(1/\Omega) = k_B \ln \Omega$, as we know for that ensemble.

For the canonical ensemble, there are two constraints, the normalization and the energy constraint, so $S = -k_B\left[-\beta\langle E\rangle - \ln Z(T, V, \mathcal{N})\right] = \langle E\rangle/T + k_B \ln Z(T, V, \mathcal{N})$, where $Z(T, V, \mathcal{N}) = \sum_i \exp[-\beta\mathcal{E}_i(V, \mathcal{N})]$ is the canonical partition function. Thus, the Helmholtz free energy $F(T,V,\mathcal{N}) = -k_B T \ln Z$. We also find $p = -\sum_i P_i \partial E_i(V,\mathcal{N})/\partial V$ as well as $\mu = \sum_i P_i \partial E_i(V,\mathcal{N})/\partial \mathcal{N}$, as in Section 19.1.3.

For the grand canonical ensemble, we have the results of the previous section.

For the pressure ensemble, which was treated in Section 21.5 by another method, we have the normalization constraint, the energy constraint, and a volume constraint of the form $\sum_{i,\ell} P_{i\ell} V_\ell =$ constant, where V_ℓ is the set of volumes on which the energy eigenstates $\mathcal{E}_i(V_\ell,\mathcal{N})$ depend. The entropy is therefore $S = -k_B\left[-\beta\langle E\rangle - \beta p\langle V\rangle - \ln Z_p(T,p,\mathcal{N})\right] = \langle E\rangle/T + (p/T)\langle V\rangle + k_B \ln Z_p(T,p,\mathcal{N})$, where the partition function

$$Z_p(T,p,\mathcal{N}) = \sum_\ell \exp(-\beta p V_\ell) \sum_i \exp\left[-\beta \mathcal{E}_i(V_\ell,\mathcal{N})\right]. \tag{22.24}$$

Thus, the Gibbs free energy $G(T,p,\mathcal{N}) = -k_B T \ln Z_p$. We also have $\mu = \sum_{i,\ell} P_{i\ell} \partial \mathcal{E}_i (V_\ell,\mathcal{N})/\partial \mathcal{N}$.

For the sake of illustration, we invent another ensemble for which the normalizing function for the probabilities can be related to a Massieu function of the system. The energies of all of the eigenstates in the ensemble will have the same energy E, just as for the microcanonical ensemble, so we have the normalization constraint but no additional energy constraint. But we will allow the members of the ensemble to have a set of volumes, V_ℓ as they did for the pressure ensemble. Thus we will have a volume constraint $\sum_{i,\ell} P_{i\ell} V_\ell =$ constant. The probabilities will be given by

$$P_{i\ell} = \exp(-\gamma V_\ell)/\Omega^*, \tag{22.25}$$

where the normalizing function[3]

$$\Omega^*(E,\gamma,\mathcal{N}) = \sum_{i,\ell} \exp(-\gamma V_\ell) = \sum_\ell \Omega(E,V_\ell,\mathcal{N}) \exp(-\gamma V_\ell). \tag{22.26}$$

Here, γ is the Lagrange multiplier for the volume constraint and $\Omega(E,V_\ell,\mathcal{N})$ is the number of eigenstates having the given energy E and particle number $\mathcal{N}$ for a state with volume V_ℓ. The probabilities $P_{i\ell}$ depend on E, γ, and $\mathcal{N}$, so to find γ we allow it to vary at fixed E and $\mathcal{N}$. In this case, the differential of the entropy is simply

$$\mathrm{d}S = \sum_{i,\ell} k_B \gamma V_\ell \frac{\partial P_{i\ell}}{\partial \gamma}\,\mathrm{d}\gamma. \tag{22.27}$$

For the average volume of the system, we have

$$\langle V\rangle = \sum_{i,\ell} P_{i\ell} V_\ell; \quad \mathrm{d}\langle V\rangle = \sum_{i,\ell} V_\ell \frac{\partial P_{i\ell}}{\partial \gamma}\,\mathrm{d}\gamma. \tag{22.28}$$

At fixed E and $\mathcal{N}$, $\mathrm{d}S = (p/T)\,\mathrm{d}\langle V\rangle$, so

$$\gamma = p/(k_B T). \tag{22.29}$$

[3]This normalizing function is the partition function for this ensemble but we give it a different notation because of its close association with the microcanonical ensemble, as clarified in the next section.

The entropy is therefore

$$S = -k_B\left[-\beta p\langle V\rangle - \ln \mathcal{Z}(T,p,\mathcal{N})\right] = (p/T)\langle V\rangle + k_B \ln \Omega^*(E,p/T,\mathcal{N}). \tag{22.30}$$

Thus

$$k_B \ln \Omega^*(E,p/T,\mathcal{N}) = S - (p/T)\langle V\rangle \equiv M_2(E,p/V,\mathcal{N}), \tag{22.31}$$

which is a Legendre transform of the entropy, ordinarily called a Massieu function. From the differential of S, we find

$$\mathrm{d}M_2(E,p/V,\mathcal{N}) = (1/T)\,\mathrm{d}E - \langle V\rangle\,\mathrm{d}(p/T) - (\mu/T)\,\mathrm{d}\mathcal{N}. \tag{22.32}$$

Thus from the partial derivatives of $\Omega^*(E,p/T,\mathcal{N})$, we are able to compute $1/T$, $-\langle V\rangle$, and $-(\mu/T)$.

22.2 Summation over Energy Levels

As pointed out by Hill [64, p. 30], the partition function for all of these ensembles can be written as sums over the extensive variables that are needed to characterize the microcanonical ensemble provided that we sum over energy levels (instead of quantum states) with an appropriate degeneracy factor for the energy eigenstates. For a single component system, that factor is $\Omega(E,V,\mathcal{N})$ which is the number of eigenstates having energy E for a system with volume V and particle number $\mathcal{N}$.

For the microcanonical ensemble, there is no summation and one has simply

$$\ln \Omega(E,V,\mathcal{N}) = S(E,V,\mathcal{N})/k_B. \tag{22.33}$$

For the canonical ensemble,

$$\ln \sum_E \Omega(E,V,\mathcal{N}) \exp(-\beta E) = -\beta F(\beta,V,\mathcal{N}). \tag{22.34}$$

For the grand canonical ensemble

$$\ln \sum_{E,\mathcal{N}} \Omega(E,V,\mathcal{N}) \exp(-\beta E + \beta\mu\mathcal{N}) = -\beta K(\beta,V,\mu). \tag{22.35}$$

For the pressure ensemble

$$\ln \sum_{E,V} \Omega(E,V,\mathcal{N}) \exp(-\beta E - \beta p V) = -\beta G(\beta,p,\mathcal{N}). \tag{22.36}$$

For the ensemble related to the Massieu function discussed above,

$$\ln \sum_{V} \Omega(E,V,\mathcal{N}) \exp(-\beta p V) = M_2(E,p/T,\mathcal{N})/k_B. \tag{22.37}$$

Note that the form of the right-hand sides of Eq. (22.37) and Eq. (22.33) depend on E which is not summed over. In all of these cases, one could use distribution functions for any of the variables that are summed over and then integrate over those variables.

This is necessary if V is to be treated as a continuous variable in Eqs. (22.36) and (22.37). Every ensemble involves a weighted sum of entropies of a microcanonical ensemble. The extensive variables that are summed over are the ones that have dispersion in the respective ensemble.

23

Unified Treatment of Ideal Fermi, Bose, and Classical Gases

In Chapter 21, we introduced the grand canonical ensemble which applies to a system having a fixed temperature and a fixed chemical potential, but not a fixed energy or a fixed number of particles. In Section 21.2.5, we discussed a unified treatment of orbitals of ideal Fermi, Bose, and classical gases for which the grand partition function $\mathcal{Z}$ factored and could be written formally in the form

$$\mathcal{Z} = \prod_{\varepsilon} \left[1 + a\lambda \exp(-\beta\varepsilon)\right]^{1/a}, \tag{23.1}$$

where the product is over all orbitals[1] having energy ε, $\lambda = \exp(\beta\mu)$ is the absolute activity with chemical potential μ, and

$$a = \begin{cases} 1 & \text{fermions} \\ -1 & \text{bosons} \\ 0 & \text{classical.} \end{cases} \tag{23.2}$$

This yields

$$\ln \mathcal{Z} = \frac{1}{a} \sum_{\varepsilon} \ln\left[1 + a\lambda \exp(-\beta\varepsilon)\right]. \tag{23.3}$$

The classical case must be interpreted as a limit $a \to 0$ to give

$$\ln \mathcal{Z} = \sum_{\varepsilon} \lambda \exp(-\beta\varepsilon) = \lambda z, \tag{23.4}$$

where z is the canonical partition function of a single particle. From Eq. (21.32) with $q = \ln \mathcal{Z}$, we obtain

$$\langle \mathcal{N} \rangle = \lambda \left(\frac{\partial q}{\partial \lambda}\right)_{\beta,V} = \sum_{\varepsilon} f(\varepsilon, a) \tag{23.5}$$

and

$$U = -\left(\frac{\partial q}{\partial \beta}\right)_{\lambda,V} = \sum_{\varepsilon} \varepsilon f(\varepsilon, a), \tag{23.6}$$

[1]As explained in Section 21.2, an orbital is a quantum state of a particle specified by all quantum numbers of its spatial wave function and its spin, which we incorporate in the single symbol ε which is also the energy of that state, usually degenerate.

Thermal Physics. http://dx.doi.org/10.1016/B978-0-12-803304-3.00023-5

where

$$f(\varepsilon, a) := \frac{1}{\lambda^{-1}\exp(\beta\varepsilon) + a} = \frac{1}{\exp[\beta(\varepsilon - \mu)] + a}. \tag{23.7}$$

Note that $f(\varepsilon, a)$ agrees with Eq. (21.88) for $a = 1$ and Eq. (21.92) for $a = -1$. From Eq. (21.38) we also obtain

$$\frac{pV}{k_B T} = \frac{1}{a}\sum_{\varepsilon} \ln\left[1 + a\lambda \exp(-\beta\varepsilon)\right]. \tag{23.8}$$

23.1 Integral Formulae

If the temperature is not too low, the sums in Eqs. (23.5), (23.6), and (23.8) can be converted to integrals because the spacings of the energy levels will be small compared with $k_B T$ and ε will be quasi-continuous. However, conversion to an integral is insufficient for bosons below the condensation temperature, which we discuss in the next chapter. If every state has a degeneracy g_0 due to spin, then $\sum_{\varepsilon} = g_0 \sum_{\varepsilon}'$, where the primed sum is over states with spin degeneracy ignored. For a free particle in a rectangular box of dimensions H, K, L, these states can be expressed in terms of the wave vector $\mathbf{k}$ given by Eq. (16.51). If one of the integers in that expression, say n_x, changes by unity, the x component of $\mathbf{k}$ changes by $\Delta k_x = 2\pi/H$, and similarly the y and z components change by $\Delta k_y = 2\pi/K$ and $\Delta k_z = 2\pi/L$. Thus we have

$$\sum_{\varepsilon} = g_0 \sum_{\varepsilon}{}' = g_0 \sum_{\mathbf{k}} = g_0 \sum_{k_x}\sum_{k_y}\sum_{k_z} = g_0 \frac{HKL}{(2\pi)^3}\sum_{k_x}\sum_{k_y}\sum_{k_z} \Delta k_x \Delta k_y \Delta k_z. \tag{23.9}$$

If we apply this to any nearly continuous function $\mathcal{F}(\mathbf{k})$ that does not vary significantly over the $\mathbf{k}$-space volume element $\Delta k_x \Delta k_y \Delta k_z$, we can replace summation by integration and obtain[2]

$$\frac{HKL}{(2\pi)^3}\sum_{k_x}\sum_{k_y}\sum_{k_z} \Delta k_x \Delta k_y \Delta k_z \, \mathcal{F}(\mathbf{k}) \to \frac{V}{(2\pi)^3}\int \mathrm{d}^3k \, \mathcal{F}(\mathbf{k}), \tag{23.10}$$

where HKL has been replaced by the volume V. Furthermore, if $\mathcal{F}(\mathbf{k})$ depends only on the magnitude of $\mathbf{k}$, as it would for an integrand of the form $\mathcal{G}(\varepsilon(|\mathbf{k}|))$, where $\varepsilon = \hbar^2|\mathbf{k}|^2/2m$, we would have

$$\frac{V}{(2\pi)^3}\int \mathrm{d}^3k \, \mathcal{G}(\varepsilon(|\mathbf{k}|)) = \frac{V}{(2\pi)^3} 4\pi \int_0^\infty k^2 \, \mathrm{d}k \, \mathcal{G}(\varepsilon(|\mathbf{k}|)) = \frac{V}{2\pi^2}\int_0^\infty \mathcal{G}(\varepsilon)\, k^2 \frac{\mathrm{d}k}{\mathrm{d}\varepsilon}\, \mathrm{d}\varepsilon. \tag{23.11}$$

Since $(1/2\pi^2)k^2\mathrm{d}k/\mathrm{d}\varepsilon = (2/\pi^{1/2})(m/2\pi\hbar^2)^{3/2}\varepsilon^{1/2}$, we finally obtain

$$\sum_{\varepsilon} \mathcal{G}(\varepsilon(|\mathbf{k}|)) = g_0 V n_Q(T) \frac{1}{\Gamma(3/2)}\int_0^\infty \mathcal{G}(\varepsilon)\, \beta^{3/2}\varepsilon^{1/2}\, \mathrm{d}\varepsilon, \tag{23.12}$$

where $n_Q(T) = (mk_B T/2\pi\hbar^2)^{3/2}$ is the quantum concentration and the gamma function $\Gamma(3/2) = (1/2)\pi^{1/2}$ has been introduced to unify subsequent notation.

[2]This would not be true for thin samples. For instance, if H were small, then Δk_x would be large.

With the use of Eq. (23.12) and the substitution $u = \beta\varepsilon$, Eqs. (23.5) and (23.6) can be written in the forms

$$n = g_0 n_Q(T)\, h_{3/2}(\lambda, a) \tag{23.13}$$

and

$$u_V = (3/2)k_B T\, g_0 n_Q(T)\, h_{5/2}(\lambda, a), \tag{23.14}$$

where $n = \langle\mathcal{N}\rangle/V$ is the average number of particles per unit volume, $u_V = U/V$ is the energy per unit volume, and the function

$$h_\nu(\lambda, a) := \frac{1}{\Gamma(\nu)} \int_0^\infty \frac{u^{\nu-1}\, du}{\lambda^{-1}\, e^u + a}. \tag{23.15}$$

Equation (23.13) determines λ, or equivalently the chemical potential μ, as a function of n and T which can then be substituted into Eq. (23.14) to determine u_V. For the classical gas, we have $h_\nu = \lambda$ for any $\nu > 0$ so Eq. (23.13) becomes simply $\lambda = n/(n_Q g_0)$ and Eq. (23.14) becomes the familiar $u_V = (3/2)nk_B T$. For a strictly classical gas, there is no spin degree of freedom so $g_0 = 1$.

Before exploring the behavior of the functions $h_\nu(\lambda, a)$, we return to Eq. (23.8) for the pressure and convert the sum to an integral to obtain

$$\frac{p}{k_B T} = \frac{1}{a} g_0 n_Q(T) \frac{1}{\Gamma(3/2)} \int_0^\infty \ln(1 + a\lambda\, e^{-u}) u^{1/2}\, du. \tag{23.16}$$

Then we use $u^{1/2} = (2/3)(d/du)u^{3/2}$ to integrate by parts and obtain

$$\frac{1}{a}\int_0^\infty \ln(1 + a\lambda\, e^{-u}) u^{1/2}\, du = \frac{2}{3a} u^{3/2} \ln(1 + a\lambda\, e^{-u})\Big|_0^\infty + \frac{2}{3}\int_0^\infty \frac{u^{3/2}\, du}{\lambda^{-1}\, e^u + a}. \tag{23.17}$$

The integrated part vanishes at both limits and we obtain

$$p = k_B T g_0 n_Q(T) h_{5/2}(\lambda, a) = (2/3)u_V, \tag{23.18}$$

where $(3/2)\Gamma(3/2) = \Gamma(5/2)$ has been used. In view of Eq. (23.3), the same integration by parts can be used to evaluate the partition function, resulting in

$$\ln \mathcal{Z} = g_0 V n_Q(T) h_{5/2}(\lambda, a). \tag{23.19}$$

Therefore, the Kramers potential is

$$K = -k_B T g_0 V n_Q(T) h_{5/2}(\lambda, a) = -(2/3)U = -pV \tag{23.20}$$

as expected.

The entropy can be determined by using Eq. (21.14) to obtain

$$\frac{S}{k_B} = \ln \mathcal{Z} - \beta\left(\frac{\partial \ln \mathcal{Z}}{\partial \beta}\right)_{V,\mu}. \tag{23.21}$$

This results in

$$\frac{S}{k_B} = (5/2)V g_0 n_Q(T) h_{5/2}(\lambda, a) - \beta V g_0 n_Q \left(\frac{\partial h_{5/2}(\lambda, a)}{\partial \beta}\right)_{V,\mu}, \tag{23.22}$$

where we have recalled that $n_Q(T) \propto \beta^{-3/2}$. Since $(\partial\lambda/\partial\beta)_{V,\mu} = \mu\lambda$, Eq. (23.22) becomes

$$\frac{S}{k_B V g_0 n_Q(T)} = (5/2)h_{5/2}(\lambda, a) - \beta\mu\lambda \frac{\partial h_{5/2}(\lambda, a)}{\partial\lambda}. \tag{23.23}$$

In the next section, we will show that $\lambda\partial h_{5/2}(\lambda, a)/\partial\lambda = h_{3/2}(\lambda, a)$. Then noting that $\beta\mu = \ln\lambda$, we obtain

$$S = k_B V g_0 n_Q(T) \left[(5/2)h_{5/2}(\lambda, a) - \ln\lambda\, h_{3/2}(\lambda, a)\right]. \tag{23.24}$$

Alternatively we could compute the entropy from $S/k_B = \beta(U - K - \mu\langle\mathcal{N}\rangle)$, which leads to the same answer.

Thus, it remains to determine the behavior of the functions $h_\nu(\lambda, a)$ which we take up in the next section.

23.2 The Functions $h_\nu(\lambda, a)$

We first derive the relation

$$\lambda\frac{\partial h_\nu(\lambda, a)}{\partial\lambda} = h_{\nu-1}(\lambda, a); \quad \nu > 1. \tag{23.25}$$

We begin with

$$\lambda\frac{\partial h_\nu(\lambda, a)}{\partial\lambda} = \lambda\frac{\partial}{\partial\lambda}\frac{1}{\Gamma(\nu)}\int_0^\infty \frac{u^{\nu-1}\,du}{\lambda^{-1}\,e^u + a} \tag{23.26}$$

and note that

$$\lambda\frac{\partial}{\partial\lambda}\frac{1}{\lambda^{-1}\,e^u + a} = -\frac{\partial}{\partial u}\frac{1}{\lambda^{-1}\,e^u + a}. \tag{23.27}$$

We integrate by parts to obtain

$$\lambda\frac{\partial h_\nu(\lambda, a)}{\partial\lambda} = -\frac{1}{\Gamma(\nu)}\frac{u^{\nu-1}}{\lambda^{-1}\,e^u + a}\bigg|_0^\infty + \frac{\nu-1}{\Gamma(\nu)}\int_0^\infty \frac{u^{\nu-2}\,du}{\lambda^{-1}\,e^u + a}. \tag{23.28}$$

The integrated term vanishes at both limits provided that $\nu > 1$. In the second term, we use $\Gamma(\nu) = (\nu - 1)\Gamma(\nu - 1)$, resulting in Eq. (23.25).

For $0 \le \lambda < 1$, we can obtain series expansions for $h_\nu(\lambda, a)$. Since $\lambda = e^{\beta\mu}$ and $\beta\mu$ is real, we certainly have $\lambda \ge 0$. Returning to Eqs. (23.5) and (23.7), we see that $f(\varepsilon, a)$ must be finite and positive for all values of ε. If we examine the ground state $\varepsilon = 0$ for bosons, we see that $f(0, -1) = \lambda/(1 - \lambda)$, which means that $\lambda < 1$, or equivalently $\mu < 0$. For fermions, $f(0, 1) = \lambda/(1 + \lambda)$ so no such restriction exists and $0 \le \lambda < \infty$. Therefore, our series expansion will cover the range of $0 \le \lambda < 1$ but we will have to examine $\lambda = 1$ carefully. For fermions, we will have to examine $\lambda > 1$ separately.

From Eq. (23.15) we obtain

$$h_\nu(\lambda, a) = \frac{1}{\Gamma(\nu)} \int_0^\infty \frac{u^{\nu-1}\lambda\, \mathrm{e}^{-u}\, \mathrm{d}u}{1 + a\lambda\, \mathrm{e}^{-u}} = \frac{1}{\Gamma(\nu)} \left(\frac{1}{-a}\right) \sum_{n=1}^{\infty} (-a\lambda)^n \int_0^\infty u^{\nu-1}\, \mathrm{e}^{-nu}\, \mathrm{d}u$$
$$= \frac{1}{\Gamma(\nu)} \left(\frac{1}{-a}\right) \sum_{n=1}^{\infty} \frac{(-a\lambda)^n}{n^\nu} \int_0^\infty v^{\nu-1}\, \mathrm{e}^{-v}\, \mathrm{d}v = \left(\frac{1}{-a}\right) \sum_{n=1}^{\infty} \frac{(-a\lambda)^n}{n^\nu}. \tag{23.29}$$

For bosons,

$$g_\nu(\lambda) := h_\nu(\lambda, -1) = \lambda + \frac{\lambda^2}{2^\nu} + \frac{\lambda^3}{3^\nu} + \cdots = \sum_{n=1}^{\infty} \frac{\lambda^n}{n^\nu} \tag{23.30}$$

and for fermions,

$$f_\nu(\lambda) := h_\nu(\lambda, 1) = \lambda - \frac{\lambda^2}{2^\nu} + \frac{\lambda^3}{3^\nu} + \cdots = \sum_{n=1}^{\infty} \frac{\lambda^n}{n^\nu} (-1)^{n+1}. \tag{23.31}$$

For classical particles, $h_\nu(\lambda, 0) = \lambda$ for $\nu > 0$ as mentioned previously.

The value $\lambda = 1$ must be handled with care. It turns out that

$$h_\nu(1, -1) = g_\nu(1) = 1 + \frac{1}{2^\nu} + \frac{1}{3^\nu} + \cdots = \sum_{n=1}^{\infty} \frac{1}{n^\nu} = \zeta(\nu), \tag{23.32}$$

where

$$\zeta(\nu) := \sum_{k=1}^{\infty} k^{-\nu}, \quad \Re\, \nu > 1, \tag{23.33}$$

is the Riemann zeta function. For $\nu = 1$, this is the well-known harmonic series and diverges. Important values for our purposes are $g_{3/2}(1) = \zeta(3/2) = 2.61238$ and $g_{5/2}(1) = \zeta(5/2) = 1.34149$. Since $g_{1/2}(1) = \infty$, Eq. (23.25) shows that $g_{3/2}(\lambda)$ approaches $g_{3/2}(1)$ with infinite slope. For fermions, nothing special happens at $\lambda = 1$ because

$$h_\nu(1, 1) = f_\nu(1) = 1 - \frac{1}{2^\nu} + \frac{1}{3^\nu} + \cdots = \sum_{n=1}^{\infty} \frac{1}{n^\nu} (-1)^{n+1}, \tag{23.34}$$

which is an alternating series with terms of decreasing size for positive ν. In fact, $f_\nu(1) = (1 - 2^{1-\nu})\zeta(\nu)$ for $\nu > 1$, as the reader may verify by writing out the series.

Figure 23–1 shows some plots of $h_\nu(\lambda, a)$ as a function of λ, including values of $\lambda > 1$ which we have not yet discussed for fermions.

For fermions and $\lambda > 1$, one can either compute the integrals for $h_\nu(\lambda, 1) = f_\nu(\lambda)$ numerically or resort to an asymptotic expansion that is valid for large λ. This expansion, known as the **Sommerfeld expansion** [65], is actually a series in $(\ln \lambda)^{-1} = (\beta\mu)^{-1} = k_{\mathrm{B}}T/\mu$ and is useful at high temperatures to treat degenerate Fermi gases. For now we only quote the first few terms [8, p. 510]:

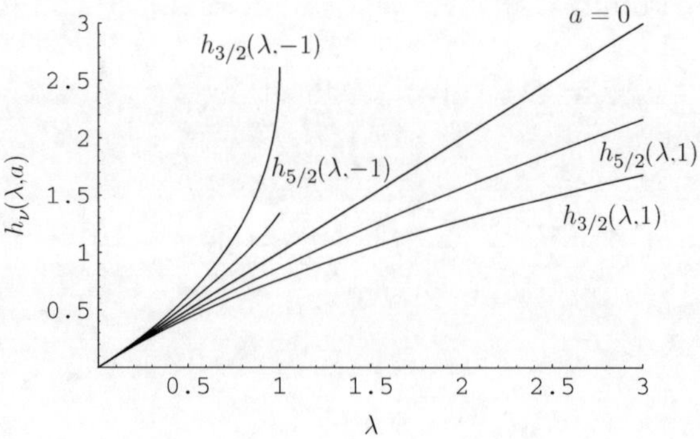

FIGURE 23–1 Plots of the function $h_\nu(\lambda, a)$ for ideal Fermi, Bose, and classical gases as a function of λ. Note that all plots merge for $\lambda \ll 1$ which is the classical limit. The upper two curves are for bosons and the lower two are for fermions. The middle line is for the classical case $a = 0$.

$$f_\nu(\lambda) \sim \frac{(\ln\lambda)^\nu}{\Gamma(\nu+1)}\left[1 + \nu(\nu-1)\frac{\pi^2}{6}\left(\frac{1}{\ln\lambda}\right)^2 + \nu(\nu-1)(\nu-2)(\nu-3)\frac{7\pi^4}{360}\left(\frac{1}{\ln\lambda}\right)^4 + \cdots\right]. \tag{23.35}$$

In Chapter 25, we shall examine a related expansion in more detail to treat the free-electron model of metals.

23.3 Virial Expansions for Ideal Fermi and Bose Gases

We digress briefly to discuss so-called virial expansions which are series expansions for $p/(nk_BT)$ in powers of n. We first discuss these expansions for very small values of λ and then present some more general results.

From Eqs. (23.14) and (23.18) we have

$$y := \frac{p}{g_0 n_Q k_B T} = h_{5/2}(\lambda, a) = \left(\frac{1}{-a}\right)\sum_{n=1}^{\infty}\frac{(-a\lambda)^n}{n^{5/2}} \tag{23.36}$$

and Eq. (23.13) becomes

$$x := \frac{n}{g_0 n_Q} = h_{3/2}(\lambda, a) = \left(\frac{1}{-a}\right)\sum_{n=1}^{\infty}\frac{(-a\lambda)^n}{n^{3/2}}. \tag{23.37}$$

Dividing Eq. (23.36) by Eq. (23.37) we obtain

$$\frac{y}{x} = \frac{p}{nk_BT} = \frac{h_{5/2}(\lambda, a)}{h_{3/2}(\lambda, a)}. \tag{23.38}$$

For a given value of a, $p/(nk_BT)$ depends only on λ and hence only on x. We can then invert the series in Eq. (23.37) by successive approximations and obtain a series expansion for $p/(nk_BT)$ in terms of x. For the classical case, $a = 0$, we have $h_{5/2}(\lambda, 0) = h_{3/2}(\lambda, 0) = \lambda$, so Eq. (23.38) becomes simply

$$\frac{p}{nk_BT} = 1. \tag{23.39}$$

This turns out to be the leading term for $a \neq 0$ for sufficiently small λ.

We illustrate the expansion procedure by calculating the next term in the series. For the Fermi and Bose gases, we have, to order λ^2, the expressions

$$y = \lambda - a\lambda^2/2^{5/2} + \cdots \tag{23.40}$$

and

$$x = \lambda - a\lambda^2/2^{3/2} + \cdots. \tag{23.41}$$

To lowest order, we have $\lambda = x$ which we substitute into the second-order term in Eq. (23.41) to obtain $\lambda = x + ax^2/2^{3/2} + \cdots$. Substitution into Eq. (23.40) gives $y = x + ax^2/2^{3/2} - ax^2/2^{5/2} \cdots$ so to this order we have

$$\frac{y}{x} = \frac{p}{nk_BT} = 1 - ax\left(-\frac{1}{4\sqrt{2}}\right) + \cdots = 1 - ax(-0.17678) + \cdots. \tag{23.42}$$

As the concentration n, and therefore x, increases, we get a positive correction (compared to a classical gas) for fermions (repulsive effect consistent with the exclusion principle) and a negative correction for bosons. This iteration process can be carried out to higher order and results in a virial expansion of the form (see [8, p. 160])

$$\frac{p}{nk_BT} = \sum_{\ell=1}^{\infty}(-a)^{\ell-1}a_\ell\, x^{\ell-1}, \tag{23.43}$$

where the first few virial coefficients are $a_1 = 1$, $a_2 = -1/(4\sqrt{2}) = -0.17678$, $a_3 = -[2/(9\sqrt{3}) - 1/8] = -0.00330$, and $a_4 = -[3/32 + 5/(32\sqrt{2}) - 1/(2\sqrt{6})] = -0.00011$. It turns out that the higher order terms in the series are not very important and Eq. (23.42) is accurate to within about 1% for fermions and about 5% for bosons even up to $\lambda = 1$.

Figure 23–2 shows a plot of $p/(nk_BT)$ versus $x = n/(g_0n_Q)$ for ideal Fermi and Bose gases up to values that correspond to $\lambda = 1$. The plot was constructed by evaluation of

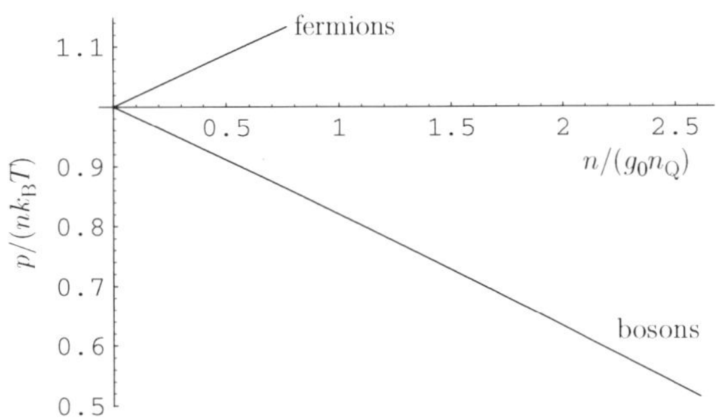

FIGURE 23–2 Plot of $p/(nk_BT)$ versus $x = n/(g_0n_Q)$ for ideal Fermi and Bose gases up to values that correspond to $\lambda = 1$. This plot was constructed by evaluation of the functions $h_\nu(\lambda, a)$ numerically and then using a parametric plotting routine. Amazingly, the deviations from linearity are only a few percent.

the functions $h_\nu(\lambda, a)$ numerically as functions of λ and then using the parametric plotting routine in Mathematica®. The plot agrees extremely well with the series expansion up to $n/(g_0 n_Q) = 0.5$. By using values of $\lambda > 1$ it can be extended to larger values than shown for fermions and deviates only slightly from a straight line. Therefore, Eq. (23.42) suffices approximately over a considerable range of x. We shall see later that Bose condensation sets in very near to $\lambda = 1$, in agreement with the limited range of the plot for bosons.

23.4 Heat Capacity

We can compute the heat capacity at constant volume by partial differentiation of the internal energy U with respect to T with V and $\langle\mathcal{N}\rangle$ held constant. From Eq. (23.14) we have

$$C_V = \frac{15}{4} V k_B g_0 n_Q h_{5/2}(\lambda, a) + \frac{3}{2} V k_B T g_0 n_Q \frac{\partial h_{5/2}(\lambda, a)}{\partial \lambda} \left(\frac{\partial \lambda}{\partial T}\right)_{V,\langle\mathcal{N}\rangle}, \tag{23.44}$$

where we have recalled that $n_Q \propto T^{3/2}$. To calculate $(\partial\lambda/\partial T)_{V,\langle\mathcal{N}\rangle}$ we differentiate Eq. (23.13) to obtain

$$0 = \frac{3}{2} V k_B \frac{1}{T} g_0 n_Q h_{3/2}(\lambda, a) + V k_B T g_0 n_Q \frac{\partial h_{3/2}(\lambda, a)}{\partial \lambda} \left(\frac{\partial \lambda}{\partial T}\right)_{V,\langle\mathcal{N}\rangle}. \tag{23.45}$$

Then after using Eq. (23.25) we solve for the required derivative to obtain

$$\left(\frac{\partial \lambda}{\partial T}\right)_{V,\langle\mathcal{N}\rangle} = -\frac{3}{2}\frac{\lambda}{T}\frac{h_{3/2}(\lambda, a)}{h_{1/2}(\lambda, a)}. \tag{23.46}$$

By substituting into Eq. (23.44) and again using Eq. (23.25), we obtain

$$C_V = \frac{3}{2} V k_B g_0 n_Q \left\{\frac{5}{2} h_{5/2}(\lambda, a) - \frac{3}{2}\frac{\left[h_{3/2}(\lambda, a)\right]^2}{h_{1/2}(\lambda, a)}\right\}. \tag{23.47}$$

Finally, we can use Eq. (23.13) for $\langle\mathcal{N}\rangle$ to obtain

$$C_V = \frac{3}{2} \langle\mathcal{N}\rangle k_B \left\{\frac{5}{2}\frac{h_{5/2}(\lambda, a)}{h_{3/2}(\lambda, a)} - \frac{3}{2}\frac{h_{3/2}(\lambda, a)}{h_{1/2}(\lambda, a)}\right\}. \tag{23.48}$$

We caution, however, that Eqs. (23.4), (23.13), and (23.48) are not valid for bosons for temperatures below the Bose condensation temperature that we treat in the next chapter.

24

Bose Condensation

An ideal Bose fluid is one composed of noninteracting bosons, which are particles having integral spin $s = 0, 1, 2, \ldots$ and orbitals ε. The partition function for a single orbital is given by Eq. (21.91) and the average number of particles occupying that orbital is given by Eq. (21.92). The average number of particles in the system is given by Eq. (21.93) but ordinarily this number is specified and Eq. (21.93) is used to find the absolute activity λ or, equivalently, the chemical potential μ. If we take the lowest energy state to be $\varepsilon = 0$, we see for systems having a finite number of bosons that $\lambda < 1$ (μ must be negative) to prevent $f_{\rm BE}(\varepsilon)$ from becoming infinite.

In Chapter 23, we gave a unified treatment of ideal Fermi, Bose, and classical gases. This treatment is applicable to bosons, for which $a < 1$, provided that the temperature is above the so-called condensation temperature $T_{\rm c}$, a critical temperature to be defined in the next section. For $T < T_{\rm c}$, λ becomes very nearly equal to one and many of the results in Chapter 23 for bosons require modification. In some cases, it will no longer be possible to convert sums over ε entirely to integrals. Instead, the ground state $\varepsilon = 0$ will have to be treated by means of a separate term and the integrals in Chapter 23 will only be applicable to the excited states.

To simplify the notation we will use

$$f_{\rm BE}(\varepsilon) := f(\varepsilon, -1) = \frac{1}{\lambda^{-1}\exp(\beta\varepsilon) - 1} = \frac{1}{\exp[\beta(\varepsilon - \mu)] - 1}; \tag{24.1}$$

$$g_\nu(\varepsilon) := h_\nu(\varepsilon, -1) = \frac{1}{\Gamma(\nu)} \int_0^\infty \frac{u^{\nu-1}\,{\rm d}u}{\lambda^{-1}\,{\rm e}^u - 1}. \tag{24.2}$$

Hereafter in this chapter, we will also write $\mathcal{N}$ instead of $\langle \mathcal{N} \rangle$ with the understanding that the average number of particles will be specified and λ (or equivalently μ) will be determined consistently as a function of particle density and temperature.

24.1 Bosons at Low Temperatures

To focus attention on the problem that occurs at low temperatures, we recall Eq. (23.13) which we now write in the form

$$\mathcal{N} = V g_0 n_{\rm Q}(T) g_{3/2}(\lambda), \tag{24.3}$$

where we have written $n_{\rm Q}(T)$ for the quantum concentration to emphasize its dependence on temperature. Here, $g_0 = 2s + 1$ accounts for degeneracy due to spin s that must be an integer for bosons. The problem with this equation becomes evident when we examine the function $g_{3/2}(\lambda)$ which is plotted in Figure 24–1. We see that $g_{3/2}(\lambda)$ is a monotonically

Thermal Physics. http://dx.doi.org/10.1016/B978-0-12-803304-3.00024-7

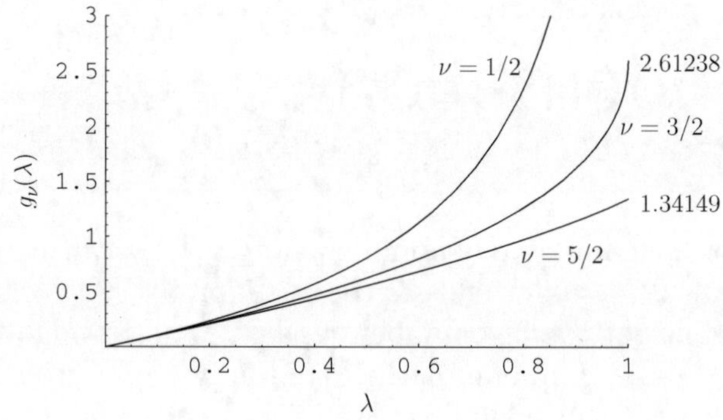

FIGURE 24–1 Plots of the functions $g_\nu(\lambda)$ given by Eq. (24.2). Recall from Chapter 23 that $\lambda g'_\nu(\lambda) = g_{\nu-1}(\lambda)$ and $g_\nu(1) = \zeta(\nu)$, the Riemann zeta function. $g_{1/2}(1) = \infty$ so $g_{3/2}(\lambda)$ has an infinite slope at $\lambda = 1$.

increasing function of λ which has its maximum value at $\lambda = 1$, namely $g_{3/2}(1) = 2.61238$. Inserting this value into Eq. (24.3) gives

$$\mathcal{N} = V g_0 n_Q(T) g_{3/2}(1). \tag{24.4}$$

Since $n_Q(T) \propto T^{3/2}$, the right-hand side of Eq. (24.4) gets smaller as T decreases. Therefore, for a given value of $n = \mathcal{N}/V$, there exists a critical temperature T_c below which all particles cannot be accommodated. This temperature satisfies

$$n = g_0 \left(\frac{m k_B T_c}{2\pi\hbar^2} \right)^{3/2} g_{3/2}(1). \tag{24.5}$$

This presents a problem because we must accommodate all particles, even at low temperatures! For $T < T_c$, Eq. (24.3) cannot be correct and must be modified.

The source of this problem is related to an approximation made in the conversion of a sum to an integral. We therefore return to Eq. (23.5) for bosons which we write in the form

$$\mathcal{N} = \sum_{\epsilon} \frac{1}{\lambda^{-1} \exp(\beta\varepsilon) - 1}. \tag{24.6}$$

Examination of this sum shows that the term arising from $\varepsilon = 0$ contributes a number of particles

$$\mathcal{N}_0 = g_0 \frac{1}{\lambda^{-1} - 1} = g_0 \frac{\lambda}{1 - \lambda}. \tag{24.7}$$

As $\lambda \to 1$ this term becomes infinite, so for λ very slightly less than 1 it cannot be ignored. If the sum in Eq. (24.7) were converted to an integral, as was done to obtain Eq. (24.4), that integral would be unaffected by leaving out a single point, so it would not properly include the nearly singular term given by Eq. (24.7). Thus, the right-hand side of Eq. (24.4) accounts only for the number $\mathcal{N}_e$ of particles in the excited states. This is no problem for $T > T_c$ but for $T < T_c$ we must account explicitly for particles that have "condensed" into the ground state.

Accordingly, we replace Eq. (24.4) by

$$\mathcal{N} = \mathcal{N}_0 + \mathcal{N}_e = g_0 \frac{\lambda}{1-\lambda} + Vg_0 n_Q(T) g_{3/2}(1); \quad T < T_c. \tag{24.8}$$

The concentration of particles in *excited* states is therefore

$$n_e := \frac{\mathcal{N}_e}{V} = g_0 n_Q(T) g_{3/2}(1) = g_0 \left(\frac{mk_B T}{2\pi\hbar^2}\right)^{3/2} g_{3/2}(1); \quad T < T_c. \tag{24.9}$$

Solving Eq. (24.7) for λ gives

$$\lambda = \left[1 + \frac{g_0}{\mathcal{N}_0}\right]^{-1}. \tag{24.10}$$

There appears at first to be an inconsistency between Eqs. (24.5) and (24.7) for the following reason: λ has been set equal to 1 in the argument of $g_{3/2}$ in both Eqs. (24.5) and (24.8). But Eq. (24.5) is based on $\mathcal{N}_0 = 0$ and this would require $\lambda = 0$ according to Eq. (24.7)! This turns out to be a false argument because $\mathcal{N}_0$ is never zero, but can still be negligible with respect to $\mathcal{N}$. All we need for Eq. (24.5) to hold for determination of T_c is $\mathcal{N}_0 \ll \mathcal{N}$. For example, suppose that $\mathcal{N}_0 = 10^{-6}\mathcal{N}$. Then Eq. (24.10) becomes

$$\lambda = \left[1 + \frac{g_0}{10^{-6}\mathcal{N}}\right]^{-1} \approx 1 - \frac{g_0}{10^{-6}\mathcal{N}} \approx 1 \tag{24.11}$$

for any reasonably large value of $\mathcal{N}$. One might expect $\mathcal{N} \sim 10^{23}$, but even for a sample so small that $\mathcal{N} \sim 10^{12}$, λ is less than 1 by a few parts in a million. Thus we can safely set $\lambda = 1$ in $g_{3/2}(\lambda)$ while letting it be a variable extremely close to 1 in Eqs. (24.7) and (24.8). In terms of the chemical potential, Eq. (24.11) would become

$$\mu \approx -k_B T \frac{g_0}{10^{-6}\mathcal{N}}, \tag{24.12}$$

which means that the chemical potential is negative and very slightly less than zero.[1] If $\mathcal{N}_0$ accounted for all of $\mathcal{N}$, equations of the form of Eqs. (24.11) and (24.12) would hold but without the factor of 10^{-6}. Then λ would be even closer to 1 and μ would be even closer to 0.

Dividing Eq. (24.9) by Eq. (24.5) and denoting the concentration of particles in the ground state by $n_0 := \mathcal{N}_0/V$, we obtain

$$n_e = n\left(\frac{T}{T_c}\right)^{3/2} \tag{24.13}$$

and

$$n_0 = n\left[1 - \left(\frac{T}{T_c}\right)^{3/2}\right]. \tag{24.14}$$

[1]The ground state energy has been set equal to 0 for convenience. Otherwise, μ would be slightly less than the ground state energy ε_0.

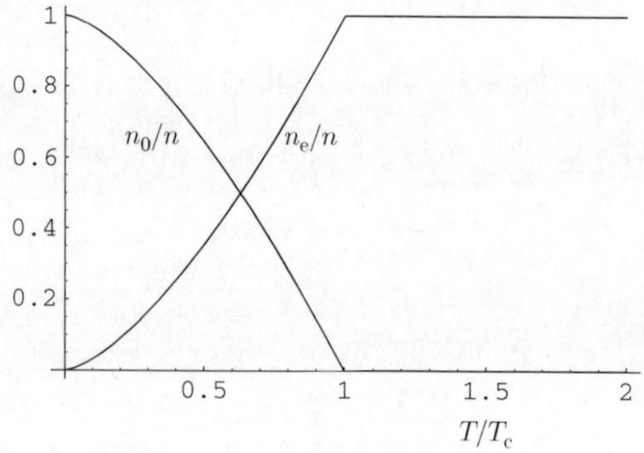

FIGURE 24–2 Plots of the fraction of condensate, n_0/n associated with the ground state, and of the normal fluid, n_e/n associated with the excited states, as a function of T/T_c according to Eqs. (24.13) and (24.14).

The concentrations (number densities) n_e and n_0 are represented graphically in Figure 24–2. For an exact treatment that agrees with the present results in the thermodynamic limit, see Pathria and Beale [9, Appendix F].

24.2 Thermodynamic Functions

To determine the thermodynamic functions for $T \le T_c$, we return to the sum in Eq. (23.3) for $a = -1$ and account explicitly for the $\varepsilon = 0$ term in $\ln \mathcal{Z}$, namely,

$$\ln \mathcal{Z}_0 = -g_0 \ln(1 - \lambda). \tag{24.15}$$

This term contributes an amount

$$K_0 = g_0 k_B T \ln(1 - \lambda) \tag{24.16}$$

to the Kramers potential K which must be added to the part of K that comes from the excited states, calculated by converting the sum to an integral. Explicitly,

$$K = K_0 + k_B T V g_0 n_Q(T) \frac{1}{\Gamma(3/2)} \int_0^\infty \ln(1 - \lambda\, e^{-u}) u^{1/2}\, du. \tag{24.17}$$

After integration by parts, as in Eq. (23.17), this becomes

$$K = K_0 - k_B T V g_0 n_Q(T) g_{5/2}(\lambda). \tag{24.18}$$

Fortunately, K_0 is negligible except for computation of $\mathcal{N}$, which we have already considered.[2] For example, since $K = -pV$, the term K_0 would appear to produce an excess pressure

$$p_0 = -\frac{k_B T}{V} g_0 \ln(1 - \lambda) \tag{24.19}$$

[2]Note that $-(\partial K_0/\partial \mu)_{T,V} = \lambda (\partial \ln \mathcal{Z}_0/\partial \lambda)_{\beta,V} = g_0\lambda/(1-\lambda) = \mathcal{N}_0$ in agreement with Eq. (24.7).

over and above the pressure given by Eq. (23.18) for bosons, namely

$$p = k_B T g_0 n_Q(T) g_{5/2}(\lambda). \tag{24.20}$$

But from Eq. (24.10) this excess pressure is

$$p_0 = \frac{k_B T}{V} g_0 \ln\left[(\mathcal{N}_0/g_0) + 1\right]. \tag{24.21}$$

Since $V \propto \mathcal{N}$, this term is of the order of $\mathcal{N}^{-1} \ln \mathcal{N}$ and should be neglected in the thermodynamic limit of large $\mathcal{N}$. Thus we have

$$p = g_0 n_Q(T) k_B T g_{5/2}(\lambda); \quad T > T_c \tag{24.22}$$

and

$$p = g_0 n_Q(T) k_B T g_{5/2}(1); \quad T \le T_c. \tag{24.23}$$

Note especially that Eq. (24.23) shows that p depends only on T, independent of n.

Care must therefore be taken in expressing the pressure in terms of n because Eq. (24.3) holds for $T > T_c$ but Eq. (24.5) holds for $T \le T_c$. Thus,

$$p = n k_B T \frac{g_{5/2}(\lambda)}{g_{3/2}(\lambda)}; \quad T > T_c. \tag{24.24}$$

but

$$p = \left(\frac{T}{T_c}\right)^{3/2} n k_B T \frac{g_{5/2}(1)}{g_{3/2}(1)} = n_e k_B T \frac{g_{5/2}(1)}{g_{3/2}(1)}; \quad T \le T_c. \tag{24.25}$$

Equation (24.25) shows that *the condensate makes no contribution to the pressure.* Those bosons in the ground state, the so-called condensate, exert no pressure. As written, Eq. (24.25) appears to depend on n, thus contradicting Eq. (24.23), but it must be recalled that $T_c^{3/2}$ is proportional to n, so n actually cancels in Eq. (24.25) (see Eq. (24.9)).

Similar considerations pertain to the internal energy and the entropy, although the calculations are more complicated because one must first take derivatives. For the internal energy, the situation is quite simple because the ground state has zero energy so the condensate does not contribute (see Eq. (23.6)). This can be seen more formally by writing $q = \ln \mathcal{Z} = -\beta K$ which leads to

$$q = q_0 + V g_0 n_Q(T) g_{5/2}(\lambda), \tag{24.26}$$

where $q_0 = -g_0 \ln(1-\lambda)$. Then from $U = -\left(\partial q/\partial \beta\right)_\lambda$, we see that q_0 makes no contribution and the second term gives

$$U = \frac{3}{2} k_B T V g_0 n_Q(T) g_{5/2}(\lambda) \tag{24.27}$$

in agreement with Eq. (23.14). Thus the energy density,

$$u_V = \frac{3}{2} n k_B T \frac{g_{5/2}(\lambda)}{g_{3/2}(\lambda)}; \quad T > T_c \tag{24.28}$$

but

$$u_V = \left(\frac{T}{T_c}\right)^{3/2} \frac{3}{2} n k_B T \frac{g_{5/2}(1)}{g_{3/2}(1)} = \frac{3}{2} n_e k_B T \frac{g_{5/2}(1)}{g_{3/2}(1)}; \quad T \le T_c. \tag{24.29}$$

We see that the *condensate makes no contribution to the internal energy* for $T < T_c$. It therefore follows that $p = (2/3)u_V$ holds at all temperatures.

Finally, the contribution of K_0 to the entropy is

$$S_0 = -\left(\frac{\partial K_0}{\partial T}\right)_{\mu,V} = -g_0 k_B \left[\ln(1-\lambda) + \frac{\lambda \ln \lambda}{1-\lambda}\right]. \tag{24.30}$$

In view of Eq. (24.10), this becomes

$$S_0 = k_B g_0 \left\{\ln\left[(\mathcal{N}_0/g_0) + 1\right] + (\mathcal{N}_0/g_0)\ln[1 + (g_0/\mathcal{N}_0)]\right\}. \tag{24.31}$$

Provided that $\mathcal{N}_0$ is any reasonable fraction of $\mathcal{N}$, we have $g_0/\mathcal{N}_0 \ll 1$ in which case $(\mathcal{N}_0/g_0)\ln[1 + (g_0/\mathcal{N}_0)] \approx 1$; the remaining term in Eq. (24.31) is of order $\ln \mathcal{N}$ and is also negligible in the thermodynamic limit. Thus,

$$S = k_B V g_0 n_Q(T)\left[(5/2)g_{5/2}(\lambda) - \ln\lambda\, g_{3/2}(\lambda)\right], \tag{24.32}$$

in agreement with Eq. (23.24). This result can also be expressed in terms of the number density, resulting in an entropy density

$$s_V = k_B n \left[\frac{5}{2}\frac{g_{5/2}(\lambda)}{g_{3/2}(\lambda)} - \ln\lambda\right]; \quad T > T_c, \tag{24.33}$$

but[3]

$$s_V = \left(\frac{T}{T_c}\right)^{3/2} n k_B \frac{5}{2}\frac{g_{5/2}(1)}{g_{3/2}(1)} = n_e k_B \frac{5}{2}\frac{g_{5/2}(1)}{g_{3/2}(1)}; \quad T \le T_c. \tag{24.34}$$

We observe that the *bosons in the condensate make no contribution to the entropy.*

For $T > T_c$ the heat capacity is still given by either Eq. (23.47) or (23.48) for $a = -1$, in which case the latter becomes

$$C_V = \frac{3}{2}\mathcal{N} k_B \left\{\frac{5}{2}\frac{g_{5/2}(\lambda)}{g_{3/2}(\lambda)} - \frac{3}{2}\frac{g_{3/2}(\lambda)}{g_{1/2}(\lambda)}\right\}; \quad T > T_c. \tag{24.35}$$

It would be wrong, however, to use Eq. (23.47) (or Eq. (23.48) which is derived from it) for $T \le T_c$ because Eq. (23.47) is based on a temperature derivative of Eq. (23.13) which is no longer valid. The correct result can be obtained by differentiating either Eq. (24.29) or (24.34) with respect to T at constant $\mathcal{N}$ to obtain

$$C_V = \left(\frac{T}{T_c}\right)^{3/2} \mathcal{N} k_B \frac{15}{4}\frac{g_{5/2}(1)}{g_{3/2}(1)} = \mathcal{N}_e k_B \frac{15}{4}\frac{g_{5/2}(1)}{g_{3/2}(1)}; \quad T \le T_c. \tag{24.36}$$

[3]Note that the term in $\ln\lambda \to 0$ for $\lambda \to 1$.

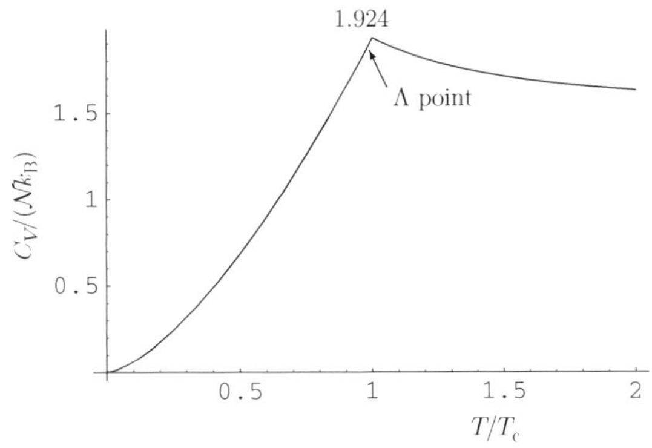

FIGURE 24–3 Heat capacity $C_V/(\mathcal{N}k_B)$ of an ideal Bose fluid as a function of T/T_c. For $T \to \infty$, $C_V/(\mathcal{N}k_B) = 3/2$, the value for a classical ideal gas. The curve resembles the letter Λ and the peak of the curve, which is about 28% higher than the classical value, occurs at the lambda point where $T = T_c$. The heat capacity of He^4 displays a similar behavior, although it is not an ideal Bose fluid.

Since $g_{1/2}(1) = \infty$, Eq. (24.35) for $T = T_c$ yields the same result as Eq. (24.36), so C_V is continuous at T_c. On the other hand, its slope is discontinuous at T_c, as illustrated in Figure 24–3. The C_V versus T curve resembles the letter Λ. Since the peak of the curve corresponds to the condensation temperature, the corresponding transition in liquid He^4 is said to occur at the "**lambda point**," even though He^4 atoms have attractive forces and are only crudely approximated by an ideal Bose gas. Evaluated at the number density of liquid He^4, $T_c \approx 3$ K; however, the **lambda transition** in liquid He^4 takes place at about 2.17 K.

Equations (24.25), (24.29), and (24.34) show explicitly that the bosons that are "condensed" in the ground state do not contribute to the pressure, the internal energy, or the entropy. This suggests that below T_c the ideal Bose fluid behaves like a mixture of two "phases," the **inactive condensate** associated with the ground state and a **normal fluid** associated with the excited states. As the temperature is lowered from T_c to $T = 0$, it is as if there is a "phase transition" from the normal fluid to the condensate. For a brief discussion of liquid helium as well as superfluidity, see Kittel and Kroemer [6, p. 20] and Pathria and Beale [9, p. 108,215].

From the Euler equation, one has

$$\mathcal{N}\mu = U - TS + pV. \tag{24.37}$$

Inserting Eqs. (24.25), (24.29), and (24.34) into Eq. (24.37) leads to $\mu = 0$ for $T \le T_c$, which is only approximately true. In fact, μ can be found from Eq. (24.10) to be

$$\mu = -k_B T \ln\left(1 + \frac{g_0}{\mathcal{N}_0}\right), \tag{24.38}$$

which is negative and very close to zero because g_0 is of order 1 and $\mathcal{N}_0$ is of order $\mathcal{N}$, even if it is negligible in comparison to $\mathcal{N}$, say $10^{-6}\mathcal{N}$. See the argument in connection with Eq. (24.12) for further detail.

24.2.1 Heat Capacity at Constant Pressure

Above T_c, the heat capacity at constant pressure can be calculated by differentiation of the enthalpy H at constant $\mathcal{N}$ and p. First of all, we have

$$H = U + pV = U + (2/3)U = (5/3)U \tag{24.39}$$

and we can use Eq. (24.28) for U to obtain

$$C_p = \frac{5}{2}\mathcal{N}k_B\left\{\frac{g_{5/2}(\lambda)}{g_{3/2}(\lambda)} + \frac{g_{3/2}(\lambda)g'_{5/2}(\lambda) - g_{5/2}(\lambda)g'_{3/2}(\lambda)}{[g_{3/2}(\lambda)]^2}T\left(\frac{\partial\lambda}{\partial T}\right)_{\mathcal{N},p}\right\}, \tag{24.40}$$

where the primes denote derivatives. Differentiation of Eq. (24.22) holding p constant then leads to

$$\left(\frac{\partial\lambda}{\partial T}\right)_{\mathcal{N},p} = -\frac{5}{2T}\frac{g_{5/2}(\lambda)}{g'_{5/2}(\lambda)}. \tag{24.41}$$

We recall that $g'_{5/2}(\lambda) = \lambda^{-1}g_{3/2}(\lambda)$ and $g'_{3/2}(\lambda) = \lambda^{-1}g_{1/2}(\lambda)$. Then substitution of Eq. (24.41) into Eq. (24.40) leads to

$$C_p = \frac{5}{2}\mathcal{N}k_B\left\{\frac{5}{2}\frac{[g_{5/2}(\lambda)]^2}{[g_{3/2}(\lambda)]^2}\frac{g_{1/2}(\lambda)}{g_{3/2}(\lambda)} - \frac{3}{2}\frac{g_{5/2}(\lambda)}{g_{3/2}(\lambda)}\right\}; \quad T > T_c. \tag{24.42}$$

Dividing by Eq. (24.35) then yields

$$\frac{C_p}{C_V} = \frac{5}{3}\frac{g_{5/2}(\lambda)g_{1/2}(\lambda)}{[g_{3/2}(\lambda)]^2}; \quad T > T_c. \tag{24.43}$$

For small λ, we recover the classical result $C_p/C_V = 5/3$ but this ratio increases with λ and for $\lambda = 1$ we obtain $C_p/C_V = \infty$. Of course we never quite reach $\lambda = 1$ as shown by Eq. (24.10), so the ratio remains finite but very large. With some algebra, Eq. (24.43) can be rewritten in the form

$$\frac{C_p}{C_V} = 1 + \frac{4}{9}\frac{C_V}{\mathcal{N}k_B}\frac{g_{1/2}(\lambda)}{g_{3/2}(\lambda)}, \tag{24.44}$$

which leads to

$$\frac{C_p - C_V}{\mathcal{N}k_B} = \frac{4}{9}\left(\frac{C_V}{\mathcal{N}k_B}\right)^2\frac{g_{1/2}(\lambda)}{g_{3/2}(\lambda)}. \tag{24.45}$$

Equation (24.45) shows that $C_p > C_V$ as expected.

For $T \le T_c$, we see from Eq. (24.25) that p depends only on T. So in the approximation $\lambda = 1$ inherent in this equation, constant p demands constant T. On the other hand, the energy U and the enthalpy $H = U + pV = (5/3)U = (5/2)pV$ depend on both T and V or, alternatively, on both p and V. Therefore, at constant p and T, H can change linearly with V. In other words, at constant p one can add or subtract heat from the system by changing V and the system remains at constant T. The system therefore behaves as if it has an infinite heat capacity C_p. The same conclusion would be reached if we relate the heat

$Q = T\Delta S$ to a change in the entropy S, since by Eq. (24.34) we see that S is also proportional to V at constant T. In any event, when heat is added to the system at constant p and T, the amount of condensate $\mathcal{N}_0$ changes. This becomes more evident if we use Eq. (24.5) to rewrite Eq. (24.14) in the explicit form

$$\mathcal{N}_0 = \mathcal{N} - Vg_0 \left(\frac{mk_B T}{2\pi\hbar^2}\right)^{3/2} g_{3/2}(1). \tag{24.46}$$

When heat is added to the system by increasing V at constant p and T, we see that $\mathcal{N}_0$ decreases linearly with V until $\mathcal{N}_0 = 0$, at which point the system will have a critical volume V_c

$$V_c = \frac{\mathcal{N}}{g_0 g_{3/2}(1)} \left(\frac{2\pi\hbar^2}{mk_B T}\right)^{3/2}. \tag{24.47}$$

For $V > V_c$ at the same T, the fluid will be entirely in the gaseous state in which virtually all of the bosons are accommodated in the excited states. See Section 24.3 for a related discussion.

24.3 Condensate Region

Except in the preceding section, we have regarded the volume V to be fixed and focused our discussion on temperature T relative to the critical temperature T_c. But T_c actually depends on V, so in this section we take a broader approach.

24.3.1 In the v, T Plane

We return to Eq. (24.4) which we now write in the form

$$\frac{1}{v} = g_0 n_Q(T) g_{3/2}(1), \tag{24.48}$$

where $v = V/\mathcal{N}$ is the volume per particle. Equation (24.48) can be rewritten in the form

$$\frac{1}{vT^{3/2}} = g_0 (mk_B/2\pi\hbar^2)^{3/2} g_{3/2}(1) =: C^*, \tag{24.49}$$

where C^* is a constant. If the quantity $vT^{3/2}$ is too small, Eq. (24.49) cannot be satisfied, and this defines the condensate region

$$vT^{3/2} < 1/C^*, \quad \text{condensate region}, \tag{24.50}$$

depicted in Figure 24–4 where the population in the ground state is so large that it must be taken explicitly into account. This zone is bounded from above by the curve $vT^{3/2} = 1/C^*$ which can be solved to give either a critical temperature $T_c(v) = (1/C^*)^{2/3}(1/v)^{2/3}$ as a function of v or a critical volume per particle $v_c(T) = (1/C^*)(1/T)^{3/2}$ as a function of T.

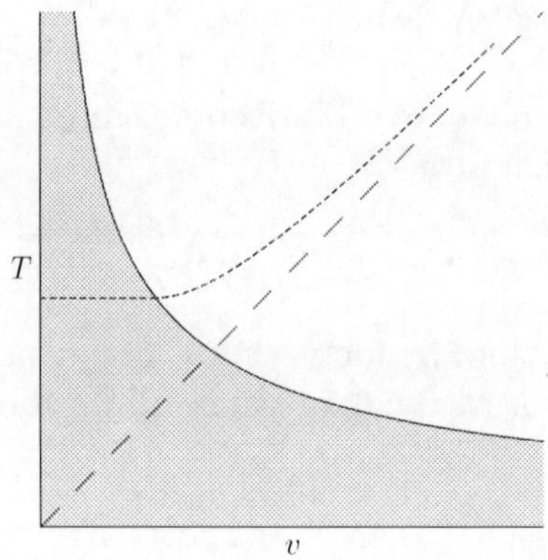

FIGURE 24–4 Condensate region (shaded) $vT^{3/2} < 1/C^*$ for a Bose fluid in arbitrary units. The dotted line is an isobar (p = constant) that depends only on T in the condensate zone and asymptotes the dashed line $T = pv/k$ for a classical ideal gas at high temperatures and large volumes.

According to Eq. (24.23), the pressure p depends only on temperature in the condensate region but in the normal region it follows a curved path that may be obtained by eliminating λ between Eqs. (24.22) and (24.48), which cannot be done analytically. This curved path may, however, be plotted parametrically by defining variables

$$\tilde{v} = g_0(m/2\pi\hbar^2)^{3/2}; \quad \tilde{t} = k_B T; \quad \tilde{p} = g_0^{-1}(m/2\pi\hbar^2)^{-3/2}, \tag{24.51}$$

which allows Eqs. (24.22) and (24.48) in the forms

$$\tilde{v}^{-1}\tilde{t}^{-3/2} = g_{3/2}(\lambda); \quad \tilde{p}\tilde{t}^{-5/2} = g_{5/2}(\lambda). \tag{24.52}$$

Then an isobar may be plotted from the parametric equations

$$\tilde{v} = \left[\frac{g_{5/2}(\lambda)}{\tilde{p}}\right]^{3/5}\left[\frac{1}{g_{3/2}(\lambda)}\right]; \quad \tilde{t} = \left[\frac{\tilde{p}}{g_{5/2}(\lambda)}\right]^{2/5} \tag{24.53}$$

by choosing some constant value of $\tilde{p}$ and letting λ range from very small values to 1. For $\lambda = 1$, such an isobar will intersect the boundary of the condensate region given by $\tilde{v}^{-1}\tilde{t}^{-3/2} = g_{3/2}(1)$. We also observe that

$$\frac{\tilde{p}\tilde{v}}{\tilde{t}} = \frac{g_{5/2}(\lambda)}{g_{3/2}(\lambda)} \to 1 \text{ as } \lambda \to 0, \tag{24.54}$$

which is the classical ideal gas law that is approached asymptotically far from the condensate region.

24.3.2 In the v, p Plane

Since Eq. (24.23) for the pressure may be rewritten in the form

$$p = k_B C^* g_{5/2}(1) T^{5/2}/g_{3/2}(1), \quad \text{condensate region,} \tag{24.55}$$

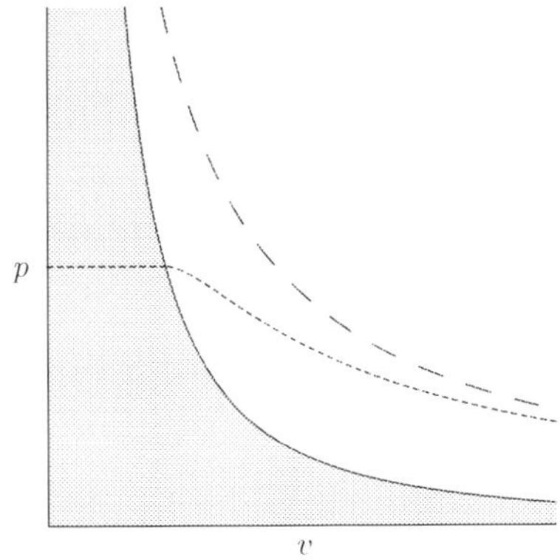

FIGURE 24–5 Condensate region (shaded) $pv^{5/3}$ < constant given by Eq. (24.56) for an ideal Bose fluid in arbitrary units. The dotted curve is an isotherm (T = constant) that depends only on p in the condensate zone and asymptotes the dashed hyperbola $p = k_B T/v$ for a classical ideal gas at low pressures and large volumes.

we can combine it with Eq. (24.50) to rewrite the condensate region in the form $pv^{5/3} < k_B(C^*)^{-2/3} g_{5/2}(1)/g_{3/2}(1)$ or explicitly

$$pv^{5/3} < \frac{2\pi\hbar^2}{m} \frac{g_{5/2}(1)}{[g_{3/2}(1)]^{5/3}} \frac{1}{g_0^{2/3}}, \quad \text{condensate region.} \tag{24.56}$$

The condensate region given by Eq. (24.56) is depicted in Figure 24–5.

Inside the condensate region, an isotherm is independent of v and is therefore a horizontal line at some value of p. Outside the condensate region, we can make a parametric plot of an isotherm by using Eq. (24.52) and solving for $\tilde{v}$ and $\tilde{p}$, resulting in

$$\tilde{v} = \tilde{t}^{-3/2} \left[g_{3/2}(\lambda)\right]^{-1}; \quad \tilde{p} = \tilde{t}^{5/2} g_{5/2}(\lambda). \tag{24.57}$$

Then for some fixed value of $\tilde{t}$, we let λ range from small values to 1. In these variables, the condensate region is bounded by $\tilde{p}\tilde{v}^{5/2} = g_{5/2}(1)\left[g_{3/2}(1)\right]^{-5/3}$. Far from the condensate region such an isotherm asymptotes the hyperbola $\tilde{p}\tilde{v} = \tilde{t}$ for a classical ideal gas.

24.3.3 Isentropic Transformation

In a reversible adiabatic transformation, the number of particles $\mathcal{N}$ and the entropy S must remain constant. For $T > T_c$, Eq. (24.33) applies, so constant $\mathcal{N}$ and S requires constant λ. Then Eq. (24.3) shows that

$$vT^{3/2} = \text{constant}, \tag{24.58}$$

where $v = V/\mathcal{N}$ is the volume per particle. Similarly, Eq. (24.22) shows that

$$p/T^{5/2} = \text{constant}. \tag{24.59}$$

By eliminating T from Eqs. (24.58) and (24.59), we obtain

$$pv^{5/3} = \text{constant}. \tag{24.60}$$

Furthermore,

$$pv/T = \text{constant}, \tag{24.61}$$

which can be obtained by multiplying Eq. (24.58) by Eq. (24.59). These equations resemble the equations for an isentropic transformation of a classical monatomic ideal gas for which the exponent $5/3 = C_p/C_V$, but Eq. (24.43) for the ideal Bose gas shows that this ratio is only equal to 5/3 in the classical limit.

For an isentropic transformation for $T \le T_c$, Eq. (24.32) for $\lambda = 1$ yields Eq. (24.58) whereas Eq. (24.23) for $\lambda = 1$ yields Eq. (24.59), so Eqs. (24.60) and (24.61) are still valid. Comparison of Eq. (24.58) with Eq. (24.49) shows that the boundary of the condensate region is an isentrope.

25

Degenerate Fermi Gas

In this chapter, we examine in more detail the behavior of an ideal Fermi gas. Even for temperatures near absolute zero, the Pauli exclusion principle forces fermions into high energy states, and the gas is said to be degenerate. Consequently, raising the temperature causes only a small change in occupation of even higher energy states. This gives rise to a heat capacity that is much smaller than for a classical gas. This and other phenomena are illustrated for a simple model of a metal in which the valence electrons are treated as an ideal Fermi gas. In the presence of a magnetic field, the two spin states of each electron have different energies which gives rise to weak magnetic behavior known as Pauli paramagnetism. The magnetic field also affects the nonspin states, which gives rise to weak Landau diamagnetism. If sufficiently heated, some electrons can overcome an energy barrier and leave the metal, a phenomenon known as thermionic emission. If an external electric field is applied, this energy barrier can be reduced and thermionic emission can be enhanced. Electron emission can also be enhanced by radiation, the photoelectric effect. Finally, we examine semiconductors that have densities of single electron quantum states separated by a forbidden region of energy known as a band gap. Such states pertain to an electron in an effective periodic potential that accounts approximately for interactions with the lattice. With increase of temperature, some electrons can be excited to states above that band gap, resulting in an overall increase in electron mobility and enhanced electrical conductivity. Adding small amounts of impurities to such a metal, a process known as doping, can cause major changes in the way electrons are thermally excited in semiconductors.

25.1 Ideal Fermi Gas at Low Temperatures

For an ideal Fermi gas the average occupancy $f_{\rm FD}(\varepsilon)$ (see Eq. (21.88)) of an orbital ε is given by the bounded quantity

$$0 \leq \frac{1}{\lambda^{-1}\,{\rm e}^{\beta\varepsilon}+1} = \frac{1}{\exp[\beta(\varepsilon-\mu)]+1} \leq 1. \tag{25.1}$$

For an ideal Bose gas, the corresponding average occupancy becomes infinite for $\varepsilon = 0$ as $\lambda \to 1$. However, for fermions, $\lambda = {\rm e}^{\beta\mu}$ can be any positive number, so $0 \leq \lambda \leq \infty$. In particular, one does not have to take the ground state into account explicitly, so conversion from a sum to an integral presents no problem. Therefore, for fermions, there is no critical temperature, such as the condensation temperature $T_{\rm c}$ for bosons.

Thermal Physics. http://dx.doi.org/10.1016/B978-0-12-803304-3.00025-9

At all temperatures (see Section 23.1 and Eq. (23.13) with $f_\nu(\lambda) = h_\nu(\lambda, 1)$), the particle density $n = \mathcal{N}/V$ can be written in the form

$$n = g_0 n_Q(T) f_{3/2}(\lambda), \tag{25.2}$$

which can be regarded as an implicit equation for $\mu(n, T)$, with n specified. In particular, the function $f_{3/2}(\lambda)$ is not bounded as $\lambda \to \infty$. As we shall see later, $f_{3/2}(\lambda) \to (\ln \lambda)^{3/2}/\Gamma(5/2) = (\beta\mu)^{3/2}/\Gamma(5/2)$ as $\lambda \to \infty$, so the product $n_Q(T) f_{3/2}(\lambda)$ becomes independent of T and proportional to $\mu^{3/2}$ as $T \to 0$. This leads to an equation for $\mu(n, 0)$, the chemical potential at zero temperature, which is known as the **Fermi energy**, $\varepsilon_F \equiv \mu(n, 0)$.

This same $T = 0$ limit may be explored in an elementary way by returning to the sum (see Eq. (21.89)) that led to Eq. (25.2), namely

$$\mathcal{N} = g_0 \sum_{\varepsilon}{}' \frac{1}{\exp[\beta(\varepsilon - \mu)] + 1}, \tag{25.3}$$

where $g_0 = 2s + 1$ is the degeneracy due to spin s that is half integral for fermions. Thus, the prime on the sum means that one should exclude the degeneracy g_0 due to spin. As $T \to 0$, $\beta \to \infty$, so $\mu \to \varepsilon_F$ that depends only on n. Thus, $f_{FD}(\varepsilon)$ becomes a step function of the form

$$\lim_{\beta \to \infty} \frac{1}{\exp[\beta(\varepsilon - \mu)] + 1} = \begin{cases} 1 & \text{if } \varepsilon < \varepsilon_F \\ 0 & \text{if } \varepsilon > \varepsilon_F. \end{cases} \tag{25.4}$$

In other words, all of the states for $\varepsilon < \varepsilon_F$ are full and all of the states for $\varepsilon > \varepsilon_F$ are empty. So for $T = 0$, Eq. (25.3) takes the simple form

$$\mathcal{N} = g_0 \sum_{\varepsilon < \varepsilon_F}{}' 1. \tag{25.5}$$

For the free particle and periodic boundary conditions, we know that $\varepsilon = \hbar^2 k^2/2m$ and that the quantum states are distributed uniformly in **k** space with a density $V/(2\pi)^3$. Therefore, we only need to compute the volume in **k** space for which $\varepsilon < \varepsilon_F$, known as the volume of the **Fermi sphere**. Specifically,

$$\mathcal{N} = g_0 \frac{V}{(2\pi)^3} \int_0^{k_F} 4\pi k^2 \, dk = g_0 \frac{V}{(2\pi)^3} \frac{4}{3} \pi k_F^3, \tag{25.6}$$

where the **Fermi wavenumber** k_F satisfies $\varepsilon_F = \hbar^2 k_F^2/2m$. We therefore obtain

$$k_F = (6\pi^2 n/g_0)^{1/3} \tag{25.7}$$

and

$$\varepsilon_F = \frac{\hbar^2}{2m} (6\pi^2 n/g_0)^{2/3}. \tag{25.8}$$

At $T = 0$ the energy is also easy to calculate because one can include a factor of ε in the sum in Eq. (25.5) to obtain

$$U_0 = g_0 \sum_{\varepsilon<\varepsilon_{\rm F}}' \varepsilon = g_0 \frac{\hbar^2}{2m}\frac{V}{(2\pi)^3}\int_0^{k_{\rm F}} 4\pi k^4 \, {\rm d}k = \frac{3}{5}\mathcal{N}\varepsilon_{\rm F}. \tag{25.9}$$

According to Eq. (23.18), the pressure is two-thirds of the energy density at all temperatures, so the pressure at $T = 0$ is given by

$$p_0 = \frac{2}{5} n\varepsilon_{\rm F}. \tag{25.10}$$

In summary, at $T = 0$, the fermions are forced by the Pauli exclusion principle to fill the states of lowest energy that can accommodate all of them. Thus, all states up to the Fermi energy $\varepsilon_{\rm F}$ are occupied and all states above that energy are unoccupied. This forced occupation of high energy states results in a cumulative energy given by Eq. (25.9) and a corresponding pressure given by Eq. (25.10).

One can also define a **Fermi temperature**

$$T_{\rm F} := \frac{\varepsilon_{\rm F}}{k_{\rm B}} = \frac{\hbar^2}{2mk_{\rm B}}(6\pi^2 n/g_0)^{2/3}. \tag{25.11}$$

This may be rewritten in the form

$$n = g_0 \left(\frac{mk_{\rm B}T_{\rm F}}{2\pi\hbar^2}\right)^{3/2} \frac{4}{3\pi^{1/2}}, \tag{25.12}$$

which greatly resembles Eq. (24.5) for the critical temperature $T_{\rm c}$ of an ideal Bose gas. We emphasize, however, that $T_{\rm F}$ is *not* a critical temperature but rather a temperature that characterizes the degree to which fermions at $T = 0$ are forced into excited states by the Pauli exclusion principle.

A word about the relative magnitudes of $T_{\rm F}$ and $T_{\rm c}$ is relevant. If we consider fermions or bosons that have comparable number densities and masses, say the masses of $\rm He^3$ (a fermion with half integral spin) and $\rm He^4$ (a boson with integral spin), the magnitudes of $T_{\rm F}$ and $T_{\rm c}$ will be comparable. As we saw previously, $T_{\rm c}$ was typically a few K degrees at the density of $\rm He^4$ near the lambda transition. But electrons are fermions and the electron mass is about 1836 times smaller than the mass of a proton. Therefore, for free electron gases in metals at their usual densities, $T_{\rm F}$ is typically 50,000 K degrees. In such cases, one has $T \ll T_{\rm F}$ for any temperature of interest. We shall see that a Fermi gas at temperature $T > 0$ but $T \ll T_{\rm F}$ displays characteristics very similar to a Fermi gas at $T = 0$ except a small fraction $\sim T/T_{\rm F}$ of electrons is now in excited states. Consequently, a Fermi gas at $T \ll T_{\rm F}$ is usually referred to as a **degenerate Fermi gas**. Equivalent conditions for a degenerate Fermi gas are therefore $\beta\mu \gg 1$, $\lambda \gg 1$, or $n/n_{\rm Q}(T) \gg 1$.

Before leaving this section, it is worth pointing out that the integrals in Eqs. (25.6) and (25.9) could equally well have been written as integrals over ε by expressing $k = \sqrt{2m\varepsilon/\hbar^2}$ and then using ${\rm d}k = (\partial k/\partial\varepsilon)\,{\rm d}\varepsilon$. This leads to an intensive density of states of the form

$$g(\varepsilon) := \frac{G(\varepsilon)}{V} := \frac{g_0}{2}\frac{m}{\hbar^2\pi^2}\left(\frac{2m}{\hbar^2}\right)^{1/2}\varepsilon^{1/2} = \frac{3}{2}\frac{n}{\varepsilon_{\rm F}}\left(\frac{\varepsilon}{\varepsilon_{\rm F}}\right)^{1/2}. \tag{25.13}$$

Here, $G(\varepsilon)\,d\varepsilon$ is the number of states, including spin, with energy between ε and $\varepsilon + d\varepsilon$ and $g(\varepsilon)\,d\varepsilon$ is the number of states per unit volume in that same interval. Then

$$n = \int_0^{\varepsilon_F} g(\varepsilon)\,d\varepsilon \tag{25.14}$$

and the energy density

$$u_V(T=0) = \int_0^{\varepsilon_F} g(\varepsilon)\varepsilon\,d\varepsilon = \frac{3}{5}n\varepsilon_F. \tag{25.15}$$

25.2 Free Electron Model of a Metal

As an example of an ideal Fermi gas with spin $s = 1/2$, we treat the free electron model of a metal. According to this model, each atom contributes z_v valence electrons to a sea of electrons that are shared by the remaining ion cores. Interactions among the valence electrons as well as interactions with the ion cores are treated only on average. Specifically, one assumes that each valence electron experiences an *effective potential* that is constant (and set equal to zero for convenience) within the metal. The potential outside the metal is assumed to be sufficiently large that the electrons are confined to the volume V of the metal. Thus, each valence electron behaves as if it were free but confined to a box of volume V. We shall see that the valence electrons constitute a very dense gas, typically 1000-10,000 times more dense than a classical gas, so quantum effects are important. Even though the free electron model is quite naive, it works rather well for some elements, especially the alkali metals.

The quantum statistics of such an electron gas are governed by the Fermi-Dirac distribution function Eq. (25.1). Quantitative details for $k_BT \ll \varepsilon_F$ are handled by a series expansion in k_BT/μ due to Sommerfeld. We shall see that μ depends very weakly on T, so ultimately results for μ and u_V can be expressed as a series expansion in $k_BT/\varepsilon_F = T/T_F$. This free electron model of a metal was the first to explain why an electron gas in a metal contributes only a small fraction of the heat capacity that it would if it were a classical gas.

We estimate the number density of an electron gas in a metal. Consider a simple cubic lattice with lattice constant $a = 2.5$ Å and only one valence electron per unit cell. The number n of free electrons per unit volume is

$$n \sim \frac{1}{a^3} \sim \frac{1}{(2.5 \times 10^{-8}\ \text{cm})^3} \sim 6.4 \times 10^{22}\ \text{cm}^{-3}. \tag{25.16}$$

This should be compared to the number density n_c of a classical ideal gas at standard temperature and pressure, where one mole occupies 22.4 l. Thus

$$n_c \sim \frac{6.02 \times 10^{23}}{22.4 \times 10^3\ \text{cm}^3} \sim 2.7 \times 10^{19}\ \text{cm}^{-3}. \tag{25.17}$$

We see that the electron gas has a number density that is about 1000 times that of a classical gas. For $T = 273$ K, we find the quantum concentration $n_Q = (mk_BT/2\pi\hbar^2)^{3/2} \sim 1.1 \times 10^{19}$ cm^{-3} for electrons and 2.4×10^{24} cm^{-3} for hydrogen. Thus, $n \gg n_Q$ for electrons, which are expected to behave like a dense quantum gas; however, for hydrogen $n \ll n_Q$ so it behaves like a classical gas. For $n \sim 6.4 \times 10^{22}$ cm^{-3}, we have $k_F \sim 1.1 \times 10^8$ cm^{-1} which corresponds to a Fermi wavelength $\lambda_F = 2\pi/k_F \sim 5.7 \times 10^{-8}$ cm which is comparable to the lattice constant a. The Fermi energy $\varepsilon_F \sim 7.7 \times 10^{-12}$ erg ~ 4.4 eV, which corresponds to a Fermi temperature $T_F = \varepsilon_F/k_B \sim 51,000$ K. These numerical estimates are typical values; for actual values for given materials, see table 2.1 of Ashcroft and Mermin [58, p. 28]. In any case, it is important to recognize that for temperatures T of practical interest for metals, one has $T \ll T_F$, so only a small fraction $\sim k_BT/\varepsilon_F = T/T_F$ of the free electrons are thermally activated with respect to their energy levels for $T = 0$. This thermal activation is governed by the Fermi-Dirac distribution function, as discussed in the next section.

For now, we shall assume that no magnetic field is present, so each state corresponding to a given value of **k** is twofold degenerate because of spin. This degeneracy has already been incorporated in $g(\varepsilon)$ given by Eq. (25.13) with $g_0 = 2$.

25.3 Thermal Activation of Electrons

The population of electronic orbitals for $T > 0$ is governed by the Fermi-Dirac distribution function (see Eq. (21.88)),

$$f_{FD}(\varepsilon) = \frac{1}{\exp[(\varepsilon - \mu)/k_BT] + 1}, \tag{25.18}$$

where μ is the chemical potential. This distribution function $f_{FD}(\varepsilon)$ gives the average number of electrons in a single orbital having energy ε. The chemical potential at number density n and temperature T is to be calculated from

$$n = \int_0^\infty g(\varepsilon) f_{FD}(\varepsilon)\, d\varepsilon. \tag{25.19}$$

The internal energy density is given by

$$u_V = \int_0^\infty \varepsilon\, g(\varepsilon) f_{FD}(\varepsilon)\, d\varepsilon. \tag{25.20}$$

Equations of the forms of Eqs. (25.19) and (25.20) would hold even if $g(\varepsilon)$ were for a more general model in which the valence electrons were subject to an effective single-electron potential due to a crystal lattice.

As $T \to 0$, Eq. (25.4) shows that $f_{FD}(\varepsilon)$ is a step function as depicted in Figure 25–1. For $T > 0$ but still $T \ll T_F$, the corners of the step function become rounded as also shown in Figure 25–1. In three dimensions, the value of μ becomes slightly less than

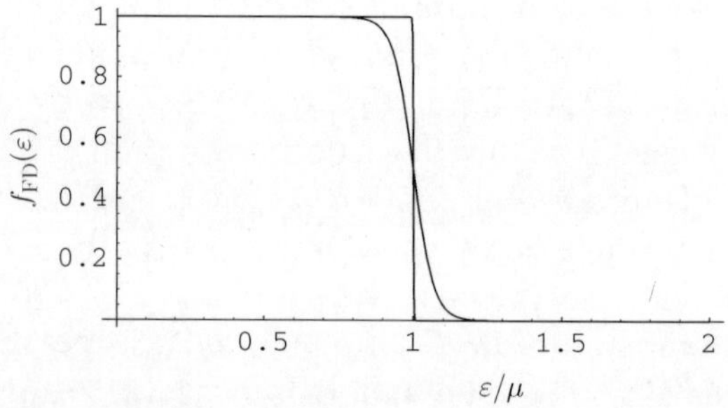

FIGURE 25–1 Plots of the Fermi-Dirac distribution function as a function of ε/μ for $T = 0$ (step function) and $T > 0$ but $T \ll T_F$ (curve). Note that μ also depends on T but is practically equal to ε_F (see Eq. (25.35)). Thus $\mu/k_B T \gg 1$, but $\mu/(k_B T) = 30$ was chosen for the sake of illustration.

ε_F in order to satisfy Eq. (25.19), for reasons to be discussed later. We note that $f_{FD}(\mu) = 1/2$ for any $T > 0$.

25.3.1 Sommerfeld Expansion

In order to treat Eqs. (25.19) and (25.20) in the general case of $T > 0$ but still $T \ll T_F$, we make use of an expansion due to Sommerfeld [65]. Either of these integrals is of the form

$$I := \int_0^\infty w(\varepsilon) f(\varepsilon)\, d\varepsilon, \tag{25.21}$$

where $w(\varepsilon)$ is either $g(\varepsilon)$ or $\varepsilon\, g(\varepsilon)$. We define an auxiliary function

$$H(\varepsilon) = \int_0^\varepsilon w(\eta)\, d\eta, \tag{25.22}$$

which has the properties $H(0) = 0$ and

$$\frac{dH(\varepsilon)}{d\varepsilon} = w(\varepsilon). \tag{25.23}$$

Substitution into Eq. (25.21) gives

$$I = \int_0^\infty \frac{dH(\varepsilon)}{d\varepsilon} f(\varepsilon)\, d\varepsilon = H(\varepsilon) f(\varepsilon)\Big|_0^\infty + \int_0^\infty H(\varepsilon)\left(-\frac{df(\varepsilon)}{d\varepsilon}\right) d\varepsilon. \tag{25.24}$$

The first term on the right-hand side of Eq. (25.24) vanishes because of the properties of $H(\varepsilon)$ at the lower limit and $f(\varepsilon)$ at the upper limit. The function $-df/d\varepsilon$ is highly peaked near $\varepsilon = \mu$ and nearly 0 elsewhere because of the shape of $f(\varepsilon)$. In fact, as $T \to 0$ it tends toward a Dirac delta function, $\delta(\varepsilon - \mu)$, which is the formal derivative of a unit step function. We therefore realize that $H(\varepsilon)$ is only important in the vicinity of $\varepsilon = \mu$, so we expand it in a power series near μ.

For convenience we make a change of variable to $x := (\varepsilon - \mu)/k_B T$ which gives

$$I = \int_{-\mu/k_B T}^\infty H(\mu + x k_B T)\left(-\frac{df}{dx}\right) dx, \tag{25.25}$$

where

$$-\frac{df}{dx} = -\frac{d}{dx}\frac{1}{e^x+1} = \frac{e^x}{(e^x+1)^2} = \frac{1}{4}\frac{1}{\cosh^2(x/2)}, \tag{25.26}$$

which is an even function of x. Then we expand H in a Taylor series

$$H(xk_BT+\mu) = H(\mu) + H'(\mu)xk_BT + \frac{1}{2!}H''(\mu)(xk_BT)^2 + \cdots, \tag{25.27}$$

where a prime denotes the derivative with respect to the argument of a function. We substitute Eq. (25.27) into Eq. (25.25) and perform the integrals over x. The lower limit in Eq. (25.25) is essentially $-\infty$, so integrals over odd powers of x are negligible to an excellent asymptotic approximation. The integrals over even powers can be done analytically,[1] resulting in

$$I = H(\mu) + \frac{\pi^2}{6}w'(\mu)(k_BT)^2 + \cdots, \tag{25.28}$$

where we have used $w'(\mu) = H''(\mu)$. Equations (25.19) and (25.20) therefore become

$$n = \int_0^\mu g(\eta)\,d\eta + \frac{\pi^2}{6}g'(\mu)(k_BT)^2 + \cdots; \tag{25.29}$$

$$u_V = \int_0^\mu \eta g(\eta)\,d\eta + \frac{\pi^2}{6}[\mu g(\mu)]'(k_BT)^2 + \cdots. \tag{25.30}$$

Unfortunately we are still not done because Eqs. (25.29) and (25.30) depend on μ which is still an unknown function of T and n. We therefore take advantage of the fact that $|\mu-\varepsilon_F|$ is small compared to ε_F and expand again to obtain[2]

$$n = \int_0^{\varepsilon_F} g(\eta)\,d\eta + g(\varepsilon_F)(\mu-\varepsilon_F) + \frac{\pi^2}{6}g'(\varepsilon_F)(k_BT)^2 + \cdots \tag{25.31}$$

and

$$u_V = \int_0^{\varepsilon_F} \eta g(\eta)\,d\eta + \varepsilon_F g(\varepsilon_F)(\mu-\varepsilon_F) + \frac{\pi^2}{6}[\varepsilon_F g(\varepsilon_F)]'(k_BT)^2 + \cdots. \tag{25.32}$$

By definition of the Fermi energy, the first integral in Eq. (25.31) is equal to n. Therefore, the remaining terms in Eq. (25.31) must vanish, resulting in

$$\mu-\varepsilon_F = -\frac{\pi^2}{6}\frac{g'(\varepsilon_F)}{g(\varepsilon_F)}(k_BT)^2 + \cdots. \tag{25.33}$$

[1]See Eq. (23.35) for additional terms. See Ashcroft and Mermin [58, appendix C] or Pathria [8, appendix E] for details and even higher order terms of the expansion.

[2]Consistent to second order in k_BT/ε_F, we do not need to expand the second-order term but simply evaluate it at $\mu=\varepsilon_F$.

Equation (25.33) shows that the chemical potential shifts from ε_F by a small amount in a direction of opposite sign to $g'(\varepsilon_F)$. Substitution of Eq. (25.33) into Eq. (25.32) gives

$$u_V = \int_0^{\varepsilon_F} \eta g(\eta)\, d\eta + \frac{\pi^2}{6} g(\varepsilon_F)(k_B T)^2 + \cdots, \tag{25.34}$$

which depends only on the value of g (not its derivative) at the Fermi energy. The first term in Eq. (25.34) is just the value of u_V at $T = 0$ given by Eq. (25.9).

For the free electron model, for which $g(\varepsilon)$ is given by Eq. (25.13), Eqs. (25.33) and (25.34) become

$$\mu = \varepsilon_F - \frac{\pi^2}{12}\varepsilon_F \left(\frac{k_B T}{\varepsilon_F}\right)^2 + \cdots; \tag{25.35}$$

$$u_V = \frac{3}{5}\varepsilon_F n + \frac{\pi^2}{4}\frac{(k_B T)^2}{\varepsilon_F} n + \cdots. \tag{25.36}$$

The chemical potential (sometimes called the Fermi *level*) is therefore different from the Fermi energy ε_F except at $T = 0$.

The shift in chemical potential relative to the Fermi energy can be understood by noting that the Fermi-Dirac function given by Eq. (25.18) can be written in the form.

$$f_{FD} = \frac{1}{2} - \frac{1}{2}\tanh[\beta(\varepsilon - \mu)/2]. \tag{25.37}$$

Thus for $T > 0$ but $T \ll T_F$, the increase in the probability of occupancy with $\varepsilon > \mu$ is exactly equal to the decrease in the probability of occupancy with $\varepsilon < \mu$. But because $g(\varepsilon) \propto \varepsilon^{1/2}$, this change in probabilities would result in a greater *number* of electrons having $\varepsilon > \mu$ with respect to the *number* lost from $\varepsilon < \mu$. Thus, μ must decrease slightly from ε_F in order to conserve the total number of electrons. The analytical result Eq. (25.33) shows that the shift from ε_F has the same sign as $g'(\varepsilon_F)$. In two dimensions, $g(\varepsilon_F)$ is a constant, so $\mu = \varepsilon_F + k_B T \ln[1 - \exp(-\varepsilon_F/k_B T)]$; thus there is no shift in chemical potential to exponential order for $\varepsilon_F/k_B T \gg 1$. In one dimension, $g(\varepsilon) \propto \varepsilon^{-1/2}$, so μ is slightly larger than ε_F.

25.3.2 Heat Capacity

We differentiate Eq. (25.36) with respect to T to get the heat capacity per unit volume

$$c_V = \frac{3}{2} n k_B \frac{\pi^2}{3}\frac{k_B T}{\varepsilon_F} + \cdots. \tag{25.38}$$

We observe that c_V depends linearly on T and is reduced from the heat capacity, $3nk_B/2$, of a classical ideal gas by the small factor $(\pi^2/3)(k_B T/\varepsilon_F)$. This factor arises because the Pauli exclusion principle forces the electrons to occupy energy levels up to ε_F at $T = 0$. Therefore, only a small fraction $\sim k_B T/\varepsilon_F$ of electrons are thermally activated for $T \neq 0$ and each of these will have energy $\sim (k_B T)$ above ε_F. They will therefore lead to a heat capacity $c_V \sim 2nk_B(k_B T/\varepsilon_F)$, in agreement with Eq. (25.38) except for a numerical factor.

At high T, the electronic heat capacity given by Eq. (25.38) is quite small compared with the heat capacity $\sim 3nk_B$ due to lattice vibrations,[3] but at sufficiently low T it dominates the heat capacity due to lattice vibrations, which is proportional to T^3. Thus at low T, we have a dependence of heat capacity on temperature of the form

$$c_V = \begin{cases} AT + BT^3, & \text{electronic conductor,} \\ BT^3, & \text{insulator,} \end{cases} \qquad \text{at low } T, \tag{25.39}$$

where A and B are constants.

25.4 Pauli Paramagnetism

In the presence of a magnetic field B, we no longer have spin degeneracy so there are two sets of states having energies:

$$\begin{aligned} &\frac{\hbar^2 k^2}{2m} - \mu^* B, \quad \text{spin up;} \\ &\frac{\hbar^2 k^2}{2m} + \mu^* B, \quad \text{spin down,} \end{aligned} \tag{25.40}$$

where μ^* is the magnetic moment, taken to be positive. For $\mathcal{N}$ electrons, we have

$$\mathcal{N} = \sum_{\mathbf{k}} \left[\frac{1}{\exp[\beta(\hbar^2 k^2/2m - \mu^* B - \mu)] + 1} + \frac{1}{\exp[\beta(\hbar^2 k^2/2m + \mu^* B - \mu)] + 1} \right], \tag{25.41}$$

where μ is the chemical potential in the presence of the magnetic field. For $T = 0$, both of the Fermi functions become step functions and μ becomes the Fermi energy ε_F in the presence of the magnetic field. The sums can then be converted to integrals and we obtain

$$\mathcal{N} = \frac{V}{(2\pi)^3} \left[\int_0^{[(\varepsilon_F + \mu^* B)2m/\hbar^2]^{1/2}} 4\pi k^2 \, dk + \int_0^{[(\varepsilon_F - \mu^* B)2m/\hbar^2]^{1/2}} 4\pi k^2 \, dk \right]. \tag{25.42}$$

These integrals are over spheres in $\mathbf{k}$ space having slightly different radii. We obtain

$$n = \frac{1}{6\pi^2} \left(\frac{2m}{\hbar^2} \right)^{3/2} \left[(\varepsilon_F + \mu^* B)^{3/2} + (\varepsilon_F - \mu^* B)^{3/2} \right]. \tag{25.43}$$

For $B = 0$, Eq. (25.43) yields

$$\varepsilon_{F0} \equiv \varepsilon_F(B = 0) = \frac{\hbar^2}{2m} (3\pi^2 n)^{2/3}, \tag{25.44}$$

which agrees with Eq. (25.8) for $g_0 = 2$. For $B \neq 0$ we can expand Eq. (25.43) in powers of B. The linear term in B cancels and we are left, to second order in B, with

$$n = \frac{1}{3\pi^2} \left(\frac{2m}{\hbar^2} \right)^{3/2} \varepsilon_F^{3/2} \left[1 + \frac{3}{8} \left(\frac{\mu^* B}{\varepsilon_F} \right)^2 + \cdots \right]. \tag{25.45}$$

[3] For lattice vibrations, n would be the number of lattice sites per unit volume, not the number of electrons per unit volume. For the monovalent alkali metals, these number densities would be the same.

Then by substitution of Eq. (25.44) to eliminate n and expansion in B we find

$$\varepsilon_{\rm F} = \varepsilon_{\rm F0}\left[1 - \frac{1}{4}\left(\frac{\mu^* B}{\varepsilon_{\rm F0}}\right)^2 + \cdots\right]. \tag{25.46}$$

Except for extremely large magnetic fields, $(\mu^* B/\varepsilon_{\rm F0})^2$ is negligible, so hereafter we will take $\varepsilon_{\rm F} = \varepsilon_{\rm F0}$ and drop the extra subscript 0.

The magnetization $m_{\rm V}$ (magnetic moment per unit volume) at $T = 0$ can now be calculated easily by recognizing that the two terms in Eq. (25.43) come from spin up and spin down electrons. We can therefore multiply the first of them by μ^* and the second by $-\mu^*$ to obtain

$$m_{\rm V} = \frac{1}{6\pi^2}\left(\frac{2m}{\hbar^2}\right)^{3/2}\left[(\varepsilon_{\rm F} + \mu^* B)^{3/2}\mu^* - (\varepsilon_{\rm F} - \mu^* B)^{3/2}\mu^*\right]. \tag{25.47}$$

We then expand in powers of B to obtain, to lowest order,

$$m_{\rm V} = \frac{3}{2} n (\mu^*)^2 \frac{1}{\varepsilon_{\rm F}} B. \tag{25.48}$$

The corresponding susceptibility per unit volume is therefore

$$\chi_0 = \frac{\partial m_{\rm V}}{\partial B} = \frac{3}{2} n (\mu^*)^2 \frac{1}{\varepsilon_{\rm F}}. \tag{25.49}$$

For high temperatures, the corresponding result for a spin 1/2 paramagnet can be calculated from Eq. (19.125) with $\mu_{\rm B} = \mu^*$, $g = 2$ and $J = 1/2$, resulting in

$$\chi_\infty = \frac{1}{V}\frac{\partial \mathcal{M}}{\partial B} = n(\mu^*)^2 \frac{1}{k_{\rm B} T}. \tag{25.50}$$

Thus the electron gas has a susceptibility that is smaller by a factor of $(3/2)(k_{\rm B}T/\varepsilon_{\rm F})$, similar to the situation for the heat capacity. This weak paramagnetism is known as **Pauli paramagnetism**.

We can give a more general treatment by returning to Eq. (25.41) and converting to integrals, which leads to

$$n = n_{\rm Q}(T)[f_{3/2}(\lambda_+) + f_{3/2}(\lambda_-)], \tag{25.51}$$

where

$$\lambda_\pm = \exp[\beta(\mu \pm \mu^* B)] = \lambda \exp(\pm \beta \mu^* B). \tag{25.52}$$

For $B = 0$, Eq. (25.51) becomes Eq. (25.2) for $g_0 = 2$. Equation (23.20) can be generalized in the same way to yield the Kramers potential

$$K = -k_{\rm B} T V n_{\rm Q}(T)[f_{5/2}(\lambda_+) + f_{5/2}(\lambda_-)]. \tag{25.53}$$

To obtain the magnetic moment $\mathcal{M}$, we note that $K = F - \mu\mathcal{N}$ and then use Eq. (19.96) to obtain

$$dK = -S\,dT - p\,dV - \mathcal{M}\,dB - \mathcal{N}\,d\mu \tag{25.54}$$

from which

$$\mathcal{M} = -\left(\frac{\partial K}{\partial B}\right)_{T,V,\mu}. \tag{25.55}$$

Thus, the magnetization $m_V = \mathcal{M}/V$ becomes

$$m_V = n_Q(T)[f_{3/2}(\lambda_+) - f_{3/2}(\lambda_-)]\mu^*. \tag{25.56}$$

To compute the susceptibility χ per unit volume, we need to take the derivative of m_V with respect to B but holding $T, V, \mathcal{N}$ constant. This yields

$$\begin{aligned}\chi &= n_Q(T)[f_{1/2}(\lambda_+) + f_{1/2}(\lambda_-)]\beta(\mu^*)^2 \\ &\quad + n_Q(T)[f_{1/2}(\lambda_+) - f_{1/2}(\lambda_-)]\mu^*\beta\left(\frac{\partial \mu}{\partial B}\right)_{T,V,\mathcal{N}}.\end{aligned} \tag{25.57}$$

From Eq. (25.51) we compute the required derivative

$$\left(\frac{\partial \mu}{\partial B}\right)_{T,V,\mathcal{N}} = -\mu^*\frac{[f_{1/2}(\lambda_+) - f_{1/2}(\lambda_-)]}{[f_{1/2}(\lambda_+) + f_{1/2}(\lambda_-)]}. \tag{25.58}$$

This results in a rather complicated expression for χ, but unless we are interested in the very weak dependence of χ on magnetic field, we can take the $B = 0$ limit in which case both $\lambda_\pm$ can be replaced by $\lambda = \exp(\beta\mu)$. Then $\partial\mu/\partial B = 0$ and Eq. (25.57) becomes

$$\chi = 2n_Q(T)\beta(\mu^*)^2 f_{1/2}(\lambda). \tag{25.59}$$

In this same $B = 0$ limit, Eq. (25.51) becomes

$$n = 2n_Q(T)f_{3/2}(\lambda). \tag{25.60}$$

Together, Eqs. (25.59) and (25.60) allow determination of χ and μ in the $B = 0$ limit at all temperatures.

At low temperatures, $T \ll T_F$, we will have $\lambda \gg 1$ and we can use the asymptotic expansion Eq. (23.35) for large λ to obtain

$$\chi = n\beta(\mu^*)^2\frac{f_{1/2}(\lambda)}{f_{3/2}(\lambda)} = n(\mu^*)^2\frac{3}{2\mu}\left[1 - \frac{\pi^2}{6}\left(\frac{k_B T}{\mu}\right)^2 + \cdots\right]. \tag{25.61}$$

Then Eq. (25.60) yields the same value of μ as given by Eq. (25.35) and we get our final answer at low temperatures,

$$\chi = n(\mu^*)^2\frac{3}{2}\frac{1}{\varepsilon_F}\left[1 - \frac{\pi^2}{12}\left(\frac{k_B T}{\varepsilon_F}\right)^2 + \cdots\right], \tag{25.62}$$

which agrees with Eq. (25.49) at $T = 0$.

At high temperatures, $T \gg T_F$, if attainable for some spin 1/2 ideal Fermi gas, we would have the classical case $\lambda \ll 1$ and we can use the expansion Eq. (23.31) to obtain

$$\chi = n(\mu^*)^2\beta\left[1 - \frac{\lambda}{2^{3/2}} + \cdots\right]. \tag{25.63}$$

It is then sufficient to estimate (see Eq. (21.101)) $\lambda = n/(2n_Q(T))$ and thus obtain the high temperature result

$$\chi = n(\mu^*)^2\beta\left[1 - \frac{n}{2^{5/2}n_Q(T)} + \cdots\right]. \tag{25.64}$$

Although we have calculated χ only for spin 1/2 particles, the same technique would work for any half integral spin s, in which case one would have $2s+1$ different λ_i to deal with.

25.5 Landau Diamagnetism

In Section 25.4, we treated Pauli paramagnetism that results from the splitting of electron spin states in a magnetic field. It turns out that a magnetic field can also influence the orbital states, which gives rise to a diamagnetic effect in which the magnetic moment opposes the applied field. In other words, the magnetic susceptibility for diamagnetism is negative. We shall see that Landau diamagnetism gives rise to a susceptibility that is $-1/3$ the susceptibility for Pauli paramagnetism, provided that the effective orbital magnetic moment is equal to that for spin.

For a magnetic field applied along the z-axis, the velocity in the z-direction of a classical particle of charge e is unaffected but its velocity in the x- and y-directions is affected by the Lorentz force[4] which acts perpendicular to z with magnitude $Bev_\perp/c$. Thus, such a classical charged particle would move in a spiral of radius $R = mv_\perp c/eB$. Setting $v_\perp = R\omega$, we see that its angular frequency would be $\omega = eB/mc$. According to quantum mechanics, however, this motion is quantized and gives rise to energy levels, in addition to those associated with the free motion in the z-direction, that are spaced by $\hbar\omega = (e\hbar/mc)B$, namely,[5]

$$\varepsilon = \frac{e\hbar B}{mc}\left(j + \frac{1}{2}\right) + \frac{\hbar^2 k_z^2}{2m}, \tag{25.65}$$

where $j = 0, 1, 2, \ldots$.

The energy levels associated with the quantum number j are strongly degenerate, which can be understood by relating them to a coalescence of states associated with free x, y motion in the absence of a magnetic field. In that case, we know for a rectangle of dimensions L_x, L_y,

$$\sum_{k_x}\sum_{k_y} \to \frac{L_xL_y}{(2\pi)^2}\int 2\pi k_\perp\,\mathrm{d}k_\perp = \frac{L_xL_y}{(2\pi)^2}\int 2\pi k_\perp \frac{\mathrm{d}k_\perp}{\mathrm{d}\varepsilon_\perp}\mathrm{d}\varepsilon_\perp = \frac{L_xL_y}{h^2}2\pi m\int \mathrm{d}\varepsilon_\perp, \tag{25.66}$$

[4]For SI units, set $c = 1$ in this and subsequent formulas in this section.

[5]The quantity $e\hbar/mc$ is twice the Bohr magneton $\mu_B = e\hbar/2mc$ that we introduced in Section 19.6.2. Except for possible corrections for effective masses, $\mu_B = \mu^*$ as used in Section 25.4. The given energy levels can be obtained by mapping the x, y motion onto the problem for a harmonic oscillator.

where we have used $\varepsilon_\perp = \hbar^2 k_\perp^2/2m$ for energy associated with motion in the directions perpendicular to z. If we then make the correspondence

$$\int \mathrm{d}\varepsilon_\perp = \sum_j \Delta\varepsilon_j = \sum_j \frac{e\hbar B}{mc}, \tag{25.67}$$

we deduce that the degeneracy associated with each j level is

$$\frac{L_x L_y}{h^2} 2\pi m \frac{e\hbar B}{mc} = L_x L_y \frac{eB}{hc}. \tag{25.68}$$

This degeneracy, exclusive of spin, turns out to be correct based on a detailed solution of the problem [66, p. 424].

We proceed to compute the grand partition function

$$\ln \mathcal{Z} = g_0 \sum_\varepsilon{}' \ln(1 + \lambda\, \mathrm{e}^{-\beta\varepsilon}), \tag{25.69}$$

where the factor of $g_0 = 2$ is due to spin degeneracy. In fact, the spin states are not degenerate in the presence of B as we know from our treatment of Pauli paramagnetism in Section 25.4, so we should really treat each spin state separately or, better yet, treat Landau diamagnetism and Pauli paramagnetism simultaneously. But here we limit ourselves to the calculation of the zero field susceptibility, so we can treat each phenomenon separately.[6]

We therefore obtain

$$\ln \mathcal{Z} = g_0 \frac{V}{2\pi} \frac{eB}{hc} \int_{-\infty}^{\infty} \mathrm{d}k_z \sum_{j=0}^{\infty} \ln(1 + \lambda\, \mathrm{e}^{-\beta\varepsilon}), \tag{25.70}$$

where we have replaced the sum over k_z with an integral over k_z and a factor of $L_z/2\pi$ as usual, recognizing that the volume $V = L_x L_y L_z$. For $e\hbar B/mc \ll k_B T$, which we assume to be the case, we might consider replacing the sum over j by an integral but this turns out to give a result independent of B, as we shall see. We must use instead a form of the **Euler-Maclaurin sum formula**

$$\sum_{j=0}^{\infty} g(j + \tfrac{1}{2}) = \int_0^\infty g(x)\, \mathrm{d}x + \frac{1}{24} g'(0) + \cdots, \tag{25.71}$$

that is derived in Appendix H, Eq. (H.25), to obtain the first term that depends on B. Thus,

$$\sum_{j=0}^{\infty} \ln(1 + \lambda\, \mathrm{e}^{-\beta\varepsilon}) = \int_0^\infty \mathrm{d}x \ln\left[1 + \lambda \exp\left(-\beta \frac{e\hbar B}{mc} x - \beta \frac{\hbar^2 k_z^2}{2m}\right)\right]$$
$$- \frac{1}{24} \beta \frac{e\hbar B}{mc} \frac{1}{\lambda^{-1} \exp(\beta\hbar^2 k_z^2/2m) + 1} + \cdots. \tag{25.72}$$

[6]The decomposition represented later by Eq. (25.72) would still be valid if λ were replaced by $\lambda_\pm$ given by Eq. (25.52).

The integral in Eq. (25.72) can be written

$$\begin{aligned}
&\frac{mc}{e\hbar B}\int_0^\infty \mathrm{d}y \ln\left[1+\lambda\exp\left(-\beta y-\beta\frac{\hbar^2 k_z^2}{2m}\right)\right] \\
&= \frac{mc}{e\hbar B}\frac{\hbar^2}{m}\int_0^\infty k_\perp\,\mathrm{d}k_\perp \ln\left[1+\lambda\exp\left(-\beta\frac{\hbar^2(k_\perp^2+k_z^2)}{2m}\right)\right] \\
&= \frac{mc}{e\hbar B}\frac{\hbar^2}{m}\frac{1}{2\pi}\int_0^\infty \mathrm{d}k_x\int_0^\infty \mathrm{d}k_y \ln\left[1+\lambda\exp\left(-\beta\frac{\hbar^2(k_x^2+k_y^2+k_z^2)}{2m}\right)\right].
\end{aligned} \tag{25.73}$$

Therefore, the contribution of Eqs. (25.70)–(25.73) is

$$\ln \mathcal{Z}_0 = g_0\frac{V}{(2\pi)^3}\int \mathrm{d}^3k \ln[1+\lambda\exp(-\beta\hbar^2k^2/2m)], \tag{25.74}$$

which is precisely the result for zero field. It is therefore only the second term in Eq. (25.72), which resulted from the discrete nature of the quantum number j, that leads to diamagnetism. The contribution of that term to Eq. (25.70) is

$$\begin{aligned}
\ln \mathcal{Z}_B = -\frac{\beta}{24}\frac{g_0 V}{(2\pi)^2}\frac{e^2B^2}{mc^2}\int_{-\infty}^\infty \mathrm{d}k_z\frac{1}{\lambda^{-1}\exp(\beta\hbar^2k_z^2/2m)+1} \\
= -\frac{1}{6}\frac{g_0 V n_Q(T)}{(k_B T)^2}(\mu^* B)^2 f_{1/2}(\lambda),
\end{aligned} \tag{25.75}$$

where $\mu^* = e\hbar/2mc$ is the Bohr magneton, provided that the mass of the free electron in the metal can be taken as the electron mass. The magnetization is

$$m_V = \frac{k_B T}{V}\left(\frac{\partial \ln \mathcal{Z}_B}{\partial B}\right)_{V,\mu} = -\frac{1}{3}\frac{g_0 n_Q(T)}{k_B T}(\mu^*)^2 B f_{1/2}(\lambda) = -\frac{n(\mu^*)^2}{3k_B T}\frac{f_{1/2}(\lambda)}{f_{3/2}(\lambda)}B \tag{25.76}$$

to lowest order in B, where Eq. (25.2) has been used. The susceptibility per unit volume is therefore

$$\chi = \frac{k_B T}{V}\left(\frac{\partial \ln \mathcal{Z}_B}{\partial B}\right)_{V,\mu} = -\frac{1}{3}\frac{g_0 n_Q(T)}{k_B T}(\mu^*)^2 f_{1/2}(\lambda) = -\frac{n(\mu^*)^2}{3k_B T}\frac{f_{1/2}(\lambda)}{f_{3/2}(\lambda)}. \tag{25.77}$$

This diamagnetic susceptibility is $-1/3$ of the Pauli paramagnetic susceptibility given by Eq. (25.61), provided of course that the values of μ^* are the same (no effective mass corrections).

Example Problem 25.1. What is the total zero-field susceptibility for $T \ll T_F$ due to Pauli paramagnetism and Landau diamagnetism if there are effective mass corrections $m \to m_{\text{eff}}$ for the translational energies $\hbar^2k^2/2m_{\text{eff}}$ for both, and also a correction for the magnetic moment for Landau diamagnetism?

Solution 25.1. For Pauli paramagnetism, we assume that $\mu^* = \mu_B$, the Bohr magneton. For the Landau diamagnetism, we need $\mu^* = r\mu_B$, where $r = m/m_{\text{eff}}$. According to Eq. (25.8), the

Fermi energy depends on mass, so we should use $r\varepsilon_F$ in place of ε_F. In view of Eq. (25.62), we have the low temperature result

$$\chi_{\rm tot} = \left(1 - \frac{r^2}{3}\right) \frac{3}{2} \frac{n(\mu_B)^2}{r\varepsilon_F} \left[1 - \frac{\pi^2}{12}\left(\frac{k_B T}{r\varepsilon_F}\right)^2 + \cdots\right]. \tag{25.78}$$

25.6 Thermionic Emission

If a metal is heated, electrons can acquire sufficient energy to escape, a process known as thermionic emission. The process is somewhat similar to effusion, treated for a classical gas in Section 20.1.1, except for effusion one calculates the slow rate of escape through a small hole in a cavity. For thermionic emission, one considers the possibility that electrons moving in a given direction, say the z-direction, can overcome a potential energy barrier W^* that keeps the otherwise free electrons in the metal to begin with. We measure W^* from the zero of energy used for free electrons inside the metal. We can think of W^* as being made up of two parts, a positive part W_0 that would be necessary to remove an electron very far from the metal in the absence of surface relaxation effects, and another positive part W_s due to surface relaxation that accounts for a layer of surface dipoles (called the double layer).[7] What we actually calculate is the flux J through an imaginary small window of area a perpendicular to the z-direction, recognizing, however, that the electrons can escape in all directions. Moreover, we assume that the rate of escape is so slow that the system remains in quasi-equilibrium. We also assume that electrons are continuously supplied to the metal by a suitable electrical circuit so that the metal remains electrically neutral.

Since we know that electrons in a metal obey Fermi-Dirac statistics, they fill energy levels up to the Fermi energy ε_F even at $T = 0$. Therefore, we anticipate that they start out with an energetic boost of approximately ε_F so that they only have a barrier $W = W^* - \varepsilon_F$ to overcome. This turns out approximately to be the case, and follows naturally from a formal treatment based on Fermi statistics. The quantity W is called the **work function** of the metal.

We assume that the flux can be written in the form

$$J = \frac{2}{V} \sum_{k_x} \sum_{k_y} \sum_{k_z > k_z^*} \frac{1}{\exp\{\beta[(\hbar^2 k^2/2m) - \mu]\} + 1} \frac{\hbar k_z}{m}, \tag{25.79}$$

[7]Surface relaxation can be quite complicated and the value of the potential experienced by an electron outside a metal can depend on surface condition and surface charges that depend on surface orientation. To remove this complication in the case where all surfaces are not equivalent, one considers the electron to be removed from the metal only to a point sufficiently far outside the double layer that the electron no longer experiences any changes due to the presence of the double layer but not so far away as to be influenced by fields external to the metal, for example due to surface charges. For a more comprehensive discussion see Ashcroft and Mermin [58, p. 354].

where $k_z^* = (2mW^*/\hbar^2)^{1/2}$ is assumed to be the threshold value of k_z needed for escape. The quantity $\hbar k_z/m$ plays the role of velocity in the z-direction and the remainder of the expression is the number density of eligible electrons. The factor of 2 is due to spin degeneracy. This is the analog of Eq. (20.25) in the case of classical effusion. We approximate summation by integration by means of a factor $V/(2\pi)^3$ to obtain

$$J = \frac{2}{(2\pi)^3}\int_{-\infty}^{\infty} \mathrm{d}k_x \int_{-\infty}^{\infty} \mathrm{d}k_y \int_{k_z^*}^{\infty} \mathrm{d}k_z \frac{1}{\exp\{\beta[(\hbar^2 k^2/2m) - \mu]\} + 1}\frac{\hbar k_z}{m}. \tag{25.80}$$

We then pass to cylindrical coordinates $\hbar k_x = p' \cos\varphi$, $\hbar k_y = p' \sin\varphi$, and $\hbar k_z = p_z$ and do the φ integral to obtain a factor of 2π. This results in

$$J = \frac{4\pi}{mh^3}\int_0^{\infty} p'\,\mathrm{d}p' \int_{p_z^*}^{\infty} p_z\,\mathrm{d}p_z \frac{1}{\exp\{\beta[(p'^2/2m) + (p_z^2/2m) - \mu]\} + 1}, \tag{25.81}$$

where $p_z^* = (2mW^*)^{1/2}$. We perform the integral over p' and change variables to $\varepsilon_z = p_z^2/2m$ to obtain

$$J = \frac{4\pi m k_B T}{h^3}\int_{W^*}^{\infty} \mathrm{d}\varepsilon_z \ln\{1 + \exp[-\beta(\varepsilon_z - \mu)]\}. \tag{25.82}$$

To proceed, we make the approximation that $W^* - \mu \gg k_B T$ which means that $\exp[-\beta(\varepsilon_z - \mu)] \ll 1$ in the range of integration. We can therefore expand the logarithm to obtain

$$J = \frac{4\pi m k_B T}{h^3}\int_{W^*}^{\infty} \mathrm{d}\varepsilon_z \exp[-\beta(\varepsilon_z - \mu)] = \frac{4\pi m (k_B T)^2}{h^3}\exp[-\beta(W^* - \mu)]. \tag{25.83}$$

The chemical potential μ is given by Eq. (25.35) but in view of prior approximations, the lowest order $\mu \approx \varepsilon_F$ will suffice. We multiply by the magnitude $|e|$ of the charge of the electron to get the magnitude of the flux of charge

$$J_q = |e|\frac{4\pi m (k_B T)^2}{h^3}\exp(-\beta W), \tag{25.84}$$

where $W = W^* - \varepsilon_F$ is the work function of the metal, introduced previously. This result is known as the **Richardson-Dushman equation** and is supported by experiment if reduced by a transmission coefficient that accounts for the surface condition of the metal and the simplifying assumptions that have been made about the barrier for escape.

As anticipated, the energy barrier W^* is reduced to $W = W^* - \varepsilon_F$ since at $T = 0$, the electrons already occupy energy levels up to ε_F. If the electron gas had behaved like a classical gas, we would have $\lambda = \exp(\beta\mu) = n/2n_Q(T) \ll 1$ which would result in a much smaller flux of charge[8]

$$J_q^{\text{class}} = |e|\left(\frac{k_B T}{2\pi m}\right)^{1/2}\exp(-\beta W^*) \tag{25.85}$$

with a higher activation energy W^* and a prefactor proportional to $T^{1/2}$ instead of T^2.

[8]We write this formula only for the sake of comparison, recognizing that it is not true because $n/n_Q(T) \gg 1$ for free electrons in metals at any reasonable temperature because of their high density and the small electron mass.

25.6.1 Schottky Effect

An electric field of strength E at the surface of a metal and directed toward the metal is known to enhance thermionic emission. This is known as the **Schottky effect**, which is reasonable to expect because an electron outside the metal, having a negative charge, would experience a force in the opposite direction of the field. If z measures distance outside the metal, the electrical potential due to the electric field is Ez and the potential energy of an electron at distance z due to the field is $-eEz$, all relative to the energy W_0. But an electron at distance z outside the metal creates an electric field of its own that must be canceled inside the metal by inducing a positive surface charge on the metal.[9] Formally, the effect of this surface charge can be handled by placing an imaginary image charge e at a distance z inside the metal, in other words at location $-z$. The force on the electron due to this image charge (really the induced surface charge) will be $-e^2/(2z)^2$, so the electron will be attracted toward the metal. The potential energy felt by the electron due to this image effect will be

$$\int_z^\infty -e^2/(2z)^2\,\mathrm{d}z = -\mathrm{e}^2/(4z), \tag{25.86}$$

relative again to W_0. The combined effect of these two potentials is $-eEz - \mathrm{e}^2/4z$ which has a maximum at $z = (eE)^{1/2}/2$ where its value is $-e(eE)^{1/2}$. The barrier for escape to far distances from the surface at which the electric field is applied therefore becomes $W_0 + W_s - e(eE)^{1/2} = W^* - e(eE)^{1/2}$, resulting in an effective work function

$$W_\mathrm{E} = W^* - e(eE)^{1/2} - \varepsilon_\mathrm{F} = W - e(eE)^{1/2} \tag{25.87}$$

instead of W in Eq. (25.84). In SI units, $-e(eE)^{1/2} \to -e(eE/4\pi\epsilon_0)^{1/2} = (1.44\times 10^{-9}E)^{1/2}$ eV, where E is measured in V/m. To reduce W by even 0.1 eV would require a large field, $E \sim 7 \times 10^6$ V/m. Typically, W is 2-4 eV.

25.6.2 Photoelectric Effect

If photons of monochromatic light of frequency ν enter a metal, they can collide with electrons and reduce the barrier for emission from W^* to $W^* - h\nu$. If $h\nu \ll W^*$, Eq. (25.84) will apply with W replaced by $W - h\mu$, analogous to the small reduction in the effective work function caused by an applied electric field. But in the case of sufficiently energetic photons, $h\nu$ can be comparable to W or even exceed W so we must return to Eq. (25.82) which now becomes

$$J = \frac{4\pi m k_\mathrm{B} T}{h^3}\int_{W^*-h\nu}^\infty \mathrm{d}\varepsilon_z \ln\{1 + \exp[-\beta(\varepsilon_z - \mu)]\}. \tag{25.88}$$

[9]Recall that we ignored the effect of surface charges and external fields in the result that led to the work function W. The presence of the field E itself also requires a positive surface charge on the metal to prevent penetration of E into the metal.

We substitute $\varepsilon_z = uk_B T + W^* - h\nu$ to obtain

$$J = \frac{4\pi m(k_B T)^2}{h^3}\int_0^\infty du \ln\{1 + \exp[\beta(h\nu - W^* + \mu) - u]\}. \tag{25.89}$$

Then approximating $W^* - \mu \approx W^* - \varepsilon_F = W$ and defining $\nu_0 := W/h$, we obtain

$$J = \frac{4\pi m(k_B T)^2}{h^3}\int_0^\infty du \ln\{1 + \exp[\beta h(\nu - \nu_0) - u]\}. \tag{25.90}$$

We introduce the notation

$$\lambda_\nu := \exp[\beta h(\nu - \nu_0)] \tag{25.91}$$

and integrate by parts to obtain

$$\int_0^\infty du \ln[1 + \lambda_\nu \, e^{-u}] = \int_0^\infty du \frac{u}{\lambda_\nu^{-1} e^u + 1} = f_2(\lambda_\nu), \tag{25.92}$$

where $f_2(\lambda_\nu) = h_2(\lambda_\mu, 1)$ is given by Eq. (23.15). We therefore have

$$J = \frac{4\pi m(k_B T)^2}{h^3} f_2(\lambda_\nu). \tag{25.93}$$

Since $f_2(\lambda_\nu)$ is a monotonically increasing function of λ_ν, we see that J increases monotonically with $\nu - \nu_0$ as one would expect. In the limit $h(\nu - \nu_0) \gg k_B T$, we can use the asymptotic form (see Eq. (23.35)) $f_2(\lambda_\nu) \sim \ln(\lambda_\nu)^2/2 = \beta^2 h^2(\nu - \nu_0)^2$, so J saturates at a value

$$J_{sat} = \frac{2\pi m}{h}(\nu - \nu_0)^2 \tag{25.94}$$

that is independent of temperature.

25.7 Semiconductors

In this section, we treat the statistical mechanics of semiconductors based on single particle states (orbitals) of an electron in an effective periodic potential due to interaction with a crystal lattice. Thus the density of states is no longer given by the free electron result, Eq. (25.13). As shown in a number of books on solid-state physics, the density of states can have the following approximate form, as sketched in Figure 25–2:

$$g(\varepsilon) = \begin{cases} g_v(\varepsilon) & \text{for } 0 \le \varepsilon \le \varepsilon_V \\ 0 & \text{for } \varepsilon_V < \varepsilon < \varepsilon_C = \varepsilon_V + \varepsilon_g \\ g_c(\varepsilon) & \text{for } \varepsilon_C < \varepsilon. \end{cases} \tag{25.95}$$

The region of width $\varepsilon_g = \varepsilon_C - \varepsilon_V$, where $g(\varepsilon) = 0$ is known as a **band gap** that separates the **valence band** $g_V(\varepsilon)$ from the **conduction band** $g_C(\varepsilon)$.

We consider a material for which the valence band is completely full and the conduction band is completely empty at $T = 0$. In this condition, each electron is in a definite state and cannot move in response to an applied electric field, so the material will behave

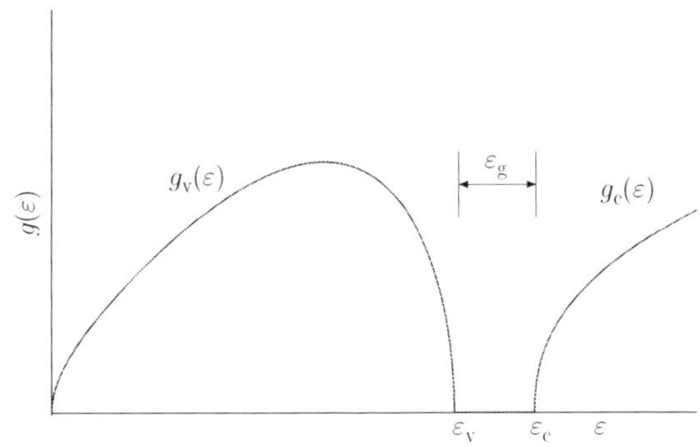

FIGURE 25–2 Sketch of density of states $g(\varepsilon)$ given by Eq. (25.95) versus electron energy ε for a simple semiconductor. The size ε_g of the band gap is exaggerated for the sake of illustration.

as an insulator. For $T > 0$, some electrons will be excited into the conduction band, leaving unoccupied states called **holes** in the valence band. Under these conditions, electrons in both the valence and conduction bands can move in response to an electric field and the material can conduct electricity. Provided that $\varepsilon_g \gg k_B T$, only a small number of electrons will be excited to the conduction band. For $T = 300$ K, we have $k_B T = 0.026$ eV. If $\varepsilon_g \geq 10$ eV, hardly any electrons will be excited into the conduction band and the material will be a good insulator. However, if $\varepsilon_g \sim 1$ eV or less, there will be a significant number of electrons excited to the conduction band, accompanied by a dramatic increase in electrical conductivity at $T = 300$ K. Such a material is called an **intrinsic semiconductor**. Certain **dopants**, which are foreign atoms of very low concentrations, can be substituted for host atoms in the material and can greatly modify this behavior. Dopants referred to as **donors** can lead to a greatly enhanced number of electrons in the conduction band whereas so-called **acceptors** can lead to a greatly enhanced number of holes in the valence band. Strongly doped materials are called **extrinsic semiconductors**. We first treat the intrinsic case and then show how dopants can be accounted for.

25.7.1 Intrinsic Semiconductors

In the absence of dopants and for a density of states given by Eq. (25.95), we assume that

$$\frac{\mathcal{N}}{V} = \int_0^{\varepsilon_v} g_v(\varepsilon)\, d\varepsilon; \quad T = 0, \tag{25.96}$$

where $\mathcal{N}$ is the number of valence electrons. This will be true if the Fermi energy ε_F is at ε_v or anywhere else within the band gap because there are no states in the gap. By extrapolation from $T > 0$, we will see later that ε_F is located near the middle of the band gap (see Eq. (25.108) for $T = 0$). For $T > 0$, the corresponding equation is

$$\frac{\mathcal{N}}{V} = \int_0^{\varepsilon_v} g_v(\varepsilon) \frac{1}{e^{\beta(\varepsilon-\mu)}+1} d\varepsilon + \int_{\varepsilon_c}^{\infty} g_c(\varepsilon) \frac{1}{e^{\beta(\varepsilon-\mu)}+1}\, d\varepsilon. \tag{25.97}$$

We subtract Eq. (25.96) from Eq. (25.97) to obtain

$$-\int_0^{\varepsilon_V} g_V(\varepsilon)\frac{1}{1+e^{-\beta(\varepsilon-\mu)}}\,d\varepsilon + \int_{\varepsilon_C}^{\infty} g_C(\varepsilon)\frac{1}{e^{\beta(\varepsilon-\mu)}+1}\,d\varepsilon = 0. \tag{25.98}$$

The second term in Eq. (25.97) is the concentration of electrons in the conduction band, namely,

$$n = \int_{\varepsilon_C}^{\infty} g_C(\varepsilon)\frac{1}{e^{\beta(\varepsilon-\mu)}+1}\,d\varepsilon \approx \int_{\varepsilon_C}^{\infty} g_C(\varepsilon)e^{-\beta(\varepsilon-\mu)}\,d\varepsilon, \tag{25.99}$$

where the second approximate form follows, provided that $\varepsilon - \mu \gg k_B T$ in the range of integration, which we assume for now to be the case. Similarly, the negative of the first term in Eq. (25.98) is defined to be the concentration p of **holes**, which are hypothetical *positive* charge carriers in the valence band. Thus

$$p = \int_0^{\varepsilon_V} g_V(\varepsilon)\frac{1}{1+e^{-\beta(\varepsilon-\mu)}}\,d\varepsilon \approx \int_0^{\varepsilon_V} g_V(\varepsilon)e^{-\beta(\mu-\varepsilon)}\,d\varepsilon, \tag{25.100}$$

where now $\mu - \varepsilon \gg k_B T$ in this range of integration. When the approximate forms in Eqs. (25.99) and (25.100) are valid, which will be the case if $\varepsilon_g \gg k_B T$ and μ remains near the center of the band gap, the semiconductor is said to be nondegenerate [6, p. 358]. See Eqs. (25.129) and (25.131) for the degenerate case when these approximate forms are not valid. Equation (25.98), which may be rewritten $n - p = 0$, can be thought of as expressing overall charge neutrality.

By using the approximate forms for n and p, we see that

$$pn = \left[\int_0^{\varepsilon_V} g_V(\varepsilon)e^{\beta\varepsilon}\,d\varepsilon\right]\left[\int_{\varepsilon_C}^{\infty} g_C(\varepsilon)e^{-\beta\varepsilon}\,d\varepsilon\right], \tag{25.101}$$

which is independent of μ. Equation (25.101) is independent of Eq. (25.98) provided that n and p are given by the approximate forms on the right-hand sides of Eqs. (25.99) and (25.100), respectively, and is known as the **law of mass action.**[10] In the degenerate case, pn is given by Eq. (25.134) of Section 25.7.3, where we also treat doped extrinsic semiconductors.

In the integral for n we substitute $\varepsilon = w + \varepsilon_C$ and in the integral for p we substitute $\varepsilon = \varepsilon_V - w$ which results in

$$n = e^{-\beta\varepsilon_C}e^{\beta\mu}\int_0^{\infty} g_C(w+\varepsilon_C)e^{-\beta w}\,dw \tag{25.102}$$

and

$$p = e^{\beta\varepsilon_V}e^{-\beta\mu}\int_0^{\varepsilon_V} g_V(\varepsilon_V - w)e^{-\beta w}\,dw. \tag{25.103}$$

[10]This is by analogy to a gaseous chemical reaction of the form AB = A + B; in the present case, we would think of an electron-hole pair dissociating into an electron and a hole. In the event that the approximate forms of Eqs. (25.99) and (25.100) do not hold, as is the case for degenerate semiconductors that result from high doping levels, modification is required because the full Fermi-Dirac distribution function must be used. See Section 25.7.3 and Kittel and Kroemer [6, p. 365] for details.

The integrals in Eqs. (25.102) and (25.103) sample the densities of state only near the band edges. They can be evaluated by following the conventional approximations of semiconductor physics, according to which these densities of state near the band gap can be approximated by those for a free electron but with the mass of the electron replaced by an effective mass, either m_n or m_p. In that case, the specific forms (see Eq. (25.13)) would be

$$g_c(w+\varepsilon_c) = 2\frac{2}{\pi^{1/2}}\left(\frac{m_n}{2\pi\hbar^2}\right)^{3/2} w^{1/2}; \quad g_v(\varepsilon_v - w) = 2\frac{2}{\pi^{1/2}}\left(\frac{m_p}{2\pi\hbar^2}\right)^{3/2} w^{1/2}. \tag{25.104}$$

Then we obtain

$$n = n^* e^{-\beta(\varepsilon_c-\mu)}; \quad p = p^* e^{-\beta(\mu-\varepsilon_v)}, \tag{25.105}$$

where

$$n^* = 2\left(\frac{m_n k_B T}{2\pi\hbar^2}\right)^{3/2}; \quad p^* = 2\left(\frac{m_p k_B T}{2\pi\hbar^2}\right)^{3/2}. \tag{25.106}$$

We note that n^* and p^* each have the form of a quantum concentration of an ideal gas times a factor of 2 for spin. Then Eq. (25.101) becomes

$$pn = p^* n^* e^{-\beta\varepsilon_g}. \tag{25.107}$$

For an intrinsic semiconductor, Eq. (25.98) requires $p_i = n_i$, where we have added the subscript "i" to denote the intrinsic case. Then from Eq. (25.105) we obtain

$$\mu_i = \frac{\varepsilon_v+\varepsilon_c}{2} + \frac{k_B T}{2}\ln\left(\frac{p^*}{n^*}\right) = \varepsilon_v + \frac{1}{2}\varepsilon_g + \frac{3k_B T}{4}\ln\left(\frac{m_p}{m_n}\right), \tag{25.108}$$

which locates the chemical potential[11] very near the middle of the band gap. If $m_n = m_p$, μ_i is exactly in the middle of the band gap. As $T \to 0$, we have $\mu_i \to \varepsilon_F$ so ε_F is regarded as being near the middle of the band gap for a nondegenerate intrinsic semiconductor. By taking the square root of Eq. (25.107), we find the individual concentrations

$$n_i = p_i = (p^* n^*)^{1/2} e^{-\beta\varepsilon_g/2}. \tag{25.109}$$

Example Problem 25.2. For silicon, $\varepsilon_g = 1.14$ eV and at $T = 300$ K one has $k_B T = 0.0259$ eV, $p^* = 1.1 \times 10^{19}$ cm^{-3}, and $n^* = 2.7 \times 10^{19}$ cm^{-3}. Silicon is diamond cubic with a cube edge of $a = 3.57 \times 10^{-8}$ cm; there are eight atoms in each cube and each has a valence of 4. Calculate n_i and compare with the total valence electron concentration $\mathcal{N}/V$. Then calculate $\mu_i - \varepsilon_v$ and compare with the middle of the band gap.

Solution 25.2. From Eq. (25.109) we calculate $(p^*n^*)^{1/2} = 1.7 \times 10^{19}$ cm^{-3} and also $\exp(-\beta\varepsilon_g/2) = 2.77 \times 10^{-10}$. Thus $n_i = p_i = 4.8 \times 10^9$ cm^{-3}. Since $\mathcal{N}/V = 32/a^3 = 7.0 \times 10^{23}$ cm^{-3}, the ratio of n_i to $\mathcal{N}/V$ is 6.8×10^{-15}. We first calculate $(k_B T/2)\ln(p^*/n^*) = -0.012$ eV

[11] In the semiconductor literature, the chemical potential μ is usually called the Fermi level, which should not be confused with the Fermi energy, which is the value of μ at $T = 0$.

whereas $\varepsilon_g/2 = 0.57$ eV. Thus, $\mu_i - \varepsilon_V = 0.56$ eV, about 2% lower than the middle of the band gap.

25.7.2 Semiconductors with Dopants

As mentioned above, dopants known as donors and acceptors can be substituted for host atoms to affect the carrier concentrations of electrons in the conduction band and holes in the valence band. In the presence of dopants, we shall see that the equation $n = p$ will be replaced by

$$n - p = \Delta, \tag{25.110}$$

where Δ, to be discussed in the subsequent paragraph, depends on dopant concentrations and temperature. Moreover, Eq. (25.107) can be written

$$pn = n_i^2, \tag{25.111}$$

where Eq. (25.109) has been used. Therefore, without yet knowing Δ, we can write

$$n - \frac{n_i^2}{n} = \Delta, \tag{25.112}$$

which can be solved to yield

$$n = \sqrt{n_i^2 + (\Delta/2)^2} + \Delta/2; \quad p = \sqrt{n_i^2 + (\Delta/2)^2} - \Delta/2. \tag{25.113}$$

For $(\Delta/2)^2 \gg n_i^2$, the semiconductor is said to be extrinsic (dominated by dopants) and

$$n \approx (\Delta + |\Delta|)/2 + n_i^2/|\Delta|; \quad p \approx (-\Delta + |\Delta|)/2 + n_i^2/|\Delta|, \tag{25.114}$$

so $n \sim \Delta$ and $p \sim 0$ if $\Delta > 0$ and $n \sim 0$ and $p \sim |\Delta|$ if $\Delta < 0$. A semiconductor is said to be n-type or p-type depending on which is the dominant species.

By using the approximate forms in Eqs. (25.99) and (25.100), we can write

$$n = n_i e^{\beta(\mu-\mu_i)}; \quad p = n_i e^{-\beta(\mu-\mu_i)} \tag{25.115}$$

from which

$$\Delta/2 = n_i \sinh[\beta(\mu - \mu_i)]. \tag{25.116}$$

For the approximate forms in Eq. (25.115) to be valid, we need $|\mu - \mu_i| \ll \varepsilon_g/2$, so

$$|\Delta/2| \ll n_i \sinh[\beta\varepsilon_g/2], \tag{25.117}$$

but this still allows $|\Delta/2n_i|$ to be fairly large. In the extrinsic limit, μ approaches either ε_C or ε_V, shown in Figure 25–3, depending on the sign of $|\Delta|$. If $|\Delta|$ becomes comparable to p^* or n^*, the chemical potential shifts so far from the center of the band gap that the approximate forms of Eqs. (25.99) and (25.100) are no longer valid. See the Example Problem 25.3 below for more detail.

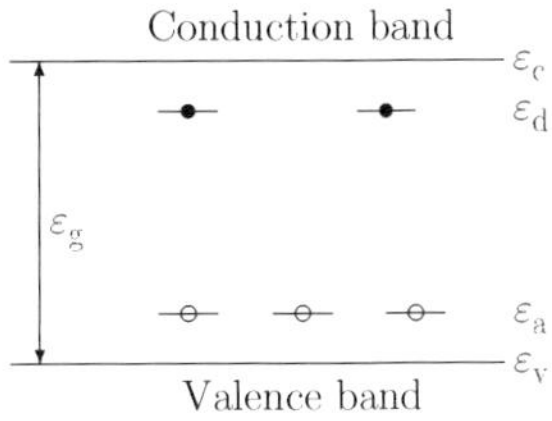

FIGURE 25–3 Schematic diagram of the valence and conduction bands with acceptor levels at ε_a just above the valence band (which lies below ε_v), and donor levels at ε_d just below the conduction band (which lies above ε_c).

We now proceed to calculate Δ. A donor[12] provides one more valence electron than a host atom, and this extra electron can possibly be added to the pool of electrons that are subject, to a first approximation, to the effective periodic potential. However, if the extra electron is not localized near the donor site, the core of the donor atom, which consists of its nucleus and other bound electrons, would appear to have a net positive charge (relative to the host atoms). By analogy with the hydrogen atom, but in a medium with a greatly altered dielectric constant, a donor can be regarded as a localized defect that has a weakly bound state at an energy ε_d that is slightly below ε_c, as illustrated in Figure 25–3. We assume that the number of donors is $\mathcal{N}_d \ll \mathcal{N}$, where $\mathcal{N}$ is still the number of valence electrons for the intrinsic case. Since these defects are quite dilute, they can be treated by means of the grand canonical ensemble in which the bulk of the system imposes a chemical potential μ, corresponding to an absolute activity $\lambda = e^{\beta\mu}$. There are three possible states: The donor can be occupied with an electron of either spin (two distinct quantum states) in which case its energy is ε_d or it can be in an unoccupied state in which case its energy is 0. At temperature T, the number of donors that are occupied by a localized electron is

$$\mathcal{N}_d \frac{2\lambda e^{-\beta\varepsilon_d}}{1+2\lambda e^{-\beta\varepsilon_d}} = \mathcal{N}_d \frac{1}{(1/2)e^{\beta(\varepsilon_d-\mu)}+1} \approx \mathcal{N}_d 2e^{-\beta(\varepsilon_d-\mu)} \ll \mathcal{N}_d, \tag{25.118}$$

provided that μ is still near the middle of the band gap and $(\varepsilon_d-\mu)/k_BT \sim (\varepsilon_g/2)/k_BT \gg 1$. The number of electrons that are not locally bound to donors, and therefore the number of positively charged (ionized) donors, is

$$\mathcal{N}_d^+ \equiv \mathcal{N}_d \frac{1}{1+2\lambda e^{-\beta\varepsilon_d}} = \mathcal{N}_d \frac{1}{1+2e^{-\beta(\varepsilon_d-\mu)}} \approx \mathcal{N}_d(1-2e^{-\beta(\varepsilon_d-\mu)}) \sim \mathcal{N}_d. \tag{25.119}$$

In other words, if there are $\mathcal{N}_d$ donors, practically all of them will donate an electron to the bands, provided that the given restrictions on μ are valid.

An acceptor[13] provides one less electron than a host atom. Its core therefore appears to have a net negative charge unless a hole is bound to the acceptor site. A hole bound

[12]For example, one could dope silicon with the donor phosphorus. For the host silicon, each Si atom, atomic number 14, provides four valence electrons ($3s^23p^2$). Each donor atom P, atomic number 15, that is substituted for Si provides five valence electrons ($3s^23p^3$).

[13]If the host atom is Si, each acceptor atom Al, atomic number 13, that is substituted for Si provides three valence electrons ($3s^23p$).

to an acceptor site essentially means that an electron is rejected from the site. If a hole is not bound to the acceptor site, meaning an electron occupies the site, it gives rise to an electronic energy level ε_a that lies slightly above ε_v. Such a state is a singlet because the occupying electron and another electron with opposite spin from a host atom constitute a bond. If a hole is bound to an acceptor site, an electron of either spin has left the site, so such an unoccupied state is doubly degenerate and has energy zero. If there are $\mathcal{N}_a$ acceptors, the number of electrons that are bound to acceptor sites, and therefore the number of negatively charged (ionized) acceptors, is

$$\mathcal{N}_a^- \equiv \mathcal{N}_a \frac{\lambda e^{-\beta\varepsilon_a}}{2+\lambda e^{-\beta\varepsilon_a}} = \mathcal{N}_a \frac{1}{2e^{\beta(\varepsilon_a-\mu)}+1} \approx \mathcal{N}_a(1-2e^{-\beta(\mu-\varepsilon_a)}) \sim \mathcal{N}_a, \tag{25.120}$$

and the number of acceptor sites that are not occupied by a valence electron is

$$\mathcal{N}_a \frac{2}{2+\lambda e^{-\beta\varepsilon_a}} = \mathcal{N}_a \frac{1}{1+(1/2)e^{\beta(\mu-\varepsilon_a)}} \approx \mathcal{N}_a(e^{-\beta(\mu-\varepsilon_a)}) \ll \mathcal{N}_a. \tag{25.121}$$

Therefore, practically all of the acceptor sites will take on an electron from the bands (i.e., create holes in the bands), provided that μ remains near the center of the band gap.

We can now establish the following balance for electrons:

$$\begin{aligned}\text{(Electrons in bands)} = {} & \text{(Valence electrons if all sites were host atoms)} \\ & + \text{(Electrons freed from donors)} \\ & - \text{(Electrons bound to acceptors)}.\end{aligned} \tag{25.122}$$

The total number of valence electrons if all sites were host atoms is just $\mathcal{N}$, so Eq. (25.122) per unit volume takes the form

$$\begin{aligned}\int_0^{\varepsilon_v} g_v(\varepsilon)\frac{1}{e^{\beta(\varepsilon-\mu)}+1}\,d\varepsilon + \int_{\varepsilon_c}^{\infty} g_c(\varepsilon)\frac{1}{e^{\beta(\varepsilon-\mu)}+1}\,d\varepsilon \\ = \frac{\mathcal{N}}{V} + \frac{\mathcal{N}_d}{V}\frac{1}{1+2e^{-\beta(\varepsilon_d-\mu)}} - \frac{\mathcal{N}_a}{V}\frac{1}{2e^{\beta(\varepsilon_a-\mu)}+1}.\end{aligned} \tag{25.123}$$

We now subtract $\mathcal{N}/V$ from both sides and use Eq. (25.96) and the definitions of n and p given by Eqs. (25.99) and (25.100) to obtain

$$\Delta = n - p = n_d^+ - n_a^-, \tag{25.124}$$

where

$$n_d^+ = \frac{\mathcal{N}_d^+}{V} = n_d \frac{1}{1+2e^{-\beta(\varepsilon_d-\mu)}}; \tag{25.125}$$

$$n_a^- = \frac{\mathcal{N}_a^-}{V} = n_a \frac{1}{2e^{\beta(\varepsilon_a-\mu)}+1}. \tag{25.126}$$

The quantity on the right-hand side of Eq. (25.124) is called the **net ionized donor concentration**. We note that Eq. (25.124) is just a statement of overall charge neutrality, namely,

$$n + n_a^- = p + n_d^+. \tag{25.127}$$

If Δ is not too large, we will have $\mu \sim \mu_{\rm i} \sim \varepsilon_{\rm v} + \varepsilon_{\rm g}/2$ which gives approximately $\Delta = n_{\rm d} - n_{\rm a}$ to lowest order. See Kittel and Kroemer [6, p. 371] for an interesting graphical solution for μ under conditions of rather large Δ for which μ becomes comparable to $\varepsilon_{\rm d}$.

Example Problem 25.3. Suppose that $n_{\rm i} \ll \Delta$ so that Eq. (25.114) holds. Then $n \sim \Delta$ and $p \sim 0$. Show that μ approaches the edge of the conduction band $\varepsilon_{\rm c}$ as Δ approaches n^*. Discuss the breakdown of the approximation that leads to the second form of Eq. (25.99). Evaluate μ under conditions for which Δ becomes sufficiently large that μ enters the conduction band. Then examine the same problem except for Δ negative but of large magnitude.

Solution 25.3. From Eq. (25.105) with $n = \Delta$ we have

$$\varepsilon_{\rm c} - \mu = k_{\rm B} T \ln(n^*/\Delta). \tag{25.128}$$

As Δ approaches n^* from below, we see that μ appears to approach $\varepsilon_{\rm c}$ from below. Examination of Eq. (25.99) shows that the second form becomes invalid under these conditions because $\exp[\beta(\varepsilon - \mu)]$ will no longer be large in the range of integration. Thus, Eq. (25.105) is no longer valid. From the first form of Eq. (25.99) and with $g_{\rm c}(\varepsilon)$ given by Eq. (25.104), we obtain

$$n = n^* \frac{1}{\Gamma(3/2)} \int_0^\infty \frac{u^{1/2}}{\lambda_{\rm c}^{-1} {\rm e}^u + 1} \, {\rm d}u = n^* f_{3/2}(\lambda_{\rm c}), \tag{25.129}$$

where $\lambda_{\rm c} = \exp[\beta(\mu - \varepsilon_{\rm c})]$ and $f_{3/2}$ is a fermion function defined in Chapter 23. If $\lambda_{\rm c}$ were small, as it would be for μ near the middle of the band gap, one would have $f_{3/2}(\lambda_{\rm c}) = \lambda_{\rm c}$ and Eq. (25.105) for n is recovered. But for heavy doping of donors, so that $n \approx \Delta \gg n^*$, $\lambda_{\rm c}$ will be large and we can use the asymptotic form $f_{3/2}(\lambda_{\rm c}) = [\beta(\mu - \varepsilon_{\rm c})]^{3/2}/\Gamma(5/2)$ to obtain

$$\mu - \varepsilon_{\rm c} = \frac{\hbar^2}{2m_n} \left(3\pi^2 \Delta\right)^{2/3}. \tag{25.130}$$

For $\Delta < 0$ but $|\Delta| \gg p^*$, one will have $p \sim |\Delta|$ and $n \sim 0$. Now (with $\beta\varepsilon_{\rm v} \approx \infty$ in the upper limit of the integral),

$$p = p^* \frac{1}{\Gamma(3/2)} \int_0^\infty \frac{u^{1/2}}{\lambda_{\rm v}^{-1} {\rm e}^u + 1} \, {\rm d}u = p^* f_{3/2}(\lambda_{\rm v}) \tag{25.131}$$

with $\lambda_{\rm v} = \exp[\beta(\varepsilon_{\rm v} - \mu.)]$ The chemical potential will move into the valence band and

$$\varepsilon_{\rm v} - \mu = \frac{\hbar^2}{2m_p} \left(3\pi^2 |\Delta|\right)^{2/3}. \tag{25.132}$$

25.7.3 Degenerate Semiconductors

Substitution of the more general expressions for n and p given by Eqs. (25.129) and (25.131) into Eq. (25.124) gives

$$n^* f_{3/2}(\lambda_c) - p^* f_{3/2}(\lambda_v) = \frac{n_d}{2\lambda_c + 1} - \frac{n_a}{2\lambda_v + 1}, \tag{25.133}$$

where $\lambda_c = \exp[\beta(\mu - \varepsilon_c)]$ and $\lambda_v = \exp[\beta(\varepsilon_v - \mu)]$. Since $\lambda_v \lambda_c = \exp(-\beta\varepsilon_g) < 1$, Eq. (25.133) can be solved to determine μ. When the doping is such that λ_c and λ_v are *both* small, the semiconductor is said to be nondegenerate. Then μ is near the middle of the band gap, $f_{3/2}(\lambda_c) \approx \lambda_c$ and $f_{3/2}(\lambda_v) \approx \lambda_v$. Then one recovers the cases treated that are based on the approximate expressions on the right-hand side of Eqs. (25.99) and (25.100). If the doping is such that *either* λ_c or λ_v is not small, the semiconductor is said to be degenerate and the $f_{3/2}$ functions associated with the dominant carrier must be used. Alternatively, one could replace Eq. (25.107) by

$$pn = p^* n^* f_{3/2}(\lambda_v) f_{3/2}(\lambda_c) \tag{25.134}$$

and then solve simultaneously with Eq. (25.124) by means of power series expansions, such as the Joyce-Dixon approximation [6, p. 366].

26

Quantum Statistics

In this chapter, we discuss several formal aspects of the statistical mechanics of quantum systems. Two types of averaging arise. The first type pertains to the intrinsically statistical nature of quantum mechanics itself and is present even when the system is in a pure quantum state $|\psi(t)\rangle$ with wave function $\psi(\mathbf{r},t) = \langle \mathbf{r}|\psi(t)\rangle$. The second type of averaging pertains to averages over many quantum states related to an ensemble used to represent a system for which complete information about its quantum state is not known. Such an ensemble might be used to represent a system in a state of thermodynamic equilibrium under some constraints, for example, near isolation or contact with a temperature reservoir. To treat such systems, it is convenient to introduce a statistical operator $\hat{\rho}$ that is known as the **density operator**. In terms of $\hat{\rho}$, we shall see that the expectation value of some observable having operator $\hat{f}$ can be written in the form of a trace, $\mathrm{tr}(\hat{f}\hat{\rho})$, which is invariant if calculated for any complete set of states of the system. This allows us to express results in a manner independent of representation and also leads to approximation methods for problems that cannot be solved exactly.

26.1 Pure States

If a quantum mechanical system is in a pure time-dependent state $|\psi(t)\rangle$, the probability density of finding the system at coordinate[1] $\mathbf{r}$ is $|\psi(\mathbf{r},t)|^2$, where the wave function is assumed to be normalized, so

$$\langle \psi(t)|\psi(t)\rangle = \int \psi^*(\mathbf{r},t)\psi(\mathbf{r},t)\,\mathrm{d}\mathbf{r} = 1. \tag{26.1}$$

The expectation value of some operator $\hat{f}$ in a **pure state** is

$$\langle f\rangle = \langle \psi(t)|\hat{f}|\psi(t)\rangle = \int \psi^*(\mathbf{r},t)\hat{f}(\mathbf{r})\psi(\mathbf{r},t)\,\mathrm{d}\mathbf{r}, \tag{26.2}$$

where $\hat{f}(\mathbf{r})$ is the corresponding operator (in general, a differential operator in the Schrödinger representation).

[1] Here, the vector $\mathbf{r}$ denotes the coordinates of the entire system; for a system composed of $\mathcal{N}$ particles, $\mathbf{r}$ would have $3\mathcal{N}$ components. The function $\psi(\mathbf{r},t)$ is also assumed to carry information about nonclassical variables such as spin, but these variables are suppressed in the interest of simplicity.

Thermal Physics. http://dx.doi.org/10.1016/B978-0-12-803304-3.00026-0

An alternative expression for $\langle f \rangle$ can be obtained by employing a complete set of states $|f\rangle$ in which $\hat{f}$ is diagonal, that is, $\hat{f}|f\rangle = f|f\rangle$, where f is an eigenvalue. From closure, the unit operator $\hat{1}$ can be expressed in the form[2]

$$\hat{1} = \sum_f |f\rangle\langle f|. \tag{26.3}$$

Thus

$$\langle f \rangle = \langle \psi(t)|\hat{f}|\psi(t)\rangle = \sum_{f,f'} \langle \psi(t)|f'\rangle\langle f'|\hat{f}|f\rangle\langle f|\psi(t)\rangle = \sum_f |\langle f|\psi(t)\rangle|^2 f. \tag{26.4}$$

The quantities $\langle f|\psi(t)\rangle$ are just the expansion coefficients for $|\psi(t)\rangle$ in the basis $|f\rangle$, that is,

$$|\psi(t)\rangle = \sum_f |f\rangle\langle f|\psi(t)\rangle. \tag{26.5}$$

By rearranging terms, Eq. (26.4) may be rewritten in the form

$$\langle f \rangle = \sum_{f,f'} \langle f|\psi(t)\rangle\langle\psi(t)|f'\rangle\langle f'|\hat{f}|f\rangle = \sum_f \langle f|\hat{\rho}\hat{f}|f\rangle = \mathrm{tr}(\hat{\rho}\hat{f}), \tag{26.6}$$

where the Hermitian operator

$$\hat{\rho} := |\psi(t)\rangle\langle\psi(t)| \tag{26.7}$$

is the **density operator** for the pure state $|\psi(t)\rangle$. It is a **projection operator** onto the state $|\psi(t)\rangle$, so

$$\hat{\rho}\hat{\rho} = |\psi(t)\rangle\langle\psi(t)|\psi(t)\rangle\langle\psi(t)| = |\psi(t)\rangle\langle\psi(t)| = \hat{\rho}. \tag{26.8}$$

Since the trace of an operator is invariant if calculated in any representation, we can calculate it with respect to an arbitrary, complete set of states $|\phi_n\rangle$. Because of the cyclic properties of the trace, we have

$$\langle f \rangle = \mathrm{tr}(\hat{\rho}\hat{f}) = \mathrm{tr}(\hat{f}\hat{\rho}) = \sum_n \langle\phi_n|\hat{f}\hat{\rho}|\phi_n\rangle \tag{26.9}$$

or in matrix form

$$\langle f \rangle = \sum_{m,n} f_{nm}\rho_{mn}, \tag{26.10}$$

where $f_{nm} = \langle\phi_n|\hat{f}|\phi_m\rangle$ and $\rho_{mn} = \langle\phi_m|\psi(t)\rangle\langle\psi(t)|\phi_n\rangle = \rho^*_{nm}$. The quantities ρ_{mn} are the elements of the **density matrix** ρ, which is the matrix representation of the density operator, in this case for a pure state. By setting $\hat{f}$ equal to the unit operator, Eq. (26.9) shows that $\mathrm{tr}(\hat{\rho}) = 1$. Alternatively, Eq. (26.10) shows that $\sum_n \rho_{nn} = 1$.

[2]It is possible to have a continuous spectrum of states as well as discrete states in which case the closure relation requires both integration over the complete spectrum as well as summation over the continuous spectrum (see Dirac [67, p. 37]). Schiff [57, p. 156] uses the symbol S_f instead of a summation sign to indicate this process. For simplicity we use only the summation sign with the implicit understanding that one must also integrate over the continuous spectrum if relevant.

26.2 Statistical States

Suppose we have incomplete knowledge of a quantum system. Instead of being sure that the system is in some pure state $|\psi(t)\rangle$, all we know is that the system has a *probability* p_i of being in the pure state $|\psi_i(t)\rangle$, where $i = 1, 2, \ldots$. Such a state is called a **statistical state**, also known as a **mixed state**. For convenience, we take the set of states $|\psi_i(t)\rangle$ to be mutually orthonormal, although not necessarily complete. From the results of the preceding section, the average value in a statistical state of some observable represented by the operator $\hat{f}$ is therefore

$$\langle f \rangle = \sum_i p_i \, \langle \psi_i(t)|\hat{f}|\psi_i(t)\rangle. \tag{26.11}$$

Note that Eq. (26.11) involves an averaging process for each quantum state $|\psi_i(t)\rangle$ as well as a weighted average over the quantum states $i = 1, 2, \ldots$, each with probability p_i, that make up the statistical state. By employing any complete set of states ϕ_n for which the unit operator is $\sum_n |\phi_n\rangle\langle\phi_n|$, this average may be written

$$\langle f \rangle = \sum_{i,n} p_i \, \langle \psi_i(t)|\hat{f}|\phi_n\rangle\langle\phi_n|\psi_i(t)\rangle = \sum_{i,n} p_i \, \langle\phi_n|\psi_i(t)\rangle\langle\psi_i(t)|\hat{f}|\phi_n\rangle = \mathrm{tr}(\hat{\rho}^S \hat{f}), \tag{26.12}$$

where the Hermitian operator

$$\hat{\rho}^S = \sum_i |\psi_i(t)\rangle p_i \langle\psi_i(t)| \tag{26.13}$$

is the **density operator for a statistical state**. For the special case in which $|\phi_n\rangle$ are chosen to be eigenstates $|f\rangle$ of $\hat{f}$ with eigenvalues f, we obtain

$$\langle f \rangle = \sum_{i,f} p_i f |\langle f|\psi_i(t)\rangle|^2, \tag{26.14}$$

which illustrates that two averaging processes are involved. One is quantum mechanical averaging with weighting factors $|\langle f|\psi_i(t)\rangle|^2$ given by the squares of the wave function for the pure state i; the second is statistical averaging with probabilities p_i of that state.

Since the p_i are probabilities, we have $p_i \geq 0$ and $\sum_i p_i = 1$. Thus,

$$\mathrm{tr}(\hat{\rho}^S) = \sum_n \langle\phi_n| \sum_i |\psi_i(t)\rangle p_i \langle\psi_i(t)||\phi_n\rangle = \sum_i p_i \sum_n \langle\psi_i(t)|\phi_n\rangle\langle\phi_n|\psi_i(t)\rangle = 1. \tag{26.15}$$

Moreover,

$$(\hat{\rho}^S)^2 = \sum_i |\psi_i(t)\rangle p_i \langle\psi_i(t)| \sum_j |\psi_j(t)\rangle p_j \langle\psi_j(t)| = \sum_i |\psi_i(t)\rangle p_i^2 \langle\psi_i(t)|. \tag{26.16}$$

Since Eq. (26.8) holds for a pure state, Eq. (26.16) show that $\hat{\rho}^S$ represents a pure state only in the special case when one of the p_i is equal to unity and the rest are zero. In general,

$$\mathrm{tr}((\hat{\rho}^S)^2) = \sum_i p_i^2 \leq 1, \tag{26.17}$$

with the equality holding only for a pure state. For an arbitrary basis $|\phi_n\rangle$ we would have

$$\langle f\rangle = \mathrm{tr}(\hat{\rho}^S\hat{f}) = \mathrm{tr}(\hat{f}\hat{\rho}^S) = \sum_n \langle\phi_n|\hat{f}\hat{\rho}^S|\phi_n\rangle = \sum_{m,n} f_{nm}\rho^S_{mn}, \tag{26.18}$$

where $\rho^S_{mn} = \sum_i \langle\phi_m|\psi_i(t)\rangle p_i\langle\psi_i(t)|\phi_n\rangle = (\rho^S_{nm})^*$.

26.3 Random Phases and External Influence

A statistical density operator of the form of Eq. (26.13) can be rationalized in several ways. We discuss two possibilities, for which the author is grateful for private discussions with R.B. Griffiths.

The first rationalization is based on the assumption of random phases, for example, see [8, p. 109] and [68, chapter 9]. First, consider a normalized pure state of the form

$$|\Psi^\alpha(t)\rangle = \sum_j \sqrt{p_j}\exp(i\alpha_j)|\psi_j(t)\rangle, \tag{26.19}$$

where the α_j are a set of phases. Normalization requires $\sum_i p_i = 1$. The projection operator for such a state is

$$|\Psi^\alpha(t)\rangle\langle\Psi^\alpha(t)| = \sum_{j,k}\sqrt{p_jp_k}\exp[i(\alpha_j-\alpha_k)]|\psi_j(t)\rangle\langle\psi_k(t)|. \tag{26.20}$$

A density operator of the form of Eq. (26.13) can be obtained by averaging over the phases α_j with the assumption that the phases corresponding to different values of j are random. Explicitly,

$$\begin{aligned}\overline{|\Psi^\alpha(t)\rangle\langle\Psi^\alpha(t)|} &= \sum_{j,k}\sqrt{p_jp_k}\,\overline{\exp[i(\alpha_j-\alpha_k)]}\,|\psi_j(t)\rangle\langle\psi_k(t)| \\ &= \sum_{j,k}\sqrt{p_jp_k}\,\delta_{jk}\,|\psi_j(t)\rangle\langle\psi_k(t)| = \sum_j p_j|\psi_j(t)\rangle\langle\psi_j(t)|.\end{aligned} \tag{26.21}$$

The second rationalization is based on a description of both the system and its environment. We can represent the total normalized wave function of a system by a state of the form

$$|\Psi^\epsilon(t)\rangle = \sum_j |\epsilon_j\rangle\otimes\exp(i\alpha_j)\sqrt{p_j}\,|\psi_j(t)\rangle, \tag{26.22}$$

where $\otimes$ represents the outer product of the subspace spanned by an orthonormal set (not necessarily complete) of external states $|\epsilon_j\rangle$ and the states $|\psi_j(t)\rangle$ of the system. The corresponding (total) projection operator is

$$|\Psi^\epsilon(t)\rangle\langle\Psi^\epsilon(t)| = \sum_{j,k}|\epsilon_j\rangle\langle\epsilon_k|\otimes\exp[i(\alpha_j-\alpha_k)]\sqrt{p_jp_k}\,|\psi_j(t)\rangle\langle\psi_k(t)|. \tag{26.23}$$

A density operator for the system of interest of the form of Eq. (26.13) can be obtained by taking the expectation value, and hence the trace, of this total projection operator with respect to any complete set of orthonormal *external* states $|\phi_\epsilon\rangle$, resulting in

$$\begin{aligned}\mathrm{tr}_\epsilon(|\Psi^\epsilon(t)\rangle\langle\Psi^\epsilon(t)|) &= \sum_{j,k}\sum_{\phi_\epsilon}\langle\phi_\epsilon|\epsilon_j\rangle\langle\epsilon_k|\phi_\epsilon\rangle\exp[i(\alpha_j-\alpha_k)]\sqrt{p_j p_k}\,|\psi_j(t)\rangle\langle\phi_k(t)| \\ &= \sum_{j,k}\langle\epsilon_k|\epsilon_j\rangle\exp[i(\alpha_j-\alpha_k)]\sqrt{p_j p_k}\,|\psi_j(t)\rangle\langle\phi_k(t)| \\ &= \sum_j p_j|\psi_j(t)\rangle\langle\phi_j(t)|. \end{aligned} \tag{26.24}$$

Either of these rationalizations demonstrates that the statistical operator describes a quantum mechanical system for which there is incomplete information. The phases of the associated quantum states are unknown but one can still average over those quantum states by knowledge of their probabilities.

26.4 Time Evolution

One can calculate the time evolution of the statistical density operator by recognizing that the probabilities p_i are independent of time and making use of the evolution equations for the states $|\psi_i(t)\rangle$ which evolve according to

$$i\hbar\frac{\mathrm{d}}{\mathrm{d}t}|\psi_i(t)\rangle = \hat{\mathcal{H}}|\psi_i(t)\rangle \tag{26.25}$$

and its Hermitian conjugate

$$-i\hbar\frac{\mathrm{d}}{\mathrm{d}t}\langle\psi_i(t)| = \langle\psi_i(t)|\hat{\mathcal{H}}, \tag{26.26}$$

where $\hat{\mathcal{H}}$ is the Hamiltonian operator (which, of course, is Hermitian). Thus

$$\begin{aligned} i\hbar\frac{\mathrm{d}}{\mathrm{d}t}\hat{\rho}^S &= \frac{\mathrm{d}}{\mathrm{d}t}\sum_i|\psi_i(t)\rangle p_i\langle\psi_i(t)| \\ &= \sum_i\left[\left(i\hbar\frac{\mathrm{d}}{\mathrm{d}t}|\psi_i(t)\rangle\right)p_i\langle\psi_i(t)| + |\psi_i(t)\rangle p_i\left(i\hbar\frac{\mathrm{d}}{\mathrm{d}t}\langle\psi_i(t)|\right)\right] \\ &= \sum_i\left[\hat{\mathcal{H}}|\psi_i(t)\rangle p_i\langle\psi_i(t)| - |\psi_i(t)\rangle p_i\langle\psi_i(t)|\hat{\mathcal{H}}\right]. \end{aligned} \tag{26.27}$$

Thus[3]

$$i\hbar\frac{\mathrm{d}}{\mathrm{d}t}\hat{\rho}^S = \hat{\mathcal{H}}\hat{\rho}^S - \hat{\rho}^S\hat{\mathcal{H}} \equiv \left[\hat{\mathcal{H}},\hat{\rho}^S\right], \tag{26.28}$$

where the latter expression is a commutator. Equation (26.28) also applies to the density operator for a pure state $|\psi_i(t)\rangle$ if $p_i = 1$ and $p_j = 0$ for $j \neq i$.

[3]Here, the operator $\hat{\rho}^S$ is in the Schrödinger representation. The result in Eq. (26.28) should not be confused with the time derivative of an operator in the Heisenberg representation, which contains a commutator with opposite sign.

If $\hat{\rho}^S$ is the *statistical operator for an equilibrium state,* we need $\mathrm{d}\hat{\rho}^S/\mathrm{d}t = 0$ in which case

$$\left[\hat{\mathcal{H}}, \hat{\rho}^S\right] = 0, \tag{26.29}$$

that is, $\hat{\rho}^S$ commutes with the Hamiltonian. Equation (26.28) is the quantum mechanical analog of the classical Liouville equation (Eq. (17.9)) for an equilibrium ensemble for which the density in phase space has no explicit dependence on time, $\partial\rho/\partial t = 0$. The lack of explicit time dependence of the classical density function ρ is the counterpart of the fact that the probabilities p_i are independent of time. The classical analog to Eq. (26.29) is Eq. (17.11), namely the vanishing of the Poisson bracket $\{\rho, \mathcal{H}\}$.

The time derivative of the average value $\langle f\rangle$ of some observable may be computed in a similar way as follows:[4]

$$\frac{\mathrm{d}}{\mathrm{d}t}\langle f\rangle = \frac{\mathrm{d}}{\mathrm{d}t}\mathrm{tr}(\hat{\rho}^S\hat{f}) = \mathrm{tr}\left(\frac{\mathrm{d}\hat{\rho}^S}{\mathrm{d}t}\hat{f}\right) + \mathrm{tr}\left(\hat{\rho}^S\frac{\partial\hat{f}}{\partial t}\right) = \frac{1}{i\hbar}\mathrm{tr}\left(\left[\hat{\mathcal{H}}, \hat{\rho}^S\right]\hat{f}\right) + \mathrm{tr}\left(\hat{\rho}^S\frac{\partial\hat{f}}{\partial t}\right). \tag{26.30}$$

If the observable is explicitly independent of time, $\partial\hat{f}/\partial t = 0$, and for an equilibrium state Eq. (26.29) applies, so $\mathrm{d}\langle f\rangle/\mathrm{d}t = 0$, as expected.

26.5 Density Operators for Specific Ensembles

In this section, we present the statistical density operators for the three main ensembles, microcanonical, canonical, and grand canonical, employed in statistical thermodynamics. These ensembles pertain to equilibrium states, so Eq. (26.29) applies and can be satisfied by choosing $\hat{\rho}^S$ to be a function of a Hamiltonian $\hat{\mathcal{H}}$ that is independent of time. $\hat{\rho}^S$ can therefore be expressed in terms of a set of probabilities and the *stationary* eigenstates $|E_n\rangle$ of $\hat{\mathcal{H}}$. It is for this reason that we only had to deal with the stationary eigenstates of $\hat{\mathcal{H}}$ in our previous description of statistical mechanics, beginning with the microcanonical ensemble.

For brevity of notation we drop the superscript S in the rest of this section, but bear in mind that we are dealing with a *statistical operator for a system in equilibrium.* The results can therefore be expressed easily in the energy representation where the matrix representations of $\hat{\mathcal{H}}$, and therefore also $\hat{\rho}(\hat{\mathcal{H}})$, are diagonal. Specifically, we employ a complete set of orthonormal stationary eigenstates $|E_n\rangle$ that satisfy $\hat{\mathcal{H}}|E_n\rangle = E_n|E_n\rangle$. Note especially that n labels states, not energies, so there can be many values of n for a given energy in the case of degeneracy. For the case of the grand canonical ensemble, we will employ states that are also eigenstates of the number operator $\hat{\mathcal{N}}$. See Appendix I for more information about number operators.

[4]The operator $\hat{f}$ is in the Schrödinger representation so its only dependence on time is explicit; we therefore use a partial derivative for its time rate of change.

26.5.1 Microcanonical Ensemble

The microcanonical ensemble applies in principle to an isolated system having constant total energy E. We recognize, however, that a truly isolated system is an impossibility because there will always be some interaction of a system with its environment, even if ever so slight. Because of the uncertainty relation $\Delta E \sim \hbar/\Delta t$, a constant energy would require isolation for an infinite time. Therefore, we actually treat a quasi-isolated system (see [66, p. 14]) for which the energy lies in a very narrow range $E - \Delta E$ to E. Within this range, the number of quantum states of the system is represented by Ω, and each is assumed to be equally probable. Then the density operator has the form

$$\hat{\rho} = \sum_n |E_n\rangle p_n \langle E_n| = \sum_{n=1}^{\Omega} |E_n\rangle \frac{1}{\Omega} \langle E_n|; \quad p_n = \begin{cases} 1/\Omega & \text{for } E - \Delta E \le E_n \le E. \\ 0 & \text{otherwise.} \end{cases} \tag{26.31}$$

The entropy is given by $S = k_B \ln \Omega$. In terms of $\hat{\rho}$, it can be calculated from the formula

$$S = -k_B \mathrm{tr}(\hat{\rho} \ln \hat{\rho}), \tag{26.32}$$

where the function $\ln \hat{\rho}$ is to be understood as the operator whose eigenvalues, in a representation where $\hat{\rho}$ is diagonal, are equal to the logarithm of the eigenvalues of $\hat{\rho}$. The quantity $-\mathrm{tr}(\hat{\rho} \ln \hat{\rho})$ in Eq. (26.32) is just the expectation value of $-\ln \hat{\rho}$ in the statistical state represented by $\hat{\rho}$; in a representation where $\hat{\rho}$ can be represented by a diagonal matrix with diagonal elements P_n, Eq. (26.32) gives the familiar result $S = -k_B \sum_n P_n \ln P_n$. For the microcanonical ensemble we can evaluate the trace in an arbitrary, complete set of states $|\phi_m\rangle$ to obtain

$$\begin{aligned} -\mathrm{tr}(\hat{\rho} \ln \hat{\rho}) &= -\sum_m \langle \phi_m| \sum_{n=1}^{\Omega} |E_n\rangle \frac{\ln(1/\Omega)}{\Omega} \langle E_n||\phi_m\rangle \\ &= \sum_{n=1}^{\Omega} \sum_m \langle E_n|\phi_m\rangle\langle\phi_m|E_n\rangle \frac{\ln \Omega}{\Omega} = \sum_{n=1}^{\Omega} \langle E_n|E_n\rangle \frac{\ln \Omega}{\Omega} = \ln \Omega. \end{aligned} \tag{26.33}$$

26.5.2 Canonical Ensemble

The canonical ensemble pertains to a system in contact with a heat reservoir that maintains the system at temperature T. The corresponding probabilities in the energy representation are just $P_n = \exp(-\beta E_n)/Z$, where $\beta = 1/(k_B T)$ and $Z = \sum_m \exp(-\beta E_m)$ is the canonical partition function. Thus we can write the density operator in the form

$$\hat{\rho} = \sum_n |E_n\rangle \frac{\exp(-\beta E_n)}{Z} \langle E_n| = \frac{\exp(-\beta \hat{\mathcal{H}})}{Z} = \frac{\exp(-\beta \hat{\mathcal{H}})}{\mathrm{tr}\left[\exp(-\beta \hat{\mathcal{H}})\right]}. \tag{26.34}$$

In this case, the sum is over all energy states, a complete set. From the last form of Eq. (26.34), it is obvious that $\mathrm{tr}\, \hat{\rho} = 1$. In this case, Eq. (26.32) leads to the familiar formula

$$S/k_B = -\mathrm{tr}(\hat{\rho}\ln\hat{\rho}) = -\sum_m \langle\phi_m|\sum_n |E_n\rangle P_n \ln P_n \langle E_n||\phi_m\rangle = -\sum_n P_n \ln P_n, \tag{26.35}$$

where P_n are the probabilities of occupation of the states. Of course the expectation value of the energy itself is the internal energy

$$U = \langle\hat{\mathcal{H}}\rangle = \mathrm{tr}(\hat{\rho}\hat{\mathcal{H}}) = \frac{\mathrm{tr}\left[\hat{\mathcal{H}}\exp(-\beta\hat{\mathcal{H}})\right]}{\mathrm{tr}\left[\exp(-\beta\hat{\mathcal{H}})\right]}. \tag{26.36}$$

If the eigenvalues of $\hat{\mathcal{H}}$ cannot be calculated, the last expression in Eq. (26.36) can be calculated, at least approximately, in any convenient representation. If the eigenvalues are known, then we retrieve the familiar result

$$U = \frac{\sum_n E_n \exp(-\beta E_n)}{\sum_m \exp(-\beta E_m)}. \tag{26.37}$$

26.5.3 Grand Canonical Ensemble

Based on the considerations of Chapter 21, the density operator in the grand canonical ensemble will be diagonal in a set of states that are simultaneous eigenfunctions of the number operator $\hat{\mathcal{N}}$ and the Hamiltonian operator $\hat{\mathcal{H}}$ for a system having $\mathcal{N}$ particles. Such states $|\mathcal{N}_s\mathcal{E}_{rs}\rangle$ satisfy

$$\hat{\mathcal{H}}|\mathcal{N}_s\mathcal{E}_{rs}\rangle = \mathcal{E}_{rs}|\mathcal{N}_s\mathcal{E}_{rs}\rangle; \quad \hat{\mathcal{N}}|\mathcal{N}_s\mathcal{E}_{rs}\rangle = \mathcal{N}_s|\mathcal{N}_s\mathcal{E}_{rs}\rangle. \tag{26.38}$$

Thus with P_{rs} being the probability of the state $|\mathcal{N}_s\mathcal{E}_{rs}\rangle$, we have

$$\begin{aligned}\hat{\rho} &= \sum_{r,s} |\mathcal{N}_s\mathcal{E}_{rs}\rangle P_{rs}\langle\mathcal{N}_s\mathcal{E}_{rs}| = \sum_{r,s} |\mathcal{N}_s\mathcal{E}_{rs}\rangle \frac{\exp[-\beta(\mathcal{E}_{rs}-\mu\mathcal{N}_s)]}{\mathcal{Z}}\langle\mathcal{N}_s\mathcal{E}_{rs}| \\ &= \frac{\exp[-\beta(\hat{\mathcal{H}}-\mu\hat{\mathcal{N}})]}{\mathcal{Z}} = \frac{\exp[-\beta(\hat{\mathcal{H}}-\mu\hat{\mathcal{N}})]}{\mathrm{tr}\left[\exp[-\beta(\hat{\mathcal{H}}-\mu\hat{\mathcal{N}})]\right]},\end{aligned} \tag{26.39}$$

where the grand partition function (see Eq. (21.21))

$$\mathcal{Z} = \sum_s \exp(\beta\mu\mathcal{N}_s)\sum_r \exp(-\beta\mathcal{E}_{rs}) = \sum_s \lambda^{\mathcal{N}_s}\sum_r \exp(-\beta\mathcal{E}_{rs}). \tag{26.40}$$

Here, $\lambda = \exp(\beta\mu)$ is the absolute activity. The expectation value of some observable having operator $\hat{f}$ is therefore

$$\langle f\rangle = (1/\mathcal{Z})\mathrm{tr}\left[\hat{f}\exp[-\beta(\hat{\mathcal{H}}-\mu\hat{\mathcal{N}})]\right] = \frac{\sum_{\mathcal{N}}\lambda^{\mathcal{N}}\langle f\rangle_{\mathcal{N}} Z_{\mathcal{N}}}{\sum_{\mathcal{N}}\lambda^{\mathcal{N}} Z_{\mathcal{N}}}, \tag{26.41}$$

where $Z_{\mathcal{N}}$ is the canonical partition function for a system of $\mathcal{N}$ particles and $\langle f\rangle_{\mathcal{N}}$ is the canonical average of $\hat{f}$ for that system. From Eq. (26.32), the entropy is just $S = -k_B\sum_{r,s} P_{rs}\ln P_{rs}$ as expected.

26.6 Examples of the Density Matrix

For the canonical ensemble, we calculate the matrix elements of the equilibrium statistical density operator for several simple systems. First, we treat a free spinless particle in a box, then a one-dimensional harmonic oscillator, and finally a spin 1/2 particle.

26.6.1 Single Free Particle

We consider a single free particle in a cubical box of dimension L and volume $V = L^3$ with periodic boundary conditions. The wave function is

$$\psi_{\mathbf{k}}(\mathbf{r}) = V^{-1/2} \exp(i\mathbf{k} \cdot \mathbf{r}), \tag{26.42}$$

which satisfies

$$\hat{\mathcal{H}}\psi_{\mathbf{k}} = \varepsilon_{\mathbf{k}}\psi_{\mathbf{k}} \tag{26.43}$$

with $\varepsilon_{\mathbf{k}} = \hbar^2 k^2/(2m)$. Here, $\hat{\mathcal{H}} = \hat{p}^2/2m$, where $\hat{p} = (\hbar/i)\nabla$ is the momentum operator. For periodic boundary conditions, $\psi_{\mathbf{k}}$ is also an eigenfunction of the momentum operator with eigenvalue $\hbar\mathbf{k}$, so we can label the energy eigenstates by $\mathbf{k}$ as well as $\varepsilon_{\mathbf{k}}$. The allowed values of $\mathbf{k}$ satisfy $k_\alpha = 2n_\alpha\pi/L$, where $\alpha = x, y, z$ and n_α are integers (positive, negative, and zero). In terms of the eigenstates $|\varepsilon_{\mathbf{k}}\rangle$, for which $\psi_{\mathbf{k}}(\mathbf{r}) = \langle \mathbf{r}|\varepsilon_{\mathbf{k}}\rangle$, the relevant matrix elements are

$$\langle \varepsilon_{\mathbf{k}}| \exp(-\beta\hat{\mathcal{H}})|\varepsilon_{\mathbf{k}'}\rangle = \exp(-\beta\varepsilon_{\mathbf{k}})\, \delta_{\mathbf{k}\mathbf{k}'}. \tag{26.44}$$

Thus

$$\begin{aligned} \operatorname{tr}(\exp(-\beta\hat{\mathcal{H}})) &= \sum_{\mathbf{k}} \exp(-\beta\varepsilon_{\mathbf{k}}) \approx \frac{V}{(2\pi)^3} \int \mathrm{d}^3k \, \exp[-\beta\hbar^2 k^2/(2m)] \\ &= \frac{V}{(2\pi)^3} \left[\left(\frac{2m}{\beta\hbar^2}\right)^{1/2} \int_{-\infty}^{\infty} \mathrm{d}x \, \exp(-x^2) \right]^3 = V/\lambda_{\mathrm{T}}^3. \end{aligned} \tag{26.45}$$

Here, λ_{T} is the thermal wavelength given by

$$\frac{1}{\lambda_{\mathrm{T}}^3} = \left(\frac{m}{2\pi\hbar^2\beta}\right)^{3/2} = n_{\mathrm{Q}}, \tag{26.46}$$

where n_{Q} is the quantum concentration. Thus, in the energy representation, the density operator for a single free particle is represented by the diagonal matrix

$$\langle \varepsilon_{\mathbf{k}}|\hat{\rho}|\varepsilon_{\mathbf{k}'}\rangle = (\lambda_{\mathrm{T}}^3/V) \exp(-\beta\varepsilon_{\mathbf{k}})\delta_{\mathbf{k}\mathbf{k}'}. \tag{26.47}$$

Since these energy eigenstates are also eigenstates of the momentum operator, Eq. (26.47) could also be written

$$\langle \mathbf{k}|\hat{\rho}|\mathbf{k}'\rangle = (\lambda_{\mathrm{T}}^3/V) \exp[-\beta\hbar^2 k^2/(2m)]\delta_{\mathbf{k}\mathbf{k}'}. \tag{26.48}$$

We proceed to calculate the matrix elements of $\hat{\rho}$ in the coordinate representation $|\mathbf{r}\rangle$ where it is not diagonal. Thus

$$\langle \mathbf{r}|\hat{\rho}|\mathbf{r}'\rangle = \sum_{\mathbf{k}\mathbf{k}'} \langle \mathbf{r}|\mathbf{k}\rangle\langle \mathbf{k}|\hat{\rho}|\mathbf{k}'\rangle\langle \mathbf{k}'|\mathbf{r}'\rangle = \frac{\lambda_T^3}{V^2}\sum_{\mathbf{k}} \exp[-\beta\hbar^2k^2/(2m)]\exp[i\mathbf{k}\cdot(\mathbf{r}-\mathbf{r}')]$$

$$\approx \frac{\lambda_T^3}{V^2}\frac{V}{(2\pi)^3}\int d^3k \exp[-\beta\hbar^2k^2/(2m)]\exp[i\mathbf{k}\cdot(\mathbf{r}-\mathbf{r}')] \tag{26.49}$$

$$= \frac{\lambda_T^3}{(2\pi)^3 V}\exp[-m|\mathbf{r}-\mathbf{r}'|^2/(2\beta\hbar^2)]\int d^3k \exp[-(\beta\hbar^2/2m)|\mathbf{k}-i(m/\beta\hbar^2)(\mathbf{r}-\mathbf{r}')|^2]$$

$$= \frac{\lambda_T^3}{(2\pi)^3 V}\exp[-m|\mathbf{r}-\mathbf{r}'|^2/(2\beta\hbar^2)]\left(\frac{2\pi m}{\beta\hbar^2}\right)^{3/2}, \tag{26.50}$$

where we have completed the square in the argument of the exponential to get a Gaussian integral. In terms of λ_T, this result can be written simply as

$$\langle \mathbf{r}|\hat{\rho}|\mathbf{r}'\rangle = \frac{1}{V}\exp[-\pi\,(|\mathbf{r}-\mathbf{r}'|/\lambda_T)^2]. \tag{26.51}$$

The diagonal element$\langle \mathbf{r}|\hat{\rho}|\mathbf{r}\rangle = 1/V$ is independent of $\mathbf{r}$ and shows that there is a uniform probability density of finding the particle anywhere in the box, as would be expected for periodic boundary conditions. Of course $\mathrm{tr}\hat{\rho} = \int_V \langle \mathbf{r}|\hat{\rho}|\mathbf{r}\rangle\, d^3r = V/V = 1$.

One could also treat this problem with boundary conditions for which the wave function vanishes on the sides of the box. In that case, $\psi_{\varepsilon_\mathbf{k}} = (8/V)^{1/2}\sin(k_x x)\sin(k_y y)\sin(k_z z)$, with $\mathbf{k}$ now given by Eq. (16.52). In that case, $\varepsilon_\mathbf{k} = \hbar^2k^2/2m$, but $\psi_{\varepsilon_\mathbf{k}}$ is no longer an eigenfunction of the momentum operator $\hat{\mathbf{p}}$. As expected, $\langle \mathbf{r}|\hat{\rho}|\mathbf{r}\rangle$ goes to zero on the sides of the box and increases to a maximum at the center of the box. For λ_T much smaller than any edge length of the box, $\langle \mathbf{r}|\hat{\rho}|\mathbf{r}\rangle \approx 1/V$ except within a distance of order λ_T near the walls of the box.

26.6.2 One-Dimensional Harmonic Oscillator

For a harmonic oscillator in one dimension, x, the energies are given by $\varepsilon_n = \hbar\omega(n+1/2)$ and the partition function $z = \exp(-\beta\hbar\omega/2)/[1-\exp(-\beta\hbar\omega)]$. Thus the probabilities are

$$p_n = \exp(-\beta\varepsilon_n)/z = \exp(-n\beta\hbar\omega)[1-\exp(\beta\hbar\omega)], \tag{26.52}$$

independent of the zero point energy. The density operator is therefore

$$\hat{\rho} = \sum_{n=0}^{\infty} |n\rangle p_n\langle n|. \tag{26.53}$$

The expectation value of x^2 is given by

$$\langle x^2\rangle = \mathrm{tr}(\hat{\rho}\hat{x}^2) = \sum_{n=0}^{\infty} p_n\langle n|\hat{x}^2|n\rangle. \tag{26.54}$$

In Appendix I, it is shown by Eq. (I.10) that the operator $\hat{x}^2$ can be expressed in terms of the raising and lowering operators $a^\dagger$ and a, resulting in

$$\hat{x}^2 = \frac{\hbar}{m\omega}\left(a^\dagger a + \frac{1}{2} + \frac{aa + a^\dagger a^\dagger}{2}\right). \tag{26.55}$$

The operators aa and $a^\dagger a^\dagger$ have no diagonal elements, so $\langle n|\hat{x}^2|n\rangle = \hbar(n + 1/2)/(m\omega)$. Therefore,

$$\langle x^2\rangle = \sum_{n=0}^{\infty} p_n \hbar(n+1/2)/(m\omega) = \sum_{n=0}^{\infty} p_n \varepsilon_n/(m\omega^2) = \langle H\rangle/(m\omega^2). \tag{26.56}$$

Here, $\langle H\rangle = \hbar\omega[1/2 + (\exp(\beta\hbar\omega) - 1)^{-1}]$ is the average energy. Thus the average potential energy is $(1/2)m\omega^2\langle x^2\rangle = (1/2)\langle H\rangle$ just as for the time average of the potential energy of a classical harmonic oscillator. The average of the kinetic energy is therefore $\langle H\rangle - (1/2)\langle H\rangle = (1/2)\langle H\rangle$, the same as the time average of the kinetic energy of a classical harmonic oscillator.

See Pathria [8, pp. 113-115] for a representation of this density matrix in the x representation, where it is also shown that $\langle x|\hat{\rho}|x\rangle$ follows a Gaussian distribution.

26.6.3 Spin 1/2 Particle

In the previous examples, we have not included spin, so we proceed here to treat electrons having spin 1/2. We will begin by treating a pure state and then proceed to discuss a statistical state. We represent the pure state by

$$|\chi\rangle = c_1|\alpha\rangle + c_2|\beta\rangle, \tag{26.57}$$

where, for some arbitrary z-axis, $|\alpha\rangle$ corresponds to spin up, $|\beta\rangle$ corresponds to spin down, and c_1 and c_2 are complex numbers. $|\chi\rangle$ is assumed to be normalized, so $|c_1|^2 + |c_2|^2 = 1$. The density operator is the projection operator

$$\hat{\rho} = |\chi\rangle\langle\chi| = |c_1|^2|\alpha\rangle\langle\alpha| + c_1c_2^*|\alpha\rangle\langle\beta| + c_1^*c_2|\beta\rangle\langle\alpha| + |c_2|^2|\beta\rangle\langle\beta|. \tag{26.58}$$

In a matrix notation, Eq. (26.57) can be written as a **spinor**

$$\chi = \begin{pmatrix} c_1 \\ c_2 \end{pmatrix} = c_1\begin{pmatrix} 1 \\ 0 \end{pmatrix} + c_2\begin{pmatrix} 0 \\ 1 \end{pmatrix} \equiv c_1\alpha + c_2\beta, \tag{26.59}$$

so the density matrix corresponding to Eq. (26.58) is

$$\rho = \chi\chi^\dagger = \begin{pmatrix} c_1 \\ c_2 \end{pmatrix}(c_1^* \; c_2^*) = \begin{pmatrix} |c_1|^2 & c_1c_2^* \\ c_1^*c_2 & |c_2|^2 \end{pmatrix}, \tag{26.60}$$

which is Hermitian.

It is usual to express ρ in terms of the **Pauli spin matrices**

$$\sigma_x = \begin{pmatrix} 0 & 1 \\ 1 & 0 \end{pmatrix}; \quad \sigma_y = \begin{pmatrix} 0 & -i \\ i & 0 \end{pmatrix}; \quad \sigma_z = \begin{pmatrix} 1 & 0 \\ 0 & -1 \end{pmatrix}. \tag{26.61}$$

These have the properties

$$\sigma_x\sigma_y = -\sigma_y\sigma_x = i\sigma_z; \quad \sigma_y\sigma_z = -\sigma_z\sigma_y = i\sigma_x; \quad \sigma_z\sigma_x = -\sigma_x\sigma_z = i\sigma_y;$$

$$\sigma_x^2 = \sigma_y^2 = \sigma_z^2 = E \equiv \begin{pmatrix} 1 & 0 \\ 0 & 1 \end{pmatrix}; \quad \text{tr}\,\sigma_x = \text{tr}\,\sigma_y = \text{tr}\,\sigma_z = 0. \tag{26.62}$$

The Pauli spin matrices can be related to fermion operators that *anticommute*, as developed in Section I.3 of Appendix I.

It is useful to define a quantity $\boldsymbol{\sigma}$, whose x, y, and z components are the matrices σ_x, σ_y, and σ_z. By studying the transformation of χ under rotations, it can be shown that the expectation value of σ_x transforms like a vector [69, pp. 261-270]. In this sense, $\boldsymbol{\sigma}$ is the matrix representation of a vector operator. Moreover, we recognize that σ_x, σ_y, σ_z, and E constitute a linearly independent set in terms of which ρ can be expanded. This results in

$$\rho = (1/2)[E + \mathbf{P}\cdot\boldsymbol{\sigma}], \tag{26.63}$$

where

$$P_x = c_1^* c_2 + c_1 c_2^* = 2\Re(c_1^* c_2), \quad P_y = (c_1^* c_2 - c_1 c_2^*)/i = 2\Im(c_1^* c_2) \quad P_z = |c_1|^2 - |c_2|^2, \tag{26.64}$$

which may be verified as follows. First, take the trace of Eq. (26.63) and recognize that $\text{tr}\,\rho = 1$, $\text{tr}\,E = 2$ and $\text{tr}\,\boldsymbol{\sigma} = 0$, which verifies the term $(1/2)E$. Then multiply Eq. (26.63) by σ_x and take the trace to obtain

$$\langle\sigma_x\rangle = \text{tr}\,(\sigma_x\rho) = (1/2)P_x\,\text{tr}\,(\sigma_x^2) = P_x, \tag{26.65}$$

where Eq. (26.62) has been used. By using the explicit form Eq. (26.60) for ρ, we can compute $\sigma_x\rho$ and take its trace, thus verifying P_x in Eq. (26.64). Proceeding in a similar way for the y and z components, we verify the expressions for P_y and P_z and obtain the result

$$\langle\boldsymbol{\sigma}\rangle = \mathbf{P}. \tag{26.66}$$

It also turns out that $\mathbf{P}$ is a unit vector, which is known as the **polarization vector** for the pure state χ under consideration. This can be seen readily by writing $c_1 = |c_1|e^{i\gamma_1}$ and $c_2 = |c_2|e^{i\gamma_2}$ in which case $c_1^* c_2 = |c_1||c_2|e^{i\gamma}$, where $\gamma = \gamma_2 - \gamma_1$. Then $P_x = 2|c_1||c_2|\cos\gamma$ and $P_y = 2|c_1||c_2|\sin\gamma$. Thus

$$P_x^2 + P_y^2 + P_z^2 = 4|c_1|^2|c_2|^2 + (|c_1|^2 - |c_2|^2)^2 = (|c_1|^2 + |c_2|^2)^2 = 1. \tag{26.67}$$

We are now in a position to relate to the magnetic moment of an electron which has spin (1/2). We associate the "spin up" state α with the spin component (1/2) and the "spin down" state β with the spin component $-(1/2)$. For simplicity, we approximate the g-factor for spin (approximately 2.0023) by 2, so the magnetic moment for spin up would be $-2\mu_B(1/2) = -\mu_B$, where $\mu_B = e\hbar/(2mc) > 0$ is the Bohr magneton and the minus sign results from the negative charge of the electron. The magnetic moment operator in matrix notation is therefore

$$\boldsymbol{\mu} = -\mu_{\rm B}\boldsymbol{\sigma} \tag{26.68}$$

and the Hamiltonian is[5]

$$\mathcal{H} = -\boldsymbol{\mu} \cdot \mathbf{B} = \mu_{\rm B}\boldsymbol{\sigma}\cdot\mathbf{B}. \tag{26.69}$$

More insight can be gained by noting that ρ given by Eq. (26.60) has eigenvalues $\lambda = 1$ and $\lambda = 0$. Moreover, χ is a normalized eigenvector of ρ corresponding to $\lambda = 1$, unique except for an overall phase factor. An eigenvector of ρ that corresponds to $\lambda = 0$ must be perpendicular to χ and can be taken to be $\chi_\perp = c_2^*\alpha - c_1^*\beta$ which is also normalized. Operating on χ given by Eq. (26.59) yields

$$\rho\chi = \chi = (1/2)[\chi + \mathbf{P}\cdot\boldsymbol{\sigma}\,\chi] \tag{26.70}$$

from which we deduce that $\mathbf{P}\cdot\boldsymbol{\sigma}\,\chi = \chi$, so χ is also an eigenvector of $\mathbf{P}\cdot\boldsymbol{\sigma}$ with eigenvector 1. Similarly, operating on $\chi_\perp$ gives

$$\rho\chi_\perp = 0 = (1/2)[\chi_\perp + \mathbf{P}\cdot\boldsymbol{\sigma}\,\chi_\perp], \tag{26.71}$$

so $\mathbf{P}\cdot\boldsymbol{\sigma}\,\chi_\perp = -\chi_\perp$. This is equivalent to saying that $\chi_\perp$ is an eigenstate of the operator $-\mathbf{P}\cdot\boldsymbol{\sigma}$ with eigenvalue 1. Therefore, for an axis along $\mathbf{P}$, χ corresponds to the "spin up" state and $\chi_\perp$ corresponds to the "spin down" state. This shows that the operator $\mathbf{P}\cdot\boldsymbol{\sigma}$ is the spin operator for the direction $\mathbf{P}$ that corresponds to σ_z for our original but arbitrary choice of the orientation of the z-axis, consistent with Eq. (26.68) for the magnetic moment operator.

For a statistical state, all we know are the probabilities p_α of being in the eigenstate $|\alpha\rangle$ and p_β of being in the eigenstate $|\beta\rangle$. The density operator is

$$\hat{\rho}^S = p_\alpha|\alpha\rangle\langle\alpha| + p_\beta|\beta\rangle\langle\beta|, \tag{26.72}$$

where $|\alpha\rangle\langle\alpha|$ and $|\beta\rangle\langle\beta|$ are projection operators for the respective states. The corresponding density matrix

$$\rho^S = \begin{pmatrix} p_\alpha & 0 \\ 0 & p_\beta \end{pmatrix} \tag{26.73}$$

is diagonal with elements equal to p_α and p_β. We can rewrite Eq. (26.73) in the form

$$\rho^S = (1/2)[E + (p_\alpha - p_\beta)\sigma_z], \tag{26.74}$$

which is quite different from ρ for a pure state given by Eq. (26.60) for the case in which $|c_1|^2 = p_\alpha$ and $|c_2|^2 = p_\beta$. For the statistical state and the pure state, the expectation value of α will be p_α and that for β will be p_β. This is easily verified by taking the

[5]We often say that the spins tend to line up with the magnetic field, but the low energy state for an electron, due to its negative charge, occurs when its spin is opposite to the magnetic field so its magnetic moment is along the magnetic field.

trace of the density matrix with the respective projection operator. For example, for the pure state

$$\mathrm{tr}(\hat{\rho}|\alpha\rangle\langle\alpha|) = \langle\alpha|\hat{\rho}|\alpha\rangle + \langle\beta|\hat{\rho}|\alpha\rangle\langle\alpha|\beta\rangle = \langle\alpha|\hat{\rho}|\alpha\rangle = |c_1|^2 = p_\alpha. \tag{26.75}$$

But at least one of the quantities $\langle\sigma_x\rangle = P_x$ or $\langle\sigma_y\rangle = P_y$ will not be zero for the pure state, except for the special values $c_1 = 0$ or $c_2 = 0$, in which cases both density matrices represent the same pure state. On the other hand, for the statistical state $\langle\sigma_x\rangle = \langle\sigma_y\rangle = 0$.

We can also construct a statistical state based on the states χ and $\chi_\perp$ with density operator

$$\hat{\rho}^S_\chi = p_\chi|\chi\rangle\langle\chi| + p_\perp|\chi_\perp\rangle\langle\chi_\perp|, \tag{26.76}$$

where p_χ and $p_\perp$ are probabilities. By using Eq. (26.63) for each pure state and recalling that $-\mathbf{P}$ corresponds to $\chi_\perp$, we deduce the corresponding density matrix

$$\rho^S_\chi = (1/2)[E + (p_\chi - p_\perp)\mathbf{P}\cdot\boldsymbol{\sigma}]. \tag{26.77}$$

Equation (26.77) resembles Eq. (26.74) but with respect to the **P**-axis, as opposed to our original arbitrary z-axis. For **P** not along the z-axis, these represent different statistical states except for the special values $p_\alpha = p_\beta = p_\chi = p_\perp = 1/2$, in which case $\rho^S = \rho^S_\chi = (1/2)E$. Such a statistical state is isotropic in the sense that the expectation value for a state of any orientation is equal to 1/2.

There is an interesting relationship for expectation values of χ and $\chi_\perp$ for ρ^S and for α and β in ρ^S_χ. Thus

$$\mathrm{tr}(\rho^S|\chi\rangle\langle\chi|) = p_\alpha|c_1^2 + p_\beta|c_2|^2; \quad \mathrm{tr}(\rho^S|\chi_\perp\rangle\langle\chi_\perp|) = p_\alpha|c_2|^2 + p_\beta|c_1|^2, \tag{26.78}$$

which sum to 1 whereas

$$\mathrm{tr}(\rho^S_\chi|\alpha\rangle\langle\alpha|) = p_\chi|c_1^2 + p_\perp|c_2|^2; \quad \mathrm{tr}(\rho^S_\chi|\beta\rangle\langle\beta|) = p_\chi|c_2|^2 + p_\perp|c_1|^2, \tag{26.79}$$

which also sum to 1. For the isotropic statistical state, each of these probabilities is equal to 1/2. Equations (26.78) and (26.79) are special cases of Eq. (26.11) for which the operator $\hat{f}$ is a projection operator for some pure state.

For a magnetic field **B**, which for convenience we can take to be along the z-axis, the Hamiltonian will be $\mathcal{H} = \mu_B B\sigma_z$. The eigenstates are then just α and β with respective energies $\mu_B B$ and $-\mu_B B$. For thermal equilibrium, the probabilities would be

$$p_\alpha = e^{-w}/(e^w + e^{-w}); \quad p_\beta = e^w/(e^w + e^{-w}) \tag{26.80}$$

with $w = \beta\mu_B B$. Then

$$\langle\sigma_z\rangle = (p_\alpha - p_\beta) = (e^{-w} - e^w)/(e^w + e^{-w}) = -\tanh w \tag{26.81}$$

and the magnetic moment is $\mu_z = -\mu_B\langle\sigma\rangle_z = \mu_B \tanh w$ in the direction of B. Of course we could have obtained this last result by elementary methods, so the use of the statistical density matrix for this simple two-state problem is overkill. Nevertheless, in this simple case, we see in detail the difference between a pure state and a statistical state.

See Schiff [57, p. 382] for a treatment of general spin s.

Example Problem 26.1. Compare the isotropic statistical state $(1/2)|\alpha\rangle\langle\alpha| + (1/2)|\beta\rangle\langle\beta|$ with the pure states $|\phi_1\rangle = (1/\sqrt{2})|\alpha\rangle + (1/\sqrt{2})|\beta\rangle$, $|\phi_2\rangle = (1/\sqrt{2})|\alpha\rangle + (i/\sqrt{2})|\beta\rangle$, $|\phi_3\rangle = (1/\sqrt{2})|\alpha\rangle + (1/\sqrt{2})e^{i\gamma}|\beta\rangle$ by computing and discussing their density matrices.

Solution 26.1. The respective density matrices are

$$\rho^S = \begin{pmatrix} 1/2 & 0 \\ 0 & 1/2 \end{pmatrix} \tag{26.82}$$

and

$$\rho_1 = \begin{pmatrix} 1/2 & 1/2 \\ 1/2 & 1/2 \end{pmatrix}; \quad \rho_2 = \begin{pmatrix} 1/2 & -i/2 \\ i/2 & 1/2 \end{pmatrix}; \quad \rho_3 = \begin{pmatrix} 1/2 & e^{-i\gamma}/2 \\ e^{i\gamma}/2 & 1/2 \end{pmatrix}. \tag{26.83}$$

For all four, the probability of finding the system in $|\alpha\rangle$ or $|\beta\rangle$ is 1/2. For the statistical state, $\langle\boldsymbol{\sigma}\rangle = 0$. For $|\phi_1\rangle$, $P_x = 1$, $P_y = 1$, and $P_z = 0$. For $|\phi_2\rangle$, $P_x = 0$, $P_y = 1$, and $P_z = 0$. And for $|\phi_3\rangle$, $P_x = \cos\gamma$, $P_y = \sin\gamma$, and $P_z = 0$. For the three pure states, the vector $\mathbf{P}$ is perpendicular to the z-axis, but spins in those eigenstates still *have probabilities* of 1/2 of being in either $|\alpha\rangle$ or $|\beta\rangle$.

Example Problem 26.2. Compare the statistical state $(1/3)|\alpha\rangle\langle\alpha| + (2/3)|\beta\rangle\langle\beta|$ with the pure state $|\phi\rangle = \sqrt{1/3}|\alpha\rangle + \sqrt{2/3}\,e^{i\gamma}|\beta\rangle$. What would be the value of $\mathbf{P}$ for a state $|\phi_\perp\rangle$ that is perpendicular to $|\phi\rangle$ and what would be its density matrix?

Solution 26.2. The respective density matrices are

$$\rho^S = \begin{pmatrix} 1/3 & 0 \\ 0 & 2/3 \end{pmatrix}; \quad \rho = \begin{pmatrix} 1/3 & e^{-i\gamma}\sqrt{2/9} \\ e^{i\gamma}\sqrt{2/9} & 2/3 \end{pmatrix}. \tag{26.84}$$

For both, the probability of finding the system in $|\alpha\rangle$ is 1/3 and in $|\beta\rangle$ it is 2/3. For the statistical state, $\langle\sigma_x\rangle = \langle\sigma_y\rangle = 0$ and $\langle\sigma_z\rangle = -1/3$. For $|\phi\rangle$, $P_x = (2\sqrt{2}/3)\cos\gamma$, $P_y = (2\sqrt{2}/3)\sin\gamma$, and $P_z = -1/3$. The value of $\mathbf{P}$ for $|\phi_\perp\rangle$ would be the negative of $\mathbf{P}$ for $|\phi\rangle$. Within an overall phase factor, one could take $|\phi_\perp\rangle = \sqrt{2/3}\,e^{-i\gamma}|\alpha\rangle - \sqrt{1/3}|\beta\rangle$ and its density matrix (which is independent of the overall phase factor) would be

$$\rho_\perp = \begin{pmatrix} 2/3 & -e^{-i\gamma}\sqrt{2/9} \\ -e^{i\gamma}\sqrt{2/9} & 1/3 \end{pmatrix}. \tag{26.85}$$

26.7 Indistinguishable Particles

Suppose we have a set of identical particles that cannot be distinguished from one another in the sense that interchange of any pair of particles will not lead to a new quantum state.[6] We shall refer to such particles as indistinguishable particles. Then quantum mechanical considerations require their quantum states to have certain symmetry properties,

[6]If identical particles were imbedded in a solid, they could be distinguished by their position in the solid, so we would not regard them as indistinguishable. On the other hand, if they shared the same volume as they would in a gas, they would be indistinguishable.

depending on whether they are bosons (integral spin) or fermions (half integral spin). If they are bosons (fermions), their wave vectors must be symmetric (antisymmetric) under interchange of any pair of particles. For a very thorough treatment of the general case, see Messiah [59, chapter XIV]. See Appendix I for an introduction to creation and annihilation operators that can be used to construct such states.

We proceed to illustrate this symmetry requirement for ideal Bose and Fermi gases for which the interaction energies of particles are assumed to be negligible. The Hamiltonian for such a system will be of the form

$$\hat{\mathcal{H}}(\xi_1, \xi_2, \ldots, \xi_{\mathcal{N}}) = \sum_{i=1}^{\mathcal{N}} \hat{h}(\xi_i), \tag{26.86}$$

where ξ_i represents the coordinates, momenta, and spin of particle i and $\hat{h}(\xi)$ is the Hamiltonian for a single particle having coordinates ξ, the same function for each particle. For each particle, we label the eigenstates of a complete set of commuting observables, including $\hat{h}$, by a single number[7] α, where $\alpha = 1, 2, \ldots$. Thus,

$$\hat{h}\, u_\alpha(\xi) = \varepsilon_\alpha u_\alpha(\xi), \qquad \text{for } \xi = \xi_1, \xi_2, \ldots, \tag{26.87}$$

or more succinctly $\hat{h}|\alpha\rangle = \varepsilon_\alpha|\alpha\rangle$, where ε_α is the energy of the state α.

Each physically distinct quantum state of a system of $\mathcal{N}$ bosons or fermions can be described by specifying a set $\{n_\alpha\}$ of occupation numbers n_α (sometimes called a distribution) for each of the states $|\alpha\rangle$. Thus there will be n_1 particles in the state $|\alpha = 1\rangle$, n_2 particles in $|\alpha = 2\rangle$, and so on, with the understanding that unoccupied states will simply have $n_\alpha = 0$. Since every particle will be in some state, we shall have

$$\sum_\alpha n_\alpha = \mathcal{N} \tag{26.88}$$

and

$$\sum_\alpha n_\alpha \varepsilon_\alpha = \mathcal{E}, \tag{26.89}$$

where $\mathcal{E}$ is the total energy.

Given a set of occupation numbers $\{n_\alpha\}$, we focus on only the subset of nonzero occupation numbers, $n_\alpha, n_\beta, \ldots, n_\gamma$, where now $\alpha, \beta, \ldots, \gamma$ represent specific eigenstates of the single particle Hamiltonian $\hat{h}$. Since each of these occupation numbers will be at least equal to 1, there can be at most $\mathcal{N}$ of them. A trial wave function of the form

$$\psi_{\{n_\alpha\}}(\xi_1, \xi_2, \ldots, \xi_{\mathcal{N}}) = \prod_{i=1}^{n_\alpha} u_\alpha(\xi_i) \prod_{j=n_\alpha+1}^{n_\alpha+n_\beta} u_\beta(\xi_j) \cdots \prod_{k=\mathcal{N}-n_\gamma}^{\mathcal{N}} u_\gamma(\xi_k), \tag{26.90}$$

[7]Generally, a set of quantum numbers is used to label each state, but we simply renumber these sets sequentially according to some scheme.

in which the first n_α factors are u_α, the second n_β factors are u_β, and so on, will be an eigenfunction of the total Hamiltonian $\hat{\mathcal{H}}$ with energy $\mathcal{E}$, but it will not be acceptable because it specifies which particles are in a given state. We can, however, obtain from it a wave function having the desired symmetry properties by summing over all permutations of the ξ_i as follows:

For bosons, we apply the symmetrization operator

$$S := \frac{1}{\mathcal{N}!}\sum_{\text{all}} P, \tag{26.91}$$

where the sum is over all permutations and P is a permutation operator that permutes the coordinates ξ_i. For fermions we apply the anti-symmetrization operator

$$A := \frac{1}{\mathcal{N}!}\sum_{\text{all}} P(-1)^p, \tag{26.92}$$

where the factor $(-1)^p$ is $+1$ or -1 according to whether the permutation p generated by P is even or odd. In applying A to $\psi(\xi_1, \xi_2, \ldots, \xi_{\mathcal{N}})$ in Eq. (26.90), we see immediately that the result is zero if any $n_\alpha > 1$. This follows because one possible permutation would involve an interchange of two particles in the same state, which would produce terms of opposite sign. To get a nonvanishing result for fermions, all of the states belonging to the subset of nonvanishing occupation numbers must be different and have occupation numbers equal to 1. Thus for fermions, the only possible occupation numbers for the single particle states are 0 and 1, which is equivalent to the Pauli exclusion principle.

Normalized wave functions can be obtained as follows:

bosons

$$\Psi^B_{\{n_\alpha\}}(\xi_1, \xi_2, \ldots, \xi_{\mathcal{N}}) = \left[\frac{\mathcal{N}!}{n_\alpha! n_\beta! \cdots n_\gamma!}\right]^{1/2} S \prod_{i=1}^{n_\alpha} u_\alpha(\xi_i) \prod_{j=n_\alpha+1}^{n_\alpha+n_\beta} u_\beta(\xi_j) \cdots \prod_{k=\mathcal{N}-n_\gamma}^{\mathcal{N}} u_\gamma(\xi_k). \tag{26.93}$$

In this case, the application of S produces the same function $n_\alpha! n_\beta! \cdots n_\gamma!$ times, so the result can also be written

$$\Psi^B_{\{n_\alpha\}}(\xi_1, \xi_2, \ldots, \xi_{\mathcal{N}}) = \left[\frac{n_\alpha! n_\beta! \cdots n_\gamma!}{\mathcal{N}!}\right]^{1/2} \sum_{\text{dis}} P \prod_{i=1}^{n_\alpha} u_\alpha(\xi_i) \prod_{j=n_\alpha+1}^{n_\alpha+n_\beta} u_\beta(\xi_j) \cdots \prod_{k=\mathcal{N}-n_\gamma}^{\mathcal{N}} u_\gamma(\xi_k), \tag{26.94}$$

where now the sum is only over $\mathcal{N}!/(n_\alpha! n_\beta! \cdots n_\gamma!)$ distinct permutations.

fermions

$$\begin{aligned}\Psi^F_{\{n_\alpha\}}(\xi_1, \xi_2, \ldots, \xi_{\mathcal{N}}) &= [\mathcal{N}!]^{1/2} A\, u_\alpha(\xi_1) u_\beta(\xi_2) \cdots u_\gamma(\xi_{\mathcal{N}}) \\ &= \left[\frac{1}{\mathcal{N}!}\right]^{1/2} \sum_{\text{all}} P(-1)^p\, u_\alpha(\xi_1) u_\beta(\xi_2) \cdots u_\gamma(\xi_{\mathcal{N}}).\end{aligned} \tag{26.95}$$

Consequently, for fermions, the wave function can be expressed as a Slater determinant:

$$\Psi^F_{\{n_\alpha\}}(\xi_1, \xi_2, \ldots, \xi_\mathcal{N}) = \left[\frac{1}{\mathcal{N}!}\right]^{1/2} \begin{bmatrix} u_\alpha(\xi_1) & u_\beta(\xi_1) & \cdots & u_\gamma(\xi_1) \\ u_\alpha(\xi_2) & u_\beta(\xi_2) & \cdots & u_\gamma(\xi_2) \\ \vdots & \vdots & \vdots & \vdots \\ u_\alpha(\xi_\mathcal{N}) & u_\beta(\xi_\mathcal{N}) & \cdots & u_\gamma(\xi_\mathcal{N}) \end{bmatrix}. \tag{26.96}$$

See Appendix I for an alternative way of representing boson and fermion states in Dirac vector space by means of creation operators.

From the point of view of statistical mechanics, the counting of the number of microstates is quite different for systems of identical bosons, fermions, or classical particles, which for brevity we will refer to as **'boltzons'** since they are the sort of particle treated by Maxwell-Boltzmann statistics. For a wave function of the type $\psi_{\{n_\alpha\}}(\xi_1, \xi_2, \ldots, \xi_\mathcal{N})$ given by Eq. (26.90), the number of independent states would be

$$\mathcal{N}!/(n_\alpha! n_\beta! \cdots n_\gamma!), \quad \text{for identical but distinguishable boltzons.} \tag{26.97}$$

For indistinguishable boltzons, this number could be reduced by $\mathcal{N}!$, as suggested by Gibbs, to give a weighting factor

$$W_\mathrm{G} = 1/(n_\alpha! n_\beta! \cdots n_\gamma!), \quad \text{for indistinguishable boltzons.} \tag{26.98}$$

This weighting factor $W_\mathrm{G} < 1$ unless $n_\alpha = n_\beta = \cdots = n_\gamma = 1$ (or 0, which means here that the state is not included) in which case $W_\mathrm{G} = 1$. If the wave functions in Eq. (26.90) were used to represent indistinguishable bosons, they would constitute only one quantum state, as represented by Eq. (26.94). If the functions in Eq. (26.90) were used to represent indistinguishable fermions, they could not represent a quantum state unless they were all different, in which case they would represent only one state represented by Eq. (26.96). Therefore, the weighting factors for any configuration set $\{n_\alpha\}$ that satisfies Eq. (26.88) are

$$W_\mathrm{B} = 1, \quad \text{for indistinguishable bosons, any } \{n_\alpha\} \tag{26.99}$$

and

$$W_\mathrm{F} = 1, \quad \text{for indistinguishable fermions, } \{n_\alpha\} = 0, 1. \tag{26.100}$$

One might ask under what circumstances systems of indistinguishable boltzons, bosons, and fermions would lead to the same number of quantum states. The answer is: under conditions for which the number of available single particle states is extremely large compared to the total number of particles. A single particle state is deemed to be accessible if its Boltzmann factor $\exp(-\beta\varepsilon)$ is not negligibly small. Thus there will be a huge number of accessible states at high temperature. Then if the system is also sufficiently dilute, the probability of multiply-occupied states will be extremely small and most states will be either unoccupied or singly occupied. Under these conditions, every significant set of occupation numbers will contain only ones and zeros, so the Gibbs-boltzon weighting factor W_G for such a state will be practically unity.

27 Ising Model

Until now we have confined most of our treatments to systems of weakly interacting particles. A number of new phenomena, generally referred to as cooperative phenomena, arise whenever particles interact. These phenomena often include phase transitions, such as liquification of a gas as discussed from a thermodynamic viewpoint in connection with the van der Waals model in Chapter 9. Another example would be an order-disorder transition in a binary alloy.

The present chapter is devoted primarily to the study of a model known as the Ising model which is a simple tractable model for a magnetic system. We begin by considering a **spin Hamiltonian** of the form

$$\mathcal{H} = -\frac{1}{2}\sum_{i,j} J_{ij}\mathbf{S}_i \cdot \mathbf{S}_\mathbf{j}, \tag{27.1}$$

where the quantities $\mathbf{S}_i$ play the role of spins situated on a lattice, the sums are over all lattice sites, and J_{ij} is a coupling constant. Such a spin Hamiltonian is a drastic simplification itself. The spins are actually pseudo-spins that might be combinations of spin and orbital angular momenta. The interaction itself is primarily due to electrostatic energies associated with the different orbital wave functions needed to construct, according to the Pauli exclusion principle, antisymmetric wave functions for each electronic spin state. For a simplified motivation of Eq. (27.1), the reader is referred to Ashcroft and Mermin [58, p. 679]. A simpler version of a spin Hamiltonian is the **Heisenberg model** for which

$$\mathcal{H} = -\frac{1}{2}J\sum_{i,j}^{nn} \mathbf{S}_i \cdot \mathbf{S}_j. \tag{27.2}$$

Here, there is just one coupling constant J and the sum is only over nearest neighbors.

An even simpler model is the **Ising model** wherein the spins are replaced by quantities $\sigma_i = \pm 1$. This results in

$$\mathcal{H} = -\frac{1}{2}J\sum_{i,j}^{nn} \sigma_i\sigma_j = -J\sum_{i,j}^{nnp} \sigma_i\sigma_j, \tag{27.3}$$

where the second sum is over nearest-neighbor pairs. This model gives rise to only two energy states for a pair of nearest-neighbor spins, aligned neighbors $(1, 1$ or $-1, -1)$ with energy $-J$ and opposite neighbors $(1, -1$ or $-1, 1)$ with energy J. Despite this drastic simplification, the Ising model still presents some challenging problems, even though it allows for exact solutions for lattices in one and two spatial dimensions.

Thermal Physics. http://dx.doi.org/10.1016/B978-0-12-803304-3.00027-2

27.1 Ising Model, Mean Field Treatment

In the presence of a magnetic field B, we write the Hamiltonian for the Ising model in the form[1]

$$\mathcal{H} = -\frac{1}{2}\sum_{i,j} J_{ij}\sigma_i\sigma_j - \mu^* B \sum_i \sigma_i, \tag{27.4}$$

where $\mu^* > 0$ is a magnetic moment (not the chemical potential) and

$$J_{ij} = \begin{cases} J > 0 & \text{if } i \text{ and } j \text{ are nearest neighbors.} \\ 0 & \text{otherwise.} \end{cases} \tag{27.5}$$

There is still only one coupling constant J as in Eq. (27.3) but this change of notation will facilitate the sums over nearest neighbors.

We denote by $\langle\sigma_i\rangle$ the ensemble average value of σ_i. Then we substitute the identity

$$\sigma_i = (\sigma_i - \langle\sigma_i\rangle) + \langle\sigma_i\rangle \tag{27.6}$$

and a similar one for σ_j to obtain

$$\begin{aligned} -\frac{1}{2}\sum_{i,j} J_{ij}\sigma_i\sigma_j &= \frac{1}{2}\sum_{i,j} J_{ij}\langle\sigma_i\rangle\langle\sigma_j\rangle - \sum_{i,j} J_{ij}\sigma_i\langle\sigma_j\rangle \\ &\quad -\frac{1}{2}\sum_{i,j} J_{ij}(\sigma_i - \langle\sigma_i\rangle)(\sigma_j - \langle\sigma_j\rangle), \end{aligned} \tag{27.7}$$

where cross terms have been combined after interchange of i and j to give a factor of 2 in the second term. We shall use periodic boundary conditions so all lattice sites are equivalent. Thus $\langle\sigma_i\rangle = \langle\sigma_j\rangle \equiv \langle\sigma\rangle$ and Eq. (27.7) takes the form

$$\begin{aligned} -\frac{1}{2}\sum_{i,j} J_{ij}\sigma_i\sigma_j &= \frac{1}{2}J\mathcal{N}q\langle\sigma\rangle^2 - Jq\langle\sigma\rangle\sum_i \sigma_i \\ &\quad -\frac{1}{2}\sum_{i,j} J_{ij}(\sigma_i - \langle\sigma\rangle)(\sigma_j - \langle\sigma\rangle), \end{aligned} \tag{27.8}$$

where $\mathcal{N}$ is the number of lattice sites and q is the number of nearest neighbors.

The term on the second line of Eq. (27.8) represents correlations between nearest-neighbor spins. This may be seen because its average would vanish if $\langle\sigma_i\sigma_j\rangle = \langle\sigma_i\rangle\langle\sigma_j\rangle = \langle\sigma\rangle^2$ for $i \neq j$. This can also be seen because the average of the first two terms is $-(1/2)Jq\langle\sigma\rangle^2$ which would not equal the average of the left-hand side unless nearest-neighbor spins were uncorrelated. The second term on the right-hand side resembles the term in Eq. (27.4) that contains the external magnetic field B. This becomes more evident if we introduce the notation

$$B_J := Jq\langle\sigma\rangle/\mu^* \tag{27.9}$$

[1]The factor of 1/2 avoids double counting of interactions. The reader is cautioned that Ising Hamiltonians are often written without this factor of 1/2 and sometimes also with a factor of 2. If one sums only over nearest-neighbor *pairs*, the factor of 1/2 should be omitted. The sign convention here is that $\mu^*\sigma_i$ is the magnetic moment for this pseudo-spin.

in which case Eq. (27.4) takes the form

$$\mathcal{H} = \frac{1}{2} J\mathcal{N}q\langle\sigma\rangle^2 - \mu^*(B+B_J)\sum_i \sigma_i - \frac{1}{2}\sum_{i,j} J_{ij}(\sigma_i - \langle\sigma\rangle)(\sigma_j - \langle\sigma\rangle). \tag{27.10}$$

The quantity B_J is seen to play the role of a mean field experienced by a given spin due to the presence of the other spins.

The **mean-field approximation**[2] consists of ignoring the correlation term, resulting in a mean field Hamiltonian

$$\mathcal{H}_\mathrm{M} = \frac{1}{2} J\mathcal{N}q\langle\sigma\rangle^2 - \mu^*(B+B_J)\sum_i \sigma_i. \tag{27.11}$$

Many books also ignore the first term in Eq. (27.11) because it depends only on average quantities and plays no role in computing the magnetization. Omitting it, however, leads to an average energy for $B = 0$ that is too large by a factor of 2; this requires "patching" by a factor of $1/2$ in a somewhat *ad hoc* manner.

By using the mean field Hamiltonian given by Eq. (27.11), we obtain a problem for which the spins are formally independent. The canonical partition function for a single spin is therefore

$$z = \exp[-\beta(1/2)Jq\langle\sigma\rangle^2]\, 2\cosh[\beta\mu^*(B+B_J)] \tag{27.12}$$

and the probabilities are

$$p_+ = \frac{\exp[\beta\mu^*(B+B_J)]}{2\cosh[\beta\mu^*(B+B_J)]}; \quad p_- = \frac{\exp[-\beta\mu^*(B+B_J)]}{2\cosh[\beta\mu^*(B+B_J)]} \tag{27.13}$$

for $\sigma_i = \pm 1$, respectively. Note that these probabilities do not depend on the exponential factor in z, which came from the first term in Eq. (27.11). The magnetization $\mathcal{M} = \mathcal{N}\mu^*\langle\sigma\rangle$, where

$$\langle\sigma\rangle = p_+ - p_- = \tanh[\beta\mu^*(B+B_J)] \tag{27.14}$$

and the average energy is[3]

$$U = \langle\mathcal{H}_\mathrm{M}\rangle = -\frac{1}{2} J\mathcal{N}q\langle\sigma\rangle^2 - \mu^*\mathcal{N}B\langle\sigma\rangle. \tag{27.15}$$

Since B_J is given by Eq. (27.9), we see that Eq. (27.14) can be rewritten in the form

$$\langle\sigma\rangle = \tanh[\beta\mu^*(B + Jq\langle\sigma\rangle/\mu^*)], \tag{27.16}$$

which is a self-consistency equation for $\langle\sigma\rangle$. We can solve this equation graphically by defining a dimensionless parameter

$$x = \beta\mu^*(B + Jq\langle\sigma\rangle/\mu^*). \tag{27.17}$$

[2]This is also known as the Bragg-Williams or the Weiss molecular field approximation.

[3]Since B_J depends on $\langle\sigma\rangle$ which in turn depends on β and B, the formulae $\mathcal{M} = -\mathcal{N}k_\mathrm{B}T\partial\ln z/\partial B$ and $U = -\partial\ln z/\partial\beta$ will only work if $\langle\sigma\rangle$ is held constant during the differentiation. This inconsistency of the mean field approximation arises because average quantities appear in the mean field Hamiltonian.

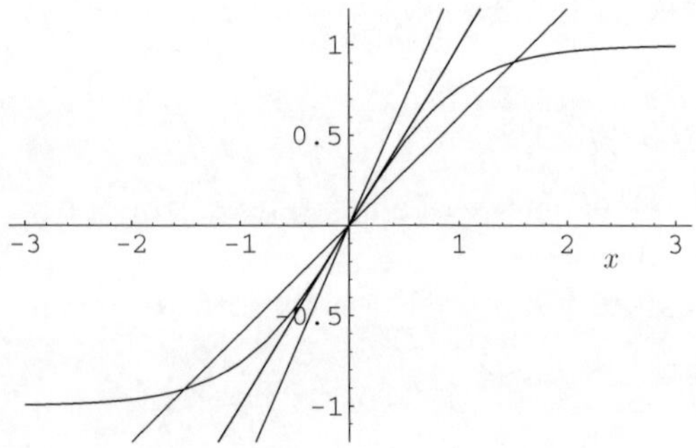

FIGURE 27–1 Graphical solution of Eq. (27.19) for $B = 0$. The curve is $\tanh x$ and the lines have slopes $k_B T/Jq$ of 1.4, 1, and 0.6. There are only solutions for $x \neq 0$ for $k_B T/Jq < 1$.

Then

$$\langle \sigma \rangle = -\frac{\mu^* B}{Jq} + \frac{k_B T}{Jq} x \tag{27.18}$$

and Eq. (27.16), which is now $\langle \sigma \rangle = \tanh x$, becomes

$$-\frac{\mu^* B}{Jq} + \frac{k_B T}{Jq} x = \tanh x. \tag{27.19}$$

Viewed as a function of x, the left-hand side of Eq. (27.19) is just a straight line of slope $k_B T/Jq$ and intercept $-\mu^* B/Jq$ and the right-hand side is a curve that can be drawn once and for all. The case $B = 0$ is of special importance and is illustrated in Figure 27–1. In that case, x is just proportional to $\langle \sigma \rangle$. Since the slope of $\tanh x$ is one at $x = 0$, we see that there are solutions for $x \neq 0$ provided that $k_B T/Jq < 1$ and otherwise no solutions. This defines a **critical temperature**

$$T_c = qJ/k_B \tag{27.20}$$

below which there is a spontaneous magnetization in the absence of an applied magnetic field.

Note that if x is a solution, $-x$ is also a solution. This degeneracy arises because $B = 0$ so there is no preferred direction for the spontaneous magnetic field. If we started with a finite positive field and then let it shrink to zero, we would create a bias for the positive solution.

Graphical solutions for $B > 0$ are illustrated in Figure 27–2. We see that positive solutions[4] for x exist for all values of $k_B T/Jq$, but those for large T correspond to small values of x and therefore to small values of $\langle \sigma \rangle = \tanh x$. Similar considerations lead to negative solutions for all T when $B < 0$.

[4]For sufficiently small positive values of $k_B T/Jq$, there can also be negative solutions. These can be shown to correspond to metastable or unstable solutions that represent cases in which a magnetic field is applied in a direction opposite to the spontaneous magnetization that occurs in zero field.

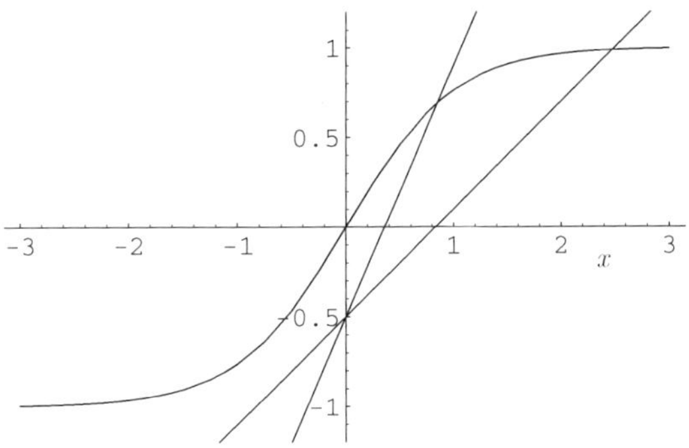

FIGURE 27–2 Graphical solution of Eq. (27.19) for $B > 0$, namely $\mu^* B/Jq = 0.5$ for the sake of illustration. The curve is $\tanh x$ and the lines have slopes $k_B T/Jq$ of 1.4 and 0.6. There are positive solutions for all values of $k_B T/Jq$ but those for large T correspond to small values of x and therefore to small values of $\langle\sigma\rangle = \tanh x$. For $B < 0$, the lines would have a positive intercept and the solutions for x would be negative.

The foregoing results in the mean field approximation are suggestive but *incorrect*. Indeed, it is possible to solve the Ising model exactly in one dimension and two dimensions and numerically in three and higher dimensions. There are also better approximate solutions for all dimensions. The most serious discrepancy occurs in one dimension where the mean field model leads to a critical temperature at $k_B T_c/J = 2$ but the exact solution displays no spontaneous magnetization. In higher dimensions, there are critical temperatures $T_c > 0$ but the numerical values of $k_B T_c/J$ are different. For instance, the exact two-dimensional solution for a square lattice, due to Onsager, gives

$$\frac{k_B T_c}{J} = \frac{2}{\ln(1+\sqrt{2})} = 2.26919, \tag{27.21}$$

whereas the mean field approximation gives $k_B T_c/J = 4$. Some comparative values of T_c are given in Table 27–1. We see that the mean field model shows the general trend with dimensionality but is certainly wrong in detail because correlations are neglected. The cluster model of Boethe (see Pathria [8, p. 329]) takes into account the correlations of a given spin with its neighbors but all other interactions (e.g., the interactions of its neighbors with other neighbors) are taken into account by a mean field. The critical temperature for the Boethe model satisfies

$$\frac{k_B T_c}{J} = \frac{2}{\ln[q/(q-2)]}. \tag{27.22}$$

Table 27–1 Values of $k_B T_c/J$ for the Ising Model for a "Simple Cubic" Lattice of Various Dimensionality According to Several Theories

Dimensionality	1	2	3	4	5	6	7
Exact	0	2.26919					
Numerical		2.26919	4.51153	6.68003	8.77739	10.8348	12.8690
Boethe	0	2.88539	4.93261	6.95212	8.96284	10.9696	12.9743
Mean field	2	4	6	8	10	12	14

Note: Numerical results are from Galam and Mauger [70].

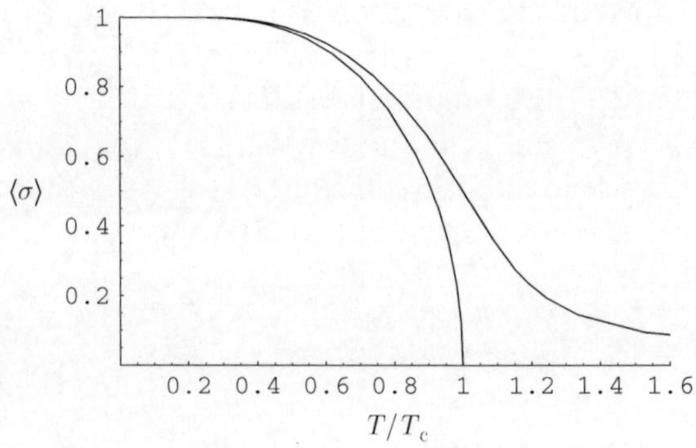

FIGURE 27–3 Dimensionless magnetization per spin, $\langle\sigma\rangle = \mathcal{M}/\mathcal{N}\mu^*$, versus dimensionless temperature $t = T/T_c$ for $B = 0$ and $b = \mu^* B/Jq = 0.05$ according to the parametric equations Eq. (27.23). For $B = 0$, the magnetization is zero for $T > T_c$ but for $B > 0$ it extends beyond T_c.

In one dimension, $q = 2$ so Eq. (27.22) gives $T_c = 0$, correctly showing that there is no phase transition for any $T > 0$.

For the mean field model, Figure 27–3 shows a plot of the dimensionless magnetization per spin, $\mathcal{M}/\mathcal{N}\mu^* = \langle\sigma\rangle$, as a function of T/T_c for $B = 0$ and $B > 0$. These plots were constructed by writing Eqs. (27.14) and (27.19) in the parametric form

$$\langle\sigma\rangle = \tanh x; \quad t = \frac{\tanh x}{x} + \frac{b}{x}, \tag{27.23}$$

where the dimensionless temperature $t := T/T_c$ and the dimensionless magnetic field $b := (\mu^* B/Jq) = (\mu^* B/k_B T_c)$. For a given value of b, one can assign values of the parameter x and construct a plot of $\langle\sigma\rangle$ versus t. For $b = 0$, we see that $\langle\sigma\rangle = 0$ for $T > T_c$ but for $b > 0$ it extends beyond T_c, although its value is small.

For the case $B = 0$ we can get a series representation for $1/t$ in terms of $m \equiv \langle\sigma\rangle$ as follows: For $B = 0$ we can eliminate x from Eq. (27.23) to obtain

$$m = \tanh(m/t) = \frac{e^{2m/t} - 1}{e^{2m/t} + 1}. \tag{27.24}$$

We can then solve for m/t, which amounts to finding the inverse of the hyperbolic tangent function to obtain

$$\frac{1}{t} = \frac{1}{2m} \ln\left(\frac{1+m}{1-m}\right) = \sum_{p=0}^{\infty} \frac{m^{2p}}{2p+1}, \tag{27.25}$$

where the series converges for $|m| < 1$. Thus

$$t = \frac{1}{1 + m^2/3 + m^4/5 + \cdots} = 1 - m^2/3 - 4m^4/45 + \cdots, \tag{27.26}$$

which can be solved to lowest order to give

$$m = \sqrt{3}(1 - t)^{1/2}. \tag{27.27}$$

Equation (27.27) shows how the magnetization rises from zero as T decreases slightly from T_c and the exponent of $1/2$ is referred to as a **critical exponent**. In one dimension, we know from the exact solution that $T_c = 0$, so there really is no critical exponent in that case. For two dimensions, the correct critical exponent is $1/8$ for a square lattice and approximately 0.313 for a simple cubic lattice in three dimensions. As was the case with T_c itself, the mean field theory shows some qualitative trends but is certainly wrong in detail.

By differentiation of U from Eq. (27.15) we can calculate the heat capacity, but we have to remember that $\langle\sigma\rangle \equiv m$ depends on T. Thus

$$C_V = \frac{\partial U}{\partial T} = -\mathcal{N}Jq(m+b)\frac{\partial m}{\partial T}. \tag{27.28}$$

To calculate the derivative of m, we write Eq. (27.16) in the form

$$m = \tanh\left[\frac{1}{t}(b+m)\right]. \tag{27.29}$$

Then

$$\frac{\partial m}{\partial T} = \operatorname{sech}^2\left[\frac{1}{t}(b+m)\right]\left[\frac{1}{t}\frac{\partial m}{\partial T} - \frac{1}{t^2 T_c}(b+m)\right], \tag{27.30}$$

which we can solve to obtain

$$\frac{\partial m}{\partial T} = -\frac{(b+m)(1-m^2)}{t^2 - t(1-m^2)}\frac{1}{T_c}, \tag{27.31}$$

where we have used $\operatorname{sech}^2\left[(b+m)/t\right] = 1 - m^2$. Combining Eqs. (27.28) and (27.31) gives

$$\frac{C_V}{\mathcal{N}k_B} = \frac{(b+m)^2(1-m^2)}{t^2 - t(1-m^2)}. \tag{27.32}$$

Equation (27.32), however, is not very enlightening, so we introduce $x = (b+m)/t$ as in Eq. (27.23) and obtain, after some algebra, the parametric equations

$$t = \frac{\tanh x}{x} + \frac{b}{x}; \quad \frac{C_V}{\mathcal{N}k_B} = \frac{x^2(\tanh x + b)\operatorname{sech}^2 x}{(\tanh x + b) - x\operatorname{sech}^2 x}. \tag{27.33}$$

Figure 27–4 shows a plot of $C_V/\mathcal{N}k_B$ versus T/T_c for $B = 0$ and $B > 0$. For $B = 0$ there is a sharp peak at $C_V/\mathcal{N}k_B = 3/2$ at $T = T_c$ and zero heat capacity for $T > T_c$. The height of this peak is not obvious from Eq. (27.32) or (27.33) because the function is discontinuous at $T = T_c$ for $b = 0$. It can be computed easily, however, by substituting Eq. (27.27), which holds for $t \approx 1$, into Eq. (27.15) for $B = 0$ to obtain

$$U/\mathcal{N} = -(1/2)Jqm^2 = -(3/2)k_B(T_c - T); \quad T \approx T_c, \tag{27.34}$$

and then differentiating with respect to T.

The same technique of implicit differentiation can be used to compute the magnetic susceptibility

$$\chi = \frac{\partial \mathcal{M}}{\partial B} = \mathcal{N}\mu^*\frac{\partial m}{\partial B} = \frac{\mathcal{N}\mu^{*2}}{k_B T_c}\frac{1-m^2}{t-(1-m^2)}. \tag{27.35}$$

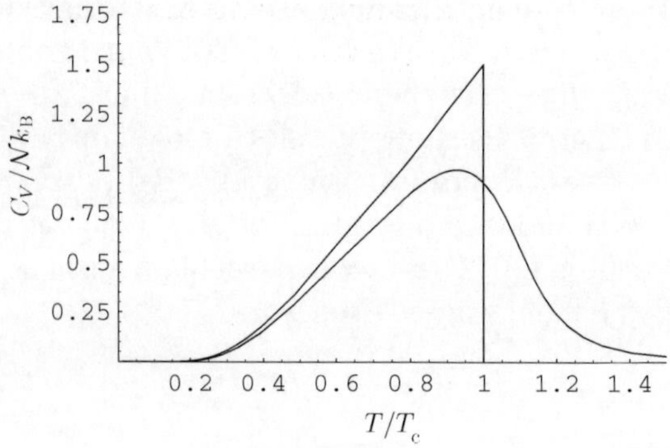

FIGURE 27–4 Dimensionless heat capacity per spin, $C_V/\mathcal{N}k_B$, versus dimensionless temperature $t = T/T_c$ for $B = 0$ and $b = \mu^*B/Jq = 0.05$ according to the parametric equations Eq. (27.33). For $B = 0$, the heat capacity rises to a sharp peak and then drops to zero for $T > T_c$, but for $B > 0$ it extends beyond T_c.

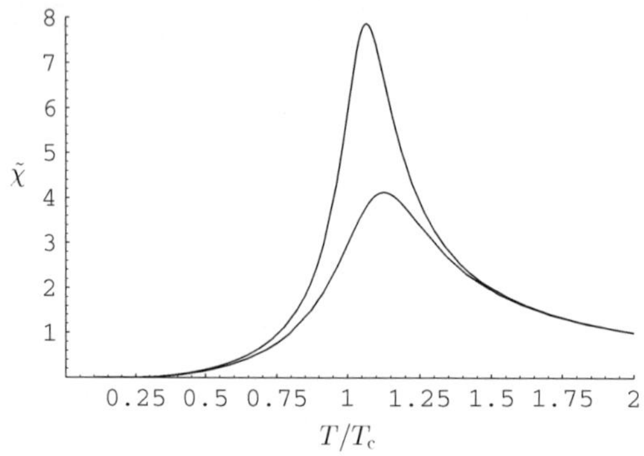

FIGURE 27–5 Dimensionless magnetic susceptibility, $\tilde{\chi}$, versus dimensionless temperature $t = T/T_c$ for $b = \mu^* B/Jq = 0.05$ (low peak) and $b = 0.02$ (high peak) according to the parametric equations Eq. (27.37). For $B = 0$, the susceptibility diverges as $T \to T_c$.

For $t \gg 1$ we have $m \ll 1$ so

$$\chi \approx \frac{\mathcal{N}\mu^{*2}}{k_B(T - T_c)}, \tag{27.36}$$

which is known as the **Curie-Weiss law**. Parametric equations for the general case are

$$t = \frac{\tanh x}{x} + \frac{b}{x}; \quad \tilde{\chi} := \frac{\chi k_B T_c}{\mathcal{N}\mu^{*2}} = \frac{x\,\mathrm{sech}^2 x}{(\tanh x + b) - x\,\mathrm{sech}^2 x}. \tag{27.37}$$

Figure 27–5 shows a plot of the dimensionless susceptibility $\tilde{\chi}$ as a function of $t = T/T_c$ for two positive magnetic fields. As the field strength is decreased, the peak in the vicinity of $t = 1$ becomes progressively higher and ultimately diverges as $B \to 0$. We can see the nature of this divergence by substituting Eq. (27.27) into Eq. (27.35) to obtain

$$\tilde{\chi} \approx \frac{1}{2}(1 - t)^{-1} \quad \text{for } t < 1 \text{ and } t \approx 1. \tag{27.38}$$

Thus χ diverges like $(1 - t)^{-\gamma}$ with a critical exponent $\gamma = 1$. Although the mean field model is incorrect, this value of γ is close to values 1.2-1.4 measured for magnetic systems [8, p. 336].

27.2 Pair Statistics

More insight about the Ising model can be gained by studying the **statistics of nearest-neighbor pairs**. To do this, we follow Pathria [8, p. 318] and introduce the following notation:

- $\mathcal{N}$ = total number of spins
- $\mathcal{N}_+$ = total number of "up" spins
- $\mathcal{N}_-$ = total number of "down" spins
- $\mathcal{N}_{++}$ = total number of "up-up" nearest-neighbor pairs
- $\mathcal{N}_{--}$ = total number of "down-down" nearest-neighbor pairs
- $\mathcal{N}_{+-}$ = total number of "opposite" nearest-neighbor pairs

In general, we certainly have $\mathcal{N} = \mathcal{N}_+ + \mathcal{N}_-$. We treat the case of periodic boundary conditions so that all lattice sites are equivalent. It follows that:

$$q\mathcal{N}_+ = 2\mathcal{N}_{++} + \mathcal{N}_{+-}; \quad q\mathcal{N}_- = 2\mathcal{N}_{--} + \mathcal{N}_{+-}. \tag{27.39}$$

If we sum these equations we obtain $q\mathcal{N}/2 = \mathcal{N}_{++} + \mathcal{N}_{--} + \mathcal{N}_{+-}$ which is a correct expression for the total number of pairs. For a given $\mathcal{N}$, we choose $\mathcal{N}_+$ and $\mathcal{N}_{++}$ to be independent variables and obtain the remaining quantities:

$$\mathcal{N}_- = \mathcal{N} - \mathcal{N}_+ \tag{27.40}$$

$$\mathcal{N}_{+-} = q\mathcal{N}_+ - 2\mathcal{N}_{++} \tag{27.41}$$

$$\mathcal{N}_{--} = q\mathcal{N}/2 + \mathcal{N}_{++} - q\mathcal{N}_+. \tag{27.42}$$

The part of the Ising Hamiltonian that is independent of the magnetic field is

$$-\frac{1}{2}\sum_{i,j} J_{ij}\sigma_i\sigma_j = -J\left(\mathcal{N}_{++} + \mathcal{N}_{--} - \mathcal{N}_{+-}\right) = -J\left(q\mathcal{N}/2 + 4\mathcal{N}_{++} - 2q\mathcal{N}_+\right) \tag{27.43}$$

and the magnetic moment for such a configuration is

$$M = \mu^*\left(\mathcal{N}_+ - \mathcal{N}_-\right) = \mu^*\left(2\mathcal{N}_+ - \mathcal{N}\right). \tag{27.44}$$

The difficulty of solving the Ising problem, even for zero magnetic field, can be appreciated by realizing that it amounts to enumerating all possible configurations of $\mathcal{N}_+$ and $\mathcal{N}_{++}$, a difficult combinatorial problem.

27.2.1 Average Pair Statistics for Mean Field

We can learn more about the nature of the mean field approximation by taking averages of the above equations. The average of Eq. (27.43) with correlations ignored, so $\langle\sigma_i\sigma_j\rangle = \langle\sigma\rangle^2$, gives

$$-q\mathcal{N}\langle\sigma\rangle^2/2 = -J\left(q\mathcal{N}/2 + 4\langle\mathcal{N}_{++}\rangle - 2q\langle\mathcal{N}_+\rangle\right) \tag{27.45}$$

and the average of Eq. (27.44) gives

$$\mathcal{N}\langle\sigma\rangle = 2\langle\mathcal{N}_+\rangle - \mathcal{N}. \tag{27.46}$$

We can therefore solve Eqs. (27.45) and (27.46) for $\langle\mathcal{N}_+\rangle$ and $\langle\mathcal{N}_{++}\rangle$ and use them to compute averages $\langle\mathcal{N}_-\rangle$, $\langle\mathcal{N}_{--}\rangle$, and $\langle\mathcal{N}_{+-}\rangle$ from Eqs. (27.40) to (27.42). Then recalling that the total number of pairs is $\mathcal{N}q/2$, we can compute the following probabilities:

$$p_+ := \frac{\langle\mathcal{N}_+\rangle}{\mathcal{N}} = \frac{1}{2}(1+\langle\sigma\rangle), \tag{27.47}$$

$$p_- := \frac{\langle\mathcal{N}_-\rangle}{\mathcal{N}} = \frac{1}{2}(1-\langle\sigma\rangle), \tag{27.48}$$

$$p_{++} := \frac{\langle\mathcal{N}_{++}\rangle}{\mathcal{N}q/2} = \frac{1}{4}(1+\langle\sigma\rangle)^2 = p_+^2, \tag{27.49}$$

$$p_{+-} := \frac{\langle\mathcal{N}_{+-}\rangle}{\mathcal{N}q/2} = \frac{1}{2}(1+\langle\sigma\rangle)(1-\langle\sigma\rangle) = 2p_+p_-, \tag{27.50}$$

$$p_{--} := \frac{\langle\mathcal{N}_{--}\rangle}{\mathcal{N}q/2} = \frac{1}{4}(1-\langle\sigma\rangle)^2 = p_-^2. \tag{27.51}$$

We observe that p_{++}, p_{+-}, and p_{--} are just the terms in the expansion of $(p_+ + p_-)^2$. This indicates that the spins are randomly distributed in the mean field approximation and further emphasizes that correlations have been ignored.

Plots of these probabilities as a function of temperature are shown in Figure 27–6 for $B \to 0$ from positive values. At the critical temperature, $\langle\sigma\rangle = 0$ so $p_+ = p_- = 1/2$, $p_{++} = p_{--} = 1/4$, and $p_{+-} = 1/2$. From the shapes of the p_{+-} and p_{--} plots, we see that the "down" spins tend to form in isolation as the temperature rises and only form a significant number of "down-down" pairs near the critical temperature.

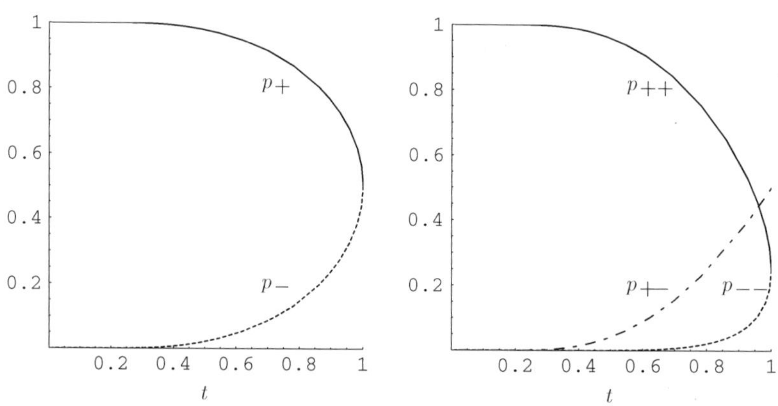

FIGURE 27–6 Probabilities p_+ and p_- of up and down spins (left) and pairs p_{++}, p_{+-}, and p_{--} as a function of dimensionless temperature $t = T/T_c$ for the Ising model in the mean field approximation for $B \to 0$ from positive values.

27.3 Solution in One Dimension for Zero Field

In one dimension with periodic boundary conditions, it is possible to solve exactly the Ising model for $B = 0$ by a rather elementary method that demonstrates explicitly the way that correlations enter. We can write the Hamiltonian in the form

$$\mathcal{H} = -\frac{1}{2}\sum_{i,j} J_{ij}\sigma_i\sigma_j = -J\sum_{i=1}^{\mathcal{N}} \sigma_i\sigma_{i+1} = -J\sum_{i=1}^{\mathcal{N}} \tau_i, \tag{27.52}$$

where the pair operators $\tau_i := \sigma_i\sigma_{i+1}$. For periodic boundary conditions, we have $\sigma_{\mathcal{N}+1} = \sigma_1$, so it is easy to see that the pair operators are correlated because

$$\prod_{i=1}^{\mathcal{N}} \tau_i = \prod_{i=1}^{\mathcal{N}} \sigma_i\sigma_{i+1} = \prod_{i=1}^{\mathcal{N}} \sigma_i^2 = 1. \tag{27.53}$$

Note that the τ_i take on the values ± 1. The canonical partition function for the whole system is therefore

$$Z = \sum_{\sigma_1}\cdots\sum_{\sigma_{\mathcal{N}}} \exp\left[y\sum_i \sigma_i\sigma_{i+1}\right] = 2\sum_{\tau_1}\cdots\sum_{\tau_{\mathcal{N}}} \exp\left[y\sum_i \tau_i\right], \tag{27.54}$$

where the sums are constrained by Eq. (27.53) and $y := \beta J$. The factor of 2 on the right-hand side arises because as the set of the σ_i range over their values once, the set of τ_i range over their values twice.[5]

At first we ignore the constraint Eq. (27.53) and evaluate the right-hand side of Eq. (27.54) to obtain

$$Z = 2\sum_{\tau_1} e^{y\tau_1}\cdots\sum_{\tau_{\mathcal{N}}} e^{y\tau_{\mathcal{N}}} = 2(e^y + e^{-y})^{\mathcal{N}} = 2(2\cosh y)^{\mathcal{N}}, \quad \text{no constraint.} \tag{27.55}$$

Equation (27.55) would be correct for a chain of length $\mathcal{N}$ with open ends, except not all spins would be equivalent for such a chain. This is, however, a very small effect for large $\mathcal{N}$ since the fraction of nonequivalent spins is $2/\mathcal{N}$.

To account for the constraint Eq. (27.54) due to periodic boundary conditions, we expand the binomial in Eq. (27.55) to obtain

$$2(e^y + e^{-y})^{\mathcal{N}} = 2\sum_{r=1}^{\mathcal{N}} \frac{\mathcal{N}!}{r!(\mathcal{N}-r)!}(e^y)^{\mathcal{N}-r}(e^{-y})^r. \tag{27.56}$$

Because of the constraint, what we should have instead is this same sum but with all of the terms for odd values of r missing. To accomplish this, we note the related series

$$2(e^y - e^{-y})^{\mathcal{N}} = 2\sum_{r=1}^{\mathcal{N}} \frac{\mathcal{N}!}{r!(\mathcal{N}-r)!}(e^y)^{\mathcal{N}-r}(-e^{-y})^r \tag{27.57}$$

[5] Note that $\sigma_i\sigma_{i+1} = 1$ when both factors are 1 and when both factors are -1; in each case, $\tau_i = 1$.

in which all of the terms with odd r enter with a minus sign. If we add these two series, we get twice the terms with even r, double what we want. The correct partition function for periodic boundary conditions is therefore

$$Z = 2\frac{(e^y + e^{-y})^{\mathcal{N}} + (e^y - e^{-y})^{\mathcal{N}}}{2} = (e^y + e^{-y})^{\mathcal{N}} + (e^y - e^{-y})^{\mathcal{N}}. \tag{27.58}$$

In terms of hyperbolic functions,

$$Z = (2\cosh y)^{\mathcal{N}} + (2\sinh y)^{\mathcal{N}}, \tag{27.59}$$

which agrees with the exact result obtained by using a transfer matrix, which we present in the next section.

27.4 Transfer Matrix

By using a computational technique known as the transfer matrix, we can obtain an exact solution to the Ising model in one dimension even in the presence of a magnetic field. For periodic boundary conditions in one dimension, we can write the Ising Hamiltonian in the form

$$\mathcal{H} = -J\sum_{i=1}^{\mathcal{N}} \sigma_i\sigma_{i+1} - (1/2)\mu^* B\sum_{i=1}^{\mathcal{N}}(\sigma_i + \sigma_{i+1}). \tag{27.60}$$

Then

$$\exp(-\beta\mathcal{H}) = \exp\left[y\sum_{i=1}^{\mathcal{N}} \sigma_i\sigma_{i+1} + (x/2)\sum_{i=1}^{\mathcal{N}}(\sigma_i + \sigma_{i+1})\right], \tag{27.61}$$

where $x := \beta\mu^* B$ and $y = \beta J$. The partition function is given by

$$Z = \sum_{\{\sigma_i\}=\pm 1} \exp(-\beta\mathcal{H}) \equiv \sum_{\sigma_1=\pm 1}\sum_{\sigma_2=\pm 1}\cdots\sum_{\sigma_{\mathcal{N}}=\pm 1} \exp(-\beta\mathcal{H}). \tag{27.62}$$

Since the exponential of a sum can be written as a product of exponentials, Eq. (27.61) can be written as a product of terms of the form

$$\exp[y\sigma_i\sigma_{i+1} + (x/2)(\sigma_i + \sigma_{i+1})]. \tag{27.63}$$

Such a term depends only on the product $\sigma_i\sigma_{i+1}$, which can take on only the values ± 1, and the sum $\sigma_i + \sigma_{i+1}$, which can take on only the values $-2, 0, 2$, irrespective of the value of i. Therefore, for *any* value of i, the expression given by Eq. (27.63) can take on only the values $\exp(x + y)$, $\exp(-y)$, and $\exp(x - y)$. Thus, the partition function can be written in the form

$$Z = \sum_{\{\sigma_i\}=\pm 1} \langle\sigma_1|\hat{P}|\sigma_2\rangle\langle\sigma_2|\hat{P}|\sigma_3\rangle\cdots\langle\sigma_{\mathcal{N}-1}|\hat{P}|\sigma_{\mathcal{N}}\rangle\langle\sigma_{\mathcal{N}}|\hat{P}|\sigma_1\rangle, \tag{27.64}$$

where $\hat{P}$ is an operator with matrix representation

$$P = \begin{pmatrix} e^{y+x} & e^{-y} \\ e^{-y} & e^{y-x} \end{pmatrix}. \tag{27.65}$$

In this matrix representation, the bra $\langle 1|$ is represented by the row vector $(1,0)$, the bra $\langle -1|$ is represented by the row vector $(0,1)$, and the kets $|1\rangle$ and $|-1\rangle$ are represented by their respective transposes, which are column vectors. For example,

$$\langle 1|\hat{P}|1\rangle = \begin{pmatrix} 1 & 0 \end{pmatrix}\begin{pmatrix} e^{y+x} & e^{-y} \\ e^{-y} & e^{y-x} \end{pmatrix}\begin{pmatrix} 1 \\ 0 \end{pmatrix} = \begin{pmatrix} 1 & 0 \end{pmatrix}\begin{pmatrix} e^{y+x} \\ e^{-y} \end{pmatrix} = e^{y+x}, \tag{27.66}$$

whereas

$$\langle -1|\hat{P}|1\rangle = \begin{pmatrix} 0 & 1 \end{pmatrix}\begin{pmatrix} e^{y+x} & e^{-y} \\ e^{-y} & e^{y-x} \end{pmatrix}\begin{pmatrix} 1 \\ 0 \end{pmatrix} = \begin{pmatrix} 0 & 1 \end{pmatrix}\begin{pmatrix} e^{y+x} \\ e^{-y} \end{pmatrix} = e^{-y}. \tag{27.67}$$

Since $\sum_{\sigma_i=\pm 1}|\sigma_i\rangle\langle\sigma_i| = |1\rangle\langle 1| + |-1\rangle\langle -1|$ is equal to the unit operator in a two-state space for any value of i, we have

$$Z = \sum_{\sigma_1=\pm 1}\langle\sigma_1|\hat{P}^{\mathcal{N}}|\sigma_1\rangle = \text{trace}\,\hat{P}^{\mathcal{N}} = \lambda_1^{\mathcal{N}} + \lambda_2^{\mathcal{N}}, \tag{27.68}$$

where λ_1 and λ_2 are the eigenvalues of the matrix P.

The partition function can therefore be calculated by diagonalizing the matrix P, which can be accomplished by solving

$$\det\begin{vmatrix} e^{y+x}-\lambda & e^{-y} \\ e^{-y} & e^{y-x}-\lambda \end{vmatrix} = 0. \tag{27.69}$$

This results in

$$\lambda^2 - 2\lambda e^{y}\cosh x + e^{2y} - e^{-2y} = 0, \tag{27.70}$$

which yields the eigenvalues

$$\lambda_{1,2} = e^{y}\cosh x \pm \sqrt{e^{2y}\sinh^2 x + e^{-2y}}, \tag{27.71}$$

where the plus sign goes with subscript 1.

We first examine limiting cases and then the general case. For $y = 0$, which is the case of noninteracting spins in a magnetic field, we obtain $\lambda_1 = 2\cosh x$ and $\lambda_2 = 0$ resulting in $Z = (2\cosh x)^{\mathcal{N}}$, which is the familiar result for a two-state paramagnetic system. In this case, the internal energy is $U/\mathcal{N} = -\mu^* B\tanh x$, the magnetization is $M/\mathcal{N}\mu^* = \tanh x$, and the entropy is $S/\mathcal{N}k_B = x - x\tanh x + \ln(1+e^{-2x})$, which goes to 0 for $T = 0$ and to $\ln 2$ for $T = \infty$ as expected.

For $x = 0$, which is the case of spin-spin interaction but in a zero magnetic field, we obtain $\lambda_1 = 2\cosh y$ and $\lambda_2 = 2\sinh y$, resulting in

$$Z = (2\cosh y)^{\mathcal{N}} + (2\sinh y)^{\mathcal{N}} = (2\cosh y)^{\mathcal{N}}\left[1 + (\tanh y)^{\mathcal{N}}\right] \tag{27.72}$$

in agreement with Eq. (27.59). The factor $[1 + (\tanh y)^{\mathcal{N}}]$ in Eq. (27.72) ranges in value between 1 and 2 and turns out not to be important in calculating the energy, although it could be kept for aesthetic reasons to get an entropy of $S = k_B\ln 2$ at $T = 0$ because of the doubly degenerate ground state (all $\sigma_i = 1$ or all $\sigma_i = -1$). However, this small entropy at $T = 0$ is not of order $\mathcal{N}$ and is an unimportant technicality, as we shall see subsequently.

The internal energy is

$$U = -\frac{\partial \ln Z}{\partial \beta} = -\mathcal{N}J\left\{\tanh y + \frac{(\tanh y)^{\mathcal{N}-1}\text{sech}^2 y}{1+(\tanh y)^{\mathcal{N}}}\right\}, \tag{27.73}$$

which may also be written

$$U = -\frac{\partial \ln Z}{\partial \beta} = -\mathcal{N}J \tanh y\left[\frac{1+a^{\mathcal{N}-2}}{1+a^{\mathcal{N}}}\right], \tag{27.74}$$

where $a := \tanh y$. Since $0 \le a \le 1$, the term in square brackets involving a is nearly equal to 1 for large $\mathcal{N}$. For $\mathcal{N} > 2$ it is equal to 1 at $a = 0$ and $a = 1$. It has a maximum[6] value of $1 + 0.55693/\mathcal{N}$ for very large $\mathcal{N}$ near $a = 1$. Therefore, the term in brackets in Eq. (27.74) can be ignored and the internal energy for $B = 0$ becomes

$$U_0 = -\mathcal{N}J \tanh y. \tag{27.75}$$

The corresponding heat capacity for $B = 0$ is therefore

$$C_0 = \frac{\partial U_0}{\partial T} = -\mathcal{N}k_B(\beta J)^2 \text{sech}^2 y. \tag{27.76}$$

As T increases, C_0 increases from zero, passes through a smooth peak, and then decreases to zero at high T. Thus, there is no sign of a phase transition at any $T > 0$.

The entropy is given by

$$S/k_B = \beta U + \ln Z = \beta U + \mathcal{N}y + \mathcal{N}\ln(1+e^{-2y}) + \ln\left[1+(\tanh y)^{\mathcal{N}}\right]. \tag{27.77}$$

For $T \to \infty$, $y \to 0$, and $S/k_B \to \mathcal{N}\ln 2$ because the populations of $\sigma_i = \pm 1$ are equal. For $T \to 0$, $y \to \infty$, $\beta U \to -\mathcal{N}y$ (which cancels the $\mathcal{N}y$ term from $\ln Z$), $\ln(1+e^{-2y}) \to 0$ and $\tanh y \to 1$, so only the last term contributes, resulting in $S/k_B \to \ln 2$, because of the doubly degenerate ground state mentioned above. Since, however, the last term in Eq. (27.77) is not of order $\mathcal{N}$, it can be dropped. Thus the entropy for $B = 0$ is given by

$$S_0/k_B = \mathcal{N}y(1-\tanh y) + \mathcal{N}\ln(1+e^{-2y}). \tag{27.78}$$

Of more interest is the magnetization which can be calculated from the full partition function given by Eq. (27.68), namely from

$$M = k_B T \frac{\partial}{\partial B}\ln[\lambda_1^{\mathcal{N}} + \lambda_2^{\mathcal{N}}], \tag{27.79}$$

[6]The maximum occurs at values of a that satisfy $\mathcal{N}a^2 + 2a^{\mathcal{N}} = \mathcal{N} - 2$. We can obtain an approximate solution by setting $a = 1 - b/\mathcal{N}$ and noting that $a^{\mathcal{N}} = e^{-b}$ as $\mathcal{N} \to \infty$. Thus a root occurs at approximately $a^2 = 1 - (2/\mathcal{N})(1+e^{-b})$ or $a = 1 - (1/\mathcal{N})(1+e^{-b})$. This gives $1 + e^{-b} = b$ whose solution is $b = 1.27846$. Then the factor in brackets in Eq. (27.74) becomes approximately $1 + (1/\mathcal{N})(2be^{-b})/(1+e^{-b}) = 1 + 0.55693/\mathcal{N}$. A numerical solution of the exact equations gives a corresponding value of $1 + 5.5684 \times 10^{-6}$ for $\mathcal{N} = 10^5$, in good agreement.

which results in

$$\frac{M}{\mathcal{N}\mu^*} = \frac{\lambda_1^{\mathcal{N}-1}\partial\lambda_1/\partial x + \lambda_2^{\mathcal{N}-1}\partial\lambda_2/\partial x}{\lambda_1^{\mathcal{N}} + \lambda_2^{\mathcal{N}}}. \tag{27.80}$$

But

$$\frac{\partial\lambda_{1,2}}{\partial x} = \pm\frac{e^y \sinh x}{\sqrt{e^{2y}\sinh^2 x + e^{-2y}}}\lambda_{1,2}, \tag{27.81}$$

where the plus sign goes with subscript 1 and the minus sign goes with subscript 2. The magnetic moment is therefore given by

$$\frac{M}{\mathcal{N}\mu^*} = \frac{\sinh x}{\sqrt{\sinh^2 x + e^{-4y}}}\frac{\lambda_1^{\mathcal{N}} - \lambda_2^{\mathcal{N}}}{\lambda_1^{\mathcal{N}} + \lambda_2^{\mathcal{N}}}. \tag{27.82}$$

The important thing to notice about Eq. (27.82) is that it is proportional to $\sinh x$ which goes to zero as $x \to 0$, while all of the other factors remain finite. Since $x = \beta\mu^* B$, this means there is no spontaneous magnetization for $B = 0$ at any $T > 0$. In other words, *this exact solution to the one-dimensional Ising model displays no phase transition, contrary to the mean field model in one dimension. In this respect, the mean field model is qualitatively incorrect.* The same conclusion would follow from neglecting the smaller eigenvalue λ_2 in comparison to λ_1, in which case the last factor in Eq. (27.82) would be unity.

In the case that the interaction between spins is very strong, such that $y = J/k_B T \gg 1$, one has approximately

$$\lambda_{1,2} \approx e^y \cosh x \pm e^y \sinh x = e^{y \pm x}. \tag{27.83}$$

In that case, as $y \to \infty$, Eq. (27.82) becomes

$$M \approx \mathcal{N}\mu^* \tanh \mathcal{N}x \tag{27.84}$$

which has a very large slope proportional to $\mathcal{N}^2$ as $x \to 0$. This is sometimes interpreted to suggest that a phase transition is about to happen at $T = 0$, that is, effectively $T_c = 0$. An alternative interpretation would be to note that *a very small magnetic field would lead to the saturation magnetization* $M = \mathcal{N}\mu^*$ as T approaches zero. Such a field would need to satisfy $\mu^* B > k_B T/\mathcal{N}$.

27.5 Other Methods of Solution

The Ising model has been solved exactly in two dimensions for several lattices and approximately by various methods in spaces of higher dimensionality. See Pathria and Beale [9, p. 488] for an extensive discussion of two-dimensional Ising models and several related models. They report critical values of $K_c = J/k_b T_c$ for several exact solutions. For the Onsager solution of the square lattice, previously mentioned, $K_c = (1/2)\sinh^{-1}(1) = (1/2)\ln(\sqrt{2} + 1) \approx 0.4407$. For a triangular lattice, $K_c = (1/2)\sinh^{-1}(1/\sqrt{3}) \approx 0.2747$ and for a honeycomb lattice, $K_c = (1/2)\sinh^{-1}(\sqrt{3}) \approx 0.6585$. In three dimensions, numerical solutions [70] yield $K_c = 0.36982$, 0.15740, and

0.10209 for diamond cubic, FCC, and BCC lattices, respectively. In many cases, associated critical exponents have been calculated.

Even though the Ising model is quite simple, it has stimulated a great deal of activity and has led to important insight that is useful in understanding more realistic models. Renormalization group (RG) methodology, discussed very briefly below, has been used extensively to simulate the Ising model and has led to major advances in the study of phase transitions in more realistic models.

27.6 Monte Carlo Simulation

As we have seen, the solution of the Ising model by means of the mean field approximation is incorrect because correlations among the spins are not taken into account. Monte Carlo (MC) simulation is an important tool that can be used to include such correlations. It can also be used to solve many other problems in statistical physics as well as other fields. It is a huge subject to which we can only give an introduction. For comprehensive treatments, the reader is referred to a number of recent books, [71–74].

27.6.1 MC Simulation of the Ising Model

We introduce computer simulation by using MC methods to treat the Ising model in two dimensions for a square lattice. The basic idea is to work with a square system having $n \times n$ spins, each of which can take on the values $s_i = \pm 1$. We define a **configuration** of the system to be a specification of the set $\{s_i\}$ of $\mathcal{N} = n^2$ spin values. It is convenient to think of $\{s_i\}$ as a vector $\mathbf{s}$ with components $s_1, s_2, \ldots, s_{\mathcal{N}}$. In the absence of a magnetic field, the energy of such a configuration is taken to be

$$E(\mathbf{s}) = -\frac{J}{2}\sum_{i,j}^{nn} s_i s_j = -J\sum_{i,j}^{nnp} s_i s_j, \tag{27.85}$$

where the first sum is over nearest neighbors and the second sum is over nearest-neighbor pairs. The objective of the simulation is to find a set of configurations such that the probability $P(\mathbf{s})$ of any given configuration is proportional to its Boltzmann factor,

$$P(\mathbf{s}) \propto \exp[-\beta E(\mathbf{s})]. \tag{27.86}$$

This can be accomplished by taking a random walk through configuration space in steps called MC steps. At the end of the kth step, we suppose the configuration to be in a state $\mathbf{s}$ and then proceed by means of a rule, to be discussed below, to establish a configuration $\mathbf{s}'$ at the next MC step, $k + 1$. This is accomplished by means of a **Markov process** [75, p. 135], according to which the conditional transition probability $W_k(\mathbf{s} \to \mathbf{s}')$ to the state $\mathbf{s}'$, given the occurrence of the state $\mathbf{s}$ at step k, depends only on the previous state $\mathbf{s}$, independent of any prior state $\mathbf{s}''$ at step $p < k$. This process is repeated a large number of times, resulting in the generation of a so-called **Markov chain**. The steps in configuration space are often referred to as MC time steps that are imagined to take place at equal intervals of some dimensionless (but not continuous) MC time, $t = k$. However, the progression through configuration space by means of MC time steps should not be confused with following the dynamics of the system in real time, as would take place in a simulation called molecular dynamics.[7]

We shall proceed to discuss a particular algorithm, usually referred to the **Metropolis algorithm** [76]. This algorithm employs MC sampling methods for a Markov process that leads to the Boltzmann distribution. It has been generalized by Hastings [77] to treat many other problems by similar methods. The Metropolis algorithm can be implemented by beginning with some arbitrary initial configuration, say $\mathbf{s}$, having energy $E(\mathbf{s})$, randomly selecting a given spin, reversing its value ($1 \rightarrow -1$ or $-1 \rightarrow 1$) and calculating the energy $E(\mathbf{s}')$ of a *trial* configuration at step 1. Such a spin can be selected by generating[8] a pseudo-random number, r between 0 and 1, and comparing $\mathcal{N}r$ with the number used to label each spin, $1, 2, \ldots, \mathcal{N}$, to see which is closest. In event that the selected spin is on the border of the $n \times n$ square, one uses periodic boundary conditions (in the x- and y-directions) to ascertain the spin of any missing nearest neighbor. Then at step 1 the trial configuration $\mathbf{s}'$ is rejected or accepted according to the following rules depending on the energy difference $\Delta E(\mathbf{s}', \mathbf{s}) = E(\mathbf{s}') - E(\mathbf{s})$:

- If $\Delta E(\mathbf{s}', \mathbf{s}) < 0$, the trial configuration $\mathbf{s}'$ is accepted and becomes the actual configuration at the next MC time step (initially, time step 1).
- If $\Delta E(\mathbf{s}', \mathbf{s}) \geq 0$, the configuration at the next MC time step is the trial configuration $\mathbf{s}'$ with probability $\exp[-\beta \Delta E(\mathbf{s}', \mathbf{s})]$, but reverts to the former configuration $\mathbf{s}$ with probability $[1 - \exp[-\beta \Delta E(\mathbf{s}', \mathbf{s})]$. This can be accomplished by comparison of a pseudo-random number r' between 0 and 1 with the Boltzmann factor $\exp[-\beta \Delta E(\mathbf{s}', \mathbf{s})]$.

This same process is then repeated to progress from step 1 to step 2, etc., until a very large number $\mathcal{N}' \gg \mathcal{N}$ of MC steps has been taken. The MC chain will begin to follow a trajectory in configuration space that corresponds approximately to the Boltzmann distribution.[9] Then by studying a correlation function between the configuration $\mathbf{s}$ at step q and $\mathbf{s}''$ at step $q - m$ for sufficiently large $q > \mathcal{N}'$, an interval of m MC time steps can established beyond which correlations become negligible. This establishes a dimensionless MC correlation time, $\tau = m$. At that stage, one can begin to store these statistically independent configurations at intervals of p steps for some $p > m$ and this set of configurations is deemed to be representative of a Boltzmann distribution of configurations. From that distribution, various quantities of interest can be computed; for example, the average value of a spin or the correlation of spins separated by a given distance. As discussed below, other considerations are necessary to obtain an efficient simulation.

[7]For a classical system, molecular dynamics would be accomplished by integrating numerically Newton's equations for a system of $\mathcal{N}$ particles, given some initial condition.

[8]A number of algorithms for generating pseudo-random numbers are readily available. See [73, chapter 16] for an extensive discussion.

[9]Theorems for MC chains [75, p. 142] exist to demonstrate some conditions for which this will occur.

So what is the physical basis of the Metropolis algorithm? It is based on a so-called master equation of the form[10]

$$P_{k+1}(\mathbf{s}) - P_k(\mathbf{s}) = \sum_{s'} \left\{-W_k(\mathbf{s} \to \mathbf{s}')P_k(\mathbf{s}) + W_k(\mathbf{s}' \to \mathbf{s})P_k(\mathbf{s}')\right\}. \tag{27.87}$$

In Eq. (27.69), the quantities $P_k(\mathbf{s})$ represent the probability of being in the state $\mathbf{s}$ at step k. However, once an equilibrium distribution has been established, $P_{k+1}(\mathbf{s}) - P_k(\mathbf{s}) = 0$, so the quantities $P_k(\mathbf{s})$ become independent of k. Specifically, we want them to tend to the Boltzmann distribution

$$P_k(\mathbf{s}) \to P(\mathbf{s}) = (1/Z)\exp[-\beta E(\mathbf{s})], \tag{27.88}$$

where Z is the partition function needed to normalize P. Then Eq. (27.87) becomes

$$0 = \sum_{s'} \left\{-W_k(\mathbf{s} \to \mathbf{s}')\exp[-\beta E(\mathbf{s})] + W_k(\mathbf{s}' \to \mathbf{s})\exp[-\beta E(\mathbf{s}')]\right\}, \tag{27.89}$$

where the partition function has been canceled.

As a guide to finding an algorithm that will lead to the desired distribution, we want to be sure that all states of the system are accessible, even though their probabilities may be small. In the language of MC simulations, this is referred to as "ergodicity," but should not be confused with the ergodic hypothesis for the microcanonical ensemble in classical statistical mechanics [14, p. 144]. We return briefly to the master equation Eq. (27.87) and note that $\sum_{\mathbf{s}'} W_k(\mathbf{s} \to \mathbf{s}') = 1$, so it can be rewritten as

$$P_{k+1}(\mathbf{s}) = \sum_{s'} W_k(\mathbf{s}' \to \mathbf{s})P_k(\mathbf{s}'), \tag{27.90}$$

which has the form of a matrix equation except the matrix is stochastic. As $k \to \infty$, we want $P_\infty(\mathbf{s})$ to approach the Boltzmann distribution. But we want to avoid a so-called limit cycle in which the system, which starts in some state $P_0(\mathbf{s}'')$, reaches a dynamic equilibrium in which only a subset of states of the system are visited [73, p. 37].

With the foregoing considerations in mind, we need to remember that we are not following the true dynamics of the system, so all we need is an algorithm that leads efficiently to the correct distribution. This can be accomplished by making use of the **principle of detailed balance**, according to which we satisfy Eq. (27.89) by making *each term in the sum equal to zero*, resulting in

$$W_k(\mathbf{s} \to \mathbf{s}')\exp[-\beta E(\mathbf{s})] = W_k(\mathbf{s}' \to \mathbf{s})\exp[-\beta E(\mathbf{s}')]. \tag{27.91}$$

[10]In the MC literature, one often writes this equation with the notation $P_{\mathbf{s}}(t) \equiv P_k(\mathbf{s})$, where $t = k$ is dimensionless MC time. Then $P_{k+1}(\mathbf{s}) - P_k(\mathbf{s}) = P_{\mathbf{s}}(t+1) - P_{\mathbf{s}}(t)$. In that case, $P_{\mathbf{s}}(t+1) - P_{\mathbf{s}}(t)$ would be the finite forward difference approximation to the derivative $dP_{\mathbf{s}}(t)/dt$ and Eq. (27.87) could be written as a differential equation with the quantities W regarded as transition rates. Although this is common, it is misleading so we avoid its use.

Although Eq. (27.91) is not necessary to satisfy Eq. (27.89), it is a sufficient condition. It can be written in the form

$$W_k(\mathbf{s} \to \mathbf{s}') = \exp[-\beta \Delta E(\mathbf{s}', \mathbf{s})] W_k(\mathbf{s}' \to \mathbf{s}), \tag{27.92}$$

where $\Delta E(\mathbf{s}', \mathbf{s}) = E(\mathbf{s}') - E(\mathbf{s})$, so we only have to deal with energy differences of configurations. Since the factor $\exp[-\beta \Delta E(\mathbf{s}', \mathbf{s})]$ is never zero, there will always be a *nonzero probability* of returning from $\mathbf{s}'$ to $\mathbf{s}$ if there is *nonzero probability* of going from $\mathbf{s}$ to $\mathbf{s}'$, so there is no possibility of a limit cycle.

The Metropolis algorithm is a convenient and efficient way of satisfying Eq. (27.92). As mentioned above, we start with a state $\mathbf{s}$ and select a state $\mathbf{s}'$ at random. Then we can choose to reject or accept that state such that the probability

$$W_k(\mathbf{s} \to \mathbf{s}') = W_0 \begin{cases} 1 & \text{for } \Delta E(\mathbf{s}', \mathbf{s}) < 0 \\ \exp[-\beta \Delta E(\mathbf{s}', \mathbf{s})] & \text{for } \Delta E(\mathbf{s}', \mathbf{s}) \geq 0. \end{cases} \tag{27.93}$$

Then evidently

$$W_k(\mathbf{s}' \to \mathbf{s}) = W_0 \begin{cases} 1 & \text{for } \Delta E(\mathbf{s}, \mathbf{s}') < 0 \Rightarrow \Delta E(\mathbf{s}', \mathbf{s}) \geq 0 \\ \exp[-\beta \Delta E(\mathbf{s}, \mathbf{s}')] & \text{for } \Delta E(\mathbf{s}, \mathbf{s}') \geq 0 \Rightarrow \Delta E(\mathbf{s}', \mathbf{s}) < 0. \end{cases} \tag{27.94}$$

For $\Delta E(\mathbf{s}', \mathbf{s}) < 0$, we can substitute the top line of Eq. (27.93) and the bottom line of Eq. (27.94) into Eq. (27.92) and see that it is satisfied. Similarly, for $\Delta E(\mathbf{s}', \mathbf{s}) \geq 0$, we can substitute the bottom line of Eq. (27.93) and the top line of Eq. (27.94) into Eq. (27.92) and see that it is satisfied. Since $W_0 \neq 0$ can be canceled after these substitutions, it can be chosen for convenience. A very efficient choice is $W_0 = 1$, which leads to the maximum probability that the new state will be accepted. With $W_0 = 1$, Eq. (27.93) gives the Metropolis algorithm.[11]

Although the above description of a MC simulation presents the basic methodology, it omits many practical considerations. For example, even for a fairly small system with $n = 50$, $\mathcal{N} = n^2 = 2500$, so there are $2^{2500} \approx 10^{753}$ possible configurations. In principle, one could calculate the Boltzmann factor for each of them, sum the results to get a partition function, and hence calculate the Boltzmann probabilities for each, but that would involve so much computation that it is absurd. Fortunately, most such configurations have much higher energies than others, and therefore much smaller Boltzmann factors, so small they are negligible. The Metropolis algorithm avoids this problem by sampling only those configurations that have a significant probability in the Boltzmann distribution (**Boltzmann sampling**). This technique is an example of **importance sampling** which makes MC simulation tractable for many other applications.

Nevertheless, one must still develop practical criteria to decide the number $\mathcal{N}'$ of iterations that are needed for the Markov chain to settle into an approximation of the Boltzmann distribution. Moreover, system size will be limited by the actual time and cost that a computer must run to accurately compute and store the equilibrium distribution.

[11] Since we are using the condition of detailed balance, only two configurations are involved in updating from MC step k to step $k+1$. So if $\mathbf{s}$ does not become $\mathbf{s}'$ at step $k+1$, it remains $\mathbf{s}$ with probability $[1 - \exp[-\beta \Delta E(\mathbf{s}', \mathbf{s})]]$.

Fortunately, the problem has been well-studied and efficient algorithms have been devised. Some of these sample the spins in some order until all $\mathcal{N}$ spins have been sampled at least once, a so-called MC sweep, and then rely on empirical rules to decide how many MC sweeps are needed to calculate a MC Boltzmann chain with reasonable accuracy [73, p. 55]. See also [78, 79] for some specialized techniques. Empirical rules can be established by carrying out the simulation for systems for which analytical solutions are available. See figure 16.1 of [9, p. 643] for a graph of the specific heat of the two-dimensional Ising model calculated by MC simulation as compared to that calculated from the exact solution. In that case, for $n = 128$, 10^5 sweeps gives good agreement except near the critical temperature where 10^6 sweeps are necessary. In general, empirical rules to decide the accuracy of a simulated equilibrium distribution must be established by running the simulation even longer and comparing with previous results. In any case, one should also run the simulation with different initial conditions to see if the results are statistically equivalent.

Just looking at the configurations produced by MC simulation can reveal patterns that are very different at high and low temperatures. At low temperatures, differences in the energies of configurations are extremely important and one can see large islands of spins of the same kind. At high temperatures, differences in energy of configurations are not so important and the resulting patterns show much smaller clusters of each spin in no particular arrangement. Results can also be analyzed quantitatively by generating a large set $\{\mathbf{s}_i^{\rm MC}\}$ of statistically independent configurations and taking the averages $\langle\cdots\rangle_{\rm MC}$ of various quantities with respect them, each weighted equally with probability $1/\mathcal{N}_{\rm MC}$. For example, one could compute the average value of an individual spin,

$$\langle s\rangle = \left\langle \frac{\sum_{i=1}^{\mathcal{N}} s_i}{\mathcal{N}} \right\rangle_{\rm MC} = \frac{1}{\mathcal{N}_{\rm MC}} \sum_{i=1}^{\mathcal{N}_{\rm MC}} \left(\frac{\sum_{j=1}^{\mathcal{N}} s_j}{\mathcal{N}} \right)_{\{\mathbf{s}_i^{\rm MC}\}}. \tag{27.95}$$

To analyze patterns, one could choose $\mathcal{N}_{ij}$ pairs of spins $(s_i s_j)_d$ that are separated by a distance d and compute a correlation function of the form

$$C(d) = \left\langle \frac{\sum_{ij} (s_i s_j)_d}{\mathcal{N}_{ij}} \right\rangle_{\rm MC} - \langle s\rangle^2. \tag{27.96}$$

Study of $C(d)$ as a function of d would help to quantify the cluster sizes viewed in patterns. It can also be used to establish a correlation length ξ beyond which $C(\xi)$ becomes negligibly small.

Near a critical point, MC simulations become difficult because the correlation length ξ becomes very large. Thus large systems and long-run times would be necessary to obtain accuracy. This problem can be alleviated by using the **renormalization group** (RG) approach. As suggested by Kadanoff in 1966 [80], the basic idea is to perform a length scaling that leads to an approximately equivalent problem with scaled coupling constants, such as $J \to J'$, now known as a **Kadanoff transformation**. The success of the technique is based on the idea that aspects of the problem, such as the existence of a phase transition, are insensitive to the lattice constant a. Specifically, for a new lattice constant $a' = \ell a$, where $\ell > 1$, there is insensitivity of results provided $\ell a \ll \xi$ and conditions are close to criticality. This scaling-up idea can also be viewed as removing spins from the system, or more generally as reducing the number of degrees of freedom of a more general system, a process known as **decimation**. A systematic way of handling transformations based on this idea was developed later by Wilson [81, 82] by means of RG theory. By using such techniques, one can begin with very weak coupling constants, for which an approximate solution is possible. Then by successive scalings, one can use a recurrence relation to step up to values of the coupling constants or other parameters that are of interest. A successful application of this technique will result in successive transformations leading to a fixed point corresponding to criticality in parameter space. A detailed presentation of RG techniques is beyond the scope of this book. For a lucid introduction see chapter 5 of Chandler [12]; for a more extensive treatment, including the RG formulation, see chapter 14 of Pathria and Beale [9].

Other types of sampling can be accomplished by doing a MC simulation for a given problem and using the configurations so obtained to simulate a different problem. We illustrate this for two cases, the first involving a different energy but the same temperature, and the second involving a change in temperature for the same energy.

In the first case, suppose that

$$E(\mathbf{s}) = E_0(\mathbf{s}) + E_1(\mathbf{s}). \tag{27.97}$$

Then for $E_0(\mathbf{s})$ we have a probability and partition function given by

$$P_0(\mathbf{s}) = (Z_0)^{-1} \exp[-\beta E_0(\mathbf{s})]; \quad Z_0 = \sum_{\mathbf{s}} \exp[-\beta E_0(\mathbf{s})]. \tag{27.98}$$

By using MC simulation, we obtain a set of configurations $\{\mathbf{s}_i^0\}, i = 1, 2, \ldots, \mathcal{N}_{\text{MC}}$ that approximate $P_0(\mathbf{s})$ if they are equally weighted with probability $1/\mathcal{N}_{\text{MC}}$. Then the average value of some quantity $R(\mathbf{s})$ is given by

$$\langle R \rangle_0 = \sum_{\mathbf{s}} P_0(\mathbf{s}) R(\mathbf{s}) \approx (\mathcal{N}_{\text{MC}})^{-1} \sum_{i=1}^{\mathcal{N}_{\text{MC}}} R(\{\mathbf{s}_i^0\}). \tag{27.99}$$

For $E(\mathbf{s})$ we have

$$P(\mathbf{s}) = Z^{-1} \exp[-\beta E(\mathbf{s})] = Z^{-1} Z_0 P_0(\mathbf{s}) \exp[-\beta E_1(\mathbf{s})], \tag{27.100}$$

where

$$Z = \sum_{\mathbf{s}} \exp[-\beta E_0(\mathbf{s})] \exp[-\beta E_1(\mathbf{s})] = Z_0 \sum_{\mathbf{s}} P_0(\mathbf{s}) \exp[-\beta E_1(\mathbf{s})]. \tag{27.101}$$

Thus,

$$P(\mathbf{s}) = \frac{P_0(\mathbf{s}) \exp[-\beta E_1(\mathbf{s})]}{\sum_{\mathbf{s}} P_0(\mathbf{s}) \exp[-\beta E_1(\mathbf{s})]} = \frac{P_0(\mathbf{s}) \exp[-\beta E_1(\mathbf{s})]}{\langle \exp[-\beta E_1(\mathbf{s})] \rangle_0}. \tag{27.102}$$

Then the average value of $R(\mathbf{s})$ is given by

$$\langle R \rangle = \sum_{\mathbf{s}} P(\mathbf{s})R(\mathbf{s}) = \frac{\langle R(\mathbf{s}) \exp[-\beta E_1(\mathbf{s})] \rangle_0}{\langle \exp[-\beta E_1(\mathbf{s})] \rangle_0}. \tag{27.103}$$

When the averages $\langle \cdots \rangle_0$ in Eq. (27.103) are computed by the right-hand member of Eq. (27.99), which is only approximate, accurate results are only expected if $E_1(\mathbf{s})$ is a small perturbation.

The second case is somewhat similar except we use MC simulation to approximate the Boltzmann distribution

$$P(\mathbf{s}, \beta) = [Z(\beta)]^{-1} \exp[-\beta E(\mathbf{s})]; \quad Z(\beta) = \sum_{\mathbf{s}} \exp[-\beta E(\mathbf{s})], \tag{27.104}$$

resulting in a set of configurations $\{\mathbf{s}_i(\beta)\}, i = 1, 2, \ldots, \mathcal{N}_{\mathrm{MC}}$. The average value of some $R(\mathbf{s})$ corresponding to β is then

$$\langle R \rangle_\beta = \sum_{\mathbf{s}} P(\mathbf{s}, \beta)R(\mathbf{s}) \approx (\mathcal{N}_{\mathrm{MC}})^{-1} \sum_{i=1}^{\mathcal{N}_{\mathrm{MC}}} R(\{\mathbf{s}_i(\beta)\}). \tag{27.105}$$

Then we change the temperature by changing β to $\beta + \Delta\beta$ and seek to evaluate $P(\mathbf{s}, \beta + \Delta\beta)$. By using steps similar to those used to treat the first case above, we find

$$P(\mathbf{s}, \beta + \Delta\beta) = \frac{P(\mathbf{s}, \beta) \exp[-\Delta\beta E(\mathbf{s})]}{\langle \exp[-\Delta\beta E(\mathbf{s})] \rangle_\beta} \tag{27.106}$$

and

$$\langle R \rangle_{(\beta+\Delta\beta)} = \frac{\langle R(\mathbf{s}) \exp[-\Delta\beta E(\mathbf{s})] \rangle_\beta}{\langle \exp[-\Delta\beta E(\mathbf{s})] \rangle_\beta}. \tag{27.107}$$

When the averages $\langle \cdots \rangle_\beta$ are evaluated from MC simulations at β, Eq. (27.107) is likely to be accurate only for small $\Delta\beta$.

Although the two cases above illustrate how the properties of the Boltzmann distribution can be used to treat changes of the Hamiltonian, or of β, by MC sampling, they should not be construed as efficient algorithms. Histogram methods such as those used by Ferrenburg and Swendsen [83, 84] are much more accurate, efficient, and versatile. These methods batch the results of MC simulation to generate **histograms** that depend on parameters of the problem. For example, for the Ising model in the presence of a magnetic field, one has

$$E_{J,B}(\mathbf{s}) = -J \sum_{i,j}^{nnp} s_i s_j - \mu^* B \sum_i s_i, \tag{27.108}$$

so the parameters $K := \beta J$ and $h := \beta \mu^* B$ enter the probability distribution. Associated with given K and H, one can use MC simulation to calculate histograms of values of the dimensionless spin-spin interaction, $S = \sum_{i,j}^{nnp} s_i s_j$, and the dimensionless magnetization, $M = \sum_i s_i$. Those histograms can then be used to generate histograms of S and M for $K + \Delta K$ and $h + \Delta h$ by methods similar to those discussed above.

Many other kinds of sampling can be used to treat specific problems. For an introduction to umbrella sampling, used to remove barriers or sample rare configurations, and path integral quantum MC techniques, see Chandler [12, p. 170].

27.6.2 MC Simulation of Classical Particles

MC simulations are also useful to treat systems of $\mathcal{N}$ classical particles of mass m with a Hamiltonian of the form

$$\mathcal{H} = \mathcal{T}(\mathbf{p}) + V(\mathbf{q}), \tag{27.109}$$

where $\mathbf{p}$ and $\mathbf{q}$ are $3\mathcal{N}$-dimensional vectors of momenta and coordinates, respectively. The quantity

$$\mathcal{T}(\mathbf{p}) = \sum_{i=1}^{3\mathcal{N}} p_i^2/2m \tag{27.110}$$

is the kinetic energy and $V(\mathbf{q})$ is the potential energy, usually taken to be a function of pairwise interaction energies of particles. The classical partition function is[12]

$$Z_{\mathrm{C}}^* = (h^{3\mathcal{N}}\mathcal{N}!)^{-1} \int \mathbf{dp}\mathbf{dq}\exp(-\beta\mathcal{H}) = \int \mathbf{dp}\exp(-\beta\mathcal{T}(\mathbf{p})) \int \mathbf{dq}\exp(-\beta V(\mathbf{q})). \tag{27.111}$$

The integrals over the momenta factor into Gaussian integrals that are easily evaluated to give

$$(h^{3\mathcal{N}}\mathcal{N}!)^{-1} \int \mathbf{dp}\exp(-\beta\mathcal{T}(\mathbf{p})) = (\mathcal{N}!)^{-1}(mk_{\mathrm{B}}T/2\pi\hbar^2)^{3\mathcal{N}/2} = (\mathcal{N}!)^{-1}n_{\mathrm{Q}}^{\mathcal{N}}, \tag{27.112}$$

where n_{Q} is the quantum concentration. The integral

$$Q := \int \mathbf{dq}\exp(-\beta V(\mathbf{q})) \tag{27.113}$$

plays the role of a partition function for the coordinates. Thus the normalized distribution function of any configuration $\{q_i\}$ of the coordinates is given by

$$P(\{q_i\}) = Q^{-1}\exp(-\beta V(\{q_i\}). \tag{27.114}$$

Equation (27.114) is the Boltzmann distribution of coordinate configurations that can be simulated by using the Metropolis algorithm, which can be done without knowledge of Q. For example, one of the coordinates q_i could be shifted by some small amount to position q_i' to give a trial configuration and then $\beta\Delta V = \beta(V' - V)$ can be evaluated to decide whether to keep the trial configuration. For short-range forces, this evaluation would involve only a small number of particles. This is particularly simple for simulation of particles that are hard spheres, since $\beta\Delta V$ is either zero or infinity (the latter occurring when the shift would cause hard spheres to overlap). Quantities such as the pair

[12]We have included a factor $(h^{3\mathcal{N}}\mathcal{N}!)^{-1}$ if appropriate to connect with quantum mechanics at high temperatures, but such a factor is irrelevant to the simulation of particle configurations.

correlation function $g(r)$ can be calculated from the configurations obtained from an MC simulation.

As discussed in Section 20.5, the virial theorem can be used to relate $g(r)$ to the equation of state of a nonideal gas, as given by Eq. (20.77). This equation contains the derivative $\partial u/\partial r$ of the potential function $u(r)$ for pairwise central-force interactions of the particles. For a hard-sphere gas of particles having diameter σ, $u(r)$ is a step function at σ, so the formal derivative of u is a delta function and must be handled with care. As shown by Widom [17, p. 126], a carefully taken limit leads to a hard-sphere gas pressure $p_{\rm hsg}$ given by

$$\frac{p_{\rm hsg}}{nk_{\rm B}T} = 1 - \frac{2}{3}\pi n\sigma^3 g(\sigma^+), \tag{27.115}$$

where $g(\sigma^+)$ is the value of the pair correlation function for a hard-sphere gas in the limit that r approaches σ from larger values. See Figure 20–1 and the surrounding discussion of $g(r)$. $g(\sigma^+)$ can be evaluated from the results of computer simulation as a function of n. An approximate analytical fit to the data can be represented by the Carnahan-Starling [85] equation of state

$$\frac{p_{\rm hsg}}{nk_{\rm B}T} = \frac{1+y+y^2-y^3}{(1-y)^3} \equiv W(y), \quad \text{hard-sphere gas,} \tag{27.116}$$

where $y = v_s n = (\pi/6)\sigma^3 n$ is the volume of hard spheres per total volume. The function $W(y)$ is in approximate agreement with an expansion of the pressure in terms of virial coefficients [9, p. 314].

One might wonder about the origin of the excess pressure, p^{xs}, contained in $p_{\rm hsg}$ in addition to that for an ideal gas. We shall proceed to show that p^{xs} is related to the change of excess configurational entropy when y changes. To do this, we write $p_{\rm hsg} = p_i + p^{xs}$, where $p_i = nk_{\rm B}T$ is the ideal gas pressure and

$$p^{xs} = nk_{\rm B}T\left[W(y) - 1\right]. \tag{27.117}$$

For the hard-sphere gas, there is no penetration of the spheres, so the molar internal energy, $u(T)$, depends only on the temperature, as is also the case for an ideal gas. Therefore, the differential of the entropy becomes

$$ds = \frac{du}{T} + \frac{p}{T}\,dv = \frac{du(T)}{dT}\frac{dT}{T} - \frac{p}{n^2T}\,dn = \frac{c_v(T)}{T}\,dT - \frac{p}{nT}\frac{dy}{y}, \tag{27.118}$$

where $c_v(T)$ is the molar heat capacity at constant volume. By integrating at constant T, we obtain

$$s(y,T) = -k_{\rm B}\ln y + \tilde{s}(T) - k_{\rm B}\int_0^y \left[\frac{W(x)}{x} - \frac{1}{x}\right] dx, \tag{27.119}$$

where $\tilde{s}(T)$ is a function of integration. Since $p = -nTy\big(\partial s(y,T)/\partial y\big)_T$, comparison of the differential of Eq. (27.119) with Eq. (27.118) shows that $d\tilde{s}(T) = c_v(T)/T$, so $\tilde{s}(T) = \int [c_v(T)/T]\,dT + \text{constant}$, as expected. Thus, the first term in Eq. (27.119) represents the configurational molar entropy $s_i(y)$ of an ideal gas and the last term represents the excess configurational molar entropy $s^{xs}(y)$ of the hard-sphere gas. The lower limit of the integral has been set equal to zero so that only $s_i(y) + \tilde{s}(T)$ remains as $y \to 0$. Evaluation of the integral gives

$$s^{xs}(y) = -k_B \int_0^y \left[\frac{W(x)}{x} - \frac{1}{x}\right] dx = -k_B \frac{y(4-3y)}{(1-y)^2}. \tag{27.120}$$

The excess pressure is given by

$$p^{xs} = -nyT \frac{\partial s^{xs}(y)}{\partial y} = nk_B T \left[W(y) - 1\right], \tag{27.121}$$

in agreement with Eq. (27.117).

Therefore, the excess pressure p^{xs} of the hard-sphere gas arises because the excess configurational molar entropy $s^{xs}(y)$ is a decreasing function of y. This decrease of $s^{xs}(y)$ must be related to the decrease of unoccupied volume as y increases. Although we have demonstrated this by using the Carnahan-Starling approximate function $W(y)$, the same conclusion would follow if a more accurate function were used.

Widom [17, p. 106] has shown that an approximate equation of state of a normal liquid can be obtained by adding to the hard-sphere gas pressure a term $-\alpha n^2$, where $-\alpha n < 0$ represents an average potential energy per atom due to binding forces. The essence of the argument is that the attractive forces between liquid atoms nearly cancel for a given atom but the associated potentials are additive and nearly uniform. The result is

$$p = k_B T n \frac{1 + y + y^2 - y^3}{(1-y)^3} - \alpha n^2, \tag{27.122}$$

in which the hard-sphere radius σ, contained in y, should be interpreted as an effective radius related to the repulsive part of the actual potential. If the right-hand side of Eq. (27.116) is expanded for small y to lowest order, the result is $1/(1-4y) = 1/(1-4v_s n)$. Then Eq. (27.122) becomes

$$p = \frac{k_B T}{(n^{-1} - 4v_s)} - \alpha n^2, \tag{27.123}$$

which is just another form of the van der Waals equation, Eq. (9.2), but in different units.[13] Equation (27.122) can be analyzed by the same method used to analyze the van der Waals fluid. The spinodal curve in the y, T plane is given by $v_s k_B T/\alpha = 2y/[yW(y)]'$, where the prime denotes the derivative with respect to y. The maximum of the spinodal curve occurs at $y = 0.1304$ and the critical temperature is given by $v_s k_B T/\alpha = 0.09433$.

By means of computer simulation, one finds that the hard-sphere model displays a phase transition between a hard-sphere gas at low volume fractions of the spheres and a hard-sphere crystalline solid phase at high volume fractions (see figure 16.3 of [9, p. 649]).

[13]The correspondence can be made by setting $n = \mathcal{N}_A/v$, where $\mathcal{N}_A$ is Avogadro's number and v is the volume per mole. Then $b = 4\mathcal{N}_A v_s$ and $a = \mathcal{N}_A^2 \alpha$.

The gas phase ends at $y = 0.491$ and the solid phase begins at $y = 0.543$, with co-existence of phases for volume fractions in between. For $y > 0.543$, Speedy [86] gives the pressure $p_{\rm hsc}$ of the hard-sphere crystal,

$$\frac{p_{\rm hsc}}{nk_{\rm B}T} = \frac{3}{(1-z)} - 0.5921\frac{(z-0.7072)}{(z-0.601)}, \quad \text{hard-sphere crystal}, \tag{27.124}$$

in terms of the relative solid fraction $z = n\sigma^3/\sqrt{2} = y/y_{\rm cp} = 1.35y$, where $y_{\rm cp} = \pi\sqrt{2}/6 = 0.7405$ corresponds to a close-packed FCC crystal. It is also possible to conduct simulations [87] that avoid the transition from the hard-sphere gas to the hard-sphere crystal and follow the disordered state into the metastable region where the pressure tends to infinity at $z = 0.644 \pm 0.005$, which corresponds well with the Bernal [88] packing fraction established experimentally.

The above simple example of the hard-sphere gas begins to illustrate the power of computer simulation in describing the liquid state, something that is very limited by using analytical methods alone. MC simulation has been used for simulation of many systems that involve other classical particles for decades. A favorite for simulation is the Lennard-Jones potential,

$$u(r) = 4\varepsilon\left[\left(\frac{\sigma}{r}\right)^{12} - \left(\frac{\sigma}{r}\right)^{6}\right], \tag{27.125}$$

where $\varepsilon > 0$ is an energy parameter and σ is a length parameter. The first term is strongly repulsive and its form is selected for convenience; the second is attractive and yields a force of the same form as that between electric dipoles. The potential minimum occurs at $r_{\rm min} = 2^{1/6}\sigma = 1.12\sigma$ at which $u = -\varepsilon$. The Lennard-Jones potential was used for simulations over 50 years ago that were compared to experimental results for argon [89]. More recent simulations using the Lennard-Jones potential have dealt with liquid-crystal phase transitions [90], including those involving several crystal phases [91].

As computing power has improved exponentially over the years, MC simulation has become a potent tool for the statistical study of models of materials with more realistic potentials, resulting in greater variety and accuracy of results. The reader is referred to several books cited above as well as the vast journal literature.

PART III

Appendices

A

Stirling's Approximation

In the process of going from statistical mechanics to thermodynamics, we will often use Stirling's approximation in the form

$$\ln(N!) \sim N \ln N - N, \tag{A.1}$$

which is a good approximation when N is a large number. A good approximation for $N!$ itself is

$$N! \sim N^N e^{-N}(2\pi N)^{1/2}. \tag{A.2}$$

Taking the logarithm of Eq. (A.2) gives

$$\ln(N!) \sim N \ln(N) - N + (1/2)\ln(2\pi N), \tag{A.3}$$

but the last term in Eq. (A.3) is quite negligible for large N. For example, for $N = 10^6$, the last term is 7.83 and the sum of the first two terms is 12815510.56. In statistical mechanics, we usually deal with $\ln N!$ and much larger values of N, so this extra term in Eq. (A.3) is completely negligible. In Eq. (A.2), however, its counterpart $(2\pi N)^{1/2}$ occurs as a multiplicative factor and must be kept to achieve reasonable accuracy.

For $N > 0$, it can be shown [92, p. 253] that

$$N! = N^N e^{-N}(2\pi N)^{1/2} e^{\theta/(12N)}, \tag{A.4}$$

where $0 < \theta < 1$.

For the particular case of a polynomial coefficient $\Omega = N!/(N_1!N_2!\cdots N_r!)$ where $N = \sum_{i=1}^{r} N_i$, Eq. (A.1) leads to

$$\begin{aligned}\ln \Omega &\sim N \ln N - \sum_{i=1}^{r} N_i \ln N_i - N + \sum_{i=1}^{r} N_i \\ &= N \ln N - \sum_{i=1}^{r} N_i \ln N_i \\ &= -\sum_{i=1}^{r} N_i \ln(N_i/N),\end{aligned} \tag{A.5}$$

which is an extensive function of the N_i. Note in this special case that the final result would have been obtained even if we had dropped the second term in Eq. (A.1). Expressions of this type arise frequently in statistical mechanics and are used to represent extensive thermodynamic functions, particularly the entropy.

One can use Mathematica® to compute numerical values of $N!$ either exactly or from Stirling's approximation and compare the results. Table A–1 gives some values of $\ln N!$ and its approximations according to Eqs. (A.1) and (A.3). Table A–2 gives some values of $N!$ and

Table A–1 Illustration of Accuracy of Stirling's Approximation for $\ln N!$

N	$\ln N!$	$N \ln N - N$	$N \ln N - N + (1/2)\ln(2\pi N)$
10	15.10441257	13.02585093	15.09608201
100	363.7393756	360.5170186	363.7385422
1000	5912.128178	5907.755279	5912.128095
10,000	82108.92784	82103.40372	82108.92783

Table A–2 Illustration of Accuracy of Stirling's Approximation for $N!$

N	$N!$	$(2\pi N)^{1/2}N^N e^{-N}$	$(2\pi N)^{1/2}N^N e^{-N}[1+1/(12N)]$
1	1	0.9221370	0.9989818
2	2	1.919004	1.998963
5	120	118.0192	119.9862
10	3,628,800	3,598,696	3,628,685

its approximation by Eq. (A.2) and its correction to next order by a factor of $[1 - 1/(12N)]$. Even for these small values of N, the results are quite reasonable. For numbers $N > 10^{10}$ typical of thermodynamic systems, Stirling's approximation is excellent.

One should still be cautious, however, in using Stirling's approximation for $\ln N!$ to evaluate complex expressions. For example, the probability p that a well-shuffled deck of cards, when cut into two equal parts, will contain an equal number of red and black cards in each part is given by $p = (26!/13!)^4/52! = 16232365000/74417546961 = 0.218126$. If Stirling's approximation equation (A.1) is used to evaluate $\ln p$, the result is $\ln p = 0$ which would give the ridiculous result $p = 1$. By using Eq. (A.3), one obtains $\ln p = -(1/2)\ln(13\pi/2)$ which results in $p = 0.221293$, correct within 1.5%. This numerical example illustrates that the use of Eq. (A.1) ignores the pre-factor $(2\pi N)^{1/2}$ in Eq. (A.2), which is fine for calculating logarithms, but leads to inaccurate results when those results are exponentiated to compute factorials themselves or ratios of them.

A.1 Elementary Motivation of Eq. (A.1)

Equation (A.1) can be motivated by elementary methods. We first note that

$$I(q) := \int_1^q \ln u \, du = u \ln u - u \Big|_1^q = q \ln q - q + 1. \tag{A.6}$$

For $q = N$, we can bound this integral from above and below by sums of rectangular areas (upper and lower staircases) as illustrated in Figure A–1 for $N = 10$. We obtain

$$\ln 1 + \ln 2 + \cdots + \ln(N-1) < I(N) < \ln 2 + \ln 3 + \cdots + \ln N, \tag{A.7}$$

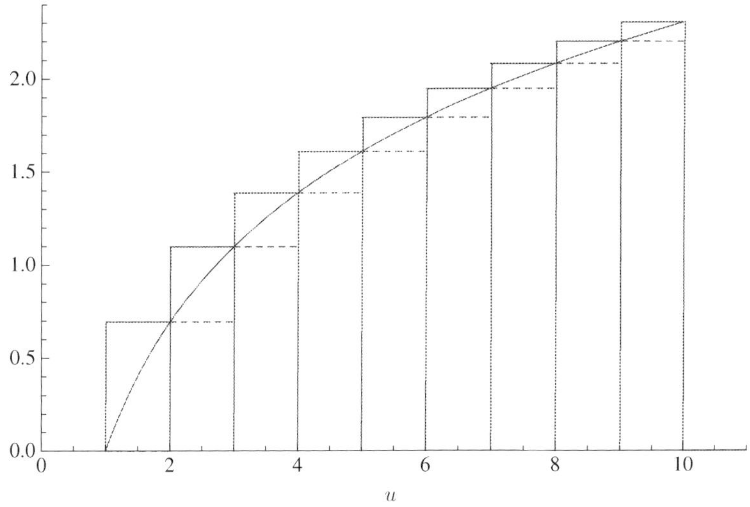

FIGURE A–1 Staircase diagram used to illustrate bounds for the area under the curve ln *u* for 1 < *u* < 10. The area under the upper staircase is larger than that under ln *u* while the area under the lower staircase (dashed) is smaller than that under ln *u*.

which can be rewritten as

$$\ln(N-1)! < N\ln N - N + 1 < \ln N!. \tag{A.8}$$

Subtracting $\ln N! + 1$ from Eq. (A.8) and dividing by $\ln N!$ we obtain

$$-\frac{(1+\ln(N))}{\ln N!} < \frac{N\ln N - N - \ln N!}{\ln N!} < -\frac{1}{\ln N!}, \tag{A.9}$$

which shows that the fractional error in Eq. (A.1) is of order $1/N$. Note also from Eq. (A.9) that Eq. (A.1) will give a slight underestimate of $\ln N!$.

A.2 Asymptotic Series

Equation (A.4) is based on Stirling's asymptotic series [92, p. 253]

$$\Gamma(x) \sim x^x e^{-x}(2\pi/x)^{1/2}\left[1 + \frac{1}{12x} + \frac{1}{288x^2} - \frac{139}{51840x^3} - \frac{571}{2488320x^4} + O\left(\frac{1}{x^5}\right)\right] \tag{A.10}$$

for the gamma function, $\Gamma(x)$. The coefficients in Eq. (A.10) are not very simple and are related to Bernoulli numbers. The gamma function is defined by the integral

$$\Gamma(x) = \int_0^\infty t^{x-1}e^{-t}\,dt \tag{A.11}$$

for the continuous variable $x > 0$. In general, $\Gamma(x+1) = x\Gamma(x)$, which may be verified for $x > 0$ by integration by parts in Eq. (A.11). For integer N, we have $\Gamma(N+1) = N!$. Another special value worth noting is $\Gamma(1/2) = \sqrt{\pi}$. The gamma function can be

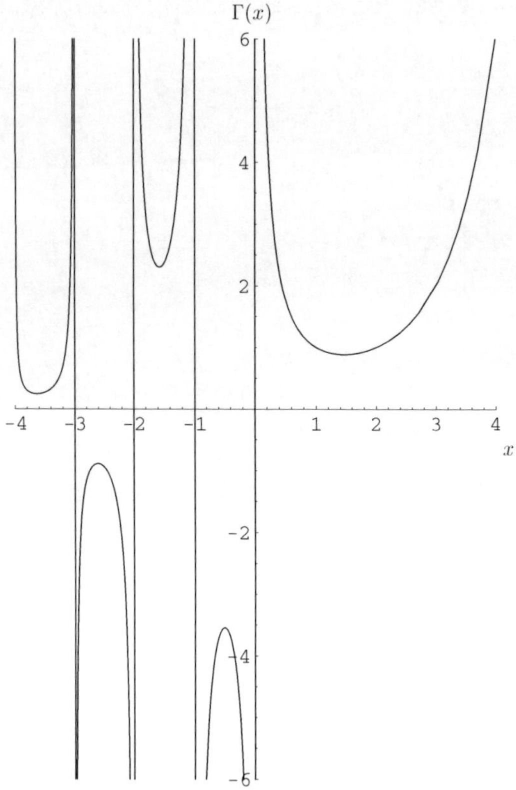

FIGURE A–2 Graph of the function $\Gamma(x)$ versus x for continuous values of x. For x equal to a positive integer $N\Gamma(N) = (N-1)!$. For N equal to zero or a negative integer, $\Gamma(N) \to \pm\infty$. Values of $\Gamma(x)$ for negative x are obtained by means of analytic continuation using $\Gamma(x) = \Gamma(x+1)/x$ and values of the function defined by Eq. (A.11). Note especially $\Gamma(1) = 0! = 1$, $\Gamma(2) = 1! = 1$, $\Gamma(3) = 2! = 2$, and $\Gamma(4) = 3! = 6$.

extended to negative values of x and to complex variables by a process known as analytical continuation. In general, $z! \equiv \Gamma(z+1) = z\Gamma(z)$, where $z = x + iy$ is a complex variable. Figure A–2 shows a graph of the function $\Gamma(x)$ versus x for real continuous values of x.

A.2.1 Asymptotic Versus Convergent Series

Asymptotic series should be contrasted with convergent series. If we speak of a convergent power series

$$f(z) = \sum_{n=0}^{\infty} a_n z^n, \tag{A.12}$$

we mean that the difference

$$\left| f(z) - \sum_{n=0}^{m} a_n z^n \right| \tag{A.13}$$

can be made as small as desired for fixed z by taking m sufficiently large. On the other hand, if the series

$$F(z) \sim \sum_{n=0}^{\infty} \frac{A_n}{z^n} \tag{A.14}$$

is asymptotic, then [92, p. 151]

$$|z|^m \left| F(z) - \sum_{n=0}^{m} \frac{A_n}{z^n} \right| \tag{A.15}$$

can be made as small as desired for fixed m by taking $|z|$ sufficiently large.[1] Thus to get more accuracy in Eq. (A.12), we take more terms; however, to get more accuracy in Eq. (A.14) we cut off the series and take larger $|z|$. In fact, for fixed z we usually *must* cut off an asymptotic series because many asymptotic series do not converge, so taking more terms might give a worse result.

A generalization of Eq. (A.14) is to say that if

$$F(z) = \frac{G(z)}{H(z)} \sim \sum_{n=0}^{\infty} \frac{A_n}{z^n}, \tag{A.16}$$

then

$$G(z) \sim H(z) \sum_{n=0}^{\infty} \frac{A_n}{z^n}. \tag{A.17}$$

We note that Eq. (A.10) is actually of the form of Eq. (A.17). Equation (A.10) can be derived by consecutive integration by parts and then proving that the remainder, after m terms, satisfies Eq. (A.15). Equation (A.4) can be proven in a similar way.

[1] As a function of a complex variable z, these results only hold in some sector $\alpha < \arg(z) < \beta$. For our purposes, we only need this sector to include z real and positive.

B

Use of Jacobians to Convert Partial Derivatives

Often in thermodynamics one is faced with the problem of converting partial derivatives with certain quantities held constant to expressions involving other partial derivatives with different quantities held constant. For example, one might want to relate the isothermal compressibility $\kappa_T = V^{-1}(\partial V/\partial P)_{T,N}$ to the isentropic (sometimes called adiabatic) compressibility $\kappa_S = V^{-1}(\partial V/\partial P)_{S,N}$. This can be done by trial and error by using the chain rule of partial differentiation together with appropriate Maxwell relations. The use of Jacobians, however, provides a systematic approach to this problem. For other treatments of this topic, see Landau and Lifshitz [7, p. 50] and the first edition of Callen [2].

B.1 Properties of Jacobians

We review briefly the definition and main properties of Jacobians. We illustrate these for three variables, but the results hold for any number of variables.

We consider the variables u, v, w that depend on x, y, z. A Jacobian is defined as a determinant of partial derivatives as follows:

$$\frac{\partial\,(u,v,w)}{\partial\,(x,y,z)} = \begin{vmatrix} \partial u/\partial x & \partial u/\partial y & \partial u/\partial z \\ \partial v/\partial x & \partial v/\partial y & \partial v/\partial z \\ \partial w/\partial x & \partial w/\partial y & \partial w/\partial z \end{vmatrix} = \begin{vmatrix} \partial u/\partial x & \partial v/\partial x & \partial w/\partial x \\ \partial u/\partial y & \partial v/\partial y & \partial w/\partial y \\ \partial u/\partial z & \partial v/\partial z & \partial w/\partial z \end{vmatrix}. \tag{B.1}$$

Interchange of two rows or two columns of a determinant gives rise to an overall minus sign. Thus, for example,

$$\frac{\partial\,(u,v,w)}{\partial\,(x,y,z)} = -\frac{\partial\,(v,u,w)}{\partial\,(x,y,z)} = \frac{\partial\,(v,u,w)}{\partial\,(y,x,z)} = -\frac{\partial\,(u,v,w)}{\partial\,(y,x,z)}. \tag{B.2}$$

If A and B are square matrices, it is well known that the determinant of their matrix product is the product of their determinants, that is, $|AB| = |A||B|$. Then by the chain rule of partial differentiation it follows that

$$\frac{\partial\,(u,v,w)}{\partial\,(x,y,z)} = \frac{\partial\,(u,v,w)}{\partial\,(r,s,t)}\frac{\partial\,(r,s,t)}{\partial\,(x,y,z)} \tag{B.3}$$

and

$$\frac{\partial\,(u,v,w)}{\partial\,(x,y,z)} = 1 \Bigg/ \frac{\partial\,(x,y,z)}{\partial\,(u,v,w)}. \tag{B.4}$$

Thus, determinants obey an algebra similar to fractions.

There is a simple connection of a determinant to a single partial derivative. Since

$$\frac{\partial(u,y,z)}{\partial(x,y,z)} = \begin{vmatrix} \partial u/\partial x & \partial u/\partial y & \partial u/\partial z \\ \partial y/\partial x & \partial y/\partial y & \partial y/\partial z \\ \partial z/\partial x & \partial z/\partial y & \partial z/\partial z \end{vmatrix} = \begin{vmatrix} \partial u/\partial x & \partial u/\partial y & \partial u/\partial z \\ 0 & 1 & 0 \\ 0 & 0 & 1 \end{vmatrix}, \tag{B.5}$$

it follows that

$$\left(\frac{\partial u}{\partial x}\right)_{y,z} = \frac{\partial(u,y,z)}{\partial(x,y,z)}. \tag{B.6}$$

B.2 Connection to Thermodynamics

One often wants to relate thermodynamic derivatives to measurable quantities such as the heat capacity at constant pressure, C_p; the isobaric coefficient of thermal expansion, α; and the isothermal compressibility, κ_T, where

$$C_p = T\left(\frac{\partial S}{\partial T}\right)_{p,N}; \quad \alpha = \frac{1}{V}\left(\frac{\partial V}{\partial T}\right)_{p,N}; \quad \kappa_T = -\frac{1}{V}\left(\frac{\partial V}{\partial p}\right)_{T,N}. \tag{B.7}$$

Example 1 We first relate the heat capacity at constant volume, namely

$$C_V = T\left(\frac{\partial S}{\partial T}\right)_{V,N}, \tag{B.8}$$

to C_p. This was done in the text (see Eq. (5.32)) by elementary methods but we now use determinants. Thus

$$C_V = T\frac{\partial(S,V,N)}{\partial(T,V,N)} = T\frac{\partial(S,V,N)}{\partial(T,p,N)}\frac{\partial(T,p,N)}{\partial(T,V,N)}. \tag{B.9}$$

We recognize that

$$\frac{\partial(T,p,N)}{\partial(T,V,N)} = 1\Big/\left(\frac{\partial V}{\partial p}\right)_{T,N} \tag{B.10}$$

and readily compute[1]

$$\frac{\partial(S,V,N)}{\partial(T,p,N)} = \left(\frac{\partial S}{\partial T}\right)_{p,N}\left(\frac{\partial V}{\partial p}\right)_{T,N} - \left(\frac{\partial S}{\partial p}\right)_{T,N}\left(\frac{\partial V}{\partial T}\right)_{p,N}. \tag{B.11}$$

This results in

$$C_V = C_p - T\left(\frac{\partial S}{\partial p}\right)_{T,N}\left(\frac{\partial V}{\partial T}\right)_{p,N}\Big/\left(\frac{\partial V}{\partial p}\right)_{T,N}. \tag{B.12}$$

From the differential $dG = -S\,dT + V\,dp + \mu\,dN$ we obtain the Maxwell relation $(\partial S/\partial p)_{T,N} = -(\partial V/\partial T)_{p,N}$, so Eq. (B.12) becomes

[1] The last line of the 3×3 determinant is 0, 0, 1 so the result is a 2×2 determinant.

$$C_V = C_p + T\left[\left(\frac{\partial V}{\partial T}\right)_{p,N}\right]^2 \bigg/ \left(\frac{\partial V}{\partial p}\right)_{T,N} \tag{B.13}$$

which may be rewritten

$$C_V = C_p - TV\alpha^2/\kappa_T. \tag{B.14}$$

A result for this same quantity that looks somewhat different can be obtained by starting with C_p. Thus, more briefly,

$$\begin{aligned} C_p = T\frac{\partial(S,p,N)}{\partial(T,p,N)} &= T\frac{\partial(S,p,N)}{\partial(T,V,N)}\frac{\partial(T,V,N)}{\partial(T,p,N)} \\ &= T\left[\left(\frac{\partial S}{\partial T}\right)_{V,N}\left(\frac{\partial p}{\partial V}\right)_{T,N} - \left(\frac{\partial S}{\partial V}\right)_{T,N}\left(\frac{\partial p}{\partial T}\right)_{V,N}\right]\left(\frac{\partial V}{\partial p}\right)_{T,N} \\ &= C_V - T\left[\left(\frac{\partial p}{\partial T}\right)_{V,N}\right]^2\left(\frac{\partial V}{\partial p}\right)_{T,N}. \end{aligned} \tag{B.15}$$

So

$$C_p = C_V + TV\left[\left(\frac{\partial p}{\partial T}\right)_{V,N}\right]^2 \kappa_T. \tag{B.16}$$

Both Eqs. (B.14) and (B.16) show that $C_p \geq C_V$ but they appear to be different. They can be reconciled, however, by noting that

$$\mathrm{d}V = \left(\frac{\partial V}{\partial T}\right)_{p,N}\mathrm{d}T + \left(\frac{\partial V}{\partial p}\right)_{T,N}\mathrm{d}p + \left(\frac{\partial V}{\partial N}\right)_{p,T}\mathrm{d}N \tag{B.17}$$

from which we readily deduce that

$$\left(\frac{\partial p}{\partial T}\right)_{V,N} = -\left(\frac{\partial V}{\partial T}\right)_{p,N} \bigg/ \left(\frac{\partial V}{\partial p}\right)_{T,N} = \alpha/\kappa_T. \tag{B.18}$$

Example 2 A more powerful use of Jacobians can be used to relate the isentropic compressibility

$$\kappa_S = -\frac{1}{V}\left(\frac{\partial V}{\partial p}\right)_{S,N} \tag{B.19}$$

to the isothermal compressibility κ_T. Thus,

$$\left(\frac{\partial V}{\partial p}\right)_{S,N} = \frac{\partial(V,S,N)}{\partial(p,S,N)} = \frac{\partial(V,S,N)}{\partial(V,T,N)}\frac{\partial(V,T,N)}{\partial(p,T,N)}\frac{\partial(p,T,N)}{\partial(p,S,N)}. \tag{B.20}$$

In this case, each Jacobian can be identified as a single partial derivative and we readily deduce

$$\kappa_S/\kappa_T = C_V/C_p. \tag{B.21}$$

From this relationship, we see that $\kappa_T \geq \kappa_S$. Furthermore, division of Eq. (B.14) by C_p, substitution of Eq. (B.21) and rearrangement leads to

$$\kappa_S = \kappa_T - TV\alpha^2/C_p. \tag{B.22}$$

Similarly, dividing Eq. (B.16) by C_V and substituting of Eq. (B.21) gives

$$\frac{1}{\kappa_S} = \frac{1}{\kappa_T} + TV\left[\left(\frac{\partial p}{\partial T}\right)_{V,N}\right]^2 \frac{1}{C_V} = \frac{1}{\kappa_T} + \frac{TV\alpha^2}{\kappa_T^2 C_V}. \tag{B.23}$$

Example 3 By analogy to Eq. (B.19), one can define an isentropic (sometimes called adiabatic) coefficient of expansion

$$\alpha_S = \frac{1}{V}\left(\frac{\partial V}{\partial T}\right)_{S,N} \tag{B.24}$$

and relate it to the isothermal coefficient of expansion α. Thus

$$\left(\frac{\partial V}{\partial T}\right)_{S,N} = \frac{\partial\,(V,S,N)}{\partial\,(T,S,N)} = \frac{\partial\,(V,S,N)}{\partial\,(V,T,N)}\frac{\partial\,(V,T,N)}{\partial\,(T,S,N)} = -\left(\frac{\partial S}{\partial T}\right)_{V,N}\left(\frac{\partial V}{\partial S}\right)_{T,N}. \tag{B.25}$$

We recognize $(\partial S/\partial T)_{V,N} = C_V/T$. From $\mathrm{d}F = -S\,\mathrm{d}T - p\,\mathrm{d}V + \mu\,\mathrm{d}N$, we obtain the Maxwell relation $(\partial S/\partial V)_{T,N} = \left(\partial p/\partial T\right)_{V,N} = \alpha/\kappa_T$, where Eq. (B.18) has been used in the last step. Putting everything together gives

$$\alpha_S = -\frac{C_V\kappa_T}{VT\alpha}. \tag{B.26}$$

This result shows unexpectedly that α_S varies inversely with α and has the opposite sign. For an ideal gas it becomes $\alpha_S = -C_V/pV = -C_V/NRT$, which follows easily from Eq. (3.56) for the entropy of one mole of an ideal gas.

Example 4 For a monocomponent system, the Kramers potential $K = U - TS - \mu N$ so we have $\mathrm{d}K = -S\,\mathrm{d}T - p\,\mathrm{d}V - N\,\mathrm{d}\mu$. The independent variables are T, V and μ. As shown in Chapter 21, this potential is related to the grand partition function $\mathcal{Z}$ by Eq. (21.13). We proceed to express the heat capacity at constant volume in terms of derivatives with respect to these independent variables as follows:

$$\begin{aligned} C_V = T\left(\frac{\partial S}{\partial T}\right)_{V,N} &= T\frac{\partial\,(S,V,N)}{\partial\,(T,V,N)} = T\frac{\partial\,(S,V,N)}{\partial\,(T,V,\mu)}\frac{\partial\,(T,V,\mu)}{\partial\,(T,V,N)} \\ &= T\left[\left(\frac{\partial S}{\partial T}\right)_{\mu,V}\left(\frac{\partial N}{\partial \mu}\right)_{T,V} - \left(\frac{\partial S}{\partial \mu}\right)_{T,V}\left(\frac{\partial N}{\partial T}\right)_{\mu,V}\right]\left(\frac{\partial \mu}{\partial N}\right)_{T,V} \\ &= T\left(\frac{\partial S}{\partial T}\right)_{\mu,V} - T\left[\left(\frac{\partial S}{\partial \mu}\right)_{T,V}\right]^2 \Big/ \left(\frac{\partial N}{\partial \mu}\right)_{T,V}, \end{aligned} \tag{B.27}$$

where the Maxwell relation $(\partial S/\partial\mu)_{T,V} = (\partial N/\partial T)_{\mu,V}$ from $\mathrm{d}K$ has been used.

Example 5 If there is a functional relationship among three variables x, y, z, then

$$\frac{\partial\,(x,y)}{\partial\,(z,x)}\frac{\partial\,(y,z)}{\partial\,(x,y)}\frac{\partial\,(z,x)}{\partial\,(y,z)} = 1. \tag{B.28}$$

Interpreting each Jacobian as a partial derivative we obtain

$$[-(\partial y/\partial z)_x][-(\partial z/\partial x)_y][-(\partial x/\partial y)_z] = 1 \tag{B.29}$$

or simply

$$(\partial y/\partial z)_x(\partial z/\partial x)_y(\partial x/\partial y)_z = -1. \tag{B.30}$$

Although the Jacobians in Eq. (B.28) behave like fractions, the corresponding partial derivatives are each accompanied by a minus sign; therefore, they do not quite behave like fractions, resulting in the net minus sign on the right of Eq. (B.30). Equation (5.31) is a relation of this type where the independent variables are p, V, T with N being constant in all derivatives and therefore irrelevant.

C Differential Geometry of Surfaces

In this appendix, we develop some formulae based on the differential geometry of surfaces that are useful in the treatment of surfaces and interfaces, as discussed in Chapters 13 and 14. We also explore some more aspects of the $\boldsymbol{\xi}$ vector used to treat anisotropic solid-fluid interfaces, as well as the calculus of variations needed to treat curved interfaces. For convenience, we give the main differential and integral formulas that involve the surface gradient ∇_s, surface divergence $\nabla_s\cdot$ and surface curl $\nabla_s\times$ operators. This is followed by a formula for $\nabla_s\cdot\boldsymbol{\xi}$ that we use to derive a generalization of Herring's formula for the chemical potential at a point on curved surface, as well as a formula for the equilibrium shape. The equilibrium shape is also calculated from a variational formulation that can be used to prove the Wulff construction for differentiable anisotropic surface free energy.

C.1 Alternative Formulae for ξ Vector

In Chapter 14 (see Eqs. (14.30) and (14.31)) we defined the vector

$$\xi_\alpha(\hat{\mathbf{n}}) := \frac{\partial \tilde{\gamma}(\mathbf{P})}{\partial P_\alpha}; \quad \boldsymbol{\xi}(\hat{\mathbf{n}}) := \nabla_P \tilde{\gamma}(\mathbf{P}), \tag{C.1}$$

where $\tilde{\gamma}(\mathbf{P}) = P\gamma(\hat{\mathbf{n}})$, $\mathbf{P} = P\hat{\mathbf{n}}$ and $\gamma(\hat{\mathbf{n}})$ is the interfacial free energy per unit area as a function of its unit normal $\hat{\mathbf{n}}$, with other variables held constant and suppressed. We also showed that $\gamma = \boldsymbol{\xi}\cdot\hat{\mathbf{n}}$, $\mathrm{d}\gamma = \boldsymbol{\xi}\cdot \mathrm{d}\hat{\mathbf{n}}$ and $\hat{\mathbf{n}}\cdot \mathrm{d}\boldsymbol{\xi} = 0$, where all derivatives are assumed to exist and be continuous. Now we develop some alternative ways of calculating $\boldsymbol{\xi}(\hat{\mathbf{n}})$ directly from derivatives with respect to $\hat{\mathbf{n}}$.

First, we simply recognize that the chain rule of differentiation can be used to compensate for the fact that the components of $\hat{\mathbf{n}}$ are not independent. Thus

$$\xi_\alpha = \frac{\partial[P\gamma(\mathbf{P}/P)]}{\partial P_\alpha} = \gamma\,\frac{P_\alpha}{P} + P\sum_{\beta=1}^{3}\frac{\partial\gamma}{\partial n_\beta}\frac{\partial(P_\beta/P)}{\partial P_\alpha}, \tag{C.2}$$

where the partial derivatives with respect to n_β are formal derivatives taken as if the n_β were independent. But

$$\frac{\partial(P_\beta/P)}{\partial P_\alpha} = \frac{\delta_{\alpha\beta}}{P} - \frac{P_\alpha P_\beta}{P^3} \tag{C.3}$$

so

$$\xi_\alpha = \gamma\, n_\alpha + \sum_{\beta=1}^{3}\frac{\partial\gamma}{\partial n_\beta}(\delta_{\alpha\beta} - n_\alpha n_\beta). \tag{C.4}$$

If we define a formal gradient operator ∇_n whose components are $\partial/\partial n_\alpha$, Eq. (C.4) can be written in the vector form

$$\boldsymbol{\xi}(\hat{\mathbf{n}}) = \gamma\,\hat{\mathbf{n}} + [\nabla_n\gamma - \hat{\mathbf{n}}(\hat{\mathbf{n}}\cdot\nabla_n\gamma)]. \tag{C.5}$$

Given the various ways that γ can be expressed in terms of the components of $\hat{\mathbf{n}}$, the quantity $\nabla_n\gamma$ is not unique but the quantity $[\nabla_n\gamma - \hat{\mathbf{n}}(\hat{\mathbf{n}}\cdot\nabla_n\gamma)]$ is unique and represents the tangential part $\boldsymbol{\xi}_t$ of $\boldsymbol{\xi}$.

Another option can be used to simplify Eq. (C.5) even further. Given a function $\gamma(n_\alpha)$ of the components n_α, one can always write it in the form

$$\gamma_h := \gamma\left(\frac{n_\alpha}{(n_x^2+n_y^2+n_z^2)^{1/2}}\right) \tag{C.6}$$

so that it is a homogeneous function of degree zero in the components of $\hat{\mathbf{n}}$. Then from Euler's theorem,

$$\sum_{\beta=1}^{3} n_\beta \frac{\partial\gamma_h}{\partial n_\beta} = 0 \tag{C.7}$$

or more succinctly $\hat{\mathbf{n}}\cdot\nabla_n\gamma_h = 0$. Then Eq. (C.5) reduces to

$$\boldsymbol{\xi}(\hat{\mathbf{n}}) = \gamma\,\hat{\mathbf{n}} + \nabla_n\gamma_h. \tag{C.8}$$

Example Problem C.1. For a crystal having cubic symmetry, the leading anisotropy is

$$\gamma(\hat{\mathbf{n}}) = \gamma_0 + \gamma_4(n_x^4+n_y^4+n_z^4), \tag{C.9}$$

where γ_0 and γ_4 are constants. Calculate $\boldsymbol{\xi}(\hat{\mathbf{n}})$ directly by differentiation with respect to the components of $\hat{\mathbf{n}}$.

Solution C.1. We write

$$\gamma_h = \gamma_0 + \gamma_4\frac{(n_x^4+n_y^4+n_z^4)}{(n_x^2+n_y^2+n_z^2)^2} \tag{C.10}$$

so

$$\nabla_n\gamma_h = 4\gamma_4\frac{(n_x^3\hat{\mathbf{i}}+n_y^3\hat{\mathbf{j}}+n_z^3\hat{\mathbf{k}})}{(n_x^2+n_y^2+n_z^2)^2} - 4\gamma_4\hat{\mathbf{n}}\frac{(n_x^4+n_y^4+n_z^4)}{(n_x^2+n_y^2+n_z^2)^3}. \tag{C.11}$$

Now that the differentiation is finished, we can set both denominators equal to one. Thus

$$\boldsymbol{\xi}_t = 4\gamma_4[(n_x^3\hat{\mathbf{i}}+n_y^3\hat{\mathbf{j}}+n_z^3\hat{\mathbf{k}}) - 4\hat{\mathbf{n}}(n_x^4+n_y^4+n_z^4)] \tag{C.12}$$

and of course $\boldsymbol{\xi}_n = \gamma\hat{\mathbf{n}}$, in agreement with Eq. (14.40).

A popular alternative is to express $\mathbf{P}$ in terms of spherical polar coordinates with radius $r = P$, where θ is the polar angle and φ is the azimuthal angle. Then the gradient operator ∇_P becomes

$$\nabla_r = \hat{\mathbf{r}}\frac{\partial}{\partial r} + \frac{1}{r}\left[\hat{\boldsymbol{\theta}}\frac{\partial}{\partial \theta} + \hat{\boldsymbol{\varphi}}\frac{1}{\sin\theta}\frac{\partial}{\partial \varphi}\right], \tag{C.13}$$

where the unit vectors

$$\begin{aligned}\hat{\mathbf{r}} &= \sin\theta\cos\varphi\,\hat{\mathbf{i}} + \sin\theta\sin\varphi\,\hat{\mathbf{j}} + \cos\theta\,\hat{\mathbf{k}};\\ \hat{\boldsymbol{\theta}} &= \cos\theta\cos\varphi\,\hat{\mathbf{i}} + \cos\theta\sin\varphi\,\hat{\mathbf{j}} - \sin\theta\,\hat{\mathbf{k}};\\ \hat{\boldsymbol{\varphi}} &= -\sin\varphi\,\hat{\mathbf{i}} + \cos\varphi\,\hat{\mathbf{j}},\end{aligned} \tag{C.14}$$

can be related to $\hat{\mathbf{i}}, \hat{\mathbf{j}}, \hat{\mathbf{k}}$ in a Cartesian space. Thus

$$\boldsymbol{\xi} = \nabla_r[r\,\gamma(\theta,\varphi)] = \hat{\mathbf{r}}\gamma + \left[\hat{\boldsymbol{\theta}}\frac{\partial \gamma}{\partial \theta} + \hat{\boldsymbol{\varphi}}\frac{1}{\sin\theta}\frac{\partial \gamma}{\partial \varphi}\right]. \tag{C.15}$$

Here, $\hat{\mathbf{r}}$ must be identified with the local normal vector $\hat{\mathbf{n}}$ at a point on the surface of the crystal, where $\hat{\boldsymbol{\theta}}$ and $\hat{\boldsymbol{\varphi}}$ are local unit tangent vectors. This representation can be confusing because $\mathbf{r}$ is not the radius vector to some point on that surface unless that surface happens to be a sphere of radius r. See Section C.3 for a representation that relates to a general surface.

C.2 Surface Differential Geometry

We present some elements of surface differential geometry that are useful in treating curved interfaces with anisotropic $\gamma(\mathbf{n})$. We also introduce the surface gradient operator ∇_s and give equations for the surface divergence $\nabla_s \cdot \mathbf{V}$ and some of its properties. We follow a straightforward treatment by Weatherburn [93, 94].

We define a surface in terms of parameters u and v by means of the parametric equations $x = x(u, v)$, $y = y(u, v)$ and $z = z(u, v)$, or briefly $\mathbf{r} = \mathbf{r}(u, v)$, where the involved functions are assumed to have continuous first and second derivatives. The vectors

$$\mathbf{r}_u := \frac{\partial \mathbf{r}(u,v)}{\partial u}; \quad \mathbf{r}_v := \frac{\partial \mathbf{r}(u,v)}{\partial v} \tag{C.16}$$

are locally tangent to the surface at the point u, v; they are not collinear but they are not necessarily orthogonal to one another. We choose the vectors $\mathbf{r}_u$, $\mathbf{r}_v$ and $\hat{\mathbf{n}}$ to form a right-handed triad, so the local unit outward normal is given by

$$\hat{\mathbf{n}} = \frac{\mathbf{r}_u \times \mathbf{r}_v}{|\mathbf{r}_u \times \mathbf{r}_v|} = \frac{\mathbf{H}}{H}, \tag{C.17}$$

where the vector $\mathbf{H} := \mathbf{r}_u \times \mathbf{r}_v$ and $H = |\mathbf{r}_u \times \mathbf{r}_v|$ is its magnitude. The vector area element is

$$\mathrm{d}\mathbf{A} = \hat{\mathbf{n}}\,\mathrm{d}A = \hat{\mathbf{n}}\,|\mathbf{r}_u \times \mathbf{r}_v|\,\mathrm{d}u\,\mathrm{d}v = \hat{\mathbf{n}}H\,\mathrm{d}u\,\mathrm{d}v = (\mathbf{r}_u \times \mathbf{r}_v)\,\mathrm{d}u\,\mathrm{d}v. \tag{C.18}$$

We note that $H = \mathbf{r}_u \times \mathbf{r}_v \cdot \hat{\mathbf{n}}$ and readily compute $H^2 = EG - F^2$ where

$$E := \mathbf{r}_u \cdot \mathbf{r}_u; \quad F := \mathbf{r}_u \cdot \mathbf{r}_v; \quad G := \mathbf{r}_v \cdot \mathbf{r}_v. \tag{C.19}$$

In order to handle the possible non-orthogonality of $\mathbf{r}_u$ and $\mathbf{r}_v$, we introduce the reciprocal vectors

$$\mathbf{r}_u^\dagger := \frac{\mathbf{r}_v \times \hat{\mathbf{n}}}{H}; \quad \mathbf{r}_v^\dagger := \frac{\hat{\mathbf{n}} \times \mathbf{r}_u}{H}, \tag{C.20}$$

which are orthogonal to $\hat{\mathbf{n}}$ and satisfy

$$\mathbf{r}_u^\dagger \cdot \mathbf{r}_u = 1; \quad \mathbf{r}_u^\dagger \cdot \mathbf{r}_v = \mathbf{r}_v^\dagger \cdot \mathbf{r}_u = 0; \quad \mathbf{r}_v^\dagger \cdot \mathbf{r}_v = 1. \tag{C.21}$$

The manner in which $\hat{\mathbf{n}}$ changes as one moves along the surface can be related to its local curvatures. We examine the derivatives[1]

$$\hat{\mathbf{n}}_u := \frac{\partial \hat{\mathbf{n}}(u, v)}{\partial u}; \quad \hat{\mathbf{n}}_v := \frac{\partial \hat{\mathbf{n}}(u, v)}{\partial v} \tag{C.22}$$

which are necessarily normal to $\hat{\mathbf{n}}$. They can therefore be resolved along $\mathbf{r}_u$ and $\mathbf{r}_v$ or alternatively along $\mathbf{r}_u^\dagger$ and $\mathbf{r}_v^\dagger$. Moreover,

$$d\hat{\mathbf{n}} = \mathrm{d}\left(\frac{\mathbf{H}}{H}\right) = \frac{\mathrm{d}\mathbf{H}}{H} - \mathbf{H}\frac{dH}{H^2} = \frac{\mathrm{d}\mathbf{H}}{H} - \left(\hat{\mathbf{n}} \cdot \frac{\mathrm{d}\mathbf{H}}{H}\right)\hat{\mathbf{n}}. \tag{C.23}$$

Then if we write $\hat{\mathbf{n}}_u = L\mathbf{r}_u^\dagger + M\mathbf{r}_v^\dagger$, we see that $L = \mathbf{r}_u \cdot \hat{\mathbf{n}}_u = \mathbf{r}_u \cdot (\partial \mathbf{H}/\partial u)/H$ and $M = \mathbf{r}_v \cdot \hat{\mathbf{n}}_u = \mathbf{r}_v \cdot (\partial \mathbf{H}/\partial u)/H$. Since $\partial \mathbf{H}/\partial u = \mathbf{r}_{uu} \times \mathbf{r}_v + \mathbf{r}_u \times \mathbf{r}_{uv}$, we readily compute $L = -\hat{\mathbf{n}} \cdot \mathbf{r}_{uu}$ and $M = -\hat{\mathbf{n}} \cdot \mathbf{r}_{uv}$. Similarly, we find $\hat{\mathbf{n}}_v = M\mathbf{r}_u^\dagger + N\mathbf{r}_v^\dagger$, where $N = -\mathbf{n} \cdot \mathbf{r}_{vv}$. These results can be summarized in matrix notation by the equation

$$\begin{pmatrix} \hat{\mathbf{n}}_u \\ \hat{\mathbf{n}}_v \end{pmatrix} = \begin{pmatrix} L & M \\ M & N \end{pmatrix} \begin{pmatrix} \mathbf{r}_u^\dagger \\ \mathbf{r}_v^\dagger \end{pmatrix} = \begin{pmatrix} P & R \\ Q & S \end{pmatrix} \begin{pmatrix} \mathbf{r}_u \\ \mathbf{r}_v \end{pmatrix}, \tag{C.24}$$

where

$$L := -\hat{\mathbf{n}} \cdot \mathbf{r}_{uu}; \quad M := -\hat{\mathbf{n}} \cdot \mathbf{r}_{uv}; \quad N := -\hat{\mathbf{n}} \cdot \mathbf{r}_{vv} \tag{C.25}$$

and

$$\begin{aligned} P &= \frac{LG - MF}{H^2}; \quad R = \frac{ME - LF}{H^2}; \\ Q &= \frac{MG - NF}{H^2}; \quad S = \frac{NE - MF}{H^2}. \end{aligned} \tag{C.26}$$

The second matrix in Eq. (C.24) is obtained by using

$$\mathbf{r}_u^\dagger = \frac{G\mathbf{r}_u - F\mathbf{r}_v}{H^2}; \quad \mathbf{r}_v^\dagger = \frac{-F\mathbf{r}_u + E\mathbf{r}_v}{H^2}. \tag{C.27}$$

As shown by Weatherburn, the mean curvature and the Gaussian curvature are equal to the trace and the determinant of the second matrix in Eq. (C.24), specifically

$$\mathcal{K} \equiv \frac{1}{R_1} + \frac{1}{R_2} = P + S; \quad \mathcal{G} \equiv \frac{1}{R_1}\frac{1}{R_2} = PS - QR = \frac{LN - M^2}{H^2}, \tag{C.28}$$

where R_1 and R_2 are the principal radii of curvature measured in principal planes that are orthogonal to each other and that contain $\hat{\mathbf{n}}$. One could transform to coordinates in

[1]Note that $\hat{\mathbf{n}}_u$ and $\hat{\mathbf{n}}_v$ are not unit vectors; they are derivatives of unit vectors.

the principal planes by means of a series of linear transformations that would ultimately result in transforming that matrix by means of a similarity transformation that preserves the trace and the determinant.

One can see the connection to curvatures easily by supposing that u and v are already orthogonal coordinates at the point of the surface under consideration and that they have been oriented so that $\mathbf{r}_{uv} = 0$. Then $F = 0$, the line element would be $\mathrm{d}s^2 = E(\mathrm{d}u)^2 + G(\mathrm{d}v)^2$, $\mathbf{r}_u^\dagger = \mathbf{r}_u/E$, $\mathbf{r}_v^\dagger = \mathbf{r}_v/G$, and both matrices in Eq. (C.24) would be diagonal. Unit tangent vectors in the principal directions would be $\hat{\mathbf{t}}_u = \mathbf{r}_u/E^{1/2}$ and $\hat{\mathbf{t}}_v = \mathbf{r}_v/G^{1/2}$. Equation (C.24) would become simply

$$\hat{\mathbf{n}}_u = P\mathbf{r}_u = (L/E)\mathbf{r}_u; \quad \hat{\mathbf{n}}_v = S\mathbf{r}_v = (N/G)\mathbf{r}_v. \tag{C.29}$$

For this special choice, the principal curvatures would be

$$\frac{1}{R_1} = \left(\frac{\mathrm{d}\theta}{\mathrm{d}s}\right)_u = \frac{\hat{\mathbf{t}}_u \cdot \hat{\mathbf{n}}_u\,\mathrm{d}u}{E^{1/2}\mathrm{d}u} = \frac{L}{E} = P; \quad \frac{1}{R_2} = \left(\frac{\mathrm{d}\theta}{\mathrm{d}s}\right)_v = \frac{\hat{\mathbf{t}}_v \cdot \hat{\mathbf{n}}_v\,\mathrm{d}v}{G^{1/2}\,\mathrm{d}v} = \frac{N}{G} = S. \tag{C.30}$$

Thus we have

$$\mathrm{d}\hat{\mathbf{n}} = \frac{1}{R_1}\mathbf{r}_u\,\mathrm{d}u + \frac{1}{R_2}\mathbf{r}_v\,\mathrm{d}v, \quad \text{principal axes,} \tag{C.31}$$

which is equivalent to the formulae of Rodrigues.

In the general case, the principal curvatures are given by

$$\frac{1}{R_{1,2}} = \frac{P+S}{2} \pm \sqrt{\left(\frac{P-S}{2}\right)^2 + QR}, \tag{C.32}$$

which can be found by determining the eigenvalues that correspond to the principal axes. An outline of this transformation is the following: If we denote the second matrix in Eq. (C.24) by $\mathcal{P}$, it may be taken into diagonal form by a transformation of the form $\mathcal{Q}^{-1}\mathcal{P}\mathcal{Q}$ where the matrix $\mathcal{Q} = \mathcal{A}\Lambda^{-1/2}\mathcal{B}$ encompasses three successive transformations. The matrix $\mathcal{A}$ is orthogonal and takes the line element into diagonal form with positive definite eigenvalues. Λ is the resulting diagonal eigenvalue matrix and $\Lambda^{1/2}$ is its square root; it provides a stretching transformation. The combination of transformations $\mathcal{A}\Lambda^{-1/2}$ takes the line element into the form $\mathrm{d}s^2 = \mathrm{d}X^2 + \mathrm{d}Y^2$ and takes $\mathcal{P}$ into a symmetric matrix. The final matrix $\mathcal{B}$ is an orthogonal matrix that rotates the already orthogonal axes into the principal axes. For future reference, we note for general parameters u, v that Eq. (C.28) can be written

$$\mathcal{K} = \mathbf{r}_u^\dagger \cdot \hat{\mathbf{n}}_u + \mathbf{r}_v^\dagger \cdot \hat{\mathbf{n}}_v; \quad \mathcal{G} = \frac{\mathbf{n} \cdot \hat{\mathbf{n}}_u \times \hat{\mathbf{n}}_v}{H}. \tag{C.33}$$

C.2.1 Surface Differential Operators

The surface gradient operator ∇_s is defined such that

$$\nabla_s\phi \cdot \mathrm{d}\mathbf{r} = \mathrm{d}\phi = \frac{\partial\phi}{\partial u}\mathrm{d}u + \frac{\partial\phi}{\partial v}\mathrm{d}v, \tag{C.34}$$

where $\phi(u, v)$ is a scalar function defined on the surface. Since $\mathbf{dr} = \mathbf{r}_u\, du + \mathbf{r}_v\, dv$ it follows that

$$\nabla_s = \mathbf{r}_u^\dagger \frac{\partial}{\partial u} + \mathbf{r}_v^\dagger \frac{\partial}{\partial v}. \tag{C.35}$$

For a vector of the form

$$\mathbf{V} = V^u \mathbf{r}_u + V^v \mathbf{r}_v + V^n \hat{\mathbf{n}}, \tag{C.36}$$

one can form a surface divergence

$$\nabla_s \cdot \mathbf{V} = \nabla_s \cdot (V^u \mathbf{r}_u + V^v \mathbf{r}_v) + V^n \nabla_s \cdot \hat{\mathbf{n}} \tag{C.37}$$

in which there is no contribution from the derivatives of the component V^n because $\mathbf{r}_u^\dagger$ and $\mathbf{r}_v^\dagger$ are perpendicular to $\hat{\mathbf{n}}$. By the first member of Eq. (C.33) we see that

$$\nabla_s \cdot \hat{\mathbf{n}} = \mathcal{K}, \tag{C.38}$$

so Eq. (C.37) becomes

$$\nabla_s \cdot \mathbf{V} = \nabla_s \cdot (V^u \mathbf{r}_u + V^v \mathbf{r}_v) + V^n \mathcal{K}. \tag{C.39}$$

Note especially that the term $V^n\mathcal{K}$ arises because the surface is curved; no such term would be present for a divergence in the x, y plane of a Cartesian coordinate system. The tangential components of $\mathbf{V}$ each lead to two terms because $\mathbf{r}_u$ and $\mathbf{r}_v$ are not constants. After some algebra one obtains

$$\nabla_s \cdot \mathbf{V} = \frac{1}{H}\left[\frac{\partial}{\partial u}(HV^u) + \frac{\partial}{\partial v}(HV^v)\right] + V^n \mathcal{K}. \tag{C.40}$$

A case of special importance occurs when $\mathbf{V} = \mathbf{r}(u, v)$, the position vector itself. Then

$$\nabla_s \cdot \mathbf{r} = \left[\mathbf{r}_u^\dagger \frac{\partial}{\partial u} + \mathbf{r}_v^\dagger \frac{\partial}{\partial v}\right] \cdot \mathbf{r} = \mathbf{r}_u^\dagger \cdot \mathbf{r}_u + \mathbf{r}_v^\dagger \cdot \mathbf{r}_v = 2. \tag{C.41}$$

One can also define a surface Laplacian and a surface curl and obtain various vector identities. The surface Laplacian is

$$\nabla_s^2 \phi = \nabla_s \cdot \nabla_s \phi = \frac{1}{H}\left[\frac{\partial}{\partial u}\left(\frac{G\phi_u - F\phi_v}{H}\right) + \frac{\partial}{\partial v}\left(\frac{G\phi_u - F\phi_v}{H}\right)\right]. \tag{C.42}$$

As shown by Weatherburn, $\nabla_s^2 \mathbf{r} = -\mathcal{K}\hat{\mathbf{n}}$ and $\nabla_s^2 \hat{\mathbf{n}} = -(\mathcal{K}^2 - 2\mathcal{G})\hat{\mathbf{n}} + \nabla_s \mathcal{K}$. The surface curl is given by

$$\nabla_s \times \mathbf{V} = \frac{\hat{\mathbf{n}}}{H}\left[\frac{\partial}{\partial u}\left(FV^u + GV^v\right) - \frac{\partial}{\partial v}\left(EV^u + FV^v\right)\right]$$
$$+ \frac{1}{H}\left[\left(MV^u + NV^v\right)\mathbf{r}_u - \left(LV^u + MV^v\right)\mathbf{r}_v\right] - \hat{\mathbf{n}} \times \nabla_s V^n. \tag{C.43}$$

A special case is $\nabla_s \times \hat{\mathbf{n}} = 0$. Moreover, $\nabla_s \times \nabla_s \phi$ can be shown to be a vector in the tangent plane, not necessarily zero; this is a significant deviation from $\nabla \times \nabla \phi = 0$ in three dimensions.

Before leaving this section we calculate the variation of $\mathbf{H} = \mathbf{r}_u \times \mathbf{r}_v$ and $\hat{\mathbf{n}} = \mathbf{H}/H$ for a normal variation of the form

$$\delta\mathbf{r}(u, v) := \mathbf{r} - \mathbf{r}_0(u, v) = \hat{\mathbf{n}}_0(u, v)\,\eta(u, v), \tag{C.44}$$

where $\mathbf{r}_0(u, v)$ is a point on some initial surface, $\mathbf{r}$ is the position on a neighboring varied surface, $\hat{\mathbf{n}}_0(u, v)$ is the unit normal on the original surface and the infinitesimal quantity $\eta(u, v)$ is arbitrary but differentiable. Evidently

$$\delta\mathbf{H} = \mathbf{r}_u \times \delta\mathbf{r}_v - \mathbf{r}_v \times \delta\mathbf{r}_u = (\mathbf{r}_{0u} \times \hat{\mathbf{n}}_{0v} - \mathbf{r}_{0v} \times \hat{\mathbf{n}}_{0u})\eta + (\mathbf{r}_u \times \hat{\mathbf{n}}_0)\eta_v - (\mathbf{r}_v \times \hat{\mathbf{n}}_0)\eta_u \tag{C.45}$$

to first order in η. The coefficient of η can be calculated by using Eq. (C.24) and carrying out the cross products; it turns out to be $\mathbf{H}_0\,\mathcal{K}_0$. The terms involving the derivatives of η can be identified in terms of the surface gradient operator Eq. (C.35) applied to η. The result is

$$\delta\mathbf{H} = \mathbf{H}_0\,\mathcal{K}_0\,\eta - H_0\nabla_s\eta. \tag{C.46}$$

Since $\mathrm{d}\hat{\mathbf{n}}$ is perpendicular to $\hat{\mathbf{n}}$, we see from Eq. (C.23) that the first term in $\delta\mathbf{H}$ makes no contribution to $\delta\hat{\mathbf{n}}$ but the second term contributes to give the important result

$$\delta\hat{\mathbf{n}} = -\nabla_s\eta. \tag{C.47}$$

C.2.2 Integral Theorems

The surface divergence theorem is similar to the Gauss divergence theorem except it applies to a surface whose curvature must be accounted for. It applies to a curved surface A having local unit normal $\hat{\mathbf{n}}$ surrounded by a closed skew curve C with vector line element $\mathrm{d}\boldsymbol{\ell}$ with the convention that positive circulation around the area is governed by the right-hand rule. The outwardly directed unit tangent vector along that curve is denoted by $\hat{\mathbf{t}}$ and points in the direction $\mathrm{d}\boldsymbol{\ell} \times \hat{\mathbf{n}}$. The theorem states that

$$\int_A \nabla_s \cdot \mathbf{V}\,\mathrm{d}A = \oint_C \mathbf{V}_t \cdot \hat{\mathbf{t}}\,\mathrm{d}\ell + \int_A V^n\mathcal{K}\,\mathrm{d}A, \tag{C.48}$$

where $\mathbf{V}_t = V^u\mathbf{r}_u + V^v\mathbf{r}_v$ is the tangential part of $\mathbf{V}$. Since $\hat{\mathbf{t}}\,\mathrm{d}\ell = \mathrm{d}\boldsymbol{\ell} \times \hat{\mathbf{n}}$, Eq. (C.48) can also be written in the form

$$\int_A \nabla_s \cdot \mathbf{V}\,\mathrm{d}A = \oint_C \hat{\mathbf{n}} \cdot \mathbf{V}_t \times \mathrm{d}\boldsymbol{\ell} + \int_A V^n\mathcal{K}\,\mathrm{d}A. \tag{C.49}$$

The term involving the curvature $\mathcal{K}$ follows directly from the last term in Eq. (C.40) so we have only to deal with

$$\int_A \nabla_s \cdot \mathbf{V}_t\,\mathrm{d}A = \int_{u,v} (\nabla_s \cdot \mathbf{V}_t)H\,\mathrm{d}u\,\mathrm{d}v = \int_{u,v}\left[\frac{\partial(HV^u)}{\partial u} + \frac{\partial(HV^v)}{\partial v}\right]\mathrm{d}u\,\mathrm{d}v, \tag{C.50}$$

where the second two integrals are taken over the corresponding domain in u, v. But

$$\int_{u,v}\left[\frac{\partial(HV^u)}{\partial u} + \frac{\partial(HV^v)}{\partial v}\right]\mathrm{d}u\,\mathrm{d}v = \oint_{u,v} HV^u\,\mathrm{d}v - \oint_{u,v} HV^v\,\mathrm{d}u, \tag{C.51}$$

where the minus sign on the second term on the right arises because of a choice of positive circulation according to the right-hand rule. The integrand in the line integral in Eq. (C.49) can be written

$$\begin{aligned}\mathbf{V}_t \cdot d\boldsymbol{\ell} \times \hat{\mathbf{n}} &= (V^u \mathbf{r}_u + V^v \mathbf{r}_v) \cdot (\mathbf{r}_u\, du + \mathbf{r}_v\, dv) \times \hat{\mathbf{n}} \\ &= (V^u \mathbf{r}_u + V^v \mathbf{r}_v) \cdot H(-\mathbf{r}_v^\dagger\, du + \mathbf{r}_u^\dagger\, dv) = HV^u\, dv - HV^v\, du,\end{aligned} \tag{C.52}$$

the same as Eq. (C.51). Therefore, the tangential part of $\mathbf{V}$ contributes the line integral in Eq. (C.48) or Eq. (C.49) and the normal component V^n generates the term containing the curvature $\mathcal{K}$. In a flat two-dimensional space, one would have only the line integral.

There is also a surface curl (Stokes) theorem, specifically

$$\int_A (\nabla_s \times \mathbf{V}) \cdot d\mathbf{A} = \oint_C \mathbf{V} \cdot d\boldsymbol{\ell}, \tag{C.53}$$

which is similar to the Stokes theorem for the three-dimensional curl.

C.3 ξ Vector for General Surfaces

We return to Eq. (C.1) and choose $\mathbf{P} = \mathbf{H}$, where $\mathbf{H} = \mathbf{r}_u \times \mathbf{r}_v$ for some crystal surface under consideration, to obtain

$$\boldsymbol{\xi} = \nabla_H [H\gamma(\mathbf{H}/H)] = \gamma \hat{\mathbf{n}} + H \nabla_H \gamma. \tag{C.54}$$

Then by using the relations from differential geometry from Section C.2, we find

$$\boldsymbol{\xi} = \xi^u\, \mathbf{r}_u + \xi^v\, \mathbf{r}_v + \xi^n \hat{\mathbf{n}}, \tag{C.55}$$

where

$$\begin{aligned}\xi^u &= \mathbf{r}_u^\dagger \cdot H\nabla_H \gamma = (\mathbf{r}_v \times \hat{\mathbf{n}}) \cdot \nabla_H \gamma; \\ \xi^v &= \mathbf{r}_v^\dagger \cdot H\nabla_H \gamma = (\hat{\mathbf{n}} \times \mathbf{r}_u) \cdot \nabla_H \gamma; \\ \xi^n &= \gamma.\end{aligned} \tag{C.56}$$

To calculate $\nabla_H \gamma$, we write

$$\gamma = \gamma^*(\alpha, \beta), \tag{C.57}$$

where $\alpha = H_x/H$, $\beta = H_y/H$, and $H_z/H = \pm\sqrt{1 - \alpha^2 - \beta^2}$ with the sign chosen locally to make $\hat{\mathbf{n}} = \mathbf{H}/H$ the outward normal of the crystal under consideration. Then

$$\boldsymbol{\xi}_t = H\nabla_H \gamma = \frac{\partial \gamma^*}{\partial \alpha}\left(\hat{\mathbf{i}} - \alpha\hat{\mathbf{n}}\right) + \frac{\partial \gamma^*}{\partial \beta}\left(\hat{\mathbf{j}} - \beta\hat{\mathbf{n}}\right), \tag{C.58}$$

which is perpendicular to $\hat{\mathbf{n}}$ as expected. Thus

$$\begin{aligned}\xi^u &= \frac{\partial \gamma^*}{\partial \alpha}\left(\mathbf{r}_u^\dagger \cdot \hat{\mathbf{i}}\right) + \frac{\partial \gamma^*}{\partial \beta}\left(\mathbf{r}_u^\dagger \cdot \hat{\mathbf{j}}\right); \\ \xi^v &= \frac{\partial \gamma^*}{\partial \alpha}\left(\mathbf{r}_v^\dagger \cdot \hat{\mathbf{i}}\right) + \frac{\partial \gamma^*}{\partial \beta}\left(\mathbf{r}_v^\dagger \cdot \hat{\mathbf{j}}\right).\end{aligned} \tag{C.59}$$

To proceed further, we adopt a specific parameterization of the surface:

$$x = u; \quad y = v; \quad z = w(u, v). \tag{C.60}$$

Then with $p := w_u$ and $q := w_v$ we have

$$\mathbf{r}_u = \hat{\mathbf{i}} + p\hat{\mathbf{k}}; \quad \mathbf{r}_v = \hat{\mathbf{j}} + q\hat{\mathbf{k}}; \quad \mathbf{H} = -p\hat{\mathbf{i}} - q\hat{\mathbf{j}} + \hat{\mathbf{k}} \tag{C.61}$$

so that

$$H = \sqrt{1 + p^2 + q^2}; \quad \alpha = -p/H; \quad \beta = -q/H. \tag{C.62}$$

We can also calculate the reciprocal vectors

$$\mathbf{r}_u^\dagger = [(1 + q^2)\hat{\mathbf{i}} - pq\hat{\mathbf{j}} + p\hat{\mathbf{k}}]/H^2; \quad \mathbf{r}_v^\dagger = [-pq\hat{\mathbf{i}} + (1 + p^2)\hat{\mathbf{j}} + q\hat{\mathbf{k}}]/H^2. \tag{C.63}$$

Then with $\gamma(p, q) = \gamma^*(\alpha, \beta)$, Eq. (C.59) becomes

$$\begin{aligned} \xi^u &= \frac{(1 + q^2)}{H^2}\frac{\partial \gamma^*}{\partial \alpha} - \frac{pq}{H^2}\frac{\partial \gamma^*}{\partial \beta} = -H\frac{\partial \gamma(p, q)}{\partial p}; \\ \xi^v &= -\frac{pq}{H^2}\frac{\partial \gamma^*}{\partial \alpha} + \frac{(1 + q^2)}{H^2}\frac{\partial \gamma^*}{\partial \beta} = -H\frac{\partial \gamma(p, q)}{\partial q}. \end{aligned} \tag{C.64}$$

We can use the general expression Eq. (C.40) to compute $\nabla_s \cdot \boldsymbol{\xi}$. To bear in mind the specific parameterization of Eq. (C.60), we now use x and y instead of u and v and write $p = \partial z/\partial x$ and $q = \partial z/\partial y$, resulting in

$$\nabla_s \cdot \boldsymbol{\xi} = -\frac{1}{H}\left[\frac{\partial}{\partial x}\left(H^2\frac{\partial \gamma}{\partial p}\right) + \frac{\partial}{\partial y}\left(H^2\frac{\partial \gamma}{\partial q}\right)\right] + \gamma\,\mathcal{K}. \tag{C.65}$$

The curvature $\mathcal{K}$ can be calculated from Eq. (C.33) which can be simplified because $\hat{\mathbf{n}}_u = \mathbf{H}_u/H - \mathbf{H}H_u/H^2$ and $\mathbf{H}$ is perpendicular to $\mathbf{r}_u^\dagger$ and $\mathbf{r}_v^\dagger$. Therefore, in general

$$\mathcal{K} = \frac{\mathbf{r}_u^\dagger \cdot \mathbf{H}_u + \mathbf{r}_v^\dagger \cdot \mathbf{H}_v}{H}. \tag{C.66}$$

In our special Cartesian representation,

$$\mathcal{K} = -\frac{(1 + p)^2 z_{xx} - 2pqz_{xy} + (1 + q)^2 z_{yy}}{H^3}. \tag{C.67}$$

Straightforward algebra allows this curvature to be written in the form

$$\mathcal{K} = -\frac{\partial}{\partial x}\frac{\partial H}{\partial p} - \frac{\partial}{\partial y}\frac{\partial H}{\partial q}. \tag{C.68}$$

Equation (C.68) can be combined with Eq. (C.65) to produce, after considerable algebra, the compact result

$$\nabla_s \cdot \boldsymbol{\xi} = -\frac{\partial}{\partial x}\frac{\partial \Phi}{\partial p} - \frac{\partial}{\partial y}\frac{\partial \Phi}{\partial q}, \tag{C.69}$$

where $\Phi := H\gamma$. This result can be obtained more easily by means of a variational calculation (see Section C.4.1) committed to our choice of a Monge representation from the outset. Note particularly the form

$$\nabla_s \cdot \boldsymbol{\xi} = -(\Phi_{pp} z_{xx} + 2\Phi_{pq} z_{xy} + \Phi_{qq} z_{yy}), \tag{C.70}$$

which displays the symmetry of the result.

C.4 Herring Formula

We proceed to derive a formula due to Herring at a point on an anisotropic surface. The partial derivatives of Φ in Eq. (C.70) are given explicitly by

$$\begin{aligned}
\Phi_{pp} &= \frac{(1+p^2)\gamma}{(1+p^2+q^2)^{3/2}} + \frac{2p\gamma_p}{(1+p^2+q^2)^{1/2}} + (1+p^2+q^2)^{1/2}\gamma_{pp};\\
\Phi_{pq} &= -\frac{pq\gamma}{(1+p^2+q^2)^{3/2}} + \frac{q\gamma_p + p\gamma_q}{(1+p^2+q^2)^{1/2}} + (1+p^2+q^2)^{1/2}\gamma_{pq};\\
\Phi_{qq} &= \frac{(1+q^2)\gamma}{(1+p^2+q^2)^{3/2}} + \frac{2q\gamma_q}{(1+p^2+q^2)^{1/2}} + (1+p^2+q^2)^{1/2}\gamma_{qq}.
\end{aligned} \tag{C.71}$$

For the case in which γ is a constant, Eq. (C.70) must give just $\gamma\mathcal{K}$, where $\mathcal{K}$ is the mean curvature, so we obtain the well known formula

$$\mathcal{K} = -\frac{(1+q^2)z_{xx} - 2pqz_{xy} + (1+p^2)z_{yy}}{(1+p^2+q^2)^{3/2}} \tag{C.72}$$

for the sum of the principal curvatures. Simplification of Eq. (C.70) for anisotropic γ can be obtained by choosing very special Cartesian axes at each point of the equilibrium shape. The z axis is chosen to be along the normal to the equilibrium shape with the x, y plane locally tangent to the shape. In that case, $p = q = 0$ when evaluated at the chosen point, which gives

$$\nabla_s \cdot \boldsymbol{\xi} = -(\gamma + \gamma_{pp})z_{xx} - 2\gamma_{pq} z_{xy} - (\gamma + \gamma_{qq})z_{yy}, \quad \text{at a point } x_0, y_0, \text{ for } z \text{ along } \hat{\mathbf{n}}_0. \tag{C.73}$$

If, in addition, the x and y axes are oriented along the principal axes of curvature of the surface, we have $z_{xy} = 0$ and Eq. (C.73) reduces to

$$\nabla_s \cdot \boldsymbol{\xi} = \frac{\gamma + \gamma_{pp}}{R_1} + \frac{\gamma + \gamma_{qq}}{R_2}, \quad \text{at a point } x_0, y_0, \text{ for } z \text{ along } \hat{\mathbf{n}}_0, \text{ principal axes,} \tag{C.74}$$

where $1/R_1 = -\partial^2 z/\partial x^2 = K_1$ and $1/R_2 = -\partial^2 z/\partial y^2 = K_2$ are principal curvatures.

In the vicinity of the surface point x_0, y_0 under consideration, the angle θ made by $\hat{\mathbf{n}}$ with $\hat{\mathbf{n}}_0 = \hat{\mathbf{k}}$ is given by $\cos\theta = \hat{\mathbf{n}} \cdot \hat{\mathbf{k}} = (1+p^2+q^2)^{-1/2}$ so $\tan^2\theta = p^2 + q^2$. For principal planes, $\tan\theta_1 = \pm p$ and $\tan\theta_2 = \pm q$. With this notation, Eq. (C.74) can be written in the Herring form

$$\nabla_s \cdot \boldsymbol{\xi} = \frac{\gamma + \gamma_{\theta_1\theta_1}}{R_1} + \frac{\gamma + \gamma_{\theta_2\theta_2}}{R_2}, \quad \text{at a point } x_0, y_0, \text{ principal planes,} \tag{C.75}$$

where the derivatives are to be evaluated at $\theta_1 = 0$ and $\theta_2 = 0$.

By incorporating Eq. (C.75) in Eq. (14.90) of the text, we obtain

$$\omega_v^F - \omega_v^s = \frac{\gamma + \gamma_{\theta_1\theta_1}}{R_1} + \frac{\gamma + \gamma_{\theta_2\theta_2}}{R_2}, \tag{C.76}$$

which is a somewhat more general version of Herring's result. The original Herring formula [38, 41] pertained to the case of a solid-vapor interface for a single component for which Eq. (14.102) of the text becomes

$$\mu = \mu_\infty + \Omega_0 \left[\frac{\gamma + \gamma_{\theta_1\theta_1}}{R_1} + \frac{\gamma + \gamma_{\theta_2\theta_2}}{R_2} \right]. \tag{C.77}$$

Although the form of Eq. (C.76) is elegant, it is not very useful computationally because it requires one to find the principal axes of curvature beforehand. In particular, it is not a differential equation for the equilibrium shape, since it applies only at a single point. A more useful expression that still applies only at a point but does not require finding the principal axes can be obtained by using Eq. (C.73), namely

$$\omega_v^F - \omega_v^s = -(\gamma + \gamma_{pp}) z_{xx} - 2\gamma_{pq} z_{xy} - (\gamma + \gamma_{qq}) z_{yy}. \tag{C.78}$$

An elegant geometrical interpretation of the terms $\gamma + \gamma_{\theta_1\theta_1}$ and $\gamma + \gamma_{\theta_2\theta_2}$ was given by Johnson [95] for the case in which Eq. (C.77) is applied to give a *local* equilibrium condition at the surface of a body that is *not* the equilibrium shape. In that case, R_1 and R_2 are principal radii of the non-equilibrium body at the point under consideration. Johnson shows that $\gamma + \gamma_{\theta_1\theta_1}$ and $\gamma + \gamma_{\theta_2\theta_2}$ are proportional to the radii of curvature ρ_1 and ρ_2 of the equilibrium shape *projected* onto principal planes of the non-equilibrium body. Since the convex part of the $\boldsymbol{\xi}$ plot is similar to the equilibrium shape, it turns out that $\gamma + \gamma_{\theta_1\theta_1}$ and $\gamma + \gamma_{\theta_2\theta_2}$ are equal to the radii of curvature of the $\boldsymbol{\xi}$ plot, calculated in the respective principal planes of the non-equilibrium body.

C.4.1 Variational Formulation

If we adopt a Monge representation $z = z(x, y)$ of the interface, one can formulate the variational problem for the equilibrium shape as follows.We minimize the interfacial free energy

$$\int_{A_{xy}} \Phi \, dx \, dy, \tag{C.79}$$

where $\Phi = \gamma H = \gamma(p, q)\sqrt{1 + p^2 + q^2}$, subject to the constraint of constant volume,

$$\int_{A_{xy}} z(x, y) \, dx \, dy. \tag{C.80}$$

Here, Φ is the free energy per unit area of the x, y plane and the integration is over A_{xy}, a fixed projected area in the x, y plane. By means of a Lagrange multiplier 2λ, we obtain the variational problem

$$\delta \int_{A_{xy}} [\Phi - 2\lambda z]\, \mathrm{d}x\, \mathrm{d}y = 0. \tag{C.81}$$

By carrying out the variation, Eq. (C.81) becomes

$$\int_{A_{xy}} \left[\frac{\partial \Phi}{\partial p} \delta p + \frac{\partial \Phi}{\partial q} \delta q - 2\lambda \delta z \right] \mathrm{d}x\, \mathrm{d}y = 0. \tag{C.82}$$

Then with $\delta p = \delta \partial z/\partial x = \partial(\delta z)/\partial x$ and $\delta q = \delta \partial z/\partial y = \partial(\delta z)/\partial y$, we obtain

$$\int_{A_{xy}} \left\{ \frac{\partial}{\partial x}\left(\frac{\partial \Phi}{\partial p} \delta z \right) + \frac{\partial}{\partial y}\left(\frac{\partial \Phi}{\partial q} \delta z \right) - \left[\frac{\partial}{\partial x}\left(\frac{\partial \Phi}{\partial p} \right) + \frac{\partial}{\partial y}\left(\frac{\partial \Phi}{\partial q} \right) + 2\lambda \right] \delta z \right\} \mathrm{d}x\, \mathrm{d}y = 0. \tag{C.83}$$

The first two terms can be integrated to the boundary where the result vanishes, either because δz vanishes or because the boundary is closed. Since δz is arbitrary within A_{xy}, its coefficient in the integral must vanish, resulting in

$$-\frac{\partial}{\partial x}\left(\frac{\partial \Phi}{\partial p} \right) - \frac{\partial}{\partial y}\left(\frac{\partial \Phi}{\partial q} \right) = 2\lambda. \tag{C.84}$$

In view of Eqs. (C.69) and (14.90), $\nabla_s \cdot \boldsymbol{\xi} = 2\lambda = \omega_F - \omega_v^s$.

For a closed body, one can find an integral of Eq. (C.84) by following a method outlined by Landau and Lifshitz [7, p. 460]. We replace these derivatives by Jacobians as follows:

$$\left(\frac{\partial(\partial \Phi/\partial p)}{\partial x} \right)_y = \frac{\partial\,(\partial \Phi/\partial p, y)}{\partial\,(x, y)}; \quad \left(\frac{\partial(\partial \Phi/\partial q)}{\partial y} \right)_x = \frac{\partial\,(x, \partial \Phi/\partial q)}{\partial\,(x, y)}. \tag{C.85}$$

We multiply the resulting expression by the Jacobian $\partial(x, y)/\partial(p, q)$ to obtain

$$-\frac{\partial\,(\partial \Phi/\partial p, y)}{\partial\,(p, q)} - \frac{\partial\,(x, \partial \Phi/\partial q)}{\partial\,(p, q)} = 2\lambda \frac{\partial\,(x, y)}{\partial\,(p, q)}. \tag{C.86}$$

Then we introduce the function

$$\phi(p, q) = z - xp - yq \tag{C.87}$$

whose differential

$$\mathrm{d}\phi = -x\, \mathrm{d}p - y\, \mathrm{d}q \tag{C.88}$$

follows because $\mathrm{d}z = p\, \mathrm{d}x + q\, \mathrm{d}y$. Thus Eq. (C.86) becomes

$$\frac{\partial\,(\partial \Phi/\partial p, \partial \phi/\partial q)}{\partial\,(p, q)} + \frac{\partial\,(\partial \phi/\partial p, \partial \Phi/\partial q)}{\partial\,(p, q)} = 2\lambda \frac{\partial\,(\partial \phi/\partial p, \partial \phi/\partial q)}{\partial\,(p, q)}. \tag{C.89}$$

An obvious integral of Eq. (C.89) is $\Phi = \lambda \phi$, so

$$\Phi(p, q)/\lambda = z - xp - yq = z - x\frac{\partial z}{\partial x} - y\frac{\partial z}{\partial y}, \tag{C.90}$$

which has the form of a Legendre transform. According to Eq. (C.88)

$$\mathrm{d}(\Phi/\lambda) = -x\, \mathrm{d}p - y\, \mathrm{d}q, \tag{C.91}$$

so

$$\left(\frac{\partial(\Phi/\lambda)}{\partial p}\right)_q = -x; \quad \left(\frac{\partial(\Phi/\lambda)}{\partial q}\right)_p = -y. \tag{C.92}$$

Therefore, the inverse of Eq. (C.90) is

$$z = (\Phi/\lambda) + px + yq = (\Phi/\lambda) - p\frac{\partial(\Phi/\lambda)}{\partial p} - q\frac{\partial(\Phi/\lambda)}{\partial q} \tag{C.93}$$

which also has the form of a Legendre transform. The transformation $X = \lambda x$, $Y = \lambda y$ and $Z = \lambda z$ gives the forms of these equations developed in Section 14.7.

The form of Eq. (C.90) can be used to obtain the Wulff construction for the equilibrium shape. It can be rewritten in the form

$$\gamma(p, q) = \lambda \frac{z - xp - yq}{\sqrt{1 + p^2 + q^2}}, \tag{C.94}$$

which is a first order nonlinear partial differential equation for $z(x, y)$. The components of the unit surface normal are

$$n_x = \frac{-p}{\sqrt{1 + p^2 + q^2}}; \quad n_y = \frac{-q}{\sqrt{1 + p^2 + q^2}}; \quad n_z = \frac{1}{\sqrt{1 + p^2 + q^2}}, \tag{C.95}$$

in agreement with Eq. (C.62). Regarding p and q to be parameters, the right-hand side of Eq. (C.94) represents a family of tangent planes to the equilibrium shape and the envelope of such planes is the integral of that nonlinear partial differential equation for $z(x, y)$. This is the basis of the Wulff construction. This becomes more obvious if we write Eq. (C.94) in the form

$$\gamma(\hat{\mathbf{n}}) = \lambda\,\mathbf{r} \cdot \hat{\mathbf{n}} \tag{C.96}$$

from which it is clear that γ is proportional to the so-called support function for the equilibrium shape. In terms of the scaled coordinates $\mathbf{R} = \lambda\mathbf{r}$, it is the support function for the shape. In fact we know from Section 14.7 that $\boldsymbol{\xi} = \lambda\mathbf{r}$ so we can also write

$$\gamma(\hat{\mathbf{n}}) = \boldsymbol{\xi} \cdot \hat{\mathbf{n}}, \tag{C.97}$$

which we know to be one of the properties of the $\boldsymbol{\xi}$ vector. There is a subtle but important difference between Eqs. (C.96) and (C.97) that is worth attention. If the equilibrium shape has missing orientations, Eq. (C.96) only gives the true $\gamma(\hat{\mathbf{n}})$ for those orientations that actually appear on the shape; for orientations that are missing it gives another function that we called $\Gamma(\hat{\mathbf{n}})$ in Section 14.4. On the other hand, if $\boldsymbol{\xi}$ were known for all orientations, including the ears that must be truncated to give the equilibrium shape, one would obtain $\gamma(\hat{\mathbf{n}})$ for all orientations from Eq. (C.97).

D Equilibrium of Two-State Systems

We use the microcanonical ensemble to make a detailed study of equilibrium of a composite system consisting of two subsystems, each having different numbers of spin 1/2 particles. This will serve as an explicit demonstration of how the composite system achieves its most probable state, as well as the approximations that lead to additivity of entropy. We follow closely a treatment of two identical spin systems by Kittel and Kroemer [6, p. 37] but allow each system to have a different number of spins and evaluate explicitly the overlap integral to determine the entropy of the combined system.

First we consider a system made up of $\mathcal{N}$ spins, each fixed in a solid and having two non-degenerate energy levels. We examine a configuration of the system in which the lower state with energy $-m_0B$ (spin up) is occupied by n_1 spins and the upper state with energy m_0B (spin down) is occupied by $n_2 = \mathcal{N} - n_1$ spins. Here, $m_0 > 0$ is the magnetic moment of a spin and B is the strength of the magnetic field. Following Kittel and Kroemer, we introduce the spin excess, $2s$, where

$$2s =: n_1 - n_2, \tag{D.1}$$

which results in

$$n_1 = \frac{\mathcal{N}}{2} + s; \quad n_2 = \frac{\mathcal{N}}{2} - s. \tag{D.2}$$

Here, s can be integral $\dots, -3, -2, -1, 0, 1, 2, 3, \dots$ or half integral $\dots, -\frac{5}{2}, -\frac{3}{2}, -\frac{1}{2}, \frac{1}{2}, \frac{3}{2}, \frac{5}{2}, \dots$, depending on whether $\mathcal{N}$ is even or odd. In any case, $2s$ will represent the excess[1] number of spins in the ground state, and $0 \le s \le \mathcal{N}/2$. The energy of this state is

$$E = n_1(-m_0B) + n_2(m_0B) = -2sm_0B. \tag{D.3}$$

We assume that these spins are identical but distinguishable by virtue of their fixed positions in a solid. Then the number of microstates of the system that corresponds to the given configuration is

$$\frac{\mathcal{N}!}{n_1!n_2!} = \frac{\mathcal{N}!}{n_1!(\mathcal{N} - n_1)!} = \frac{\mathcal{N}!}{(\frac{\mathcal{N}}{2} + s)!(\frac{\mathcal{N}}{2} - s)!} =: \tilde{g}(\mathcal{N}; s), \tag{D.4}$$

where $\tilde{g}(\mathcal{N}; s)$ is a multiplicity function that plays the same role as the multiplicity function $g(\mathcal{N}, \mathcal{M})$ in Section 16.2 except in terms of a different variable, the correspondence being $s = \mathcal{N}/2 - \mathcal{M}$. Thus, the entropy

[1] Kittel and Kroemer take $\mathcal{N}$ to be even, so s can be considered to be the number of spin flips with respect to equally populated states. Negative values of s correspond to states of the whole system in which the upper spin state has a higher population than the lower one, and hence formally to negative temperatures, which we do not allow. At infinite temperature, the upper and lower spin states are equally populated and $s = 0$.

$$S = k_B \ln \Omega(\mathcal{N}, E) = k_B \ln \tilde{g}(\mathcal{N}; s), \tag{D.5}$$

where k_B is Boltzmann's constant.

We proceed to illustrate explicitly what happens to the entropy when two spin systems, one of size $\mathcal{N}_1$ and the other of size $\mathcal{N}_2$, combine to form a system of size $\mathcal{N} = \mathcal{N}_1 + \mathcal{N}_2$. To do this, we note that the coefficients of t^n in the binomial expansion

$$(1+t)^{\mathcal{N}} = \sum_{n=0}^{\mathcal{N}} \frac{\mathcal{N}!}{n!(\mathcal{N}-n)!} t^n \tag{D.6}$$

are the same as those that enter into Eq. (D.4). In view of the relation

$$(1+t)^{\mathcal{N}_1}(1+t)^{\mathcal{N}_2} = (1+t)^{\mathcal{N}}, \tag{D.7}$$

we seek to relate the multiplicity functions for the system with $\mathcal{N}$ spins to the multiplicity functions of the systems having $\mathcal{N}_1$ and $\mathcal{N}_2$ spins by expanding each binomial and equating the coefficients of like powers of t. Thus

$$\sum_{r_1=0}^{\mathcal{N}_1} \frac{\mathcal{N}_1!}{r_1!(\mathcal{N}_1 - r_1)!} t^{r_1} \sum_{r_2=0}^{\mathcal{N}_2} \frac{\mathcal{N}_2!}{r_2!(\mathcal{N}_2 - r_2)!} t^{r_2} = \sum_{r=0}^{\mathcal{N}} \frac{\mathcal{N}!}{r!(\mathcal{N}-r)!} t^r. \tag{D.8}$$

Equating the coefficient of t^r results in

$$\sum_{r_1} \frac{\mathcal{N}_1!}{r_1!(\mathcal{N}_1 - r_1)!} \frac{\mathcal{N}_2!}{r_2!(\mathcal{N}_2 - r_2)!} = \frac{\mathcal{N}!}{r!(\mathcal{N}-r)!}, \tag{D.9}$$

where the sum over r_1 is restricted by[2] the set of constraints $r_1 + r_2 = r$, $0 \le r_1 \le \mathcal{N}_1$ and $0 \le r_2 \le \mathcal{N}_2$, which also guarantees $0 \le r \le \mathcal{N}$. In terms of the multiplicity functions $\tilde{g}$, Eq. (D.9) can be written

$$\sum_{s_1} \tilde{g}(\mathcal{N}_1; s_1)\tilde{g}(\mathcal{N}_2; s - s_1) = \tilde{g}(\mathcal{N}; s); \quad \mathcal{N} = \mathcal{N}_1 + \mathcal{N}_2, \tag{D.10}$$

where the sum over s_1 has the additional restrictions $0 \le s_1 \le \mathcal{N}_1/2$ and $0 \le s_2 = s - s_1 \le \mathcal{N}_2/2$.

We know that $s = s_1 + s_2$ because of conservation of energy. We are interested in systems having a huge number of spins, say of order 10^{22}, in which case Eq. (D.10) can be simplified greatly because the sum on the left will be dominated by its largest terms. To see this, one can use Stirling's approximation[3] which leads to

$$\tilde{g}(\mathcal{N}; s) = \tilde{g}(\mathcal{N}; 0)\, \mathrm{e}^{-2s^2/\mathcal{N}}, \tag{D.11}$$

[2]An equivalent way of restricting the sum over r_1 is to require $r_1 + r_2 = r$ and replace the factorials with gamma functions according the relation $n!\Gamma(n+1)$. Then since $\Gamma(m) = \pm\infty$ when m is zero or a negative integer, one can sum over all non-vanishing terms.

[3]We use $\mathcal{N}! \sim \mathcal{N}^{\mathcal{N}} \mathrm{e}^{-\mathcal{N}} (2\pi\mathcal{N})^{1/2}$ for better accuracy, since we deal here with the factorial rather than its logarithm. To obtain Eq. (D11) one must expand formally in the small variable $2|s|/\mathcal{N}$; however, this Gaussian approximation is accurate to quite large values of s as shown by the local DeMovire-Laplace theorem. See Gnedenko [75, p. 94] for details.

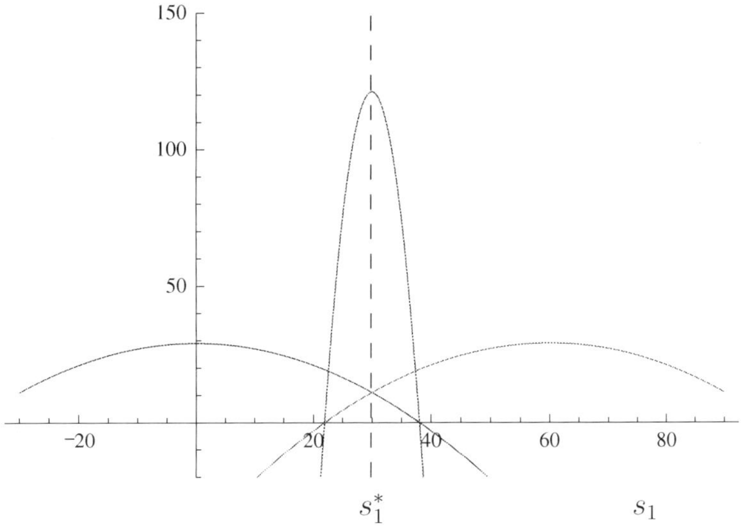

FIGURE D–1 Illustration of the high and narrow peak resulting from the product of two Gaussian peaks of equal height as a function of s_1 for $\mathcal{N}_1 = \mathcal{N}_2 = 100$ and $s = 60$. Since the peaks are so high we have plotted their logarithms to the base 10, specifically $\log_{10} \tilde{g}(\mathcal{N}_1; s_1)$, $\log_{10} \tilde{g}(\mathcal{N}_2; s - s_1)$, and $\log_{10}[\tilde{g}(\mathcal{N}_1; s_1)\tilde{g}(\mathcal{N}_2; s - s_1)]$. Even for these small numbers, we see that the Gaussian peak due to overlap has a height of about 10^{120} and a width of about 10 at half height.

where

$$\tilde{g}(\mathcal{N}; 0) := \sqrt{\frac{2}{\pi \mathcal{N}}}\, 2^{\mathcal{N}}. \tag{D.12}$$

This is often referred to as the **Gaussian approximation** and applies also to $\tilde{g}(\mathcal{N}_1; s_1)$ and $\tilde{g}(\mathcal{N}_2; s_2)$. For huge $\mathcal{N}$, the function $\tilde{g}(\mathcal{N}; s)$ is highly peaked near $s_1 = 0$ and is a quasi-continuous function of s. The same is true for $\tilde{g}(\mathcal{N}_1; s_1)$ and $\tilde{g}(\mathcal{N}_2; s_2)$ as functions of s_1 and s_2. We can therefore approximate the sum in Eq. (D.10) by an integral to obtain

$$\int ds_1 \tilde{g}(\mathcal{N}_1; 0)\tilde{g}(\mathcal{N}_2; 0) e^{-2s_1^2/\mathcal{N}_1} e^{-2(s-s_1)^2/\mathcal{N}_2} = \tilde{g}(\mathcal{N}; 0) e^{-2s^2/\mathcal{N}}. \tag{D.13}$$

As illustrated in Figure D–1, the integral in Eq. (D.13) is over the region of overlap of two Gaussians, one centered at $s_1 = 0$ and the other centered at $s_2 = s - s_1$. The product of these overlapping Gaussian peaks forms an even higher and narrower Gaussian peak. For huge numbers of spins typical of a thermodynamic system, the overlap peak is so high and narrow that it dominates the integral in Eq. (D.13).

The overlap peak occurs at the maximum of the product $\tilde{g}(\mathcal{N}_1; s_1)\tilde{g}(\mathcal{N}_2; s_2)$, with $s_2 = s - s_1$. We can find the position of this peak by differentiation of $\tilde{g}(\mathcal{N}_1; s_1)\tilde{g}(\mathcal{N}_2; s_2)$ with respect to s_1 or, more simply, by differentiating its logarithm with respect to s_1 to obtain

$$\frac{\partial \ln \tilde{g}(\mathcal{N}_1; s_1)}{\partial s_1} = \frac{\partial \ln \tilde{g}(\mathcal{N}_2; s_2)}{\partial s_2}; \quad s_2 = s - s_1. \tag{D.14}$$

Equation (D.14) determines values s_1^* and $s_2^* = s - s_1^*$ that correspond to the overlap peak. Therefore, s_1^* and s_2^* correspond to the dominant contributions of the thermodynamic macrostates of the subsystems. The total energy is divided between the two subsystems so that Eq. (D.14) is satisfied, which is equivalent to equalizing their temperatures.

In terms of the explicit representations of $\tilde{g}(\mathcal{N}_1; s_1)$ and $\tilde{g}(\mathcal{N}_2; s_2)$ (see Eqs. (D.11) and (D.12)), we can write Eq. (D.14) in the form

$$\frac{\partial}{\partial s_1}\left[\ln \tilde{g}(\mathcal{N}_1; 0) - 2s_1^2/\mathcal{N}_1\right] = \frac{\partial}{\partial s_2}\left[\ln g(\mathcal{N}_2; 0) - 2s_2^2/\mathcal{N}_2\right] \tag{D.15}$$

which results in

$$\frac{s_1^*}{\mathcal{N}_1} = \frac{s_2^*}{\mathcal{N}_2} = \frac{s}{\mathcal{N}}, \tag{D.16}$$

where the last equality follows because $s_1^* + s_2^* = s$ and $\mathcal{N}_1 + \mathcal{N}_2 = \mathcal{N}$.

We have yet to demonstrate the additivity of entropy when these subsystems are combined. To do this, we return to Eq. (D.13) and introduce the variable $\delta = s_1 - s_1^*$ such that $\delta = 0$ corresponds to the peak of the product of the Gaussians. After some algebra and the use of Eq. (D.16), the integrand in Eq. (D.13) can be written

$$\tilde{g}(\mathcal{N}_1; s_1)\tilde{g}(\mathcal{N}_2; s - s_1) = (\tilde{g}_1\tilde{g}_2)_{\max} e^{-\delta^2/\delta_0^2}, \tag{D.17}$$

where

$$(\tilde{g}_1\tilde{g}_2)_{\max} := \tilde{g}(\mathcal{N}_1; s_1^*)\tilde{g}(\mathcal{N}_1; s_2^*) = \tilde{g}(\mathcal{N}_1; 0)\tilde{g}(\mathcal{N}_2; 0) e^{-2s^2/\mathcal{N}} \tag{D.18}$$

and

$$\delta_0 := \sqrt{\frac{\mathcal{N}_1\mathcal{N}_2}{2\mathcal{N}}}. \tag{D.19}$$

Therefore Eq. (D.13) becomes approximately

$$(\tilde{g}_1\tilde{g}_2)_{\max} \int e^{-\delta^2/\delta_0^2}\, d\delta = \tilde{g}(\mathcal{N}; s). \tag{D.20}$$

Since the Gaussian peak represented by Eq. (D.17) is so high and narrow, the range of integration of the integral in Eq. (D.20) can be taken to be $-\infty$ to ∞, in which case it becomes $\sqrt{\pi}\delta_0$. Thus Eq. (D.20) becomes

$$\sqrt{\pi}\delta_0(\tilde{g}_1\tilde{g}_2)_{\max} = \tilde{g}(\mathcal{N}; s). \tag{D.21}$$

From Eq. (D.21), we see with the help of Eqs. (D.11) and (D.19) that $(\tilde{g}_1\tilde{g}_2)_{\max}$ is *not* equal to $\tilde{g}(\mathcal{N}; s)$ because of the multiplicative factor $\sqrt{\pi}\delta_0$. But if we take the *logarithm* of both sides to relate to the entropy, we see that

$$\ln(\tilde{g}_1\tilde{g}_2)_{\max} + \frac{1}{2}\ln\pi + \frac{1}{2}\ln\left(\frac{\mathcal{N}_1\mathcal{N}_2}{2\mathcal{N}}\right) = \ln \tilde{g}(\mathcal{N}; s). \tag{D.22}$$

In view of Eqs. (D.12) and (D.18), we see that the first term in Eq. (D.22) is of order $\mathcal{N}$, the second is of order 1, and the third is of order $\ln \mathcal{N}$. The second two terms are negligible compared to the first, so we have

$$\ln(\tilde{g}_1\tilde{g}_2)_{\max} = \ln \tilde{g}(\mathcal{N}_1; s_2^*) + \ln \tilde{g}(\mathcal{N}_2; s_1^*) = \ln \tilde{g}(\mathcal{N}; s), \tag{D.23}$$

which demonstrates the additivity of the entropy.

In other words, in the thermodynamic limit of large numbers of spins, each of the spin subsystems can be regarded as being in its most probable state, consistent with a common temperature that governs how they share the total energy of their combined equilibrium state. This is a general property, believed to be true of all thermodynamic systems. Here we have only demonstrated it explicitly in a simple case.

E

Aspects of Canonical Transformations

We present some aspects of canonical transformations that are used in classical mechanics to transform from one set of generalized coordinates $q = q_1, q_2, \ldots, q_N$ and their conjugate momenta $p = p_1, p_2, \ldots, p_N$ to another independent set $Q = Q_1, Q_2, \ldots, Q_N$ and $P = P_1, P_2, \ldots, P_N$ according to relations of the form

$$q_i = q_i(Q, P, t); \quad p_i = p_i(Q, P, t). \tag{E.1}$$

In this somewhat compressed notation, we regard q, p, Q, and P to be N-dimensional vectors which we do not write in bold face in order to avoid cumbersome expressions. For $\mathcal{N}$ particles each moving in three dimensions, we would have $N = 3\mathcal{N}$ and the entire phase space for the system would have dimension $6\mathcal{N}$, but we retain the more general notation which could be applicable in a two-dimensional world, where $N = 2\mathcal{N}$, or if certain degrees of freedom are suppressed.

We shall treat the general case in which the Hamiltonian $\mathcal{H}(q, p, t)$ as well as the transformation equations depend on time, even though our primary interest will be applications to conservative systems for which there is no explicit dependence on time. As is well known, the dynamical equations are given in the original variables by Hamilton's equations

$$\dot{q}_i = \frac{\partial \mathcal{H}}{\partial p_i}; \quad \dot{p}_i = -\frac{\partial \mathcal{H}}{\partial q_i}. \tag{E.2}$$

Here, a dot above a variable denotes its total time derivative, $\mathrm{d}/\mathrm{d}t$. For a canonical transformation, dynamical equations are given in terms of the new variables by equations of the same form

$$\dot{Q}_i = \frac{\partial \mathcal{K}}{\partial P_i}; \quad \dot{P}_i = -\frac{\partial \mathcal{K}}{\partial Q_i}, \tag{E.3}$$

where $\mathcal{K}(Q, P, t)$ is the new Hamiltonian.

Our treatment of this general case follows closely a treatment by Courant [96, p. 248] but in the modern notation of classical mechanics, as in Goldstein [97, p. 378]. It also includes a demonstration that the necessary and sufficient conditions for a canonical transformation, to be derived below, lead explicitly to Hamilton's equations for the new variables. Courant proceeds to show that canonical transformations belong to the so-called **symplectic group**, from which many properties follow easily.[1] In particular, it will be shown that the Jacobian

$$J = \frac{\partial\,(q,p)}{\partial\,(Q,P)} \equiv \frac{\partial\,(q_1,q_2,\ldots,q_N,p_1,p_2,\ldots,p_N)}{\partial\,(Q_1,Q_2,\ldots,Q_N,P_1,P_2,\ldots,P_N)} = \pm 1. \tag{E.4}$$

Since the absolute value of this Jacobian $|J| = 1$, the volume element in phase space takes the same form $\mathrm{d}Q_1\mathrm{d}Q_2\cdots\mathrm{d}Q_N\mathrm{d}P_1\mathrm{d}P_2\cdots\mathrm{d}P_N$ in terms of the transformed variables as it did in terms of the original variables, namely $\mathrm{d}q_1\mathrm{d}q_2\cdots\mathrm{d}q_N\mathrm{d}p_1\mathrm{d}p_2\cdots\mathrm{d}p_N$. *This fact can sometimes be used to simplify the calculation of the classical partition function.*

E.1 Necessary and Sufficient Conditions

We begin by recalling that Hamilton's equations can be derived by means of the variational principle

$$\delta\int_{t_1}^{t_2}\left[\sum_i p_i\dot{q}_i - \mathcal{H}(q,p,t)\right]\mathrm{d}t = 0, \tag{E.5}$$

where δ denotes virtual synchronous[2] variations of the actual trajectory that connects a fixed point in phase space at time t_1 to another fixed point in phase space at time t_2. The resulting Euler-Lagrange equations, obtained by considering variations in coordinates and momenta to be independent, are just the $2\mathcal{N}$ first order Hamilton equations, Eq. (E.2). The transformation to the new variables will have the same form if a similar variational principal holds, namely

$$\delta\int_{t_1}^{t_2}\left[\sum_k P_k\dot{Q}_k - \mathcal{K}(Q,P,t)\right]\mathrm{d}t = 0. \tag{E.6}$$

We are, of course, free to add the total time derivative of some function $F(Q,P,t)$ to the integrand in Eq. (E.6) to obtain

$$\delta\int_{t_1}^{t_2}\left[\sum_k P_k\dot{Q}_k - \mathcal{K}(Q,P,t) + \frac{\mathrm{d}F(Q,P,t)}{\mathrm{d}t}\right]\mathrm{d}t = 0 \tag{E.7}$$

because the end points are fixed, so

$$\delta\int_{t_1}^{t_2}\frac{\mathrm{d}F(Q,P,t)}{\mathrm{d}t}\,\mathrm{d}t = \delta\,F(Q,P,t)|_{t_1}^{t_2} = 0. \tag{E.8}$$

[1]The author would like to acknowledge David Kinderlehrer for bringing this to his attention and for introducing him to the relevant literature.

[2]Here, synchronous means that the independent variations δq and δp are at a fixed time. For details, see Goldstein [60, p. 225].

By comparison of Eqs. (E.5) and (E.8), we deduce that a canonical transformation is possible if functions $\mathcal{K}$ and F can be found such that the equation[3]

$$\sum_i p_i \dot{q}_i - \mathcal{H}(q,p,t) = \sum_k P_k \dot{Q}_k - \mathcal{K}(Q,P,t) + \frac{\mathrm{d}F(Q,P,t)}{\mathrm{d}t} \tag{E.9}$$

holds *identically* as a function of the variables $Q, P, \dot{Q}, \dot{P}, t$, where it is understood that the left-hand side is to be evaluated by substitution of Eq. (E.1).

By carrying out the substitution and differentiation in Eq. (E.9), one obtains the following set of equations from the coefficients of $\dot{Q}_k$, of $\dot{P}_k$ and remaining terms:

$$\frac{\partial F}{\partial Q_k} = \sum_i p_i \frac{\partial q_i}{\partial Q_k} - P_k; \tag{E.10}$$

$$\frac{\partial F}{\partial P_k} = \sum_i p_i \frac{\partial q_i}{\partial P_k}; \tag{E.11}$$

$$\frac{\partial F}{\partial t} = \mathcal{K}(Q,P,t) - \mathcal{H}(q,p,t) + \sum_i p_i \frac{\partial q_i}{\partial t}. \tag{E.12}$$

Since these equations determine the partial derivatives of F, they will be solvable if and only if all second mixed partial derivatives are independent of the order of differentiation. We first deal with Eqs. (E.10) and (E.11) and then return later to Eq. (E.12) which can be satisfied by a suitable choice of $\mathcal{K}(Q,P,t)$. We obtain:

$$\frac{\partial^2 F}{\partial P_j \partial P_k} - \frac{\partial^2 F}{\partial P_k \partial P_j} = \sum_i \left(\frac{\partial q_i}{\partial P_k}\frac{\partial p_i}{\partial P_j} - \frac{\partial q_i}{\partial P_j}\frac{\partial p_i}{\partial P_k} \right) = 0; \tag{E.13}$$

$$\frac{\partial^2 F}{\partial P_j \partial Q_k} - \frac{\partial^2 F}{\partial Q_k \partial P_j} = \sum_i \left(\frac{\partial q_i}{\partial Q_k}\frac{\partial p_i}{\partial P_j} - \frac{\partial q_i}{\partial P_j}\frac{\partial p_i}{\partial Q_k} \right) - \delta_{jk} = 0; \tag{E.14}$$

$$\frac{\partial^2 F}{\partial Q_j \partial Q_k} - \frac{\partial^2 F}{\partial Q_k \partial Q_j} = \sum_i \left(\frac{\partial q_i}{\partial Q_k}\frac{\partial p_i}{\partial Q_j} - \frac{\partial q_i}{\partial Q_j}\frac{\partial p_i}{\partial Q_k} \right) = 0. \tag{E.15}$$

Equations (E.13) to (E.15) are the necessary and sufficient conditions for a canonical transformation. They can be written in a compact form in terms of **Lagrange brackets**

$$[S,T]_{qp} := \sum_i \left(\frac{\partial q_i}{\partial S}\frac{\partial p_i}{\partial T} - \frac{\partial q_i}{\partial T}\frac{\partial p_i}{\partial S} \right), \tag{E.16}$$

where S and T are any two members of the set Q, P. In that case these conditions become

$$[P_k, P_j]_{qp} = 0; \quad [Q_k, P_j]_{qp} = \delta_{kj}; \quad [Q_k, Q_j]_{qp} = 0. \tag{E.17}$$

[3]This treatment is different from treatments that involve generating functions that are functions of both the old and new variables because the $2\mathcal{N}$ variables Q, P in Eq. (E.1) are always independent. Thus if $F(Q,P,t)$ were replaced by $F_1(Q,q,t)$, one could obtain a canonical transformation only if the $2\mathcal{N}$ variables Q, q were independent, which would not be the case for a coordinate transformation alone. The present approach therefore leads to general conditions that are necessary and sufficient.

Of course we could carry out everything by interchanging the roles of the original and new variables, in which case equivalent conditions would be

$$[p_k, p_j]_{QP} = 0; \quad [q_k, p_j]_{QP} = \delta_{kj}; \quad [q_k, q_j]_{QP} = 0. \tag{E.18}$$

We now return to consider the mixed second derivatives involving time. From Eq. (E.12) we compute

$$\begin{aligned}\frac{\partial^2 F}{\partial P_j \partial t} &= \frac{\partial \mathcal{K}}{\partial P_j} - \sum_i \left(\frac{\partial \mathcal{H}}{\partial p_i}\frac{\partial p_i}{\partial P_j} + \frac{\partial \mathcal{H}}{\partial q_i}\frac{\partial q_i}{\partial P_j} \right) + \sum_i \frac{\partial p_i}{\partial P_j}\frac{\partial q_i}{\partial t} + \sum_i p_i \frac{\partial^2 q_i}{\partial P_j \partial t} \\ &= \frac{\partial \mathcal{K}}{\partial P_j} - \sum_i \left(\dot{q}_i \frac{\partial p_i}{\partial P_j} - \dot{p}_i \frac{\partial q_i}{\partial P_j} \right) + \sum_i \frac{\partial p_i}{\partial P_j}\frac{\partial q_i}{\partial t} + \sum_i p_i \frac{\partial^2 q_i}{\partial P_j \partial t}\end{aligned} \tag{E.19}$$

and from Eq. (E.11)

$$\frac{\partial^2 F}{\partial t \partial P_j} = \frac{\partial q_i}{\partial P_j}\frac{\partial p_i}{\partial t} + \sum_i p_i \frac{\partial^2 q_i}{\partial t \partial P_j}. \tag{E.20}$$

Equating these mixed partials and solving for $\partial \mathcal{K}/\partial P_j$, we obtain

$$\begin{aligned}\frac{\partial \mathcal{K}}{\partial P_j} &= \sum_i \left(\dot{q}_i \frac{\partial p_i}{\partial P_j} - \dot{p}_i \frac{\partial q_i}{\partial P_j} \right) - \sum_i \frac{\partial p_i}{\partial P_j}\frac{\partial q_i}{\partial t} + \frac{\partial q_i}{\partial P_j}\frac{\partial p_i}{\partial t} \\ &= \sum_k [Q_k, P_j]_{qp} \dot{Q}_k + \sum_k [P_k, P_j]_{qp} \dot{P}_k = \sum_k \delta_{jk} \dot{Q}_k = \dot{Q}_j.\end{aligned} \tag{E.21}$$

Similarly, from $\partial^2 F/\partial t \partial Q_j = \partial^2 F/\partial Q_j \partial t$ we obtain

$$\begin{aligned}\frac{\partial \mathcal{K}}{\partial Q_j} &= \sum_i \left(\dot{q}_i \frac{\partial p_i}{\partial Q_j} - \dot{p}_i \frac{\partial q_i}{\partial Q_j} \right) - \sum_i \frac{\partial p_i}{\partial Q_j}\frac{\partial q_i}{\partial t} + \frac{\partial q_i}{\partial Q_j}\frac{\partial p_i}{\partial t} \\ &= \sum_k [Q_k, Q_j]_{qp} \dot{Q}_k + \sum_k [P_k, Q_j]_{qp} \dot{P}_k = -\sum_k \delta_{jk} \dot{P}_k = -\dot{P}_j.\end{aligned} \tag{E.22}$$

Equations (E.21) and (E.22) show explicitly that the conditions Eq. (E.17) lead to Hamilton's equations in the new variables.

E.1.1 Symplectic Transformation

We now demonstrate that the conditions Eq. (E.17) can be written in the form of a symplectic transformation. This can be accomplished by introducing two $2\mathcal{N} \times 2\mathcal{N}$ matrices

$$M = \begin{pmatrix} \partial q/\partial Q & \partial q/\partial P \\ \partial p/\partial Q & \partial p/\partial P \end{pmatrix}; \qquad S = \begin{pmatrix} 0 & 1 \\ -1 & 0 \end{pmatrix}, \tag{E.23}$$

where each entry is, itself, an $\mathcal{N} \times \mathcal{N}$ matrix. For example, $\partial q/\partial Q$ has matrix elements $(\partial q/\partial Q)_{ij} = \partial q_i/\partial Q_j$. In particular, the Jacobian of the transformation, given by Eq. (E.4), is just $J = \det M$. In the matrix S, 1 is understood to be the $\mathcal{N} \times \mathcal{N}$ unit matrix and 0 is the $\mathcal{N} \times \mathcal{N}$ null matrix. We observe that

$$S^2 = \begin{pmatrix} 0 & 1 \\ -1 & 0 \end{pmatrix} \begin{pmatrix} 0 & 1 \\ -1 & 0 \end{pmatrix} = \begin{pmatrix} -1 & 0 \\ 0 & -1 \end{pmatrix} = -\begin{pmatrix} 1 & 0 \\ 0 & 1 \end{pmatrix}, \tag{E.24}$$

so S plays the role of $i = \sqrt{-1}$ in this space. We also observe that

$$(\det S)^2 = \det S^2 = (-1)^{2\mathcal{N}} = 1, \tag{E.25}$$

so $\det S = \pm 1$. Inspection shows that $\det S = -1$ if $\mathcal{N}$ is odd and $\det S = 1$ if $\mathcal{N}$ is even. Evidently the inverse $S^{-1} = \tilde{S}$, the transpose of S. We shall also need the transpose of M, namely

$$\tilde{M} = \begin{pmatrix} \widetilde{\partial q/\partial Q} & \widetilde{\partial p/\partial Q} \\ \widetilde{\partial q/\partial P} & \widetilde{\partial p/\partial P} \end{pmatrix}. \tag{E.26}$$

Then from the conditions given by Eq. (E.17) it follows that

$$\tilde{M} S M = S \tag{E.27}$$

and M is said to be a symplectic matrix. To see this, first compute

$$SM = \begin{pmatrix} 0 & 1 \\ -1 & 0 \end{pmatrix} \begin{pmatrix} \partial q/\partial Q & \partial q/\partial P \\ \partial p/\partial Q & \partial p/\partial P \end{pmatrix} = \begin{pmatrix} \partial p/\partial Q & \partial p/\partial P \\ -\partial q/\partial Q & -\partial q/\partial P \end{pmatrix}. \tag{E.28}$$

Then $\tilde{M}SM$ is given by

$$\begin{pmatrix} \widetilde{\partial q/\partial Q} & \widetilde{\partial p/\partial Q} \\ \widetilde{\partial q/\partial P} & \widetilde{\partial p/\partial P} \end{pmatrix} \begin{pmatrix} \partial p/\partial Q & \partial p/\partial P \\ -\partial q/\partial Q & -\partial q/\partial P \end{pmatrix} = \begin{pmatrix} ((QQ)) & ((QP)) \\ ((PQ)) & ((PP)) \end{pmatrix}, \tag{E.29}$$

where the symbols in double parentheses are $\mathcal{N} \times \mathcal{N}$ matrices given by the Lagrange brackets as follows:

$$((PP))_{kj} = [P_k, P_j]_{qp}; \quad ((QP))_{kj} = [Q_k, P_j]_{qp}; \quad ((QQ))_{kj} = [Q_k, Q_j]_{qp} \tag{E.30}$$

with $((QP)) = -((PQ))$. Then by Eq. (E.17), we see that the right-hand side of Eq. (E.29) is equal to S.

Having established Eq. (E.27), we can now easily compute the Jacobian for any canonical transformation. We have

$$\det \tilde{M}SM = \det \tilde{M} \det S \det M = (\det M)^2 \det S = \det S. \tag{E.31}$$

Since $\det S = \pm 1$, it can be canceled and we obtain $(\det M)^2 = 1$ from which

$$J = \det M = \pm 1. \tag{E.32}$$

Thus, as stated above, for a canonical transformation the volume element in phase space takes the same form $\mathrm{d}Q_1\mathrm{d}Q_2 \cdots \mathrm{d}Q_{\mathcal{N}}\mathrm{d}P_1\mathrm{d}P_2 \cdots \mathrm{d}P_{\mathcal{N}}$ in terms of the new variables as it did in terms of the original variables, namely $\mathrm{d}q_1\mathrm{d}q_2 \cdots \mathrm{d}q_{\mathcal{N}}\mathrm{d}p_1\mathrm{d}p_2 \cdots \mathrm{d}p_{\mathcal{N}}$. This may seem counterintuitive because one is so familiar with the fact that the volume element in real Cartesian space is $\mathrm{d}x\mathrm{d}y\mathrm{d}z$, whereas in cylindrical coordinates it is $r^2 \sin\theta \mathrm{d}r\mathrm{d}\theta\mathrm{d}\phi$, which contains scale factors. But for canonical transformations, we must remember that *both* coordinates and their conjugate momenta are transformed. It often happens that after integration over conjugate momenta, familiar scale factors for the coordinates appear.

Symplectic matrices form a group. Since $\det M = \pm 1$, the inverse matrix M^{-1} exists. Multiplication of Eq. (E.27) from the right by M^{-1} and from the left by $\tilde{M}^{-1}$ gives

$$S = \tilde{M}^{-1} S M^{-1} = \widetilde{(M^{-1})} S M^{-1}, \tag{E.33}$$

so M^{-1} is also symplectic. Furthermore, if M and W are symplectic, their product $A = MW$ is symplectic because

$$\tilde{A} S A = \tilde{W} \tilde{M} S M W = \tilde{W} S W = S, \tag{E.34}$$

which establishes the group property. Multiplication pertains to successive canonical transformations, which generate yet another canonical transformation.

E.2 Restricted Canonical Transformations

An important special case is a restricted canonical transformation in which the transformation from q, p to Q, P does not involve the time explicitly, namely,

$$q_i = q_i(Q, P); \quad p_i = p_i(Q, P). \tag{E.35}$$

Under these circumstances, the terms containing $\partial q_i/\partial t$ on the right-hand side of Eq. (E.12) will vanish and it will be possible to choose the function F to have no explicit dependence on time, that is, $F = F(Q, P)$. In that case, one can obtain a canonical transformation in which

$$\mathcal{H}(q, p, t) = \mathcal{K}(Q, P, t), \tag{E.36}$$

so the new Hamiltonian can be obtained by substituting $q(Q, P)$ and $p(Q, P)$ in the original Hamiltonian.

Of course the transformation will only be canonical if the conditions given by Eq. (E.17) or, alternatively, Eq. (E.18) apply. For this restricted situation, however, we can follow Goldstein [97, p. 391] to derive a simpler set of conditions that will guarantee that the transformation will be canonical. Thus

$$\frac{\partial \mathcal{K}}{\partial P_i} = \sum_k \left(\frac{\partial \mathcal{H}}{\partial q_k} \frac{\partial q_k}{\partial P_i} + \frac{\partial \mathcal{H}}{\partial p_k} \frac{\partial p_k}{\partial P_i} \right) \tag{E.37}$$

and

$$\dot{Q}_i = \sum_k \left(\frac{\partial Q_i}{\partial q_k} \dot{q}_k + \frac{\partial Q_i}{\partial p_k} \dot{p}_k \right) = \sum_k \left(-\frac{\partial \mathcal{H}}{\partial q_k} \frac{\partial Q_i}{\partial p_k} + \frac{\partial \mathcal{H}}{\partial p_k} \frac{\partial Q_i}{\partial q_k} \right). \tag{E.38}$$

For the transformation to be canonical, we need $\partial \mathcal{K}/\partial P_i = \dot{Q}_i$ and for this to be true for any function $\mathcal{H}(q, p, t)$, we see from comparison of Eq. (E.37) with Eq. (E.38) that

$$\frac{\partial q_k}{\partial P_i} = -\frac{\partial Q_i}{\partial p_k}; \quad \frac{\partial p_k}{\partial P_i} = \frac{\partial Q_i}{\partial q_k}. \tag{E.39}$$

Similarly

$$\frac{\partial \mathcal{K}}{\partial Q_i} = \sum_k \left(\frac{\partial \mathcal{H}}{\partial q_k}\frac{\partial q_k}{\partial Q_i} + \frac{\partial \mathcal{H}}{\partial p_k}\frac{\partial p_k}{\partial Q_i} \right) \tag{E.40}$$

and

$$-\dot{P}_i = -\sum_k \left(\frac{\partial P_i}{\partial q_k}\dot{q}_k + \frac{\partial P_i}{\partial p_k}\dot{p}_k \right) = \sum_k \left(\frac{\partial \mathcal{H}}{\partial q_k}\frac{\partial P_i}{\partial p_k} - \frac{\partial \mathcal{H}}{\partial p_k}\frac{\partial P_i}{\partial q_k} \right). \tag{E.41}$$

So requiring $\partial \mathcal{K}/\partial Q_i = -\dot{P}_i$ leads to

$$\frac{\partial q_k}{\partial Q_i} = \frac{\partial P_i}{\partial p_k}; \quad \frac{\partial p_k}{\partial Q_i} = -\frac{\partial P_i}{\partial q_k}. \tag{E.42}$$

In Eqs. (E.39) and (E.42) it is important to bear in mind that the variable set for each of these partial derivatives is either q, p or Q, P. Thus, in a somewhat expanded notation,

$$\frac{\partial p_k}{\partial P_i} \equiv \left(\frac{\partial p_k}{\partial P_i}\right)_Q; \qquad \text{whereas} \quad \frac{\partial P_i}{\partial p_k} \equiv \left(\frac{\partial P_i}{\partial p_k}\right)_q, \tag{E.43}$$

so $\partial p_k/\partial P_i$ is not the reciprocal of $\partial P_i/\partial p_k$.

By using Eqs. (E.39) and (E.42) it is easy to see that the general conditions for a canonical transformation are satisfied. For example, for $[P_k, P_j]_{qp}$ we have

$$\sum_i \left(\frac{\partial q_i}{\partial P_k}\frac{\partial p_i}{\partial P_j} - \frac{\partial q_i}{\partial P_j}\frac{\partial p_i}{\partial P_k} \right) = \sum_i \left(\frac{\partial q_i}{\partial P_k}\frac{\partial Q_j}{\partial q_i} + \frac{\partial Q_j}{\partial p_k}\frac{\partial p_i}{\partial P_k} \right) = \frac{\partial Q_j}{\partial P_k} = 0, \tag{E.44}$$

where Eq. (E.39) has been used. Similarly, for $[Q_k, P_j]_{qp}$,

$$\sum_i \left(\frac{\partial q_i}{\partial Q_k}\frac{\partial p_i}{\partial P_j} - \frac{\partial q_i}{\partial P_j}\frac{\partial p_i}{\partial Q_k} \right) = \sum_i \left(\frac{\partial q_i}{\partial Q_k}\frac{\partial Q_j}{\partial q_i} + \frac{\partial Q_j}{\partial p_i}\frac{\partial p_i}{\partial Q_k} \right) = \frac{\partial Q_j}{\partial Q_k} = \delta_{jk}. \tag{E.45}$$

And finally for $[Q_k, Q_j]_{qp}$,

$$\sum_i \left(\frac{\partial q_i}{\partial Q_k}\frac{\partial p_i}{\partial Q_j} - \frac{\partial q_i}{\partial Q_j}\frac{\partial p_i}{\partial Q_k} \right) = \sum_i \left(\frac{\partial P_k}{\partial p_i}\frac{\partial p_i}{\partial Q_j} + \frac{\partial q_i}{\partial Q_j}\frac{\partial P_k}{\partial Q_i} \right) = \frac{\partial P_k}{\partial Q_j} = 0. \tag{E.46}$$

F

Rotation of Rigid Bodies

There is no such thing as a rigid body but under many circumstances, solid bodies can often be treated to a good approximation as if they were rigid. Moreover, molecules can be approximated by rigid bodies composed of point masses provided that vibrational modes are not excited. We therefore summarize some useful properties of such bodies.

The formulae below pertain to bodies whose center of mass is at rest. We know that the total kinetic energy of a body is the sum of the kinetic energy of the center of mass and the kinetic energy with respect to the center of mass. Similarly, the total angular momentum is the sum of the angular momentum with respect to the center of mass plus the angular momentum of the center of mass with respect to the origin of coordinates. For this and other reasons that afford simplification, we treat only bodies whose centers of mass are at rest.

We denote the coordinate of a point of such a body by the vector $\mathbf{r}$. We shall write a number of formulae for the continuum case for which mass is distributed according to a density $\rho(\mathbf{r})$. To obtain formulae for the case of discrete masses, one only needs to write the density as a sum of delta functions, for example, $\rho(\mathbf{r}) = \sum_i m_i \delta(\mathbf{r} - \mathbf{r}_i)$ in which case the integrals are replaced by sums.[1]

F.1 Moment of Inertia

The **moment of inertia** of a body with respect to some axis passing through its center of mass is defined by the formula

$$\mathcal{I} := \int \rho(\mathbf{r})\, r_\perp^2 \,\mathrm{d}V, \tag{F.1}$$

where the integral is over the volume of the body and $r_\perp$ is the distance to a point in the body measured perpendicular to the specified axis. We can specify the axis by supposing it to lie along a unit vector $\hat{\mathbf{a}}$ in which case $r_\perp^2 = |\hat{\mathbf{a}} \times \mathbf{r}|^2 = r^2 - (\mathbf{r} \cdot \hat{\mathbf{a}})^2$. Thus Eq. (F.1) can be written more explicitly as

$$\mathcal{I}(\hat{\mathbf{a}}) = \int \rho(\mathbf{r})\,(r^2 - (\mathbf{r} \cdot \hat{\mathbf{a}})^2)\,\mathrm{d}V = \mathcal{I}(-\hat{\mathbf{a}}) = \sum_{\alpha,\beta} \hat{a}_\alpha \mathcal{I}_{\alpha\beta} \hat{a}_\beta, \tag{F.2}$$

[1]The notation $\rho(\mathbf{r})$ is applicable if the body is at rest. In Section F.1 we discuss some cases of rotating bodies. This is handled by treating discrete masses located at positions $\mathbf{r}_i(t)$ that depend on time. In that case we should write $\rho(\mathbf{r}(t))$ but we will suppress the dependence on t for simplicity. In Section F.5 we use a rotating coordinate system $\mathbf{r}'$ in which the body is at rest, so in that case we write $\rho(\mathbf{r}')$.

where α and β denote Cartesian coordinates and the symmetric **moment of inertia tensor**

$$\mathcal{I}_{\alpha\beta} := \int \rho(\mathbf{r})\,(r^2\delta_{\alpha\beta} - x_\alpha x_\beta)\,\mathrm{d}V. \tag{F.3}$$

For the case of a rigid body made up of discrete masses m_i located at positions $\mathbf{r}_i$, as mentioned previously, Eq. (F.3) becomes[2]

$$\mathcal{I}_{\alpha\beta} := \sum_i m_i\,(r_i^2\delta_{\alpha\beta} - x_{i\alpha}x_{i\beta}). \tag{F.4}$$

In dyadic notation, this tensor could be written

$$\boldsymbol{I} = \sum_i m_i\,[r_i^2\mathbb{1} - \mathbf{r}_i\mathbf{r}_i], \tag{F.5}$$

where $\mathbb{1}$ corresponds to the unit tensor.

From its definition, it is clear that the actual components of $\mathcal{I}_{\alpha\beta}$ will depend on the orientation of the body with respect to the chosen axes. If the body is at rest with respect to these axes, these components will be constants. We observe that $\mathcal{I}_{\alpha\beta} = \mathcal{I}_{\beta\alpha}$ so this tensor can be diagonalized by means of a choice of axes, rotated with respect to some original choice of axes. Such a transformation can be accomplished by means of an orthogonal transformation. It is worth noting, however, that the quantity $\mathcal{I}(\hat{\mathbf{a}})$ with respect to any fixed axis $\hat{\mathbf{a}}$ is unchanged if the body is rotated about that axis because it only depends on $r_\perp$.

If the rigid body is in motion with respect to the axes of the chosen reference frame, the *components* of the tensor $\mathcal{I}_{\alpha\beta}$ will generally depend on time. For reasons just mentioned, however, the value of $\mathcal{I}(\hat{\mathbf{a}})$ will not depend on time as the body rotates about any *fixed* axis. This provides some simplification of some of the formulae given below but also leads to some complications in formulae in which time derivatives of $\mathcal{I}_{\alpha\beta}$ occur. In such cases, one must either evaluate such time derivatives explicitly or employ a reformulation in which two coordinate systems are used, one at rest with respect to the body and in which the components of $\mathcal{I}_{\alpha\beta}$ will be constants, and another with respect to which the body can rotate.

F.1.1 Diatomic Molecule

The moment of inertia tensor for a diatomic molecule consisting of two point particles can be calculated from Eq. (F.3) by replacing the density by a sum of two delta functions,

$$\rho(\mathbf{r}) = m_1\delta(\mathbf{r} - \mathbf{r}_1) + m_2\delta(\mathbf{r} - \mathbf{r}_2), \tag{F.6}$$

where one particle of mass m_1 is located at $\mathbf{r}_1$ and the other particle of mass m_2 is located at $\mathbf{r}_2$. Thus

[2]If an origin other than the center of mass is used, it is apparent from Eq. (F.4) that one must add to $\mathcal{I}_{\alpha\beta}$ the quantity $M(R^2\delta_{\alpha\beta} - R_\alpha R_\beta)$, where M is the total mass and $\mathbf{R}$ is the vector from the new origin to the center of mass. Cross terms vanish because of the definition of the center of mass. This tensor would contribute an additional term $M|\hat{\mathbf{a}} \times \mathbf{R}|^2$ to $\mathcal{I}(\hat{\mathbf{a}})$.

$$\mathcal{I}_{\alpha\beta} = m_1\left[r_1^2\delta_{\alpha\beta} - x_{1\alpha}x_{1\beta}\right] + m_2\left[r_2^2\delta_{\alpha\beta} - x_{2\alpha}x_{2\beta}\right]. \tag{F.7}$$

We recall that the origin is to be located at the center of mass and proceed to calculate $\mathcal{I}_{\alpha\beta}$ in a principal axis system x, y, z in which the coordinates of the particles are $x_1 = x_2 = y_1 = y_2 = 0$, and

$$z_1 = \frac{m_2}{m_1 + m_2}\ell_0; \quad z_2 = -\frac{m_1}{m_1 + m_2}\ell_0, \tag{F.8}$$

where $\ell_0 = |z_1 - z_2|$ is the distance of separation between particles. In this coordinate system, $r_1 = z_1$ and $r_2 = z_2$ so we see immediately that $\mathcal{I}_{zz} = 0$ and

$$\mathcal{I}_{xx} = \mathcal{I}_{yy} = m_1 z_1^2 + m_2 z_2^2 = \frac{\ell_0^2}{(m_1 + m_2)^2}(m_1 m_2^2 + m_2 m_1^2) = \ell_0^2 \frac{m_1 m_2}{m_1 + m_2}. \tag{F.9}$$

The quantity multiplying ℓ_0^2 is known as the reduced mass, familiar from mechanics.

If the particles were not point particles, but spheres of radii a_1 and a_2 respectively, there would be a small value of $\mathcal{I}_{zz} = (2/5)(m_1 a_1^2 + m_2 a_2^2)$ and each of $\mathcal{I}_{xx}$ and $\mathcal{I}_{yy}$ would be increased by this same amount. Since most of the mass of an atom resides in its nucleus, say of radius r_0, the ratio $\mathcal{I}_{zz}/\mathcal{I}_{xx}$ will be of the order of magnitude of $(r_0/\ell_0)^2 \sim 10^{-8}$. See Section F.8 for a related discussion.

F.2 Angular Momentum

The **angular momentum** with respect to the center of mass is defined by

$$\mathbf{L} := \int \rho(\mathbf{r})\, \mathbf{r} \times \mathbf{v}\, \mathrm{d}V, \tag{F.10}$$

where $\mathbf{v}$ is the velocity of the body at the point $\mathbf{r}$. The quantity

$$\hat{\mathbf{a}} \cdot \mathbf{L} = \int \rho(\mathbf{r})\, \hat{\mathbf{a}} \cdot \mathbf{r} \times \mathbf{v}\, \mathrm{d}V = \int \rho(\mathbf{r})\, \hat{\mathbf{a}} \times \mathbf{r} \cdot \mathbf{v}\, \mathrm{d}V = \int \rho(\mathbf{r})\, \mathbf{r} \times \mathbf{v} \cdot \hat{\mathbf{a}}\, \mathrm{d}V = \mathbf{L} \cdot \hat{\mathbf{a}} \tag{F.11}$$

is the angular momentum with respect to the axis $\hat{\mathbf{a}}$. This follows because $\hat{\mathbf{a}} \times \mathbf{r}$ has magnitude $r_\perp$ and a direction perpendicular to the plane made by $\hat{\mathbf{a}}$ and $\mathbf{r}$. Its dot product with $\mathbf{v}$ selects the component of $\mathbf{v}$ perpendicular to this plane in a direction related to the $\hat{\mathbf{a}}$ axis in accordance with the right-hand rule.

We define a vector $\boldsymbol{\omega} := \hat{\mathbf{a}}\omega$, where ω is an angular velocity. Then for rotation of a rigid body about an axis $\hat{\mathbf{a}}$, in the sense of the right-hand rule, we can write $\mathbf{v} = \boldsymbol{\omega} \times \mathbf{r}$ in which case $\mathbf{r} \times \mathbf{v} = \mathbf{r} \times (\boldsymbol{\omega} \times \boldsymbol{r}) = [r^2\boldsymbol{\omega} - (\mathbf{r} \cdot \boldsymbol{\omega})\mathbf{r}] = [r^2\mathbb{1} - \mathbf{rr}] \cdot \boldsymbol{\omega}$, where $\mathbb{1}$ is the unit dyadic and $\mathbf{rr}$ is a tensor product (dyadic). Substitution into Eq. (F.10) gives

$$\mathbf{L} = \boldsymbol{I} \cdot \boldsymbol{\omega}, \tag{F.12}$$

where $\boldsymbol{I}$ is the moment of inertia tensor with components given by Eq. (F.3). We also observe that $\hat{\mathbf{a}} \cdot \mathbf{L} = \hat{\mathbf{a}} \cdot \boldsymbol{I} \cdot \boldsymbol{\omega} = \hat{\mathbf{a}} \cdot \boldsymbol{I} \cdot \hat{\mathbf{a}}\,\omega = \mathcal{I}(\hat{\mathbf{a}})\omega$.

F.3 Kinetic Energy

The kinetic energy with respect to the center of mass is[3]

$$T = \frac{1}{2}\int \rho(\mathbf{r})\,\mathbf{v}\cdot\mathbf{v}\,\mathrm{d}V. \tag{F.13}$$

For rigid rotation, $\mathbf{v}\cdot\mathbf{v} = (\boldsymbol{\omega}\times\mathbf{r})\cdot(\boldsymbol{\omega}\times\mathbf{r}) = \boldsymbol{\omega}\cdot[r^2\mathbb{1} - \mathbf{rr}]\cdot\boldsymbol{\omega}$, so

$$T = \frac{1}{2}\boldsymbol{\omega}\cdot\boldsymbol{I}\cdot\boldsymbol{\omega} = \frac{1}{2}\boldsymbol{\omega}\cdot\mathbf{L} = \frac{1}{2}\mathcal{I}(\hat{\mathbf{a}})\,\omega^2. \tag{F.14}$$

If we use a coordinate system in which $\boldsymbol{I}$ is momentarily diagonal, we would have[4]

$$T = \frac{1}{2}\left(\mathcal{I}_{xx}\,\omega_x^2 + \mathcal{I}_{yy}\,\omega_y^2 + \mathcal{I}_{zz}\,\omega_z^2\right) \tag{F.15}$$

and Eq. (F.12) becomes

$$L_x = \mathcal{I}_{xx}\,\omega_x; \quad L_y = \mathcal{I}_{yy}\,\omega_y; \quad L_z = \mathcal{I}_{zz}\,\omega_z. \tag{F.16}$$

Then Eq. (F.15) can be written in the form

$$T = \frac{L_x^2}{2\mathcal{I}_{xx}} + \frac{L_y^2}{2\mathcal{I}_{yy}} + \frac{L_z^2}{2\mathcal{I}_{zz}}. \tag{F.17}$$

F.4 Time Derivatives

As remarked toward the end of Section F.1, the components of $\mathcal{I}_{\alpha\beta}$ will depend on time if a body is rotating with respect to the coordinates of the reference frame. We first calculate the time derivative of Eq. (F.5) at a moment when the body is rotating with angular velocity $\boldsymbol{\omega}$ with respect to that reference frame. A vector $\mathbf{r}_i(t)$ that locates point i of the body will have a velocity $\mathbf{v}_i = \boldsymbol{\omega}\times\mathbf{r}_i$ in that frame. Thus $\mathrm{d}r_i^2/\mathrm{d}t = 2\mathbf{v}_i\cdot\mathbf{r}_i = 0$. On the other hand, the time derivative of the dyadic $\mathbf{r}_i\mathbf{r}_i$ is

$$\frac{\mathrm{d}}{\mathrm{d}t}(\mathbf{r}_i\mathbf{r}_i) = \mathbf{v}_i\mathbf{r}_i + \mathbf{r}_i\mathbf{v}_i = (\boldsymbol{\omega}\times\mathbf{r}_i)\mathbf{r}_i - \mathbf{r}_i(\mathbf{r}_i\times\boldsymbol{\omega}) = \boldsymbol{\omega}\times\mathbf{r}_i\mathbf{r}_i - \mathbf{r}_i\mathbf{r}_i\times\boldsymbol{\omega}, \tag{F.18}$$

where the parentheses can be omitted without ambiguity. Substituting these results into the time derivative of Eq. (F.5) gives

$$\frac{\mathrm{d}}{\mathrm{d}t}\boldsymbol{I} = -\sum_i m_i[\boldsymbol{\omega}\times\mathbf{r}_i\mathbf{r}_i - \mathbf{r}_i\mathbf{r}_i\times\boldsymbol{\omega}]. \tag{F.19}$$

It also turns out that $\boldsymbol{\omega}\times\mathbb{1} = \mathbb{1}\times\boldsymbol{\omega}$, which can be shown by straightforward algebra by writing $\mathbb{1} = \hat{\mathbf{i}}\hat{\mathbf{i}} + \hat{\mathbf{j}}\hat{\mathbf{j}} + \hat{\mathbf{k}}\hat{\mathbf{k}}$ in terms of the unit vectors of a Cartesian coordinate system. Hence Eq. (F.19) can be written in the form

$$\frac{\mathrm{d}\boldsymbol{I}}{\mathrm{d}t} = \boldsymbol{\omega}\times\boldsymbol{I} - \boldsymbol{I}\times\boldsymbol{\omega}. \tag{F.20}$$

[3] In Chapter 1 we called this quantity T^i to distinguish it from the total kinetic energy.
[4] If the axis of rotation remains fixed, $\mathcal{I}(\hat{\mathbf{a}})$ would not change with time, as shown by Eq. (F.21).

For $\boldsymbol{\omega}$ along a *fixed* axis $\hat{\mathbf{a}}$, we see from Eq. (F.2) that

$$\frac{\mathrm{d}}{\mathrm{d}t}\mathcal{I}(\hat{\mathbf{a}}) = \hat{\mathbf{a}} \cdot \frac{\mathrm{d}\boldsymbol{I}}{\mathrm{d}t} \cdot \hat{\mathbf{a}} = \hat{\mathbf{a}} \cdot (\boldsymbol{\omega} \times \boldsymbol{I} - \boldsymbol{I} \times \boldsymbol{\omega}) \cdot \hat{\mathbf{a}} = 0 \tag{F.21}$$

because we can interchange the dot and the cross on either side of $\boldsymbol{I}$ and the cross product of $\hat{\mathbf{a}}$ with $\boldsymbol{\omega}$ vanishes. This further supports our statement at the end of Section F.1 that $\mathcal{I}(\hat{\mathbf{a}})$ is independent of time even for a body that rotates about a fixed axis $\hat{\mathbf{a}}$.

On the other hand, from Eq. (F.12) we see that

$$\frac{\mathrm{d}\mathbf{L}}{\mathrm{d}t} = \boldsymbol{I} \cdot \frac{\mathrm{d}\boldsymbol{\omega}}{\mathrm{d}t} + \frac{\mathrm{d}\boldsymbol{I}}{\mathrm{d}t} \cdot \boldsymbol{\omega} = \boldsymbol{I} \cdot \frac{\mathrm{d}\boldsymbol{\omega}}{\mathrm{d}t} + \boldsymbol{\omega} \times \boldsymbol{I} \cdot \boldsymbol{\omega} \tag{F.22}$$

or finally

$$\frac{\mathrm{d}\mathbf{L}}{\mathrm{d}t} = \boldsymbol{I} \cdot \frac{\mathrm{d}\boldsymbol{\omega}}{\mathrm{d}t} + \boldsymbol{\omega} \times \mathbf{L}. \tag{F.23}$$

The "extra" term $\boldsymbol{\omega} \times \mathbf{L}$ comes from the dependence of $\boldsymbol{I}$ on time when calculated in a fixed frame in which the body is rotating.

Turning to the kinetic energy given by Eq. (F.14) we see that

$$\frac{\mathrm{d}T}{\mathrm{d}t} = \frac{1}{2}\frac{\mathrm{d}\boldsymbol{\omega}}{\mathrm{d}t} \cdot \boldsymbol{I} \cdot \boldsymbol{\omega} + \frac{1}{2}\boldsymbol{\omega} \cdot \boldsymbol{I} \cdot \frac{\mathrm{d}\boldsymbol{\omega}}{\mathrm{d}t} + \frac{1}{2}\boldsymbol{\omega} \cdot \frac{\mathrm{d}\boldsymbol{I}}{\mathrm{d}t} \cdot \boldsymbol{\omega}. \tag{F.24}$$

The last term vanishes after substitution of Eq. (F.20) and the other two combine to give

$$\frac{dT}{dt} = \boldsymbol{\omega} \cdot \boldsymbol{I} \cdot \frac{\mathrm{d}\boldsymbol{\omega}}{\mathrm{d}t} = \mathbf{L} \cdot \frac{\mathrm{d}\boldsymbol{\omega}}{\mathrm{d}t}, \tag{F.25}$$

from which we see that the dependence of $\boldsymbol{I}$ on time does not result in an additional term.

We now return to Eq. (F.23) and recognize that the left-hand side is equal to the torque $\mathbf{N}$ that may be applied to the body. If this torque vanishes, $\mathbf{L}$ is just a constant, and angular momentum is conserved. If it does not, we can take the dot product with $\boldsymbol{\omega}$ to obtain

$$\boldsymbol{\omega} \cdot \mathbf{N} = \boldsymbol{\omega} \cdot \boldsymbol{I} \cdot \frac{\mathrm{d}\boldsymbol{\omega}}{\mathrm{d}t} = \frac{\mathrm{d}T}{\mathrm{d}t} \tag{F.26}$$

from which we recognize that the left-hand side is the power supplied by the torque, which is equal to the time rate of change of the kinetic energy. If the torque vanishes, the kinetic energy is constant, as expected.

F.5 Rotating Coordinate System

Some of the complications of Section F.4 can be avoided by using two coordinate systems, the unprimed system with coordinates x, y, z as dealt with above, which we here take to be an inertial frame, and a primed system with coordinates x', y', z' having the same origin and coordinate axes imbedded in the rigid body. The axes of the primed system rotate with the body. A point i in the body can be described by either a vector $\mathbf{r}_i = x_i\hat{\mathbf{i}} + y_i\hat{\mathbf{j}} + z_i\hat{\mathbf{k}}$ or a vector $\mathbf{r}'_i = x'_i\hat{\mathbf{i}}' + y'_i\hat{\mathbf{j}}' + z'_i\hat{\mathbf{k}}'$, and since it is the same point, we have $\mathbf{r}_i = \mathbf{r}'_i$. As the body rotates, the coordinates x_i, y_i, z_i change with time but the unit vectors $\hat{\mathbf{i}}, \hat{\mathbf{j}}, \hat{\mathbf{k}}$ remain constant in time; on the other hand, the coordinates x'_i, y'_i, z'_i remain constant in time but the unit vectors $\hat{\mathbf{i}}', \hat{\mathbf{j}}', \hat{\mathbf{k}}'$ rotate with time.

It follows that the moment of inertia tensor, evaluated with respect to the primed system, will have *components* that are independent of time, although the dyadic that represents the tensor will depend on time through the unit vectors $\hat{\mathbf{i}}', \hat{\mathbf{j}}', \hat{\mathbf{k}}'$. In particular, one could choose the orientation of the primed axes with respect to the body, once and for all, such that the moment of inertia tensor is diagonal, with diagonal elements $\mathcal{I}_1 = \mathcal{I}_{x'x'}$, $\mathcal{I}_2 = \mathcal{I}_{y'y'}$, and $\mathcal{I}_3 = \mathcal{I}_{z'z'}$ and with off diagonal elements $\mathcal{I}_{x'y'} = \mathcal{I}_{y'z'} = \mathcal{I}_{z'x'} = 0$. In such a representation, the moment of inertia dyadic would be

$$\boldsymbol{I}' = \sum_i m_i [r_i'^2 \mathbb{1} - \mathbf{r}'_i \mathbf{r}'_i] = \hat{\mathbf{i}}' \mathcal{I}_1 \hat{\mathbf{i}}' + \hat{\mathbf{j}}' \mathcal{I}_2 \hat{\mathbf{j}}' + \hat{\mathbf{k}}' \mathcal{I}_3 \hat{\mathbf{k}}'. \tag{F.27}$$

In this notation, the prime on $\boldsymbol{I}'$ is only a reminder that it is the moment of inertia dyadic expressed in terms of the unit vectors $\hat{\mathbf{i}}', \hat{\mathbf{j}}', \hat{\mathbf{k}}'$ rather than the unit vectors $\hat{\mathbf{i}}, \hat{\mathbf{j}}, \hat{\mathbf{k}}$ that are independent of time. In fact, ironically, $\boldsymbol{I} = \boldsymbol{I}'$, just as $\mathbf{r}_i = \mathbf{r}'_i$, but the tensor *components* of these dyadics are different, those for $\boldsymbol{I}'$ being independent of time.

Similarly, the angular momentum can be represented by $\mathbf{L} = L_x\hat{\mathbf{i}} + L_y\hat{\mathbf{j}} + L_z\hat{\mathbf{k}}$ or alternatively $\mathbf{L}' = L_{x'}\hat{\mathbf{i}}' + L_{y'}\hat{\mathbf{j}}' + L_{z'}\hat{\mathbf{k}}'$, with $\mathbf{L} = \mathbf{L}'$. Since $\mathbf{L} = \boldsymbol{I} \cdot \boldsymbol{\omega}$ we also have

$$\mathbf{L}' = \boldsymbol{I}' \cdot \boldsymbol{\omega}' = \mathcal{I}_1 \omega_{x'} \hat{\mathbf{i}}' + \mathcal{I}_2 \omega_{y'} \hat{\mathbf{j}}' + \mathcal{I}_3 \omega_{z'} \hat{\mathbf{k}}', \tag{F.28}$$

where $\boldsymbol{\omega} = \omega_x\hat{\mathbf{i}} + \omega_y\hat{\mathbf{j}} + \omega_z\hat{\mathbf{k}}$ or alternatively $\boldsymbol{\omega}' = \omega_{x'}\hat{\mathbf{i}}' + \omega_{y'}\hat{\mathbf{j}}' + \omega_{z'}\hat{\mathbf{k}}'$, with $\boldsymbol{\omega} = \boldsymbol{\omega}'$.

F.5.1 Time Derivatives Revisited

We first consider a general vector $\mathbf{G} = G_x\hat{\mathbf{i}} + G_y\hat{\mathbf{j}} + G_z\hat{\mathbf{k}}$ or alternatively $\mathbf{G}' = G_{x'}\hat{\mathbf{i}}' + G_{y'}\hat{\mathbf{j}}' + G_{z'}\hat{\mathbf{k}}'$, with $\mathbf{G} = \mathbf{G}'$, and where $G_{x'}$, $G_{y'}$, and $G_{z'}$ are not necessarily independent of time. Its time derivative is

$$\begin{aligned}\frac{\mathrm{d}\mathbf{G}}{\mathrm{d}t} &= \frac{\mathrm{d}G_x}{\mathrm{d}t}\hat{\mathbf{i}} + \frac{\mathrm{d}G_y}{\mathrm{d}t}\hat{\mathbf{j}} + \frac{\mathrm{d}G_z}{\mathrm{d}t}\hat{\mathbf{k}} \\ &= \frac{\mathrm{d}G'_x}{\mathrm{d}t}\hat{\mathbf{i}}' + \frac{\mathrm{d}G'_y}{\mathrm{d}t}\hat{\mathbf{j}}' + \frac{\mathrm{d}G'_z}{\mathrm{d}t}\hat{\mathbf{k}}' + G_{x'}\frac{\mathrm{d}\hat{\mathbf{i}}'}{\mathrm{d}t} + G_{y'}\frac{\mathrm{d}\hat{\mathbf{j}}'}{\mathrm{d}t} + G_{z'}\frac{\mathrm{d}\hat{\mathbf{k}}'}{\mathrm{d}t} = \frac{\mathrm{d}\mathbf{G}'}{\mathrm{d}t}.\end{aligned} \tag{F.29}$$

But $\mathrm{d}\hat{\mathbf{i}}'/\mathrm{d}t = \boldsymbol{\omega} \times \hat{\mathbf{i}}'$ and similarly for $\mathrm{d}\hat{\mathbf{j}}'/\mathrm{d}t$ and $\mathrm{d}\hat{\mathbf{k}}'/\mathrm{d}t$, so

$$\frac{\mathrm{d}\mathbf{G}}{\mathrm{d}t} = \frac{\mathrm{d}G'_x}{\mathrm{d}t}\hat{\mathbf{i}}' + \frac{\mathrm{d}G'_y}{\mathrm{d}t}\hat{\mathbf{j}}' + \frac{\mathrm{d}G'_z}{\mathrm{d}t}\hat{\mathbf{k}}' + \boldsymbol{\omega} \times \mathbf{G}'. \tag{F.30}$$

Equation (F.30) can be written in the form

$$\left(\frac{\mathrm{d}\mathbf{G}}{\mathrm{d}t}\right)_{\text{fixed}} = \left(\frac{\mathrm{d}\mathbf{G}'}{\mathrm{d}t}\right)_{\text{rotating}} + \boldsymbol{\omega} \times \mathbf{G}', \tag{F.31}$$

but this notation can be easily misinterpreted because in $(\mathrm{d}\mathbf{G}'/\mathrm{d}t)_{\text{rotating}}$ one only differentiates the *components* of $\mathbf{G}'$, not the unit vectors. Another word of caution concerns the interpretation of the expressions

$$\boldsymbol{\omega} \times \mathbf{G} = \boldsymbol{\omega} \times \mathbf{G}' = \boldsymbol{\omega}' \times \mathbf{G} = \boldsymbol{\omega}' \times \mathbf{G}'. \tag{F.32}$$

The first and the last are easy to evaluate in terms of the usual rule for cross products and result in expressions in terms of unprimed and primed unit vectors, respectively; the middle two are hybrid expressions and one of their members must be resolved along the coordinates of the other in order to compute the cross product. For instance, the use of $\boldsymbol{\omega}' \times \mathbf{G}'$ in Eq. (F.30) would lead directly to an expression for $\mathrm{d}\mathbf{G}/\mathrm{d}t$ resolved along the primed vectors.

For the special case of $\mathbf{G} = \mathbf{r} = \mathbf{r}'$, we recall that the *components* of $\mathbf{r}'$ are constant in the rotating frame, so

$$\frac{\mathrm{d}\mathbf{r}}{\mathrm{d}t} = \boldsymbol{\omega} \times \mathbf{r} = \boldsymbol{\omega}' \times \mathbf{r}' = \frac{\mathrm{d}\mathbf{r}'}{\mathrm{d}t}. \tag{F.33}$$

For the case of the angular momentum $\mathbf{L}$, we see from Eq. (F.28) that

$$\frac{\mathrm{d}\mathbf{L}}{\mathrm{d}t} = \mathcal{I}_1 \frac{\mathrm{d}\omega_{x'}}{\mathrm{d}t}\hat{\mathbf{i}}' + \mathcal{I}_2 \frac{\mathrm{d}\omega_{y'}}{\mathrm{d}t}\hat{\mathbf{j}}' + \mathcal{I}_3 \frac{\mathrm{d}\omega_{z'}}{\mathrm{d}t}\hat{\mathbf{k}}' + \boldsymbol{\omega}' \times \mathbf{L}' \tag{F.34}$$

which should be compared to Eq. (F.23). On the right-hand side of Eq. (F.34), $\mathrm{d}\mathbf{L}/\mathrm{d}t = \mathrm{d}\mathbf{L}'/\mathrm{d}t$ is resolved along the components of the primed basis vectors that depend on time, but the *components* $\mathcal{I}_1$, $\mathcal{I}_2$, and $\mathcal{I}_3$ are independent of time. Since the torque $\mathbf{N} = \mathrm{d}\mathbf{L}/\mathrm{d}t$, the *components* of torque in the primed system are

$$\begin{aligned} N_1 &= \mathcal{I}_1\dot{\omega}_1 + \omega_2\omega_3(\mathcal{I}_3 - \mathcal{I}_2); \\ N_2 &= \mathcal{I}_2\dot{\omega}_2 + \omega_3\omega_1(\mathcal{I}_1 - \mathcal{I}_3); \\ N_3 &= \mathcal{I}_3\dot{\omega}_3 + \omega_1\omega_2(\mathcal{I}_2 - \mathcal{I}_1), \end{aligned} \tag{F.35}$$

where $x', y', z' \rightarrow 1, 2, 3$ and $\dot{\omega}_1 = \mathrm{d}\omega_1/\mathrm{d}t$, etc. Equations (F.35) are known as the **Euler equations of motion** of a rigid body. The power delivered by the torque is then

$$\boldsymbol{\omega}' \cdot \mathbf{N}' = \mathcal{I}_1\omega_1\dot{\omega}_1 + \mathcal{I}_2\omega_2\dot{\omega}_2 + \mathcal{I}_3\omega_3\dot{\omega}_3 = \frac{\mathrm{d}T'}{\mathrm{d}t} \tag{F.36}$$

which should be compared to Eq. (F.26) with

$$T' = \frac{1}{2}(\mathcal{I}_1\omega_1^2 + \mathcal{I}_2\omega_2^2 + \mathcal{I}_3\omega_3^2) = T. \tag{F.37}$$

In the unprimed frame, T is given by Eq. (F.14). The advantage of Eq. (F.37) is that the *components* $\mathcal{I}_1$, $\mathcal{I}_2$ and $\mathcal{I}_3$ are independent of time.

By using the same reasoning that led to Eq. (F.30), we can deduce that

$$\frac{\mathrm{d}\mathbf{I}'}{\mathrm{d}t} = \boldsymbol{\omega}' \times \mathbf{I}' - \mathbf{I}' \times \boldsymbol{\omega}' \tag{F.38}$$

which, of course, has the same form as Eq. (F.20) and holds whether or not the principal axes are used in the rotating frame. Thus,

$$\frac{d\mathbf{L}'}{dt} = \frac{d}{dt}(\mathbf{I}' \cdot \boldsymbol{\omega}') = \mathbf{I}' \cdot \dot{\boldsymbol{\omega}}' + \frac{d\mathbf{I}'}{dt} \cdot \boldsymbol{\omega}' = \mathbf{I}' \cdot \dot{\boldsymbol{\omega}}' + \boldsymbol{\omega}' \times \mathbf{I}' \cdot \boldsymbol{\omega}' \tag{F.39}$$

which agrees with Eq. (F.34) if principal axes are used. Similarly,

$$\frac{dT'}{dt} = \frac{1}{2}\frac{d}{dt}(\boldsymbol{\omega}' \cdot \mathbf{I}' \cdot \boldsymbol{\omega}') = \boldsymbol{\omega}' \cdot \mathbf{I}' \cdot \dot{\boldsymbol{\omega}}' \tag{F.40}$$

which agrees with Eq. (F.36) if principal axes are used.

F.6 Matrix Formulation

Our principal interest as far as classical statistical mechanics is concerned is to express the Hamiltonian in terms of canonical coordinates and momenta. Since the energy is provided by Eq. (F.37), it remains to express $\boldsymbol{\omega}'$ in terms of the transformation from the fixed to the rotating coordinate system. This can be done by writing the transformation in the matrix form

$$x' = Ax \tag{F.41}$$

with inverse

$$x = A^{\mathrm{T}}x', \tag{F.42}$$

where x and x' are column vectors. A is an orthogonal matrix that depends on time and A^{T} is its transpose, which is also its inverse. Following the notation and convention of Goldstein [60, p. 107], we can write A in terms of the Euler angles ϕ, θ, and ψ as a product of three successive rotations in the form

$$A = BCD, \tag{F.43}$$

where

$$D = \begin{pmatrix} \cos\phi & \sin\phi & 0 \\ -\sin\phi & \cos\phi & 0 \\ 0 & 0 & 1 \end{pmatrix}; \quad C = \begin{pmatrix} 1 & 0 & 0 \\ 0 & \cos\theta & \sin\theta \\ 0 & -\sin\theta & \cos\theta \end{pmatrix}; \quad B = \begin{pmatrix} \cos\psi & \sin\psi & 0 \\ -\sin\psi & \cos\psi & 0 \\ 0 & 0 & 1 \end{pmatrix}. \tag{F.44}$$

When the matrix A multiplies the column vector x, matrix D causes a rotation by ϕ around the z-axis; then C causes a rotation by θ about the rotated x-axis, resulting in the rotated z-axis becoming the z'-axis; and finally B causes a rotation by ψ about the z'-axis to establish the x'- and y'-axes. See Figures 4-6 of Goldstein [60].

Since the coordinates x' are independent of time, differentiation of Eq. (F.42) with respect to time gives

$$\dot{x} = \dot{A}^{\mathrm{T}}x' = \dot{A}^{\mathrm{T}}Ax. \tag{F.45}$$

Since $A^{\mathrm{T}}A = E$, where E is the unit matrix whose time derivative is zero, we have

$$\dot{A}^{\mathrm{T}}A = -A^{\mathrm{T}}\dot{A} = -(\dot{A}^{\mathrm{T}}A)^{\mathrm{T}}, \tag{F.46}$$

so $\dot{A}^{\mathrm{T}}A$ is an antisymmetric matrix that we can write in the form

$$\dot{A}^{\mathrm{T}}A \equiv \omega = \begin{pmatrix} 0 & \omega_{xy} & \omega_{xz} \\ -\omega_{xy} & 0 & \omega_{yz} \\ -\omega_{xz} & -\omega_{yz} & 0 \end{pmatrix} = \begin{pmatrix} 0 & -\omega_z & \omega_y \\ \omega_z & 0 & -\omega_x \\ -\omega_y & \omega_x & 0 \end{pmatrix}, \tag{F.47}$$

where ω_x, ω_y and ω_z are components of a pseudovector $\boldsymbol{\omega}$. With this notation, we note that Eq. (F.45) is the matrix representation of $\mathbf{v} = \omega \times \mathbf{r}$. Furthermore,

$$v^2 = \mathbf{v} \cdot \mathbf{v} = \dot{x}^{\mathrm{T}} \dot{x} = x^{\mathrm{T}} A^{\mathrm{T}} \dot{A} \dot{A}^{\mathrm{T}} A x = x^{\mathrm{T}} \omega^{\mathrm{T}} \omega x. \tag{F.48}$$

Matrix multiplication shows readily that $\omega^{\mathrm{T}}\omega$ is a symmetric matrix with components

$$(\omega^{\mathrm{T}}\omega)_{\alpha\beta} = \delta_{\alpha\beta} - \omega_\alpha \omega_\beta, \tag{F.49}$$

where $\delta_{\alpha\beta}$ is the Kronecker delta (elements of the unit matrix). Inserting this result in Eq. (F.48) with x now corresponding to the ith point of a rigid body, multiplying by the mass m_i and a factor of 1/2 and summing over all points of the body, one obtains the kinetic energy

$$T = \frac{1}{2} \sum_{\mu\nu} \omega_\mu \mathcal{I}_{\mu\nu} \omega_\nu \tag{F.50}$$

which is equivalent to Eq. (F.14).

A similar equation can be obtained in terms of the primed coordinates by returning to Eq. (F.45) to obtain $v^2 = x'^{\mathrm{T}} \omega'^{\mathrm{T}} \omega' x'$, where the antisymmetric matrix

$$A\dot{A}^{\mathrm{T}} \equiv \omega' = \begin{pmatrix} 0 & -\omega_{z'} & \omega_{y'} \\ \omega_{z'} & 0 & -\omega_{x'} \\ -\omega_{y'} & \omega_{x'} & 0 \end{pmatrix}. \tag{F.51}$$

Here, $\omega_{x'}$, $\omega_{y'}$, and $\omega_{z'}$ are to be regarded as the components of a pseudovector $\boldsymbol{\omega}'$. Then by the same reasoning as in the unprimed case,

$$T = \frac{1}{2} \sum_{\mu'\nu'} \omega_{\mu'} \mathcal{I}_{\mu'\nu'} \omega_{\nu'}, \tag{F.52}$$

where the components of the moment of inertia tensor $\mathcal{I}_{\mu'\nu'}$ are evaluated in the primed frame, which rotates with the rigid body, and are therefore independent of time. If the axes in the primed frame are chosen as principal axes of the body, then

$$T = \frac{1}{2}(\mathcal{I}_1 \omega_1^2 + \mathcal{I}_2 \omega_2^2 + \mathcal{I}_3 \omega_3^2) \tag{F.53}$$

as in Eq. (F.37).

The transformation from the matrix ω to the matrix ω' is a similarity transformation because

$$\omega' = A\dot{A}^{\mathrm{T}} = A\dot{A}^{\mathrm{T}} A A^{\mathrm{T}} = A\omega A^{\mathrm{T}}. \tag{F.54}$$

By expressing the antisymmetric matrices ω and ω' in terms of the Levi-Cavita symbols $\epsilon_{\alpha\beta\gamma}$, a rather lengthy calculation shows that

$$\omega'_\alpha = \det A \sum_\lambda A_{\alpha\lambda} \omega_\lambda \tag{F.55}$$

which defines the transformation of a pseudovector. For the matrix A given by Eq. (F.43), $\det A = 1$ so ω transforms as a vector.

It remains to express the components of ω and ω' in terms of the Euler angles and their time derivatives. This can be done in a straightforward way by appealing to the matrix formulation just described. Thus we have

$$\omega = \dot{A}^{\mathrm{T}} A = (\dot{D}^{\mathrm{T}} C^{\mathrm{T}} B^{\mathrm{T}} + D^{\mathrm{T}} \dot{C}^{\mathrm{T}} B^{\mathrm{T}} + D^{\mathrm{T}} C^{\mathrm{T}} \dot{B}^{\mathrm{T}}) BCD \tag{F.56}$$

and

$$\omega' = A\dot{A}^{\mathrm{T}} = BCD(\dot{D}^{\mathrm{T}} C^{\mathrm{T}} B^{\mathrm{T}} + D^{\mathrm{T}} \dot{C}^{\mathrm{T}} B^{\mathrm{T}} + D^{\mathrm{T}} C^{\mathrm{T}} \dot{B}^{\mathrm{T}}). \tag{F.57}$$

After an exercise in matrix algebra, we deduce

$$\omega_x = \sin\theta \sin\phi\, \dot{\psi} + \cos\phi\, \dot{\theta}; \quad \omega_y = -\sin\theta \cos\phi\, \dot{\psi} + \sin\phi\, \dot{\theta}; \quad \omega_z = \cos\theta\, \dot{\psi} + \dot{\phi} \tag{F.58}$$

and

$$\omega_{x'} = \sin\theta \sin\psi\, \dot{\phi} + \cos\psi\, \dot{\theta}; \quad \omega_{y'} = \sin\theta \cos\psi\, \dot{\phi} - \sin\psi\, \dot{\theta}; \quad \omega_{z'} = \cos\theta\, \dot{\phi} + \dot{\psi}. \tag{F.59}$$

The results in the current section are used in Section 20.7 to calculate the classical partition functions of polyatomic molecules.

F.7 Canonical Variables

For a free rotator, we are now in the position to write the Hamiltonian in the form

$$\mathcal{H} = \frac{1}{2}(\mathcal{I}_1 \omega_1^2 + \mathcal{I}_2 \omega_2^2 + \mathcal{I}_3 \omega_3^2), \tag{F.60}$$

where $\omega_1 \equiv \omega_{x'}, \omega_2 \equiv \omega_{y'}, \omega_3 \equiv \omega_{z'}$ are given by Eq. (F.59) with x', y', z' understood to correspond to the *principal* axes in the rotating frame of reference. The canonical momenta p_ϕ, p_θ, p_ψ conjugate to the three Euler angles ϕ, θ, ψ may be found by differentiation[5]:

$$\begin{aligned} p_\phi &= \frac{\partial \mathcal{H}}{\partial \dot{\phi}} = \mathcal{I}_1 \omega_1 \sin\theta \sin\psi + \mathcal{I}_2 \omega_2 \sin\theta \cos\psi + \mathcal{I}_3 \omega_3 \cos\theta \\ &= L_1 \sin\theta \sin\psi + L_2 \sin\theta \cos\psi + L_3 \cos\theta. \end{aligned} \tag{F.61}$$

Here, L_1, L_2, L_3 are the principal angular momenta, which are components of the vector $\mathbf{L}'$ given by Eq. (F.28). Similarly,

$$p_\theta = \frac{\partial \mathcal{H}}{\partial \dot{\theta}} = \mathcal{I}_1 \omega_1 \cos\psi - \mathcal{I}_2 \omega_2 = L_1 \cos\psi - L_2 \sin\psi \tag{F.62}$$

and

$$p_\psi = \frac{\partial \mathcal{H}}{\partial \dot{\psi}} = \mathcal{I}_3 \omega_3 = L_3. \tag{F.63}$$

In terms of the principal angular momenta, the Hamiltonian may be written

$$\mathcal{H} = \frac{L_1^2}{2\mathcal{I}_1} + \frac{L_2^2}{2\mathcal{I}_2} + \frac{L_3^2}{2\mathcal{I}_3}. \tag{F.64}$$

[5] Since we are only dealing with the kinetic energy, $\mathcal{H} = \mathcal{L}$, where $\mathcal{L}$ is the Lagrangian.

F.8 Quantum Energy Levels for Diatomic Molecule

The quantum energy levels associated with a Hamiltonian of the form of Eq. (F.64) are rather tricky to calculate because the angular momenta L_1 and L_2 relate to a rotating coordinate system and because the vanishing of $\mathcal{I}_3$ is only an approximation based on the assumption that each atom can be considered to be a point particle. To better understand this problem, we examine the more general case in which the kinetic energy is given by Eq. (F.64) with $\mathcal{I}_1 = \mathcal{I}_2 \equiv \mathcal{I}$ and $\mathcal{I}_3 \ll \mathcal{I}$ but $\mathcal{I}_3 \neq 0$. This is the problem of a symmetrical top and is treated by Landau and Lifshitz [66, p. 383]. The Hamiltonian is

$$\mathcal{H} = \frac{L_1^2 + L_2^2}{2\mathcal{I}} + \frac{L_3^2}{2\mathcal{I}_3} = \frac{L^2}{2\mathcal{I}} + \frac{L_3^2}{2}\left(\frac{1}{\mathcal{I}_3} - \frac{1}{\mathcal{I}}\right), \tag{F.65}$$

where $L^2 = L_1^2 + L_2^2 + L_3^2$. They proceed to show that the commutation relations among the L_i are the same as for a non-rotating coordinate system except for complex conjugation, which has no effect on the energy eigenvalues. Thus the eigenvalues of L^2 are $\hbar^2 j(j+1)$ and those of L_3^2 are $\hbar^2 k^2$, where $k = -j, \ldots, -1, 0, 1, \ldots, j$. The energy eigenvalues are therefore

$$\varepsilon_{jk} = \frac{\hbar^2}{2\mathcal{I}} j(j+1) + \frac{\hbar^2}{2}\left(\frac{1}{\mathcal{I}_3} - \frac{1}{\mathcal{I}}\right) k^2. \tag{F.66}$$

Since $\pm k$ lead to the same energy, all levels except for $k = 0$ are at least two-fold degenerate. *But for $\mathcal{I}_3 \approx 0$, all levels except for $k = 0$ are extremely large! They are so large, in fact, that they are not excited at any reasonable temperature before the molecule dissociates. So one ordinarily just ignores these levels and deals only with*

$$\varepsilon_{j0} = \frac{\hbar^2}{2\mathcal{I}} j(j+1), \tag{F.67}$$

which is the result quoted in the text, Eqs. (18.82) and (21.150). Moreover, with only $k = 0$, these levels would appear to be non-degenerate. But here, the rotating coordinate system comes into play. It introduces a degeneracy of $2j + 1$ associated with the orientation of these angular momenta with respect to a fixed coordinate system [66, footnote on p. 384]. This is, finally, in agreement with the results quoted in the text and often related to a strictly two-dimensional analysis.

G

Thermodynamic Perturbation Theory

For most problems in statistical mechanics, an exact analytical solution is impossible to obtain, so we need methods to obtain approximate solutions. Fortunately, there are many problems of interest in which the Hamiltonian can be expressed in the form

$$\mathcal{H} = \mathcal{H}_0 + V \tag{G.1}$$

in which an exact solution is known for the unperturbed Hamiltonian $\mathcal{H}_0$ and where V is a small correction, in a sense to be clarified below. Under these circumstances, it is possible to obtain an approximate solution in terms of averages of powers of V with respect to a Boltzmann distribution for the unperturbed Hamiltonian.

This technique is called thermodynamic perturbation theory, which is a mixture of perturbation theory and thermodynamic ensemble averaging. In this appendix, we discuss this topic in the context of the canonical ensemble. We first take up the classical case and then the quantum case, which requires slightly different considerations. We follow closely a treatment by Landau and Lifshitz [7, p. 93].

G.1 Classical Case

We write

$$\mathcal{H}(p,q) = \mathcal{H}_0(p,q) + V(p,q) \tag{G.2}$$

and express the classical[1] canonical partition function in the form

$$Z_C(\beta) = \int \mathrm{e}^{-\beta\mathcal{H}(p,q)}\,\mathrm{d}\omega = \int \mathrm{e}^{-\beta[\mathcal{H}_0(p,q)+V(p,q)]}\,\mathrm{d}\omega, \tag{G.3}$$

where p and q are each $3\mathcal{N}$-dimensional vectors and $d\omega$ represents the volume element in this $6\mathcal{N}$-dimensional phase space. We then expand the second exponential in powers of V to second order to obtain

$$Z_C(\beta) = \int \mathrm{e}^{-\beta\mathcal{H}_0(p,q)}\left[1 - \beta V(p,q) + \frac{(\beta V(p,q))^2}{2}\right]\mathrm{d}\omega. \tag{G.4}$$

[1]To agree with quantum mechanics, we need to divide Z_C by $h^{3\mathcal{N}}$ for $\mathcal{N}$ particles or by $h^{3\mathcal{N}}\mathcal{N}!$ for identical indistinguishable particles, as in the case of a dilute ideal gas. Here we omit those factors for simplicity.

We define an averaging process of any quantity $B(p, q)$ with respect to the unperturbed distribution as follows:

$$\langle B \rangle_0 := \frac{\int e^{-\beta \mathcal{H}_0(p,q)} B(p,q)\, d\omega}{\int e^{-\beta \mathcal{H}_0(p,q)}\, d\omega}. \tag{G.5}$$

Then Eq. (G.4) takes the form

$$Z_C(\beta) = Z_0(\beta)\left[1 - \beta\langle V\rangle_0 + (1/2)\,\beta^2\langle V^2\rangle_0\right], \tag{G.6}$$

where the unperturbed partition function

$$Z_0(\beta) := \int e^{-\beta \mathcal{H}_0(p,q)}\, d\omega. \tag{G.7}$$

We now use the formula for the Helmholtz free energy $F = -k_B T \ln Z_C$ to obtain, again to second order in V,

$$F = F_0 - k_B T \ln\left[1 - \beta\langle V\rangle_0 + (1/2)\,\beta^2\langle V^2\rangle_0\right] = F_0 + \langle V\rangle_0 - (1/2)\,\beta\langle V^2\rangle_0 + (1/2)\,\beta^2\langle V\rangle_0^2, \tag{G.8}$$

where $F_0 = -k_B T \ln Z_0$ is the unperturbed Helmholtz free energy. This result can be written in the form

$$F = F_0 + \langle V\rangle_0 - \frac{\langle (V - \langle V\rangle_0)^2\rangle_0}{2k_B T}. \tag{G.9}$$

Equation (G.9) shows that the free energy is first changed by the average value of V and then diminished by a second-order term proportional to the variance of V. Both the mean and the variance are proportional to $\mathcal{N}$, so the condition for the validity of the expansion is that V per particle be small compared to $k_B T$.

G.2 Quantum Case

According to second-order perturbation theory, the eigenvalues of the Hamiltonian in Eq. (G.1) are

$$E_n = E_n^0 + V_{nn} + \sum_{m \neq n} \frac{|V_{nm}|^2}{E_n^0 - E_m^0}, \tag{G.10}$$

where n stands for a set of quantum numbers that label the unperturbed states of the system, E_n^0 are the unperturbed energies and $V_{nm} \equiv \langle n|V|m\rangle$ are matrix elements of V with respect to the unperturbed states. We expand the partition function

$$Z(\beta) = \sum_n e^{-\beta E_n} = \sum_n e^{-\beta E_n^0}\left[1 - \beta V_{nn} - \beta \sum_{m \neq n} \frac{|V_{nm}|^2}{E_n^0 - E_m^0} + (1/2)\beta^2 V_{nn}^2\right] \tag{G.11}$$

to second order in V and define a quantum averaging process

$$\langle T_n \rangle_0 := \frac{\sum_n e^{-\beta E_n^0} T_n}{\sum_n e^{-\beta E_n^0}}. \tag{G.12}$$

Then the partition function becomes

$$Z(\beta) = Z_0(\beta)\left[1 - \beta\langle V_{nn}\rangle_0 - \beta\left\langle \sum_{m\neq n}\frac{|V_{nm}|^2}{E_n^0 - E_m^0} + (1/2)\beta^2 V_{nn}^2\right\rangle_0\right], \tag{G.13}$$

where

$$Z_0(\beta) = \sum_n \mathrm{e}^{-\beta E_n^0}. \tag{G.14}$$

The Helmholtz free energy then becomes

$$F = F_0 + \langle V_{nn}\rangle_0 + \left\langle \sum_{m\neq n}\frac{|V_{nm}|^2}{E_n^0 - E_m^0}\right\rangle_0 - (1/2)\beta\langle V_{nn}^2\rangle_0 + (1/2)\beta\langle V_{nn}\rangle_0^2, \tag{G.15}$$

where $F_0 = -k_\mathrm{B}T\ln Z_0$. This last result can be rewritten in the form

$$F = F_0 + \langle V_{nn}\rangle_0 + \left\langle \sum_{m\neq n}\frac{|V_{nm}|^2}{E_n^0 - E_m^0}\right\rangle_0 - \frac{\langle(V_{nn} - \langle V_{nn}\rangle_0)^2\rangle_0}{2k_\mathrm{B}T} \tag{G.16}$$

which has an extra term compared to its classical equivalent Eq. (G.9). This extra term can be written in the form[2]

$$\begin{aligned}\left\langle \sum_{m\neq n}\frac{|V_{nm}|^2}{E_n^0 - E_m^0}\right\rangle_0 &= \frac{1}{Z_0}\sum_n\sum_{m\neq n}\frac{|V_{nm}|^2}{E_n^0 - E_m^0}\mathrm{e}^{-\beta E_n^0} = \frac{1}{Z_0}\sum_m\sum_{n\neq m}\frac{|V_{nm}|^2}{E_m^0 - E_n^0}\mathrm{e}^{-\beta E_m^0}\\ &= -\frac{1}{2Z_0}\sum_n\sum_{m\neq n}\frac{|V_{nm}|^2}{E_n^0 - E_m^0}\left(\mathrm{e}^{-\beta E_m^0} - \mathrm{e}^{-\beta E_n^0}\right).\end{aligned} \tag{G.17}$$

Since $\mathrm{e}^{-\beta E_m^0} - \mathrm{e}^{-\beta E_n^0}$ and $E_n^0 - E_m^0$ have the same sign, this extra term is negative, so *both second order terms in Eq. (G.16) are negative.*

If the *unperturbed* energy splittings $E_n^0 - E_m^0$ happen to be small compared to $k_\mathrm{B}T$, we can further simplify Eq. (G.17) by expanding the exponentials. In that case

$$\begin{aligned}-\frac{1}{2}\sum_n\sum_{m\neq n}\frac{|V_{nm}|^2}{E_n^0 - E_m^0}\left(\mathrm{e}^{-\beta E_m^0} - \mathrm{e}^{-\beta E_n^0}\right) &= -\frac{1}{2}\sum_n\sum_{m\neq n}\frac{|V_{nm}|^2}{E_n^0 - E_m^0}\mathrm{e}^{-\beta E_n^0}\left(\mathrm{e}^{-\beta(E_m^0 - E_n^0)} - 1\right)\\ &\approx -\frac{1}{2kT}\sum_n\sum_{m\neq n}|V_{nm}|^2\mathrm{e}^{-\beta E_n^0}.\end{aligned} \tag{G.18}$$

Then Eq. (G.16) becomes

$$F = F_0 + \langle V_{nn}\rangle_0 - \frac{1}{2kT}\left[\left\langle \sum_{m\neq n}|V_{nm}|^2 + (V_{nn} - \langle V_{nn}\rangle_0)^2\right\rangle_0\right]. \tag{G.19}$$

[2] In making this identification, note that $\sum_n\sum_{m\neq n}A_{mn} = \sum_m\sum_{n\neq m}A_{nm}$ = the sum of all off-diagonal elements of the matrix A.

The term in square brackets in Eq. (G.19) can be written

$$\frac{1}{Z_0}\sum_n \mathrm{e}^{-\beta E_n^0}\left[\sum_{m\neq n}|V_{nm}|^2 + V_{nn}^2 - \langle V_{nn}\rangle_0^2\right]. \tag{G.20}$$

But since V is Hermitian, the rules of matrix multiplication lead to

$$\sum_{m\neq n}|V_{nm}|^2 + V_{nn}^2 = \sum_{m\neq n} V_{nm}V_{nm}^* + V_{nn}^2 = \sum_{m\neq n} V_{nm}V_{mn} + V_{nn}^2 = \sum_m V_{nm}V_{mn} = (V^2)_{nn}. \tag{G.21}$$

Thus Eq. (G.20) reduces to

$$\frac{1}{Z_0}\sum_n \mathrm{e}^{-\beta E_n^0}\left[(V^2)_{nn} - \langle V_{nn}\rangle_0^2\right] = \langle (V^2)_{nn}\rangle_0 - \langle V_{nn}\rangle_0^2 \tag{G.22}$$

and Eq. (G.16) takes the form

$$F = F_0 + \langle V_{nn}\rangle_0 - \frac{\langle (V - \langle V_{nn}\rangle_0)_{nn}^2\rangle_0}{2k_{\mathrm{B}}T} \tag{G.23}$$

which is similar to Eq. (G.9) for the classical case.

Finally, we remark that Eq. (G.23) can be expressed in an invariant form by using properties of the trace. The unperturbed partition function is

$$Z_0(\beta) = \sum_n \mathrm{e}^{-\beta E_n^0} = \mathrm{tr}\,\mathrm{e}^{-\beta\mathcal{H}_0}. \tag{G.24}$$

Furthermore, we can introduce the unperturbed density operator

$$\hat{\rho}^0 = \frac{\mathrm{e}^{-\beta\mathcal{H}_0}}{\mathrm{tr}\,\mathrm{e}^{-\beta\mathcal{H}_0}}. \tag{G.25}$$

Then the averaging process expressed by Eq. (G.12) can be written in the form

$$\langle\mathcal{O}\rangle_0 = \mathrm{tr}\,(\hat{\rho}^0\mathcal{O}) \tag{G.26}$$

for any operator $\mathcal{O}$. Thus

$$F = F_0 + \mathrm{tr}\,(\hat{\rho}^0 V) - \frac{\mathrm{tr}\,\left\{\hat{\rho}^0[V - \mathrm{tr}\,(\hat{\rho}^0 V)]^2\right\}}{2k_{\mathrm{B}}T}. \tag{G.27}$$

In this approximate form of Eq. (G.16) it is more obvious that the second-order correction to F is negative, as in the classical case.

H Selected Mathematical Relations

In this appendix we develop and summarize selected mathematical relations that are employed in the text. We concentrate on simplicity of presentation and refer the reader to the mathematical literature for rigorous proofs. We first define Bernoulli numbers and Bernoulli polynomials to clarify conventions used in the literature. Then we state the Euler-Maclaurin sum formula in general terms, specialize it to approximate infinite sums with integrals, and give examples of its use in computing partition functions.

H.1 Bernoulli Numbers and Polynomials

Bernoulli numbers frequently appear as coefficients in the representation of integrals by asymptotic series as well as in formulae for the correction terms used to approximate infinite sums by integrals. Unfortunately there are alternative definitions and conventions used to define Bernoulli numbers and the associated Bernoulli polynomials.

H.1.1 Bernoulli Numbers

An infinite set of numbers known as Bernoulli numbers B_n for $n = 1, 2, \ldots$ may be defined as the expansion coefficients in the series for $|z| < 2\pi$ of the even function [92, p. 125])

$$\begin{aligned} \frac{z}{2}\cot\frac{z}{2} &= 1 - B_1\frac{z^2}{2!} - B_2\frac{z^4}{4!} - B_3\frac{z^6}{6!} - \cdots \\ &= 1 - \sum_{n=1}^{\infty} B_n \frac{z^{2n}}{(2n)!}. \end{aligned} \tag{H.1}$$

By setting $z = t/i$, it is easily seen for $|t| < 2\pi$ that

$$\begin{aligned} \frac{t}{e^t - 1} = -\frac{t}{2} + \frac{t}{2i}\cot\frac{t}{2i} &= 1 - \frac{t}{2} + B_1\frac{t^2}{2!} - B_2\frac{t^4}{4!} + B_3\frac{t^6}{6!} - \cdots \\ &= 1 - \frac{t}{2} + \sum_{n=1}^{\infty}(-1)^{n+1} B_n \frac{t^{2n}}{(2n)!}, \end{aligned} \tag{H.2}$$

where the alternations in sign result from $i^2 = -1$. Although it is not obvious from these series expansions, it turns out that all B_n are positive numbers which can be established by the integral representations [92, p. 126][1]

[1]Whittaker and Watson [92] leave the last form as an exercise. It can be obtained by substituting $x = 2\pi t$ and then $x = \pi t$ in the first integral in Eq. (H.3) to obtain alternative expressions for B_n, solving for $(2^{2n} - 1)B_n$ and simplifying the integrand.

$$B_n = 4n\int_0^\infty \frac{t^{2n-1}}{e^{2\pi t}-1}\,dt = \frac{2n}{\pi^{2n}(2^{2n}-1)}\int_0^\infty \frac{x^{2n-1}}{\sinh x}\,dx > 0. \tag{H.3}$$

We use these positive B_n in this book, for example, $B_1 = 1/6$, $B_2 = 1/30$, $B_3 = 1/42$, $B_4 = 1/30$, $B_5 = 5/66$, $B_6 = 691/2730$, $B_7 = 7/6$, etc., from which we note that the sequence is not monotonic.

An alternative set of numbers $\tilde{B}_n$, also called Bernoulli numbers, can be defined by [98, p. 804]

$$\frac{t}{e^t-1} = \sum_{n=0}^{\infty} \tilde{B}_n \frac{t^n}{n!}. \tag{H.4}$$

In this case, $\tilde{B}_0 = 1$, $\tilde{B}_1 = -\frac{1}{2}$, $\tilde{B}_{2p+1} = 0$, and $\tilde{B}_{2p} = (-1)^{p+1}B_p$ for $p \geq 1$. Another convention is $B_n^* = (-1)^{n+1}B_n$.

H.1.2 Bernoulli Polynomials

Bernoulli polynomials $\phi_n(z)$ for $n \geq 1$ can be defined by [92]

$$\frac{t(e^{zt}-1)}{e^t-1} = \sum_{n=1}^{\infty} \phi_n(z)\frac{t^n}{(n)!}; \quad |t| < 2\pi. \tag{H.5}$$

Evidently $\phi_n(0) = 0$ for all $n \geq 1$, $\phi_n(1) = 0$ for $n > 1$, but $\phi_1(1) = 1$. From the defining equation one can establish the difference equation

$$\phi_n(z+1) - \phi_n(z) = nz^{n-1}; \quad n \geq 1. \tag{H.6}$$

Alternative Bernoulli polynomials $B_n(z)$ for $n \geq 0$ can be defined by [98]

$$\frac{te^{zt}}{e^t-1} = \sum_{n=1}^{\infty} B_n(z)\frac{t^n}{(n)!}; \quad |t| < 2\pi. \tag{H.7}$$

In this case, $B_0(z) = 1$ and the remaining polynomials can be shown to satisfy $\phi_n(z) = B_n(z) - \tilde{B}_n$, $n \geq 1$, so there is no difference between these polynomials for odd $n > 1$.

H.2 Euler-Maclaurin Sum Formula

As shown by Whittaker and Watson [92, p. 128] a very general form of the Euler-Maclaurin sum formula can be obtained by using Bernoulli polynomials and a formula due to Darboux. That sum formula is

$$\begin{aligned} f(z) - f(a) = &\frac{1}{2}(z-a)\left[f'(z) + f'(a)\right] \\ &+ \sum_{k=1}^{p-1}(-1)^k (z-a)^{2k}\frac{B_k}{(2k)!}\left[f^{(2k)}(z) - f^{(2k)}(a)\right] \\ &+ \frac{(z-a)^{2p+1}}{(2p)!}\int_0^1 \phi_{2p}(t) f^{(2p+1)}(a + t(z-a)), \end{aligned} \tag{H.8}$$

where $f(z)$ is analytic on a line from a to z, $f'(z) = \mathrm{d}f/\mathrm{d}z$ and $f^{(2k)}(z) = \mathrm{d}^{2k}f/\mathrm{d}z^{2k}$ denotes higher derivatives of even order. The last term is the remainder of the finite sum and involves an integral over a Bernoulli polynomial. Following their procedure, we substitute $F(z) = f'(z)$ and $w = z - a$ into Eq. (H.8) to obtain

$$\begin{aligned}\int_a^{a+w} F(x)\,\mathrm{d}x =& \frac{1}{2} w\,[F(a+w) + F(a)] \\ &+ \sum_{k=1}^{p-1} (-1)^k w^{2k} \frac{B_k}{(2k)!}\left[F^{(2k-1)}(a+w) - F^{(2k-1)}(a)\right] \\ &+ \frac{w^{2p+1}}{(2p)!}\int_0^1 \phi_{2p}(t) F^{(2p)}(a+tw).\end{aligned} \tag{H.9}$$

Then by letting a take the values $a + w, a + 2w, \ldots, a + (r - 1)w$ and adding, we obtain

$$\begin{aligned}\int_a^{a+rw} F(x)\,\mathrm{d}x =& w\left[\frac{1}{2}F(a+rw) + F(a+(r-1)w) + \cdots + F(a+w) + \frac{1}{2}F(a)\right] \\ &+ \sum_{k=1}^{p-1} (-1)^k w^{2k} \frac{B_k}{(2k)!}\left[F^{(2k-1)}(a+rw) - F^{(2k-1)}(a)\right] \\ &+ \frac{w^{2p+1}}{(2p)!}\int_0^1 \phi_{2p}(t) \sum_{s=0}^{r-1} F^{(2p)}(a+sw+tw).\end{aligned} \tag{H.10}$$

Equation (H.10) may be used to approximate integrals by sums or *vice versa*. Under suitable conditions, the remainder term on the last line vanishes as $p \to \infty$ and the middle term becomes an infinite sum:

$$\begin{aligned}\int_a^{a+rw} F(x)\,\mathrm{d}x =& w\sum_{\ell=0}^{r} F(a+\ell w) - \frac{w}{2}\,[F(a+rw) + F(a)] \\ &+ \sum_{k=1}^{\infty} (-1)^k w^{2k} \frac{B_k}{(2k)!}\left[F^{(2k-1)}(a+rw) - F^{(2k-1)}(a)\right].\end{aligned} \tag{H.11}$$

If we let $a + rw = b$, we see that w is obtained by dividing the interval $b - a$ into r equal parts. For evaluation of sums, an important special case occurs when a and b are integers and $w = 1$ which results in

$$\begin{aligned}\sum_{\ell=a}^{b} F(\ell) =& \int_a^b F(x)dx + \frac{1}{2}\,[F(a) + F(b)] \\ &+ \sum_{k=1}^{\infty} (-1)^k \frac{B_k}{(2k)!}\left[F^{(2k-1)}(a) - F^{(2k-1)}(b)\right].\end{aligned} \tag{H.12}$$

For the frequently occurring case $a = 0$ and $b = \infty$ when both F and its derivatives vanish at infinity, we obtain

$$\sum_{\ell=0}^{\infty} F(\ell) = \int_0^{\infty} F(x)dx + \frac{1}{2}F(0) + \sum_{k=1}^{\infty} (-1)^k \frac{B_k}{(2k)!} F^{(2k-1)}(0). \tag{H.13}$$

This formula is only justified if the integral and both series converge. Alternatively, it might give an asymptotic result in some parameter on which F depends.

H.2.1 Approximate Evaluation of Infinite Sums

In statistical mechanics, especially in calculating partition functions, one frequently needs to evaluate sums of the form $\sum_{n=0}^{\infty} f(n)$, where $f(n)$ is some function of an integer n. Sometimes these sums can be done exactly, for example for a geometrical series, but in many cases they cannot and approximation methods are needed. We demonstrate how the Euler-Maclaurin sum formula in the form given by Eq. (H.13) can be used for this purpose. With the change of notation $F \to f$, it becomes explicitly

$$\begin{aligned}\sum_{n=0}^{\infty} f(n) &= \int_0^{\infty} f(n)\,\mathrm{d}n + \frac{1}{2}f(0) + \sum_{j=1}^{\infty}(-1)^j \frac{B_j}{(2j)!} f^{(2j-1)}(0) \\ &= \int_0^{\infty} f(n)\,\mathrm{d}n + \frac{1}{2}f(0) - \frac{1}{12}f'(0) + \frac{1}{720}f'''(0) - \frac{1}{30240}f^{(5)}(0) + \cdots \end{aligned} \tag{H.14}$$

As noted above, this expansion is only justified if the series converges. Otherwise, it might give an asymptotic approximation.

Example Problem H.1. Evaluate approximately the partition function for a rigid rotator (see Eq. (18.83))

$$z = \sum_{j=0}^{\infty} (2j+1)\exp[-j(j+1)x], \tag{H.15}$$

where $x := \varepsilon_0/k_{\mathrm{B}}T$, correct to order $(\varepsilon_0/k_{\mathrm{B}}T)^2$ at high temperatures. Then compute the corresponding heat capacity.

Solution H.1. The lowest order term is given by the integral and is just $1/x = k_{\mathrm{B}}T/\varepsilon_0$ (see Eq. (18.85)). We compute $f(0) = 1, f'(0) = 2 - x, f'''(0) = -12x + 12x^2 + O(x)^3, f^{(5)}(0) = 120x^2 + O(x^3), f^{(7)}(0) = O(x^3)$. Thus

$$z = \frac{1}{x} + \frac{1}{3} + \frac{x}{15} + \frac{4x^2}{315} + O(x^3). \tag{H.16}$$

$$\ln z = \ln\left[\frac{1}{x}\left(1 + \frac{x}{3} + \frac{x^2}{15} + \frac{4x^3}{315} + O(x^4)\right)\right] = -\ln x + \frac{x}{3} + \frac{x^2}{90} + \frac{8x^3}{2835} + O(x^4). \tag{H.17}$$

Therefore,

$$c = \frac{\partial}{\partial T}\left(-\frac{\partial \ln z}{\partial \beta}\right) = k_{\mathrm{B}}x^2\frac{\partial^2 \ln z}{\partial x^2} = k_{\mathrm{B}}\left(1 + \frac{x^2}{45} + \frac{16x^3}{945} + O(x^4)\right), \tag{H.18}$$

which shows clearly that c asymptotes k_{B} from larger values as $T \to \infty$, as is evident from Figure 18–12. Although the trend is clear, the accuracy is poor unless x is very small, so the series appears to be asymptotic in $\varepsilon_0/k_{\mathrm{B}}T$ as $T \to \infty$. It has no hope of representing c near $T = 0$ where the leading term is $c = x^2\mathrm{e}^{-2x}$ as $x \to \infty$.

In some applications the function $f(n) = g(n+\alpha)$, where $0 \leq \alpha < 1$, with $\alpha = 1/2$ of special importance. In that case, Eq. (H.14) becomes

$$\sum_{n=0}^{\infty} g(n+\alpha) = \int_0^{\infty} g(n+\alpha)\,\mathrm{d}n + \frac{1}{2}g(\alpha) + \sum_{j=1}^{\infty} \frac{B_j}{(2j)!} g^{(2j-1)}(\alpha)$$

$$= \int_0^{\infty} g(n+\alpha)\,\mathrm{d}n + \frac{1}{2}g(\alpha) - \frac{1}{12}g'(\alpha) + \frac{1}{720}g'''(\alpha) - \frac{1}{30240}g^{(5)}(\alpha) + \cdots. \quad \text{(H.19)}$$

With the substitution $x = n + \alpha$, the integral term can be written

$$\int_0^{\infty} g(n+\alpha)\,\mathrm{d}n = \int_{\alpha}^{\infty} g(x)\,\mathrm{d}x = \int_0^{\infty} g(x)\,\mathrm{d}x - \int_0^{\alpha} g(x)dx. \quad \text{(H.20)}$$

In the last integral, we expand $g(x)$ in a series about $x = 0$ and integrate term by term to obtain

$$\int_0^{\alpha} g(x)dx = \int_0^{\alpha} \sum_{r=0}^{\infty} g^{(r)}(0)\frac{x^r}{r!} = \sum_{r=0}^{\infty} g^{(r)}(0)\frac{\alpha^{r+1}}{(r+1)!}. \quad \text{(H.21)}$$

We also expand the other terms, $g^{(j)}(\alpha)$, in Eq. (H.19) about $\alpha = 0$ and combine results to obtain

$$\sum_{n=0}^{\infty} g(n+\alpha) = \int_0^{\infty} g(x)dx + \left(\frac{1}{2} - \alpha\right)g(0) + \left(-\frac{1}{12} + \frac{\alpha}{2} - \frac{\alpha^2}{2}\right)g'(0)$$

$$+ \left(-\frac{\alpha}{12} + \frac{\alpha^2}{4} - \frac{\alpha^3}{6}\right)g''(0) + \left(\frac{1}{720} - \frac{\alpha^2}{24} + \frac{\alpha^3}{12} - \frac{\alpha^4}{24}\right)g'''(0) + \cdots. \quad \text{(H.22)}$$

In the special case $\alpha = 1/2$, the coefficients of $g(0)$ and $g''(0)$ vanish and we are left with

$$\sum_{n=0}^{\infty} g(n+\tfrac{1}{2}) = \int_0^{\infty} g(x)\,\mathrm{d}x + \frac{1}{24}g'(0) - \frac{7}{5760}g'''(0) + \cdots. \quad \text{(H.23)}$$

Example Problem H.2. Use Eq. (8.12) to evaluate approximately

$$\sum_{n=0}^{\infty} \exp[-y(n+\tfrac{1}{2})] = \frac{\mathrm{e}^{-y/2}}{1-\mathrm{e}^{-y}} \quad \text{(H.24)}$$

for $y \ll 1$ and compare with the exact result (right-hand side) which was obtained by summing the geometric series.

Solution H.2. The leading term is

$$\int_0^{\infty} \mathrm{e}^{-yx}\,\mathrm{d}x = \frac{1}{y}. \quad \text{(H.25)}$$

$g''(0) = -y$ and $g'''(0) = -y^3$ so

$$\sum_{n=0}^{\infty} \exp[-y(n+\tfrac{1}{2})] = \frac{1}{y} - \frac{y}{24} + \frac{7y^3}{5760} + \cdots . \tag{H.26}$$

Expanding the exact result gives

$$\frac{e^{-y/2}}{1-e^{-y}} = \frac{1}{y} - \frac{y}{24} + \frac{7y^3}{5760} - \frac{31y^5}{967680} + O(y^7). \tag{H.27}$$

I Creation and Annihilation Operators

In this appendix we derive some properties of creation and annihilation operators a and $a^\dagger$ that are useful in quantum mechanics and statistical mechanics. They are also used to express the operators of quantized fields in terms of expansions. We first motivate the operators used to treat bosons by appealing to the quantum treatment of the simple harmonic oscillator in a straightforward way. Then, having established the commutation relations for a and $a^\dagger$, we derive by purely algebraic operations their properties as they act on vectors in a Hilbert space. We then relate back to the quantum harmonic oscillator and derive a few useful expressions for the coordinate and momentum operators, x and p. Finally, we introduce the corresponding operators used to create bosons and fermions as well as eigenstates for systems containing many of them.

I.1 Harmonic Oscillator

We consider a simple one-dimensional harmonic oscillator of mass m and frequency ω described by the Hamiltonian

$$\mathcal{H} = \frac{p^2}{2m} + \frac{m\omega^2}{2}x^2, \tag{I.1}$$

where p and x are operators that satisfy the well-known commutation relations

$$[p, x] \equiv (px - xp) = -i\hbar. \tag{I.2}$$

Of course $p = p^\dagger$ and $x = x^\dagger$ are Hermitian, where the superscript $\dagger$ denotes the Hermitian conjugate.

We introduce the dimensionless creation operator

$$a = \left(\frac{m\omega}{2\hbar}\right)^{1/2}\left(x + \frac{i}{m\omega}p\right) \tag{I.3}$$

and its Hermitian conjugate

$$a^\dagger = \left(\frac{m\omega}{2\hbar}\right)^{1/2}\left(x - \frac{i}{m\omega}p\right) \tag{I.4}$$

which will play the role of an annihilation operator. Then the commutator

$$[a, a^\dagger] = \frac{i}{m\omega}\frac{m\omega}{2\hbar}([p, x] - [x, p]) = 1 \tag{I.5}$$

and of course $[a, a] = 0$ and $[a^\dagger, a^\dagger] = 0$. Then a little algebra shows that

$$\hbar\omega\, aa^\dagger = \frac{p^2}{2m} + \frac{m\omega^2}{2}x^2 + \hbar\omega \tag{I.6}$$

and

$$\hbar\omega\, a^\dagger a = \frac{p^2}{2m} + \frac{m\omega^2}{2}x^2 - \hbar\omega, \tag{I.7}$$

which can be added to deduce

$$\mathcal{H} = \frac{1}{2}\hbar\omega(aa^\dagger + a^\dagger a) = \hbar\omega(aa^\dagger - \frac{1}{2}) = \hbar\omega(a^\dagger a + \frac{1}{2}). \tag{I.8}$$

As will be shown below, the eigenvalues of $a^\dagger a$ are the integers $n = 0, 1, 2, 3, \ldots$ and the eigenvalues of $aa^\dagger$ are also integers $q = 1, 2, 3, \ldots$, starting at one instead of zero. The eigenvalues of $\mathcal{H}$ are therefore $\hbar\omega(n + 1/2)$ as is well known. Note that this result is based on purely algebraic properties of the operators and did not require any discussion of the Schrödinger wave functions.

Before leaving this section, we note a few useful relations. The inverses of Eqs. (I.3) and (I.4) are

$$x = \left(\frac{2\hbar}{m\omega}\right)^{1/2} \frac{a + a^\dagger}{2}; \quad p = \frac{m\omega}{i}\left(\frac{2\hbar}{m\omega}\right)^{1/2} \frac{a - a^\dagger}{2} \tag{I.9}$$

and a little algebra shows that

$$x^2 = \frac{\hbar}{m\omega}\left(a^\dagger a + \frac{1}{2} + \frac{aa + a^\dagger a^\dagger}{2}\right) \tag{I.10}$$

and

$$p^2 = m\hbar\omega\left(a^\dagger a + \frac{1}{2} - \frac{aa + a^\dagger a^\dagger}{2}\right) \tag{I.11}$$

which, of course, are Hermitian. These results are used in Section 26.6.2.

I.2 Boson Operators

We proceed to find the eigenvalues of the Hermitian operators $aa^\dagger$ and $a^\dagger a$. From the commutator Eq. (I.5), we see that $aa^\dagger = a^\dagger a + 1$, so multiplication from the left and then from the right by $a^\dagger a$ shows that the operators $a^\dagger a$ and $aa^\dagger$ commute and can be simultaneously diagonalized. Furthermore, if $|\psi\rangle$ is an eigenvector of $a^\dagger a$ with eigenvalue λ, we have

$$a^\dagger a|\psi\rangle = \lambda|\psi\rangle, \tag{I.12}$$

where λ is some real number. But then

$$aa^\dagger|\psi\rangle = (a^\dagger a + 1)|\psi\rangle = (\lambda + 1)|\psi\rangle, \tag{I.13}$$

so $|\psi\rangle$ is also an eigenvector of $aa^\dagger$ with eigenvalue $\tilde{\lambda} = \lambda + 1$.

It is easy to see that λ can never be negative. Let $a|\psi\rangle = |\phi\rangle$ so that $\langle\phi| = \langle\psi|a^\dagger$. Then

$$\langle\phi|\phi\rangle = \langle\psi|a^\dagger a|\psi\rangle = \langle\psi|\lambda|\psi\rangle = \lambda\langle\psi|\psi\rangle. \tag{I.14}$$

We therefore deduce

$$\lambda = \frac{\langle\phi|\phi\rangle}{\langle\psi|\psi\rangle} \geq 0 \tag{I.15}$$

with $\lambda = 0$ possible only if $a|\psi\rangle \equiv 0$. Thus, the eigenvalues of $aa^\dagger$ will satisfy $\tilde{\lambda} = \lambda + 1 \geq 1$.

Next, we apply the operator $a^\dagger a$ to $a|\psi\rangle$ to obtain

$$(a^\dagger a)a|\psi\rangle = (aa^\dagger - 1)a|\psi\rangle = (a(a^\dagger a) - a)|\psi\rangle = a(\lambda - 1)|\psi\rangle = (\lambda - 1)a|\psi\rangle. \tag{I.16}$$

From this we deduce that $a|\psi\rangle$ is an eigenstate of $a^\dagger a$ with eigenvalue $\lambda - 1$. Furthermore, by continued application of a, we deduce that $a^m|\psi\rangle$ is an eigenstate of $a^\dagger a$ with eigenvalue $\lambda - m$. If λ were not an integer, we could choose m to be large enough to produce a negative eigenvalue of $a^\dagger a$, which is impossible. Therefore, λ must be an integer, say $\lambda = n$, in which case n applications of a will give

$$(a^\dagger a)a^n|\psi\rangle = (\lambda - n)a^n|\psi\rangle = (n - n)a^n|\psi\rangle = 0. \tag{I.17}$$

Therefore, there exists an eigenvector $|0\rangle$ proportional to $a^n|\psi\rangle$ such that

$$a^\dagger a|0\rangle = 0|0\rangle. \tag{I.18}$$

Furthermore, since the state $a|0\rangle$, if it existed, would be an eigenvector of $a^\dagger a$ with eigenvalue -1, which by Eq. (I.15) is impossible, it must be true that

$$a|0\rangle \equiv 0 \tag{I.19}$$

to prevent such a state from existing.

At this stage, we have shown that the eigenvalues of $a^\dagger a$ are $n = 0, 1, 2, \ldots$, and we shall denote their corresponding normalized eigenvectors by $|n\rangle$ so that

$$a^\dagger a|n\rangle = n|n\rangle; \quad \langle n|n\rangle = 1. \tag{I.20}$$

The eigenvalues of $aa^\dagger$ will then be $n + 1 = 1, 2, 3, \ldots$ and we will have

$$aa^\dagger|n\rangle = (n + 1)|n\rangle. \tag{I.21}$$

Next, we apply the operator $a^\dagger a$ to the state $a^\dagger|n\rangle$ to obtain

$$(a^\dagger a)a^\dagger|n\rangle = a^\dagger(aa^\dagger)|n\rangle = (n + 1)a^\dagger|n\rangle, \tag{I.22}$$

which shows that $a^\dagger|n\rangle$ is an eigenstate of $a^\dagger a$ with eigenvalue $n + 1$, so it is proportional to $|n + 1\rangle$. Since $\langle n|aa^\dagger|n\rangle = \langle n|n + 1|n\rangle = n + 1$, we see that

$$|n + 1\rangle = \frac{1}{(n + 1)^{1/2}}a^\dagger|n\rangle. \tag{I.23}$$

We can therefore generate all eigenvectors of $a^\dagger a$ by successive application of $a^\dagger$ to $|0\rangle$ to obtain

$$|n\rangle = \frac{1}{(n!)^{1/2}} (a^\dagger)^n |0\rangle. \tag{I.24}$$

Similarly, we note that $\langle n|a^\dagger a|n\rangle = n$, so

$$|n-1\rangle = \frac{1}{n^{1/2}} a|n\rangle. \tag{I.25}$$

It follows that repeated application of a to $|n\rangle$ gives

$$|n-m\rangle = \left(\frac{(n-m)!}{n!}\right)^{1/2} a^m|n\rangle; \quad m \le n, \tag{I.26}$$

so

$$|0\rangle = \frac{1}{(n!)^{1/2}} a^n|n\rangle \tag{I.27}$$

and $a^{n+1}|n\rangle = 0$ because $a|0\rangle \equiv 0$. Given Eqs. (I.23) and (I.25), $a^\dagger$ and a are often referred to as raising and lowering operators, respectively.

I.3 Fermion Operators

In this section, we formally introduce operators that we shall call fermion operators. We shall still denote them by a and $a^\dagger$, as is usual, despite the possible confusion with a and $a^\dagger$ for bosons. These operators obey the relations

$$\{a, a^\dagger\} = 1; \quad \{a, a\} = 0; \quad \{a^\dagger, a^\dagger\} = 0, \tag{I.28}$$

where the *anticommutator* $\{A, B\} = AB + BA$. The second two members of Eq. (I.28) actually require

$$aa = -aa \equiv 0; \quad a^\dagger a^\dagger = -a^\dagger a^\dagger \equiv 0. \tag{I.29}$$

First, we note from Eq. (I.29) that $a^\dagger aaa^\dagger = 0$ and $aa^\dagger a^\dagger a = 0$, so the operators $a^\dagger a$ and $aa^\dagger$ commute and have a common set of eigenvectors. Let $|\psi\rangle$ be an eigenvector of $a^\dagger a$ with eigenvalue λ so that

$$a^\dagger a|\psi\rangle = \lambda|\psi\rangle. \tag{I.30}$$

Applying $a^\dagger a$ again and using the anticommutation relations we obtain

$$\lambda^2|\psi\rangle = a^\dagger aa^\dagger a|\psi\rangle = (1 - aa^\dagger)a^\dagger a|\psi\rangle = a^\dagger a|\psi\rangle = \lambda|\psi\rangle. \tag{I.31}$$

Therefore, $\lambda^2 = \lambda$ so the only possible eigenvalues of $a^\dagger a$ are $\lambda = 0$ and $\lambda = 1$. We denote the corresponding eigenvectors by $|0\rangle$ and $|1\rangle$, respectively. Similarly, if $|\tilde{\psi}\rangle$ is an eigenvector of $aa^\dagger$ with eigenvalue $\tilde{\lambda}$, we have

$$\tilde{\lambda}^2|\tilde{\psi}\rangle = aa^\dagger aa^\dagger|\tilde{\psi}\rangle = (1 - a^\dagger a)aa^\dagger|\tilde{\psi}\rangle = aa^\dagger|\tilde{\psi}\rangle = \tilde{\lambda}|\tilde{\psi}\rangle, \tag{I.32}$$

so $\tilde{\lambda}^2 = \tilde{\lambda}$ and $\tilde{\lambda} = 0, 1$ are the only possible eigenvalues of $aa^\dagger$.

Next, we apply $aa^\dagger$ to an eigenvector $|\psi\rangle$ of $a^\dagger a$ to obtain

$$aa^\dagger|\psi\rangle = (1 - a^\dagger a)|\psi\rangle = (1 - \lambda)|\psi\rangle. \tag{I.33}$$

It therefore follows that

$$a^\dagger a|0\rangle = 0|0\rangle \quad a^\dagger a|1\rangle = 1|1\rangle, \tag{I.34}$$

$$aa^\dagger|0\rangle = 1|0\rangle \quad aa^\dagger|1\rangle = 0|1\rangle. \tag{I.35}$$

Therefore, if $a^\dagger a$ is regarded as a number operator, then $aa^\dagger$ behaves like an antinumber operator.

Now we apply $a^\dagger a$ to the state $a^\dagger|0\rangle$ to obtain

$$(a^\dagger a)a^\dagger|0\rangle = (1 - aa^\dagger)a^\dagger|0\rangle = a^\dagger|0\rangle, \tag{I.36}$$

from which we conclude that $a^\dagger|0\rangle = |1\rangle$, but of course $a^\dagger|1\rangle = a^\dagger a^\dagger|0\rangle \equiv 0$. Then we apply $aa^\dagger$ to the state $a|1\rangle$ to obtain

$$(aa^\dagger)a|1\rangle = (1 - a^\dagger a)a|1\rangle = a|1\rangle, \tag{I.37}$$

and use the left member of Eq. (I.35) to conclude that $a|1\rangle = |0\rangle$. Then we see that $a|0\rangle = a^2|1\rangle \equiv 0$.

Summarizing, for fermions there are only two states, $|0\rangle$ and $|1\rangle$, so $a^\dagger|0\rangle = |1\rangle$ and $a^\dagger|1\rangle \equiv 0$ whereas $a|1\rangle = |0\rangle$ and $a|0\rangle \equiv 0$.

I.4 Boson and Fermion Number Operators

Having now established in detail the allowed eigenvalues and eigenvectors of $a^\dagger a$ and $aa^\dagger$ for both bosons and fermions, we focus attention on the number operator

$$\hat{N} = a^\dagger a, \tag{I.38}$$

which in both cases has the property $\hat{N}|n\rangle = n$. The only difference is that $n = 0, 1, 2, 3, \ldots$ for bosons but $n = 0, 1$ only for fermions. For *both* bosons and fermions, $\hat{N}$ satisfies the commutation relations

$$[\hat{N}, a] = -a \quad \text{and} \quad [\hat{N}, a^\dagger] = a^\dagger. \tag{I.39}$$

Therefore

$$\hat{N}a|n\rangle = (a\hat{N} - a)|n\rangle = (n - 1)a|n\rangle \tag{I.40}$$

and

$$\hat{N}a^\dagger|n\rangle = (a^\dagger\hat{N} + a^\dagger)|n\rangle = (n + 1)a^\dagger|n\rangle. \tag{I.41}$$

For both bosons and fermions, $a|0\rangle \equiv 0$, but for fermions, we also have $a^\dagger|1\rangle \equiv 0$. If we regard n as being the number of particles in a state, then a applied to $|n\rangle$ results in a state, if such a state $|n - 1\rangle$ exists, having one less particle. Therefore, a is called an annihilation operator. Similarly, since $a^\dagger$ applied to $|n\rangle$ results in a state $|n + 1\rangle$, if such a state exists, having one more particle, $a^\dagger$ is called a creation operator. In the case of fermions, and in accordance with the Pauli exclusion principle, a state can have only zero or one particle.

These ideas can be generalized to a number of identical particles whose single-particle states we denote by Greek subscripts. For bosons, the corresponding commutation relations become

$$[a_\alpha, a_\beta^\dagger] = \delta_{\alpha,\beta}; \quad [a_\alpha, a_\beta] = 0; \quad [a_\alpha^\dagger, a_\beta^\dagger] = 0. \tag{I.42}$$

The number operator for the single-particle state α is $\hat{N}_\alpha = a_\alpha^\dagger a_\alpha$ and obviously $\hat{N}_\alpha$ and $\hat{N}_\beta$ commute and can have a common set of eigenstates. The counterpart to Eq. (I.24) becomes

$$|n_\alpha, n_\beta, n_\gamma, \ldots\rangle = \frac{1}{(n_\alpha! n_\beta! n_\gamma! \cdots)^{1/2}} (a_\alpha^\dagger)^{n_\alpha} (a_\beta^\dagger)^{n_\beta} (a_\gamma^\dagger)^{n_\gamma} \cdots |0, 0, 0, \ldots\rangle, \tag{I.43}$$

where the ground state $|0, 0, 0, \ldots\rangle$ is usually called the **vacuum state**.

For fermions, the anticommutation relations become

$$\{a_\alpha, a_\beta^\dagger\} = \delta_{\alpha,\beta}; \quad \{a_\alpha, a_\beta\} = 0; \quad \{a_\alpha^\dagger, a_\beta^\dagger\} = 0. \tag{I.44}$$

In that case, $a_\alpha a_\alpha = 0$ and $a_\alpha^\dagger a_\alpha^\dagger = 0$ for all α, but for $\alpha \neq \beta$, we have $a_\alpha a_\beta = -a_\beta a_\alpha$ and $a_\alpha^\dagger a_\beta^\dagger = -a_\beta^\dagger a_\alpha^\dagger$. For fermions as well, $\hat{N}_\alpha$ and $\hat{N}_\beta$ commute, although the fact that they do is not as obvious as for bosons. The relation corresponding to Eq. (I.43) is simply

$$|n_\alpha, n_\beta, n_\gamma, \ldots\rangle = (a_\alpha^\dagger)^{n_\alpha} (a_\beta^\dagger)^{n_\beta} (a_\gamma^\dagger)^{n_\gamma} \cdots |0, 0, 0, \ldots\rangle, \tag{I.45}$$

where the only allowed values of the n_α are zero and one. Since these operators anticommute, one can order them [8, p. 268] with increasing subscripts ($\alpha < \beta < \gamma < \cdots$) to prevent an uncertainty of ± 1 in the phase.

References

[1] E. Fermi, Thermodynamics, Dover, New York, 1956.

[2] H.B. Callen, Thermodynamics, second ed., John Wiley, New York, 1985.

[3] J.W. Gibbs, The Scientific Papers of J. Willard Gibbs, vol. 1, Dover, New York, 1961.

[4] J.W. Gibbs, Elementary Principles in Statistical Mechanics, Ox Bow Press, Woodbridge, CT, 1891.

[5] C.H.P. Lupis, Thermodynamics of Materials, North Holland, New York, 1983.

[6] C. Kittel, H. Kroemer, Thermal Physics, W.H. Freeman, New York, 1980.

[7] L.D. Landau, E.M. Lifshitz, Statistical Physics, Addison-Wesley, Reading, MA, 1958.

[8] R.K. Pathria, Statistical Mechanics, second ed., Pergamon Press, New York, 1996.

[9] R.K. Pathria, P.D. Beale, Statistical Mechanics, third ed., Elsevier, New York, 2011.

[10] G. Holton, Y. Elkana, Albert Einstein: Historical and Cultural Perspectives, Dover, Mineola, NY, 1997.

[11] W.H. Cropper, Great Physicists, Oxford University Press, New York, 2004.

[12] D. Chandler, Introduction to Modern Statistical Mechanics, Oxford University Press, New York, 1987.

[13] H.D. Young, R.A. Freedman, University Physics, ninth ed., Addison-Wesley, New York, 1996.

[14] W. Greiner, L. Neise, H. Stöcker, Thermodynamics and Statistical Mechanics, Springer, New York, 2000.

[15] M. Planck, Treatise on Thermodynamics, third ed., Translation of Seventh German, Dover, Mineola, NY, 1926.

[16] D. Kondepudi, I. Prigogine, Modern Thermodynamics, John Wiley, New York, 1999.

[17] B. Widom, Statistical Mechanics, Cambridge University Press, Cambridge, UK, 2002.

[18] K. Denbigh, Principles of Chemical Equilibrium, third ed., Cambridge University Press, London, 1971.

[19] L.S. Darken, R.W. Gurry, Physical Chemistry of Metals, McGraw-Hill, New York, 1953.

[20] R.B. Griffiths, A proof that the free energy of a spin system is extensive, J. Math. Phys. 5 (1964) 1215-1222.

[21] R.T. DeHoff, Thermodynamics in Materials Science, McGraw Hill, New York, 1993.

[22] G.N. Lewis, M. Randall, Thermodynamics, McGraw-Hill, New York, 1961 (revised by K.S. Pitzer, L. Brewer).

[23] R.A. Swalin, Thermodynamics of Solids, John Wiley, New York, 1967.

[24] T.B. Massalski, Binary Phase Diagrams, vol. 1-3, ASM, Metals Park, OH, 1990.

[25] H. Margenau, G.M. Murphy, The Mathematics of Physics and Chemistry, second ed., D. Van Nostrand, Princeton, NJ, 1956.

[26] R. Courant, D. Hilbert, Methods of Mathematical Physics, vol. I, seventh printing ed., Interscience, Princeton, NJ, 1966.

[27] J.S. Rowlinson, B. Widom, Molecular Theory of Capillarity, McGraw-Hill, New York, 1982.

[28] J.W. Cahn, Thermodynamics of solid and fluid interfaces, in: W.C. Johnson, J.M. Blakely (Eds.), Interfacial Segregation, ASM International, International Materials Park, OH, 1979, pp. 3-23 (reprinted in The Selected Works of John W. Cahn, pp. 379-399).

[29] W.C. Carter, W.C. Johnson (Eds.), The Selected Works of John W. Cahn, TMS, Warrendale, PA, 1998.

[30] H.M. Princen, The equilibrium shape of interfaces, drops, and bubbles. Rigid and deformable particles at interfaces, in: E. Matijević (Ed.), Surface and Colloid Science, Wiley-Interscience, New York, 1969.

[31] F. Larché, J.W. Cahn, A linear theory of thermochemical equilibrium of solids under stress, Acta Metall. 21 (1973) 1051-1063.

[32] F. Larché, J.W. Cahn, A nonlinear theory of thermochemical equilibrium of solids under stress, Acta Metall. 26 (1978) 53-60.

[33] W.W. Mullins, R.F. Sekerka, On the thermodynamics of crystalline solids, J. Chem. Phys. 82 (1985) 5192-5202.

[34] R.F. Sekerka, J.W. Cahn, Solid-liquid equilibrium for non-hydrostatic stress, Acta Mater. 52 (2004) 1663-1668.

[35] D.C. Wallace, Thermodynamics of Crystals, John Wiley, New York, 1972.

[36] D.W. Hoffman, J.W. Cahn, A vector thermodynamics for anisotropic surfaces, Surf. Sci. 31 (1985) 368-388.

[37] J.W. Cahn, D.W. Hoffman, A vector thermodynamics for anisotropic surfaces—II. Curved and faceted surfaces, Acta Metall. 22 (1974) 1205-1214.

[38] C. Herring, Surface tension as a motivation for sintering, in: W.E. Kingston (Ed.), The Physics of Powder Metallurgy, a Symposium Held at Bayside, L.I. New York, August 24-26, 1949, McGraw-Hill, New York, 1951, pp. 143-179.

[39] G. Wulff, Zur Frage des Geschwindigkeit des Wachstums und der Auflösung der Kristallflächen, Z. Kristallogr. Mineral. 34 (1901) 449-530.

[40] W.W. Mullins, Solid surface morphologies governed by capillarity, in: W.D. Robertson, N.A. Gjostein (Eds.), Metal Surfaces: Structure, Energetics and Kinetics, ASM International, International Materials Park, OH, 1963, pp. 17-66 (papers presented at a joint symposium of the American Society of Metals and the Metallurgical Society of AIME, October 1962).

[41] C. Herring, The use of classical macroscopic concepts in surface-energy problems, in: R. Gomer, C.S. Smith (Eds.), Structure and Properties of Solid Surfaces, a Conference Arranged by the National Research Council, Lake Geneva, WI, September 1952, University of Chicago Press, Chicago, IL, 1953, pp. 5-81.

[42] F.C. Frank, Solid surface morphologies governed by capillarity, in: W.D. Robertson, N.A. Gjostein (Eds.), Metal Surfaces: Structure, Energetics and Kinetics, ASM International, International Materials Park, OH, 1963, pp. 1-15, (papers presented at a joint symposium of the American Society of Metals and the Metallurgical Society of AIME, October 1962).

[43] R.F. Sekerka, Analytical criteria for missing orientations on three-dimensional equilibrium shapes, J. Cryst. Growth 275 (2005) 77-82.

[44] C. Herring, Some theorems on the free energies of crystal surfaces, Phys. Rev. 82 (1951) 87-93.

[45] W.W. Mullins, R.F. Sekerka, Application of linear programming theory to crystal faceting, J. Phys. Chem. Solids 23 (1962) 801-803.

[46] A.P. Sutton, R.W. Ballufi, Interfaces in Crystalline Materials, Clarendon Press, Oxford, 1955.

[47] N. Reingold, Science in America: A Documentary History 1900-1939, University of Chicago Press, Chicago, 1981.

[48] C.E. Shannon, A mathematical theory of communication, Bell Syst. Tech. J. 27 (1948) 379-423; 623-856.

[49] C.E. Shannon, W. Weaver, The Mathematical Theory of Communications, University of Illinois Press, Urbana, IL, 1949.

[50] A.I. Khinchin, The entropy concept in probability theory, in: R.A. Silverman, M.D. Friedman (Trans., Eds.), Mathematical Foundations of Information Theory, Dover, New York, 1957, pp. 2-28.

[51] L. Boltzmann, Weitere Studen über das Warmegleichgewicht unter Gasmolekölen, Wien Berichte 66 (1872) 275, reprinted in Boltzmann's *Abhandlungen*, vol. 1, Barth, Leipzig, 1909, p. 316.

[52] F. Reif, Fundamentals of Statistical and Thermal Physics, second ed., Waveland Press, Long Grove, IL, 2009.

[53] C.J. Thompson, Mathematical Statistical Mechanics, Princeton University Press, Princeton, NJ, 1972.

[54] D. McQuarrie, Statistical Mechanics, University Science Books, Sausalito, CA, 2000.

[55] M. Planck, Zur Theorie des Gesetzes der Energieverteilung im Normalspectrum, Verh. Dtsch. Phys. Ges. 2 (1900) 237-245.

[56] M. Planck, Über eine Verbesserung der Wien'schen Spectral-gleichung, Verh. Dtsch. Phys. Ges. 2 (1900) 202-204.

[57] L.I. Schiff, Quantum Mechanics, third ed., Clarendon Press, Oxford, 1968.

[58] N.W. Ashcroft, N.D. Mermin, Solid State Physics, Saunders College Publishing, New York, 1976.

[59] A. Messiah, Quantum Mechanics, vol. 2, John Wiley, New York, 1962.

[60] H. Goldstein, Classical Mechanics, Addison-Wesley, Reading, MA, 1959.

[61] A.M. Guénault, Statistical Physics, Routledge, London, 1988.

[62] D.M. Dennison, A note on the specific heat of the hydrogen molecule, Proc. R. Soc. Lond. 115 (1927) 483.

[63] L.A. Girifalco, Statistical Physics of Materials, John Wiley, New York, 1973.

[64] T.L. Hill, An Introduction to Statistical Thermodynamics, Dover, Mineola, NY, 1986.

[65] A. Sommerfeld, Zur Elektronentheorie der Metalle auf Grund der Fermischen Statistik, Z. Phys. 47 (1928) 1-3.

[66] L.D. Landau, E.M. Lifshitz, Quantum Mechanics—Non-relativistic Theory, second ed., Addison-Wesley, Reading, MA, 1965.

[67] P.A.M. Dirac, The Principles of Quantum Mechanics, fourth ed., Oxford University Press, London, 1958.

[68] K. Huang, Statistical Mechanics, John Wiley, New York, 1963.

[69] E. Merzbacher, Quantum Mechanics, John Wiley & Sons, New York, 1961.

[70] S. Galam, A. Mauger, A quasi-exact formula for Ising critical temperatures on hypercubic lattices, Physica A 235 (1997) 573-576.

[71] K. Binder, D.W. Heermann, Monte Carlo Simulation in Statistical Physics, fifth ed., Springer, New York, 2013.

[72] D.P. Landau, K. Binder, Monte Carlo Simulations in Statistical Physics, third ed., Cambridge, New York, 2009.

[73] M.E.J. Newman, G.T. Barkema, Monte Carlo Methods in Statistical Physics, Oxford University Press, New York, 1999.

[74] L.M. Sander, Equilibrium Statistical Physics: With Computer Simulations in Python, first ed., Leonard M. Sander, Middletown, DE, 2013.

[75] B.V. Gnedenko, The Theory of Probability and the Elements of Statistics, fifth ed., Chelsey, New York, 1989.

[76] N. Metropolis, A.W. Rosenbluth, A.H. Teller, E. Teller, Equations of state calculations by fast computing machines, J. Chem. Phys. 21 (1953) 1087-1092.

[77] W.K. Hastings, Monte Carlo sampling methods using Markov chains and their applications, Biometrica 57 (1970) 97-109.

[78] R.H. Swendsen, J.-S. Wang, Nonuniversal critical dynamics in Monte Carlo simulations, Phys. Rev. Lett. 58 (1987) 86-88.

[79] U. Wolff, Collective Monte Carlo updating for spin systems, Phys. Rev. Lett. 62 (1989) 361-363.

[80] L.P. Kadanoff, Scaling laws for Ising models near Tc, Physics 2 (1966) 263.

[81] K.G. Wilson, Renormalization group and critical phenomena. I. Renormalization group and Kadanoff scaling picture, Phys. Rev. B 4 (1971) 3174.

[82] K.G. Wilson, Renormalization group and critical phenomena. II. Phase space cell analysis of critical behavior, Phys. Rev. B 4 (1971) 3184.

[83] A.M. Ferrenberg, R.H. Swendsen, New Monte Carlo technique for studying phase transitions, Phys. Rev. Lett. 61 (1988) 2635-2637.

[84] A.M. Ferrenberg, R.H. Swendsen, Optimized Monte Carlo data analysis, Phys. Rev. Lett. 63 (1989) 1195-1197.

[85] N.F. Carnahan, K.E. Starling, Equation of state for nonattracting rigid spheres, J. Chem. Phys. 51 (1969) 635-636.

[86] R.J. Speedy, Pressure of the metastable hard sphere fluid, J. Phys. Condens. Matter 9 (1997) 8591-8599.

[87] M.D. Rintoul, S. Torquata, Computer simulations of dense hard sphere systems, J. Chem. Phys. 105 (1996) 9258.

[88] J.D. Bernal, J. Mason, Co-ordination of randomly packed spheres, Nature 385 (1960) 910-911.

[89] W.W. Wood, F.R. Parker, Monte Carlo simulation of state of molecules interacting with the Lennard-Jones potential, J. Chem. Phys. 27 (1957) 720.

[90] J.-P. Hansen, L. Verlet, Phase transitions of the Lennard-Jones system, Phys. Rev. 184 (1969) 151.

[91] A. Travesset, Phase diagram of power law and Lennard-Jones systems: crystal phases, J. Chem. Phys. 141 (2014) 164501.

[92] E.T. Whittaker, G.N. Watson, A Course in Modern Analysis, fourth ed., Cambridge University Press, London, UK, 1969.

[93] C.E. Weatherburn, On differential invariants in geometry of surfaces with some applications to mathematical physics, Q. J. Pure Appl. Math. 50 (1925) 230-269.

[94] C.E. Weatherburn, On families of curves and surfaces, Q. J. Pure Appl. Math. 50 (1927) 350-361.

[95] C. Johnson, Generalization of the Gibbs-Thomson equation, Surf. Sci. 3 (1965) 429-444.

[96] R. Courant, Calculus of Variations, Courant Institute of Mathematical Sciences, New York University, NY, 1962 (revised and amended by J. Moser).

[97] H. Goldstein, Classical Mechanics, second ed., Addison-Wesley, Reading, MA, 1981.

[98] M. Abramowitz, I.A. Stegun (Eds.), Handbook of Mathematical Functions, ninth ed., National Bureau of Standards, Washington, DC, 1970.

[99] E. Schrödinger, Statistical Thermodynamics, second ed., Cambridge University Press, Cambridge, 1962.

[100] J.H. van Vleck, The Theory of Electric and Magnetic Susceptibilities, first ed., Oxford University Press, London, 1932.

Index

Note: Page numbers followed by *f* indicate figures, by *t* indicate tables, and by *np* indicate footnotes.

C

F

J

K

N

T

U

V

Values of Selected Physical Constants

Name and symbol	SI value and units	cgs value and units
Magnitude of electronic charge, e	1.602177×10^{-19} C	4.80324×10^{-2} esu
Electron volt, eV	1.602177×10^{-19} J	1.602177×10^{-12} erg
Boltzmann's constant, k_B	1.380649×10^{-23} J K^{-1}	1.380649×10^{-16} erg K^{-1}
Boltzmann's constant, k_B	8.6173×10^{-5} eV K^{-1}	
Planck's constant, h	6.626070×10^{-34} J s	6.626070×10^{-27} erg s
Planck's constant, h	4.135668×10^{-15} eV s	
Planck's constant h-bar, $\hbar = h/2\pi$	1.054572×10^{-34} J s	1.054572×10^{-27} erg s
Planck's constant h-bar, $\hbar = h/2\pi$	6.582120×10^{-16} eV s	
Constant in $\hbar\omega/k_BT$, $\hbar/k_B$	7.638234 K s	7.638234 K s
Avogadro's number, $\mathcal{N}_A$	6.022141×10^{23} mol^{-1}	6.022141×10^{23} mol^{-1}
Measure of heat, cal	1.05587×10^{3} J	1.05587×10^{7} erg
British thermal unit (mean), Btu	4.184 J	4.184×10^{7} erg
Gas constant, $R = \mathcal{N}_A k_B$	8.31446 J mol^{-1}	8.31446×10^{7} erg mol^{-1}
Gas constant, $R = \mathcal{N}_A k_B$	5.189×10^{19} eV mol^{-1}	
Gas constant, $R = \mathcal{N}_A k_B$	1.987 cal mol^{-1}	1.987 cal mol^{-1}
Measure of pressure, Pa	1 N m^{-2}	10 dyne cm^{-2}
Standard atmosphere of pressure, atm	1.01325×10^{5} Pa	1.01325×10^{6} dyne cm^{-2}
cm of mercury, cmHg	1.333224×10^{3} Pa	1.333224×10^{4} dyne cm^{-2}
Electron rest mass, m	9.109384×10^{-31} kg	9.109384×10^{-28} g
Proton rest mass, m_p	1.6726×10^{-27} kg	1.6726×10^{-24} g
Neutron rest mass, m_n	1.674920×10^{-27} kg	1.674920×10^{-24} g
Ratio of proton mass to electron mass, m_p/m	1836.153	1836.153
Atomic mass unit amu, u	1.660539×10^{-27} kg	1.660539×10^{-24} g
Speed of light, c	2.99792458×10^{8} m s^{-1}	2.99792458×10^{10} cm s^{-1}
Bohr magneton, $\mu_B = e\hbar/2m$	9.2740×10^{-24} J T^{-1}	
Bohr magneton, $\mu_B = e\hbar/2m$	5.788382×10^{-5} eV T^{-1}	5.788382×10^{-9} eV G^{-1}
Bohr magneton, $\mu_B = e\hbar/2mc$		9.274×10^{-21} erg G^{-1}
Nuclear magneton, $\mu_N = e\hbar/2m_p$	5.050784×10^{-27} J T^{-1}	
Nuclear magneton, $\mu_N = e\hbar/2m_pc$		5.050784×10^{-24} erg G^{-1}
Steffan-Boltzmann constant, $\sigma = \pi^2 k_B^4/(60\hbar^3c^2)$	5.670×10^{-8} W m^{-2} K^{-4}	5.670×10^{-5} erg s^{-1} cm^{-2} K^{-4}
Reciprocal fine structure constant, $\alpha^{-1} = \hbar c/e^2$	137.036	137.036
Electron Compton wavelength, $\lambda_e = \hbar^2/mc$	3.86159×10^{-15} m	3.86159×10^{-13} cm
Electron radius, $r_e = e^2/mc^2$	2.817940×10^{-13} m	2.817940×10^{-11} cm
$k_BT = 1$ eV	$T = 1.16\times10^{4}$ K	$T = 1.16\times10^{4}$ K
$h\nu = \hbar\omega = 1$ eV	$\nu = 2.42\times10^{14}$ Hz	$\omega = 2\pi\nu = 15.2\times10^{14}\ s^{-1}$
Faraday constant, $\mathcal{F} = e\mathcal{N}_A$	9.648670×10^{4} C mol^{-1}	9.648670×10^{4} C mol^{-1}
Universal gravitational constant, G	6.674×10^{-11} N m^2 kg^{-2}	6.674×10^{-8} dyne cm^{2}g^{-2}

Avogadro's number is also known as Lodschmidt's number, L. See http://physics.nist.gov/cuu/constants for the latest recommended values. C= coulomb, cal = calorie, Pa = N m^{-2} = pascal, W = J/s = watt, G = gauss, T = tesla = 10^4 G, esu = electrostatic units.

Printed in the United States
By Bookmasters